K. Grassie, E. Teuckhoff, G. Wegner, J. Haußelt, H. Hanselka (Editors)

Functional Materials

EUROMAT 99 – Volume 13

EUROMAT 99 was the biannual meeting of the Federation of European Materials Societies (FEMS) with its 21 member societies across Europe.

The program of the EUROMAT 99 congress was divided into 12 topics. The scientific coordination was managed by topic coordinators. The responsible experts for the individual topics were:

Topic A – Materials for Information Technology
K. Grassie, Philips GmbH, Aachen (D)
Topic B – Materials for Transportation Technology
P. J. Winkler, DaimlerChrysler AG, München (D)
Topic C – Materials for Electrical Power Generation and Energy Conversion
E. Tenckhoff, Siemens AG, Erlangen (D)
Topic D – Materials for Medical Engineering
H. Stallforth, Aesculap AG, Tuttlingen (D)
P. Revell, University College London (UK)
Topic E – Materials for Buildings and Structures
F. H. Wittmann, ETH Zürich (CH)
Topic F1 – Characterization Methods and Procedures
Y. Bréchet, INPG, Grenoble (F)
Topic F2 – Metals
R. Wagner, FZ. Jülich (D)
Topic F3 – Ceramics
G. Müller, Fraunhofer-Inst. f. Silicatforschung ISC, Würzburg (D)
Topic F4 – Polymers
G. Wegner, MPI für Polymerforschung, Mainz (D)
P. J. Lemstra, Univ. of Eindhoven (NL)
Topic F5 – Interface Controlled Materials
M. Rühle, MPI für Metallforschung, Stuttgart (D)
Topic G – Surface Technology
H. Dimigen, Fraunhofer-Inst. f. Schicht- und Oberflächentechnik IST, Braunschweig (D)
Topic H – Nano- and Microtechnology
J. Haußelt, Forschungszentrum Karlsruhe (D)

K. Grassie, E. Teuckhoff, G. Wegner, J. Haußelt,
H. Hanselka (Editors)

Functional Materials

EUROMAT 99 – Volume 13

Deutsche Gesellschaft
für Materialkunde e.V.

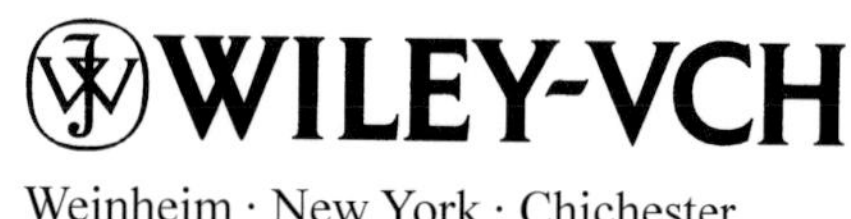

Weinheim · New York · Chichester
Brisbane · Singapore · Toronto

Editor:
Dr. K. Grassie
Philips Forschungslaboratorium
Postfach 500145
52085 Aachen
Germany

Prof. Dr. E. Teuckhoff
Siemens AG
Postfach 3240
91050 Erlangen
Germany

Prof. Dr. G. Wegner
Max-Planck-Institut für Polymerforschung
Ackermannweg 10
55128 Mainz
Germany

Prof. Dr. J. Hausselt
Forschungszentrum Karlsruhe
Postfach 3640
76201 Karlsruhe
Germany

Prof. Dr. H. Hanselka
Institut für Mechanik
Otto-von-Guericke-Universität Magdeburg
Universitätsplatz 2
39160 Magdeburg
Germany

Library of Congress Card No. applied for.

A catalogue record for this book is available from the British Library.

Deutsche Bibliothek Cataloguing-in-Publication Data:
A catalogue record for this publication is available from Die Deutsche Bibliothek
ISBN 3-527-30254-9

Printed on acid-free and chlorine-free paper.

Composition: WGV Verlagsdienstleistungen GmbH, Weinheim
Printing: betz-druck, Darmstadt
Bookbinding: Buchbinderei Osswald, Neustadt/Wstr.
Printed in the Federal Republic of Germany

Preface

Engineering progress essentially depends on the availability and the intelligent use of materials. For many key industry areas, Europe constitutes a premier place for the development of new materials and their applications. EUROMAT 99, the biannual meeting of the Federation of European Materials Societies with its 21 member societies across Europe set out to become the most comprehensive European event to demonstrate the wide range of the interdisciplinary performance of materials.

EUROMAT was essentially focused on applications of materials with high innovation potential. At the same time, fundamental approaches and processing related aspects for unconventional materials were addressed. In the frame of the 12 conference topics, 1650 papers were originally submitted to the 52 symposia. A total of 655 have been selected as oral presentation by the international group of chairpersons and were presented in 161 sessions. Further, the chairpersons have selected 65 renowned experts for keynote lectures in the frame of their symposium. Roughly 700 papers were displayed as posters.

The scope of EUROMAT was truly international. Papers originated from 57 countries. Among them the major industrial countries of the world have contributed considerably to the wealth of the programme. An overwhelming Eastern European contingent shows that there is a strong interest of these countries in international cooperation.

EUROMAT 99 represents a showcase of the competence of the European materials societies. Various European sister societies and federations act as cosponsors of the event. Joining with FEMS, they are about to establish the network MatNet in order to promote and facilitate their communication and cooperation. They have started a dialogue with the European Commission in order to discuss programme goals and priorities for maintaining Europe´s global competitiveness. In view of this promising international perspective, the European Community has agreed to sponsor EUROMAT 99 generously for which we are very grateful. EUROMAT 99 was focused to a large extent on the aims of the closing 4th Framework Programme many projects of which were presented.

EUROMAT 99 was hosted by WERKSTOFFWOCHE, a multisociety joint conference project established in Germany in 1996. Among its initiators is the Deutsche Gesellschaft für Materialkunde, one of the founding member societies of FEMS and technical organiser of this year´s EUROMAT.

EUROMAT 99 represented an outstanding success. As the President of FEMS, I would hope that it will serve as a model for future meetings, both in terms of organisation and international cooperation. I would like to extend my gratitude to the scientists, chairpersons and coordinators as well as to the various organisations and particularly to the Messe München who have made this success possible.

Dr. Paul Costa
President of the Federation of European Materials Societies

Contents

I. Materials for Information Technology

III. Polymers

IV. Sensors and Actuators

V. Adaptronics

I Materials for Information Technology

Integration of Sensor/ Actuator Films in LTCC Packages

U. Partsch *, P. Otschik *, L. Seffner *, C. Kretzschmar *, G. Reppe **
* Fraunhofer Institut für Keramische Technologien und Sinterwerkstoffe
** Radeberger Hybridelektronik GmbH

1 Abstract

The subject of the article is the integration of screen-printed sensor/actuator layers in LTCC multilayer substrates. These layers have piezoresistive (thick-film resistors) or piezoelectric (PZT) characteristics. Aspects of the material selection (characterization of thick-film resistors) and development (PZT paste) are treated. Diaphragm structures in the LTCC technique were prepared and screen printed with the appropriate functional layers. The manufactured modules were characterized.

2 Introduction

Sensors have achieved a prominent status in modern industry. Apart from the temperature sensor technique the pressure measuring technique has a special importance. Ceramic pressure sensors are characterized by high acid and base resistance, aging and temperature stability. In many cases conventional solutions are not applicable because they are too expensive or not stable or intelligent enough. The combination of Low Temperature Cofired Ceramics (LTCC) and the thick-film technique provides special advantages for small and medium-sized companies.
It was the aim of the work to combine sensors and actuators with LTCC to increase the functionality of such substrates. Therefore diaphragm structures were realized in LTCC technology.

3 Overview

3.1 LTCC

The production of the diaphragm structures follows the usual LTCC technology. Unfired LTCC layers for the ceramic base are provided with punched openings and stacked. An unpunchend LTCC layer covers the openings. Laminating and firing are the following process steps. The evenness of the diaphragm structures is strongly influenced by the production process. During the laminating process it is important to compress all areas, including the diaphragm, in the same way to avoid different shrinkage during sintering. Isostatic laminating procedures are suitable for this aim. Firing parameters also strongly influence the diaphragm configuration. Sensor or actuator layers were screen printed on the presintered diaphragm.

This technology is quite variable with regard to the dimensions of the diaphragm structures because the layers are structured with the punching tool in an unfired state. This avoids a costly finish handling. To vary the diaphragm thickness, layers with different thicknesses can be combined. Fired diaphragm thicknesses of >110 μm are realizable.

3.2 Piezoresistive Films

The characteristics of TFR's such as temperature coefficient (TCR), strain sensitivity (K-factor) and relative noise index (RNI) are defined by the conducting mechanisms. For this reason thick-film resistors are suitable as temperature, pressure, force and acceleration sensors.

Because the firing temperatures of LTCC and TFR's are practically equal the interaction of the glass constituents is probable, whereby the resistor characteristics are influenced. A thermal misfit between the TFR and the substrate material has the same effect. Therefore special attention has to be given to the selection of the strain-sensitive TFR paste with respect to the application on LTCC. Five different resistor pastes have been tested for application as pressure sensor. They were selected on the basis of the following properties: high strain sensitivity, small TCR in the application temperature range, small noise index and high aging stability.

3.3 Piezoelectric Films

The combination of the electromechanical properties of lead titanate zirconate (PZT) with the mechanical characteristics of a substrate can be used for sensors and actuators. Examples were recently described in the literature /1/. The preparation of a PZT layer is difficult because of the hindered lateral shrinkage during sintering, the PbO evaporation and the compatibility of the system substrate/electrode layer which is necessary. More recently, much progress was made with respect to the preparation of PZT layers on Al_2O_3, in particular to the development of low-sintering materials and suitable sintering technologies /2,3/. With the use of LTCC substrates one fact is added to the problems stated before. The silicate phases in the substrate can react very easily with the PbO of the PZT to lead silicate, thereby drastically decreasing the piezoelectric characteristics of the PZT /4/. Therefore the bottom electrode of such a sytem fulfills not only the function as an electrical contact but also as a barrier to avoid unwanted chemical reactions between substrate and layer. In the case of actuators, the electrode must also adhere very well to the substrate, because it transmits mechanical stresses. Attractive sensor and actuator applications are expected with the solution of these problems.

4 Results

4.1 Piezoresistive Films (TFR)

4.1.1 Film characterization

Different test structures were prepared with the selected pastes to determine the resistor characteristics (Figures 1-3).

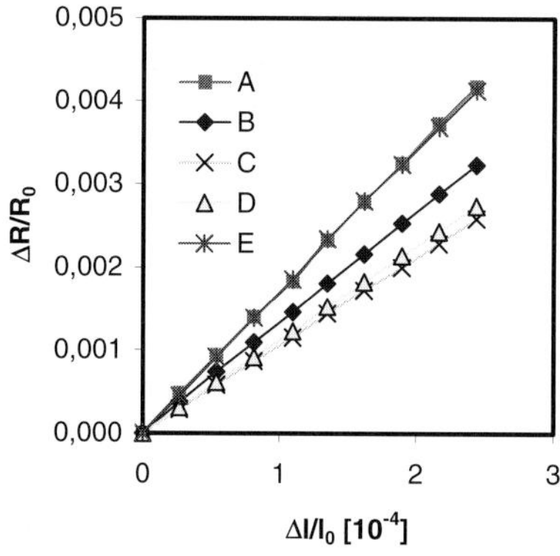

Figure 1. Strain sensitivity of the examined TFR.

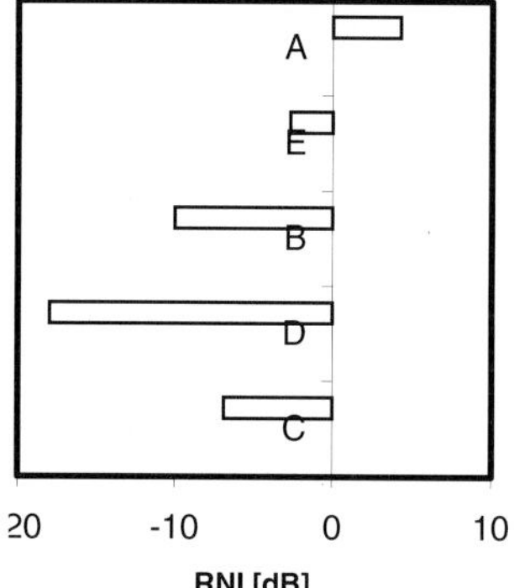

Figure 2. Relative noise index (RNI) of the examined TFR (MIL standard 308).

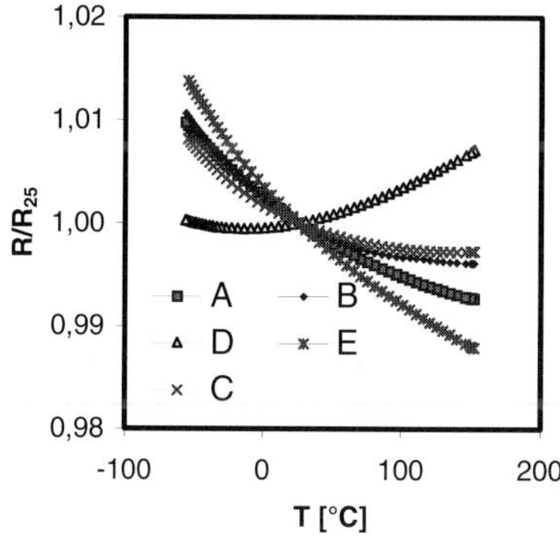

Figure 3. Thermal behaviour of resistivity of the examined TFR.

The connections between strain sensitivity and the noise index of TFR's on Al_2O_3 which were described in the literature /5/ were also acknowledged on LTCC. Thereafter a high strain sensitivity correlates with an increased noise index. These circumstances are explained by the microstructure of the TFR and the conducting mechanisms /5,6,7/.

The TFR from composition D is adapted best to LTCC concerning thermal coefficients of expansion. All other TFR's did not achieve their resistance minimum in the temperature range up to 150°C. For further investigations the resistor composition D was selected because of the best combination of properties. This paste has only an avarage strain sensitivity (K-factor 11) but a high aging stability, a very small current noise and a good TCE adjustment on LTCC.

4.2.2 Pressure sensor piezoresistive

Pressure sensors were constructed using the selected resistor paste. In each case four TFR were interconnected to a Wheatstone Bridge.

Using FEM calculations, suitable dimensions for diaphragm thickness and diameter as well as the most favorable areas for the arrangement of the TFR were found. Measurements to determine the static sensor characteristics were performed. Table 1 shows the properties of a sensor version with a nominal pressure of 10 bar.

Table 1: Properties of a sensor

P_{nom}	Nominal Pressure	bar	10
p_{burst}	Burst Pressure	bar	> 30
S	Sensitivity	mV/V*bar	> 0.3
H	Hysteresis	% FS	< 0.1
L	Linearity	% FS	< 0.25
U_{ofs}	Offset Voltage	mV/V	< 50
TC S	TC Sensitivy	% S_0/K	-0.01
TC U_{ofs}	TC Offset	% FS/K	0.05

The development of prototypes (Fig. 4) took place in cooperation with the companies Rhe (Radeberger Hybrid Electronik GmbH) and ADZ Sensortechnik GmbH Dresden.

Figure 4. Prototype of piezoresistive LTCC pressure sensor.

4.2 Piezoelectric films (PZT)

4.2.1 Film characterization

The IKTS technology /2,3/ was used for the preparation of test samples of PZT thick films. The layer thickness of the sintered PZT was 45-50 μm. Bottom and top electrodes complete the layer structure (Fig. 5).

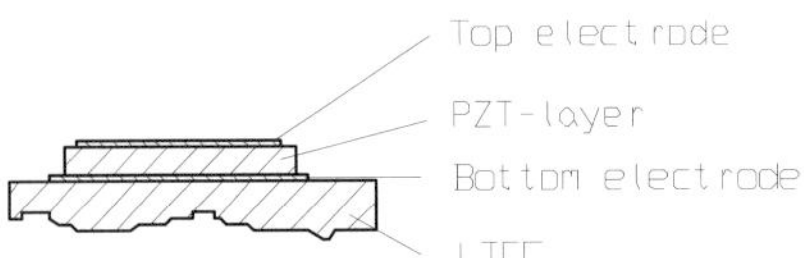

Figure 5. Layer structure of the resonant piezoelectrical LTCC pressure sensor.

With these samples the ferrouselectrical and electromechanical characteristics were measured. The realized PZT layers have a remanent polarization (P_r) of approx. 13 μC/cm^2 and a coercivity (E_c) of approx. 1,1 kV/mm.

A Pressure sensor piezoelectric

The sensors were characterized by an impedance analyzer to test the sensor characteristics. The shift of the frequency of resonance (of impedance) of the total component (layer, diaphragm, electrodes) in response to an applied pressure (Fig. 6) was measured.

The shift of the frequency of resonance amounted to 120 Hz/bar for the examined samples. Thus the resonant-piezoelectric measuring principle is suitable to detect low pressures.

B Actuator piezoelectric

The deflection of the diaphragm, which is proportional to an applied electrical field, represented in Figure 7, shows the effectiveness of the layers as an actuator.

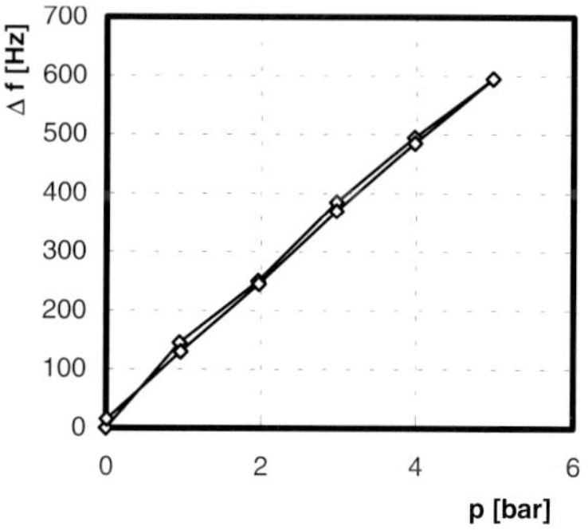

Figure 6. LTCC pressure sensor resonantly piezoelectrically. Shift of the resonant frequency.

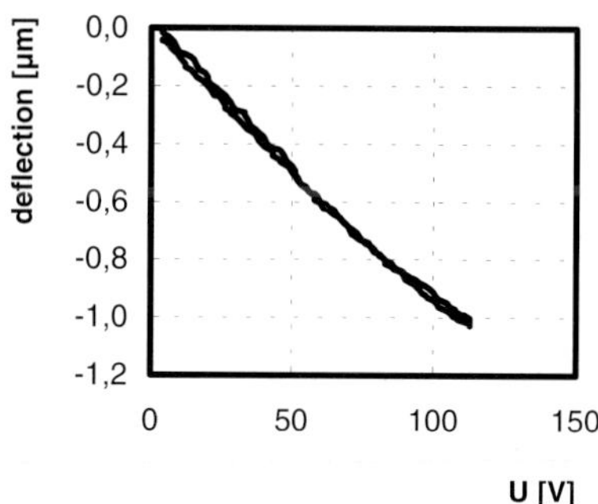

Figure 7. LTCC actuator piezoelectrically. Diaphragm deflection in a continuous electrical field.

Conceivable applications are diaphragm pumps for the active cooling of LTCC power modules.

5 Summary

The functionality of LTCC packages is increased with the presented possibilities of the integration of piezoresisitive or piezoelectric layers. Thus possibilities exist for the use of LTCC packages for sensor and actuator applications. This offers a substantial potential for the development of highly integrated, reliable and multi-functional LTCC multilayers.

The sensor/actuator ceramic modules implemented at the IKTS could be developed further as components of the Fraunhofer component system "Modular microsystem technique".

6 References

/1/ De Cicco, G.; B. Morten; Prudenziati, M. „Piezoelectric thick-film sensors Handbook of Sensors and Actuators 1, Thick Film Sensors“, Amsterdam, Lausanne, New York, Oxford, Shannon, Tokyo, Elsevier 1994

/2/ Seffner, L.; Gesemann, H.-J.; Voelker, K.: „Verfahren zur Herstellung von PZT-Schichten aus niedrig sinterndem PZT-Pulver“, Patent: DE4416245 C1 (1995)

/3/ Gesemann, H.-J.; Seffner, L.; Schönecker, A.: PZT-Dickschichten auf Al_2O_3“, Fortschrittsber. Dtsch. Ker. Ges., 10(1995)4, 213-24

/4/ Seffner, L.; Partsch, U.: „PZT-Dickschichten auf LTCC-Substraten“, Jahresbericht Fh-IKTS, Dresden 1998

/5/ Thoß, D.: „Beitrag zur Konstruktion und Technologie von Dickschichtmehrebenen-schaltungen", Dissertation, Ilmenau 1984

/6/ Pike, G. E.; Seager, C. H.: „Electrical Properties and Conduction Mechanisms of Ru-basesd Thick Film (Cermet) Resistors“, J. Appl. Phys; 12/1977

/7/ Forlani, F.; Prudenziati, M.: „Electrical Conduction by Percolation in Thick Film Resistors“, Electrocomponent Science and Technology, 3/1976

Active Media for Optical Data Processing

G.Vâle and M.Lubâne
Institute of Solid State Physics, University of Latvia, 8.Kengaraga St., Riga, LV1063, Latvia e-mail: gvale@cfi.lu.lv

1 Abstract

Optical absorption and photostimulated luminescence (PSL) have been measured in thin mono- and polycrystalline KBr and KI films doped with Tl and In. The size of grains in microstructures varied from 0.1 to 1.0 μm. Special interest is paid to the unrelaxed processes in microstructures because of their possible application as optical chips. Specifics of the photostimulated luminescence mechanisms in these systems are analysed from the viewpoint of their application in miniaturized optoelectronic and photonic devices.

PACS: 78.00; 78.40; 78.50; 78.55

Keywords: Doped alkali halide microstructures; Photostimulable storage materials; optical chips

2 Introduction

Recently X-ray storage phosphors ($BaFX:Eu^{2+}$, X=Cl, Br) have been studied extensively because they exhibit photostimulated luminescence (PSL) and offer an alternative to convencial X-ray imaging technology. It was found that powdered KBr:In also has an X-ray storage efficiency comparable to that of BaFBr:Eu [1].

We are looking for sensitive storage materials to be suitable for miniaturised optoelectronic and photonic devices for multifunctional optical data processing and transferring.

For this purpose the thin films of doped alkali halides could be very attractive photostimulable media in two aspects: as simple ionic crystals they could serve as model systems for other ionic systems, from the one hand, and also as the best physically investigated crystals in which it is possible to take advantage of unrelaxed processes to ensure high-speed and sensitive optical information storage with optical data transferring (of several terabit/s), from the other hand.

The doped micro- and nano-structures of alkali halides exhibited novel optical properties not only for technological applications but also as materials for studying physics in low dimensions.

In connection with that, it is actual to examine peculiarities of the radiation induced defect production (the storage of information) and their photostimulable recombination (the retrieval and erasure of information).

3 Experimental

In the present report the peculiarities of colour centres production and their photostimulable recombination in KBr and KI systems doped with Tl and In in concentrations of $5x10^{17}$ cm^{-3} have been investigated.

Irradiation was performed by monochromatic UV light, χ-rays (from a ^{60}Co source) and 5 MeV electrons. The F-centre concentration produced varies from 10^{10} to 10^{18} cm^{-3}, and is assumed to be proportional to the PSL intensity. Spectral measurements were made on a standard apparatus.

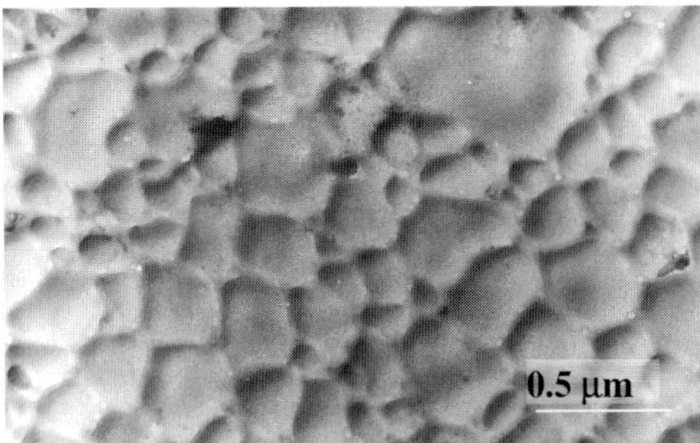

Figure 1. Structure of KBr:Tl .

Thin films, some μm thick, were produced by thermal evaporation of salts on the crystalline substrate in the vacuum of $1.0x10^{-4}$ Pa. The average grain size in polycrystalline films varied from 0.1 to 1.0 μm. The structure of films was examined by the carbon replica technique on the JEOL 100S electron microscope.

4 Physical Processes and Aspects of Application

It was found out that in doped alkali halide monocrystalline thin films like in single crystals the close genetic donor-acceptor pairs {F-centre...dopant hole-centre} are produced in a ps-time range as a result of the unrelaxed H-centre capture by the dopant ions [2].

Absorption spectra of microstructures (Figure 2) shows an additional absorption band from the high-energy side of the common F-centre absorption (shifted about 0.6-0.7 eV).

The F-centre production in microstructures also acquires an additional temperature-independent stage (Figure 3). There are observed suppressed TSL peaks related to the H-centre release (Figure 4), too. All these facts cannot be explained by the most popular opinion according to that in polycrystalline films the unrelaxed electronic excitations (electrons and holes, excitons) are able to move to the surface of the grains and, therefore, the further radiation defect production occurs at the surface [3]. It means that there must be no colour centre production in the films with fine grains (less than 1 μm). Our experimental data oppose that conclusion.

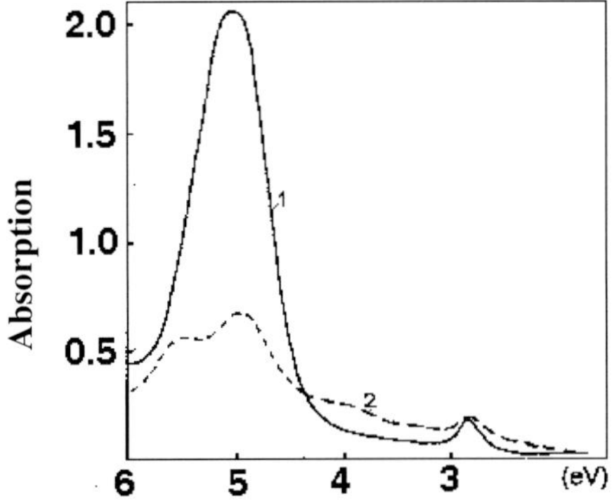

Figure 2. Absorption spectra of irradiated LiF. (1) single crystal; (2) microstructure

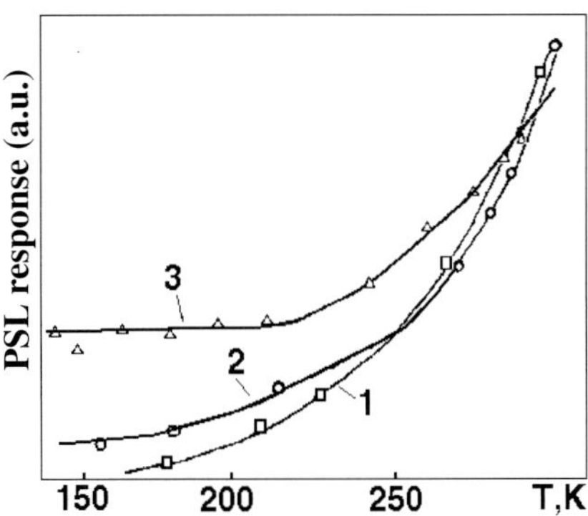

Figure 3. Temperature dependence of the PSL response for monocrystalline KBr:Tl (1) and KBr:In (2) films, and for microstructure of KBr:In (3) by X-ray excitation.

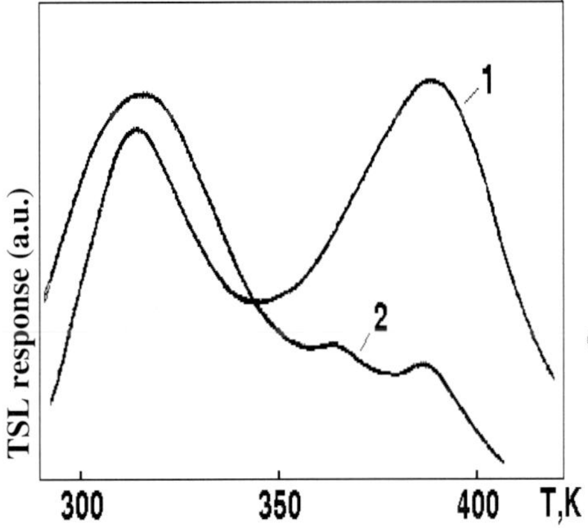

Figure 4. Thermostimulated luminescence glow curves for KBr:In monocrystalline film (1) and (2) - for microstructure after X-ray excitation at RT.

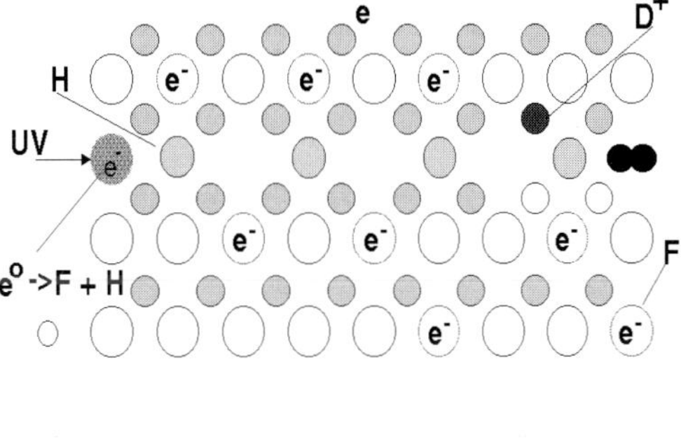

Figure 5. Schematic picture of unrelaxed H-centre moving through the F-centre channel and subsequent capture by the dopant ion.

We started from the assumption that in alkali halides the colour centres production is determinated by the primary anion exciton decay into a pair of {F,H}- centres:

$$e^{o} \rightarrow F + H^{unrel} \quad (1)$$

in the bulk of the grains and the unrelaxed H-centre is able to move to the surface and escape from it, producing in such a way in the grains of film an F-centre concentration which is greater than that of the dopant hole centres. If the F-centre concentration exceeds $5x10^{17}$ cm^{-3}, an interaction between centres begins, and a quasi F-centre chain (or a channel) is formed near the surface (Figure 5). We suppose that observed additional absorption band could be attributed to the absorption of F-centres in the chain.

This chain plays an important role in the fast temperature independent transport processes of electronic excitations and unrelaxed H-centres. It is in accord with the theoretical findings that a repulsive force between an interstitial atom and an F-centre arises if the H-centre axis is changed [4].

We also hope that the observed decorated surface picture of irradiated NaCl (Figure 6) observed by Distler [6] also confirmed our model of production the F-centre chain near the irradiated surface.

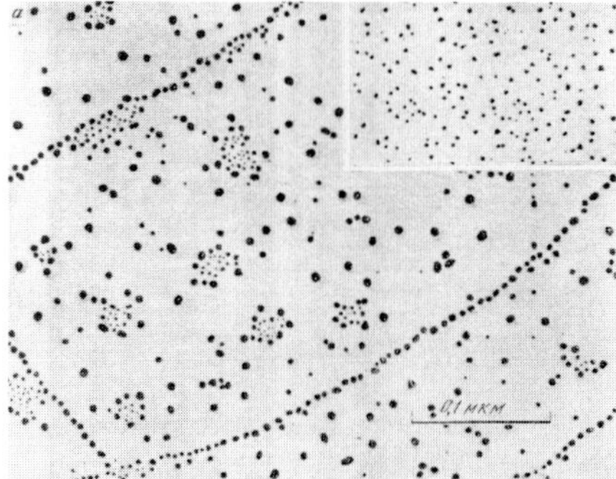

Figure 6. The surface of irradiated NaCl crystal decorated by gold [6]. At inset – unirradiated crystal

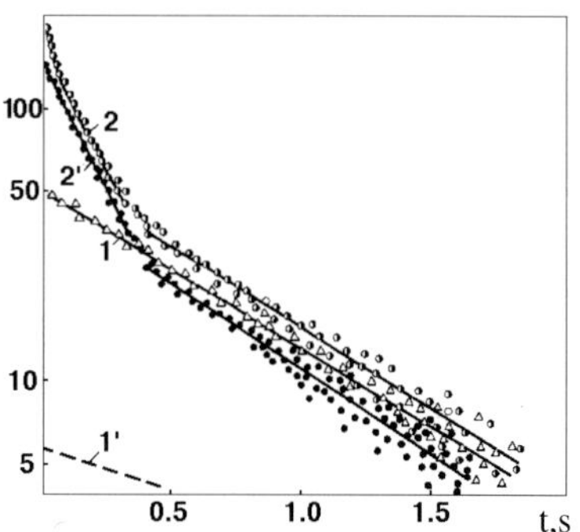

Figure 7. PSL kinetics in KBr:In systems after irradiation by UV-light at RT for monocrystalline thin film (1) and microstructure (2); (1') and (2') - stimulated at LNT.

If at the end of the path there is a dopant ion, the unrelaxed H-centre is captured by the dopant, forming the dopant hole centre. So the distance between the place of the exciton decay and the dopant hole centre production may be greater than 4-5 lattice distances, as usual in crystals.

In a similar way the F-centre chain (or the channel) affects energy transfer from the F-centre to the dopant ion. Besides the normal bleaching stage according to a simple exponential PSL decay, there appears a new, more rapid and temperature-independent stage (Figure 7). The latter could be interpreted as a fast-resonant F'-centre movement along the F-centre chain, up to the dopant hole centre where the recombination takes place with an intrinsic dopant ion luminescence:

$$h\nu_F + \{F...F\} \rightarrow \{F^*..F'\} \rightarrow \{F...F'\} \quad (2)$$

$$F' + D^{++} \rightarrow \alpha + D^* \rightarrow h\nu_{D+} + \alpha \quad (3)$$

The ends of the chains (or the channels) can be connected along the <110> directions through the squares of active medium thus forming a micro webs (optical chips) having desirable properties for multifunctional fast optical data processing and transfer with the information flux of some terabit/s (Figure 8). The variety of functional parameters is determined on the specific transport of unrelaxed electronic excitations via the F-centre chain or the channel with high quantum yield, and on the specific interaction of unrelaxed halogen atoms with the dopant in different valence and electronic states, like [5]:

$$H^{unrel} + D^* \text{ (or } D^0) \rightarrow D^+ + I \quad (4)$$

$$H^{unrel} + F...D^+ \rightarrow e^o...D^+ \rightarrow h\nu_{D+} \quad (5)$$

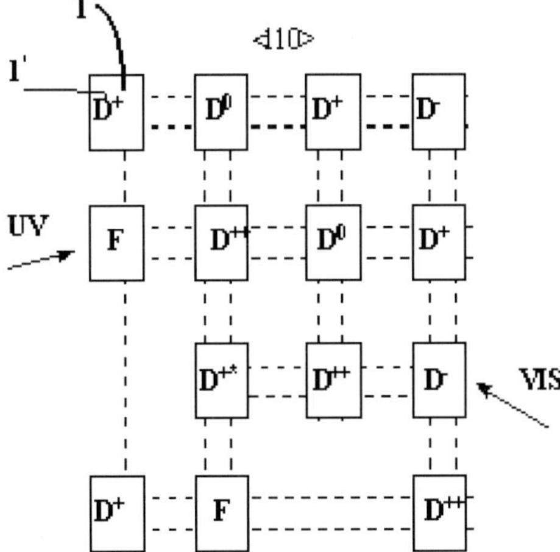

Figure 8. Schematic micro-network picture of F-centre channels along the <110> direction: - the microsquare of the active medium; - - - - the F-centre channel; D^0,D^+,D^{++} - the dopants in different valence states; D^{+*} - the dopant ion in the excited state; 1 - the optical output (optical fibre).

5 Conclusions

In the doped microstructures of alkali halides:

- An interacting F-centre chain (or a channel) is formed near the surface of the crystalline grains. The chain plays an important role in the fast movement of unrelaxed electronic excitations at any distances, and, therefore, to produce donor-acceptor pairs at these distances, too.
- The new functional abilities determined by the specific movement of unrelaxed electronic excitations via the F-centre chain or the channel with high quantum yield, on the one hand, and by the specific interaction of unrelaxed halogen atoms with the dopant in different valence and electronic states, on the other, allow to regard them as active photostimulable storage materials for high-speed optical data processing with a wide scale of functional abilities in various optical converters and storage devices, including also the optical chips.

6 References

[1] A.Kalninsh, I.Plavina and A.Tale, Nucl. Instr. and Meth.B , 1994, 84, 95.
[2] G.Vâle, Phys.stat.sol.*(b)*, 1996, 197, 293.
[3] M.Elango, Radiation Effects and Defects in Solids **,** 1994, 128 Nos.1-2, 1.
[4] A.M.Monti and E.J.Savino, Phys.stat.sol.(b), 1979 , 92, K39.
[5] G.Vâle, in Proceedings of Int.Conf. Inorganic Scintillators and Their Applications, Delft, The Netherlands, 1995, p. 449.
[6] Distler, The decorated surface of solid materials, Moskva, 1976, p.42 (in Russia).

Selected Material Aspects of GMR-Sensors used in Hard Disk Drives

Jan Marien, C. Schug, C. Waldenmaier and H.-R. Blank
IBM Deutschland Speichersysteme GmbH, Mainz

1 Introduction

Every year the increase in areal recording density of hard disk drives is about 60%. As a result, the track width becomes narrower and the transversal recording density also increases. To obtain still a high enough read signal, conventional anisotropic magneto resistive (AMR) read sensors can no longer be used at increasing recording densities. As a consequence, IBM has introduced the giant magnetoresistive (GMR) sensor, also called the spin valve device. This device consists of a stack of ultra-thin metallic layers, which exhibit a remarkable change in resistivity when the direction of an external magnetic field changes. This effect is based on the spin dependent reflectivity and the spin dependent scattering of conduction electrons. This scattering or reflection occurs at the interfaces between two ferromagnetic layers seperated by a nonmagnetic conducting spacer. Typically, the conducting spacer is made of copper and the magnetic layers are NiFe (permalloy) or Co. Used in magnetic recording applications, one layer has to follow the stray field of the media. This layer is called the free layer and due to the weak fields of 40 - 80 Oe typical of thin film media the main design issue is to make this layer magnetically as soft as possible.

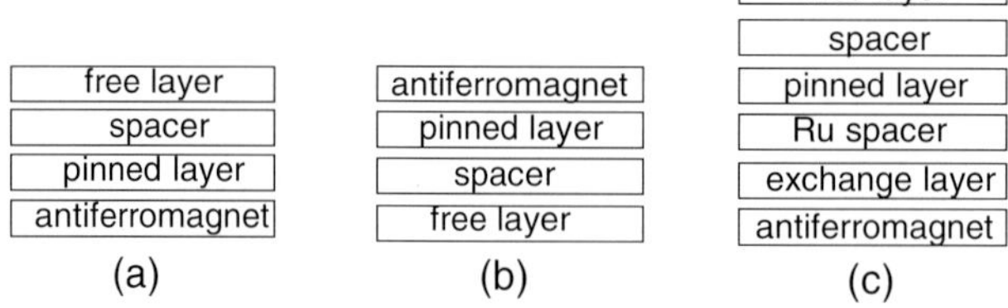

Figure 1. schematic stacking of a bottom (a) and a top spin-valve. In addition, (c) shows a bottom spin-valve stabilized with an artificial antiferromagnet

The other magnetic layer is called the reference or pinned layer and it has to be fixed in the magnetization direction regardless of the external field. This is done by using exchange bias materials, typically consisting of a very thin ferromagnetic material pinned by an antiferromagnetic oxide or metal.

Two possible device configurations are shown in Fig. 1. Fig. 1(a) shows a bottom spin-valve, where the free-layer is at the top of the metallic stack. The top spin-valve is given in Fig. 1(b), where the antiferromagnetic material is at the top of the deposition stack. Both devices are used in current read-head designs.

This paper is structured as follows. Section 2 will discuss design issues of the antiferromagnetic pinning and describe the potential antiferromagnetic materials. A special configuration of exchange coupled ferromagnetic layers can be used as an artificial antiferromagnet and help to form a more stable pinned layer. The application of X-Ray reflectometry (XRR) to measure the layer thicknesses is explained in Section 3. In section 4 the longitudinal magnetic stabilization with a hard bias film is discussed and TEM images are shown to illustrate the complex geometric design of a real device.

2 Antiferromagnetic Pinning

The choice of a proper antiferromagnet is a key design issue of a magnetic recording sensor. The ideal material would exhibit a strong coupling with the adjacent ferromagnetic layer, a low electrical conductivity to minimize shunting and should have a good corrosion resistance. However, by far the most important property is a good thermal stability as a sensor in a hard disk drive application will be used at 100°C- 120°C. Possible materials can be classified into three groups:

The first group are antiferromagnetic oxides like NiO or CoNiO. These oxides are typically insulators, and thus show no shunting. However, the magnitude of the exchange bias field is generally low and especially for NiO the blocking temperature is relatively low (190-200 °C).

The second group are localized metallic antiferromagnets with fct (CuAu-I) crystal structure and typical materials are NiMn, PtMn and PdPtMn. These materials are exceptional in that they have very high blocking temperatures and a temperature independent pinning strength up to 200°C. However, they require a high temperature anneal after deposition to transform the fcc lattice into the fct structure. During this anneal interdiffusion at the free layer/Cu or pinned layer/Cu interfaces leads to a degradation of the device performance and thus has to be minimized. In addition, although the resistivity of these alloys is high (> 200 μΩcm) compared to the other materials in the sensor (Co ~ 10 μΩcm, NiFe ~ 22 μΩcm, Cu ~ 2 μΩcm), they need to be around 20-30 nm thick to obtain good coupling and thus there is always some shunting present.

The third group is formed by itinerant metallic antiferromagnets with fcc crystal structure like FeMn, FeMnRh, IrMn, RhMn and RuMn. These materials show good coupling even at relatively low thicknesses (7nm - 10nm) which results in less shunting compared to the fct alloys. Furthermore, no anneal step is required which makes manufacturing much easier. However, these alloys generally show very poor corrosion resistance and a device design is required which carefully protects the sensor from a corrosive environment.

As only the fct metallic antiferromagnets exhibit good enough coupling up to 200 °C, special designs are necessary when using devices with either oxidic or fcc metallic antiferromagnets. In such a case, a special multilayer consisting of two ferromagnetic materials and a spacer like Ru lead to a better pinned layer. The Ru layer thickness has to be adjusted to a value where the ferromagnetic interlayer coupling between the two Co layer favours an antiparallel orientation of these two layers. This exchange coupling is known as RKKY (Rudermann, Kittel, Kasuya, Yoshida) interaction and a high external magnetic field of 450 - 650 Oe is required to flip the two layers into parallel orientation. This artificial antiferromagnetic design has the additional advantage of minimizing the remaining magnetostatic interaction of the pinned layer with the free layer.

3 Thickness Control by XRR

From the preceding it is obvious that manufacturing of spin-valve devices requires the precise control of layer thicknesses. The appropriate method to measure individual layer thicknesses on test chips is X-Ray reflectometry (XRR). At IBM, a Siemens D5000 X-Ray diffractometer equipped with a graded multilayer mirror is used for the measurement, all data evaluation procedures are performed by the software package REFS Mercury by Bede Scientific. By directly fitting the experimental XRR curve, the thicknesses, roughnesses and densities of all individual sublayers involved in a multilayer system can be extracted. Besides the monitoring of full structures from ongoing production samples also a fine tuning of individual layer thicknesses and deposition rate measurement is possible. A typical example is given in the following paragraph.

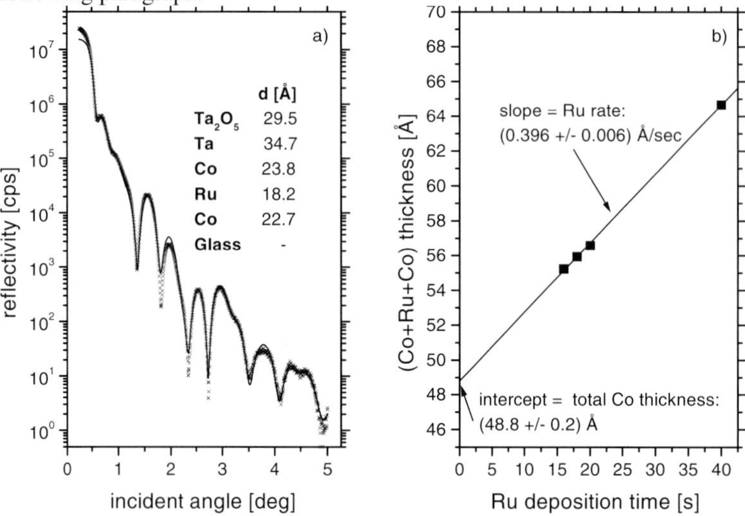

Figure 2. (a) Experimental XRR data (crosses) and fitted curve (streight line) for a Co/Ru/Co/Ta test structure, Ru deposition time 40s. (b) Plot of the total (Co+Ru+Co) thickness versus Ru deposition time.

As mentioned before spin-valve devices may utilize an artificial antiferromagnet for the stabilization of the pinned layer. To achieve best exchange coupling between the two Co layers a very precise control of both thickness and interface roughness of the Ru spacer is required. Thus, Co/Ru/Co/Ta test structures were deposited on glass subtrates to precisely determine the individual deposition rates. For this experiment 4 samples with different Ru deposition times (16s, 18s, 20s, 40s) were produced, the deposition times for all other layers remained constant. Figure 2(a) shows the experimental XRR curve for the sample with a Ru deposition time of 40s and its best fit, which was obtained by optimizing individual layer thicknesses and roughnesses. The positions of the XRR interference fringes are accurately reproduced by the fit curve. We note that for this system XRR is most sensitive to the integral (Co+Ru+Co) thickness. As a consequence, Figure 2(b) shows a plot of the resulting total (Co+Ru+Co) thicknesses as determined by XRR for all 4 samples versus the Ru deposition time. From the slope of this curve the Ru deposition rate was determined with an accuracy of better than 0.001 nm/s. The y-axis intercept is identical with the total thickness of the two Co layers. By dividing this value by the total Co deposition time, the Co deposition rate could be calculated.

4 Magnetic Stabilization and Device Geometry

The quality of the electrical read-signal of the sensor is strongly influenced by the magnetic design of the sensor. Especially spin cluster movement in the free layer is the source of electrical (Barkhausen) noise and should be avoided. A solution to this problem is a longitudinally biased free layer. In this case, the free layer is forced into a single-domain state by an external magnetostatic field.

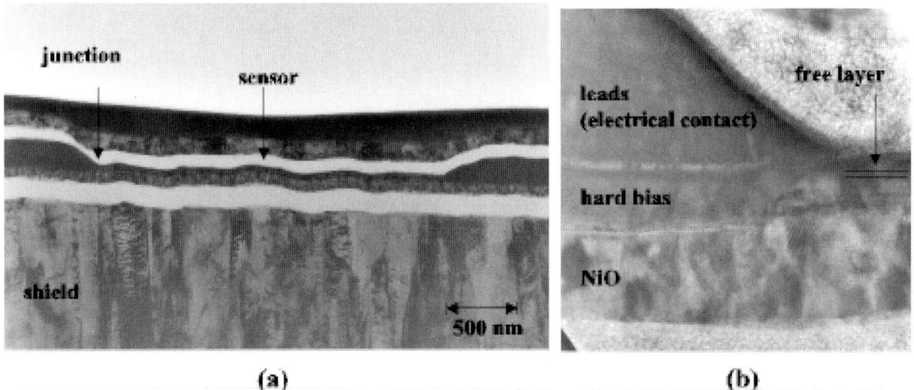

Figure 3. (a) TEM micrograph showing an overview of the sensor with the hard bias / leads contacts. (b) High magnification micrograph of the hard bias/ sensor junction. The two lines mark the region where the free layer is located.

Typically, thin layers of a ferromagnetic alloy like CoPtCr are deposited just beneath the sensor and this so-called hard bias is magnetically oriented by a proper magnetic field. However, this biasing of the free-layer has to be fine-tuned carefully. First, the layer thickness and therefore the magnetic moment has to be high enough to bias the sensor. On the other hand, the stronger the biasing the less mobile the sensor is with respect to the stray field of the media. Second, a geometry is required which directs most of the magnetic flux into the free layer. This requires a good adjustment of the corresponding photolithographic process step. A direct imaging of the resulting device geometry to control this process requires high resolution TEM imaging.

A TEM sample has been prepared by focused ion beam (FIB) and investigated in a JEM 4000EX (JEOL) high resolution TEM. Figure 3(a) shows a TEM micrograph of a cross-section of a spin-valve. The sensor is visible ion the middle of this micrograph. The hard bias junctions are located at both sides of the sensor. A higher magnification micrograph of the hard bias/sensor junction is given in figure 3(b). The two thin lines indicate roughly the location of the free layer. The tip shapeof the hard bias material at the sensor junction is clearly visible. This can be considered as the ideal geometry because the density of magnetic flux is very high and most of the flux is located at the free layer, such that the interaction of the hard bias with the pinned layer is minimized.

5 Conclusions

We have discussed different antiferromagnetic materials to build a stable and reliable pinned reference layer in a spin-valve sensor. Weak coupling oxides like NiO can be used in devices if an artificial antiferromagnet is used for further stabilization. It has been demonstrated that the XRR technique can be of use for a very precise adjustment of layer thicknesses and deposition rates. Furthermore, the design of hard biasing the sensor for magnetic stabilization has been discussed and TEM images of real structures show a well shaped hard bias / sensor junction.

6 Acknowledgement

The authors acknowledge the discussions with J. Ziermann, B. Christoffer and G. Robbel and a careful reading of the manuscript by J. Windeln.

7 References

1) M. Baibich et al., Phys. Rev. Lett. 1988, 61 (21), 2472
2) P. Grunberg, U.S. Patent 4,949,039
3) M. Lederman, IEEE Trans. Mag. 1999, 35 (2), 794
4) A. Devasahayam et al., IEEE Trans. Magn. 1999, 35 (2), 649
5) D. Heim et al., IEEE Trans. Magn. 1994, 30, 316-321
6) B. Dieny et al., J.Appl.Phys. (1991), 69 (8), 4774

Hard Disk Substrate Cleaning Using Low Energy Ions

D. Ochs, B. Cord, J. Scherer
Balzers Process Systems, Siemensstraße 100, D-63755 Alzenau, Germany

1 Abstract

We investigated the cleaning of hard disk substrates (NiP/Al, glass) using low energy ion impact. Two different sources, an inductively coupled plasma source and a DC discharge based multi cell ion source, have been studied and characterized. Ion cleaning can be used in addition to the standard heating process. The main advantages of ion impact based surface cleaning are shorter desorption times for surface contaminants and better surface cleaning compared to substrate heating.

In a first series of experiments the surface composition of NiP/Al and glass substrates was studied in an ultra high vacuum setup using X-ray photoelectron spectroscopy (XPS) as function of maximum heating temperatures.

Disk cleaning ion source characterization has been performed in a single station high vacuum test tool which is similar to a standard process station of a disk production sputtering system. Ion energies, discharge induced substrate heating up and substrate charging up was measured. Discharge induced desorption of disk contaminants was studied using a high pressure mass spectrometer. Etching rates were measured for Cr/glass layers using the optical density measurement method. Possible ion etching induced surface changing of NiP/Al and glass substrates was investigated for both sources using atomic force microscopy (AFM).

In a second step the ion source has been adapted to a disk production sputter system. The effect of surface ion treatment on the magnetics of Cr/CoCr18Pt10Ta3/C films prepared on glass and NiP/Al substrates was investigated. While a pure Ar discharge only leads to a decrease of coercivity the addition of oxygen to the discharge increases the coercivity.

Keywords: hard disk, etching, NiP/Al, glass, magnetics

2 Introduction

In recent time the surface cleanness of Hard Disk substrates as NiP/Al and glass becomes more important for Hard Disk production in order to achieve better areal densities [1]. Current substrate cleaning in production tools is done by heating the disk to about 250°C. After this procedure the amount of anorganic/organic contamination on the disk surface should be drastically reduced. Nevertheless there is some contamination left on the surface. This has to be further reduced in order to pay credit to the fact that the deposited films are getting thinner and the importance of the substrate surface – film interface increases.

To shorten cleaning process times and to achieve a better surface cleaning we investigated an additional ion bombardment process which will be performed before or after the heating process.

3 Experimental

Measurements have been done at three different experimental tools. XPS measurements have been performed at an UHV apparatus which is described in Ref. [2].

The characterization of the ion sources (ion energy, heating and charging up of substrate, CO_2 desorption, surface etch effect) has been made at a single station high vacuum test tool with a typical base pressure of about $1x10^{-6}$ mbar. Ion energies have been measured using a standard Faraday Cup set-up. For substrate temperature measurements a thermocouple was used. Partial pressure measurements have been done using a high pressure residual gas analyser (Leybold Inficon).

To investigate the influence of substrate soft etch cleaning on the magnetic performance magnetic layers were deposited onto the disks using a commercial sputter system (CIRCULUS M12) for Hard Disk production. In this tool the layer system Cr/ CoCr18Pt10Ta3/C is deposited on the substrate by magnetron sputter cathodes. Base pressure is about $5x10^{-7}$ mbar. Distance between substrate and cleaning source amounts in all cases to about 60mm.

All measurements have been performed with standard glass and NiP plated Al substrates with a diameter of 95mm and a thickness of 0.8mm as used for hard disk production. [3,4].

4 Measurements and Results

To study the efficiency of the standard heating process for cleaning the contamination of both substrate types was investigated. Using X-ray photoelectron spectroscopy (XPS) the surface composition of NiP/Al and glass substrates have been measured after heating to temperatures in the range of 20°C to 400°C. For details see Ref. [2]. The experiments showed that after heating to substrate temperatures of about 250°C a big amount of contaminating molecules is still left. For NiP/Al substrate this is about 7at% C and 20at% O, for glass substrate about 2.5at% C. Even heating to 400°C where normal surface composition is destroyed (in the case of NiP/Al substrate by diffusion of Mg from the bulk to the surface or in the case of glass substrate by ion exchange between surface and bulk) does not remove contaminants completely. There is still 15at% O and 1.5at% C (NiP/Al) and 1 at% C (glass) left on the surface.

To get a cleaner surface an additional process is needed. For this reason we investigated the cleaning efficiency of low energy ion bombardment of the surface. Two different ion sources have been investigated. The first source is a DC discharge based multi cell ion source (MCI). The working pressure of the source is $6x10^{-4}$ to $2x10^{-3}$mbar. The source consists of about 55 cells with 10mm diameter each. In each cell a DC discharge is burning. This type of source has already been used for CVD deposition of diamond like carbon. For details see ref. [5]. The other source is based on an inductively coupled plasma (ICP). The working pressure is $7x10^{-4}$ to 10^{-2}mbar. The source is operated with mid frequency (400kHz) alternating power. Sources of this type are standardly used for semiconductor applications (for example SiO removal) but with the difference that in this case the substrate is biased in order to get a higher etch effect. In our case biasing is not possible because of the necessity to etch non conductive substrates as glass simultaneously on both sides.

Measurement of the ion energies of the two different sources show major differences. In the case of the MCI source 50% of the ions have energies in the range from 0 to 40eV, the second half of ions are relatively constant distributed up to ion energies of 300 to 400eV. In contrast the ICP source produces only slow ions up to about 30eV ion energy.

Substrate ion bombardment leads to a substrate heating of about 1-2°C/s for both sources depending on the used discharge power. The floating substrate charges up to 100-300V for the MCI and 5-10V for the ICP source.

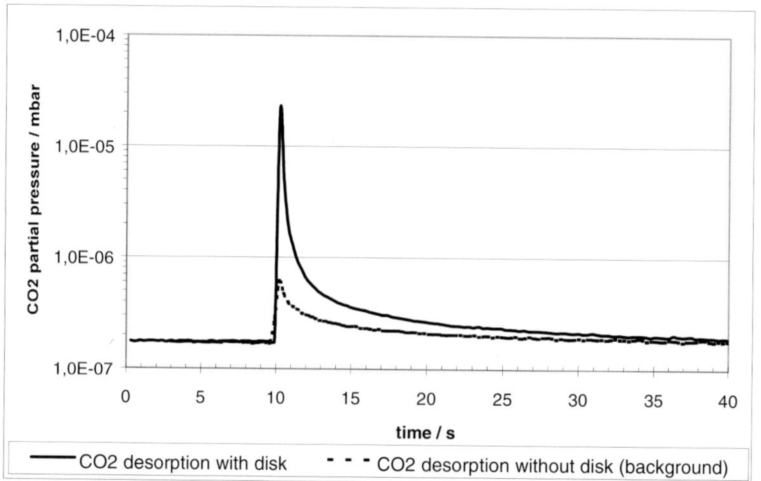

Figure 1. CO_2 partial pressure before and during surface ion bombardment using MCI source.

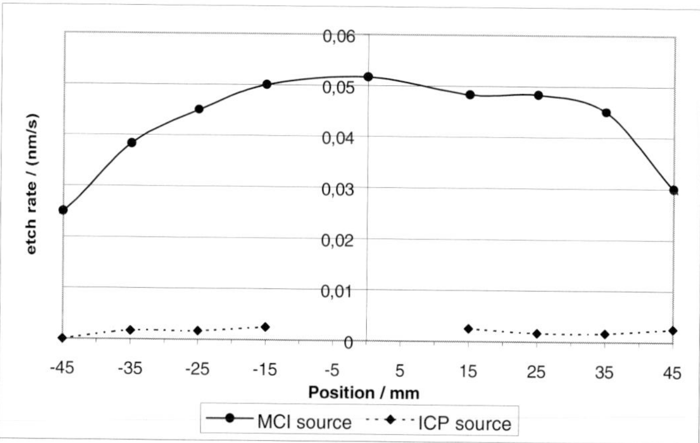

Figure 2. Ar ion etch rates for Cr films deposited on glass substrate.

CO_2 desorption from a substrate was measured before and during ion bombardment using a high pressure residual gas analyser. Fig. 1 shows the CO_2 partial pressure in the test tool chamber. The dotted line represents the measurement without disk. The curve between t= 0 and 10s corresponds to the CO_2 base pressure. At t=10s the discharge was started. With starting the discharge the CO_2 partial pressure rises quickly from about $2x10^{-7}$ mbar to $2x10^{-5}$

mbar showing that desorption starts directly. This maximum partial pressure decreases in a few seconds to about background level. After 2s the pressure is about $6x10^{-7}$ mbar. These results show that short desorption times can be realized using ion bombardment cleaning.

The Ar ion physical etching effect of both sources on Cr and C films deposited on glass was studied, too. Etch rate was calculated by measuring the change in the optical density of the films which is a linear function of the film thickness. Fig. 2 shows the etch rate for both the MCI and ICP source as function of the radial position on the Cr precoated glass disk. Both sources were operated with 100%Ar discharge gas. The etch rate using the MCI source is around 0.05 nm/s. The etch rate distribution shows a strong decrease towards the outer diameter. This is caused by the multicell structure of the source. For the ICP source we measured a much smaller etch rate (about 0.003nm/s) which corresponds to the lower ion energies of the source.

Etch rates of carbon-films deposited on glass are shown in Fig. 3. Using the MCI source with pure Ar leads only to a small carbon removal which is less than 0.01nm/s caused by the low sputter yield of carbon. In contrast adding some oxygen (15%-20%) to the discharge strongly enhances the C-removal. In the case of the MCI source this chemical etch rate amounts to 0.1nm/s, for the ICP it is even higher (0.6 nm/s). This is a sign for a higher oxygen radical production in the ICP discharge compared to the MCI source. The C removal most likely is done by the following reaction: Activated oxygen is bonded to the carbon surface. In a second step CO is desorbed from the surface [6]. For this Ar/O discharge no decrease of the etch rate towards the outer diameter of the substrate was observed.

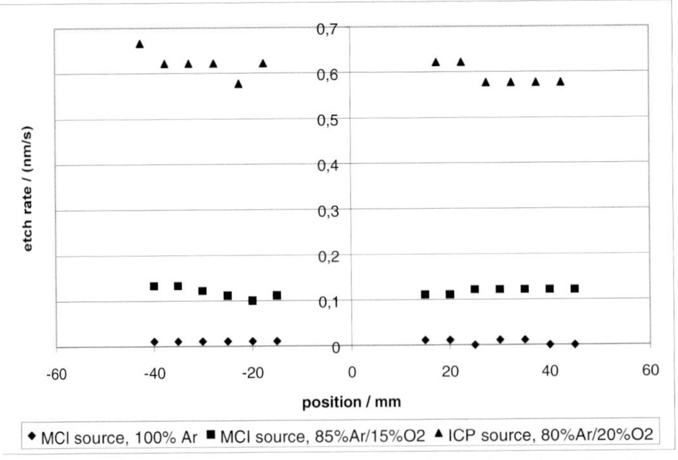

Figure 3. Comparison of physical and chemical etch rates for carbon films deposited on a glass substrate.

The influence of ion bombardment on surface roughness was studied with Atomic Force Microscopy (AFM). The measurements show that glass as well as NiP/Al surfaces are getting smoother with etch time for both sources. This effect seems to be stronger for the MCI source.

To investigate the influence of an ion etch on the magnetic properties a complete layer stack (substrate/Cr/CoCr18Pt10Ta3/C) was sputtered on NiP/Al and glass disks which have been etched before layer deposition. Coercivity was measured as function of the etch time.

Fig. 4 shows the results on NiP/Al substrates for 100%Ar and 85%Ar/15%O_2 discharge gas whereby the source was mounted after the heating stations. With 100%Ar discharge gas the coercivity decreases with etch time from 2700 Oe without etching to about 2400 Oe after 9s etch time. Using 85%Ar/15%O_2 as discharge gas coercivity first increases from 2700 Oe without etching to 2900 Oe after 2s etching and then drops to 2800 Oe after longer etch times. The etching process on glass disks shows no major change in coercivity for both gases (Ar and 85%Ar/15%O_2). Further measurements are needed to understand this dependence on the discharge gas composition.

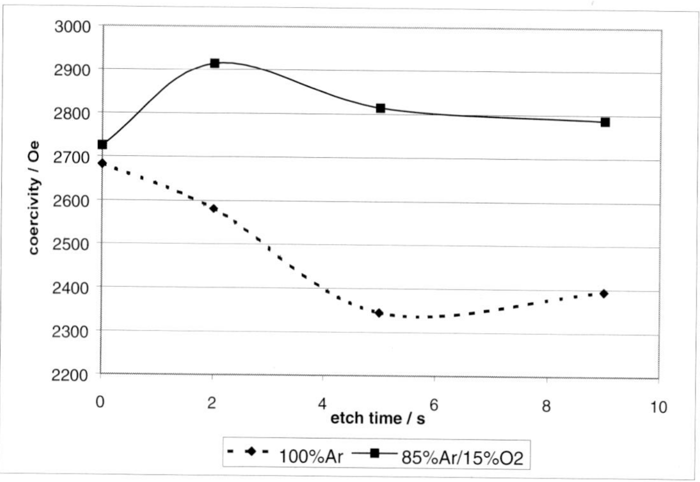

Figure 4. Coercivity of Cr/CoCr18Pt10Ta3/C layer stack on NiP/Al and glass disks with and without etching process using the MCI source at CIRCULUS M12.

5 Acknowledgements

We would like to thank Advanced Energy corporation for making the MCI source available for testing purpose.

6 References

[1] M. Takahashi, A. Kikuchi, J. Nakai, H. Shoji, J. of Magnetism and Magn. Mat. 1999, 193, p.79-84

[2] D. Ochs, S. Dieckhoff, B. Cord, ECASIA-8, Sevillia (1999), submitted to Surface and Interface Analysis

[3] K.G.Ashar, Magnetic Disk Drive Technology, IEEE Press, 1997, p163-196, ISBN 0-8703-1083-7

[4] C.D. Mee, E.D. Daniel, Magnetic Recording Technology, Mc Graw-Hill, 1995, p. 4.23-4.26, ISBN 0070412766

[5] K.L. Enisherlova, Y.A. Kontsevoi, E.N. Chervyakova, E.A. Mitrofanov, Y.P. Maiskov, Mat. Sci. and Eng. 1997, B46, p.137-140

[6] M.A. Lieberman, A.J. Lichtenberg, Principles of plasma discharges and materials processing, Wiley, New York, 1994, p. 481-485

Thermal Stability of Film Systems for Spin-Valve Applications

Andreas Paetzold, Christian Loch, Klaus Röll
FB Physik, Universität Gh Kassel, 34109 Kassel, Germany

1 Introduction

The recording density of hard disk devices for magnetic storage of information increases steadily. This is due to the introduction of giant magnetoresistive (GMR) readout heads and new developments in hard disk materials. The advantage of GMR heads is the higher sensitivity on magnetic fields. As a consequence the bit structure of the written information can be smaller. This is only one of many possible technical applications of GMR systems (more in [1]).

Spin-valve devices are a special kind of thin film systems which apply the GMR effect. A simple spin-valve system consists of two thin ferromagnetic layers (e.g. NiFe) separated by a nonferromagnetic spacer layer (e.g. Cu). The electrical resistance of the system depends on the relative magnetization directions of the two magnetic layers, where parallel alignment results in low and antiparallel in high resistance. This effect is based on the different scattering of electrons at the layer interfaces. In order to achieve the antiparallel alignment for high resistance, one layer is magnetically coupled (exchange biased) to an adjacent antiferromagnet with uniaxial anisotropy. The magnetization of the ferromagnetic layer is now 'pinned' to the preferred direction of the antiferromagnet, but the magnetization of the other ferromagnetic 'free' layer can easily follow an external magnetic field. The relative change in resistivity $\Delta R/R$ is called magnetoresistive ratio (MR) and attains values of about 5–20% depending on the quality of the spin-valve system.

Nickel oxide (NiO) is an antiferromagnetic 'pinning' material that promises some suitable characteristics for spin-valve applications. There is no shunting of the current in the film system due to the 'pinning' layer, because NiO is an insulator. Other advantages are the high corrosion resistance of NiO [2] and the relatively high Néel temperature of about 520 K. But there are also some disadvantages using NiO as 'pinning' layer. The deposition of NiO is usually done by a reactive sputtering process. This process can cause a contamination of the other metallic layers with oxygen. These oxygen atoms reduce the MR of the GMR system [3]. Another disadvantage is the strong increasing coercivity H_C with decreasing thickness of the adjacent ferromagnetic 'pinned' layer [4]. In our work we examined the influence of the O_2 flow during reactive NiO sputter deposition on the stability of the 'pinning' in NiO/NiFe bilayers under an external magnetic field. The stability of the exchange coupling in GMR spin-valve systems is a necessary condition for every technical application.

2 Experiments and Results

All thin film systems were deposited in a Leybold Z400 rf diode sputtering system with a residual gas pressure of $3*10^{-6}$ mbar. Three metal targets with 75 mm in diameter are available in the sputtering chamber. The argon and oxygen flows are controlled by two MKS mass flow controllers. In order to induce a unidirectional anisotropy into the ferromagnetic layers, a magnetic field of ≈3 kA/m is present at the substrate site during deposition. All layer systems were deposited onto glass substrates with 12 mm in diameter. The distance between target and substrate was 60 mm.

Electrically excited atoms in the rf glow discharge are emitting characteristic radiation. The optical component of the emitted plasma radiation can be examined by optical emission spectroscopy (OES). The emitted light is monochromized in a ¼ meter grating monochromator. The intensity of the monochromized light is detected by a photomultiplier tube (PMT). The output signal of the PMT, which is determined by the intensity of the emitted radiation, is recorded in dependence on the wavelength by a personal computer. The computer system also sets the gas flows and controls the stepping motor of the monochromator.

2.1 Reactive Sputter Deposition of NiO

The antiferromagnetic NiO 'pinning' layers were reactively sputtered in an Ar/O_2 gas mixture from a pure Ni target. During deposition the Ar flow was kept constant at 100 sccm, which results in a pressure of $9.7*10^{-3}$ mbar. The O_2 flow into the chamber was varied between 0 and 1.8 sccm which corresponds to O_2/Ar ratios of 0% and 1.8%. The deposition rates were determined by calibration films.

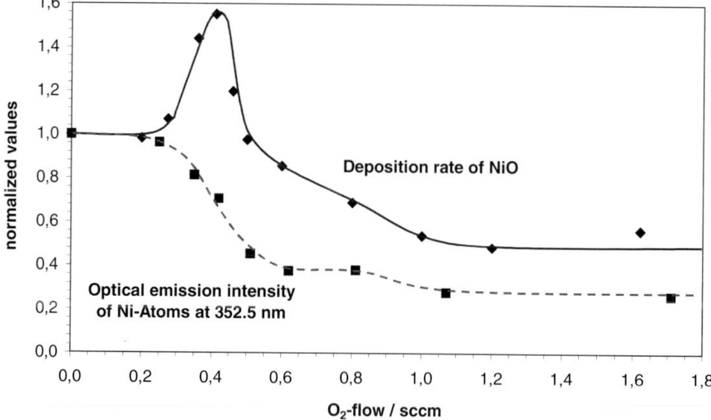

Figure 1. Deposition rate of NiO at the substrate site and plasma emission intensity at 352.5 nm of excited Ni atoms during reactive sputter deposition of NiO as a function of the oxygen flow. All values are normalized to the initial value at 0 sccm O_2 flow.

The upper values and the solid line in Fig. 1 trace the normalized deposition rate for reactively sputtered NiO at various O_2 flows. Around 0.4 sccm O_2 flow, a sharp maximum in the deposition rate is observed. This peak is caused by the formation of NiO at the substrate. Sputtered Ni atoms from the target are being oxidized at the substrate surface. The bigger NiO particles cause a faster growth of the film. A further increase in O_2 flow leads to an oxidation of the target surface. Because of the lower sputtering yield of NiO, the deposition rate drops down and reaches its saturation level at 1.1 sccm O_2 flow. The dashed line in Fig. 1 marks the normalized PMT output signal for plasma radiation at 352.5 nm of excited Ni atoms measured by OES for increasing O_2 flow. The OES signal provides information about the relative amount of Ni atoms in the glow discharge.

Comparing the two curves points out that the maximum in the deposition rate is not caused by an increasing material flow of sputtered Ni atoms from the target. The increasing deposition rate can only be explained by assuming the formation of NiO just at the substrate surface. Hence, the observation of the optical emission spectroscopy can be a useful instrument to get information about the sputtering conditions at the target surface (see also [5]).

2.2 NiO/NiFe systems

The stability of the exchange coupling in NiO/NiFe bilayers was investigated. Therefore glass/NiO(30 nm)/NiFe(4 nm)/Cu(10 nm) layer systems were deposited as described above. The magnetic field at the substrate site points parallel to the substrate plane and induces the unidirectional anisotropy into the magnetic layers. The 10 nm Cu capping layer should prevent the oxidation of the 'pinned' NiFe layer at elevated temperatures.

A. Exchange Biasfields

After deposition the hysteresis loops (parallel to the easy axis) of the NiO/NiFe/Cu systems were recorded at room temperature by using a vibrating sample magnetometer (VSM). The hysteresis loops of the exchange coupled NiFe layers show a shift from zero field axis. The corresponding exchange biasfields H_{EB} and the coercivity H_C were measured as a function of O_2 flow during reactive NiO deposition. The resulting data of the VSM measurements after deposition are shown in Fig. 2 (dashed lines). A maximum in H_{EB} and a local minimum in H_C appears between 0.6 and 0.8 sccm.

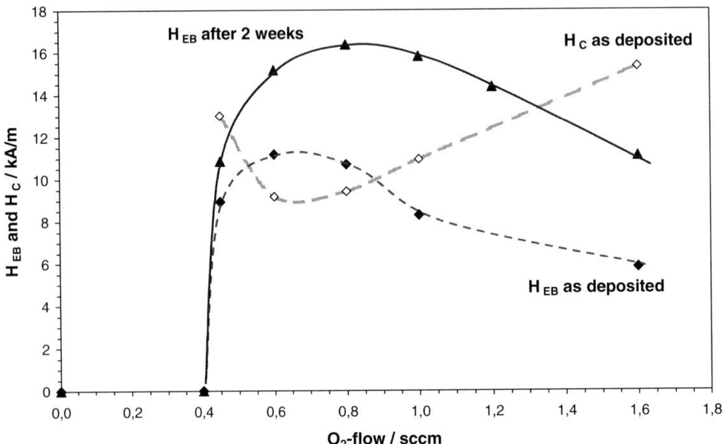

Figure 2. Results of the VSM measurements of H_{EB} and H_C as a function of O_2 flow during NiO deposition. The dashed lines represent the values just after deposition and the solid line the biasfields two weeks later.

Two weeks later the VSM measurements were repeated at the same samples and a surprising result came out. Figure 2 (solid line) shows the exchange biasfields H_{EB} of the NiO/NiFe/Cu systems two weeks after deposition. During this time the samples have been stored in air and at room temperature. The observed biasfields H_{EB} were much higher than just after the deposition while the coercivity H_C of the NiFe layers has decreased about 10% in every system.

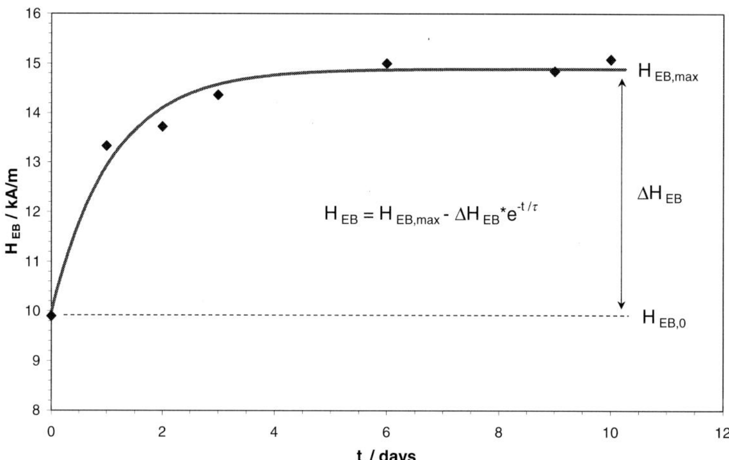

Figure 3. Development of the exchange biasfield H_{EB} after deposition in a glass/NiO(50 nm)/ NiFe(4 nm)/Cu(2.5 nm)/NiFe(4nm) spin-valve system.

Further investigations of this surprising effect were made on glass/NiO(50 nm)/NiFe(4 nm)/Cu(2.5 nm)/NiFe(4 nm) spin-valve systems. The exchange bias shift of the 'pinned' layer hysteresis loop was measured. Figure 3 shows the development of H_{EB} as a function of time t after deposition. Starting from an initial value $H_{EB,0}$ just after deposition the exchange bias field H_{EB} moves toward a saturation value $H_{EB,max}$. The increasing biasfield can be approximated by an exponential law denoted in Fig. 3 with a time constant τ.

Assuming polycrystalline NiO 'pinning' layers an explanation for this curious effect could be the delayed alignment of randomly oriented NiO grains which are being directed by the magnetic moment of the adjacent NiFe layer. But in contrast to this suggestion we found no subsequent increase of H_{EB} in NiO/Co systems after deposition.

B. Stability of the Exchange Coupling

The stability of the antiferromagnetic exchange coupling ('pinning') under an external magnetic field was investigated in NiO(30 nm)/NiFe(4 nm)/Cu(10 nm) systems. The O_2 flow during reactive NiO sputtering was varied. For these measurements a magnetic field of 4.7 kA/m was applied antiparallel to the direction of the intrinsic magnetic moment m_0 of the 'pinned' NiFe layer. Then the change in the magnetic moment m (parallel to the easy axis) of the NiFe was measured as a function of time t. In order to accelerate the "aging" process all these measurements were made at elevated temperature (85 °C) and in a N_2 atmosphere to prevent the oxidation of the NiFe.

Figure 4 displays the reversal of the normalized magnetic moment m/m_0 as a function of time t for the NiO/NiFe/Cu layer systems sputtered at different O_2 flows. Starting from the initial direction ($m = +m_0$) the resulting magnetic moment turns around until it is totally reversed ($m = -m_0$) and points into the direction of the external field with the same magnitude m_0. The measured data can be approximated by an exponential law of the form given in Fig. 4 with a time constant τ and a "form factor" β. The curves show that there is only a small range in O_2 flow around 0.6 and 0.8 sccm where the stability of the 'pinning' is high. The time constant τ of these two systems is about 200 times larger than for the other systems. This clearly shows that the stability of the 'pinning' in NiO/NiFe systems depends strongly on the deposition process of the NiO. Further characterization (stoichiometry, surface roughness, crystallography etc.) of the NiO 'pinning' layer is needed to find the origin for this behavior.

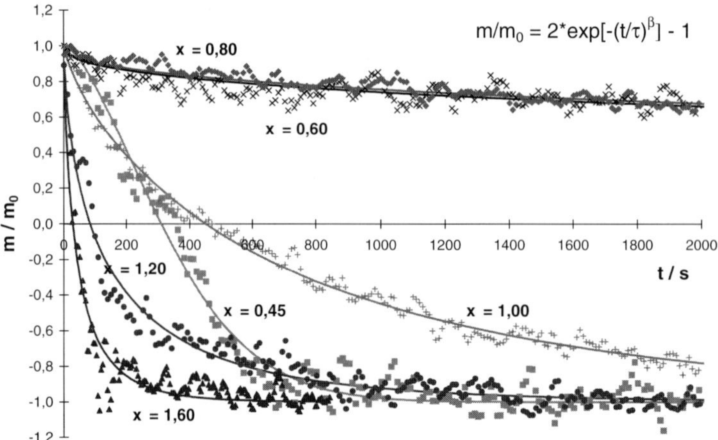

Figure 4. Reversal of the 'pinning' direction in NiO/NiFe/Cu systems as a function of time caused by an external magnetic field at elevated temperature (85 °C). The parameter x denotes the O_2/Ar flow ratio in % during reactive NiO sputter deposition.

The hysteresis loop after the reversal process shows that the NiFe layer is then 'pinned' to the opposite direction. The exchange biasfield H_{EB} changes its sign from $+H_{EB}$ to $-H_{EB}$ by keeping the magnitude. The original state of the 'pinning' direction can be restored by applying the external field in the opposite direction. This is an evidence for the reversibility of the process. The observed changes in the magnetic moment is only a magnetic reorientation of the 'pinned' system.

3 Conclusion

Antiferromagnetic NiO is formed in reactive sputtering process at very low oxygen flows. Only 0.45% O_2 in the Ar working gas is needed to achieve exchange coupling at the chosen sputtering power. The optical emission intensity of sputtered Ni atoms indicates that the target surface is not completely oxidized at this O_2 flow rate.

Our experiments show that in NiO/NiFe systems the exchange biasfield H_{EB} reaches its maximum and saturation level only after several days. The measured relative growth ΔH_{EB} /

$H_{EB,0}$ was up to 90% after 6 days of storage in air and at room temperature. This effect could not be observed in NiO/Co bilayers. The maximum H_{EB} was achieved by a reactive sputter deposition of NiO when the Ni target is between metallic and compound sputtering mode.

Also the observed stability of the 'pinning' in NiO/NiFe bilayers in an external magnetic field depends on the O_2 flow during reactive sputter deposition of the antiferro-magnetic NiO layer. There is only a small range (near to the target transition point from metallic to compound mode) where the 'pinning' direction of the adjacent NiFe layer exhibits good resistance against a manipulation by an external magnetic field. Simultaneously the coercivity H_C of the 'pinned' NiFe layer passes a local minimum. Higher biasfields H_{EB} with smaller H_C seem to be desirable for better stability of the 'pinning' against magnetic fields. Our experiments show that H_{EB} and H_C of the 'pinned' layer are strongly depending on the O_2 flow during reactive NiO deposition.

4 References

[1] J. M. Daughton, J. Magn. Magn. Mater. 1999, 192, 334–342.
[2] S. L. Burkett et al., J. Appl. Phys. 1997, 81, 4912–4914.
[3] W. F. Egelhoff et al., J. Appl. Phys. 1996, 79, 5277–5281.
[4] D.-H. Han et al., J. Appl. Phys. 1997, 81, 4996–4998.
[5] C. D. Tsiogas et al., Vacuum 1994, 45, 1181–1186.

Magnetic Interactions in CoCr/Cr Thin Films Sputtered with Different Gases

H. Z. Jia / J. Veldeman / M. Burgelman
University of Gent, Electronics and Information Systems (ELIS), B-9000 Gent, Belgium

Abstract

Magnetic properties are studied within the thin $Co_{88}Cr_{12}$/Cr films sputtered either with pure argon gas or with a helium-argon(10:3) gas mixture. Through comparison of coercivities, coercive squarenesses, $\Delta M(H)$ and SFD_{dc} plots of these thin films, the effects of variable underlayer and magnetic layer thickness are studied. There is a strong correlation between magnetic interactions in films and sputter gas types. These results indicate the crystal texture of the thin films sputtered in different types of gas is different. With XRD measurement, the crystal orientation ratio in the films sputtered in pure argon is different from that in the gas mixture. AFM images prove that the thin films sputtered in the mixture have smaller grain sizes than those sputtered in pure argon.

Keywords: Magnetic films, DC magnetron sputtering, AFM topography, Magnetic measurement, Cobalt Chromium, PET substrate

1 Introduction

Our previous work [1,2,3] has shown that a mixture of He and Ar sputter gas is beneficial for the magnetic properties of $Co_{88}Cr_{12}$/Cr films on a PET substrate. In our DC magnetron sputter equipment, we have explored a quite high pressure range with both pure argon and the gas mixture. In this work, two kinds of sputter gas are compared: pure argon and an argon-helium (3:10) gas mixture, especially their effect on the magnetic properties and on the microstructure of thin films. This ratio of the mixture has been found to be optimum elsewhere [1]. The nucleation in the films sputtered with the mixture is expected to be different. The mobility of the sputtered atoms on the substrate can be improved, which will make films with better magnetic properties.

2 Experimental

A pure Cr target is used for sputtering the underlayer, and a CoCr target with 12 at. % of Cr for the magnetic layer. The distance between the target and the substrate is 5 cm. The thickness of PET substrate is 175 μm, and its roughness is 5nm Ra. For all layers, the sputtering current is 1 A and the substrate is electrically floating. For all Cr underlayers, the sputter pressure is 50 mTorr, independent of the sputter gas type. Both the Cr underlayer and the CoCr magnetic layer of thin films are sputtered with either pure argon or the gas mixture (Ar: He=3:10). There are similar deposition rates, which are about 2 to 3 nm/s, for both types of sputter gas around 50 mTorr.

To study the magnetic properties of the films, the hysteresis curves, and the normalized isothermal (I_r) and the normalized DC demagnetization (I_d) remanence curves are measured with a vibrating sample magnetometer (vector VSM). The intrinsic coercivity, H_c and the remanence coercivity H_r are defined by respectively $M(H_c) = 0$ and $I_d(H_r) = 0$. The coercivity squareness S^* is defined as $S^* = H_c/H_r$.

The definition of both remanence curves has been explained elsewhere[3]. Afterwards, the switching field distribution, SFD_{dc}, is evaluated from the first differential of the DCD remanence curve (I_d). Magnetic interactions are characterized by $\Delta M(H)$:

$$\Delta M(H) = -I_d(H) - (1 - 2I_r(H)) \tag{1}$$

Positive values of $\Delta M(H)$ are due to the stabilizing interactions (exchange coupling) which promote magnetization of the sample. Conversely, negative values of $\Delta M(H)$ are due to demagnetizing interactions between grains (dipolar coupling). When $\Delta M(H)$ is equal to zero, the function is an expression of Wohlfarth relation [4], which holds for samples in the absence of interactions.

The topography, the grain size and the roughness of the films are studied with contact AFM. The crystal structure of the films is measured with an X-ray diffraction equipment.

3 Results and Discussion

3.1 Influence of CoCr magnetic thickness on magnetic properties

Fig.1 illustrates both H_c and H_r as a function of the sputter time of magnetic layers either sputtered with pure argon or with the gas mixture. The sputter time of every Cr underlayer is 60s. The sputter pressure of all layers is 50 mTorr whatever the sputter gas is. For both types of sputter gas, there exists an optimal thickness of magnetic layers though the optimum is different. When the thickness of CoCr layer sputtered in pure argon is too high, the coercive force goes down suddenly. But in the gas mixture, even when the film is thicker than the thickest film sputtered in pure argon, the film still has a larger coercive force. This is the case for H_r, too. Around the optimum thickness, H_r of the films sputtered in gas mixture is larger than that sputtered in pure argon.

From the $\Delta M(H)$ plots of the films sputtered in pure argon, as the thickness of CoCr layer increases, the films initially present a strong dipolar coupling and a weak exchange coupling and turn to exchange coupling afterwards. This is consistent with the literature [4]. The thinnest film, compared with the others, has the strongest dipolar coupling. The peak of the SFD_{dc} of the thickest film is the sharpest and has the largest value. The peak height values of SFD_{dc} go down monotonously as the thickness of the films decreases. The films become more dense as the thickness of magnetic layer is increasing. The density of thin films influences the magnetic interactions [5,6]. Thicker films show a more exchange coupled character than a dipolar coupled character.

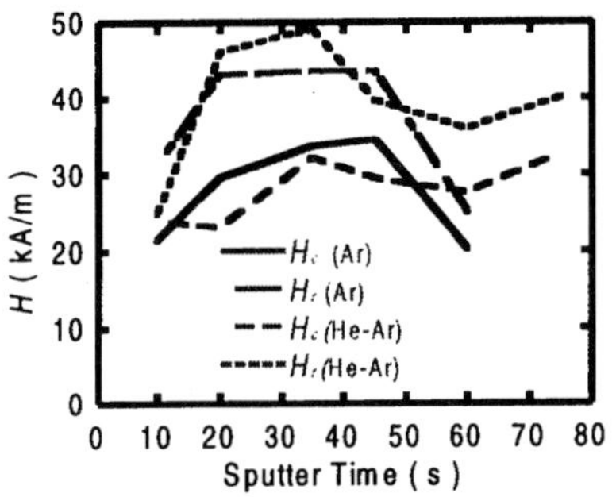

Figure 1. The coercive force varied with the sputtering time of CoCr magnetic layer

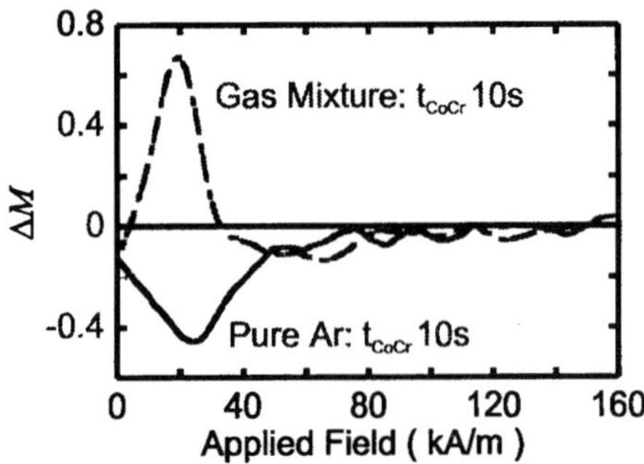

Figure 2. Comparison of ΔM plots of two films with thin magnetic layer sputtered with different gasses

Both the $\Delta M(H)$ plots and the SFD_{dc} of the thin films sputtered in the gas mixture are more complicated than those sputtered in pure argon. From the $\Delta M(H)$ plots, the thinnest film has a rather strong exchange coupling which is quite different from the film sputtered in pure argon (Fig.2). However, its coercive force is similar with that of the second thinnest film (Fig.1) which has the smallest exchange coupling. The nucleation at the initial state of thin films sputtered in the gas mixture can be different from that sputtered in pure argon. Except for the thinnest film, the minimum value of the $\Delta M(H)$ plots tends to increase as the thickness increases. The films sputtered in pure argon show a similar behavior. But the peak value of SFD_{dc} sputtered in the gas mixture is irregular, compared with the films sputtered in pure argon (see above paragraph) even if the thinnest film was not taken into account. These effects lead to the conclusion that the microstructure of the films sputtered in the gas mixture is different from those sputtered in pure argon.

The coercive squareness, S^*, has been studied as a function of the thickness of CoCr layers. For both types of sputter gas, S^* becomes larger as the thickness of CoCr layers increases. But the thinnest film sputtered in the gas mixture has the largest squareness, which consists with the largest exchange coupling indicated by $\Delta M(H)$ plots in Fig.2.

3.2 Influence of Cr underlayer thickness on magnetic properties

Fig.3 illustrates the coercive forces of the films, sputtered either with pure argon or with a gas mixture, as a function of the thickness of Cr underlayer. The sputter pressure of pure argon is 50 mTorr and 55 mTorr for the mixture. The sputter time of magnetic layer is 45s for pure argon and 40s for the mixture. The coercive force is minimum when there is no Cr underlayer for both cases. Then, it steeply goes up as the Cr underlayer becomes thicker until the films reach their maximal H_c, 37 kA/m for pure argon and 39 kA/m for the gas mixture. After that, the coercivity goes down slightly. The two curves almost coincide for small thickness of the Cr underlayer. For thick underlayers, slight differences occur. The remanence coercive force H_r of the thin films sputtered in the gas mixture is substantially larger than that sputtered in pure argon when the Cr underlayer is thick.

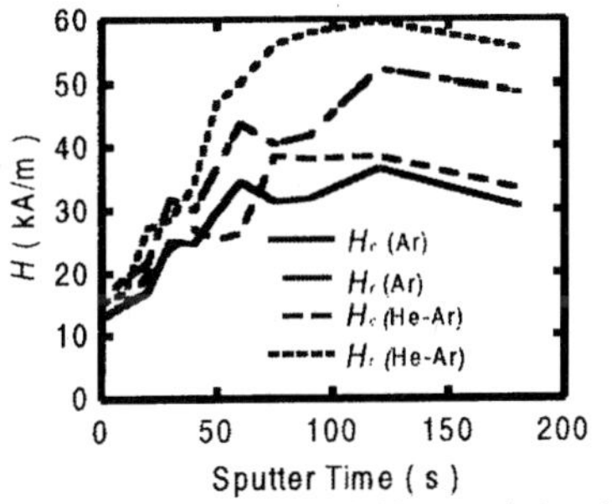

Figure 3. The coercivities varied with the sputtering time of Cr underlayer

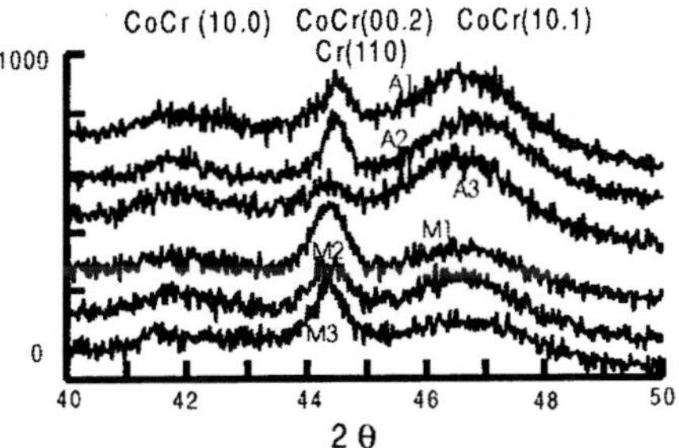

Figure 4. XRD measurement of six thin films in Table 1

According to the $\Delta M(H)$ plots, the exchange coupling in the films decreases and the dipolar coupling increases as the Cr underlayer is becoming thicker whatever the sputter gas is. This is consistent with the literature [4]. This effect is more pronounced in the films sputtered in the gas mixture than in pure Ar. The films sputtered in the mixture have a smaller exchange coupling and a larger dipolar coupling than those sputtered in pure argon. This effect is in agreement with the remanence coercive force difference in Fig.3.

For both types of sputter gas, the peak heights of SFD_{dc} plots go down when the underlayer thickness increases. If the underlayer is thick enough, the peak height is smaller and the full-width-of- half-heights of the peak is wider when the films are sputtered in the gas mixture than in pure argon.

As the thickness of underlayers increases, S^* tends to decrease. This is strictly the case for the thin films sputtered in pure argon. However, the minimum of S^* of the film sputtered in the gas mixture occurs when the sputter time is around 50 to 60s. These thin films with the minimum coercive squareness can have the maximum signal-to-noise ratio [7]. The thin films sputtered in the gas mixture have smaller coercive squareness than those sputtered in pure argon. The signal-to-noise ratio in the thin films sputtered with the gas mixture can be improved.

The crystal structure of the thin films is studied with XRD, for varying underlayer thickness and sputter gas (Fig.4). The peaks at 41.8°, 44.2° and 47.6° are identified as

$(10.0)_{CoCr}$, $(110)_{Cr}$ or $(00.2)_{CoCr}$ and $(10.1)_{CoCr}$, respectively. Table 1 gives the normalized peak intensity of XRD measurements; hereby, the sum of the intensities of these three peaks is normalized to 1. Since the peak intensity of $(110)_{Cr}$ or $(00.2)_{CoCr}$ does not increase when the thickness of the magnetic layer decreases, this peak is mainly $(00.2)_{CoCr}$. For the peak of $(10.0)_{CoCr}$, the normalized intensities are similar and independent of sputter gas types. However, for the films sputtered in the gas mixture, the peak intensities of $(10.1)_{CoCr}$ are smaller than those of $(110)_{Cr}$ or $(00.2)_{CoCr}$. But, for the films sputtered in pure argon, the trends are opposite. It indicates that the crystal orientation ratio inside thin films has been changed by the sputter gas type.

Table 1. Magnetic properties of six films and their normalized peak intensities in XRD measurement

Films	Sputter gas	Sputter time (Cr), s	H_c kA/m	CoCr (10.0)	CoCr(00.2) Cr(110)	CoCr (10.1)
M1	He-Ar	120	39	0.22	0.51	0.27
M2	He-Ar	75	39	0.24	0.41	0.35
M3	He-Ar	30	24	0.21	0.51	0.38
A1	Ar	120	37	0.22	0.37	0.41
A2	Ar	75	31	0.20	0.39	0.41
A3	Ar	30	25	0.24	0.31	0.45

3.3 AFM measurements and the ion flux on the substrate

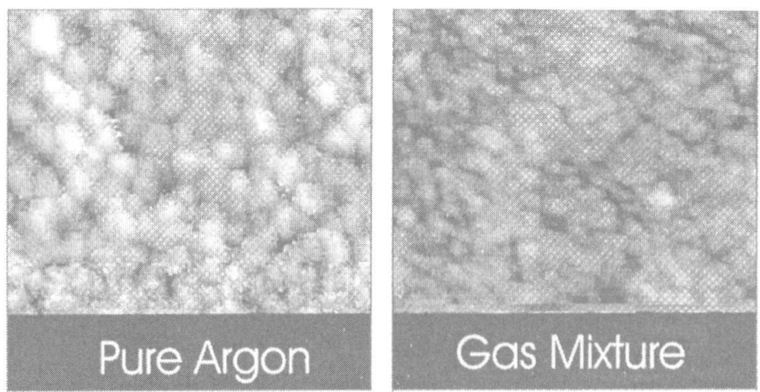

Figure 5. AFM topography of films A1 and M1; the area of every film is 1000 nm × 1000 nm

Fig.5 shows AFM images of two thin films A1 and M1. The thin films sputtered in pure argon have larger grain sizes than those sputtered in the gas mixture. We showed [1] that the presence of He enhances the ion current. The ion flux sputtered with the gas mixture is 3 to 5 times larger than that with pure argon. High flux of ion bombardment on the substrate can influence the nucleation in films. This is consistent with the result of section 3.1 (Fig.2). Thus, thin films sputtered in the gas mixture have smaller grain sizes (Fig.5) than those sputtered in pure argon. Grain size differences produce the differences of the exchange coupling and dipolar coupling between the thin films sputtered in different types of gas (see ΔM plots in 3.1 and 3.2).

4 Conclusions

The magnetic properties of thin CoCr/Cr films sputtered with pure argon and the gas mixture have been investigated by means of the measurements of coercive forces and remanence curves. $\Delta M(H)$ and SFD_{dc} plots have been applied to study the interactions in thin films. The exchange and dipolar couplings in films are quite dependent on the type of sputter gas. These measurements have shown the distinguished differences in interactions between the thin films sputtered in pure argon and in the mixture. According to AFM images and XRD measurements, the interaction differences are originated from grain size and crystal texture differences. Thin films sputtered in pure argon have larger grain sizes than those sputtered in the gas mixture. High flux of ion bombardment on the substrate with the gas mixture as a sputter gas influences the nucleation in films, thus produces small size of grains and adjusts the crystal orientation ratio.

5 References

[1] H. Jia, J. Veldeman and M. Burgelman, *J. Magn. Magn. Mat.*, **193(1-3)**, 124-127 (1999)

[2] J. Veldeman, H. Jia and M. Burgelman, *J. Magn. Magn. Mat.*, **193(1-3)**, 128-131 (1999)

[3] Huizong Jia, Joeri Veldeman and Marc Burgelman, "AFM Study of CoCr/Cr Thin Films Sputtered on PET Substrate", *ECOSS99*, Viena, Austria, 21-24 Sept. 1999

[4] P.I. Mayo, K. O'Grady, R.W. Chantrell, et al., *J. Magn. Magn. Mat.*, **95** , 109-117 (1991)

[5] A. Werner, H. Hibst and H. Mannsperger *IEEE Trans. Magn.* **26(1)**, 115-117 (1990)

[6] B.R. Acharya, E.N. Abarra, G.N. Philips, et al., *IEEE Trans. Magn.*, **34(4)**, 1594-1596 (1998)

[7] I.L. Sanders, J.K. Howard, S.E. Lambert and T. Yogi, *J. Appl. Phys.*, **65(3)**, 1234-1237 (1989)

Ti-6Al-4V with Laser Embedded SiC Particles: An Electron Microscopy Study

H. Vrenken, B.J. Kooi, A.B. Kloosterman, J.Th.M De Hosson
Dept. Applied Physics, Materials Science Centre and Netherlands Institute for metals, Research, University of Groningen, Nijenborgh 4, 9747AG Groningen, The Netherlands

1 Introduction

Titanium exhibits several excellent properties, e.g. a high strength-to-weight ratio and an excellent corrosion resistance[1]. However, the tribological properties of Ti are relatively poor. In order to improve the wear resistance of Ti, SiC particles were embedded in a top layer (about 600 μm thick) of Ti-6Al-4V employing a high-power laser. The wear rate for abrasive wear diminished by a factor 7 for the SiC embedded Ti-6Al-4V compared to the untreated material[2]. Apparently the SiC particles are bonded well to the Ti matrix contributing to the improved wear resistance. The purpose of the present work is to provide a description of the microstructure of the top layer of Ti-6Al-4V with embedded SiC particles derived on the basis of SEM, as well as conventional, analytical and high-resolution TEM. Because of the high affinity of Ti for C an exothermic reaction SiC+Ti→TiC+Si occurs during laser processing. Particularly, the influence of Si that is released during this reaction on the microstructure formed is scrutinized.

2 Experimental

Single crystal 6H-SiC particles with a typical diameter of 80 μm were embedded in Ti-6Al-4V by employing a 2 kW continuous wave Rofin Sinar Nd-YAG laser. Details about the laser process with controlled injection of particles in the melt pool are given in Ref.2. Specimens for TEM were prepared by grinding, dimpling and ion-milling round 3 mm discs to electron transparency. For TEM a JEOL 4000EX/II operating at 400 kV with a point-to-point resolution of 0.17 nm equipped with an on-axis Gatan 666 Parallel Electron Energy Loss Spectrometer located in Groningen was used. For Analytical TEM a JEOL 200 CX operating at 200 kV equipped with a Gatan 666 PEELS spectrometer as well as two Kevex system 8000 Energy Dispersive X-ray Spectroscopy (EDS) detectors located at the NCEM of the Lawrence Berkeley Laboratory (USA) was used. The EDS detectors comprise a high-angle detector and an ultra-thin window detector with a resolution of 109 eV for FKα radiation allowing quantitative analysis for elements with $Z \geq 6$. Although the used quantification procedures are standard-less for both PEELS and EDS, the procedures for quantification were calibrated internally on the embedded SiC particles. This allowed an accurate quantification of the chemical composition of TiC.

3 Results and Discussion

3.1 TiC reaction layers and Si rejection into the melt pool

An SEM image of the top part of the Ti-6Al-4V melt pool in which SiC particles were embedded is shown in Fig.1. Some different phases present have been indicated, where the large blackest parts correspond to SiC. Around the SiC particles a reaction layer can be observed which is mainly composed of TiC. The higher affinity of Ti for C than Si gives rise to an exothermic reaction between Ti and SiC resulting in this TiC layer. The Si that is released during this exothermic reaction is rejected into the melt pool.

Near the bottom of the melt pool the dissolution of SiC is modest and the amount of Si rejected into the melt gives rise to a small amount of eutectic Ti_5Si_3/α-Ti on the grain boundaries of the α-Ti grains. Approaching the top of the melt pool the amount of Si rejected increases considerably and finally gives rise to large facetted pre-eutectic Ti_5Si_3 grains with eutectic Ti_5Si_3/α-Ti in-between (see Fig.1).

Actually two types of TiC reaction layers around the SiC particles were observed using SEM: (a) A cellular layer in which the TiC grains seem to have been nucleated heterogeneously at the solid SiC/melt interface with subsequent cellular growth of the TiC into the melt. (b) A cloud-like layer of spherical TiC grains pointing at "homogeneous" nucleation of the TiC within the melt very near to the solid SiC surface.

TEM analysis of the cloud-like reaction layer revealed that besides a predominant fraction of spherical TiC grains also Ti_3SiC_2 plates appeared to be present. The Ti_3SiC_2 plates have dominant (0001) facets because they grow most easily along directions present within the basal plane.

TEM analysis of the cellular reaction layer did not reveal the presence of Ti_3SiC_2. The different way in which the Si is rejected from the TiC grains in the two types of reaction layers can be held responsible for the presence or absence of Ti_3SiC_2. In the cloud-like reaction layer volumes in-between the TiC grains can become enclosed, preventing the rejection of an excess Si into the melt pool. In the enclosed volumes the Si concentration is subsequently raised sufficiently to nucleate Ti_3SiC_2 at the interface between TiC and SiC, in accordance with the ternary Ti-Si-C phase diagram[3]. This sequence of a primary TiC nucleation followed by a nucleation of Ti_3SiC_2 can be expected because the melting point of TiC is probably highest, that is to say: 3061 ^{0}C for TiC, 2545 ^{0}C for SiC and a melting point for Ti_3SiC_2 probably in-between. In the case of the cellular reaction layer areas in the material do not become trapped and accordingly Ti_3SiC_2 is not observed. The excess of Si rejected into the "open" melt results in accordance with the ternary Ti-Si-C phase diagram in the nucleation of Ti_5Si_3 at the interface between TiC and Ti.

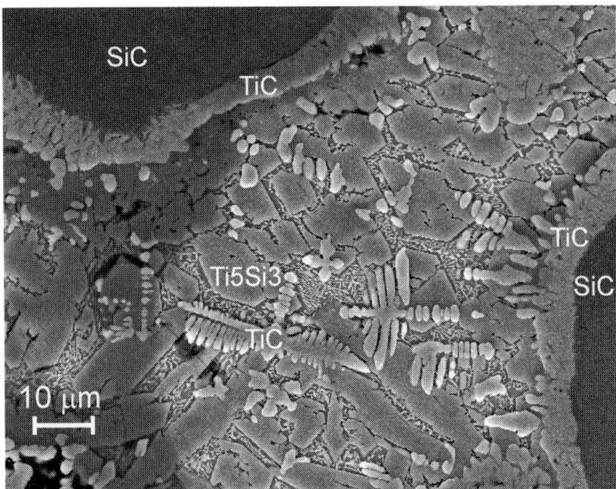

Fig.1 SEM image of SiC particles laser embedded in Ti. Around the SiC particles a TiC reaction layer is formed due to the reaction with Ti. Region at the top of the melt pool where the released Si results in large faceted pre-eutectic Ti_5Si_3 grains surrounded by eutectic Ti_5Si_3/Ti. The reaction layer is of cloud-like type.

3.2 Precipitation on stacking faults in TiC of the reaction layers

In a considerable fraction of TiC grains, predominantly in the cellular layer, a high density of widely extended stacking faults (SFs) on TiC {111} planes was observed with TEM; i.e. observations typical for (deformed) material with a very low SF energy. Since the SF energy of TiC is very high and SFs are not observable after severe deformation[5-7], no directly observable SFs were expected in TiC. Depending on the C/Ti ratio the SF energy lies in the range between 130 mJ/m^2 and 300 mJ/m^2 [4]. This first would suggest that the high SF density grains were composed of β-SiC instead of TiC, because the SF energy of SiC is much lower than of TiC and the crystallography of TiC and β-SiC is almost identical. However, subsequently performed EDS and PEELS analyses in TEM clearly indicated that these originally supposed β-SiC grains were in fact TiC grains. A second possible explanation for the SFs in TiC is the presence of impurities. Planar defects with extrinsic stacking-fault character observed before in TiC were ascribed to the effect of boron impurities[5-8]. Triangular planar defects nucleated at dislocation nodes of networks with ½<110> Burgers vectors on {111} TiC were identified as very thin TiB_2 platelets starting possibly as impurity stabilized stacking faults[5-8]. Dissociation of isolated dislocation lines resulting in (widely extended) stacking faults is less frequently observed than the triangular planar faults centered on the dislocation nodes[5,8]. The impurity content of the present TiC grains is locally probably substantially higher than in the above TiC material containing traces of boron. Hence a strong reduction of the SF energy is thus well-conceivable explaining the observed high density of widely extended SFs. Particularly Si, released during the reaction SiC+Ti→ TiC+Si, may be trapped in the TiC grains and may reduce the SF energy.

The observed faults are generally not of simple single-plane intrinsic or extrinsic stacking-fault type as can be clearly observed in the HRTEM image depicted in Fig.2. A high impurity content in these faults is probably responsible for their deviation from pure stacking-fault character. Detailed 2-beam bright- and dark-field images and (many-beam) HRTEM images revealed the presence of small precipitates on many SFs in the TiC grains with an extension perpendicular to the SF plane usually below 20 nm. The regularly spaced bright and dark fringes associated with "clean" SFs imaged in 2-beam BF and DF as often observed, become largely distorted by the presence of these precipitates at the SFs. Also SFs with nearly edge-on orientation with respect to the viewing direction, which in the case of a clean SF corresponds to a single hardly observable line, show due to the presence of the precipitates strain-field contrast extending long distances away from the defect plane. These long-range strain fields indicate at least partial coherence of the precipitates with the TiC.

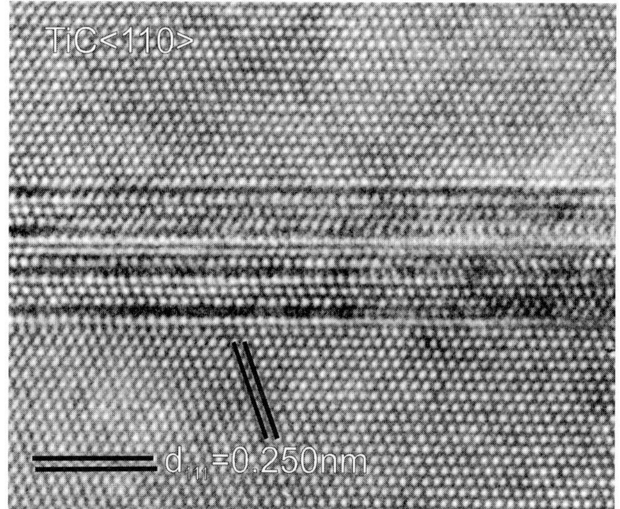

Fig.2 HRTEM image of TiC in a cellular reaction layer viewed along <110> showing an edge-on observed complex Stacking Fault containing small twinned regions of 2 to 5 {111} planes. Within these faults a strong Si enrichment is probably present.

Due to the limited size of the precipitates the identification of their phase is difficult. One relatively large lenticular-shaped precipitate (length 250 nm and maximum width of 25 nm) could be identified unambiguously on the basis of an SAED pattern as Ti_5Si_3 viewed along the $<11\bar{2}3>$ direction. Two examples of precipitates nucleated on SFs as imaged by HRTEM are depicted in Fig.3. Using the known interplanar spacings of TiC as reference, the lattice spacings in the precipitates could be determined accurately in the HRTEM images. In Fig.3a the precipitate corresponds to Ti_5Si_3 viewed along the $<11\bar{2}3>$ direction and appears being nucleated on the crossing of two edge-on seen stacking faults in the TiC. The orientation relation (OR) between the precipitate and matrix is such that $\{1\bar{2}12\}Ti_5Si_3//\{200\}TiC$ and $<1\bar{2}1\bar{3}>Ti_5Si_3//<011>TiC$. In Fig.3b the precipitate corresponds to Ti_5Si_3 viewed along the $<14\bar{5}0>$ direction and the OR and interface orientation is: $(0001)Ti_5Si_3//\{11\bar{1}\}TiC$ and $<14\bar{5}0>Ti_5Si_3//<011>TiC$. The precipitate has a large aspect ratio with very large (0001) facets parallel to the original stacking fault plane of TiC on which it is nucleated. The combined SAED and HRTEM information clearly indicates that the precipitates nucleated on SFs in TiC correspond to Ti_5Si_3. The precipitates shown in Fig.3 are already well developed and it should be realized that all the stages in-between stacking faults and these well-developed precipitates are actually present in the TiC.

The nucleation of the Ti_5Si_3 precipitates on SFs in TiC indicates that the TiC is supersaturated with Si. Apparently, apart from the rejection of the excess Si in the molten matrix, Si also appears to become trapped in the TiC when it is formed at high temperatures (~3000 ^{0}C). At 1200 ^{0}C the solubility is very small according to the Ti-Si-C phase diagram[3].

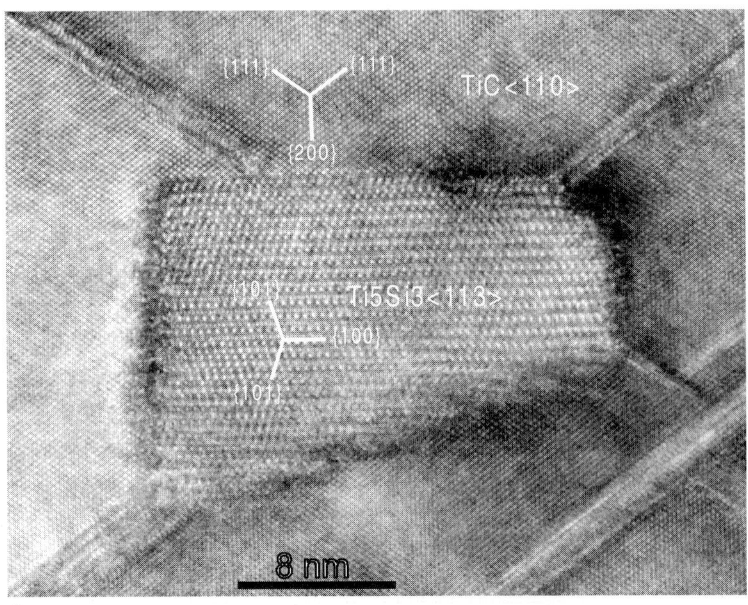

a)

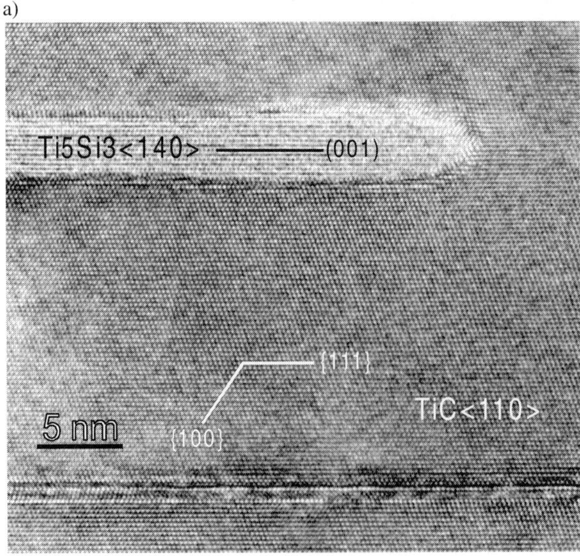

b)

Fig.3 HRTEM images of Ti_5Si_3 precipitates nucleated on stacking faults in TiC present in the reaction layer. The TiC is viewed along <110>. **a.** Ti_5Si_3 viewed along $<11\,\bar{2}\,3>$ on the crossing of two edge-on observed stacking faults **b.** Ti_5Si_3 precipitate viewed along $<14\,\bar{5}\,0>$ having its (0001) plane parallel to the formerly stacking faulted {111}TiC.

So, during cooling the supersaturated Si likes to segregate out of the TiC. A possible mechanism to achieve this within the TiC grains is that Si induces SFs in the TiC and is enriched at these SFs. This mechanism explains the presence of the unusually wide extension and high density of SFs in the TiC grains. SFs in face-centered cubic TiC correspond to locally close-packed hexagonal stacking and all at present relevant phases containing Si (Ti_5Si_3, Ti_3SiC_2 and all but one of the SiC polytypes) are based on this stacking as well. Thus, Si segregated to SFs is expected to be the most stable state for Si in TiC. If the local enrichment of Si to the SFs is sufficiently strong, new phases containing Si tend to nucleate on the SFs during cooling down, explaining the occurrence of the Ti_5Si_3 precipitates on the SFs in TiC as observed. According to the ternary phase diagram at 1200 ^{0}C, excess Si in TiC for Ti/C ratios in-between 0.33 and 0.46 will result in the formation of Ti_5Si_3. The experimental concentration of C in the TiC of the reaction layers, quantified using EDS and PEELS in the TEM, appeared to be 0.402±0.022 and 0.386±0.011, respectively and is in accordance with the observed Ti_5Si_3 precipitation on SFs in TiC.

4 Conclusions

Laser embedding of SiC particles in Ti-6Al-4V results in reaction layers of TiC around SiC with either a cellular or cloud-like structure. The Si released during the reaction SiC+Ti→TiC+Si is rejected into the melt pool and results in Ti_5Si_3 formation. Due to the extent of Si rejection, pre-eutectic Ti occurs at the bottom and pre-eutectic Ti_5Si_3 at the top of the melt pool. If in the cloud-like reaction layer regions in-between the TiC grains become enclosed, the rejected Si content increases locally and Ti_3SiC_2 plates with dominant (0001) facets nucleate. In the TiC grains particularly of the cellular reaction layer, a high density of widely extending stacking faults is observed and on these faults in many instances small Ti_5Si_3 precipitates are present. Firstly the Si supersaturating the TiC induces the SFs and secondly causes the nucleation of Ti_5Si_3 precipitates on the SFs. Thirdly, parts of the SFs become unfaulted because the precipitates withdraw Si from the SFs and hence strongly increase the SF energy locally towards the high energy value pertaining to (pure) TiC.

5 References

1. Metals handbook, 9th ed. Vol.3 p.353, Amer. Soc. for Metals, Metals Park, Ohio 44073, 1980.
2. A.B. Kloosterman, B.J. Kooi, J.Th.M De Hosson, *Acta Mater.* **46**, 6205 (1998).
3. W. J.J. Wakelkamp, F.J.J. van Loo and M. Metselaar, J.Europ.Ceram.Soc., 8, 135 (1991).
4. S. Tsurekawa, H. Yoshinaga, J. Jap. Inst. Metals 58, 390-396 (1994).
5. Q. H. Zhao et al. J, J. Mater. Res. 9, 2096- 2101 (1994).
6. F.R. Chien, S.R. Nutt, D. Cummings, Phil. Mag. A 68, 325-348 (1993).
7. G. E. Hollox, R.E. Smallman, J. Appl. Phys. 37, 818-823 (1968).
8. J. Venables, Phil. Mag , 873-890 (1967), Metall. Trans. 1, 2471-2476 (1970).

Electroless Metal Deposition Used as Underbump Metallization (UBM) for High Temperature Application - a Comparison to Other Bumping Techniques

Sabine Anhöck*, Andreas Ostmann*, Hermann Oppermann*, Rolf Aschenbrenner+, Herbert Reichl+

*Technical University of Berlin, Center of Microperipherik, TIB 4/2-1, Gustav-Meyer-Allee 25, D-13355 Berlin, email: anhoeck@izm.fhg.de, phone +49 (0)30 314 72 873 fax: +49 (0) 30 314 72 835

+ Fraunhofer Institute for Reliability and Microintegration, Gustav-Meyer-Allee 25, D-13355 Berlin

1 Abstract

Electronic components are more and more applied at higher operation temperatures. One reason is increasing density of electronic power. Another reason is the application of electronic components directly at high temperature modules especially for automotive application.

For the reliability of such high-temperature assemblies the choice of metallurgical system is very important. Critical issues for the high-temperature reliability are the phase formation and phase growth of intermetallics between the solder and the end-metallization. The electroless Ni seems to be a suitable metallization for high temperature solders. Due to the nature of the electroless deposition process the deposits consist of Ni and P and shows an amorphous structure. Compared to other materials such as Cu, Cr, Pt, Pd the phase growth of the intermetallics at the interface between solder and Ni is slow. The phosphorous seems to inhibit the Ni diffusion into the solder by formation of a phosphor-rich layer close to the interface of solder and final metallization.

The focus of the paper is to summarize the results from aging of AuSn20 solder preforms on different metallizations such as Cu, Cr, Ni, Pd and Pt to find a suitable metallization for high temperature application. Furthermore the reaction of PbSn63 solder preforms on different Ni metallizations such as electroless deposited Ni, electroplated Ni and pure Ni bulk metallization will be discussed .Differences in the phases growth due to the electroless deposition will be described.

2 Introduction

Especially for reliable packagings at high operation temperatures one of the most difficult problems is to find a suitable metallization system. Typical difficulties which limit the life time of electronic packagings include the altered electrical behavior caused by the interdiffusion of solder and UBM (under bump metallization = UBM) as well as of solder and final metallization of the substrate. Dimensional changes or brittleness caused by intermetallic phase formation play also an important role for the life time. So, the choice of the solder joint metallurgy is a key issue for the reliability for e.g. of Flip-Chip assemblies /1,2/.

A lot of solders and metallizations are widely used and well understood and the goal is to choose the best combination. Besides the issues of reliability the cost aspect is an other important factor. Different technologies offer the possibility for cost reduction. For low-cost flip chip assembly the electroless Ni deposition in combination with stencil printing of solder paste is one of the most promising technologies. For standard applications the eutectic lead-tin solder (PbSn63) is used providing an excellent reliability /3/. For special fluxless applications in opto-electronics AuSn20 solder is used /4/.

The required operation temperatures for electronic components are increasing constantly. Especially for high temperature application a Ni metallization could be an alternative compared to other materials used as final metallizations such as Cu, Cr, Pd and Pt /4/. Furthermore differences in structures and behavior between electroplated Ni and chemically deposited Ni, which is an alloy of Ni and P, can be observed.

The properties of all Ni/P films strongly depend on the phosphorous content in the deposits. Electroless Ni layers with more than 8 wt.-% P show an amorphous structure /5,6/. During the reaction between the electroless Ni and the PbSn63 it seems that the phosphorous does not react with the solder material. The formed intermetallic phase is the well known Ni_3Sn_4 phase without any phosphorous inside. The phase growth of Ni_3Sn_4 layer is less than phase growth at the interface of electroplated Ni and PbSn63.

3 Techniques for Solder Deposition

The bumping including the under bump metallization is an important issue for flip-chip assembly. The conventional C4 process of IBM uses sputtering, etching, vacuum evaporation and other processes to achieve a high melting Pb/Sn bumps. This has been the only accepted technology for a long time. Today, electroplating of lead-tin solders and gold-tin solders and other alloys is also used. The eutectic Au-Sn solder offers the possibility for fluxless application especially for opto-electronic devices and high-temperature applications /4/.

For a low-cost approach the application of solder paste by stencil printing is one of the most promising technologies. The printing technology is widely used e.g. in SMT assembly. It is also suitable for mass production. No photolithography and sputtering processes are necessary. After stencil printing of solder paste the wafer is reflowed in a reflow oven. The liquid contents of the paste (flux activators and more) are evaporated and a solderable bump is achieved. Different solder pastes are available. Usually eutectic lead-tin solder is used for low-cost assemblies and mass production /3/.

4 Under Bump Metallizations

4.1 Electroplating

For processing of an electroplated UBM several steps are necessary. In a first step organic residues and the Al oxide are remove by sputtering. After this step follows a thin film metallization of 100 nm Ti:W on the wafer to enhance the adhesion strength between the Al and the Cu which is sputtered in the next step. The Cu layer is in the range of 300 nm. A 10 μm photo resist layer is spun on the wafer and prebaked following by a exposure and development step. After post bake the electroplating of Ni or Cu is possible. To prevent the oxidation of Ni and to improve the wettability a thin Au layer (about 100 nm) is coated on the UBM by an immersion Au plating. In a last step the photo resist is stripped and the plating base is etched chemically. The typical UBM height is 5 μm /7/.

4.2 Electroless Deposition

The advantage of the chemical bumping process is the selective autocatalytic metal deposition on activated Al pads without any costly equipment for sputtering and photo resist imaging. Different electrolytes for deposition of Ni, Cu Au and Pd are available. Most suitable is the chemical nickel deposition which is extensively applied in printed circuit boards as a diffusion barrier on Cu as well as UBM on Al bond pads on wafers. In figure 1 the sequence of the Ni bumping process flow is shown. The first step of the nickel bumping process is the back side coating for protection of free Silicon from Ni plating. A cleaning step (passivation cleaning) follows to remove passivation residues from Al pads. A second Al cleaning process removes thick Al oxides and prepares the surface for metal deposition. The following zincating step activates the Al surface for nickel deposition. A thin zinc layer is deposited on Al which is substituted by Ni in the Ni bath. On this thin Ni layer the autocatalytic deposition of Ni/P starts. The temperature of the Ni bath is 90 °C with a pH 4,5 and a hypophosphite bath is used. Finally a thin gold layer is deposited on the Ni from an immersion gold bath to prevent oxidation of Ni before soldering /3/.

The chemical Ni/Au bumping process has been developed by Technical University of Berlin in cooperation with the Fraunhofer Institute for Reliability and Microintegration. It is capable to deposit Ni/Au bumps/UBMs on most wafer types.

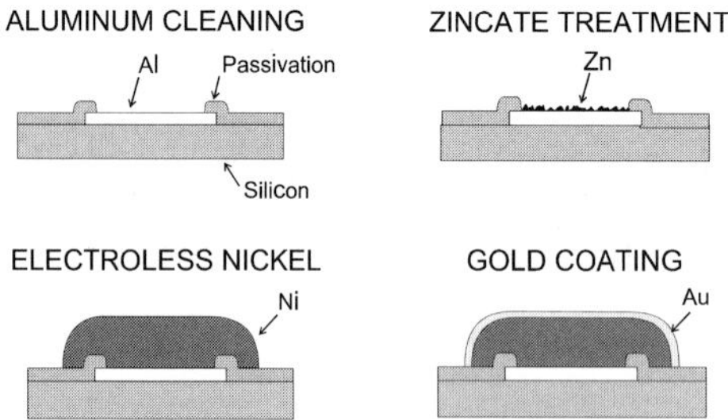

Fig.1: Process flow of the Ni bumping process /3/

5 Properties of Electroless Ni Bumps

Due to the nature of the chemical deposition process used at TU Berlin/Fraunhofer IZM the electroless Ni layer consists of Ni and P with a phosphorous content of about 9 wt.-%. Using the scheme in figure 2a the deposition flow can be explained. For the electroless Ni deposition an activation with Zn grains is essential. The Zn atoms will be replaced by Ni atoms while Ni deposition. A reducing agent (sodium hypophosphite) is required for this reaction. The hypophosphite is decomposed into H_2, H_20, P and other components. The P is built into the Ni bump in layers /6/.

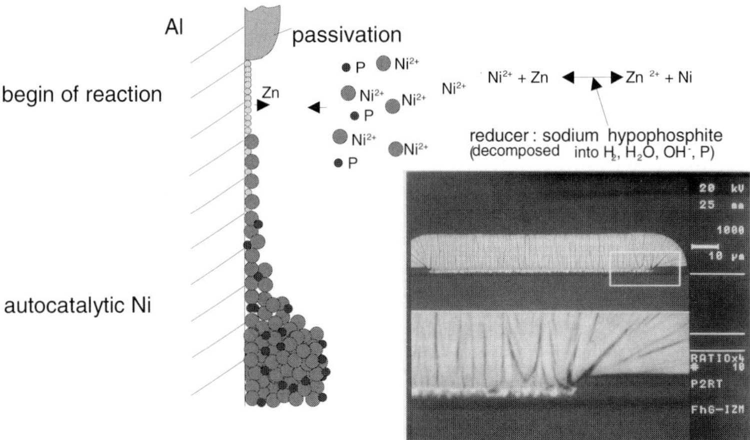

Fig. 2: a) principle of the electroless Ni deposition b) the structure of a Ni bump after etching /6/

In figure 2b the structure of a Ni bump can be observed after polishing and etching. A small variation of the phosphorous content of about 1 wt.-% can be measured. The vertical lines separate different growth columns of metallic Ni/P starting from the thin Zn seeding layer on the Al surface. Even after 4 µm Ni bump height the Ni/P layer is a compact layer because the grains are grown together. The deposition structure of chemical Pd is similar. A layer consists of needles can be observed / 6/.

6 Investigations of Reaction with Solders

Eutectic PbSn solder and AuSn20 solder were wetted on different metallizations. The samples were heat treated at different times and different temperatures. After this the test samples were prepared as cross sections and the thickness of the formed intermetallic phases were measured optically.

The combinations of solders and metallizations are given in table 1.

Tab. 1: overview of investigated combinations solders/metallizations

AuSn20	PbSn63
chromium bulk material	electroless nickel layer
nickel bulk material	electroless nickel bump
palladium bulk material	electroplated nickel layer
platinum bulk material	pure nickel bulk material
titanium bulk material	

7 Results

7.1 Au/Sn on different metallizations

The phase growth of investigated intermetallic phases as a function of square root of time for 200 °C is shown in figure 3. It could be observed that the phase growth of the investigated intermetallic phases is determined by a root-of-square rule as expected. The intermetallic phase between the Ni metallization and the AuSn20 starts with a small initial thickness after reflow and shows the smallest growth rate compared to the other. Platinum can also be a suitable metallization for AuSn20. Using the arrhenius relation the activation energies for the phase growth can be estimated. The results are given in table 2. A formed intermetallic phase between the Ti and the AuSn20 could not be observed and the wettability is very low.

The formed phases between Ni and AuSn20 and Pt and AuSn20 can be attributed to Ni-Sn /Pt-Sn phases. It can be exposed that the behavior of PbSn solder with these metallizations is similar because the lead does not react with Ni or Pt.

Tab.2: overview of activation energies of different intermetallic phases

Phase	activation energy (eV)
PtSn(Au)	0,52
Au-Cr (α´)	0,39
Pd_3Sn_2(Au)	0,28
Ni_3Sn_2(Au)	0,23
Au-Cu(Sn)	0,5 /8/

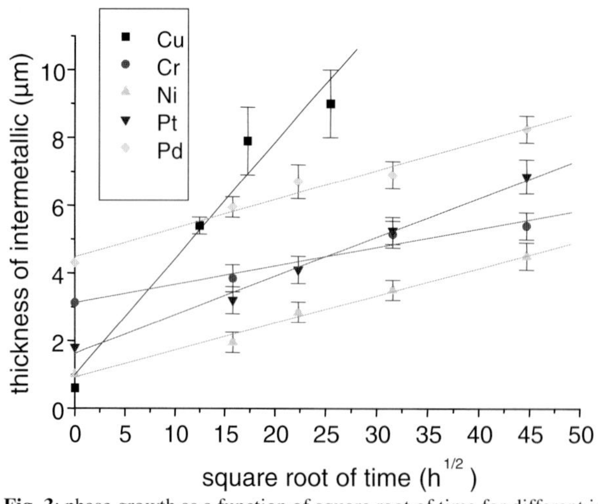

Fig. 3: phase growth as a function of square root of time for different intermetallic phases at 200 °C

7.2 Comparison of phase formation and phase growth of electroplated Ni and PbSn63 as well as chemical deposited Ni and PbSn63

Using the results of the pre-investigations for the best UBM for solders the reaction of special Ni metallizations with the usually used eutectic PbSn solder were observed. In figure 4 a comparison of the phase growth (all formed phases are added) of the intermetallic phase between the PbSn63 and several Ni types such as pure Ni bulk, electroplated Ni and chemically deposited Ni is given for 150 °C.

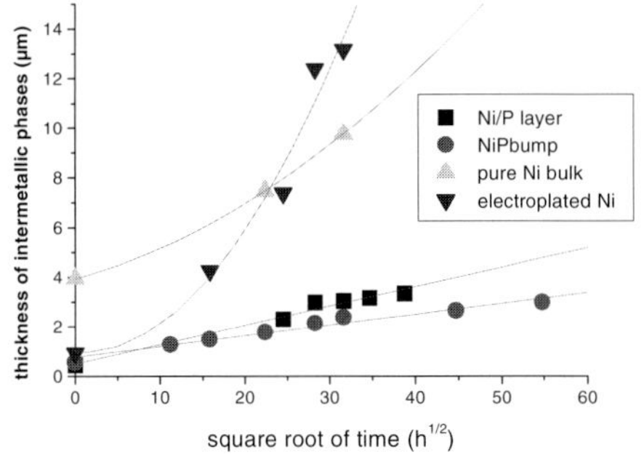

Fig. 4a: relationship between phase growth and time at 150 °C (PbSn63 and different Ni metallizations)

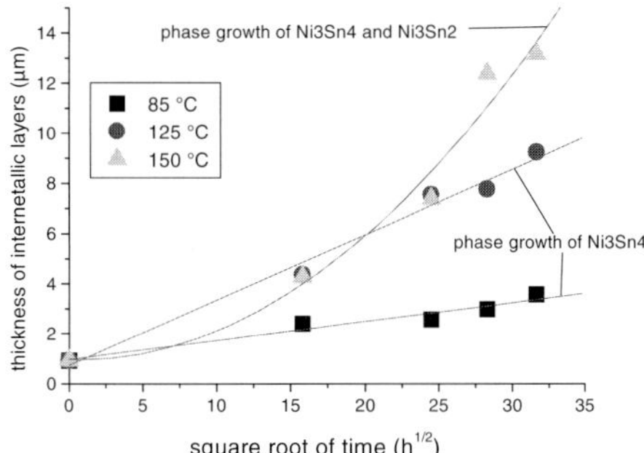

Fig. 4b: influence of time at phase growth for 85 °C, 125 °C and 150 °C for reaction of PbSn63 and electroplated Ni

In figure 4a two growth types can be observed. On one side the phases between Ni/P bumps/PbSn63 and the Ni/P layer/PbSn63 seems to react similar. The formed phase can be attributed to Ni_3Sn_4 without any phosphorous inside. The phase growth is proportional to square root of time. The growth constant can be estimated with 0,08 $\mu m/\sqrt{h}$ for Ni/P layers and 0,044 $\mu m/\sqrt{h}$ for Ni/P bumps. Differences in geometry (layer and bumps) show a small influence on the phase growth. The Ni/P layers were wetted with PbSn63 preforms which had a texture of the eutectic structure. A faster diffusion is possible.

On the other hand the electroplated Ni as well as the pure Ni bulk reacts with PbSn63 in a similar way. For temperatures above 150 °C the phase growth is not proportional to square root of time due to the growth of 2 different phases (Ni_3Sn_4 and Ni_3Sn_2). Using the figure 4b a change in phase growth for 150 °C can be observed. For 85 °C and 125 °C the phase growth is still proportional to the square root of time. At these temperatures the Ni_3Sn_2 phase could not be observed. It seems that the second phase needs a higher activation temperature for growth.

Figure 5 shows the differences in intermetallic phase formation of electroplated Ni and chemical Ni in reaction with PbSn63. As mentioned before two phases (Ni_3Sn_2 and Ni_3Sn_4) can be observed at the electroplated Ni sample. Compared to this only one phase results from the reaction between chemical Ni and PbSn63 (Ni_3Sn_4). This phase is very small. Furthermore between the Ni_3Sn_4 and the chemical Ni a small phosphorous-rich layer (dimension about 400 nm) can be observed. It seems that the phosphorous inhibits the Ni diffusion into the solder.

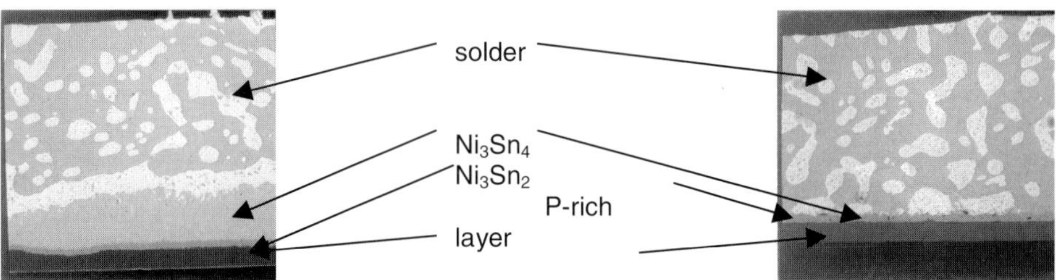

Fig. 5: cross sections of intermetallic phase formation between PbSn63/electroplated Ni and PbSn63/chemical Ni after1000h @ 150 °C (magnification of cross sections 5000x)

8 Conclusion

The phase growth and the phase formation of different metallizations soldered with AuSn20 were investigated. The results show that Ni seems to be the best metallization especially for high temperature applications. In most cases tin-rich intermetallic phases have been formed. Using this result different Ni metallizations (chemically deposited, electroplated, pure bulk) were soldered with PbSn63 and heat treated. Due to the chemical deposition process the electroless Ni deposits consist of Ni and P. This constitution seems to be a promising metallization especially for higher temperature application. The phosphorous inhibits the Ni diffusion into the solder and the phase growth of the intermetallic phase compared to electroplated and pure Ni bulk.

9 References

1. Wiley, J.D.; Perepezko, J.H.; Nordman, J.E.; Guo, K.-J.: Amorphous Metallizations for high-temperature Semiconductor Device Applications IEEE transactions on industrial electronics, Vol. IE-29, no. 2; 1982,p. 154-157
2. Kallmayer, Ch; Oppermann, H.; Anhöck, S.;. Klein, M.; Kalicki, R.; Aschenbrenner, R.; Reichl, H.: Reliability investigations of Flip Chip Solder Bumps on Pd. 49 th ECTC, San Diego, 1999, p. 135-140
3. Ostmann, A.; Kloeser, J.; Reichl, H.: Implementation of a Chemical Wafer Bumping Process. IEPS Conf. 1995, San Diego, p. 354-366
4. Anhöck, S.; Oppermann, H.; Kallmayer, Ch.; Aschenbrenner, R.; Thomas, L.; Reichl H.: First Investigations of Au-Sn Alloys on Different End-Metallizations. Third European Conference on Electronic Packaging Technology (EuPac´98), June 15-17 Nürnberg, p. 43-46
5. Müller, F.: *Herstellung, Mikrostruktur und Eigenschaften von außenstromlos abgeschiedenen Nickel-Phosphor-Schichten mit nanokristallinen Aufbau.* VDI-Fortschrittberichte, Reihe5:Grund- und Werkstoffe, Nr. 427, VDI-Verlag, p. 8ff
6. Anhöck, S.; Ostmann, A.; Oppermann, H.; Aschenbrenner, R.; Reichl H.: Reliability of Electroless Nickel for High Temperature Application. International Symposium on Advanced Packaging Materials, Braselton, 1999,p. 256-261

7. Reichl, H.: Direktmontage- Handbuch für die Verarbeitung ungehäuster IC Springer Verlag, 1998, p. 18ff
8. Zakel, E.; Reichl, H.: Au-Sn Bonding Metallurgy of TAB Contacts and its Influence on the Kirkendall Effect in the ternary Cu-Au-Sn IEEE Transactions on Compo., Hybrids and Manufacturing Technol. Vol. 16, No. 3, 1993, p. 323-332

10 Acknowledgement

The authors like to thank Ms. Christine Jauernig and Mrs. Bing Hang for preparation of diffusion samples and cross sections.

New Joining Technique for Silicon Components at Low Temperature

Jens Mehlich[1], Hans-Gerd Busmann[1], Kai Burdorf[2], Wolfgang Benecke[2]
1: Fraunhofer Institute for Manufacturing and Advanced Materials, Bremen, Germany
2: Institute for Microsensors, -actuators and –systems, University of Bremen, Germany

1 Abstract

Conventionally, silicon parts are bonded to substrates at fairly high temperatures.
This does not satisfy the demand of future micro devices. Therefore, a new joining process at low temperature (< 150 °C) has been developed using highly sinter-active gold deposits as binding agent.

The surfaces have been coated with nanoscaled Au-powder. Subsequently, the parts to be joined were pressed at room temperature for a few minutes. At 300 MPa pressure, interfacial layer with nearly 100 % density are formed that connect both parts and ensure high electric and thermal conductivity across the bond. The bonds are mechanically stable during dicing and may carry high electric current.

2 Introduction

Bonding electronic parts and components becomes increasingly important in modern sensor and actuator techniques. Most bonding techniques require relatively high temperatures, e.g. eutectic- and silicon fusion bonding. Others require additional electric fields, e.g. anodic bonding, to obtain bonds with good mechanical strength and high electrical conductivity. Bonding techniques that leave metalized patterns and structures of the parts unaffected are highly desirable.

A new technique is desirable, that allows mechanical junction and electric connection of thermal sensitive components, like the silicon parts of gas sensors.

The demands on such a process are

- applicable for fully processed devices
- low process temperatures
- high electrical and thermal conductivity
- chemically inert under corrosive conditions
- no organic solvents (gas sensor)
- surface roughness up to 1 μm tolerable

3 Experimental

Silicon bonding using nanopowder deposits was tested with 15x15 mm² sample size. For improved adherence, a thin layer of chromium and gold with a thickness of 8 and 80 nm, respectively, were deposited under normal sputtering conditions.

The raw material of the interfacial layer is highly porous gold. The synthesis of nanostructured materials was pioneered by the inert gas condensation and compaction technique (1, 2). In this investigations highly pressure sputtering is used for evaporating Au at 100 .. 300 Pa background pressure. An Au target with an effective diameter of 80 mm is used. At pressures of 10 Pa and above, the mean free path is less than 0.5 mm, leading to nucleation and growth of Au nanoparticles (3, 4).

Figure 1 is a schematic diagram of the bonding apparatus with the magnetron sputtering source and anvils of the press inside a vacuum chamber. The substrates to be bonded are adjusted and placed on the upper and lower anvils. The substrates are embedded in silicon pads in order to ensure a quasi-hydrostatic pressure and to avoid pressure peaks (5). The masks are adjusted on top of the wafers for selective Au deposition.

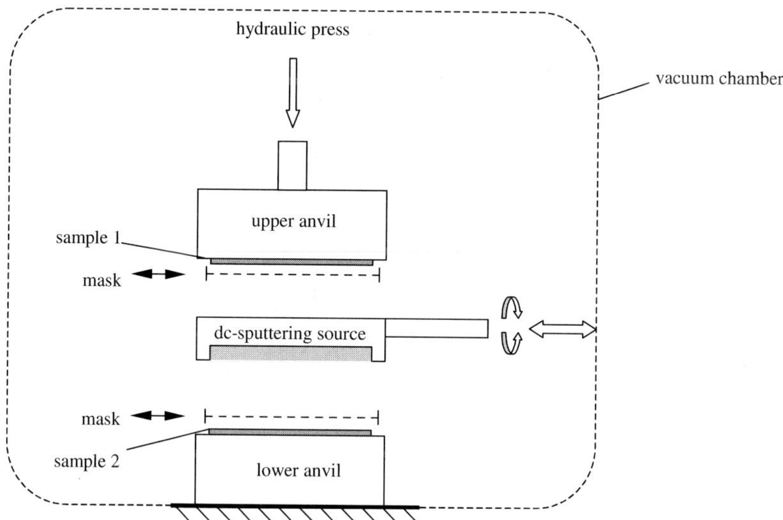

Figure 1: Schematic arrangement for the low temperature metal joining process

During sputtering, the Ar-background pressure is kept constant. In succession, both samples (1, 2) are coated with the powder consecutively. After nanopowder deposition, the device is evacuated down to 10^{-7} Pa and the substrates are pressurized with a constant load for a certain period of time at room temperature.

After the experiments, the sample stack was removed from the press and scratched with a diamond to cut it perpendicular to the interfaces into two pieces.

4 Results and Discussion

With regards to nanopowder morphology, best results were archieved with an Ar- background pressure between 190 and 250 Pa and a DC-sputtering power of 500 W. Under these conditions, particles with diameters less than 50 nm could be formed at acceptable deposition rates (200 .. 900 nm/min).

Figure 2 and 3 show a micrograph of a sample taken with a high resolution scanning electron microscope (HRSEM) before pressing.

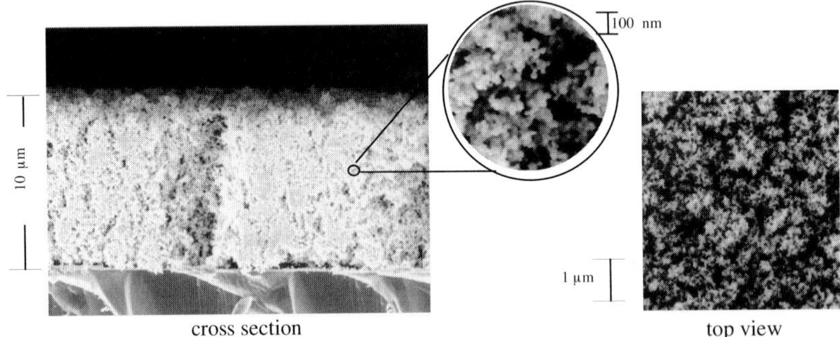

Figure 2: HRSEM micrographs of the joint before pressing (DC-power = 500 W, Ar-pressure = 250 Pa; pressed with 300 MPa, 5 minutes)

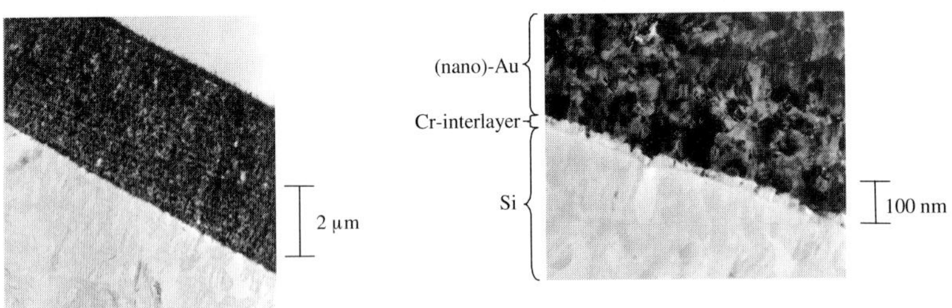

Figure 3: TEM micrographs of the interfacial area shown in figure 2 at different magnification

The silicon of the sample stacks was removed by etching with SEF 67-25-08 before it could be prepared for transmissions electron microscopy (TEM). Figure 3 shows TEM bright-field images of exactly the same sample after pressing at different magnifications.

During the compaction stage, the gold powder is compressed to about one third of its original thickness, and the porosity is almost eliminated. Some grain growth occurs, but the gold-layer still keeps fine grained.

Usually, after the integration of microsensors and systems by bonding processes, the wafers are cut by a wafer saw to obtain single sensors and systems. This step exhibits the highest external stress for this type of bondings. Therefore, also the Au-bonded stacks were diced by a wafer saw into pieces of 1x1 mm² (figure 4).

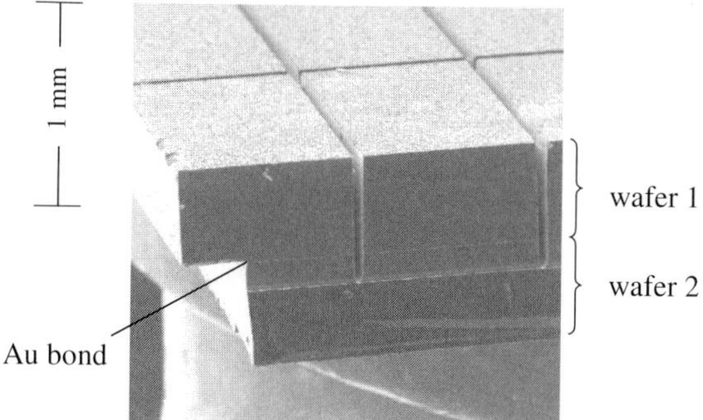

Figure 4: Dicing of a bonded stack (Si/ Cr/ Au/ Cr/ Si) with a wafer saw (DC-power = 500 W, Ar-pressure = 250 Pa; pressed with 300 MPa, 5 minutes)

Obviously, the bonded samples withstand this severe mechanical load during the cutting procedure.

The results show that bonding at low temperature with nanoscaled gold is feasible. Of course, the mechanical behavior under tensile and in particular under shear stress loading still have to be tested in detail. Such type of mechanical load occurs during thermal cycling often experienced in sensor devices.

Further work also aims at joining dissimilar materials at low temperatures. This is particularly useful for mounting electronic power devices to heatsink discs made from tungsten or molybdenum. In this case, low process temperature would decrease internal stress that is caused by the thermal mismatch between silicon and heatsink material.

5 Conclusion

Large areas of silicon parts can be joined with sufficiently strength, high electrical and high thermal conductivity by sintering nanoporous gold as an intermediate layer. The bonded stacks can be cut with a diamond wafer saw without determination of the substrates.

Even at room temperature grain growth of nanoporous Au is recognizable and applicable to this novel technique for interconnection. The required pressure at 25 °C is too high to obtain an interconnection without affecting the silicon with a high dislocation density and internal stress.

Future work aims at the decrease of required pressure for sintering and the quantification of mechanical properties under tensile and shear stress. Furthermore, bonds of dissimilar materials, e. g. Si-W, will be investigated to join backing plates and heat sinks of electronic power devices by low temperature metal bonding.

6 Acknowledgments

This work is supported by the Bremer Forschungsverbund Materialwissenschaften, MATEC Bremen, Germany.

7 References

1. C. G. Granqvist, R. A. Buhrmann, J. Appl. Phys., 47, 1976, 2200
2. H. Hahn, R.S. Averback, J. Appl. Phys., 67, 1990, 1113
3. Deng Zhaojing, Wang Qiao, Li Jian, Sun Qiang, Journal of colloid and interface science, 165, 1994, 346
4. H. H. Kaatz, G. M. Chow, A. S. Edelstein, J. Mater Research, Vol 8, 5 1993, 995
5. S. Klaka, R. Sittig, Proc. 6th int. Symposium Power Semiconductor Devices & IC's, CH, 1994, 259

Leaded and Lead-Free Solders: Bulk Mechanical Properties and the Performance and Reliability of Model Joint Assemblies

B. J. Foroodian, P.M. Mummery, B. Roebuck
National Physical Laboratory, UK

1 Abstract

A comparison of the bulk mechanical properties of a lead-tin eutectic and a lead-free (tin-silver-copper) solder was made on a novel miniaturized test system, the ETMT. Room and elevated temperature tensile tests were performed to determine the effects of material microstructure and strain rate on the flow properties. The lead-free solder had far superior mechanical properties in all material conditions and temperatures, a trend repeated in the creep and thermo-mechanical fatigue performance. Model symmetrical lap joints were assembled. Thermal and mechanical cycling, and creep tests to failure were performed. These accelerated reliability tests allowed a direct measure of the thermal cycling behaviour of the solders within an afternoon. The lead-free solder joints had two orders of magnitude greater lifetimes (cycles to failure) than the lead-tin eutectic joints. The ETMT shows great promise for generating reliable data in a short time-scale.

2 Introduction

Environmental legislation and changes in technology are urgently forcing industry to consider lead-free solders for a wide range of engineering applications, but especially in the electronics sector [1, 2]. An important new development at the National Physical Laboratory now offers the possibility of generating reliable data on properties governing service performance of solders. The system can be used on small test-pieces (in both solids and joints) in a unique miniaturized electrothermo-mechanical test (ETMT) instrument [3, 4].

3 Experimental System

The system comprises:

- An environmental chamber with comprehensive electrical lead throughs, water cooling and inert gas supply facilities.
- Computer controlled DC heating power supply (200A), with test-piece resistance and thermocouple temperature measurement facility. Heating rates up to 200 °Cs^{-1} are possible, dependent on the thermal characteristics of the test-piece.

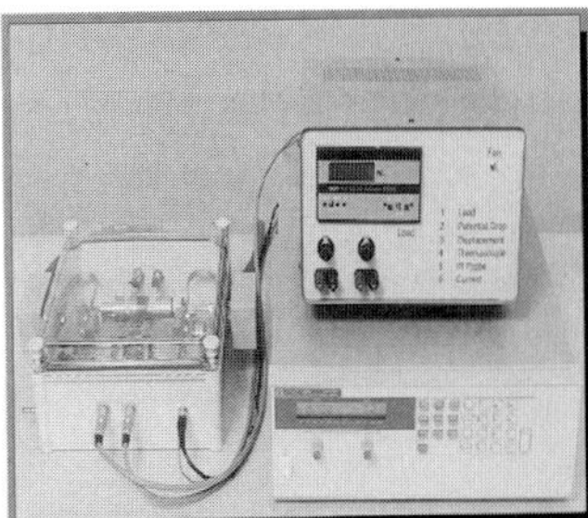

Figure 1. Electrothermo-mechanical test system (ETMT).

- Mechanical loading assembly (± 1.0 kN maximum) including a flexible grip system, load cell and displacement transducer, with computer controlled motor for null, mean or fatigue load application.
- System software to monitor and control each test type. Test data is stored in a computer (PC), and can then be analyzed with conventional PC programs.

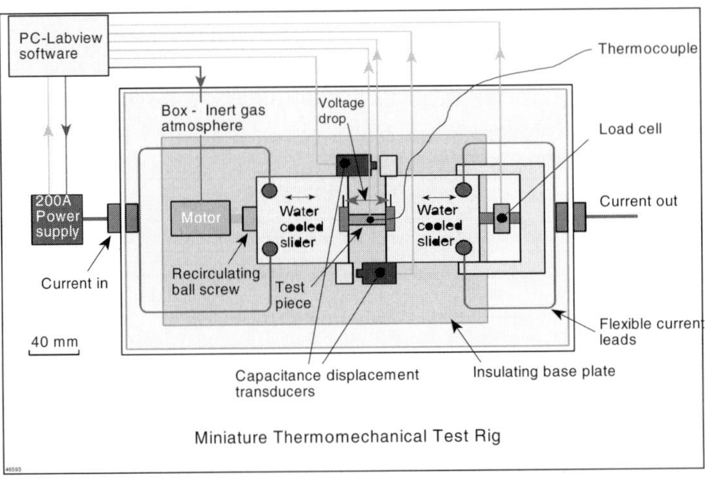

Figure 2. Schematic diagram of test system.

4 Materials and Test Pieces

4.1 Bulk Solders

Two solders were provided in the form of as-extruded bars 12x25x250 mm by Multicore Solders Ltd. Samples were tested in both the as-received (AR) and re-melted and water-quenched (WQ) conditions. WQ samples were kept in a fridge freezer at -10°C to maintain a stable microstructure. The nominal compositions of the solders were 37% Pb - 63% Sn for the lead-tin eutectic, and 95.8% Sn - 3.5% Ag - 0.7% Cu for the lead-free ternary alloy. Test-pieces were manufactured using EDM (electro discharge machining) and were 40x3x2 mm in dimensions.

4.2 Solder Joints

Solder pastes of the same composition as the bulk solders were supplied by Multicore Solders Ltd. Model symmetric lap joints were then designed. The joints were fabricated by reflow soldering techniques. A stainless steel tubular resistance furnace was designed and constructed to allow elevated temperature testing to be performed on these non-conducting test specimens.

5 Test Types: Bulk Solders and Solder Joints

Bulk Solders. Three types of test were performed:		
Uniaxial Strength Tests	at 3 deformations rates at 4 temperatures	0.25, 1 and 4 Ns^{-1} room temperature, 50, 75 and 100 °C
Creep Tests	at 4 temperatures at 5 stresses	60-120 °C lead-tin; 80-120 °C lead-free 15, 13.3, 11.7, 10, 8.3 MPa lead-tin 21.7, 20, 16.7, 15, 13.3 MPa lead-free
Thermo-mechanical Fatigue (TMF) Tests	30-75 °C cycle; load step-ramped to failure	

Solder Joints. Two types of test were performed:		
Creep Tests	at 75 °C	10 N lead-tin; 20 N lead-free
TMF Tests	at 30-75 °C cycle constant 75 °C	5, 10, 15 N lead-tin; 15, 20, 30 N lead-free Load step-ramped to failure

6 Results and Discussion

6.1 Bulk Solders

All tests on the bulk solders were performed on material in both the as-received (AR) and water-quenched (WQ) conditions. Because of the marked instability of the microstructures and sensitivities of the mechanical properties on prior thermo-mechanical treatment, data will only be presented, and comparisons with large specimens only made, for the material in the WQ condition.

6.1.1 Uniaxial Tests

Typical stress-strain curves for the lead-tin and lead-free material at 75 °C were obtained. Both materials showed the expected increase in flow stress with load rate, however, the mechanical properties of both materials are similar. The lead-free material had superior strength at lower temperatures (Figure 3). Data on the mechanical properties of lead-tin eutectic were supplied by Prof. W J Plumbridge at the Open University Materials Department. The strain rates used in the ETMT are considerably lower than those in conventional mechanical testing.

6.1.2 Creep Tests

Constant load creep tests were performed at four temperatures and three stresses appropriate for each material (Figure 4). The expected linear relationship between inverse temperature and logarithm of strain rate was observed. The lead-free solders had lower creep rates than the lead-tin materials at equivalent temperatures and stresses despite having similar strengths. The values of stress exponent calculated from those data are consistent with conventional creep tests.

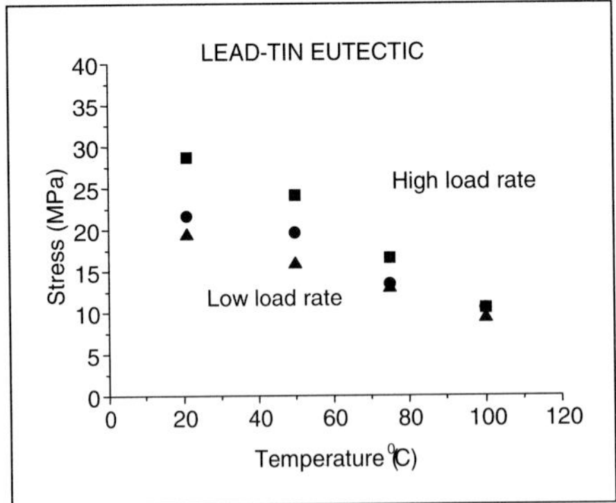

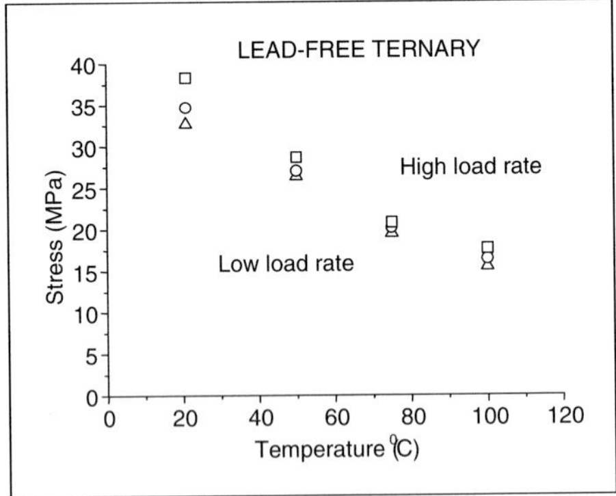

Figure 3. Flow stress as a function of temperature for both materials and 3 loading rates.

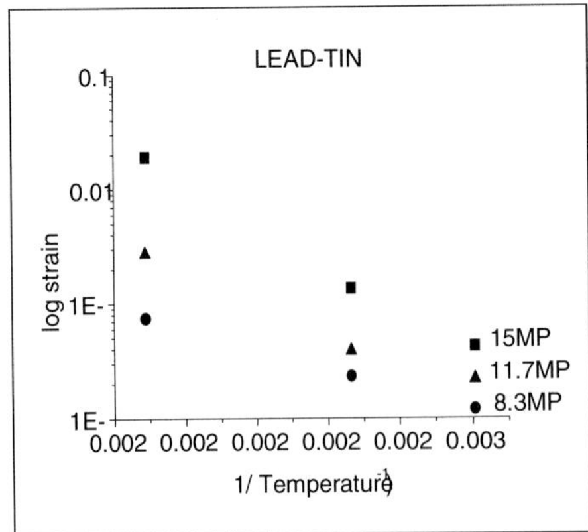

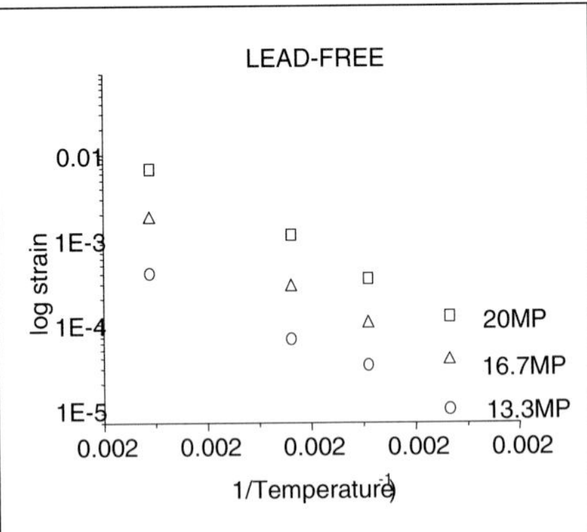

Figure 4. Creep data for both materials.

6.1.3 Thermo-mechanical Fatigue (TMF) Tests

In phase thermo-mechanical tests were performed on both materials. The load was cycled between 0 N and a fixed upper limit, initially set at 4 N and incremented by 4 N after each 50 cycles. The temperature was cycled between 30°C and 75°C. There was a dwell of 55s at the upper temperature and load, and 5s at the lower temperature and load giving a one minute cycle. The results are shown in Figure 5.

Note that the time axis can be considered to be equivalent to a stress axis as there is a linear relationship between time and load. The lead-tin eutectic extends at much lower peak loads than the lead-free solder and exhibits far fewer cycles to failure, lower by approximately a factor of two, This shows that the lead-tin eutectic is extremely sensitive to cyclic load and temperature variation. The lead-free material, however, survives to approximately the same peak stress as its failure stress under uniaxial monotonic loading at 75°C. Thus, the TMF performance of the lead-free material is significantly superior than the lead-tin eutectic. To the knowledge of the authors, equivalent data have not been generated by any other workers on conventional specimens.

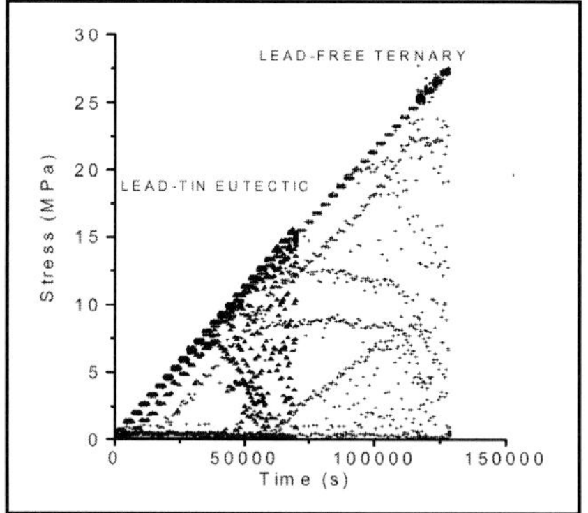

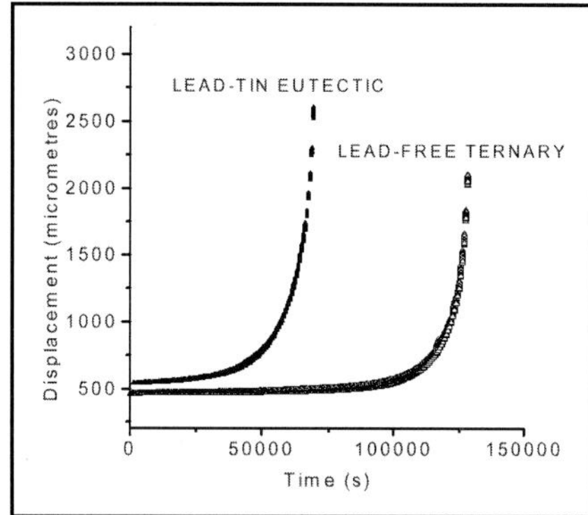

Figure 5. Thermo-mechanical fatigue data for both materials.

6.2 Solder Joints

The solder was supplied in the form of paste and used without any subsequent processing. Some difficulty in optimizing the reflow conditions to produce void-free joints was encountered. Thus, the results may indicate poorer properties than may be achievable in manufacturing conditions.

6.2.1 Creep Tests

Constant load creep tests to failure were performed on both solders (Figure 6). The superior creep performance of the bulk lead-free solder was repeated in the joint tests with loading now principally in shear. Despite being subjected to half the load, the steady state creep rate is an order of magnitude higher for the lead-tin eutectic joint.

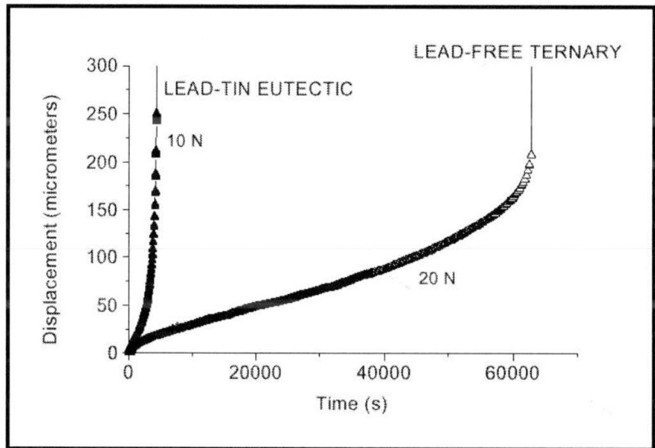

Figure 6. Constant load creep data for solder joints.

6.2.2 TMF Tests

Two types of TMF tests were performed: a constant temperature, step-ramped load cycle; and a constant load, temperature cycle. These were designed to mimic the service conditions experienced by electronic components. The constant temperature, load cycle tests (Figure 7) show the same poor creep behaviour of the lead-tin eutectic, with significantly greater extension at a given load. The load at failure, however, is only 5N higher for the lead-free solder joint.

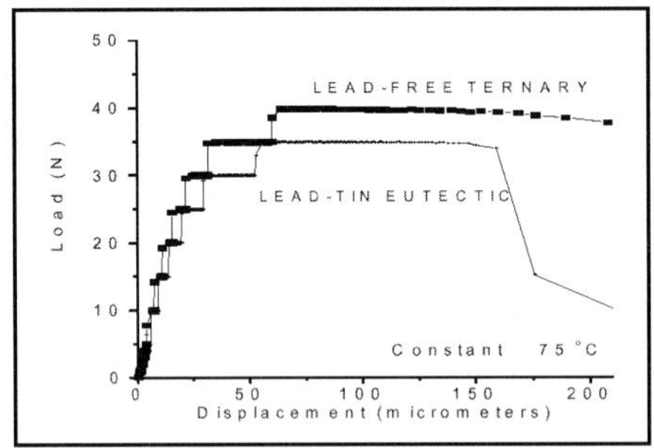

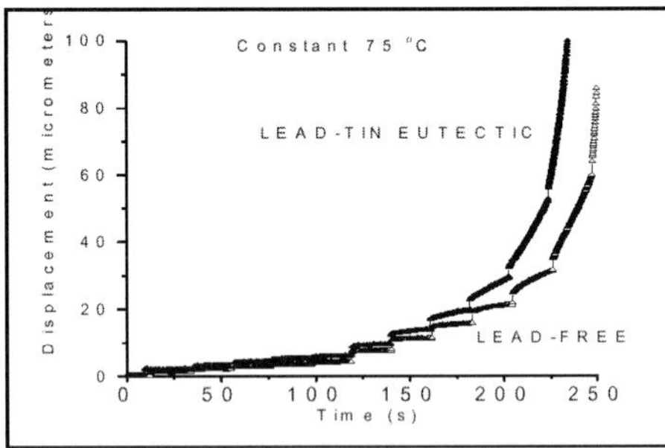

Figure 7. Constant temperature, step-ramped load cycle for both materials.

The difference between the behaviour of the materials is more clearly demonstrated in the thermal cycle tests (Figure 8). The lead-tin eutectic has a far higher baseline creep rate on cycling than the lead-free material and survives an order of magnitude fewer cycles to failure. Fractography on both systems suggests that there was significant prior voiding in the joints due to processing.

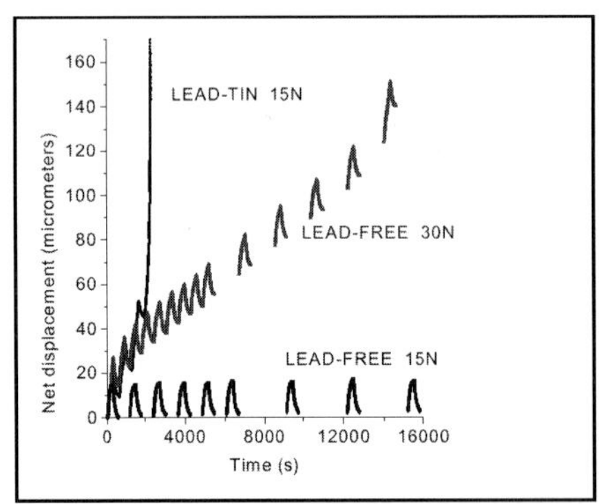

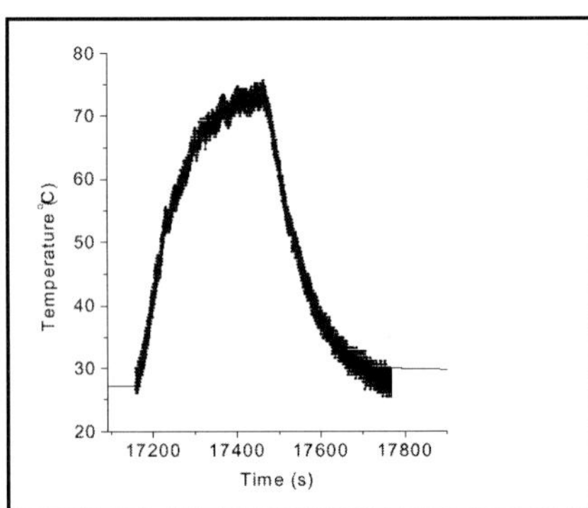

Figure 8. Constant load, thermal cycle displacement data for joints made from both solders.

7 Conclusions

It has been demonstrated that the ETMT produces credible data on bulk solder alloys that compare favorably with data produced by conventional testing [5]. The tests on the model joints enabled a rapid comparison of the TMF behavior of the two alloy systems. Data were generated in a matter of hours rather than months. The ETMT was clearly able to discriminate successfully between the solders in both bulk and joint forms. The lead-free materials exhibited superior tensile, creep and TMF behavior. The most marked improvement was noted in TMF which most closely mimics in-service conditions.

8 Acknowledgments

This work was performed with the support of the UK Department of Trade and Industry. Thanks are due to the NPL Electronics Interconnection Group for advice on solder joint design and construction, Multicore Solders Limited for supply of material and the UK Open University's Group for guidance on test configuration and analysis.

9 References

[1] NPL-ITRI Report on Lead-free Soldering for the DTI, 1999.
[2] K Nimmo, ITRI. 1998, 71.
[3] M. Miller *et al.*, JOM. 1994, 23, 595 - 601.
[4] B. J. Foroodian, P. M. Mummery, Z. Fan, B. Ralph, EngD Conf. 1999, 67 - 76.
[5] B. Roebuck, M. G. Gee, Materials Science and Engineering. 1996, A209, 358 - 365.

MoCu and WCu Alloys for MM-Wave Packaging

M. Knüwer
Fraunhofer-Institut für Angewandte Materialforschung (IFAM), Bremen (Germany)
T. Wiefel, K.-H. Wichmann
DaimlerChrysler Aerospace AG, Ulm (Germany)

1 Introduction

On ground of increasing demand for communication and multimedia technology more and more transmitting frequencies are needed. There is not any decrease of band width of these frequencies. Therefore we make way for higher frequencies which are in GHz range. 100 GHz are up to date at these so called maximum frequency applications. Systems until about 300 GHz are already in development. Thus these frequencies are in the range of vibrations of molecules. If the covibrations of the molecules are reached, they will take up the energy and it comes to absorption of the electromagtnetic radiation. On transmission paths e.g. from the surface of the earth to satellites especially absorption by water vapor is sensible so that certain frequencies cannot be used for those transmission paths. On the other hand the effect can be utilized to realize local area networks for e.g. local limited networks on airports. At this we are anxious to let electromagnetic waves be absorbed after relatively short distances.

Nevertheless high frequencies also involves many problems which have to be mastered and which lastly leads to high requirements on packagings for maximum frequency curcuits. For example it comes to the skin effect by which the conduction of electomagnetic waves is restricted on the surface area of electrical conductors. Usage of so called tubular conductors which replaces conventional cables takes this into account. Geometry of tubular conductors has crucial influence on frequencies of electromagnetic radiation. From this yields the demand for high accuracy to size.

Beyond it maximum frequency applications requires changed permittivities of substrate and curcuit materials. Therefore we superseded silicium the common material for electronical curcuits by gallium arsenic (GaAs) which is remarkable sensitive. This material is more brittle than silicium and has to be protected against humidity moreover. For this reason elecronical curcuits are placed in housings. Beyond it every conductor inside the housing represents a little transmitter for which the question is to screen against other mm-wave modules. From these properties again result the high demands on hermeticity and mechanical capacitance of the mm-wave packagings. Furthermore thermal deformations in elongation of the concerned materials have to be harmonized to each other. Thus meanwhile soldering the ceramic substrate into the packaging which has to extend more than the substrate a compressive not a tensile stress is exerted on the last one meanwhile cooling.

Since density of integration is getting higher with simultaneously increasing power a great amount of heat due to energy losses is formed on very small area. This have to be dissipated and carried off by packaging material for the safety of the curcuits. So thermal conductivity of the packaging material should be as high as possible.

Of course the curcuits inside the packaging should interact with outer world. Therefore it is necessary to provide for an interface for data transfer. As a rule this is obtained by coaxial lead-in wires, tubular conductor openings and others. At present the coaxial lead-in wires have to be realized as a glass-to-metal sealing for hermeticity of the packagings.

Pseudo alloys of the systems Mo-Cu and W-Cu includes the required physical properties concerning thermal expansion, thermal conductivity and mechanical capacitance. However, these materials cannot be produced by common metallurgical processes that only powder metallurgy comes into question. Moreover, compared with shaping by machine tool with removal of ships metal injection moulding technique as primary shaping method offers the advantage of a nearly free shaping of geometry with very low material waste. Compared with machining this renders considerable savings of costs at producing of millimeter-wave packagings possible.

To realize those mm-wave packagings the research project ‚New metallic material concepts for millimeter-wave packagings‘ was initialized by the DaimlerChrysler Aerospace AG in Ulm. The following firms respectively institutes could be secured as project partners:

- H.C. Starck GmbH & Co. KG, Laufenburg
- FhG/IFAM, Bremen
- GKN Sinter Metals GmbH & Co. KG, Radevormwald
- DaimlerChrysler AG, Forschung, Ulm

Responsable body of the project is the Kernforschungszentrum Jülich (PLR). The processing period began at 01.04.1997 and ends at 31.03.2000.

2 Results

The choosed packaging material Mo60Cu should fulfill our expectations concerning thermomechanical properties. This was checked by FE-calculations. The behaviour of a Mo60Cu-packaging in which a LTCC-substrate was soldered by means of L-Sn63Pb was simulated. This packaging is closed with a top consisitng of CuNi20. A further packaging was simulated for comparison. However, this consisted of FeNi2. Both packagings had been simulated as if they were mounted on a heat sink consisting of copper.

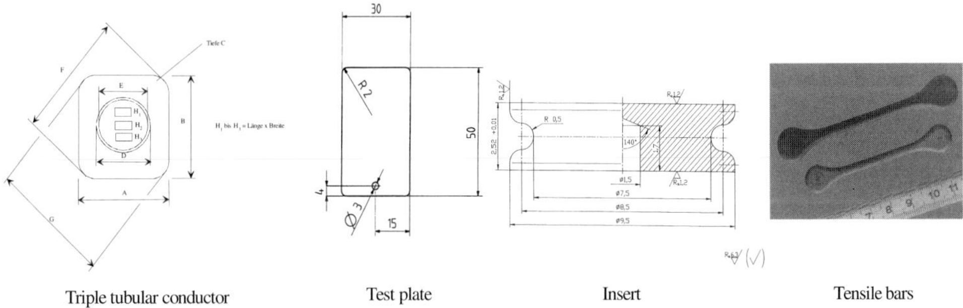

Triple tubular conductor   Test plate   Insert   Tensile bars

Fig. 1

The simulation showed that operation temperature inside the packaging can be lowered by the choosed material Mo60Cu up to more than 10°C compared with iron materials. Furthermore compressive stresses in the substrate in joint condition are only half as much as in iron materials. This verifies impressively the hitted material choice.

Several specimen which finally ended up in a so called demonstrator were produced meanwhile the running time of the projekt. On these specimen the examinations were intended to optimize materials and the manufacturing method.

Tungsten, molybdenum and copper powder were placed at disposal as raw materials. In doing so for the present upon qualitites have fallen back which can be produced by standard methods. These powders have been worked up to a mass which is able to be injected, the so called feedstock. In doing so molybdenum respectively tungsten powder has been mixed with copper powder and the mixture has been compounded with a polymer binder. Then the feedstock has been worked out in a double-sigma masticator.

The optimization of the Mo60Cu-Feedstock concerning injection moulding and properties of the sintered specimen made a high energy grinding of the powder necessary. Since this represents an expensive additional working step at feedstock production a compound powder with a low size of primary particles and microhomogenous dispersion as much as possible were intensively tried to be created parallel to that. Thereby the method of co-reduction of metal oxides appeared as a practicable way for the powder production that the high energy grinding of powders during feedstock production can fall out. The demonstrators made out of that feedstock exhibits a more homogenous microstructure and a higher dimensional stability and in addition it can be sintered at lower temperatures.

The optimized feedstock has been subsequently worked into tensile bars on a Klöckner Ferromatik FM 20 die-casting machine. Thereby parameters of injection moulding have been varied in wide but significant limits for optimization of process parameters.

Sintering of the tensile bars took place after a two-step debinding first in a solvent then thermal in a hydrogen atmosphere. Sintering temperatures and thermal retardation were varied in the following sintering process. The tensile bars produced like this were judged by density and microstructure. In spite of wide variety of injection and sintering parameters densities of MoCu alloys laid between 95,5 % and 97 % of theoretical density. In the case of WCu alloys a little bit lower densities were reached. At higher sintering temperatures we have to intend on a evaporation of copper which changes the analysis of the sintered tensile bars. Therefore residual porosity was suggested to lie a little bit higher as the determined densities let suppose.

The tensile tests of the differently produced Mo60Cu tensile bars results in the following strength factors:

- Modulus of elasticity (E): ca. 180 GPa
- Yield strength ($R_{p0.2}$): ca. 250 MPa
- Tensile strength (R_m): ca. 500 MPa

Breaking elongations varies between about 5 % and 10 %.

The tungsten powders produced after standard methods are already fine enough so that the high energy grinding can be abandoned at their processing into W-Cu-feedstock whereby dimensional stability was better than the one by using Mo60Cu-feedstock with high power grinded molybdenum powder.

The following four different W-Cu-feedstocks were produced:

P75: W90Cu10: W: 1.8μm Cu: 6.0μm
P76: W90Cu10: W: 3.6μm Cu: 13.6μm
P77: W70Cu30: W: 3.6μm Cu: 6.0μm
P78: W70Cu30: W: 1.8μm Cu: 13.6μm

There was no difficulty in debinding the demonstrators of the material system W-Cu in solvent. However, at thermal debinding in chamber oven cracks appeared in specimen which can only be avoided by using a hydrogen atmosphere. Four charges were sintered with different sintering programs. The specimen with more fine W powder generally showed a higher density as well as low roughness of the surface.

Inserts, test bars, test plates and demonstrators were produced from optimized Mo60Cu-feedstock. First over 50 % of the binder was eliminated in a solvent. After drying the specimen rest of the binder was removed thermally. The burnout process was formed so that a slight sintering of the powder particles occurs after subtilization of the binder. This guarantees sufficient stability of the specimen in the ‚brown condition'. With sintering not only a batch oven but also a walking beam furnace was used to realize production of specimen in industrial standard.

Specimen sintered in the batch oven have had higher densities than those which were sintered in the walking beam furnace in general. The test plates sintered in the batch oven have had the highest density of all specimen which was 97 %. The density reached with the walking beam furnace lay with 93 % a little lower. It is not clear for what reason the density of the demonstrator sintered in the batch oven reached only 91 %.

The inserts should attend to an improved heat flow in plastic packagings. Therefore some inserts have been surrounded by injection moulding with different plastics filled with glass fibers and the contact between metal and plastic have been examined with an elecron scan microscope. Hence it followed that the contact between Mo60Cu-inserts and PEI respectively ABS was substantially satisfactory. Only on the edge of the contact little gaps between plastic and insert can be recognized. Those can take a disadvantageous effect on the following metallization and machanical properties since dirt and liquids can be accumulated in them.

Before metallization of the plastic surrounded inserts the influence of primary treatment on the MoCu-inserts which is necessary for plastics has been examined. Roughnesses of specimen consisting of Mo40Cu were measured after conditioning for the metallization of PES. Thereby conditioning requirements for the metallization of PES, liquid cristalline polymers (LCP) and PEI were used. Erosion of material has been determined by use of density and weight of the specimen before and after the treatment.

The lowest influence on roughness has the conditioning for LCP materials at which no considerable erosion occured after 13 minutes of treatment. On the other hand a steady increase of roughness with simultaneous loss of weight had taken place at the conditioning for PEI after a delay of 3 minutes. Thereby corrosion succeded selectively at the molybdenum that the specimen became coloured like copper. The loss of weight at conditioning for PES was the highest one and was proportional to duration of treatment. The weight has decreased about 3 % after 7 minutes. At this conditioning method the highest rise of roughness has also been determined. However, it starts not till a duration of treatment of about 7 minutes. The average peak-to-valley height R_Z came to about 75 μm after 13 minutes of treatment.

The uncoated demonstrator packagings have been grey-rose-coloured which is reducable to isles consisting of copper on the surface of specimen. Before galvanical metallization in a copper bath of commercial size combined with sufuric acid the demonstrator packagings have been commonly cathodic degreased and have been activated in a 10 per cent sulfuric acid. At the following copper deposition of 10 µm it came to a rough copper surface with plentiful bud-like peaks. A preferred deposition of copper has obviously taken place on surface inhomogeneity like the copper isles. Only by a previous treatment in chromosulfuric acid in which the above mentioned inhomogeneities have been removed a deposition of copper without bud-like peaks could have been guaranteed. The demonstrator packagings coated with 10 µm copper have been additionally coated with 1 µm nickel and 2 µm gold. Thereby the nickel plating serves as barrier against diffusion and the gold plate serves for preparing an electrically conductive survace as well as it serves to cover against corrosive influence.

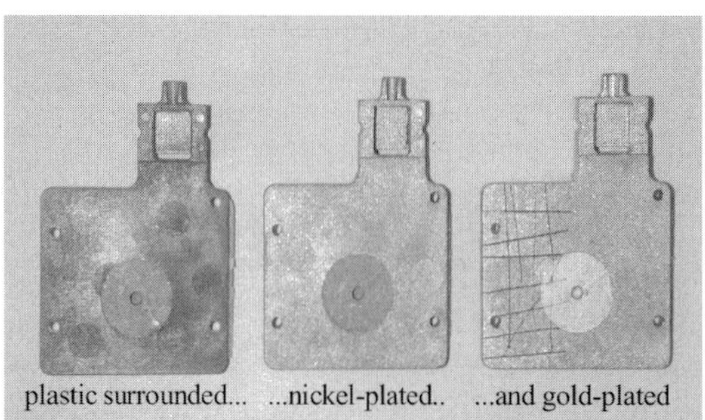

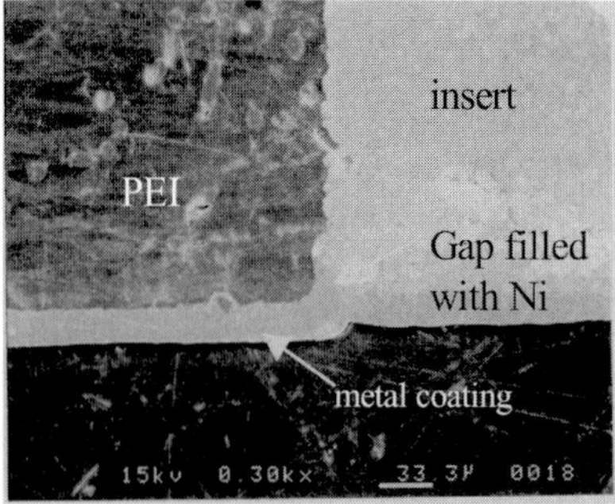

Fig. 2: Mo60Cu-insert surrounded with PEI and metallized

First the plastic surrounded inserts have been chemically nickel-plated and subsequently being galvanically gold-plated after appropriate conditioning. Adhesive strength has been examined by cross hatch method and is very satisfactory.

Thermoanalytical and electron scan micrographic examinations of the triple tubular conductors have been carried out. Objective of these examinations was to determine the homogenity and surface quality as well as to characterize the feedstock as a reference pattern. The amount of binder has been determined to 12 % by thermal analysis whereby the dispersion wasn't homogen inside this specimen.

However, microstructure of triple tubular conductors consisting of not optimized feedstock has proved to be very homogen in general compared with that of the tensile bars. Only a light porosity has been near the edges. Dimensional stability of the sintered triple tubular conductors has been determined unsafe because of burrs adhered to the edges. The surfaces of green bodies showed relatively intense structure which is reduced to the injection moulding tool as well as to inhomogeneities of powder dispersion of the feedstock which was not jet been optimized. At powder dispersion first of all molybdenum agglomerates attract attention on the micrograph.

For a better quality rating of the specimen produced in this project mircostructural examinations were made with optical microghaps of several dry powdered metal pattern which are available on the market. Compared with specimen produced in this project they showed a substantially higher inhomogeneity. The two following micrographs such patterns. Especially a lot of big pores and relatively big accumulations of copper.

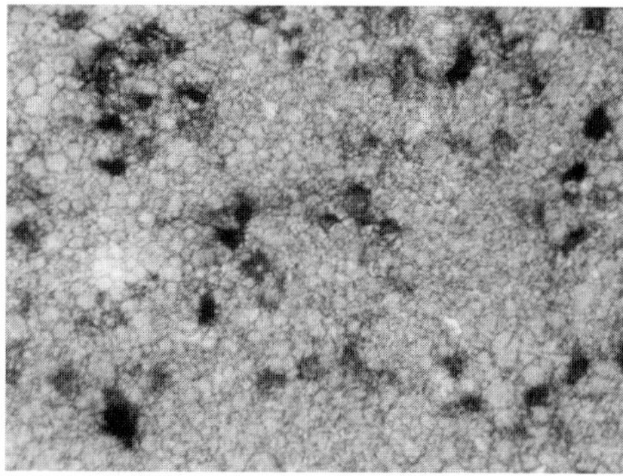

Fig. 3: Mo60Cu-pattern (density = 70%)

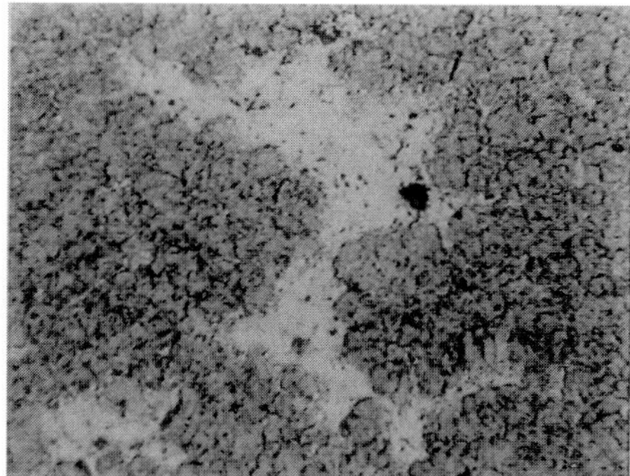

Fig. 4: W75Cu-pattern (density = 83%)

The first tests concerning the integration of substrates has been taken place when no MIM-specimens were at disposal. Because of that some packagings have been manufactured in conventional machining technique first. LTCC substrates were tried to be soldered into these packagings. Vapor phase soldering was used as method which promptly proved to be usable. The fusion of solder with substrate and packaging is very good. The solder has distributed like intensed and formed a layer with minimum thickness. The fusion also stayed optically unchanged after 100 temperature cycles from –40°C to +70°C.

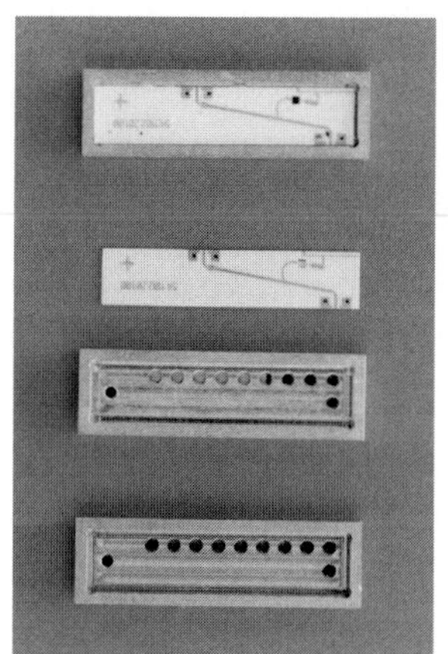

Fig 5: Pattern packagings for integration tests with the demonstrator (upwards: cutted packaging, gold-plated packaging, substrate, packaging with substrate)

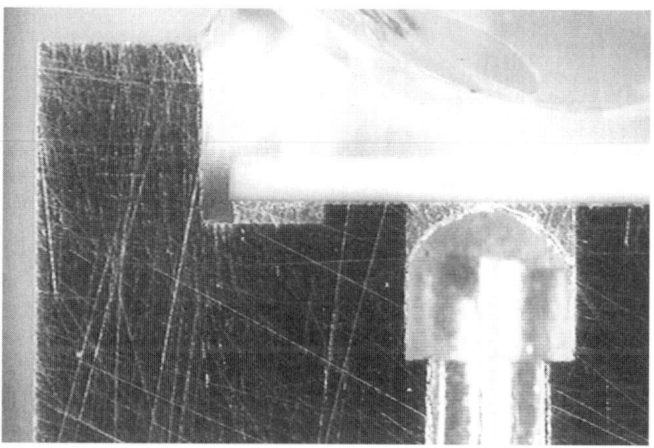

Fig 6: Soldered joint between substrate and demonstrator

Beside the works mentioned til now further characteristic values of the produced materials have been determined, whereby only the important ones shall be mentioned in this text. The chemical analysis (F-AAS) of weight proportion of molybdenum and copper in MIM tensile bars resulted in a MoCu-relation of 62.3 % to 36.4 %. The missing part of 1.3 % turned out in a brown colour of the means of attack which is a reference to a rest organic binder material. A comparison of the strength factors determined in tensile tests and the ones of pure copper verifies a moderate til good sintering of the tested specimen. The portions of elements, density, relative density and hardeness have been determined on test plates consisting of Mo60Cu respectively W75Cu and compared with specimen commonly available at the market.

Table 1: Determined values of MoCu- and WCu-materials

Specimen	Percentage of Cu	Percentage of Mo/W	Density	Relative density	Hardeness
P2-02	37 %	63 % Mo	9.49 g/cm³	98 %	387 HV 10
P2-03	37 %	63 % Mo	8.73 g/cm³	90 %	364 HV 10
M2	40 %	60 % Mo	6.8 g/cm³	70 %	52 HV 10
M3	36 %	64 % Mo	9.42 g/cm³	97 %	204 HV 10
M4	25 %	75 % W	13.83 g/cm³	83 %	186 HV 10
M5	37 %	63 % Mo	9.65 g/cm³	99 %	324 HV 10

The patterns M2 to M5 are materials commonly available at the market meanwhile specimen called P2 were selfmade. The relatively high hardenesses and high relative densities of the selfmade specimen are significant. Since the specimen M2 to M5 weren't injection moulded but pressed this is a particular circumstantial evidence for the quality of the specimen P2. The very low hardeness of the injection moulded pattern M2 is caused by its high porosity. Furthermore thermal coefficients of expansion of some of the above mentioned specimen have been measured. The average values of the injecion moulded specimen are lying over that of the pressed ones available at the market (cp. table 2). The ceramics coming into question as substrate have thermal coefficients of expansion between about $5 * 10^{-6}$ K^{-1} and $7 * 10^{-6}$ K^{-1}. The above mentioned demand for compressive stress inside the substrate after its joining is fulfilled by thermal coefficients of expansion of the examined MIM materials.

Table 2: Thermal coefficients of expansion (10^{-6} K^{-1}) of different MoCu- and WCu-materials in X- and Y-direction of the test plates

Specimen	α_X	α_Y
P2-02	9.07	9.22
P2-03	9.42	9.73
M3	7.56	7.99
M4	8.97	9.78
M5	8.93	8.95

The works are supported and continued in a current project with means of the Bundesministerium für Bildung, Wissenschaft, Forschung und Technologie in the MaTech program under the benefit registering mark 03N1030A-E.

Thermodynamic Database for Micro-Soldering Alloys

Ikuo Ohnuma, Xing Jun Liu, Hiroshi Ohtani and Kiyohito Ishida
Department of Materials Science, Graduate School of Engineering, Tohoku University, Aoba-yama 02, Sendai 980-8579, JAPAN

1 Introduction

Even though lead bearing solders have been in use for 5000 years, the drive for the fundamental study of solder metallurgy has begun only recently since 1980 with the advent of new methods for interconnecting electronic components [1]. Lead-tin solders have been, and are the most popular materials for electronic packing because of their excellent material properties and low cost. However, in view of the environmental and health issues concerning the toxicity of Pb, there has been a significant increase in the impetus for design and development of Pb-free solders. Consequently a need has arisen for thermodynamic data for reliable predictions of liquidus, solidus, phase fractions, equilibrium compositions etc in multi-component solder alloys. Such information would also provide the basic input data for understanding the reaction between the substrate and solders, solidification behavior and various thermodynamic properties relating to the solders and substrates. Recently, the author's group has reported the development of a CALPHAD (CALculation of PHAse Diagrams) [2] based thermodynamic database for micro-soldering alloys containing the elements Pb, Bi, Sn, Sb, Cu, Ag and Zn [3]. Some calculations and applications are presented in this paper to demonstrate the validity of the database.

2 Thermodynamic Database

In the CALPHAD method, the thermo-dynamic parameters for describing the Gibbs energy of the liquid and solid phases in a chosen system are evaluated by optimizing the experimental data pertaining to the phase boundary compositions and thermochemical properties such as activity, heat of mixing and enthalpy of formation. For the present evaluation, experimental determinations of phase equilibria in several binary, ternary and quaternary systems were conducted using DSC, EDX, X-ray diffraction and metallographic techniques. The parameter assessments for the binary and ternary systems were made, based on these results as well as on previous data on phase diagrams and thermochemical properties. Table 1 lists the thermodynamic assessments of binary systems that were carried out by our group [4-7] and also those previously reported [8-15]. All the ternary phase diagram assessments were conducted by our group and the thermodynamic parameters evaluated were arranged within the framework of the Thermo-Calc software [15]. Figure 1 shows the variety of information relating to the liquidus and solidus surfaces, isothermal and vertical section diagrams, mole fractions of the phase constitutions and thermodynamic properties of multi-component soldering alloys that can be derived from the present database. Surface tension [17] and viscosity [18] of the liquid phase can also be predicted from a knowledge of the Gibbs energy of liquid combined with the use of appropriate models.

Table 1: Survey of thermodynamic assesments

System	Reference	System	Reference
Ag-Bi	8	Cu-Pb	O
Ag-Cu	9	Cu-Sb	O
Ag-In	O	Cu-Sn	O
Ag-Pb	10	Cu-Zn	15
Ag-Sb	10	In-Pb	O
Ag-Sn	8	In-Sb	6,14
Ag-Zn	O	In-Sn	15
Bi-In	O	In-Zn	15
Bi-Cu	11	Pb-Sb	4
Bi-Pb	12	Pb-Sn	4
Bi-Sb	O	Pb-Zn	O
Bi-Sn	5	Sb-Sn	O
Bi-Zn	O	Sb-Zn	O
Cu-In	O	Sn-Zn	7

O Present work

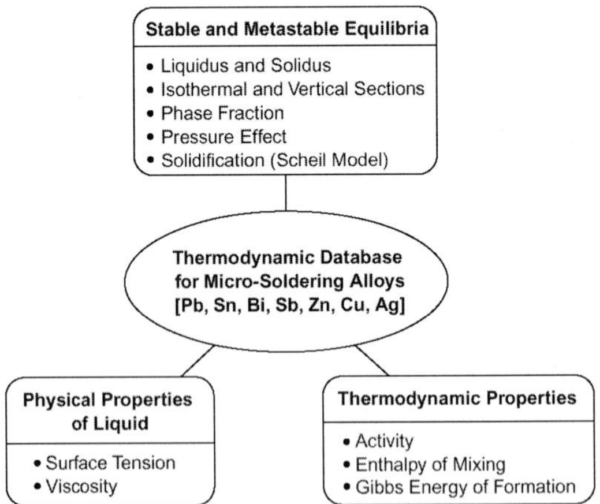

Figure 1. Derivable information from the thermodynamic database for micro-soldering alloys.

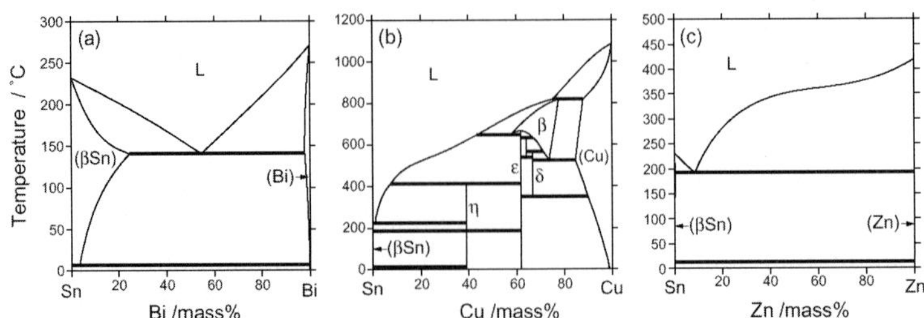

Figure 2. Calculated Sn-X binary phase diagrams of (a) Sn-Bi, (b) Sn-Cu, (c) Sn-Sb and (d) Sn-Zn systems.

3 Examples of Calculation

3.1 Phase diagrams of Pb-bearing and Pb-free solders

Some examples of phase diagram calculations in the Pb-bearing and Pb-free solder alloy systems are described in this section. Figure 2 shows the calculated phase diagrams for the Sn base binary alloys, which are potentially the alternatives to Pb-Sn solders. Figure 3 shows the isothermal section diagrams of the Pb-Sn-Bi and Pb-Bi-Sb ternary systems at 150°C. Figure 4 shows the calculated vertical section diagrams of the Pb-20mass%Sn-Bi and Pb-10at.%Bi-Sb alloys, which are compared with the DSC measurements. These analyses as well as the Pb-Sn-Sb and Sn-Sb-Bi ternary assessment have enabled reliable calculations of phase equilibria in the Pb-Sn-Bi-Sb quaternary system [3]. Sn-Ag base alloy [19] is one of the candidates in the group of lead-free tin-based alloys that are of interest as replacements for Pb-Sn eutectic alloy solders. Figure 5(a) shows the calculated vertical section diagrams of Sn-Ag-Zn system at Ag:Sn=3:1 with the superimposed experimental liquidus temperatures by Karlhuber et al.[20]. The calculated activity of Zn in the ternary liquid phase is compared with experimental data from [20] in Fig. 5 (b). The agreement between the calculation and the experimental data is quite satisfactory. Figures 6 (a) and 6 (b) show the calculated liquidus surface in the Sn-Ag-Cu ternary alloy and the vertical section diagram at 20 mass% Sn in the same system with superimposed experimental data [21] respectively.

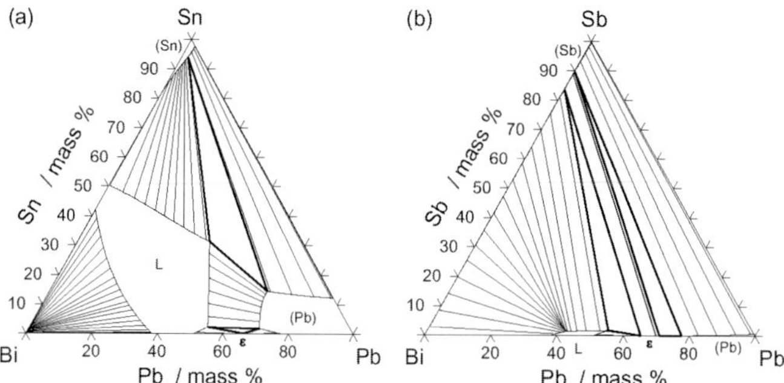

Figure 3. Isothremal section diagrams of (a) Pb-Sn-Bi and (b) Pb-Bi-Sb systems at 150°C.

3.2 Solidification

One of the ways of checking the validity of any database is by the simulation of the solidification sequence in a given alloy. The Scheil model, generally used for the simulation of the solidification process, has been applied to describe both the equilibrium and non-equilibrium solidification behaviours in the binary Sn-Bi system. The non-equilibrium case has been included because it has been previously observed that non-equilibrium solidification occurs in many solder alloy systems. The Scheil model, assumes that local equilibrium exists at the liquid/solid interface and diffusion is absent in the solid phase. Figure 7 (a) and (b) show the calculated phase fraction vs. temperature variation in the Sn-Bi binary alloys under equilibrium and non-equilibrium solidification conditions respectively. It can be seen that the liquid phase remains in most alloys even at 139°C of eutectic temperature, suggesting that the freezing temperature is very low even in alloys with very low Bi content. The presence of liquid phase at these temperatures due to non-equilibrium solidification is considered to be one of the reasons for the observed lift-off of the solder from the Cu substrate [22].

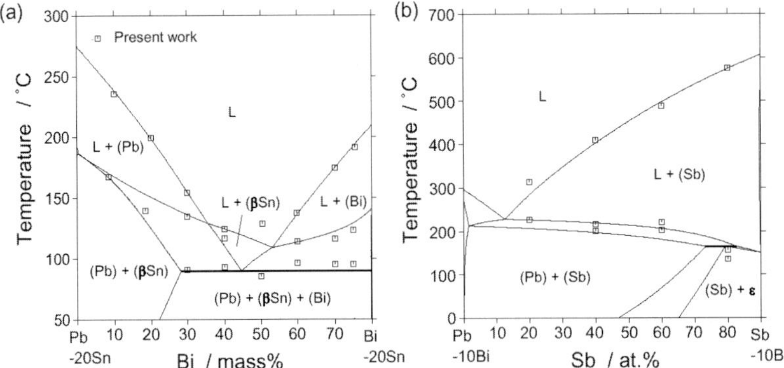

Figure 4. (a) Pb-20mass%Sn-Bi and (b) Pb-10mass%Bi-Sb vertical section diagrams.

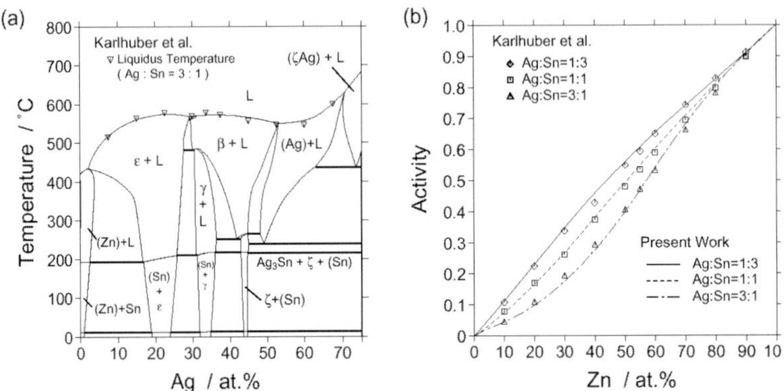

Figure 5. (a) Vertical section diagrams of the Sn-Ag-Zn system at Ag:Sn=3:1 and (b) activity of Zn in the ternary liquid phase compared with experimental data [20].

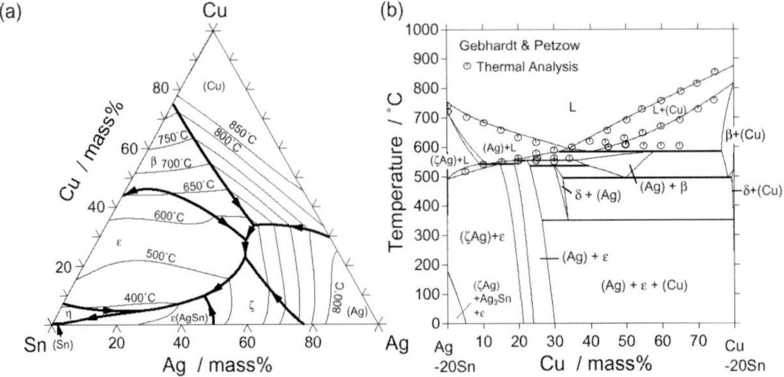

Figure 6. (a) Liqiudus surface in the Sn-Ag-Cu ternary system and (b) vertical section diagram at 20 mass% Ag compared with experimental data [21].

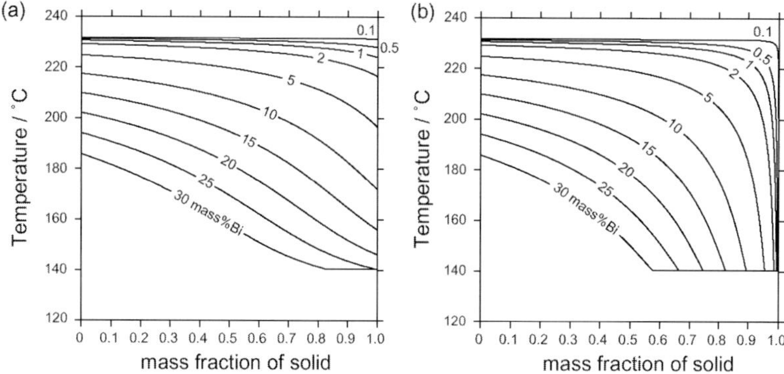

Figure 7. Calculation of phase fraction of solid vs. temperature in Sn-X binaries: (a) equilibrium solidification and (b) non-equilibrium solidification by Scheil model.

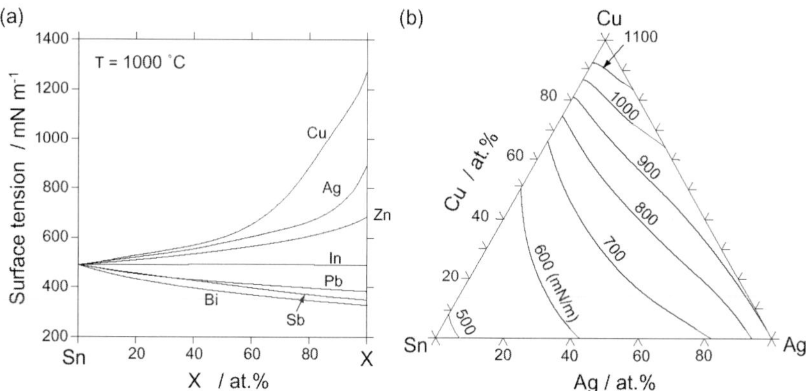

Figure 8. Calculated liquid surface tension in alloys of (a) the Sn-X binary systems and (b) the Sn-Ag-Cu ternary system at 1000°C.

3.3 Surface tension

Surface tension and viscosity in the liquid state can be estimated by combining the recently proposed calculation models [17,18] and the present data base. The surface tension calculation has been carried out assuming that (i) an equilibrium exists between the bulk phase and the monolayer at the surface, (ii) the composition of surface monolayer can be estimated from the Gibbs energy of liquid phase and (iii) the ratio of coordination number between the surface and the bulk is 3/4 [17]. Figure 8 (a) shows the calculated surface tension vs. solute content X in the Sn-X binary systems at 1000°C. Figure 8 (b) shows the iso-surface-tension contour in the Sn-Ag-Cu ternary system. This kind of information is expected to be useful for assessing the melting behavior and the manufacturability of solders from this alloy system.

4 Summary

A thermodynamic database for micro-soldering alloy systems, consisting of the elements Pb, Bi, Sn, Sb, Cu, Ag and Zn has been developed using the CALPHAD method. It is shown that the database can be used 1) to obtain the liquidus and solidus surfaces, isothermal and vertical section diagrams, the mole fraction of the phase constitutions etc. for multi-component soldering alloys, 2) to simulate the equilibrium and non-equilibrium solidification processes and 3) to estimate the physical properties such as surface tension and viscosity in solder alloys. This database will be very useful for developing Pb-bearing and Pb-free solders.

5 Acknowledgements

The authors wish to thank Dr L Chandrasekaran of DERA, UK for helping in the final stages of the preparation of the manuscript. The support from Casio Science Promotion Foundation is also acknowledged.

6 References

1. P. T. Vianco and D. R. Frear, J. Metals, 1993, 45, (7), 14-19.
2. N. Saunders and A. P. Miodownik, CALPHAD, Pergamon, 1998.
3. I. Ohnuma, X. J. Liu, H. Ohtani and K. Ishida, J. Electron. Mater., 1999, in press.
4. H. Ohtani, K. Okuda and K. Ishida, J. Phase Equilibria, 1995, 16, 416-429.
5. H. Ohtani and K. Ishida, J. Electron Mater., 1994, 23, 747-755.
6. S. Ishihara, H. Ohtani, T. Saito and K. Ishida, J. Jpn. Inst. Met., 1999, 63, 695-701.
7. H. Ohtani, M. Miyashita and K. Ishida, J. Jpn. Inst. Met. 1999, 63, 685-694.
8. U. R. Kattner and W. Boettinger, J. Electron Mater., 1994, 23, 603-610.
9. F. H. Hayes, L. Lukas, G. Effenberg and G. Petzow, Z. Metallkd., 1986, 77, 749-754.
10. B. Z. Lee, C. S. Oh and D. N. Lee, J. Alloys Comp., 1994, 215, 293-301.
11. I. Niemela, G. Effenberg, K. Hack and P. J. Spencer, CALPHAD, 1986, 10, 77-89.
12. H. L. Lukas, SGTE Report, 1991.
13. M. Kowalski and P. J. Spencer, J. Phase Equil., 1993, 14, 432-438.
14. I. Ansara, C. Chatillon, H. L. Lukas, T. Nishizawa, H. Ohtani, K. Ishida, H. Hillert, B. Sundman, B. B. Argent, A. Watson, T. G. Chart and T. Anderson, CALPHAD, 1994, 18, 177-222.
15. B. J. Lee, C. S. Oh and J. H. Shin, J. Electron Mater., 1996, 25, 983-990.
16. B. Sundman, B. Jansson and T. O. Anderson, CALPHAD, 1985, 9, 153-190.
17. T. Tanaka and I. Iida, Steel Research, 1994, 65, 21-28.
18. S. Seetharaman and D. D. Sichen, Metall. Mater. Trans. B, 1994, 25B, 589-595.
19. J. Glazer, Int. Mater. Rev., 1995, 40, 65-92.
20. S. Karlhuber, K. L. Komarek and A. Mikula, Z. Metallkd., 1994, 85, 307-311.
21. E. Gebhardt and G. Petzow, Z. Metallkd., 1959, 50, 597-605.
22. K. Suganuma, J. Jpn. Inst. Electro. Packaging, 1999, 2, 116-120.

Ageing Effects of Wire Bonds on Plasma Cleaned Flash-Gold Surfaces

Volkmar von Arnim, Jürgen Feßmann, Heinz Osterwinter, Mechtilde Schäfer
Fachhochschule Esslingen
Hans-Ulrich Völler
Transferzentrum Mikroelektronik Göppingen (TZM)

1 Introduction

Wire bond connecting techniques are widely used in electronics manufacturing. They are highly automatable and meet requirements for increasing miniaturization of electronic circuits. Only few material combinations out of many wire materials and surface metallizations became generally accepted in practice. Mainly the ultrasonic (US) wedge-wedge bonding using Al-wire on Ni/flash-gold surfaces and the thermosonic (TS) ball-wedge bonding with Au-wire on thick Au-surfaces are established in application (fig. 1) [1,2,3]. It is of great economical interest, to replace the rather slow US bond procedure on cheap and solderable Ni/flash-gold layers by the fast TS bond process [4,5]. However, a reliable TS bonding on Ni/flash-gold is only possible in laboratory or small manufacturing scale so far [6]. Ni/Pd/flash-gold layers have been developed as an economical or technical alternative for the deposition of thick reductive or galvanic gold layers. Although this rather new system has proved its suitability for TS bonding, only little practical experience towards the reliability of the bond formation and the bond strength exists [7].

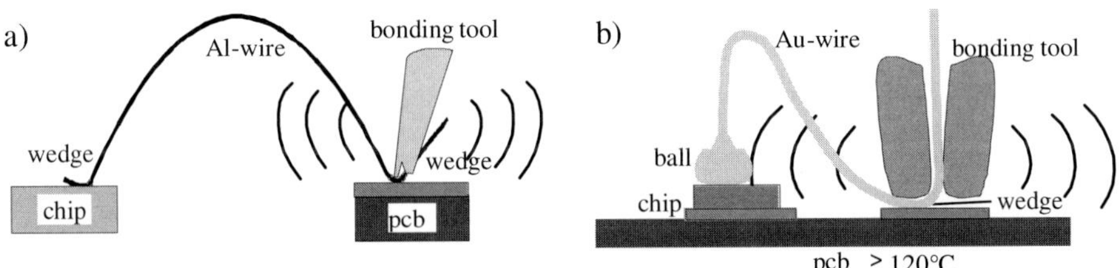

Figure 1: Ultrasonic wedge-wedge bonding (a) and thermosonic ball-wedge bonding (b)

Yield and bond reliability are influenced by many processes before bonding during manufacturing, transport, and handling of the printed circuit board (pcb). Further influences are bonding parameters during the bond procedure as well as manufacturing steps and application conditions afterwards. All these processes have got a potential for further optimization. This investigation focuses on the influence of a plasma cleaning before bonding in connection with varied bonding parameters and ageing experiments towards bond reliability. Results of reliability tests and failure mechanisms will be presented.

2 Experimental Part

US bonding experiments were carried out on a 4 µm Ni / 0.1 µm flash-gold layer on Cu (brushed). Base material of the pcb was an epoxy FR4 resin. 25 µm Al wire (Heraeus AlSi1, ≥ 11 cN) was used on a Kulicke&Soffa 4123 wedge wedge bonder. For TS bond experiments either the same 4 µm Ni / 0.1 µm flash gold layer on Cu (brushed) / FR4 or 5 µm Ni / 0.6 µm Pd / 0.05 µm flash-gold on Cu (polished) was used. Wire bonds were made with 25 µm gold wire (Heraeus Au-beta, ≥ 8 cN) on Ni/Pd/flash-gold with a Kulicke&Soffa 4124 ball wedge bonder. The temperature was 120°C. The bonds were tested with a Dage BT 22 pull tester in a destructing pull test. Average bond strengths are mean values of 30-100 bonds.

All samples were cleaned with dionized (DI) water after delivery before usage. Additional plasma treatment was performed with H_2-, O_2- or Ar-plasma for 4 minutes in a self modified plasma chamber at 0.3 to 0.6 hPa and an antenna power of 2 W/cm² with an applied frequency of 40 kHz.

For reflow soldering experiments, three different solder pastes were used and applied by dispenser to the pcb (Litton R 292-3, No. 1; Heraeus F 360, No. 2; Degussa 66110323/D, No. 3). Peak temperature during reflow soldering was 250°C.

Humidity storage was performed in a Heraeus-Vötsch HL 7015 climatic chamber at 85°C and 85% atmospheric humidity. Annealing tests were done at 150°C in air.

3 Results and Discussion

3.1 Ultrasonic aluminum wire bonding on Ni/flash-gold

Three important damage possibilities for wire bonds during ageing are damage of the layer system, changing in the wire and damage of the welding zone. The objective of the first experiment was to investigate the ageing stability of the Ni/flash-gold layer and to examine whether a plasma treatment is able to improve the bonding ability before and after humidity storage. Therefor the samples went through the following process cycle: *1. ageing of the pcb, 2. H_2-Plasma, 3. bonding, 4. pull test.* Some layer systems fail this test loosing their bonding ability. A reductive H_2-plasma was used because rather surface damage by oxidation than organic contamination was expected by the humidity. The average bond strengths in figure 2 demonstrate that for Al-wire bonding neither the pcb lost much of its bonding ability during humidity storage nor H_2-plasma was able to increase the bond strengths. This permits to refer the observed ageing effects of the wire bonds to changes in the welded joint and in the wire.

In the following experiments the reliability of US-wedge-bonded Al wires on Ni/ flash-gold was examined under various bonding and ageing conditions. The process cycle was: *1. bonding, 2. ageing, 3. pull test.* The pcbs were bonded with high or low US-power respectively bond force regarding a prior verified process window: (30 respectively 22.5 % of full power, 30 respectively 17.5 % of max. force). The bonded samples were then either annealed at 150°C or exposed to a climatic treatment at 85°C and 85% humidity. Afterwards the bond strengths were measured. Figures 3 and 4 contain the data from the pull tests. In the initial state before ageing, the gently bonded samples had significantly higher bond strengths than the cruel bonded pcbs. This is probably due to wire damage at the wedge because only wire fracture at the wedge but no lift-off from the pcb metallization occurred. However during humidity storage a slight decrease in bond strength occurred especially for the gently bonded samples.

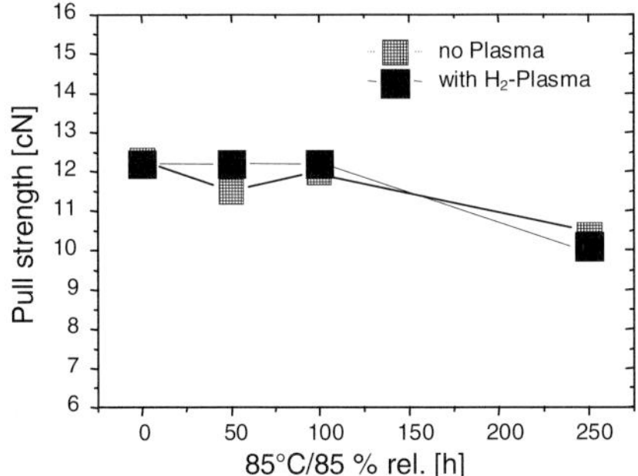

Figure 2. Bonding ability after humidity storage

After 250 hours of humidity storage all samples showed nearly the same average bond strength at 80 % of the average wire breaking load. A substantial decrease of the pull forces occurred during annealing. After 10 hours at 150°C only 60 % of the initial bond strength remained. The decrease of bond strength during annealing can be explained by diffusion processes within the 1 % of silicon containing aluminum wires which decrease the average wire breaking load [1]. This is a temperature effect so that the bond strengths decrease faster at 150°C than at 85°C. No bond damage by the humidity could be detected. In all cases the differences in strength between gently and cruel bonded samples proved to be significant in the initial but little in the aged state.

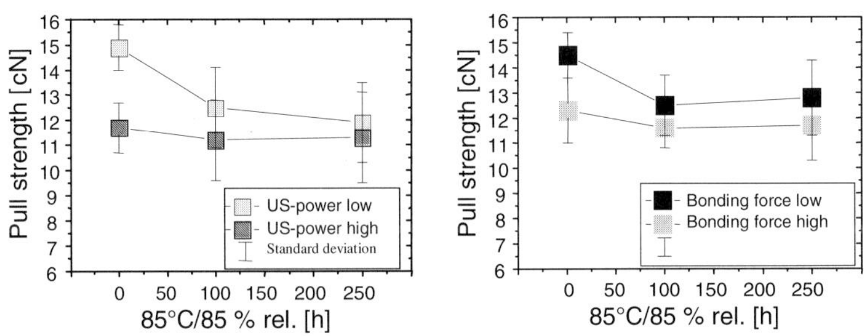

Figure 3. Al-wire bond strength after humidity storage

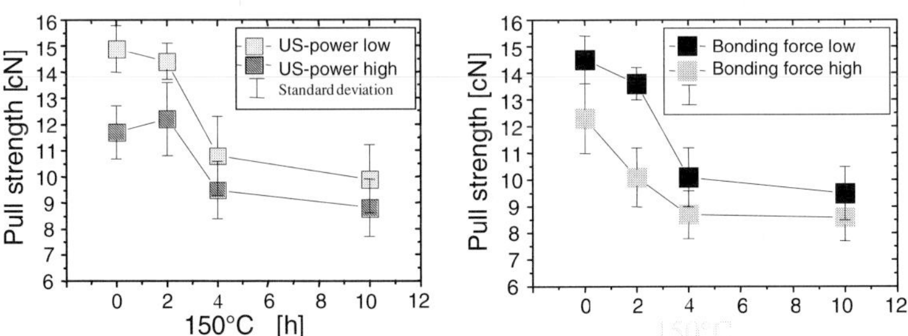

Figure 4. Al-wire bond strength after annealing

With regard to the bonding conditions the US-wedge-wedge bond process on a well manufactured Ni/flash-gold surface is a reliable process and long term reliability is mainly controlled by changes in the wire. A plasma treatment seems to be not necessary to achieve sufficient bond reliability. The influence by bonding parameters is of minor importance as long as reasonable limits are kept. Therefore the use of plasma in order to achieve particularly gentle bonding conditions for US Al wire bonding on Ni/flash-gold is not recommended because the long term reliability is not significantly affected.

3.2 Thermosonic gold wire bonding on Ni/Pd/flash-gold

In earlier investigations it was shown that the TS gold wire bonding ability of Ni/ flash-gold layers could be significantly improved by plasma treatment [8,9]. Especially H_2-Plasma proved to reduce surface nickel and nickel oxide concentration resulting in better bond adhesion. This chapter describes the reliability of TS gold wire bonds on Ni/Pd/ flash-gold layers with respect to prior plasma treatment and following ageing experiments. In the first experiments the samples were bonded with and without prior H_2-plasma treatment applying three different US-energies. The average bond strengths were then measured in the freshly bonded state and after either humidity storage or annealing of the bonds at 150°C. The results of this investigation can be seen in figures 5 and 6.

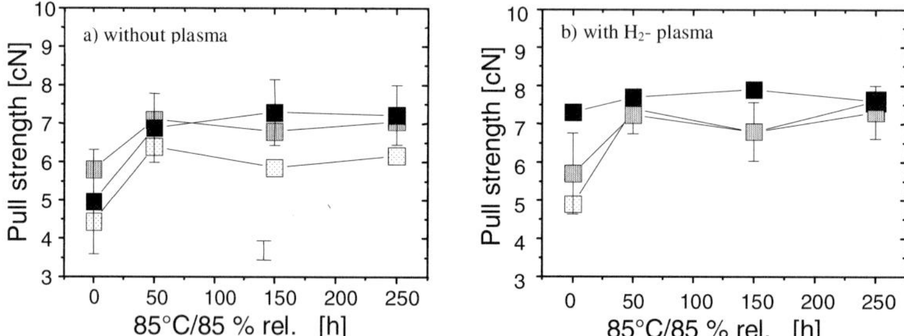

Figure 5. Au-wire bond strengths after humidity storage on Ni/Pd/Au with and without prior plasma treatment. US-energy : ☐ low (20 %.), ▒ medium (60 %), ■ high (100 %)

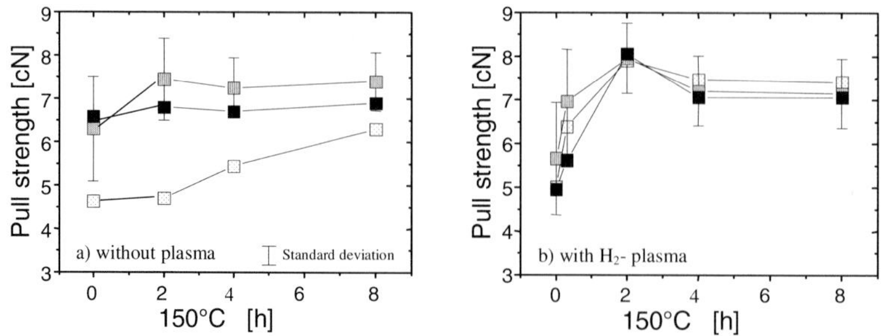

Figure 6. Au-wire bond strengths after annealing on Ni/Pd/Au with and without prior plasma treatment. US-energy : ☐ low (20 %.), ▒ medium (60 %), ■ high (100 %)

Two characteristic behaviours are generally observed. First, the bond strengths in the freshly bonded state are poor, especially for low and medium US-power. Second, bond strengths increase significantly during humidity storage and annealing. This can be most clearly seen in the annealing experiments of the plasma treated sample (fig. 6 b). The initial bond strength amounts to only 60 % of the theoretical wire breaking load. After 20 minutes annealing on the hot stage during the bond process a significant increase in the bond strength occurred and after 2 hours of annealing for all three applied US-powers the optimum bond strength with 100 % of the wire tearing strength was reached. No significant decrease of the bond strength could be detected after long annealing times. The effect of the plasma treatment expresses itself in higher bond strength after annealing, smaller standard deviations and better initial bond strengths for low US-powers. Without a plasma treatment, the low US-energy bonded bonds didn´t even show after 8 hours of annealing sufficient bond strength.

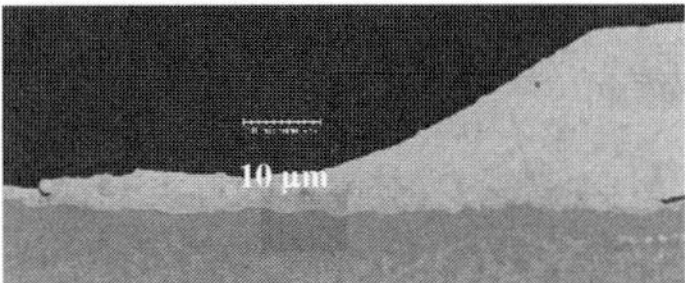

Figure 7. Micrograph of a 25 µm gold wire wedge bond on a Ni/Pd/Au layer

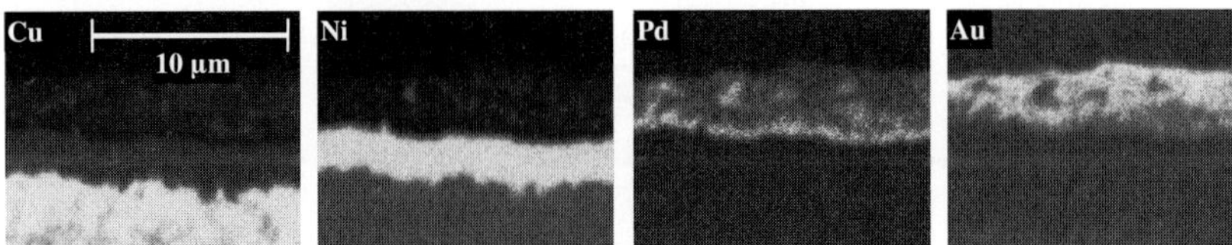

Figure 8. EDX analysis of the welding interface from fig. 7.

For a better understanding of this bonding behaviour the wedge bond formation was investigated in more detail. In figure 7 a micrograph of a 25 µm gold wire on a Cu/Ni/Pd/ flash-gold layer is presented. The wire seems to be perfectly welded over the whole contact area with the pad metallization. Figure 8 displays the atomic distribution of Cu, Ni, Pd and Au at the most intensive welded area by EDX analysis. The Cu- and Ni-layers are not affected by the welding process and an intact Pd-layer remains in the interface between gold wire and Ni-layer. That indicates relatively small mechanical changes of the layers during the bond process. Figure 9 shows a number of wedge bonds which were bent over without considerable application of force. It turns out that the low bond strength on the just di-water cleaned sample correlates to a poor wedge bond adhesion. Only the strongly deformed front part of the wedge is welded whereas the back part can be lifted off easily. On the other side the high bond strengths on the plasma cleaned sample correlate with good wedge bond adhesion. Nearly the whole wedge has a reliable contact to the pad metallization.

Figures 10 and 11 display the topographies of the wedge bottom side and the corresponding pad surface after bond failure. For a no-bond (no initial bond adhesion, fig. 10) an almost perfect imprint of the pad relief can be seen on the wedge bottom side. That proofs that wire and pad had been in intensive contact with each other. Nevertheless the pad surface stayed almost intact where the welding tried to take place. Even at high US-energies and subsequent wedge failure during pull test, only few marks of a welding remained on the pad surface besides the strongly deformed front part of the wedge bond (fig. 11).

Figure 9. poor (without plasma, left) and good (with plasma, right) wire bond adhesion.

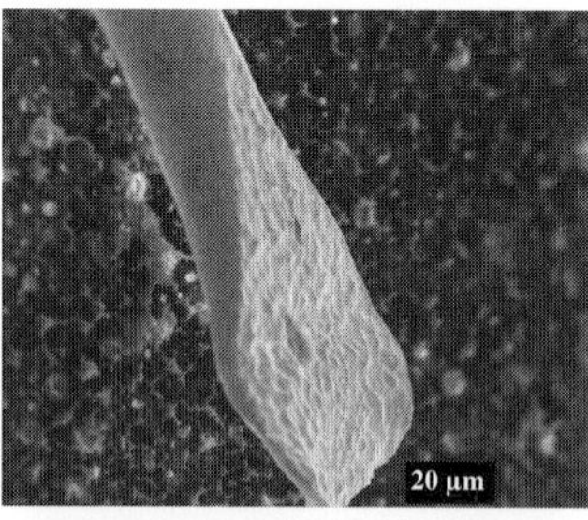

Figure 10. Failure: no-bond. wedge bottom side (top) and two bowl-shaped imprints on corresponding pad surface from bond attempt (bottom)

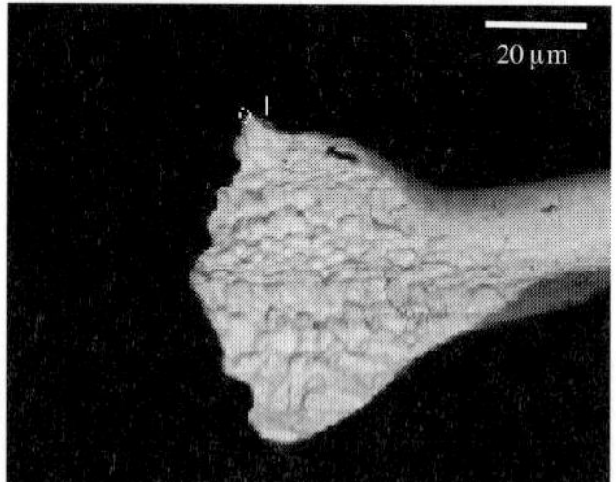

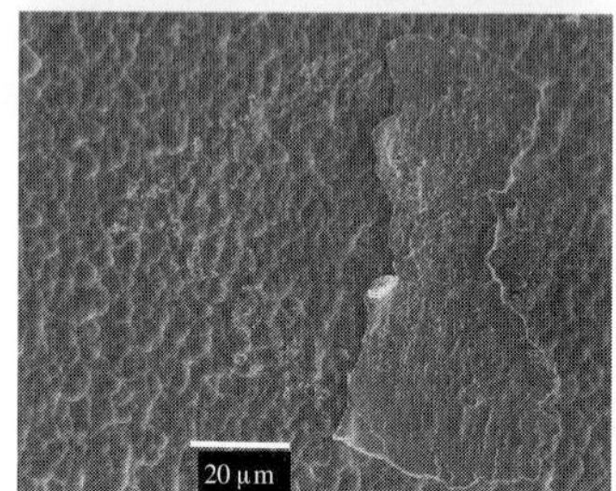

Figure 11. Failure: wedge fracture. wedge bottom side and corresponding pad surface

These observations permit the following conclusions: The welding of the back part of the wedge bond is essential for a strong and reliable adhesion. However, this part seems to undergo mainly thermocompression with less support by ultrasonic movement than on the strongly deformed front part of the wedge bond. The low initial bond adhesion and the increasing bond strength during annealing can possibly be explained by relaxation of internal stresses which are built up by the thermocompression. The ultrasonic movement should wipe the wire through surface contaminations to achieve an intensive welding. However, if the compression is predominant and US support is little, even thin disturbing surface contamination will remain in the interface between wire and pad metallization resulting in weak adhesion. The increasing bond strength after a plasma treatment can be explained by the suitability of a plasma to remove thin disturbing organic or even oxidic surface contaminations.

The process cycle for the following experiments was: *1. soldering, 2. cleaning, 3. bonding, (4 "reflow annealing"), 5. pull test.* The effect of different plasma gases on the bonding ability was investigated. Further the bond strengths of TS gold wire bonds after a reflow heat profile were examined. Figure 12 displays the average bond strength on soldered, Ar-plasma cleaned and gold wire bonded samples before and after a second reflow soldering step. The short but intense heat treatment with a peak temperature of 230°C raised the initial average bond strengths by 60 - 90 %.

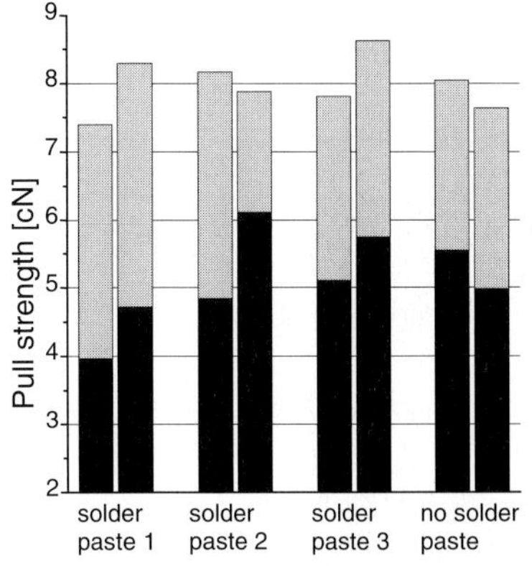

Figure 12. Bond strengths on soldered and Ar-plasma cleaned samples. ■ initial state, ▒ „reflow annealed"

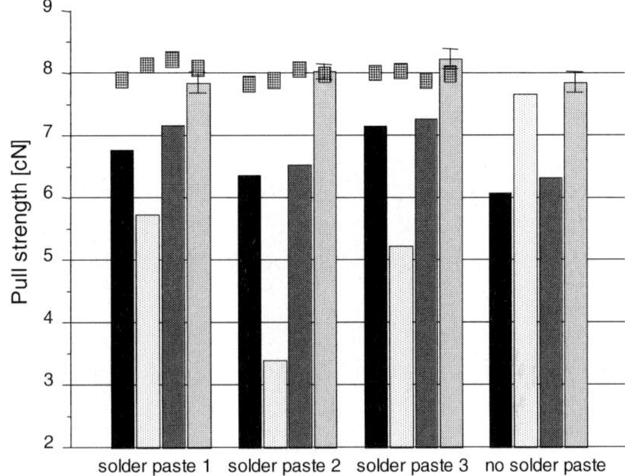

Figure 13. bond strengths on soldered and differently cleaned samples. All bonds „reflow annealed" after bonding. ■ no cleaning, □ O_2-plasma, ■ H_2-plasma, ▒ Ar-plasma, ▦ pure unsoldered

In order to investigate the effect of different plasma gases on the bonding properties of soiled surfaces after a reflow soldering process, the samples were soldered with three different solder pastes. The pcbs were then bonded without further cleaning and after plasma cleaning with H_2-, O_2- or Ar-plasma. From figure 13 can be seen that only Ar-plasma was able to restore the bondability of the soldered samples, giving the same average bond strength as the pure unsoldered samples. H_2-plasma gave nearly the same bond strength as no cleaning at all. O_2-plasma treatment however resulted in very poor bond adhesion in most cases. All average bond strengths displayed in fig. 13 are measured on "reflow annealed" bonds.

4 Conclusion

Ultrasonic aluminum wire wedge-wedge bonding and thermosonic gold wire ball- wedge bonding differ not only in the bonding tool but also in the welding mechanism and the reliability with respect to heat application and humidity storage. A well manufactured Ni/flash-gold layer can be reliably ultrasonic bonded without the need of a plasma treatment. Particularly gentle bonding conditions are not recommended because the long term reliability depends strongly on internal wire processes which are not affected by the bonding conditions. For TS gold wire bonds on Ni/Pd/flash-gold layers the wedge bond adhesion could be improved by plasma cleaning before bonding and by annealing after bonding. Both effects are probably due to a dominant thermocompression process during the wedge bond formation. Ar-plasma cleaning proved as suitable to restore good bonding properties after a reflow soldering process. A liquid cleaning was not necessary. O_2-plasma however can not be recommended for TS bonding on Ni/Pd/flash-gold as already shown for Ni/flash-gold layers.

5 References

1. G. G. Harman, Wire Bonding in Microelectronics: Materials, Processes, Reliability and Yield, 2nd edition, McGraw-Hill, 1997
2. Handbuch für die Chip-on-Board Technologie, VDI|VDE-IT, 1997
3. H. Osterwinter, Galvanotechnik 1998, 89, No. 6, 2051-2060
4. R. Bierwirth, K.D. Lang, A. Licht, F. Rudolf, V. Tiederle, VTE 1997, 9, No. 3, 127-135
5. K. Bünning, Galvanotechnik 1998, 89, No. 5, 1678-1686
6. H. Osterwinter, J. Feßmann, V. v. Arnim, U. Völler, J. Eisenlohr, M. Böhm, PLUS 1999, 6, 837 – 843
7. R. W. Johnson, M. J. Palmer, M. J. Bozack, T. Isaacs-Smith, IEEE Transactions on Electronic Packaging Manufacturing 1999, 22, No. 1, 7 - 15
8. V. v. Arnim, J. Feßmann, L. Psotta, Surface and Coatings Technology 1999, 116-119, 501-507
9. V. v. Arnim, J. Feßmann, L. Psotta 1999, PLUS, 8, 1160 – 1166

Investigation of Interface Cracking in Electronic Packages

E. Kieselstein, B. Seiler, Th. Winkler
Chemnitzer Werkstoffmechanik GmbH, Chemnitz
R. Dudek, J. Auersperg, A. Schubert, B. Michel
Fraunhofer Institute for Reliability and Microintegration , Berlin

1 Introduction

Mechanical and interfacial properties are particularly important for the prediction of reliability. A significant proportion of failures in packaging and subsystems is due to fracture or delamination. The presence of microcracks in adhesives and encapsulants is almost inevitable. Microcracks can be found in the form of surface defects, irregularities, voids, or separation between the epoxy resin and filler interface. Under excessive thermal-mechanical loading or residual stresses, it is possible that the microcracks will propagate.

Cracking or delamination in various interfaces is a significant concern for yield loss and reliability for advanced packages. There are many reasons for a interface damage, for example, low adhesion due to incompatible interfaces and loss of adhesion due to contamination or moisture absorption. Delamination at the encapsulant-chip or encapsulant-substrate interface can lead to cracking of the interconnects. To assess the integrity of the interface, one must at least have the following information: the interfacial adhesion, a pertinent numerical model, and an estimate of the maximum length of a permissible initial crack. In an interfacial fracture toughness test, the crack resistance of an interface is signified by a single parameter called the critical interfacial energy release rate. The critical interfacial energy release rate, G_C, represents the failure envelope of the interface similar to the way a yield surface represents the onset of plastic deformation. That is, once the driving force (energy release rate, G) exceeds a critical amount, an interfacial crack will propagate.

Any modification or change of the material set or process history are likely to alter the bonding strength as well as the residual stresses remaining in the interface, and therefore, affect the in-situ strength of the interface. Unfortunately, not all tests are applicable after assembly is completed. The interfacial fracture toughness test has the advantage of including all (physical, chemical, mechanical) mechanisms contributing to adhesion. It can also be used to assess the in-situ degradation or improvement of the interfacial adhesion after process flow and environmental testing are completed.

The investigations of local displacements and displacement fields by optical methods in combination with Finite Element Analysis lead to a better understanding of crack intiation and crack propagation and consequently to detailed knowledge about package stability.

2 Investigation of the Interface Adhesion of Chip-Encapsulants (Glob Tops) on Board Materials

2.1 Conditions of the Experiment

The adhesion properties of chip-encapsulants (Glob Tops) on different board materials are investigated using the optical measurement technique ***UNIDAC*** (**Uni**versal **D**eformation **A**nalysis by means of **C**orrelation) in combination with a microscopic bending device.

UNIDAC is a measurement technique supporting the processing of images, which is instrumental in the analysis of deformation modes of all kinds [1, 2].

Figure 1 is a flow-chart of the experimental set-up arranged for the investigation of the interface adhesion of different materials compounds.

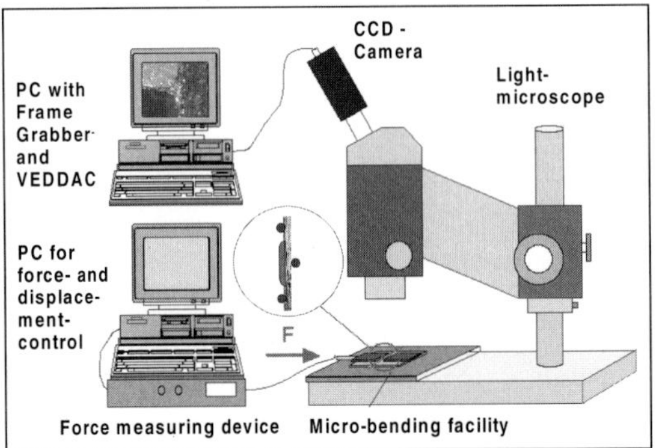

Figure 1: Measuring equipment and devices set-up for the experimental investigation of interface adhesion with the optical measurement technique ***UNIDAC***

The specimen are subjected to load by three-point bending inside the microscopic bending device, which is connected to a optical microscope. During this process, the specimen is automatically (time-controlled) photographed by a CCD-camera in situ and transferred to the personal computer where the data is immediately processed.

The results detailed in the following were obtained from such materials compounds as mentioned with varying glob-top materials on board material FR-4, and with and without the use of solder masks.

2.2 Fundamentals of Deformation Analysis by *UNIDAC*

One of the methods to present and describe the deformation of materials due to thermal and mechanical load is the correlation analysis of gray scale images, which is a special method of digital image processing.

The result of the correlation analysis is a graph of the in-plane displacement field described by vectors or overlapped lattice *(grids).* The vectors show the amount and the direction of the displacements. These results are available for both qualitative and quantitative analysis.

For the successful performance of an optical deformation analysis, it is requisite to have a sequence of photographs of the specimen under different loading conditions taken by a CCD-camera or another imaging techniques. From the *digitalized* gray-pixel images thus obtained, the in-plane displacements between two loading states, which occur locally, are determined using the two-dimensional correlation analysis.

As a result of this correlation analysis, a graph illustrating the displacement field is obtained, in which the displacements are indicated as vectors or overlapped lattices *(grids),* as well as a protocol in ASCII format with appertaining measured data, which can be used for further qualitative and quantitative analyses [3].

Thus, these results, can, for example, serve to determine the curves of local bending lines or strains in the description of interfacial phenomena in materials compounds, through which cracks and other failures in the compound can be characterized and quantified.

2.3 Investigation of Interface Adhesion of Different Chip-Encapsulants

2.3.1 Comparison of Different Materials Compounds with and without Solder Masks

Figure 2 shows the impact force versus bending deflection graphs of the different materials compounds altogether in one illustration. The description of each graph is characteristic for the three specimen used here with and without solder masks, which enables us to deduce from them the typical damage patterns which occur under different conditions for interface adhesion.

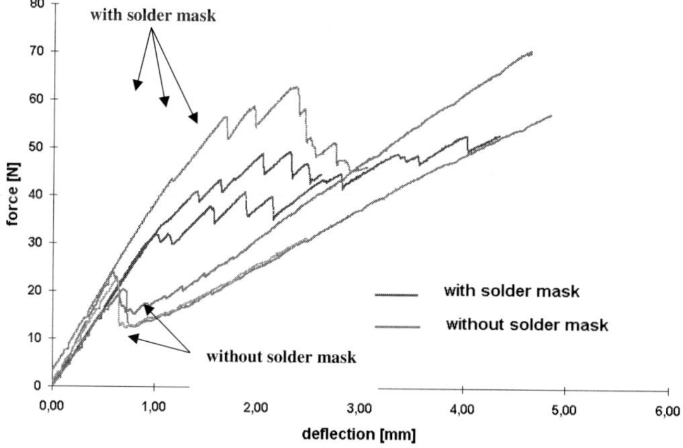

Figure 2: Impact Force versus Bending Deflection Graphs of different "glob-top - board" materials compounds, with and without solder masks

The in situ produced pictorial sequences provide information about the qualitative differences in delamination, which is exemplified in Figure 3 and Figure 4.

All of the specimens which were investigated *with* solder mask showed delamination in the interfacial area of the solder mask and the board (Figure 3). The delamination starts very early and it occurs very rapidly. The interface solder mask/glob-top, however, showed no occurrence of delamination, at all. The correlation analysis of a specimen thereof *with* solder mask (see pos. 2.3) will exemplify the extent of damage in a more precise and concrete manner.

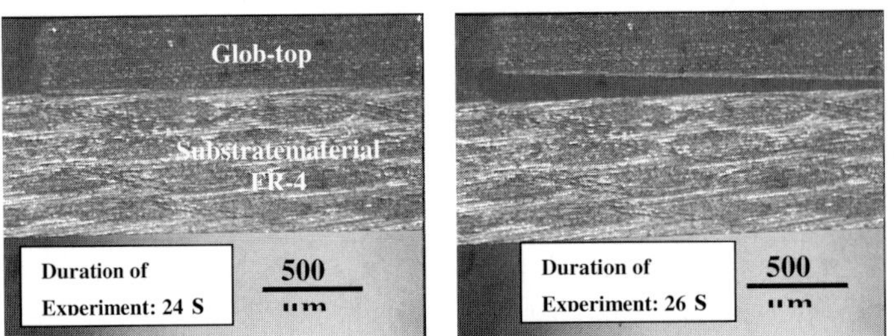

Figure 3: Specimen *with* solder mask under bending load; damage: interface delamination

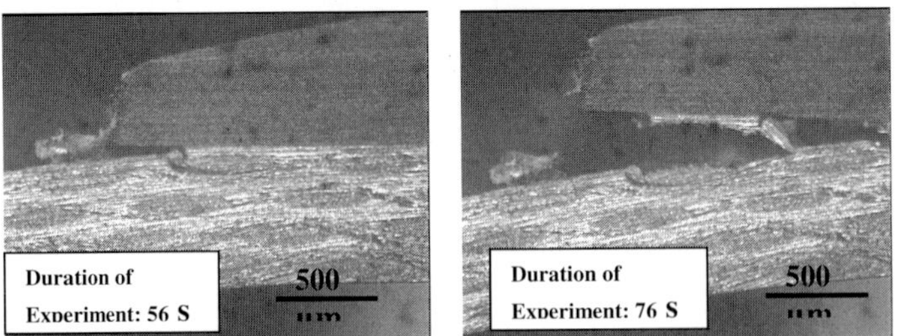

Figure 4: Specimen *without* solder mask under bending; damage: cracking of the board

With the compounds *without* solder mask, cracks pass initially into the board, but are then obstructed by the real structure (glass fibres) of the local environment, or redirected, respectively (Figure 4). In contrast to the specimen with solder mask, damages here can first be observed when impact force, which is exerted, is increased drastically. The damage process takes much more time. The specimen under investigation showed very good adhesive properties between epoxy-based glob-top material and the epoxy-based board material, and the interfaces remained intact.

2.3.2 Quantitative Characterization of the Interface Delamination on Glob-Top Compounds with solder mask

Local deformations are determined taking each image separately from the sequence which was obtained in situ in the bending experiment and using them for the gray-scale pattern correlation (see 2.2). They are then exemplified by overlapping lattices (grids) (bright: no deformation; dark: deformation) in one graph (Figure 5). The data can then be processed for quantification of the damage initiated on the compound by the Microsoft Excel calculation program. In the interface, the bending graphs (y-shift = f (x-position)) are calculated for the glob-top and the board material, by the means of which the approximate crack length can be determined (Figure 5) showing one selected loading state.

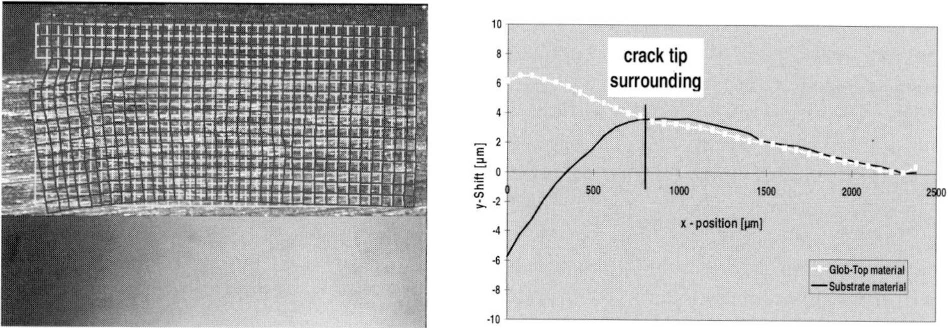

Figure 5: Correlation result and graphic evaluation for determination of the crack length; deformation state after the loading duration of 22 s

The change of the crack tip length resulting from the subjection to 3-point bending-loading is shown with remarkable clarity in Figure 6. The measured data that could be obtained is directly and immediately incorporated into the numerical models for the simulation of the delamination processes.

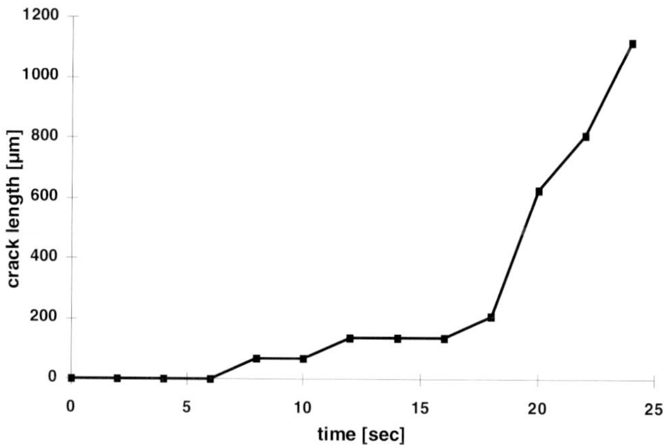

Figure 6: The crack length as a time function, determined through *gray-scale* correlation

2.4 Modeling and Numerical Simulation for the Investigation of the Interface Adhesion of Glop Top-Assemblies

Analysis, modeling and prognosis of the reliability of plastic packages do not only require profound expertise on the materials properties of the filled polymers to be used, but also about their interface adhesion on different substrate materials, as well as the extensive knowledge of their mechanical stress limits and deformation patterns, acquired through practical knowledge of production processes and stress factors.

It is no novelty that the resistance of packages to mechanical stress is essentially determined by local stress concentrations. Complicated spatial stress states, which are often of rather singular character, occur most prominently on interface boundaries, interface edges/corners and interfacial crack tips.

In the package of our example, such behavior can be expected to show at the interface between solder mask and board, no matter whether an existing delamination is progressing or prior damage does not exist.

Resistance to cracking as well as to progression of cracks are therefore parameters, which must be accounted for in any reliability evaluation. Next to the character of the stress state (as loading parameter), they are also determined by the material properties of the interface (as material parameter) itself and are therefore subject to extensive technology-induced variations.

Through direct interlinking of all experimental investigations conducted on the adhesiveness of packages and organic boards, numerical simulation via FEM could be performed almost in combination with above described specimen assemblies.

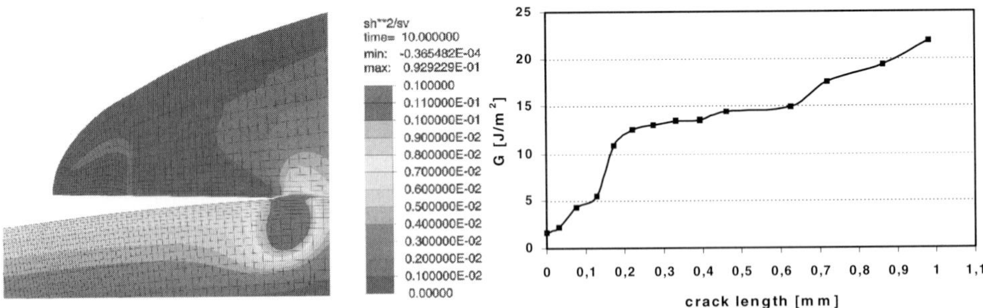

Figure 7: Numerical simulation of crack propagation and energy release rate for quantification of interface delamination

The above investigation result gives evidence of the complexity indicated in the theoretical descriptions of the interface adhesion states of glob-top assemblies. It has become evident that thermal stress states immanent in the material cannot be neglected, and that the local geometry in (interface) edges/corners also exerts a strong influence on the stress state.

With the close coupling, however, between numerical and experimental investigation, requisite parameters for the description of the interface adhesion can be obtained. Below graph, Figure 7 shows such numerically calculated parameter, the calculation basis for which were the experimental results as detailed earlier under pos. 2.3. This parameter is a representation of the **energy release rate G_c** describing the interface adhesion of the material compounds.

3 Literature

[1] VEDDAC 2.6 for WINDOWS, User's Manual, Chemnitzer Werkstoffmechnik GmbH, Chemnitz, 1998

[2] Kieselstein, E., Seiler, B., Penno, M., Wielage, B., Dost, M., Michel, B.: Einsatz der Grauwertkorrelation zur Charakterisierung von Faserverbundwerkstoffen, Werkstoffprüfung 1996, 5./6.12.1996, Bad Nauheim

[3] Kieselstein, E., Dost, M., Noack, E., Michel, B. : Charakterisierung des Verformungsverhaltens von Kunststoffen durch Korrelation an digitalisierten Grauwertbildern, TECHNOMER '97, 13.-15. November 1997, Chemnitz

[4] Auersperg, J., Vogel, D., Simon, J., Schubert, A., Michel, B.: Reliability Evaluation of Chip Scales Packages by FEA and microDAC, Proc. Symp. on Design and Reliability of Solders and Solder Interconnects held during the TMS Annual Meeting, Orlando, USA, Feb. 9-13, 1997

[5] Vogel, D., Caers, J., Auersperg, J., Schubert, A., Michel, B.: Advanced deformation and failure analysis on flip chip assemblies, Proc. of Micro Materials `97, April, 16-18,1997, Berlin

[6] Schubert, A., Dudek, R., Auersperg, J., Vogel, D., Michel, B., Reichl, H.: Thermo-Mechanical Reliability of Flip Chip Structures Used in DCA and CSP, Proc. of 4th Int. Symp. and Exhib. on Advanced Packaging Materials, Braselton, Georgia, USA, March 15-18, 1998

4 Acknowledgement

Funding of parts of this research work by the German ministry BMBF (project SIMIKO No. 16 SV 625/0) and the Saxon state ministry SMWA (No. 3908/616 and 3909/616) is gratefully appreciated. The authors would like to thank W. Schneider, MPD Dresden GmbH, for his support and the manufacturing of polymer specimen. They express their thanks to the Technical University Chemnitz, Chair Composite Materials, for providing the microscopic bending device.

Microlaminography for High-Resolution BGA and Flip-Chip Inspection

Alexander Sassov and Filip Luypaert
SkyScan, Aartselaar, Belgium

1 Introduction

Laminographical methods were developed initially for medical application as a "non-computerized" layer-by-layer visualization of human body [1]. In this case an inclined initial X-ray beam project image of specific layer of object to the detector surface with defocusing of other layers during synchronous coplanar rotation of object and photoplate.

In medical area this method recently completely changed to computerized tomography because of advantages in image sharpness and possibilities to get analytical information about local densities. In the same time all computerized tomographical (CT) methods have one strong limitation. During investigation an object should be completely displaced inside field of view. This requirement can be satisfied in medical applications, but most electronic devices and boards are the planar structures and cannot be completely rotated around axle in the device plane. By another hand big multilayer assemblies cannot be transmitted by X-ray through the direction in parallel to the object surface. Both listed CT limitations not exist in the classical laminography scanning geometry. For laminography the internal 3D-structure can be visualized layer by layer during rotation around axle orthogonal to the object surface. Combination of classical laminography approach with modern digital acquisition technique allows improving spatial resolution and image quality.

2 Industial Laminography

Classical approach for industrial coplanar rotational laminography is shown in Fig.1[2]. An X-ray source and X-ray detector rotate synchronously around the testing point in the specimen to produce sharp image from one corresponding layer in the object plane. All images during rotation integrated in X-ray CCD-camera. In real instruments the emission point can be turned by the physical rotation of the X-ray source or by the circular scanning of electron beam through the target inside the source. To change projection plane for visualization in different depths, the camera should be shifted up or down. This classical approach for industrial laminography realized in the instruments from the HP-4Pi company [3].

Other equipment existing in the market from Digiray company [4] based on the "reverse geometry" with raster scans of emission point inside the X-ray tube and data acquisition by several pointer detectors. These approach more close to X-ray stereo imaging and cannot obtain enough good depth resolution. There are also several other systems with linear movement (1-dimensional) through the conical beam [5]. In this case usable depth and spatial resolution can be achieved for specifically oriented parts of object only.

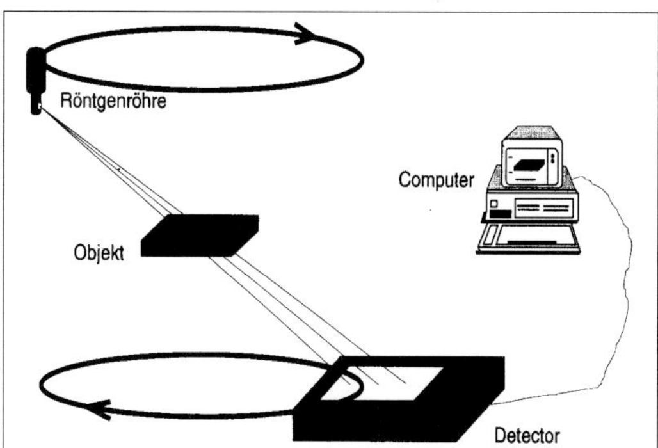

Figure 1. Coplanar rotational laminography

All recently commercially available X-ray systems for laminography have a spatial resolution in the range of hundred of microns. The spatial resolution mainly limited in accuracy of complicated coplanar movements of two heavy parts of the system. Because of modern multilayer PCBs, flip-chips, BGA-connections etc. contains structures and soldering points of 10-20 microns in size, the spatial resolution in existing systems is not enough for such applications.

There are several main limitations for spatial resolution in the listed laminography systems. In the x-ray source the spatial resolution limited by focal spot size. In the sources with power of hundreds watts the spot size cannot be smaller than 0.5 mm because of heat dissipation in the target. Modern technology open possibilities to build the microfocus X-ray sources with spot size of <10 microns. But such sources have the power in the range of 2-10 watts. That means the corresponding acquisition time should be increased to get reasonable signal-to-noise ratio.

Other limitation for spatial resolution can be found in the detector. Limited number of pixels in the camera array can be a reason for pure resolution in the case of big field of view. For example, if field of view should be 10 by 10 mm with camera division 512x512 pixels the pixel size will be approximately 20 microns. To improve relation of field of view to the spatial resolution the megapixel sensors can be used.

But a main limitation for spatial resolution can be found in complicated coplanar mechanical movements of object, camera and source. All displacements and rotations for heavy x-ray sources and cameras should be done with accuracy better than spatial resolution for all possible object positions. In the case of big size assembles and PCBs it's difficult to avoid vibrations, axle play and object non-planarity during testing.

3 Digital Microlaminography

Digital acquisition technique allows improving several important details in laminographical methods. Two-dimensional digital X-ray camera can collect images to computer memory. That means camera rotation can be changed to the image rotation into the memory by special program. Image rotation in the memory can be done very accurate and, in compare to coplanar mechanical camera movement, don't need complicated and precision mechanics. Main principles for digital multilayer microlaminography are shown in the Fig.2. During object rotation different planes inside the object produce images with rotations around different points into detector. If all projection images during one object turn will be collected into the memory the multilayer separation can be done in once just by re-rotation of initial images around corresponding points.

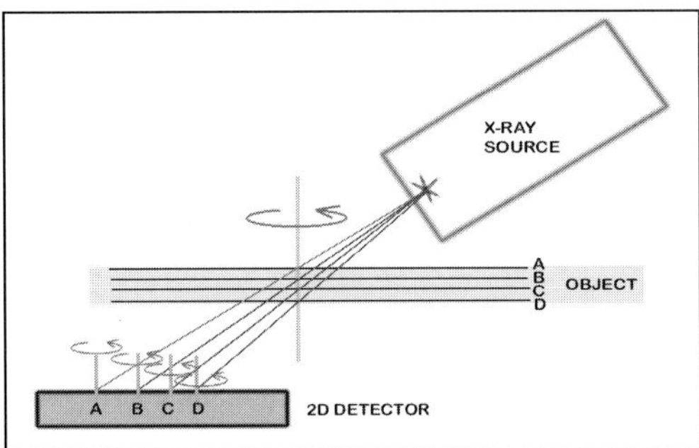

Figure 2. Principles of digital multilayer laminography

With digital technique we don't need to rescan object for visualization from another depth plane. In this case all acquired images can be rotated around corresponding points for simultaneous separation for all layers of the object structure. Digital image processing during depth separation also allows correcting most geometrical disalignments and distortions in acquired data. Using this approach the Digital Microlaminography system SkyScan-1080 has been developed. Digital microlaminography allows minimized mechanical movements and reach very good spatial resolution. In the same time the system allows to visualized full 3D-structure just after one fast scan with powerfull post-processing and visualization possibilities.

4 Desktop Microlaminography System

SkyScan-1080 is a compact desktop microlaminography instrument (Fig.3). As an x-ray source we use a microfocus sealed X-ray tube (Hamamatsu, Japan) operated at 80kV with <8um spot size, integrated with power supply. Source with 130kV sealed tube can be supplied as an option. Images acquired by immovable intensified X-ray CCD-camera with 560x768 or 1280x1024 pixels. This camera have internal compensation of thermal noise accumulation for acquire images with integration time up to 10 sec. Object stage can hold the objects up to 150x150 mm in size with precision rotation by stepping motor. For microlaminographical depth visualization any 5x5 mm area inside 150x150mm object stage can be selected. Operator can do this selection with looking to the object surface by visual camera or by programming of coordinates from the reference point. One object turn with collecting information about 3D internal structure and separation of 20-80 layers during scanning takes 0.5-5 minutes. System control and depth separation done by internal industrial single or dual Pentium III computer operated under MS Windows NT. Pixel size in any final image of the separated layer inside the object is 10 microns. Depth separation can be done with step size in the range of 10 microns to 2mm.

Figure 3. Digital desk-top microlaminography system SkyScan-1080

Powerful library of functions for post-processing allows improving image quality and depth separation. Nonlinear (logarithmically) restoration in initial images is necessary for correct independent separation of different layers. Big library of filters (Laplas, Sobel, median, unsharp mask high frequency…) can be applied to improve image quality and depth separation inside the object. Another set of functions (emboss, color transformation, zoom…) improve visualization possibilities for detection of small defects.

After processing in microlaminography system we will get a set of separated depth sections inside the object. This set can be shown on to the screen layer by layer. Special program inside the software package allows creating realistic 3D image of the object internal structure. After 3D rendering one can select the surface properties for the objects (color, reflection, diffusion…), illumination properties (color, direction, kind of the source…) and move / rotate / zoom / pan the 3D scene onto the screen. Operator can even cut this model of the internal 3D microstructure in any alternative direction or fly through the internal microstructure.

5 Application Examples

Main application areas for high-resolution microlaminography are the electronic assemblies and devices (PCBs, BGA, Flip-Chip…), multilayer structures (laminates), precision mechanics (watch…) etc. We will show several application examples for using microlaminography system SkyScan-1080 for inspection of electronic devices and assemblies.

Fig.4 shows investigation of multilayer PCB. Left image is a conventional microradiographical image where information from all 4 layers mixed together. Microlaminograpohy allows separation of information from different PCB layers, as shown in the right side of the Fig.4.

Microlaminography investigation for BGA-inspection is shown in the Fig.5. Left image shows conventional microradiography information. Right images obtained from microlaminography system. In this case the set of separated cross sections through BGA under IC shown as the realistic 3D images with different cuts by software. We can easily found several voids in balls and check all 3D details in the interconnections under IC.

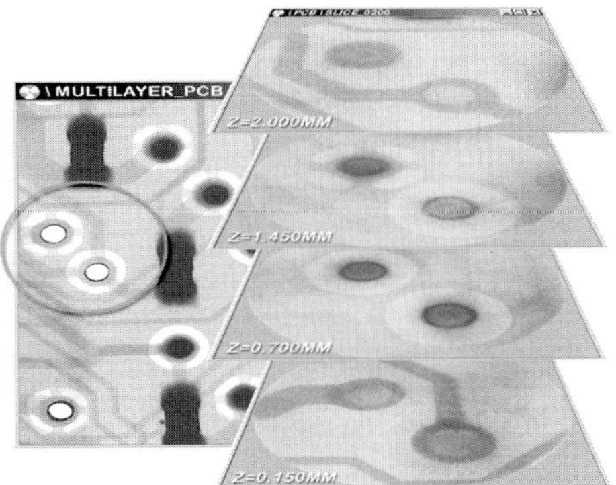

Figure 4. Microlaminography of a multilayer PCB

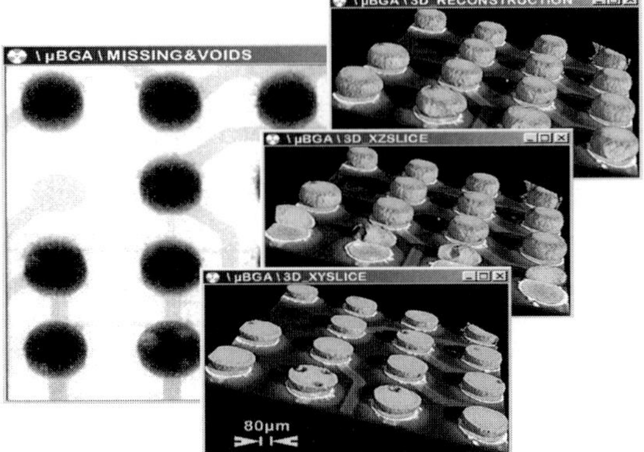

Figure 5. Microlaminography application for BGA inspection

Last example shown in Fig.6. demonstrate possibilities of microlaminography for Flip-Chip inspection. Transmission X-ray image, shown in the left side, not allows making any defect inspection for such devices. In the right separated layers one can easily see all details in interconnection under the Flip-Chip body as well as several voids with 10-30 micron in size.

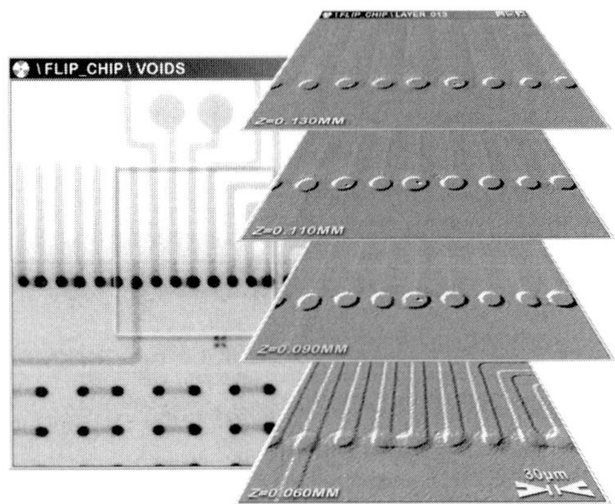

Figure 6. Microlaminography application for Flip-Chip inspection

6 Conclusions

New approach in digital microlaminography was used for high-resolution layer-by-layer separation of internal microstructure in any place of the big planar objects (multilayers, PCBs etc.), which cannot be reconstructed by computerized tomography because of limited possibilities in rotation. Depth and lateral spatial resolutions in the micron range can be reached. Microfocus X-ray sources and digital acquisition techniques with nonrotatable cameras allow improving performances.

7 References

1. G.T.Herman, Image Reconstruction from Projections, Academic Press, 1980
2. U.Ewert et al,. In: International Symp. on Computerized Tomography for Industrial Applications, Berlin, June 1994, pp.148-159
3. Hewlett-Packard Press Reliases.
4. In: web-site http://www.hp48.com/pressrel/feb97/25feb97b.htm
5. Digiray® Digital X-Ray Systems.
6. In: web-site http://www.digiray.com
7. M.Maisl et al. In: International Symp. on Computerized Tomography for Industrial Applications, Berlin, June 1994, pp.226-233

Fully Additive Copper Metallization on BCB

Thomas Stolle[1], Beatrice Schwencke[2] and Herbert Reichl[2]
[1] FhG-IZM Berlin [2]TU Berlin, Gustav Meyer-Allee 25, 13355 Berlin

1 Summary

A fully additive copper metallization process on benzocyclobutene Cyclotene™ (BCB) has been investigated for application in MCM-D technology. The process consists of surface pretreatment of the BCB basic layer by reactive ion etching (RIE), spin-coating and photopatterning of an organic seed layer by broad-band I-line photolithography followed by developing and activation steps. The metallization of the seed patterns is performed by a 2-step process by means of electroless copper baths.

A height of about 5 μm selectively deposited copper can be achieved. The electrical conductivity of patterns is in the range of 80% - 85% of the bulk conductivity of pure copper. Adhesive strength tests during accelerated aging show good adhesion of copper to the BCB surface, which is influenced by RIE pretreatment, exposure dose and thermal load. Shear experiments performed with optimal treated 200x200 μm bumps show shear forces > 150 cN. Design rules have to take into account the lateral growth of copper patterns, which is nearly equal to the vertical growth. Real spaces of ≥ 30 μm between copper lines are possible.

The process is considered as a low cost technology because of replacing of sputter technique, few process steps and waste reduction.

2 Introduction

Fully additive and selective metal deposition processes on dielectric substrates for microelectronic applications have been investigated for a long time. The generation of a seed layer, addressing the following metal deposition by catalytic reactions from electroless baths may be performed by different photoredox processes involving precious metal compounds [1, 2]. Because of its characteristic ability to give very smooth amorphous thin layers from volatile organic solvents, palladium acetate was found to be suitable for thin film applications, using incident light of short wavelength, usually vacuum UV, UV-A or UV-B [3, 4]. After mask exposure and development (dissolving unexposed parts of the layer), the amplification of the seeds generated in this way is performed by means of electroless metal deposition baths. For electrical applications the selective deposition of copper is most important because of its electrical conductivity. Up to now, controlling the reproducibility of copper baths seems to be the crucial point in an industrial process.

The aim of this work is the application of a new photoactivated process of seed layer generation based on palladium acetate (PdAc) for selective copper deposition on BCB. The seed layer is generated by means of broad-band I-line photolithography and other standard technologies used in MCM-D [5 - 8].

3 Process Overview

The process consists of the following steps:
1. RIE pretreatment of the BCB surface to ensure sufficient adhesion of the copper layer,
2. spin coating of a precursor layer of photoactivated palladium acetate from volatile organic solvents,
3. mask exposure by means of common photolithographic devices (broad band I-line photolithography),
4. development (dissolving of unexposed areas with organic solvents),
5. activation of the seed layer (reduction of Pd^{2+} to Pd^{0}) applying an alkaline KBH_4 solution in water,
6. start metallization (electroless copper to a height of about 0.3 μm),
7. growth metallization (electroless copper to a height of about 5 μm).

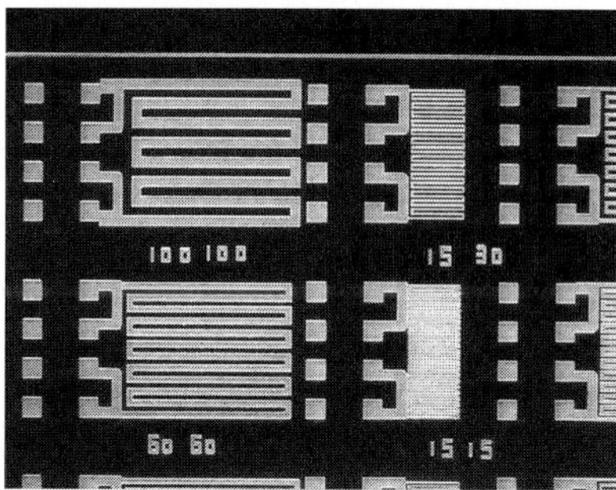

Fig. 1: Test patterns on BCB, h = 4.8 μm

The issue of this process is the realization, that definite photoactivators of the type of aromatic azides or bisazides in combination with palladium acetate are able to generate catalytically active particles (seeds) of Pd in an organic matrix after mask exposure by incident light of 360 – 410 nm. The roughening of the BCB surface by reactive ion etching (RIE) in an O_2/SF_6 plasma under optimized conditions is an important step in the process. The last steps of the electroless metallization process are crucial to avoid stress in the copper layer, which would result in a lack of adhesion to the BCB layer underneath. In the following some important results of our investigations are presented. Fig. 1 shows a test pattern of electroless deposited Cu in a height of 4.8 μm.

4 BCB Surface Pretreatment

The BCB surface as spun on the wafer and partially cured to about 70% is very plain and chemically resistant. Atomic force measurements (AFM) yielded small (< 10 Å) rms-roughness values. Therefore the adhesion of electrolessly deposited copper is very poor. The shear force of 100 x 100 μm test bumps was measured to be about 40 – 50 cN [7].

Attempts to roughen the surface by wet chemical methods failed up to now. Plasma etching (RIE) applying different plasma compositions shows very different effects on the generation of the seed layer and on the metallization results depending on the kind of plasma and on the starting activity of the copper bath. The effects include phenomena of no deposition of copper as a result of adhesion failures of the developed precursor layer in case of applying a pure O_2 plasma, as well as full-area deposition of copper in case of applying N_2 plasma.

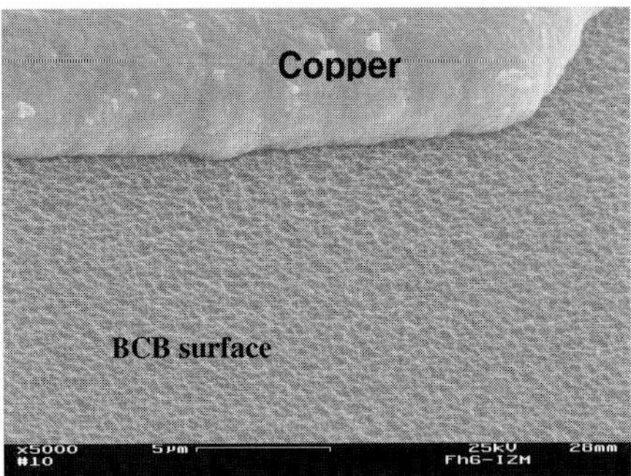

Fig. 2: BCB surface treated by RIE, after processing (magn. 5000)

Best results to roughen the BCB surface without disturbing effects were achieved by means of an O_2/SF_6 plasma for 4 – 6 min. In this case, the surface roughness is improved to 460 – 480 Å (rms roughness by AFM). This roughness becomes visible by SEM as shown in fig. 2, which presents the roughened BCB surface after performing the process steps 1 – 7.

5 Seed Layer Generation

Manufacturing of an activated precursor layer for the seed layer generation is possible by spin coating of solutions of the photoactivator 2.6-bis-(azido-bencylidene)-4-methyl-cyclohexanone and palladium acetate in a volatile organic solvent [6, 7]. Preliminary experiments using different solvents for the seed layer formulation showed best results with ethyl acetate concerning regularity and optimal layer thickness. So the drying of the wet layer was completed within a few seconds after spin coating. At this stage the layer thickness was shown to be about 800 - 1000 **Å** by profilometer measurements on quartz plates, that had been coated applying the same spin parameters.

Mask exposure was performed by means of usual photolithographic equipment to a dose of 1500 mJ/cm^2 (measured at 366 nm). Afterwards the development can be performed by organic solvents like ketones, short-chain esters or alcohols. In this work a 1 : 1 mixture of acetone and i-propanole was applied for 5 min. Cleaning of the surface was completed by a short i-propanole rinse. The patterns of this pre-seed layer might be inspected already at this stage by light-microscopy. To fabricate a catalytically active seed layer a chemical reduction must be performed to receive Pd-clusters from Pd^{2+} applying a solution of KBH_4 and KOH in water for 3 min. The thickness of the seed layer containing Pd-clusters is measured to be in the range of 150 – 250 Å . The patterns show a typical light-gray, bright-yellow color and sharp edges.

6 Selective Electroless Copper Plating

In order to receive clean and good adhering copper patterns a reliable bath control turned out to be the main problem of selective patterning. The properties of electroless copper baths do not depend on the analytic concentrations of the main components alone, which were determined and adjusted by recipes given by the suppliers, but also on the content of O_2 and special stabilizers [9, 10]. The concentration of the latter changes continously as a result of metal deposition and bath storage. Usually the concentration of O_2 and the stabilizers are only controlled by bath handling and replenishing procedures.

An O_2 control in copper baths is possible by means of membrane sensors, although not very practicable due to a lack of stability in the strong alkaline bath solution. Some stabilizers show a strong influence on the mixed potential (MP), i. e. the electrochemical potential of a copper sheet immersed in the bath solution vs. a reference electrode [9,10].

A metal deposition procedure for selective copper patterning using the catalytic properties of the seed layer has been worked out [7, 8]. This procedure involves two metallization steps. The first one we denote as start metallization, which ends up with a copper layer of about 0.3 μm thickness. The second one we denote as growth metallization to receive the final copper layer thickness of up to 5 μm in a very slow growth process. The kind of used copper baths and typical bath controlling parameters are given in table 1. The very slow growth process is necessary to avoid too high stress in the copper patterns, which would result in poor adhesion or delamination of the patterns during metal deposition.

Table 1: Working parameters for selective copper deposition

Bath type supplier	Working temp. [°C]	Work load [dm^2/l]	Plating rate [μm/h]	Layer thickness [μm]	MP [mV] vs. Ag/ AgCl ref. electr.	Used for ...
XD-6157T Mac Dermid	59 - 63	0.1	0.9 – 1.2	0.3	-640 to -550	Start metalli-zation
Doduprint 505 (Dodu- co GmbH)	20	< 0.1	0.2 – 0.4	≈ 5 (no plating on the seed layer)	-750 to -770	Growth metalli-zation

7 Results

7.1 Size and shape of structure elements

Size and shape of the deposited structure elements are influenced by the size effect caused by the O_2 content of the bath solution [11] and by the lateral growth. On one hand, the size effect is helpful to suppress extraneous deposition of copper, on the other hand it can result in suppressing the growth of very small structure elements or in generating of rounded edges on bigger ones. Examples are shown in figs. 3 (cross section of a deposited line) and 5 (bump, bird's view). Generally the contact angle to the base layer is smaller than 90° depending on the O_2 content. Taking into account the deviation from the area of a corresponding rectangle, the electrical conductivity of deposited copper lines can be considered to be 4.92 $\Omega^{-1}cm^{-1}$, which is 83% of the bulk conductivity of pure copper.

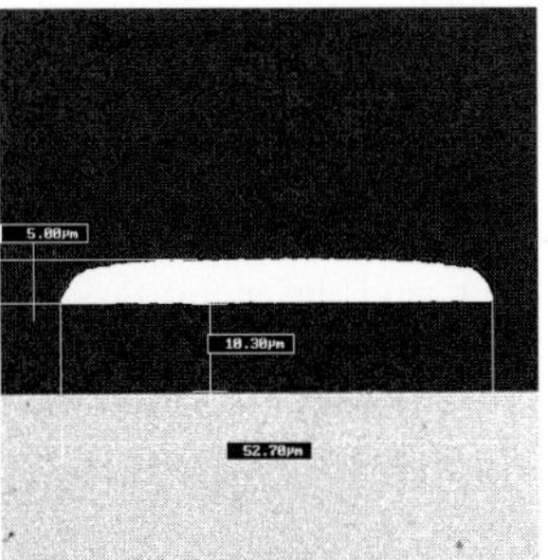

Fig. 3: Cross section of a copper line

The lateral growth causes a deviation of the sizes of structure elements from the corresponding designed dimensions, which depends on the structure height. The lateral growth is amounted to be nearly equal to the vertical growth. Assuming a structure height of 5 µm, the dimensions of a designed meander structure with 20/40 µm line/space sizes will grow up to about 30/30 µm line/space sizes. Up to now this is the limit of the resolution power of the selective deposition technique. Hence, the lateral growth of the structure elements must be taken into account for the design rules [8].

7.2 Adhesion and accelerated aging

Adhesive strength testing by shear force measurements during accelerated aging shows strong influences of the surface pretreatment by RIE as well as of the conditions of corrosion experiments. The measurements were performed with 200 x 200 µm sized bumps located on the BCB surface. Fig. 4 shows the influence of the duration of RIE, the influence of corrosion load, and of thermal cycling load on the shear force necessary for the rupture of the interface Cu/BCB. The shear force increases with both the RIE time and the corrosion load (85°C, relative humidity 85%, 300 h). An impressive result as observed in some cases is shown in fig. 5. After performing the corrosion load the shear tool leaves a track on the bump surface and pulls off a shaving with a shear force of > 200 cN without destruction of the BCB/Cu interface.

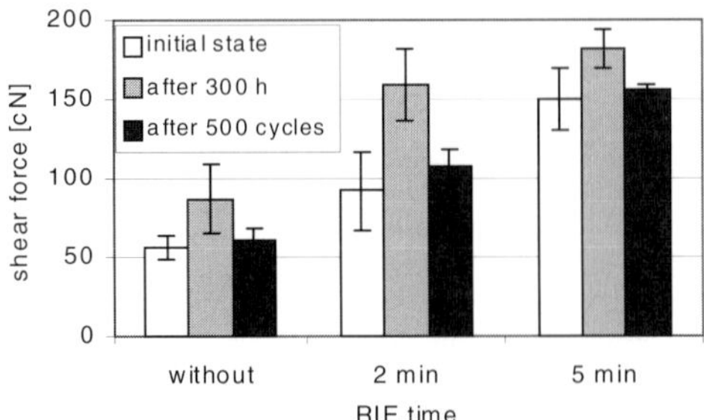

Fig. 4: Influence of RIE time, corrosion load and thermal cycling on the shear force

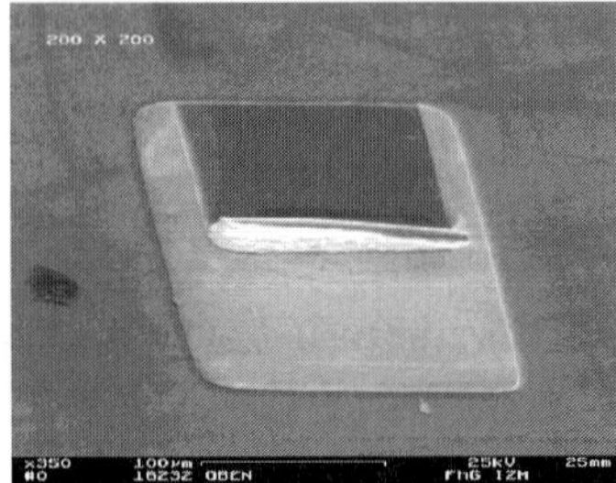

Fig. 5: Track of the shear tool on the bump surface

The influence of thermal cycling load (cycling 500 times from –55°C to +125°C) is small (third bar of each group of fig. 4). Former investigations with 100 x 100 μm bumps showed an improvement of the adhesion after performing 200 cycles [8]. This observation might be related to the increase of tension between BCB and copper during thermal cycling with increasing the structure size.

8 Conclusion

A fully additive and selective process for copper deposition on BCB is possible. The very small deposition rate is the limiting factor for an application of this process up to now. However, it involves possibilities of simultaneous metallization of large numbers of samples. The application of this process is intended for electronic packaging in the areas of MCM/D and MCM/L.

9 Acknowledgment

Funding for this research was provided by The DOW Chemical Company, by Deutsche Forschungsgemeinschaft and by Analytik-Pool at FhG-IZM, and is gratefully acknowledged. The authors would like to thank Mr. F. Krause and Mr. Dr. W. Faust from FhG-IZM, who put the results of AFM and cross-cuttings to the authors disposal.

10 References

1. M. Paunovic et al., J. Electrochem. Soc. 127 (9) 441C – 447C (1980)
2. Gemmler et al., Galvanotechnik 74 1523 - 28 (1983
3. D. J. Macauley, P. V. Kelly, Micro Mats. '97, Proceedings p. 139
4. J.-Y. Zhang, I. W. Boyd, Thin Solid Films 318 234 – 238 (1998)
5. M. Töpper, Th. Stolle et al., DE 195 18 512 Kl. G 03F 7/004
6. Th. Stolle, M. Töpper et al., J. Inf. Recording 23 209 – 221 (1996)
7. M. Töpper, Th. Stolle et al., Micro System Technologies '96, Proc. p.199-203
8. M. Töpper, Th. Stolle et al., Internat. Symp. on Advanced Packaging Materials Braselton, G., March 1999, Proceedings p. 202 - 208
9. Shiro Haruyama and Izumi Ohno, Proc. Electrochem. Soc. 88-12 p. 20 – 36 (1988)
10. D. P. Seraphim, R. Larsky and Che-Yu Li, "Principles of electronic Packaging", ed. by A. E. Elken and J. M. Morris, McGraw-Hill Book Comp.. N. Y. 1989
11. M. T. van der Putten et al., J. Electrochem. Soc. 140 (8) 2221 – 2228 (1993)

A Sintering Behavior of Glass/Ceramic Composite for Using in High Frequency Range

Chan-Joo Lee, Hyeong-Jun Kim, Sung-Churl Choi
Dept. of Ceramic Eng. Hanyang University, Seoul 133-791,KOREA

1 Abstract

Recently low temperature co-fired ceramic (LTCC) made of glass-ceramic is replaced with alumina substrate to coincide with the diversity of microwave electronic device.

The objective of this study was to investigate the sintering behavior, crystallization characteristic of glass-ceramic with high-K(dielectric constant) and optimal sintering condition on the glass/ceramic composite for fabrication LTCC. Glass/ceramic composite was made from alumina powder and glass frit , which was composed of SiO_2-TiO_2-RO/Al_2O_3(R: Ba, Sr, Ca) with PbO as additive, and was sintered for 0, 30, 60minutes in the temperature range from 700 °C to 1000 °C.

Properties of frit and glass/ceramic composite were analyzed by DTA, XRD, SEM and Network Analyzer and so on.

2 Introduction

Recently low temperature co-fired ceramic (LTCC), which was based on glass-ceramics, replaced with alumina substrate to coincide with the diversity of microwave electronic device[1)]. The properties of LTCC make it possible to fabricate the direct mounting for high density of bare chips as well as high speed signal propagation. So generally glass ceramic substrate was developed specially to reduce the resistivity of the lines and the dielectric constant of the ceramic. It has the effect that lowers the signal delay in the substrate. Furthermore the lowering of sintering temperature through the use of glass ceramic, it is possible to co-fire with low resistance conductors such as Ag and Cu.

SiO_2-TiO_2-BaO system glass ceramic have developed for using in condenser in low frequency range[3)], and Koo et al have studied that SiO_2-TiO_2-RO glass ceramic can be fabricated in low temperature in high frequency[4)]. It also is known that PbO, TiO_2, and BaO in glass are components for high dielectric constant.

Thus, SiO_2-TiO_2-BaO-PbO system glass was selected for low temperature sintering, and the sintering behavior, crystallization characteristic of glass-ceramic with high-K(dielectric constant) and optimal sintering condition on the glass/ceramic composite for fabrication LTCC was investigated.

3 Experimental Procedure

The composition of glass frit was 34 SiO_2, 25 TiO_2, 11 BaO, 15 CaO, 12 SrO, 3 PbO(mol%).

Glass was melted in a Pt-crucible for 60min. at 1400 °C and then quenched by pouring into a diluted water. To obtain glass powder with 2μm, quenched glass was crushed by attrition mill using zirconia jar and balls. The glass powder and alumina with average particle size of 3.8μm were mixed by wet ball milling using isopropylalcohol as solvent for 24h. The mixed powder was pressed into disk type and sintered for 0, 30, 60min. in the temperature range between 700 °C and 1000 °C then polished both sides to measure dielectric constant and Q-factor.

Glass transition temperature(Tg) and crystallization temperature were acquired by DTA. Crystalline phases in glass were identified by XRD. Network Analyzer was used to measure dielectric constant and Q-factor. Density was measured by Archimedean method.

4 Result and Discussion

4.1 Glass frit characterization and particle size distribution

Tg, dilatometric softening point and crystallization temperature of frit were 726 °C, 840 °C and 920 °C, respectively (Figure 1.).

Figure2. shows the particle size distribution of glass frit and Al_2O_3. Each mean particle size of frit and Al_2O_3 is 2.0 and 3.8μm.

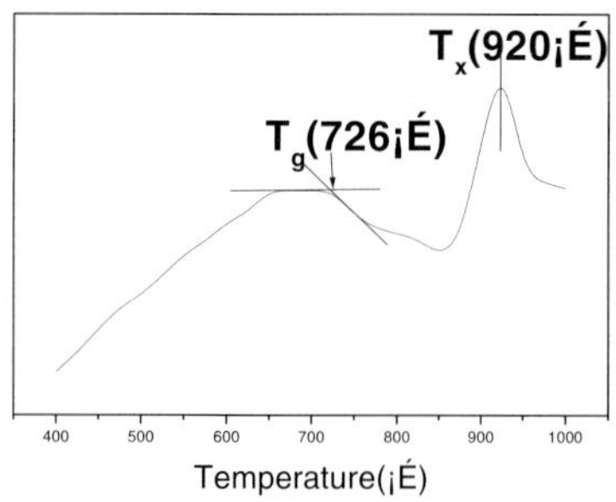

Figure 1. Differential thermal analysis curve for frit

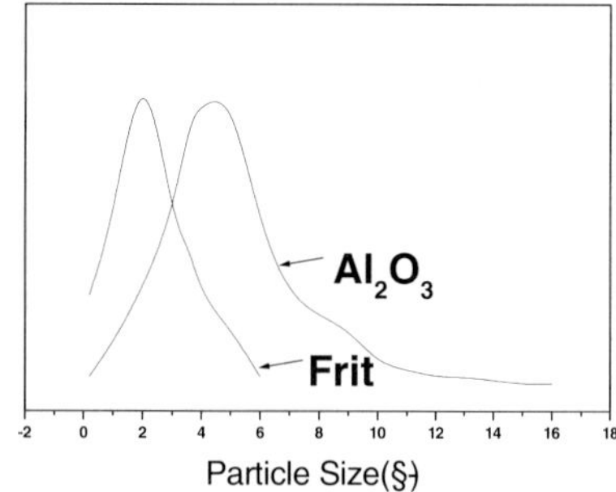

Figure 2..Particle size distribution of frit and alumina

4.2 The Sintering behavior of glass/ceramic composite

4.2.1 Densification and identification of crystalline phases

Fig 3(a) represents the change of volumetric shrinkage as duration time and (b) does the change of density at various temperatures. As duration time increased at Tg(726 °C), the change of volumetric shrinkage appeared drastically, but above Tg, it appeared to be about constant. Thus duration time did not affect shrinkage. In Figure3(b), fast increase of density appeared between 730 °C and 830 °C, which corresponded to increase of volumetric shrinkage in the temperature range.

The sintering mechanism of LTCC is that of glass sintering caused by viscous flow of the matrix glass which then penetrates into pores between frit and Al_2O_3 powders[6]. It is called NLPS(Non reactive liquid sintering) without eutectic melting.[7].

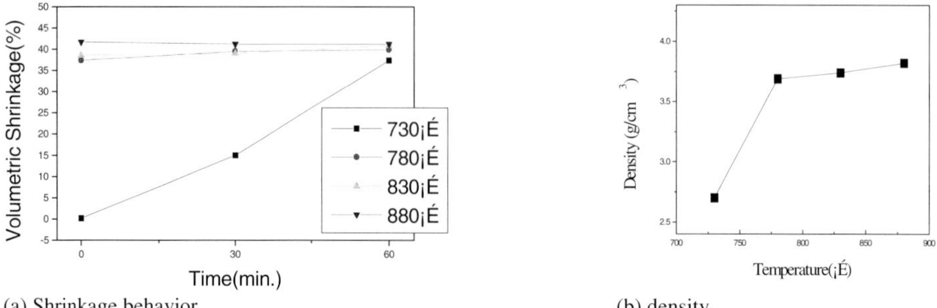

(a) Shrinkage behavior (b) density

Figure 3. Shrinkage behavior and density of glass/ceramic composite as a function of sintering temperature.

From these results, above Tg the redistribution of liquid phase and densification occurred instantly and fast but after this more change did not appeared, it was guessed that some change happened inside under sintering process. It was known that typical mechanism to impede viscous sintering was crystallization which occurred in course of the sintering of glass[8], so each of them was analyzed by XRD to identify crystalline phases in glasses (Figure 4).

In Figure 4(a), when duration time is 0 min. there is only alumina peak at 730 °C. But as sintering temperature increase, crystalline $BaTiSi_2O_7$ appeared at 780 °C and crystalline $SrTiSiO_7$ appeared at 830 °C. In Figure4 (b) and (c), as duration time increase, crystalline phases was observed at low temperatures.

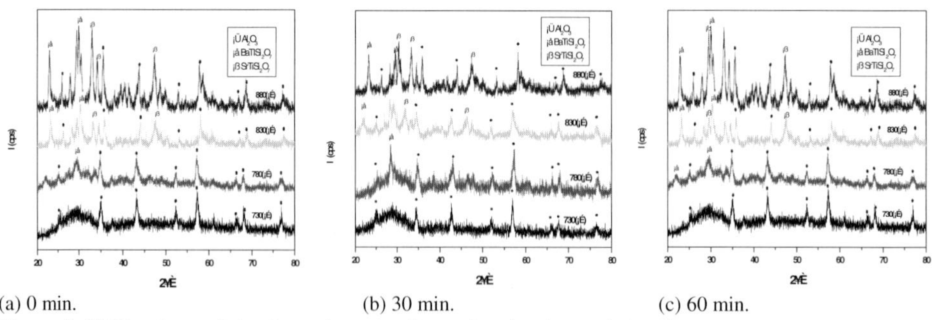

(a) 0 min. (b) 30 min. (c) 60 min.

Figure 4.. XRD patterns of glass/ceramic composites as duration time and sintering temperature

As shown Figure3 and Figure4, Densification was restraint in the moment of crystallization from shrinkage rate's being slow and the creation of crystalline phase. According to J,F,Macdowell[5], if particle size of frit is large and sintering rate is fast, the degree of densification is high but crystallization is low. On the other hand if that of frit is small and sintering rate is slow, the degree of crystallization is high but densification is low. [5]

4.2.2 Microstructure Analysis

Figure 5 presents the microstructures of surface and fracture surface at each sintering temperature for 30 minutes. Figure 5(a) is different from others and shows that sintering and densification did not progress just like results of density and shrinkage. Though shrinkage stopped above Tg(726 °C), pores remained in the fracture microstructures of glass/ceramic composite sintered at 780 and 830 °C in Figure 5(b) and (c). And surface microstructures were denser than fracture microstructures, which was thought that glass particles in surface fused faster than inner them and fused surface obstructed extrication of gas in pore.

Thus, it was known that the saturation of volume shrinkage and density at higher temperature than Tg was caused by crystallization and difference of fusing rate between surface and inside from XRD and microstructures analysis.

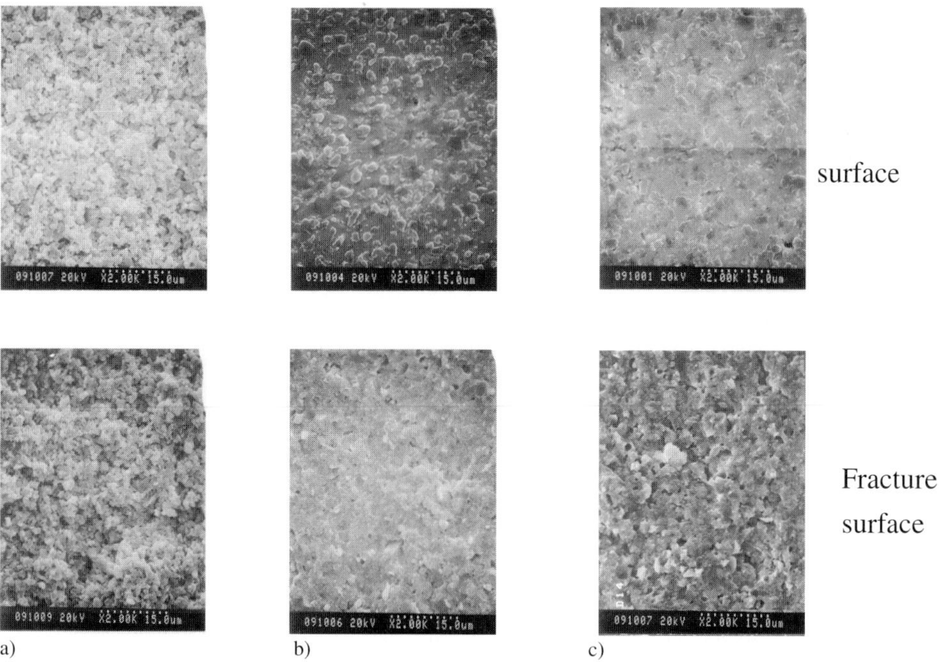

Figure 5. Microstructure changes of glass/ceramic composite of sample sintered at (a)730, (b)780, (c) 830 °C.

4.2.3 Dielectric Constant and Q-factor

Figure6 is the results of change of dielectric constant and Q as changing duration time at 830 °C. The center frequency is 12 □. The change of dielectric constant was a little in regardless of crystallization after densification as sintering temperature changes. Dielectric constant of this glass/ceramic composite was lower than general material known as having high-K, because resonant frequency is higher. But Q value was affected by crystallization from increasing from 6500 at 830 °C to 22,000 at 880 °C.

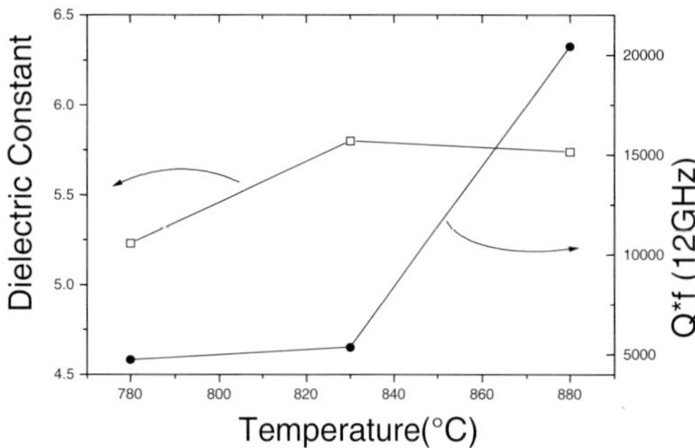

Figure 6. Dielectric Constant and Q*f of Sample sintered at various temperature for 30min.

5 Conclusion

High frequency glass/ceramic composite can be fabricated at less than 1000 °C, the melting point of Ag, Cu. At higher temperature than Tg, sintering progressed instantly but there was the end of shrinkage. It was thought that the saturation of volume shrinkage and density at higher temperature than Tg was caused by crystallization and difference of fusing rate between surface and inside from XRD and microstructures analysis.

The dielectric constant was affected a little but Q-factor was affected by crystallization after densification as sintering temperature changes.

6 Reference

1. Daniel I . Amey ," Overview of MCM technology: MCM-C" ISHM'92 , 225 – 234(1992)
2. K. Watanabe, " Coalescence and crystallization in powder high- Cordierit glass," J. Am.Cer.Soc ,68(4), 102-103 (1985)
3. Herczog," Microcrystalline $BaTiO_3$ by crystallization from glass" J. Am. Cer. Soc, 47[3],107-115(1964)
4. Koo et. al.," A study on the development of SMD-type ceramic for PCS Handyset", KETI- RD-98062
5. J.F Macdowell," Low K glass-ceramics for microelectronic packaging" Ceramic Transaction vol15, 133-151(1990)
6. W.D. Kingery, "Introduction to Ceramics", Wiley&Sons, New York,1960
7. R.M.German, " Sintering Glass-filled Ceramic Composite: Effect of Glass Properties" ,969-977, Ceramic Transaction Vol.1(B)
8. Jose Zlmmer," In situ investigation of sintering and crystallization of Lithium Aluminosillicate glass-ceramics",
9. Glastech.Ber. Glass.Sci.Tec,70(1997). No 6

A New Retarder Film Improving the Temperature Dependent Performance of STN Displays

Eckhard Hanelt, Thomas Kammel
Consortium für Elektrochemische Industrie GmbH, Central Research Company of Wacker Chemie GmbH, München, Germany
Masato Kuwabara
Tsukuba Research Laboratory, Sumitomo Chemical Co., Ltd., Tsukuba, Japan

1 Abstract

A new retarder film which shows a reversible retardation change with temperature was developed to reduce the loss of contrast of film-compensated STN-mode LCDs with increasing temperature. The retarder consists of a stretched film of polycarbonate blended with liquid-crystalline Silicone. The resulting high contrast in a wide temperature range and the stability of the film satisfy the requirements for car equipment and other mobile applications.

2 Introduction

The performance of STN (supertwisted nematic) mode LCDs has been improved so far that STN displays are used for many applications with low and high informational content [1]. In price-sensitive applications, they have proven to be competitive to the progressive TFT technology. However, the usage of reasonable film-compensated STN displays has almost been restricted to environments with rather stable surrounding temperatures which still is an obstacle for a widespread use e.g. for car equipment or mobile applications.

In order to obtain black and white images and to reduce the viewing-angle dependence, STN displays need at least one retarder film which compensates the optical retardation of the switchable cell filled with a low molecular-weight liquid-crystalline mixture. The conventional polymer retardation films used in these displays have a nearly constant optical retardation which usually is adapted to the retardation of the STN cell at room temperature. However, the cell's retardation decreases with increasing temperature due to the thermal motion of the LC molecules in the cell. This leads to a mismatch to the retarder film and a loss of the display's contrast at higher temperature.

A known technical solution to obtain retardation compensation in a wider temperature range is adding a second STN cell filled with LC molecules of low molecular weight showing a temperature dependence of the optical retardation similar to the switchable cell [2]. However, these DSTN-mode displays are heavier and more expensive.

Both disadvantages are removed by a new retarder film which shows a reversible retardation change with temperature adapted to the given cell's retardation change. The film is made from a new material containing a mixture of a high-T_g polymer such as polycarbonate

(PC) or polyvinylalcohol with a nematic liquid-crystalline Silicone. After casting and drying of the film, an orientation of the polymer chains and the mesogenic groups of the LC-Silicone is obtained by stretching. Thus, an optical anisotropy is induced in the film. The contribution of the polycarbonate shows a constant optical retardation independent of temperature up to about 130°C. The retardation of the liquid crystalline component decreases with increasing temperature down to zero above its clearing point which is set to fit the retardation curve of the switchable cell.

This new type of temperature-compensating retarder film (TCR) is commercially available now. Its first applications already have shown that the temperature dependent performance of displays for mobile applications can be considerably improved with only low additional costs. In this paper we report our results on the production and properties of the TCR films.

3 Sample Preparation

Different types of liquid-crystalline side-chain polymers based on siloxane backbones were synthesized by hydrosilylation reactions as described in [3, 4]. The resulting nematic LC-Silicones have a broad chain length distribution. An example of a LC-Silicone used for TCR films is shown in fig. 1. Its degree of polymerization is about 14 and the phase transition temperatures measured by a differential scanning calorimeter are a glass transition temperature T_g of 18°C and an isotropic transition temperature T_c of 68°C.

A polycarbonate was mixed with 2 to 15 wt% LC-Silicone and dissolved in methylene chloride. The solutions were cast and dried to produce films. The films were stretched at about 180°C and subsequently cooled down to room temperature. The result were about 100 μm thick films with low haze of less than 3%. Their optical retardation as a function of temperature was measured by the Senarmont method using a laser beam and a hot stage. Typical retardation values of the TCR films produced with the LC-Silicone of fig. 1 are 400 nm to 600 nm at 30°C and 200 nm to 400 nm at 80°C.

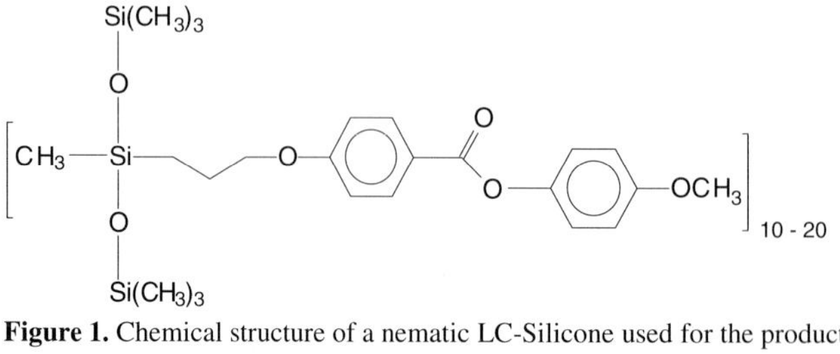

Figure 1. Chemical structure of a nematic LC-Silicone used for the production of TCR films.

4 Optical Retardation Properties

The result of the measurement of the temperature dependence of a TCR film's optical retardation is shown in fig. 2. The same film is compared to a conventional stretched polycarbonate (PC) film in a wider temperature range in fig. 3. The contributions of the LC-Silicone and the PC to the total retardation can clearly be distinguished. The retardation change in the temperature range up to 100°C is reversible and corresponds to the temperature dependence of the order parameter of the liquid crystalline phase.

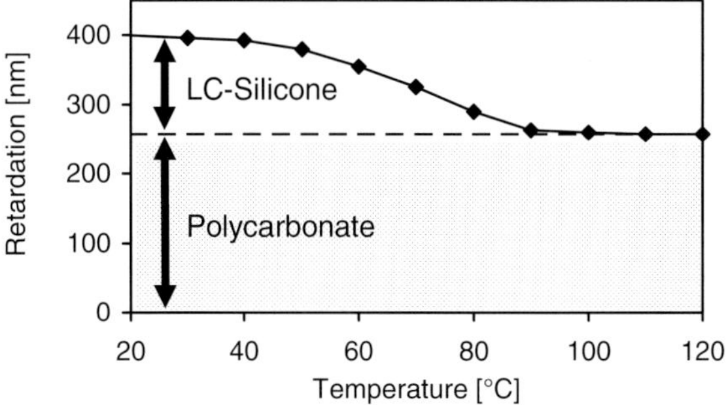

Figure 2. Temperature dependence of the optical retardation of a TCR film. The contributions of the polycarbonate (dashed line) and the LC-Silicone are schematically indicated. Measured data are marked by diamonds.

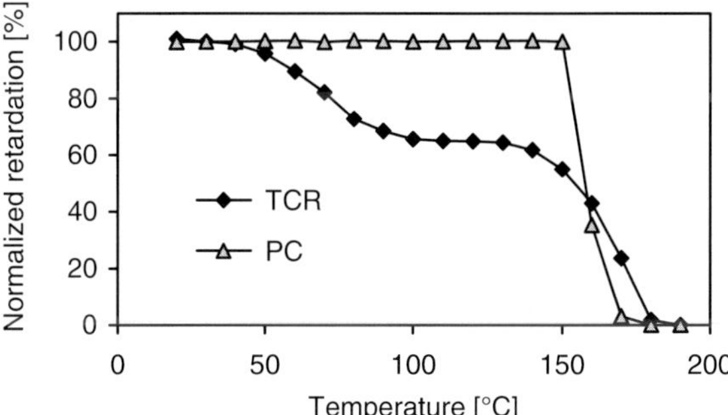

Figure 3. Temperature dependence of the optical retardation of a TCR film and a pure polycarbonate retarder film. The retardation was normalized against its value at 30°C.

The second decrease starting above 120°C is irreversible and finally leads to destruction of the film above 180°C. This decrease which is observed in both films is caused by the relaxation of the oriented PC chains at the PC glass transition around 150°C. The constant retardation in the temperature range from 100°C to 120°C is only due to the PC, because the LC material is in the isotropic phase whereas the PC is in the glassy state still keeping its retardation induced by the stretching process.

The temperature dependence of the TCR film's retardation can be adapted to the given retardation curve of any STN cell. The controllable parameters are the film thickness, the content of LC-Silicone and the isotropic transition temperature T_c of the LC. In fig. 4, the influence of the LC content is demonstrated. The slope of retardation vs. temperature increases with increasing LC content. However, there is an upper limit of the LC content given by the requirement of a high mechanical stability of the film. Another way to control the slope or the curvature of the retardation change is adjusting the T_c of the LC-Silicone. Some examples covering the range of interest for most applications between room temperature and 80°C are given in fig. 5. Finally , the film thickness is used to control the magnitude of the optical retardation.

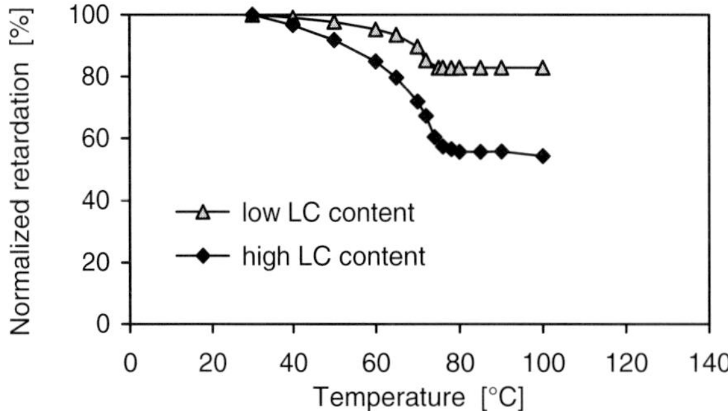

Figure 4. Comparison of the retardation of TCR films with low and high content of the LC-Silicone of Fig. 1. The measured retardation curves were normalized against their values at 30°C.

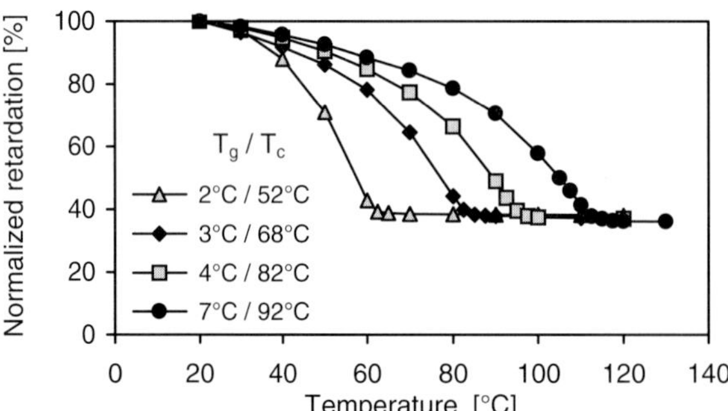

Figure 5. Comparison of TCR films containing LC-Silicones with different isotropic transition temperatures T_c. The measured retardation curves were normalized against the retardation at 20°C.

The performance of a TCR film in a STN display is demonstrated in fig. 6. The TCR film was mounted on the left-hand side of the screen and a conventional PC retarder film with corresponding retardation at room temperature on its right-hand side. Both parts of the screen show a good contrast at room temperature. At 80°C, the readability is still sufficient on the left-hand side but it diminishs on the right-hand side of the display.

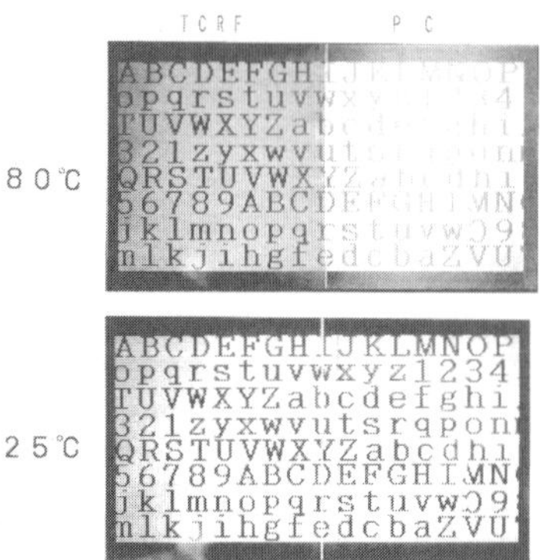

Figure 6. Pictures of a STN display containing a TCR film on the left-hand side of the screen and a conventional PC retarder film on its right hand side at 25°C and 80°C (photograph by courtesy of Optrex Co., Tokyo, Japan).

5 Film durability

The film is mechanically stable up to temperatures of about 130°C. To test the durability of its optical properties a sample was kept in an oven at 100°C. The film was taken out of the oven and the retardations at 30°C and 80°C were remeasured after the time intervals shown in fig. 7. The change of retardation was negligible during 1,000 hours of heat treatment at 100°C.

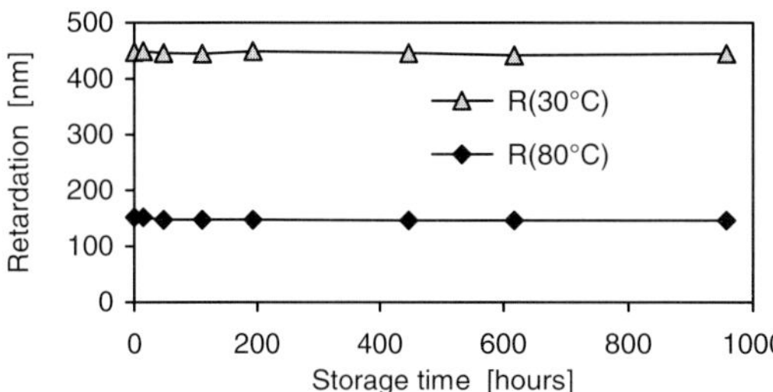

Figure 7. Stability of the optical retardation of a TCR film after storage at 100°C. The diamonds and triangles mark the retardation values measured at 80°C and 30°C, respectively.

The UV stability of the TCR film was tested in an environment similar to a LCD. The TCR was laminated with a polarizer and exposed to a high luminous intensity in a sunshine tester. It showed sufficient stability after 265 hours of irradiation with a total light dose of 67 kWh/m² which corresponds e.g. to 30 days of sunshine in Tokyo at August.

6 Conclusion

The concept of using a second layer containing LC molecules like in a DSTN display to compensate for the temperature-dependent retardation of STN cells is applicable also to film-compensated displays now. The TCR films can be produced to improve the performance of any STN display in a wide range from room temperature up to 80°C. The durability of the TCR films is comparable to conventional PC retarder films. The advantage of applying the well established and reliable production process for retarder films from polycarbonate by adding a low content of LC-Silicone is reasonable also from the economic point of view. Thus, the field of application of STN displays, especially STN for car equipment and mobile applications, is expected to be considerably expanded by using these films.

7 References

[1] T.J. Scheffer, J. Nehring in SID Seminar Lecture Notes San Jose 1999, Society for Informational Displays, Santa Ana, CA, USA, 1999, Seminar M-6.

[2] K. Katoh, Y. Endo, M. Akatsuka, M. Ohgawara, K. Sawada, Jpn. J. Appl. Phys. 1987, 26, L1784 – L1786.

[3] H. Finkelmann, G. Rehage in Advances in Polymer Science 60/61 (Ed.: M. Gordon), Springer-Verlag, Berlin Heidelberg, 1984, p. 99 – 172.

[4] H. Finkelmann, G. Rehage, Makromol. Chem., Rapid Commun. 1980, 1, 31 – 34.

Complex Defects in ZnSe-Based Phosphors

Vello Valdna, Jaan Hiie, Urve Kallavus
Tallinn Technical University, Tallinn
Roger Durst
Bruker AXS, Inc., Madison
Larry Jones
Iowa State University, Ames

1 Abstract

The influence of different thermal treatments, codopants and impurities upon the defect structure, luminescence characteristics and stability of the ZnSe-based phosphors are investigated. It was found that the brightness of annealed phosphors increases at 300 K if the flux residue is previously removed. The highest instability exhibit doped only with Cl phosphors. Dissociation of Cl-related complexes as a possible cause of instability is proposed.

2 Introduction

ZnSe is known as a well starting material for electroluminescent phosphors [1]. ZnSe-based phosphors have a high X-ray conversion efficiency [2] and well radiation stability [3]. ZnSe is intrinsically n-type [4]. Group VII elements are superior over group III elements as donor dopants in ZnSe [5] and chlorine is the most often used donor dopant. The electron concentration of Cl doped ZnSe saturates at about $2x10^{19}$ cm^{-3} [6]. Relatively little is known about the electronic structure of intrinsic defects in II-VI semiconductors [7]. The zinc interstitial in ZnSe is the only documented case of a metal interstitial in II-VI semiconductor [8]. The activation energies of supposed defects in ZnSe are given in [9]: complex $V_{Zn}Cl_{Se}$ or A-centre E_V + 0.35 eV, V_{Zn} E_V + 0.66 eV, V_{Se} E_C – 0.34 eV, V_{Se} related complexes with Cl E_C – 0.3 eV and E_C – 0.8 eV, Zn_i E_C – 0.9 eV. D.J. Chadi [10] gives E_C – 0.19 eV for antisite defect Se_{Zn}. Se_i and antisite Zn_{Se} are disregarded from the possible species as the supercell energies of these defects are quite high in all charge states [9].

The aim of this work was to investigate the influence of different thermal treatments, codopants and impurities upon the defect structure, luminescence characteristics and stability of the ZnSe-based phosphors.

3 Experimental Procedure

The samples were prepared in the way described in [11]. Ultra-pure semiconductor grade starting materials were used. Thermal treatments of the samples were carried out in the Lindberg/Blue M three-zone tube furnace. Photoluminescence (PL) spectra of samples were measured in the Janis CCS-150 He cryostat at 8 K. The 441.6 nm line of Omnichrome He-Cd laser with CC8 and 1M $CuSO_4$ solution filter was used as the excitation source. The PL signal was dispersed using the monochromator SPM-2. A chopped signal from the Hamamatsu PbS photodetector was amplified, recorded and corrected on a computer. Samples and phases condensed after thermal treatments were investigated by the scanning electron microscope (SEM), X-ray diffraction patterns (XRD) and X-ray microanalysis (XMA). The concentration of dopants in samples was determined by the glow discharge mass spectrometry (GDMS) analysis.

4 Results and Discussion

The flux residue was removed from the recrystallized with $ZnCl_2$ flux phosphor samples using the vacuum annealing before thermal treatments were carried out with the samples. GDMS analysis detected the Cl concentration around 0.07 at% in samples after a flux residue was removed from the samples. PL spectra and integral intensity at X-ray excitation of 10 keV (Mo-anode) were measured using different samples for PL and X-ray measurements.

Table 1. Stability versus thermal treatment of ZnSe-based phosphors*

No	Phosphor composition	Thermal treatment	U_F [mV]	U_{FI} [mV]	ΔU [%]
1	ZnSe:Cl	Flux residue removed	9.9	14.3	49.5
		Under Zn vapour pressure	39.3	44.4	12.9
		Vacuum annealed	21.9	25.8	17.7
		Quenched to 300 K	22.1	26.4	19.5
		Zn vapour, vac.annealed	21.5	25.4	18.2
		Zn vapour, quenched	22.9	26.4	15.4
2	ZnSe:Cl:Ag	Flux residue removed	19.3	22	14.2
		Under Zn vapour pressure	41.7	46	10.3
		Vacuum annealed	23.4	26.1	11.7
		Quenched to 300 K	25.7	28.4	10.6
		Zn vapour, vac annealed	25	25.8	3.1
		Zn vapour, quenched	24.6	29.1	18.2
3	ZnSe:Cl:Ag:Cu	Flux residue removed	46.6	50.8	9.1
		Under Zn vapour pressure	55.1	59	7.2
		Vacuum annealed	47.9	50.5	5.4
		Quenched to 300 K	46.5	51.2	10.1
		Zn vapour, vac. annealed	48.7	53.6	10
		Zn vapour, quenched	45.4	49.3	8.6

*U_F is a photovoltage of photodiode S1336-8BK with the scintillator cap; U_{FI} was measured after 9 days storage of phosphors at 300 K.

Measured under X-ray excitation integral intensities of samples are shown in Table 1. The lowest conversion efficiency have the doped only with Cl samples, the highest the doped with Cl and Cu samples. Cu is known as the most effective acceptor dopant in ZnSe. After 9 days storage at room temperature all samples exhibited increased conversion efficiency. The highest efficiency increasement show doped only with Cl samples, the smallest doped with Cu and Cl samples. PL measurements of independent samples confirmed these results.

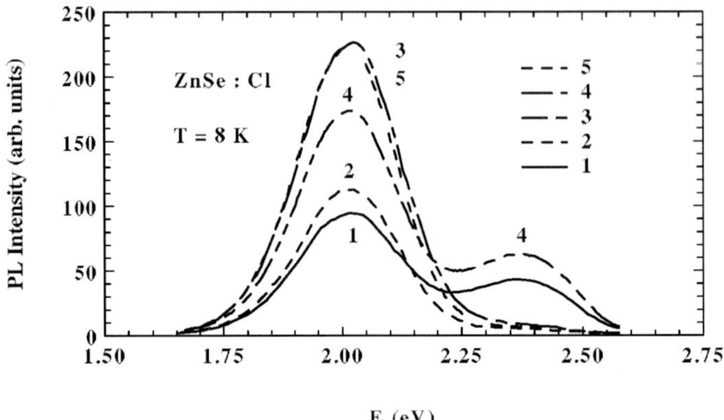

Figure 1. 1 – PL spectrum of ZnSe:Cl. 2 – Sample 1, annealed 1 hour under Zn vapour pressure of 49 kPa, slow cooled. 3 – Sample 1, vacuum annealed 30 minutes at 850 ^{0}C, quenched to 300 K. 4 – Sample 1, after 2 weeks storage at 300 K. 5 – Sample 3, after 2 weeks storage at 300 K.

In Figure 1 are shown PL spectra of doped only with Cl samples. PL band at 2.025 eV is a well known SA band, caused of transitions between A-centre and donor levels as Cl_{Se}, Zn_i and others. An origin of the 2.37 eV PL band is unknown yet. As in Table 1, the highest intensity increasement show a sample where the flux residue was remowed by the help of a low-temperature vacuum annealing, slow cooled. Higher increasement has a 2.025 peak, smaller 2.37 eV peak. Treatment under Zn vapour pressure supress the 2.37 eV PL band.

In Figure 2 are shown PL spectra of doped with Ag and Cl samples. Ag forms a new peak at 2.23 eV, SA and 2.37 eV PL bands are detected also. Treatment under Zn vapour pressure supress the 2.23 eV and 2.37 eV PL bands, high-temperature vacuum annealing restores the 2.23 eV PL band. SA and 2.23 eV PL bands show an intensity increasement, the SA band some more.

Figure 3 shows PL spectra of doped with Cu, Ag and Cl samples. These samples have intense PL band at 1.97 eV. The 2.23 eV PL band exists also. Supposedly, the 1.97 eV PL band consists of the SA and Cl – Cu donor – acceptor pair (DAP) bands. The PL spectra of Cu, Ag and Cl doped samples had the smallest drift, not shown in Figure 3.

Our results oppose the results achieved earlier on a similar materials but using a different fabrication technology [12]. Quenched II-VI phosphors where a flux residue is not previously removed have a high initial brightness after quenching that quickly decreases at room temperature. Our quenched and slow cooled samples have rather uniform characteristics.

Accounting the results, the brightness increasement of our samples at room temperature is supposedly caused of Cl-related complexes. A neutral complex $V_{Zn}2Cl_{Se}$ that can form an A-centre and donor Cl_{Se} is a possible candidate. Increased concentration of the A-centers increases the SA PL band intensity. Cl atom occupies the site of V_{Se} that are formed during the annealing process.

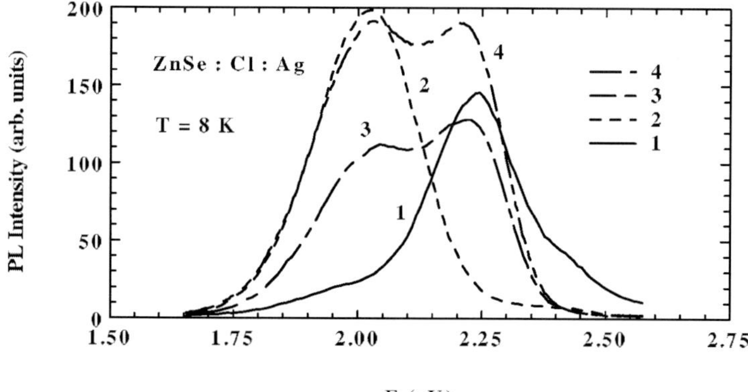

Figure 2. 1 - PL spectrum of ZnSe:Cl:Ag. 2 – Sample 1, annealed 1 hour under Zn vapour pressure of 49 kPa, slow cooled. 3 – Sample 2, vacuum annealed 30 minutes at 850 ^{0}C, quenched to 300 K. 4 – Sample 3, after 2 weeks storage at 300 K.

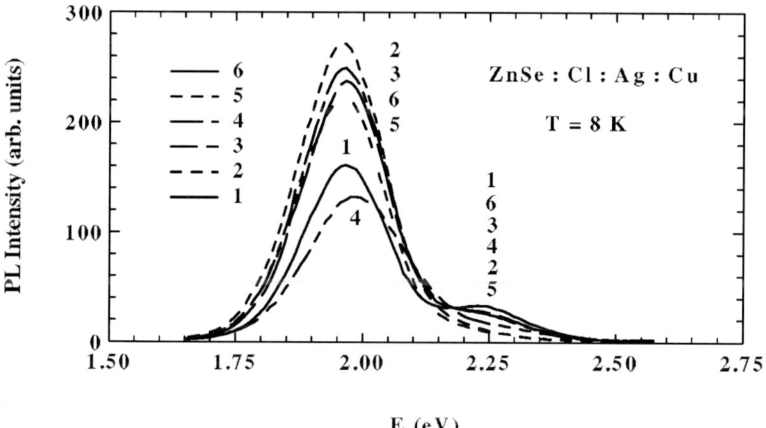

Figure 3. 1 – PL spectrum of ZnSe:Cl:Ag:Cu. 2 – Sample 1, vacuum annealed 30 minutes at 850 ^{0}C, quenched to 300 K. 3 – Sample 1, vacuum annealed 30 minutes at 850 ^{0}C, slow cooled. 4 – Sample 1, annealed 1 hour under Zn vapour pressure of 49 kPa, slow cooled. 5 – Sample 4, vacuum annealed 30 minutes at 850 ^{0}C, quenched to 300 K. 6 – Sample 4, vacuum annealed 30 minutes at 850 ^{0}C, slow cooled.

5 Summary

We are investigated the influence of different thermal treatments upon the defect structure and luminescence properties of ZnSe-based phosphors doped with different dopants. We have detected an increasement of the brightness at ZnSe-based phosphors storaged at room temperature if a flux residue is removed from phosphor before annealing. The highest brightness increasement show doped only with chlorine phosphors. We suppose that this phenomenon is caused of Cl-related complexes as $V_{Zn}2Cl_{Se}$ that dissociate at 300 K and form active centers $V_{Zn}Cl_{Se}$ and Cl_{Se}.

6 Acknowledgements

This work was supported by the Bruker Analytical X-ray Systems, Inc. and by the Estonian Science Foundation, grant No. 2826/99. The authors would like to acknowledge Dr. J. Krustok for his participation in the spectral measurements.

7 References

[1] W. G. Gelling and J. H. Haanstra, Philips Res. Repts. 1961, 16, 371 – 375.
[2] L. P. Galchinctskii, O. N. Zelenskaja, K. A. Katrunov and V. D. Ryzhikov, Atomic Energy. 1994, 76, 390 – 392.
[3] V. D. Ryzhikov, V. I. Silin and N. G. Starzhinsky, Nucl. Tracks Radiat. Meas. 1993, 21, 53 – 54.
[4] B. C. Cavenett, S. Y. Wang and K. A. Prior, Materials Science Forum. 1995, 182 –184, 5 – 10.
[5] A. Kamata, T. Uemoto, M. Okajima, K. Hirahara, M. Kawachi and T. Beppu, Journal of Crystal Growth. 1988, 86, 285 – 289.
[6] K. Akimoto, T. Ogawa, T. Maruyama, Y. Kitajima, Journal of Crystal Growth. 1996, 159, 350 – 353.
[7] M. Illgner and H. Overhof, Materials Science Forum. 1995, 196 – 201, 372 – 332.
[8] G. D. Watkins, Journal of Crystal Growth. 1996, 159, 338 – 344.
[9] G. F. Neumark, Materials Science and Engineering. 1997, R21, 1 – 46.
[10] D. J. Chadi, Materials Science Forum. 1997, 258 – 263, 1321 – 1328.
[11] V. Valdna, E. Mellikov and A. Mere, Physica Scripta. 1997, T69, 312 – 314.
[12] A. M. Gurvitch, Physical Chemistry of Luminophores (in rus.), VS, Moscow, 1982, p. 376.

Experimental Studies of ZnS-Based Phosphors

Vello Valdna, Arvo Mere
Tallinn Technical University, Tallinn
Roger Durst
Bruker AXS, Inc., Madison

1 Abstract

The photoluminescence spectra and the influence of codopants and impurities upon the luminescence efficiency and afterglow of ZnS-based phosphors were investigated. $ZnCl_2$ flux, chlorine as a donor dopant and different acceptor dopants were used. It was found that an intense blue luminescence band of doped only with chlorine ZnS has a short decay time whereas the green donor – acceptor pair band has a much higher afterglow value. Rare earth complexes with oxygen decrease the decay time whereas impurities in starting material can dramatically increase the afterglow of phosphor.

2 Introduction

Zinc sulfide and ZnS-based solid solutions are one of the most efficient and widely used host materials for electroluminescent displays [1]. Due to a wide bandgap of 3.7 eV ZnS is highly transparent to a visible light from blue to red. ZnS can be readily recrystallized and doped at temperatures less than 1000 ^{0}C if halogen flux is used. Compared with III-V semiconductors, II-VI semiconductors have been much less studied [2], and information about interstitials in these materials is scarce [3]. Most effective dopants in ZnS are group VII (donors) and IB (acceptors) elements, especially Cl and Cu that preferably occupy a substitutional position in II-VI compounds [4]. Interstitial Cu is supposed to be a luminescence killer in ZnS [5]. Produced semiconductor grade II-VI materials have a purity level around 6N whereas a needed for phosphors and detectors starting material must have a purity level of 8N. Doping level of ZnS is around 0.05 at%. Group VIII elements Fe, Ni, Co cause a detected luminescence quencing in ZnS at concentrations 10^{-5} – 10^{-4} at% [6].

The aim of this study was to investigate the influence of codopants and impurities upon the luminescence efficiency and afterglow of ZnS-based phosphors.

3 Experimental

The samples were prepared and doped in the way described in [7]. Semiconductor grade ZnS of ELMA and 4N – 5N purity chemicals for doping were used. Thermal treatments of the samples were carried out in the Lindberg/Blue M three-zone tube furnace. Photoluminescence (PL) spectra of samples were measured in the Janis CCS-150 He cryostat at 8 K. The 325 nm line of Omnichrome He-Cd laser with UFS6 and 1M $CuSO_4$ solution filter was used as the excitation source. The PL signal was dispersed using the monochromator SPM-2. A chopped signal from the photomultiplier FEU-79 was amplified, recorded and corrected on a computer. The integral intensity of samples at soft X-ray excitation of 10 keV was measured using the X-ray tube 0.8BSV2-Mo with a molybdenum anode. The photovoltage of the photodiode S1336-8BK coated with a scintillator cap was registered with the Keithley 616 digital electrometer.

4 Results and Discussion

In Table 1 are presented the measured under X-ray excitation integral intensity and an afterglow after 5 seconds of phosphors. The results confirm earlier visually detected data that a Cl – Cu donor – acceptor pair (DAP) PL band has higher afterglow value in ZnS than SA PL band caused of transitions between A-centre and donor levels. Rare earth – oxygen complex significantly decreases the afterglow whereas impurities in starting material can increase the value of afterglow (samples 10 and 11). Starting material of samples 10 and 11 was ZnS residue formed after 2 weeks sublimation of ZnS at 1100 ^{0}C in two-zone evacuated quartz ampoule, one end of ampoule at room temperature. In this residue were concentated impurities that have a small vapour pressure. A poor result with Eu:O doping (sample 7) is probably caused of impuritues in Eu dopant. We used rare earth dopants with a purity level as follows: Ce – 4N+, Eu and Yb – 4N. In Eu-doped phosphors we visually detected a weak red afterglow, not detected in Ce or Yb-doped phosphors.

Table 1. Afterglow versus composition of ZnS-based phosphors*

No	Phosphor composition	U_F [mV]	A [%]
1	ZnS : Cl	45.1	14.4
2	ZnS : Cl : Ce	40.7	9.6
3	ZnS : Cl : Ce : O	35.7	5
4	ZnS : Cl : Ce : Cu	51.7	29.2
5	ZnS : Cl : Ce : Cu : Te	44.2	23.9
6	ZnS : Cl : Eu	43.7	19
7	ZnS : Cl : Eu : O	42	14.3
8	ZnS : Cl : Yb	43.1	20.9
9	ZnS : Cl : Yb : O	32.7	7.3
10	ZnS residue : Cl : Ce	48.9	33.7
11	ZnS residue : Cl : Ce : O	40.6	34.5

*U_F is a photovoltage of photodiode S1336-8BK with the scintillator cap; A is afterglow after 5 seconds.

In Figure 1 are shown PL spectra of doped with Cl, Ce, Cu and O ZnS samples. J. W. Allen [3] has given a SA PL band peak 2.60 eV at 4 K in ZnS. The SA band peak shift to 2.58 eV can be caused of higher Zn_i concentration in our samples. V_S-related 2.88 eV PL band [3] is also well detected in Cl doped and annealed samples. G. F. Neumark gives energy level of E_V + 0.95 eV for A-centre in ZnS and SA PL band peak at 2.70 eV [8]. Zinc interstitial – zinc vacancy pair (a Frenkel pair) is annealed out at about 80 K [3]. Copper dopant supress the SA PL band and forms a peak at about 2.35 eV (curve 3). Note that in ZnS a SA PL band has higher intensity than Cl – Cu DAP PL band whereas in ZnSe a higher intensity has Cl – Cu band. A higher registered U_F value of sample 4 over sample 1 is caused of Cl – Cu band better matching to Si photodiode peak sensitivity.

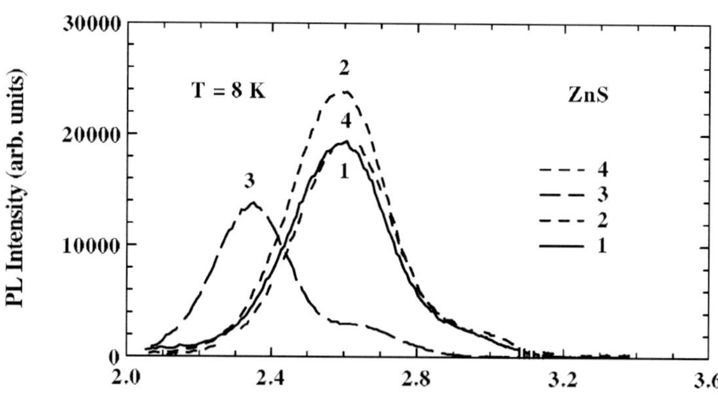

Figure 1. Photoluminescence spectra of ZnS. 1 – doped with Cl; 2 – doped with Cl and Ce; 3 – doped with Cl, Cu and Ce; 4 – doped with Cl, Ce and O.

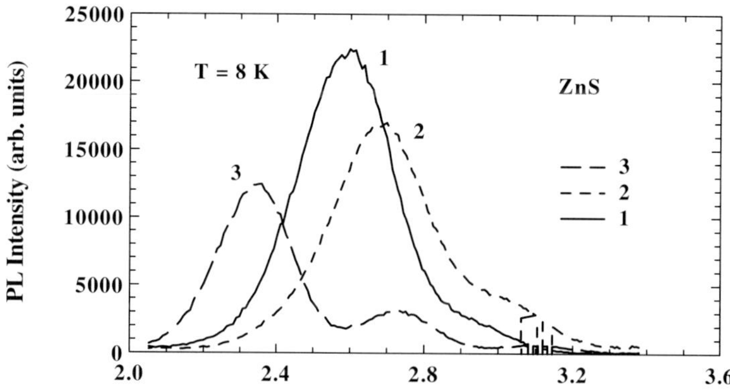

Figure 2. Photoluminescence spectra of ZnS. 1 – doped with Cl and Yb; 2 – doped with Cl, Yb and O; 3 – doped with Cl, Ce, Cu and Te.

Figure 2 shows the PL spectra of samples where codopants Yb and Te are used. Oxygen codopant shifts the band peak of Yb and Cl doped sample to the higher energy (curve 2), not detected at Ce or Eu doped samples, and promotes the band peak at about 3 eV. Isovalent Te in S site supress the SA PL band and causes a weaker peak at about 3.1 eV. M. Taguchi et al. [9] suppose that oxygen incorporation enlarges the concentration of zinc vacancies and as a result increases the SA PL band intensity. We not detected the line spectra of rare earths in our samples as the concentration of rare earths was small. In oxyde-based phosphors a line spectrum is detected at rare earths concentrations 5 at% and more.

The results of this work confirm that industrially produced at least 8N purity grade starting materials, 7N flux and 5N dopants are wanted to realize the full capabilities of II-VI materials. 6N purity grade starting material is probably well enough for the rocket eyes but not for the radiation detectors or downconwerting phosphors.

5 Summary

We are investigated an influence of the different dopants and impurities upon the luminescence efficiency and afterglow of ZnS-based phosphors. We have found that the SA PL band has a smaller decay time than the DAP PL band, and can be further decreased if the rare earths and oxygen codopants are used. We conclude that the impurities in 6N purity grade starting material can deteriorate the luminescence properties of II-VI based phosphors and that industrially produced at least 8N purity grade raw materials, 7N flux and 5N dopants are wanted to realize the full capabilities of II-VI materials.

6 Acknowledgements

This study was supported by the Bruker Analytical X-ray Systems, Inc. and Estonian Science Foundation, grant No 2826/99. The authors acknowledge Dr. J. Krustok for his assistance in the spectral measurements.

7 References

[1] S. Shionoya, Materials Forum. 1991, 15, 132 – 142.
[2] H. Luo and J. K. Furdyna, Semicond. Sci. Technol. 1995, 10, 1041 – 1048.
[3] J. W. Allen, Semicond. Sci. Technol. 1995, 10, 1049 – 1064.
[4] W. Szeto and G. A. Somorjai, The Journal of Chemical Physics. 1966, 44, 3490 – 3495.
[5] H. Kawai, S. Kuboniwa and T. Hoshina, Japanese Journal of Applied Physics. 1974, 13, 1593 – 1603.
[6] M. Tabei, S. Shionoya and H. Ohmatsu, Japanese Journal of Applied Physics. 1975, 14, 240 – 247.
[7] V. Valdna, T. Gavrish, J. Hiie, E. Mellikov, A. Mere, Mat. Res. Soc. Symp. Proc. 1997, 450, 463 – 466.
[8] G. F. Neumark, Materials Science and Engineering. 1997, R21, 1 – 46.
[9] T. Taguchi, T. Yokogawa, S. Fujita, M. Satoh and Y. Inuishi, Physica. 1983, 116B, 503 – 507.
[10] Y.-K. Park, J.-I. Han, M.-G. Kwak, H. Yang, S.-H. Ju and W.-S. Cho, Applied Physics Letters. 1998, 72, 668 – 670.

Characterizing Antiferroelectric Liquid Crystal Materials for Display Applications

R. Dabrowski[a], H. Zhang[b], H. Pauwels[b], J. L . Gayo[c], V. Urruchi[c], X. Quintana[c], J. M. Otón[c]
[a]Military University of Technology, Warsaw, Poland
[b]University of Gent, Gent, Belgium
[c]Universidad Politécnica de Madrid, Madrid, Spain

1 Introduction

Since the discovery of the tristate switching [1] in a chiral smectic liquid crystalline phase in 1989, which was later assigned to an antiferroelectric phase [2,3] (SmC_a^*), a large number of compounds exhibiting this alternating tilted phase have been synthesized. Only one year later in 1990, a first prototype AFLC display device utilizing this type of switching was reported [4]. The main characteristics of AFLC display device are fast response, possibility of analogue gray levels, wide viewing angle, wide operating temperature range, good contrast ratio and toughness against stress or layer deformation. As a result, AFLCs open the possibility of preparing high-definition full-color displays on a passive matrix. In this work, the physical and electro-optical properties of several AFLC mixtures showing tristable switching behavior have been investigated. Special attention has been paid to properties concerning their capability as high-end display materials: grey scale capability, contrast ratio, hysteresis etc.

2 AFLC Mixtures and Physical Properties

In this paper, four AFLC mixtures are characterized for display applications. They are formulated from recently prepared liquid crystalline antiferroelectric (S) 4-(1-methylheptyloxycarbonylphenyl 4'-(alkanoyloxyalkoxy)biphenyl-4-carboxylates or 4'-(perfluoroalkanoyloxyalkoxy)biphenyl-4-carboxylates[5,6,7] as main components. The composition and phase transition temperatures of these four mixtures are given in table 1 and table 2, respectively.

They can be classified into three groups. W-103 contains both compounds without fluorinated terminal chain and compounds with the fluorinated terminal chain. W-108b and W-108c contain only non-fluorinated compounds and W-110 contain only fluorinated compounds. To our knowledge, the interlayer spacing of a semifluorinated compound in SmA^* as well as in SmC_A^* phase is shorter than that of its hydrogenous analogues. As a result, the molecules are much more tilted and therefore the material shows a higher spontaneous polarisation. The similar relations are protected also in their multicomponent mixtures, which can be seen when comparing the phase transition temperatures of W-103 and W-110 to those of W-108b and W-108c in table 2. The lower transition temperatures of the last two are also due to the higher concentrations of two ring dopants.

Table 1. Chemical components in the mixtures.

	W-103	W-108b	W-108c	W-110
fluorinated compounds	*	-	-	*
non-fluorinated compounds	*	*	*	-

Table 2. Phase transition temperatures of the AFLC mixtures

temperature (°C)	W-103	W-108b	W-108c	W-110
SmA^* - I	100	72.8 - 84	90	120
SmC^* - SmA^*	90.5	no SmC*	no SmC*	107.5
SmC_A^* - SmC^*	79.5	56.7	61.6	89.7

Figure 1 compares the spontaneous polarization and the threshold voltage of two of the mixtures. As we can see, mixtures with higher content of fluorinated compounds have a higher value of the spontaneous polarization and hence need a higher threshold voltage to switch the molecules. Figure 2 compares the switching angle as a function of the temperature. It seems that fluorinated compounds have a higher tilt angle of molecules, which leads to a higher switching angle. In the ideal case, the tilt angle should be 45° as this is the only way to assure the maximum contrast demand. This is an angle calculated for a bookshelf geometry because the crossed polarizers for an AFLC are placed perpendicular and along the layer direction, respectively. By using fluorinated compounds, we are approaching this large tilt angle requirement.

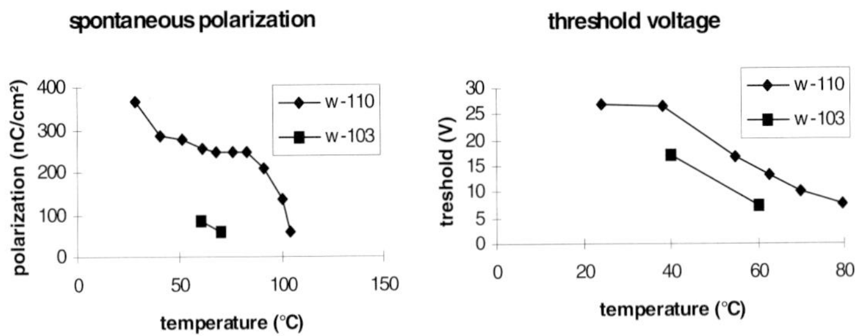

Figure 1. Comparison of spontaneous polarization and threshold voltage as a function of temperature.

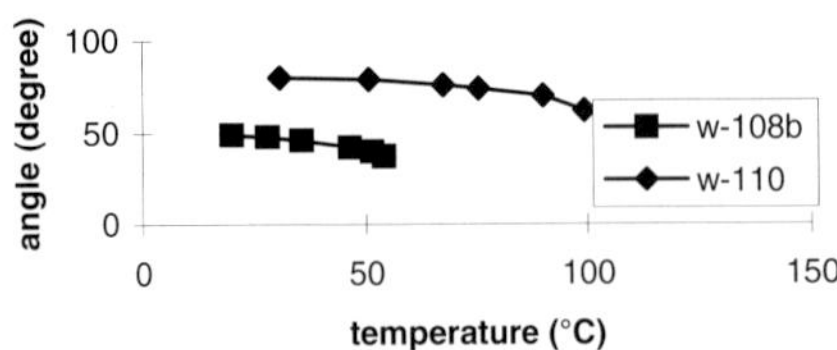

Figure 2. Comparison of the switching angle of AFLC mixtures with fluorinated and non-fluorinated compounds.

3 Electro-Optical Characteristics

We have prepared some test cells filled with these mixtures. The alignment layer used is nylon 6,6, anti-parallel rubbed with a velvet. The measurement setup is illustrated in figure 3. The sample is held in a hot plate connected to a programmable temperature controller. The hot plate is fixed on the rotating stage of a polarizing microscope. The transmitted light is transferred from the microscope to a photo detector via an optical fiber bundle.

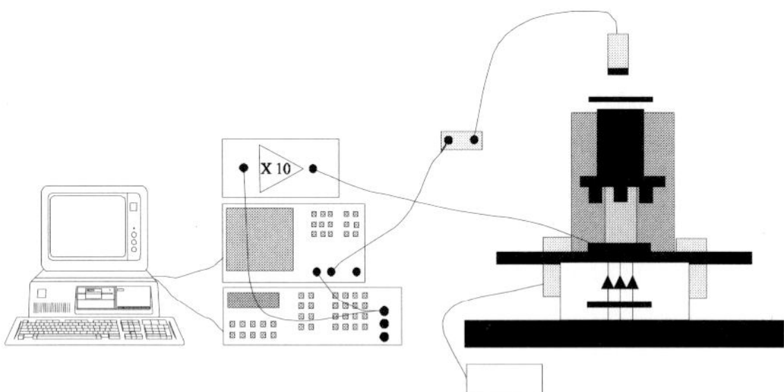

Figure 3. Measurement setup for electro-optical characterization of AFLCs.

3.1 Static behavior

Figure 4 shows the double hysteresis loop measured for the four materials, respectively. They are obtained by applying a triangular waveform of 1Hz. From these double hysteresis curves, static behavior of AFLCs can be studied and compared.

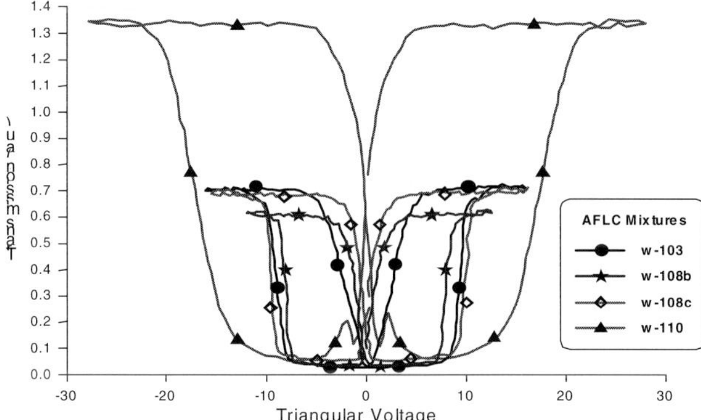

Figure 4. Static double hysteresis of AFLC mixtures.

From these curves, we see that for W-103, W-108b and W-108c, the threshold voltages are about 8V, while for W-110, the threshold almost doubles. This is due to the high value of spontaneous polarization of W-110, as the dipoles need a higher ferroelectric torque to switch from the antiferroelectric state to the ferroelectric state. Additionally, all the four mixtures show a pre-transitional effect below the threshold voltage. This is very troublesome as the transmission will only be minimum with zero voltage. A small deviation from zero will lead to the leakage of light, which will increase the dark state transmission. However, W-110 shows a much higher bright-state transmission. This is quite logical since in comparison with three other mixtures, the tilt angles of which are around 30°, it is increased to approaching 45°, which is the ideal case for AFLCDs [8]. Unlike the other three materials, W-110 does not fully relax to the antiferroelectric dark state at zero voltage. This may be attributed to the large spontaneous polarization of this material, which gives rise to a too high polar interaction between the alignment layer and the liquid crystal. This will hence result in unexpected problems in alignment and molecular configuration.

3.2 Dynamic behavior

For the dynamic behavior, we are mainly interested in analogue grey levels. They are generated by applying an addressing waveform for multiplexed passive devices, consisting of selection, bias, well, and reset time. An example of the addressing scheme and the measured optical transmission evolution is given in figure 5.

For all the four AFLC materials tested, grey levels have been obtained by varying the amplitude of the selection pulse. An appropriate bias voltage has been selected so that the optical transmission keeps more or less constant during the frame time. In the reset time, the AFLCs relax to the dark state so that the influence of the previous state can be minimized. In order to fasten the relaxation process, a voltage pulse of the opposite polarity is applied before the reset time. The measured optical transmission is then averaged through the whole frame and the transmission-selection voltage curves are obtained.

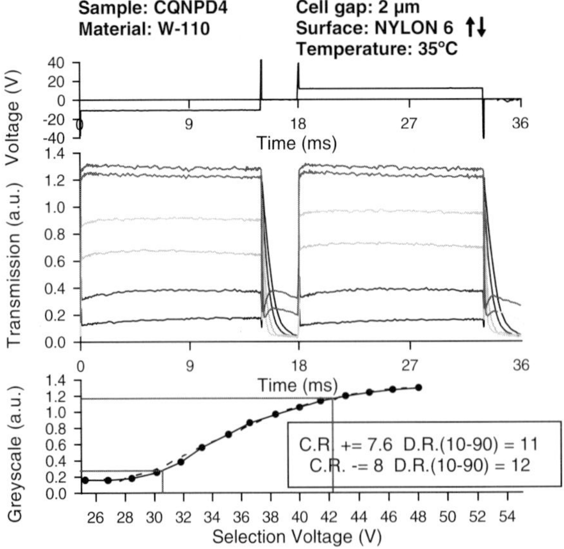

Figure 5. Addressing schemes of grey level generation by pulse amplitude modulation and optical transmission.

Table 3 summarizes the dynamic behavior of these four AFLC mixtures obtained from the experiments. For W-103, W-108b and W-108c, grey levels can be obtained by varying the amplitude of the selection voltage within a range of 4V. In a real display, this will be realised by the data signals on the column electrodes. However, in the case of W-110, this range becomes 11V. This value seems too high, since this will introduce too much crosstalk disturbance when the passive matrix addressing is applied. The pre-transitional effect will then lower the contrast ratio significantly. On the other hand, despite the high bright state transmission, the contrast ratio is even lower than the other three mixtures. This is caused mainly by the fact that this material does not fully relax to the antiferroelectric dark state when the applied voltage becomes zero, as already seen in the static double hysteresis curve. Moreover, in the application point of view, the bias voltage should be kept as low as possible. In this case, W-108b and W-108c, containing only non-fluorinated compounds, seem to be most suitable.

Table 3. Comparison of dynamic behavior of the four AFLC materials.

AFLC mixtures	contrast ratio (+/-)	grey scale dynamic range (+/-) V	threshold voltage (+/-) V	relaxation time (+/-) ms	bias voltage/cell thickness (V/μm)
W-103	14/14	3.8/5.5	14.4/14.1	1.3/3.2	6.3
W-108b	13.5/13.4	2.4/2.3	14.9/14.4	3.2/8.5	3
W-108c	13.1/13.5	3.8/3.7	17.5/17.5	12.6/14.8	3.6
W-110	7/8	11/11	30/30	8.1/7.4	5.6

4 Conclusions

Various experiments have been performed to characterize several AFLC materials. For display applications, AFLC mixtures must be optimized to meet various requirements, such as a large tilt angle of 45°, a medium value of spontaneous polarization, a fast response etc. We have found that chain fluorinated crystalline antiferroelectric compounds enable to formulate mixtures with an tilt angle almost equal 45°. Furthermore, alignment technique also needs to be improved to achieve a higher contrast ratio and lower selection and bias voltage levels.

5 Acknowledgment

This work has been carried out in the framework of the European Network program ORCHIS.

6 References

1. D. L. Chandani, T. Hagiwara, Y. Suzuki, Y. Ouchi, H. Takezoe and A. Fukuda, Jpn. J Appl Phys., 1989, L729, 27.
2. D. L. Chandani, E. Gorecka, Y. Ouchi, H. Takezoe and A. Fukuda, Jpn. J Appl. Phys, 1989, L1256, 28.
3. Fukuda, Y. Takanishi, T. Isozaki, K. Ishikawa and H. Takezoe, J. Mater. Chem., 1994, 997, 4(7).
4. Y. Yamada, N. Yamamoto, K. Mori, K. Nakamura, T. Hagiwara, Y. Suzuki, I. Kawamura, H. Orihara and Y. Ishibashi, Jpn. J Appl Phys., 1990, 1757, 29.

5. W. Drzewinsk, R. Dabrowski, K. Czuprynski, J. Przedmojski, M. Neubert, Ferroelectrics, 212,281, 1998.
6. W. Drzewinski, K. Czuprynski, R. Dabrowski, M. Neubert, Mol. Cryst. Liq. Cryst. (in press), 1999.
7. W. Drzewinski, K. Czuprynski, R. Dabrowski, J. Przedmojski, M. Neubert, S. Yakovenko, Mol. Cryst. Liq. Cryst. (in press), 1999.
8. The Optics of Thermotropic Liquid Crystals (Ed.: S. Elston and R. Sambles), Taylor & Francis Ltd. Gunpowder Square; London EC4A 3DE, UK.

II Materials for Electrical Energy Conversion

High – Speed Wheel Systems with HTS Bearings for Electric Power Applications

Frank N. Werfel
Adelwitz Technologiezentrum GmbH (ATZ), Rittergut Adelwitz, D-04886 Adelwitz

Abstract: Fast rotating wheel assemblies with passive high temperature superconducting magnetic bearings (SMB) have been developed with magnetic suspension and radial centering of the rotors. Due to air friction the speed of a 4 cm rotor is limited to about 90 000 rpm and higher than 10^5 rpm under vacuum conditions. Measurements of bearing properties and wheel performance are presented. Among several advantages of SMB, this concept does not limit the rotation speed at the level of the bearings. With v = 560 m/s circumference velocity the rotors are designed to accept high stresses to maximize the profit of the high rotation speed and to open the possibility of kinetic energy storage using flywheel energy storage systems. The double – bearing construction increases the maximum wheel imbalance relative to commercial one by a factor of five.

It is concluded that the superconducting magnetic bearings with integrated cooling devices are capable of providing reliable, long –life operation in high-speed rotating machines .

1 Introduction

Due to the improved bulk high - temperature superconducting (HTS) material fabrication the use of magnetic bearings for high – speed rotational applications is increasingly attractive. The principal benefit and the main advantages stem from low-drag torque and the self - centering, unlubricated, wear – free, vacuum - compatible operation. In contrast to ball bearings, superconducting magnetic bearings (SMB) are an integrated part of the overall design and cannot be specified in terms of simple mechanical interfaces. Control of the SMB properties such as levitation pressure, restoring forces (stiffness) and damping could be of special importance in high – speed applications. Hence, and because of the variability new concepts of the complete SMB equipment has to be designed. The present problem that now exists is besides of demonstrators and the cooling to develop effective designs in terms of the specific parameters weight, speed, power and costs.

Recent developments have shown the way for production of large-bore YBCO rings and cylinders with strong flux – pinning force and improved mechanical stability [Werfel 97, 98]. Principally, the axial or radial combination of the superconducting bulks and permanent magnets (PM) makes it feasible to construct a SMB device. The experimental superconducting magnetic bearing configurations for flywheel energy storage applications have led to a clearer outlook of the expected performance and requirements [Ishigawa 95, Bornemann 96, Coombs 98]. A flywheel with conventional bearings loses approximately 1.5 % of its stored energy per hour in bearing loss. Active magnetic bearings can reduce the losses by one order of magnitude. The used superconducting configurations eliminates the need of active feedback systems. Demonstrators mostly are designed in axial symmetry between superconductors and permanent magnet rings. However axial symmetry is known to have several drawbacks compared to a radial bearing (bearing size, limited size of PM, homogeneity problems). The objectives of this paper are to discuss the SMB application to levitate and rotate high - speed wheels safely. We describe the design of passive magnetically stabilized rotor systems and test results.

2 Background

We compared the two basic design of SMB, axial and radial configuration. In contrast to the axial bearing geometry a radial configuration has a higher variability in PM arrangement and allows a more compact design.

The price of a radial - shaped SMB is a superconducting ring produced so far by glued precision formed sector - like bulks. We have been able to fabricate polycrystalline monolithic cylinders up to a diameter of 140 mm (**Figure 2**) The polycrystalline melt textured YBCO rings with a radial - like texture together with a PM rotor provides both high stiffness, rapid damping of the critical rotor vibrations and very low rotational energy losses. With respect to the high - speed rotation magnetic field variations due to non- ideal rotors causes energy loss and eddy currents in the YBCO rings.

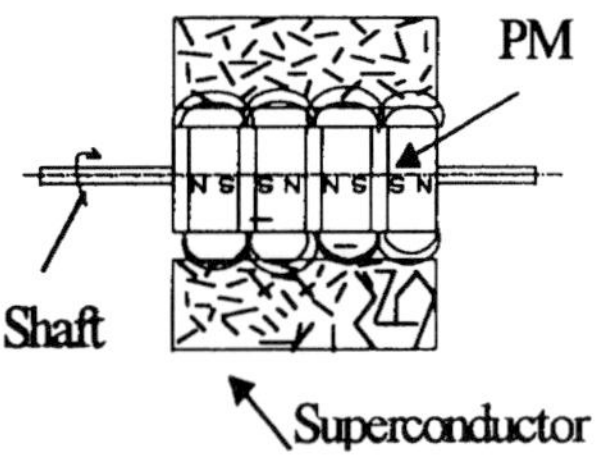

Figure 1. Radial superconducting magnetic bearing (schematically)

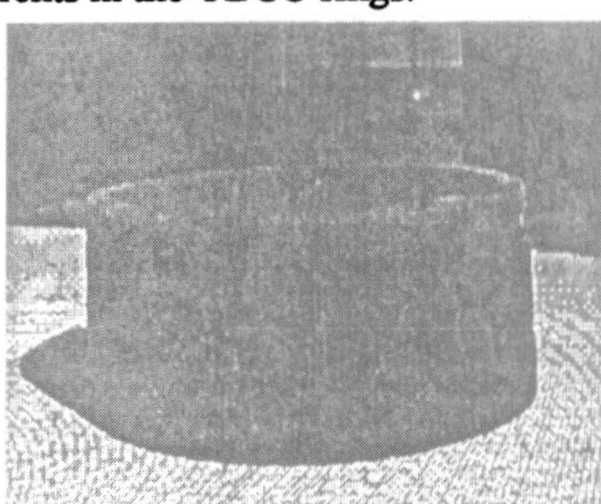

Figure 2. Melt textured YBCO superconducting stator ring, d = 80 mm (ATZ)

The magnetic force F_r between the superconductor and the PM in a radial geometry is:

(i) $$F_r = (m / \mu_0)\, dB_r / dr$$

m - magnetic moment of the superconductor

dB_r / dr - radial field gradient produced by the magnet

For the use of polycrystalline material with variable grain size D of typically 5 – 20 mm in order to obtain excellent cryo - magnetic properties we have to taken into account conditions of a high critical current density J_c and a grain overlap with respect to the magnetic pole distance.

Polycrystalline YBCO rings or cylinders with grain sizes between 1-2 cm and high J_c ($3x10^4$ A/cm² at 77 K) are excellent devices for radial SMB design. The forces and stiffnesses are strongly dependent on the magnet configuration. Sandwiched PM rings (**Figure 1**) on a shaft under repulsive force with opposing polarity and metallic shims between are for small gap bearings of 1-2 mm the best solution (mixed μ configuration).The multiplicity of opposing polarity magnets as shown in Fig.1 causes multiple flux density gradients and hence increased bearing stability. In the case of large gap operation due to isolation walls between magnet and superconductor we did observe for attracted magnets where the magnetic flux reaches a longer distance better stability conditions.

3 Design of high – speed wheels

Table 1. System technique of magnetic suspended wheel assembly

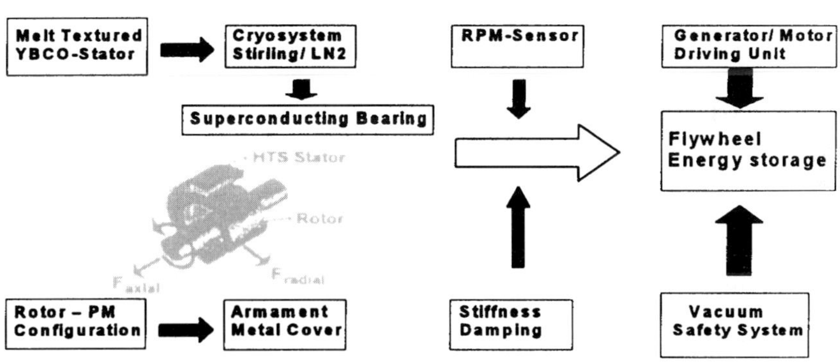

The primary function of a high speed wheel assembly is to serve as test bed for the superconducting magnetic suspension system and to demonstrate the special features that are unique to high speed bearings. Consequently in a first step no major effort was made to minimize size, weight or sound to couple a reliable cooling system with bearing. The basic features of a functional SMB system operating at the high – rotational speed level are schematically described in **Table 1**.

As previously discussed the strongest drawback of a wide application of the superconducting magnetic bearings is necessity of cooling. LN_2 based cryogenic devices with T=77K are common.

Increasingly we have designed bearing systems with Stirling - cooler devices providing 60–80 K. The driving units are usually ac motors at a frequency level of 2 – 3 kHz. The PM rings cannot withstand the extremely dynamical forces and are usually armed with a carbon fiber composite cover which has the highest specific tensile strength σ/ρ equivalent to a maximum circumference velocity of the rotors of more than 1000 m/s. If metallic covers are necessary we utilize the high – performance mechanical behavior of Maraging steel (X2NiCoMoTi) with excellent strength properties (σ_t up to 4000 MPa). Especially critical are the dynamical forces for rotating rings at high speeds. At the inner hole the highest stress occur with the possibility of material fracture.

Table 2: Prototype radial superconducting magnetic wheel asssemblies

Load [kg]	Ø PM [mm]	Stiffness (Gap) [N/mm]	T [K]	rpm [1/min]	Prototype Parameter
10	80		77	5 000	flywheel
3. 8	3x60	140 (3)	77	12 000	flywheel
2.1	4x20	68 (4)	77	50 000	unbalance
0. 2	3x24	36 (2)	60	88 000	Split – Stirling Cooler
0. 1	3x15	24 (1)	77	160 000	high-speed 2.6 kHz

4 Wheel dynamics

Centrifugal stresses in rotating wheels limit the maximum rotational speed. In the high-speed tests with a small 4 cm rotor and a weight of 80 g we have been achieved about 130 000 rpm in reduced air conditions (1 mbar). The axial stiffness for the single SMB in LN_2 was about 20 N/mm and the PM rings with a carbon fiber bandage had 16 mm diameter. The velocity at the outermost rotor face at the maximum speed of 3 kHz (180 000 rpm) of a 6 cm rotor is about 560 m/s which is 25 % lower than the critical fracture stress of the material. We used a special high quality Al alloy. At the maximum speed the 6 cm gyroscope has a elastic material deformation of about 60 μm in the diameter.

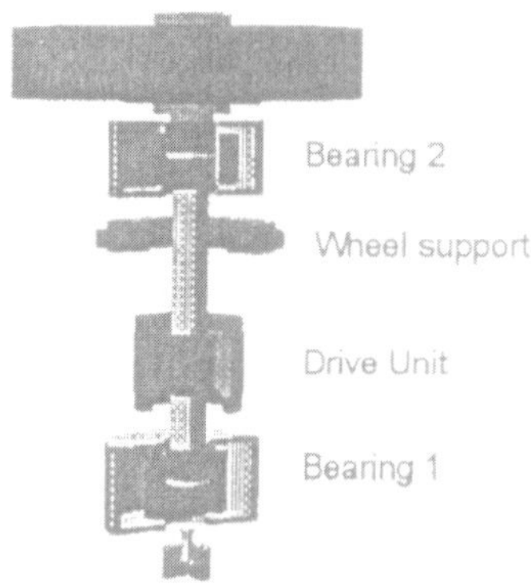

Figure 3. Prototyp magnetic double-bearing assembly

A larger double - bearing assembly we designed,

constructed and tested in order to investigate wheel damping and critical behavior (**Figure 3**). The 1.4 kg rotor was accelerated to 25 000 rpm and loaded by unbalanced mass distribution up to 5 g at the maximum radius. We calculated and measured the displacement of the center of gravity of the rotor in the case of switching from sub - critical to over - critical rotation. At the critical rpm a wheel displacement of almost 1 mm were measured. Due to the high damping the rotor can be driven safely through the critical rpm region.

With the unbalance U = m e (mass times displacement) one may calculate an engineering damping quality factor Q = e ω for rotor systems. The quality factor for the magnetic bearing configuration is with Q = 4 m/s more than 5 times higher compared to equivalent mechanical systems. The new 15 W / 80 K Stirling cryo-cooler operates the bearings after 1 hour standby time in the superconducting state. The final temperature of about 60 K increases the bearing stiffness and allows safe operation conditions.

The axial stiffness of a Ø 60 mm radial superconducting magnetic bearing is demonstrated in **Figure 4**. The gap width between the PM and the superconductor influences substantially the restoring forces (stiffness).

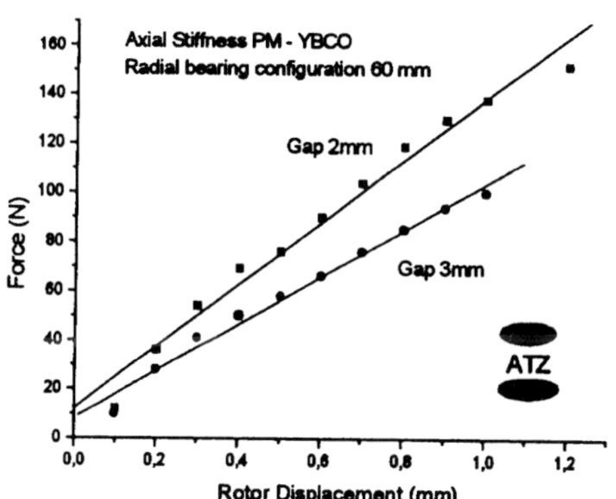

Figure 4. Stiffness measurement of a Ø 60 mm radial magnetic bearing

Generally, the passive magnetic bearing design utilizing the peculiar properties of the superconducting properties provides the capability for reliable operation over a much greater speed range than ball bearings and limited only by stress considerations of the rotors. The most significant design parameters of an SMB are the radial stiffness S_r and the damping c. There are constraints on S_r which exist within the SMB design itself, such as critical current density Jc of the superconductors and the magnetic performance of the PM rings (B_r).

The SMB program would need to be optimized to determine wheel speed and diameter as well as design improvements for the general SMB structure design. The program itself is part of different single SMB projects and follows an "HSI" philosophy (Handle Superconductors Intelligent)

5 Summary

The design, construction and test of engineering high – speed wheel assemblies with superconducting magnetic bearings indicates that passive magnetic suspension is applicable to various wheels and fast rotating machines. The construction of radial bearings with high restoring forces(stiffness) and excellent damping properties over a large frequency range of SMB meet the requirements in the technique. Because of the low drag - torque of the bearings rotations of more than 10^5 rpm`s are achieved. Consequently, we are approaching stress limiting speeds for the material of the rotors used. The wheel demonstrators described here are in touch however to real mechanical systems for exchange and improvement. The SMB solutions will not merely replace the conventional bearings, but are designed to take the full advantage of the extremely exciting properties, e. g . for rotating energy storage systems.

Acknowledgements

The work is supported by the German BMBF under the contract No. 13N7247

The cooperation with the Institut für Luft- und Kältetechnik (ILK), Dresden is gratefully acknowledged.

References

F. N. Werfel, U. Flögel-Delor, D. Wippich, Inst. Phys. Ser. No **158**, 1997, EUCAS Netherlands

F.N. Werfel, U. Flögel-Delor, D. Wippich, R. Rothfeld, Applied Superconductivity Conf.1998, Palm Desert, USA

F. Ishikawa, H. Higasa, VDI Berichte Nr. 1187, 1995

H. J. Bornemann, M. Sander Applied Superconductivity Conference 1996, Pittsburgh, USA

T. Coombs, A.M. Campbell, R. Storey, R. Weller, Applied Superconductivity Conf., 1998 Palm Desert, USA

Manufacturing and Test of a 100 kVA HTS Transformer

Peter Kummeth[1], Reinhard Schlosser[2], Cord Albrecht[1] and Heinz-Werner Neumüller[1]
[1]Siemens AG, Corporate Technology, P.O. Box 3220, D-91050 Erlangen
[2]Siemens AG, Power Transmission and Distribution, D-90027 Nuremberg

1 Abstract

The main advantages of high temperature superconducting (HTS) transformers are reduced size, weight and a better efficiency compared to conventional transformers. High critical current densities and low ac losses of HTS tape conductors are reasons therefor. As in HTS transformers nonflammable liquid nitrogen can be used as coolant, the potential fire and environmental hazards, which are present in conventional transformers, are drastically reduced. We have designed, manufactured and tested a 100 kVA HTS power transformer as a functional model operated at 77 K. Iron core and the HTS windings were mounted inside of G-FRP cryostat and cooled by liquid nitrogen. The nominal primary current of the transformer is 18 A (5.6 kV) and the secondary nominal current is 92 A (1.1 kV). Untwisted silver sheathed 2223 BPSCCO tapes with 55 filaments were used. While the primary winding (high-voltage) is made of a stack of 30 pancake coils which are connected in series, the two secondary windings (low-voltage) are solenoid coils with five tape conductors wound in parallel and transposed in situ. Primary and secondary windings are arranged concentrically and the secondary windings are connected in series. We performed repeatedly no-load tests, short-circuit tests and load tests which have proved the rated capacity of the transformer. The loss measurements resulted in HTS winding losses of 20.6 W and iron losses of 403 W.

2 Introduction

The introduction of superconductivity into electrical power equipment leads to an improvement of existing components. Several design studies on superconducting transformers have been performed in the last years. The use of superconductors instead of copper conductors promises various advantages. Beside an improved efficiency compared to conventional transformers additional benefits are reduced size and weight [1]. These advantages make superconducting transformers to very promising candidates for application in electrical power engineering and railway applications. Especially in the case of railway transformers a reduction of size and weight is desired. Low temperature superconducting (LTS) transformers were already constructed and operated in liquid helium [2]. Due to the low operation temperature of LTS transformers, which need liquid helium (T = 4.2 K) as coolant, the benefit of reduced conductor losses is easily compensated by the required cooling power. A break-through was the discovery of high temperature superconductivity in 1986. After the development of HTS tape conductors with a typical length of several hundred meters and a typical critical current density j_c = 20-25 kA/cm^2 (T = 77 K, B = 0 T) superconducting transformers can be realized with an operating temperature of T = 77 K. This leads to dramatically reduced cooling efforts. Further the use of nonflammable nitrogen as coolant minimizes potential fire hazards and environmental hazards caused by oil in conventional transformers. In the recent years several HTS transformer development projects have been carried out [3-5]. Among the known HTS materials up to now $(Bi,Pb)_2Sr_2Ca_2Cu_3O_{10+\delta}/Ag$ tape conductors are the most promising candidates for applications at T = 77 K. We used untwisted multifilamentary silver sheathed Bi2223 tape conductors with 55 filaments manufactured by Vacuumschmelze (Hanau, Germany) in our transformer project. The tapes typical width and thickness is 3.6 mm and 0.26 mm, respectively. At T = 77 K and B = 0 T the tape conductors have critical currents of $I_c \geq 45$ A (criterion: 1 μV/cm).

3 HTS Transformer Design

A 100 kVA HTS transformer was designed for operation in liquid nitrogen at T = 77 K. The superconducting windings and also the iron core were mounted inside a cryostat made of glass-fibre-reinforced-plastics (G-FRP). G-FRP cryostats are already used for various cryogenic applications [6], because the non-magnetic and non-conducting material avoids eddy-current losses. In the case of a cold iron core inside the cryostat it would also be possible to use metallic cryostats for transformers, since the cryostat is exposed only to a very low leakage flux. On the one hand manufacturing of a cryostat is much easier, if the iron core is housed inside the cryostat, but on the other hand a cold iron core dissipates its losses directly to the coolant (T = 77 K). So this set-up requires about 15 times the dissipated power for cooling. Design parameters of our single phase 100 kVA HTS transformer are primary and secondary currents (voltages) of 18A (5.6 kV) and 92A (1.1 kV), respectively. The nominal flux density inside the iron core is B = 1.7 T.

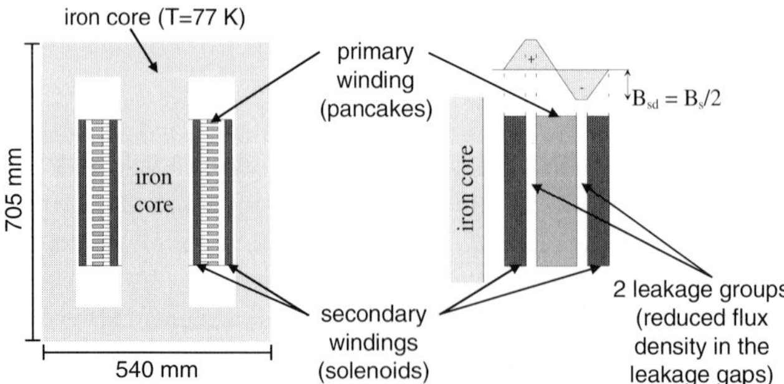

Figure 1. Sketch of 100 kVA HTS transformer.

The flux density in the leakage gaps was minimized to appr. 50% by arranging the primary winding concentrically between two secondary sub-windings as shown in figure 1. These sub-windings – designed as solenoids - are connected in series, while the primary winding consists of a stack of 30 pancake coils, also connected in series. A total length of 889 m of Bi2223/Ag tape conductor was used for the HTS transformer. 449 m and 440 m were used for the primary and secondary windings, respectively.

Manufacture of the pancake coils was performed on a winding machine. During the winding process we monitored the conductor tension permanently to avoid degradation. The winding tension was limited to values below 10 N. Each of the pancake coils with 18 turns is insulated by Kapton films with a total thickness of 100 μm, which were cowound with the Bi2223/Ag tape. Vacuum impregnation of the individual pancake coils was performed after the winding process with epoxy resin. Preparation of the secondary windings was carried out by a manual winding process. Each of the two secondary windings consists of five tape conductors which were wound in parallel and insulated with Kapton film. During the winding process the conductors were impregnated with epoxy resin. In order to obtain a homogeneous current distribution the five parallel tape conductors were transposed four times in situ, i.e. during winding.

4 Test of Components

After fabrication of windings the critical current of the manufactured coils was measured in liquid nitrogen at T = 77 K. The tests proved no degradation of conductor properties due to the winding and impregnation processes. Before assembly of the transformer we additionally investigated the AC losses of the coils. A testing facility to perform electrical ac loss measurements using a lock-in amplifier was developed especially for this purpose. The very small voltage generated in phase with the transport current in the coils was measured. Due to the inductance of the coils a compensation coil was necessary in our set-up. Figure 2 represents the results of ac loss measurements on several pancake coils. We measured typical losses $P_{ac} \approx 0.5\text{-}0.6$ W for a nominal current of I = 18 A at T = 77 K. An estimation of the magnetic field at pancake coils inside and outside the transformer revealed, that the magnetic induction in an individual pancake coil is similar to the magnetic induction inside the coils of the assembled transformer. So the primary winding losses of the stack of 30 pancake coils were estimated to $P_{prim.,ac} \approx 18$ W. Theoretical calculations of the primary transformer winding losses resulted in $P_{prim,calc} \approx 16.5$ W. The theoretical and experimental data are in quite good accordance. For loss measurements on the secondary coils the two windings were connected against one another to generate similar magnetic conditions like in the transformer. They were operated without iron core in liquid nitrogen. The total losses P_{ac} of both secondary coils and the copper wiring were measured. Afterwards dc losses P_{dc} were measured to get the wiring losses. The secondary winding losses were determined to $P_{sec.,ac} \approx 4.5$ W at a nominal current of I = 92 A. This value coincides nicely with theoretical loss estimations. In figure 3 the experimental results of these loss measurements are displayed.

HTS transformer assembly was carried out after the successfully performed component tests. In figure 4 one can regard the primary winding (30 pancake coils) and the 2 secondary windings concentrically arranged and mounted at the iron core.

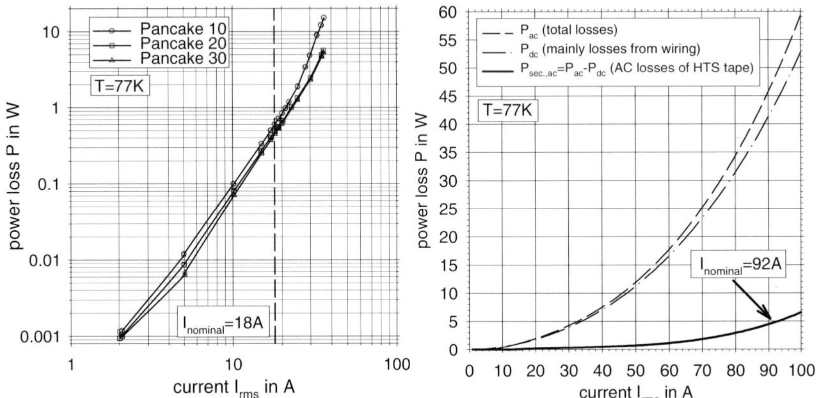

Figure 2. Power loss of pancake coils.

Figure 3. Power loss of the solenoid coils.

5 Operation and Tests

Afterwards the assembled transformer was put inside a G-FRP cryostat. The first cool down was performed very carefully. It took about 6 hours. to reach a temperature of 77 K. The maximum axial temperature gradients were 3.2 K/cm in HTS windings ($\Delta T \approx 84$ K between top and bottom) and 3.8 K/cm in the iron core during this procedure. At the typical power frequency of 50 Hz various transformer tests were performed to measure the power losses caused by the iron core and the HTS windings. In no-load tests the secondary transformer winding was connected to a power source and the power loss was measured for different voltages. At the nominal voltage $U_{sec} = 1091$ V a power loss of $P_{core} = 403$ W was measured. This power loss reflects the losses due to the iron core. Evaluation of the HTS winding losses was performed by short-circuit tests. The transformers secondary side was short-circuited and the primary winding was connected to a power supply. During the tests the primary winding voltage was adjusted to reach exactly the nominal current in the primary winding. As the secondary windings were shorted with copper wire, we had to measure the input power (primary) and the output power (secondary) dissipated in the copper. The power loss of the transformers HTS windings is given by the difference of input and output power. At the transformers nominal primary current $I_{prim} = 18$ A we measured HTS power losses of $P_{windings} = 20.6$ W. A short-circuit voltage of 48 V was found out and therefore the core losses during these short-circuit measurements are negligible. Figure 5 and figure 6 show the results of the loss measurements.

Figure 4. Transformer ready for test.

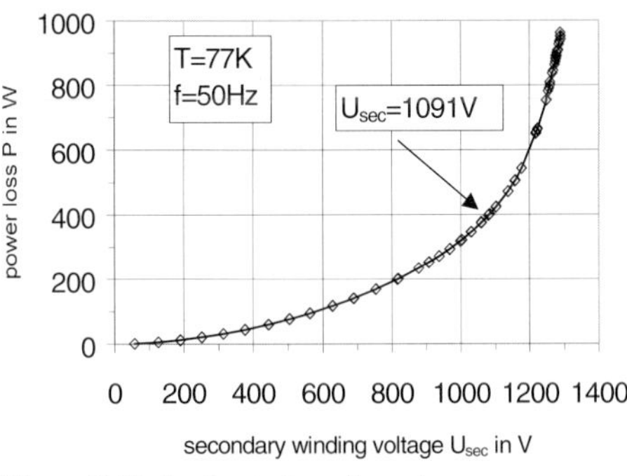

Figure 5. No-load tests (core losses).

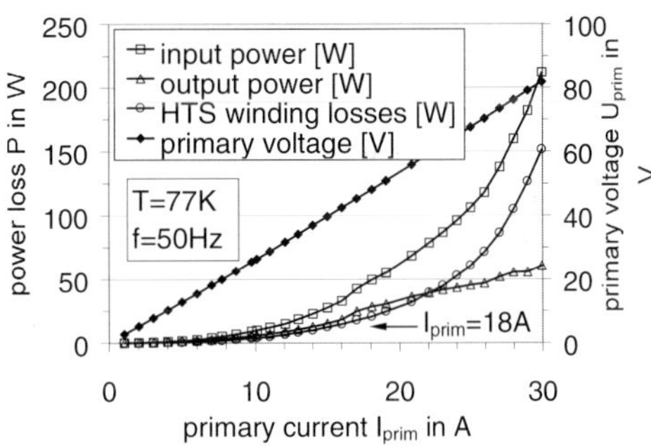

Figure 6. Short-circuit tests (HTS losses).

During short circuit tests the temperature of the HTS windings was only elevated by ΔT = 0.1-0.2 K above the liquid nitrogen temperature. A continuous operation of the transformer at the rated power was performed with an inductive load for more than 1 h and resulted in stable operation without further temperature increase. After warming up the transformer was cooled down several times and the transformer tests were performed repeatedly. The cool down rate was doubled and the transformer reached its operation temperature after about 3 hours. During all these tests stable operation without degradation of the transformer properties was proven.

6 Results and Conclusions

We developed and constructed a 100 kVA single phase HTS transformer. The transformers iron core and the Bi2223/Ag HTS windings were mounted in liquid nitrogen. Transformer tests for a frequency of 50 Hz were performed successfully in liquid nitrogen at T = 77 K. Determination of the iron losses was accomplished by no-load tests and resulted in iron core losses of 403 W. Short-circuit tests revealed HTS winding losses of 20.6 W. Stable operation without degradation of the windings was proved repeatedly at the rated power. A conventional 100 kVA transformer has typical winding losses of about 2 kW. If we assume a need of 15 W of power at a temperature of 293 K to remove 1 W at T = 77 K, the winding losses of a conventional 100 kVA transformer are about six times as high as the winding losses of our HTS transformer. Beside these drastically reduced winding losses HTS transformers have several further advantages compared to conventional transformers. Especially the reduced size and weight play an important role for railway applications. At present we are engaged in the development of a 1 MVA HTS transformer for railway applications.

7 References

1. S.P. Mehta, N. Aversa and M.S. Walker, IEEE Spectrum July 1997, 43-49.
2. T. Ise, Y. Marutani, Y. Murakami, E. Yoneda and R. Sugawara, Advances in Superconductivity VIII 1995, 1295-1298.
3. H. Zueger, Cryogenics 1998, 38, 1169-1172.
4. K. Funaki, M. Iwakuma, K. Kajikawa, M. Takeo, J. Suehiro, M. Hara, K. Yamafuji, M. Konno, Y. Kasagawa, K. Okubo, Y. Yasukawa, S. Nose, M. Ueyama, K. Hayashi and K.Sato, Cryogenics 1998, 38, 211-220.
5. S.W. Schwenterly, B.W. McConnell, J.A. Demko, A. Fadnek, J. Hsu, F.A. List, M.S. Walker, D.W. Hazelton, F.S. Murray, J.A. Rice, C.M. Trautwein, X. Shi, R.A. Farrell, J. Bascunan, R.E. Hintz, S.P. Mehta, N. Aversa, J.A. Ebert, B.A. Bednar, D.J. Neder, A.A. McIlheran, P.C. Michel, J.J. Nemec, E.F. Pleva, A.C. Swenton, W. Swets, R.C. Longsworth, R.C. Johnson, R.H. Jones, J.K. Nelson, R.C. Degeneff and S.J. Salon, IEEE Transactions on Applied Superconductivity 1999, 9(2), 680-684.
6. T. Okada and S. Nishijima, Cryogenics 1995, 35, 801-804.

Nanocrystalline Metal Hydrides for Hydrogen Storage

Wolfgang Oelerich, Thomas Klassen, Nico Eigen, Rüdiger Bormann
GKSS-Forschungszentrum Geesthacht GmbH, D-21502 Geesthacht

1 Introduction

Hydrogen is the ideal means of storage, transport and conversion of energy for a comprehensive clean-energy conception. However, appropriate storage facilities, both for stationary and for mobile applications, are complicated, because of the very low boiling point of hydrogen (20.4 K at 1 atm) and its low density in the gaseous state (0.0071 g/cm^3). Furthermore the storage of hydrogen in liquid or gaseous form imposes safety problems, in particular for mobile applications, e.g. the future zero-emission vehicle. Metal hydrides are a safe alternative for H-storage and, in addition, have a higher volumetric energy density [1]. Mg hydride has a high storage capacity by weight and is therefore favored for automotive applications. However, a potential drawback seems to be the high hydrogen desorption temperature of about 300 °C, as for most applications temperatures of about 200 °C are more suitable. Moreover, light metal hydrides have not been considered competitive because of their rather sluggish sorption kinetics. Filling a tank could take several hours. A breakthrough in hydride technology was achieved by preparing nanocrystalline hydrides using high energy ball milling [2]. These new materials show very fast ab- and desorption kinetics within few minutes, thus qualifying Mg-based hydrides for application. In this paper, we present recent results on the sorption behaviour of Mg and Mg-based alloys. The sorption kinetics of nanocrystalline Mg have been further enhanced by catalyst additions. Furthermore, different transition metals have been added to Mg to achieve a thermodynamic destabilization of the hydride, thus lowering the desorption temperatures.

2 Experimental

The milling experiments were performed with a Fritsch P5 planetary ball mill using hardened Cr-steel milling tools and an initial ball-to-powder weight ratio of 400 g : 40 g. The different initial powders (99+% for elemental powders and 95+% for MgH_2) were blended in the desired overall ratio. All handling of the powders, including milling, was performed inside a glove box under a continously purified Ar atmosphere. For characterization of the sorption properties, specimen holders were sealed inside the box and attached to a hydrogen titration apparatus (Hydro-Quèbec), that was especially designed for fast data acquisation. Kinetics and pressure-composition-temperature (pcT) measurements were performed at 300 °C (unless otherwise mentioned). The hydrogen pressure was 10 bar for absorption and up to 20 bar for pcT measurements. For measurements of the desorption kinetics the chamber was evacuated. The weight of the samples was about 150 mg.

3 Results and Discussion

Figure 1 shows the hydrogen absorption kinetics of Mg with 5 mol% catalyst in comparison to conventional pure Mg. By using MgH_2 instead of Mg for the milling process (patent pending [3]), usual activation procedures are unnecessary. During the first absorption 95% of the full capacity are already reached within only half a minute at a temperature of 300 °C and a hydrogen pressure of 10 bar. The comparison with conventional Mg clearly demonstrates the breakthrough in sorption kinetics. Another advantage is the possibility to absorb hydrogen already at lower temperatures at similar rates. Although at a lower rate hydrogen is even absorbed at room temperature, which is impossible for conventional material. Correspondingly, also the release of hydrogen is significantly enhanced, as shown in Fig. 2. While conventional material does not desorb any significant amounts at 300 °C into vacuum, hydrogen is released completely within 4 minutes at 300 °C and 15 minutes at 250 °C for ball-milled material with catalyst. The desorption rate between 80 and 20% of the capacity at 300 °C is determined to 0.022 wt.% hydrogen per second. With a net caloric power value of 120 kJ/g for hydrogen, this is equivalent to 29 kW per kg of powder.

The advantage of the ball milled material over conventionally processed alloys can be explained by the nanocrystalline microstructure and the high fraction of grain boundaries, that provide fast diffusion paths for hydrogen [2]. In addition, the splitting of H_2-molecules on the surface of the powder particles is supported by the catalyst.

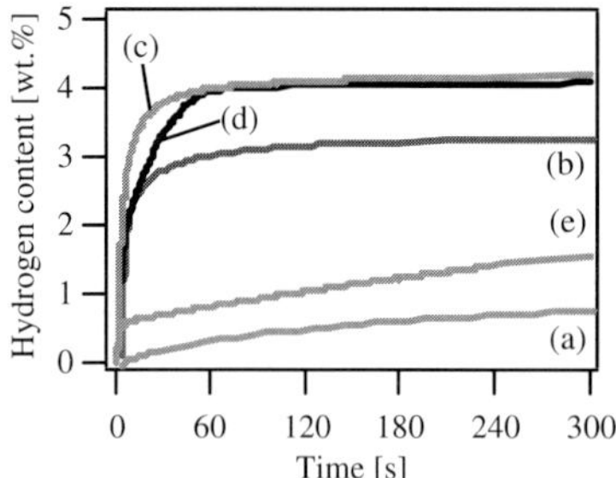

Figure 1. Absorption kinetics of nano-crystalline Mg with 5 mol% catalyst at a temperature of 40 °C (a), 100 (b), 200 (c) and 300 °C (d) in comparison to conventional Mg at 300 °C (e).

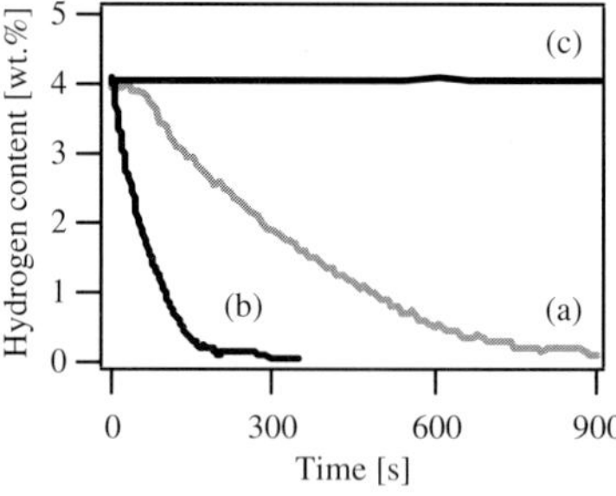

Figure 2. Desorption kinetics of nano-crystalline Mg with 5 mol% catalyst at a temperature of 250 (a) and 300 °C (b) in comparison to conventional Mg at 300 °C (c).

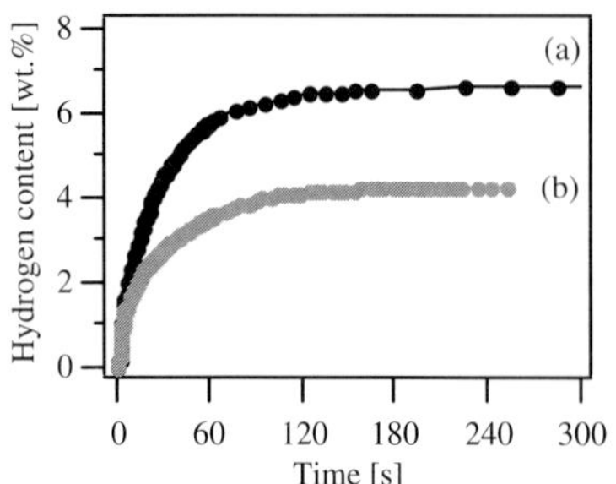

Figure 3. Absorption kinetics of nano-crystalline Mg with 0.2 mol% catalyst milled for 100 h (a) in comparison to 5 mol% (b) at 300 °C.

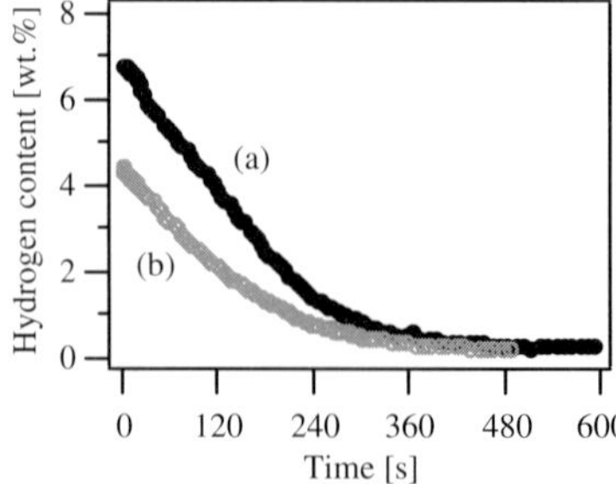

Figure 4. Desorption kinetics of nano-crystalline Mg with 0.2 mol% catalyst milled for 100 h (a) in comparison to 5 mol% (b) at 300 °C.

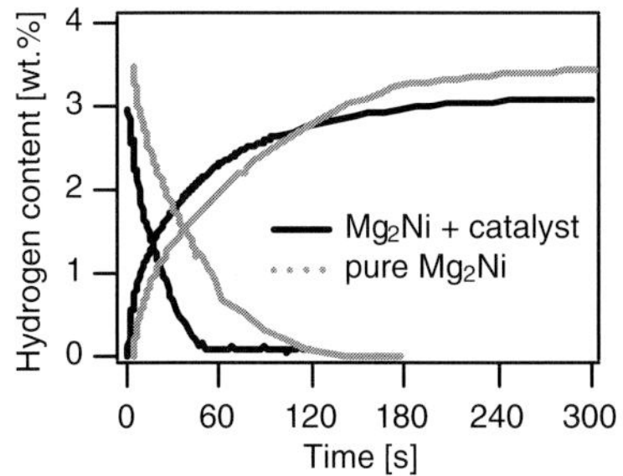

Figure 5. Absorption and desorption kinetics of nanocrystalline Mg_2Ni without and with 1 mol% catalyst at 300 °C.

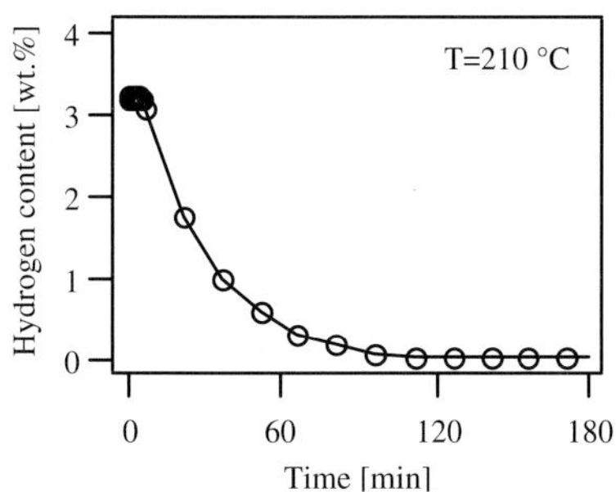

Figure 6. Desorption kinetics of nano-crystalline Mg_2Ni with 1 mol% catalyst at a temperature of 210 °C.

By lowering the amount of catalyst, the capacity can be greatly enhanced without reducing the sorption rates (Figs. 3 and 4). 5.9 wt.% of hydrogen are absorbed within 60 seconds and the full capacity of now even 6.6 wt.% is reached after 5 minutes. Upon desorption, 5.5 wt.% of hydrogen are released within 240 s, corresponding to the same desorption rate as for higher catalyst contents. Full desorption is possible within 8 minutes.

To investigate the influence of the catalyst on other Mg-based alloys, Mg_2Ni was chosen as a model system. Figs. 5 and 6 show the kinetics of nanocrystalline Mg_2Ni with and without catalyst in comparison. While absorption and desorption for conventional Mg_2Ni takes several hours [4], for nanocrystalline Mg_2Ni 80% of the full capacity are already reached after only 3 minutes and hydrogen is completely released again after 2 minutes in vacuum. The desorption kinetics of Mg_2Ni are further accelerated by small additions of catalyst. As it can be seen in Fig. 5, Mg_2NiH_4 with 1 mol% catalyst desorbs hydrogen completely in about half the time, i.e. within 1 minute at a temperature of 300 °C. In contrast to nanocrystalline Mg_2Ni without catalyst, ball-milled Mg_2Ni with 1 mol% catalyst releases hydrogen even at a temperature as low as 210 °C (Fig. 6).

In general, the lower rate of desorption in comparison to absorption can be explained by the lower driving force $|p\text{-}p_{eq}|$ [5], where p is the hydrogen pressure for absorption and desorption respectively. The equilibrium pressure p_{eq} depends on thermodynamic properties of the metal-hydrogen-reaction (see [6] for details), and is equal to 1 bar for Mg at 300 °C [7].

Many attempts have been made to weaken the Mg-H bonds and thus to increase the equilibrium pressure by adding different transition metals to Mg [8]. In this way, the desorption rate at a given temperature could be increased. More important even, the temperature for desorption could be lowered, while still reaching a usable pressure of hydrogen from the storage tank. Ni is known to reduce the stability of MgH_2, e.g. in Mg_2NiH_4 [4]. The pcT-diagram in Fig. 8 shows the equilibrium hydrogen pressure for absorption and desorption. The plateau pressure at 4 bar in the absorption cycle indicates the formation of the Mg_2NiH_4 hydride phase, that decomposes again upon desorption at a plateau pressure of about 2 bar. These pressures are significantly higher than for MgH_2, with 2 and 1 bar respectively (Fig. 7), which is important for application. Further investigations reveal that a desorption pressure of 1 bar is already achieved at 240 °C [4,9] instead of 300 °C for MgH_2 [7].

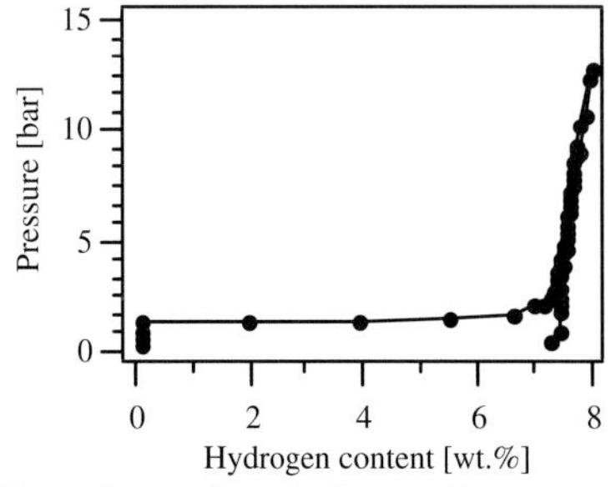

Figure 7. pcT diagram of Mg at 300 °C.

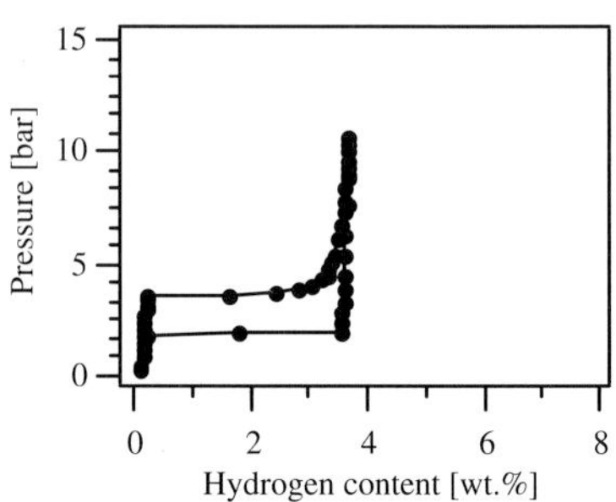

Figure 8. pcT diagram of Mg_2Ni at 300 °C.

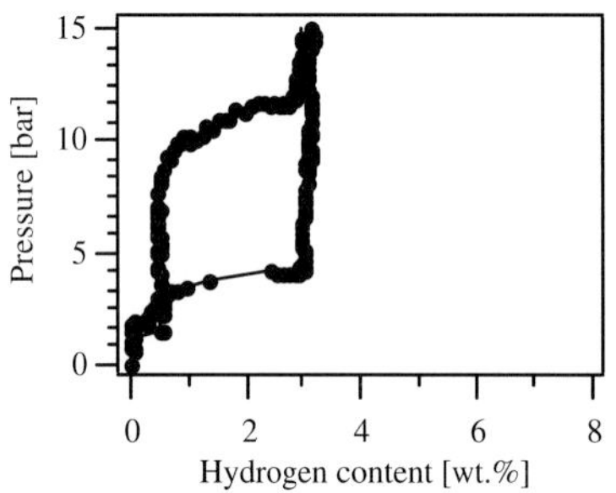

Figure 9. pcT diagram of $Mg_2Ni_{0.5}Cu_{0.5}$ at 300 °C.

Based on the promising results on the Mg_2Ni phase, half of the Ni was substituted with Cu [10]. After a milling time of 100 h a single ternary phase $Mg_2Ni_{0.5}Cu_{0.5}$ similar to Mg_2Ni is formed during high energy ball milling of metal components [1]. The initial part of the pcT-diagram (Fig. 9) reveals, that some residual pure Mg is present and alloying was obviously not yet complete after 100 h. However, it also clearly shows single absorption and desorption plateaus at pressures of about 11 and 4 bar, respectively, for the $Mg_2Ni_{0.5}Cu_{0.5}H_4$ phase. These pressures are significantly higher than those for Mg_2NiH_4 and indicate a lower thermodynamic stability of the hydride containing Cu. Further investigations indeed demonstrate, that 1 bar of hydrogen pressure is released already at 230 °C, which represents another step towards the desired application temperature of less than 200 °C. The overall capacity is slightly less than for Mg_2Ni and amounts to about 3.05 wt.% hydrogen.

5 Conclusions

Our investigations on nanocrystalline Mg-based hydrides prepared by ball milling show very fast absorption and desorption kinetics at 300 °C within few minutes, thus qualifying these light weight hydrides for application. The improved kinetics originate from explained by the high fraction of grain boundaries, that provide fast diffusion paths for hydrogen. Absorption and desorption kinetics could be further enhanced by using additional catalysts and a new processing route [3]. Thus, absorption is possible already at room temperature and desorption could be obtained even at temperatures around 200 °C. The thermodynamic stability of Mg-hydrides can be significantly reduced by alloying additions. For Mg_2NiH_4, desorption at a hydrogen pressure of 1 bar is possible at 240 °C [9] instead of 300 °C for pure Mg. Partial substitution of Ni with Cu leads to a further decrease of the desorption temperature to 230 °C.

6 Acknowledgments

This work is part of a cooperation between Hydro-Quèbec, Montreal (Canada), GfE Metals and Materials mbH, Nürnberg (Germany), and GKSS, Geesthacht (Germany). Discussions with V. Güther, R. Schulz, J. Huot, S. Boily, A. van Neste, and G. Liang are gratefully acknowledged. This project is supported by the Bavarian Government.

7 References

1. J.J. Reilly, G.D. Sandrock: *Scientific American*, **242**(5118), (1980).
2. R. Schulz, S. Boily, L. Zaluski, A. Zaluska, P. Tessier, J.O. Ström-Olsen: *Innovations in Metallic Materials*, 529-535 (1995).
3. T. Klassen, W. Oelerich, R. Bormann, V. Güther: *Deutsche Offenlegungsschrift: DE 197 58 384 A 1*, (1999).
4. J.J. Reilly, R.H. Wiswall: *Inorg. Chem.* **7**, 2254 (1968).
5. P. S. Rudman: *J. Less-Common Metals*, **89**, 93-110 (1983).
6. W.M. Mueller, J.P. Blackledge, G.G. Libowitz: *Metal Hydrides* (Academic Press, New York, London, 1968).
7. K. Zeng, T. Klassen, W. Oelerich, R. Bormann: *Int. J. Hydrogen Energy*, **24**(10), 989-1004 (1999).
8. P. Selvam, C.S. Viswanathan, C.S. Swamy, V. Srinivasan: *Int. J. Hydrogen Energy*, **11**(3), 169-192 (1986).
9. K. Zeng, T. Klassen, W. Oelerich, R. Bormann: *J. Alloys and Compounds*, **283**, 213-224 (1999).
10. J.P. Darnaudary, B. Darriet, M. Pezat: *Int. J. Hydrogen Energy*, **8**, 705 (1983).
11. T. Klassen, W. Oelerich, K. Zeng, R. Bormann: in *Magnesium Alloys and their Applications* (eds. B.L. Mordike and K.U. Kainer), p. 307-311 (Werkstoff-Informationsgesellschaft mbH, 1998).

The Effect of Zn^{2+} Substitution for A-site on Electrical Properties and Microstructure of $Pb(Mn_{1/3}Sb_{2/3})O_3$ -PZT

Min-Jae Kim, Kyong-Sop Han*, Sung-Churl Choi
Dept of Inorg. Mat. Eng., Hanyang University, Seoul 133-791, Korea
*Div. of Ceramics, Korea Institute of Science & Technology, Seoul, Korea

1 Abstract

Microstructure and piezoelectric properties of ZnO-doped (0~5mol%) $0.05Pb(Mn_{1/3}Sb_{2/3})O_3$ - $0.95Pb(Zr_{0.52}Ti_{0.48})O_3$ [PMS-PZT] ceramics were investigated. Due to eutetic reaction between PbO and ZnO, sintering temperature was lowered to 1000 °C. Grain –size increased up to 1mol% ZnO and then decreased and density increased with ZnO addition, max. 7.84 at 2 mol% ZnO. Eectromechanical coupling factor(k_p) was about 0.5 and relative dielectic constant($\varepsilon_{33}{}^{T}/\varepsilon_o$) was about 1000 and tetragonality(c/a) indicated maximum value at 3mol% ZnO. Mechanical quality facotr(Q_m) decreased with ZnO addition.(1500→760). Piezoelectric constant were d_{31}=max.-299x10^{-12}C/N and g_{33}=max.63.4x10^{-2}Vm/N. Also to investigate temperature stability and frequency stability of material, measured Curie temp.(T_c), Max. band width (Δf , $f_a - f_r$)

2 Introduction

It is known that the piezoelectric properties of Pb(Zr, Ti)O_3 solid solution can be remarkably improved by composional modifications. In order to satisfy the requirement of practical applications, many tenery solid-solution ceramics have been synthesized over binary ceramics, consisting of complex perovskite compounds[1)], and a number of minor additives have been added to the ceramics.[2),3)]

Lead zirconate-titanite ceramics are used extensively for piezoelectric devices and are promising material for piezoelectric transformer, filter, resonator, SAW filter, ultrasonic motor and piezoelectric actuator have been studied in recent years.[4)]

In one of above applications, a piezoelectric transformer is attractive component for electric device as LCD backlight inverter[5)], DC high power supplies[6)], fluorescent lamp ballast and so on. Because it has advantages over a magnetic transformer, such as a high set-up ratio of out-put to in-put voltage, high efficiency at high frequency, high energy density of an acoustic wave stored in a piezoelectric transformer, the possibility of high density mounting and so on. And many report on compositions of piezoelectric transformer including Pb(Mg,Te)O_3-PZT[7)], (Pb,Sr)[(Mn,Sb)(Zr,Ti)]O_3[8)], Pb[(Mg,Nb)(Zr,Ti)]O_3.[9)]

The aim of this work is to study the effect of Zn^{2+} substitution for Pb^{2+} on the piezoelectric properties of Pb(Mn,Sb)O_3-PZT ceramics, promising material for piezoelectric transformer and used extensively for piezoelectric device. ZnO was chosen as a element having possibility, the eutetic reaction with PbO(>860 °C) for lowering sintering temperature by formation of liquid phase and improvement of electric properties.[10), 11)]

3 Experimental Procedures

Starting materials were regent grade PbO, ZrO_2, TiO_2, Sb_2O_3 and ZnO. Chemical composition was $0.05(Mn_{1/3}Sb_{2/3})O_3$ - $0.95Pb(Zr_{.52}Ti_{0.48})O_3$ + xZnO[x = 0~5mol%]. In advance MnO_2 and Sb_2O_3 were wet-milled for 20h, calcined 600 °Cfor 2h and synthesized $MnSb_2O_5$(columbine precursor) in order to obtain stable perovskite. An excess PbO(2mol%) was added to prevent non-stoichiometry resulting from PbO evaporation and to facilitate the densification. Weighed oxides were wet-milled for 20h and calcined 850 °C for 2h. After grinding, the powder was uniaxially pressed at 1.5ton/cm^2 into pellet (diameter = 12mm). Embedded with $PbZrO_3$ powder to compensate the PbO evaporation, the green bodies were sintered at 1000 °C for 2h. Sintered specimens were chemically etched with 10%HNO_3-2%HF solution for several secs and the microstructure was observed with a scanning electron microscope(SEM). Density was measured according to conventional hydrostatic method(ASTM Std. C372-72). Crystallographic structure, lattice parameter were identified and calculated by an X-ray diffraction technique. For each sample the electrical impedance was measured as a function of frequency by using a HP-4194A impedance analyzer. From the value of resonance and anti-resonance frequency of certain slected modes, together with the minimum impedance, capacitance, density and geometrical dimensions, the piezoelectric, dielectric and elastic coeffcients, as well as the electromechanical coupling factor(k_p), mechanical quality factor(Q_m) have been determined and set-up ratio was obtained from calculation. The electrical properties of the disk were evaluated, in according to IRE standard 24 h after poling at 150 °Cfor 10min under a field of 30kV/mm.

4 Result and Discussion

4.1 Microstructure and Sintering properties

Fig 1 showed the density, apparent porosity variation with ZnO addition of sinterd PMS-PZT. Sinterd density increased up to ZnO 2mol% and then decreased.

In this result, ZnO improved sintering behavior of PMS-PZT. From Fig 1 increase of sintered density ascribed to liquid phase eutetic reaction between PbO and ZnO that stimulated sintering process. Grain-size was measured from scanning electron photographs (SEM, Fig 2, 3) fracture surface that were chemically etched by intercept linear method. Average grain size was about 0.83 μm at no addition of ZnO and about 2 μm at Zn-rich precipitate phase was detected at ZnO 5mol%, the effect of grain-growth inhibitor therefore decrease of average grain-size was seen up to ZnO 5mol%. From these result of co-existence Pb, Mn, Sb, Zr and Ti peak precipitate phase was expected to involve not only Zn^{2+} but also other components. 1mol% ZnO, over 2mol% ZnO average grain size decreased to about 0.91 μm. According to the fact that precipitate phase as grain-growth inhibitor in gain-boundary was seen.(Fig 4)

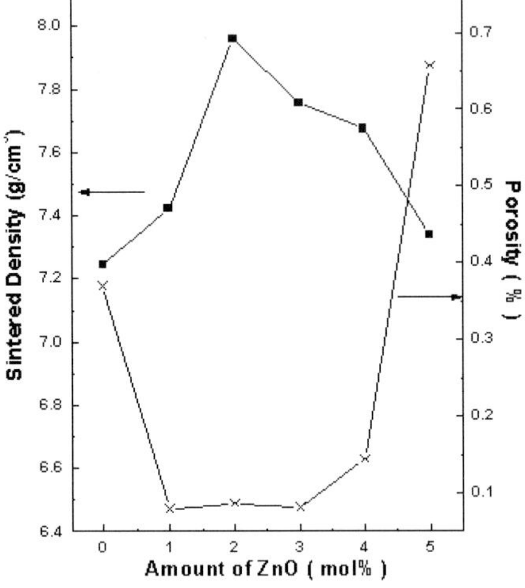

Fig 1 Variation of density of PMS- PZT-xZnO after sintering at 1000 °C for 2h

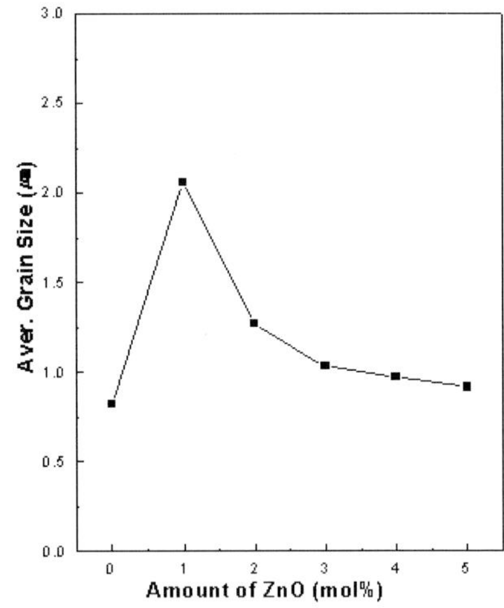

Fig 2. Variation of average grain size as a function of ZnO content in PMS-PZT-xZnO after sintering 1000 °C 2h

According to Atkin[12], it is believed that the doped ions concentrate near the grain boundaries and substantially reduce the their mobility by fluxing effect.[13] When the boundary moves, it must drag the excess(adsorbed) impurity with it, as shown schematically in Fig 5. This mechanism for reducing grain-boundary mobility and average grain-size. Fig 4 showed that result of wide wavelength dispersion spectrum analysis(WDS).

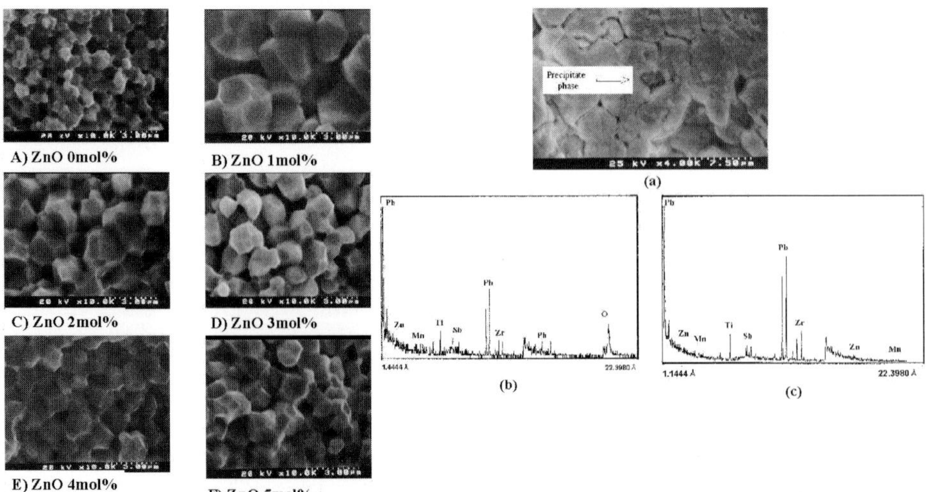

Fig 3. SEM photographs of fractured surface of PMS-PZT-xZnO after sintering at 1000 °C for 2h

Fig 4. (a) SEM photograph of ZnO 5mol% and WDS spectra of (b) 4mol% and (c) 5mol% at grain boundary (precipitate phase) of PMS-PZT after sintering for 2h

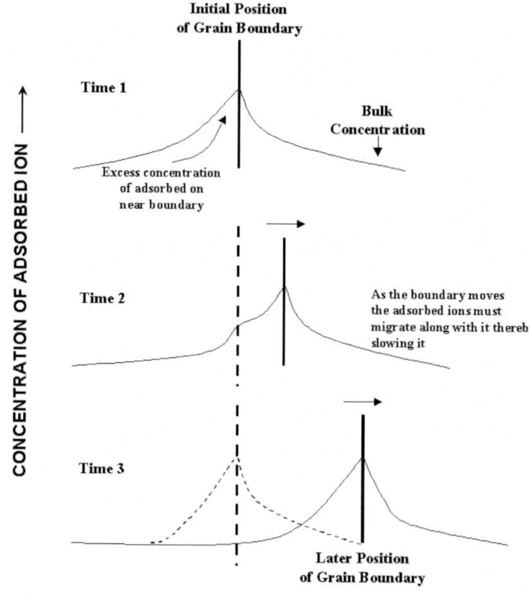

Fig 5. Mechanism by which adsorbed impuries impede grain boundary motion[13)]

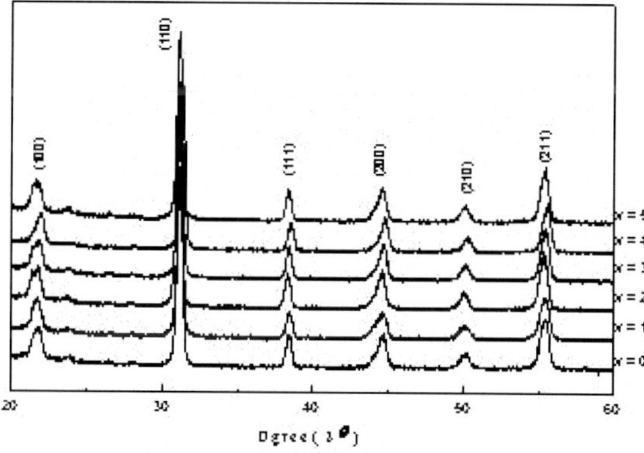

Fig 6. XRD pattern PMS-PZT-xZnO sintered at 1000 °C

4.2 X-Ray Diffraction

Fig 6 showed that X-ray diffraction(XRD) pattern of 0.05PMS-0.95PZT+xZnO (x = 0~5mol%). It obtained relatively stable perovskite structure(ABO3) over all composition and its structure is clarified rhombohedral structure by existence only (200) peak over range 2θ= 42~46 degree with addition of ZnO. Lattice parameter decreased up to 4mol% and then increased slightly. This result indicated that decrease of lattice parameter was due to substitution Zn^{2+}(0.74Å) for A-site, Pb^{2+}(1.2Å). If Zn^{2+} substituted for B-site, Mn^{2+}(0.53Å) or Sb^{3+}(0.76Å) lattice parameter must be increased of remain unchanged.(Fig 7) From above result, possibility of substitution Zn^{2+} for B-site is very low. Tetragonality(c/a) increased up to

3mol% with ZnO addition and then decreased. Accordingly chap3.1, the result of WDS analysis of ZnO 4mol%, 5mol% showed that ZnO was not detected in grain-boundary at 4mol% but was detected at 5mol% with lattice parameter increased up to 4mol%. From above result, solid-solution limit of ZnO for 0.05PMS-0.95PZT is expected 4mol% respectively.

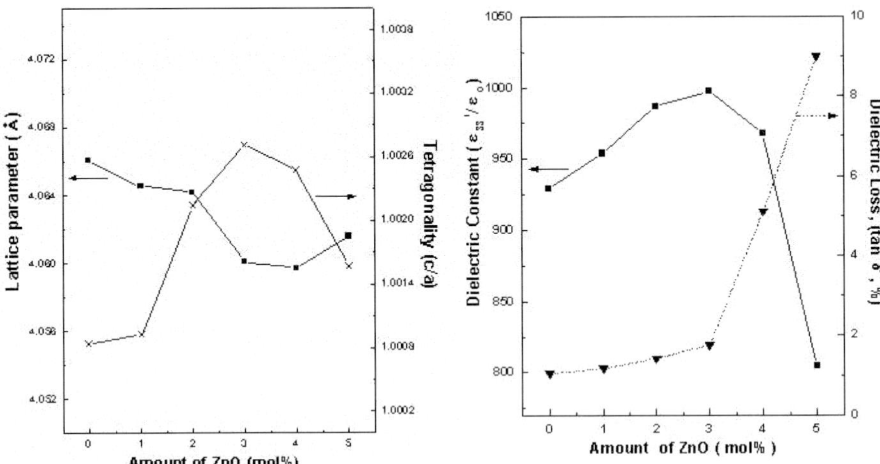

Fig 7. Variation of the lattice parameter and tetragonality as a functin of ZnO content PMS-PZT-xZnO after sintering 1000 °C 2h

Fig 8. Variation of dielectric constant and dielectric loss as a function of ZnO content in PMS-PZT-xZnO after sintering 1000 °C 2h

4.3 Electrical Properties

Fig 8 indicated that change behavior of relative dielectric constant($\varepsilon_{33}^T/\varepsilon_o$) with ZnO addition. Relative dielectric constant increased up to about 1000 at 3mol% ZnO and dielectric loss(tanδ) did vice versa. Because relative dielectric constant and dielectric loss were influenced by grain-size, porosity and A-site substitution(or vacancy)[14] therefore increase of relative dielectric constant means easiness of dipole movement due to increase of density, decrease of porosity and substitution of Zn^{2+} for Pb^{2+}(A-site). In other words, increase of relative dielectric constant($\varepsilon_{33}^T/\varepsilon_o$) up to 3mol% attributed to transfer of porosity within grain and impurity ions to grain-boundary by fluxing effect(chap.3.1), therefore formed uniform grain. After 4mol%, transferred impurity and porosity decreased density, increased porosity.

As a result, relative dielectric constant decreased with increase of porosity. Fig 9 showed the change behavior of piezoelectric strain constant(d_{33}) and electromechanical coupling factor(k_p) with addition ZnO. Piezoelectric strain constant(d_{33}) tends to decrease at ZnO 1mol% due to increase of grian-size. Maximum of d_{33} at 4mol% was regarded as solubility limit. It related to improvement of sinterbility and increase of tetragonality(c/a). Increase of tetragonality attributed to increase of dipole movement, strain easily to electric field or stress and decrease of degradation in piezoelectric properties.[15] But both mechanical quality factor(Q_m, reciprocal to internal stress) and Curie temperature(T_c) decreased. (Fig 10) A-site substitution makes easy move domain wall, which manifest decrease Qm and Tc tend to accompany to decrease of lattice parameter. Accordingly improvement of k_p, d_{33}, g_{33} and decrease of Q_m, Tc due to substitution Zn^{2+} for Pb^{2+}(A-site).

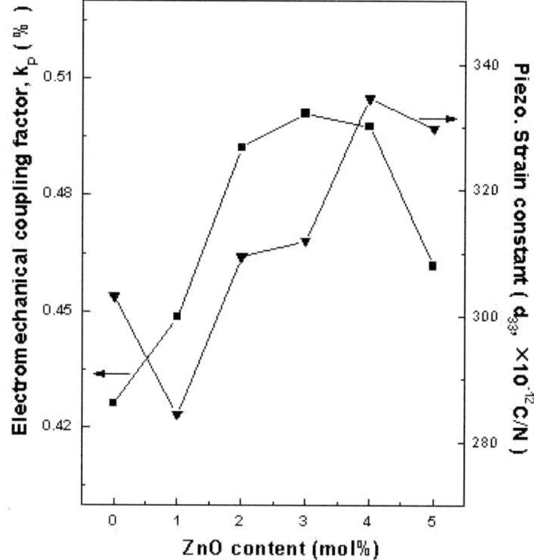

Fig 9. Variation of Piezoelectric strain(d_{33}) and electromechanical coupling factro(k_p) as a function of ZnO

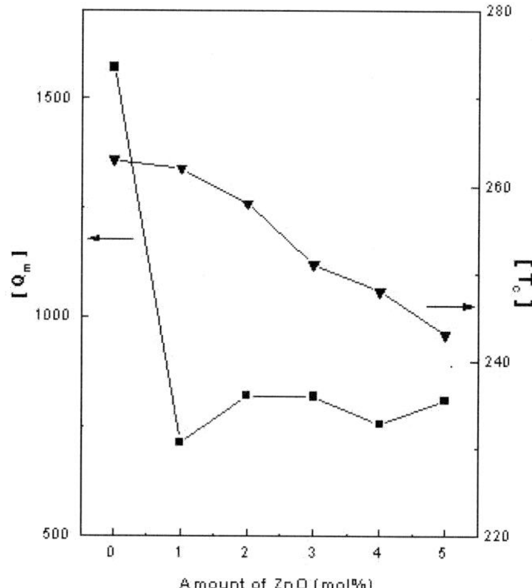

Fig 10. Variation of Curie temperature(T_c) and mechanical quality factor(Q_m) in PMS-PZT-xZnO after sintering at 1000 °C 2h

5 Conclusions

Addition of ZnO to $0.05Pb(Mn_{1/3}Sb_{2/3})O_3$-$0.95Pb(Zr_{0.52}Ti_{0.48})O_3$ system, through eutetic reaction between PbO and ZnO, it formed liquid phase and improved sintering behavior, density at 1000 °C.

Along with improvement of sinterbility, relative dielectric constant($\varepsilon_{33}{}^{T}/\varepsilon_o$) increase up to 3mol% (997) and then decreased. Doping of ZnO increased electromechanical coupling factor(k_p) from 0.43(0mol%) to 0.5 (3mol%). Maximum of piezoelectric constant d_{33} & g_{33} were $335\text{x}10^{-12}$ C/N, $63\text{x}10^{-3}$ Vm/N. From decrease of lattice parameter, mechanical quality factor(Q_m), Curie temp.(T_c), it was estimated that Zn substituted for Pb-site(A-site) respectively.

6 References

1. G.A.Smolenski and A.I.Agranoskaya, Fizika Tverdogo Tela, 10(10), pp. 1562-1572(1959)
2. M.Takahashi, Jap.J.Appl.Phys., 9(10), pp. 1236-1246 (1970)
3. R.Gerson, J.Appl.Phys, 3(1), pp. 188-194 (1960)
4. L.E.Cross, Ceram.Bull, 63(4), pp. 586-590 (1984)
5. G.Bayer, J.Amer.Ceram.Soc., 46(2), pp. 604–605 (9163)
6. L.L.Tu, Y.Y.Jin and M.Z.Han, Ferroelectrics, 28, pp. 403–406 (1980)
7. A.X.Knang, L.Y.Chai, S.N.Pan and T.S.Zhou, IEEE Proc., pp. 689–692 (1986)
8. S.Takahashi, Y.Sasaki, S.Hirose and K.Uchio, Jpn.J.Appl.Phys, 34, No.9B, pp. 5328-5331 (1995)

9. H.Kawai, Y.Sasaki, T.Inoue, T.Inoi, K.Zama and S.Takahashi, NEC Tech. Rep., 49(10),
10. pp. 135-138 (1996)
11. H.Sasaki, K.Iguchi and S.Kawashima, NEC Tech. Rep., 50(10), pp. 80-84 (1997)
12. R.S.Roth, J.R.Dennis and H.F.McMurdle, "Phase Diagram For Ceramists", Am. Ceram. Soc. 2, Fig. No.2326 (1969)
13. Z.Gui, H.Hu, L.Li and X.Zhang, Solid-State Phenomena, Vol. 25&26, pp. 309-316 (1992)
14. R.B.Atkin and R.M.Fulrath, J.Am.Ceram.Soc., 54(5), pp. 265-270 (1971)
15. J.Chan, Acta. Mat., 10, pp. 787-797 (1962)
16. Y.Xu, Ferroelectric Material and Their Applications, pp. 130-143, Elsvier Science Publishing Co., New-York (1991)
17. B.Jaffe, W.R.Cook Jr. and H.Jaffe, "Piezoelectric ceramics", pp. 91, Academic Press.
18. London & New-York (1971)

Improvement of Properties of Y- and Nd-base Melt Textured High Temperature Superconductors by High Pressure - High Temperature Treatment

Tatiana Prikhna*, Wolfgang Gawalek[&], Viktor Moshchil*, Sergey Dub*, Tobias Habisreuther[&], Vladimir Melnikov[#], Felip Sandiumenge[&], Valery Kovylayev[$], Alexey Surzhenko[&], Peter Nagorny*, Peter Shaetzle[@], Alexandr Borimsky*

* - Institute for Superhard Materials., Kiev

[&] - Institut für Physikalische Hochtechnologie, Jena

[#] - Institute of Geochemistry, Mineralogy and Ore-Formation., Kiev

[@] - Instituto de Ciéncia de Materiales de Barcelona, Barcelona

[$] - Institute of Materials Science Problems, Kiev

1 Abstract

Comparison studies of the effect of high pressure - high temperature (HP-HT) conditions (2-5 GPa, 800-1100 °C for 10-30 min), created in high - pressure apparatuses (HPA) of different types (recessed-anvil and cube) on Y- and Nd-base melt-textured high-temperature superconductors (or MT-MeBCO, where Me=Y, Nd) shows the advantages of using the cube-type apparatuses in order to reduce or avoid sample cracking, to improve their microhardness and fracture toughness as well as to attain almost theoretically-possible density and some increase in critical current density.

2 Introduction

Melt-textured $MeBa_2Cu_3O_{7-\delta}$ (Me=Y, Nd)-based or MT-MeBCO high temperature superconductive materials possess the highest level of cryomagnetic properties among the known HTSC, that makes them suitable for flying wheels, energy-storage systems, devices working on principles of levitation. Controlled seeding and melt-textured growth of YBCO results in trapped fields up to 0.9 T at 77 K and up to 6.9 T at 12 K. For example, MT-YBCO-based HTSC electromotors with an output power of about 30 kW at 3000 rpm working at liquid nitrogen temperature level, which are smaller than conventional ones and have more than 50%-higher energy density on the rotor surface have been developed in co-operation between IPHT (Jena, Germany), MAI (Moscow, Russia) and Oswald Electromotors (Miltenberg, Germany). Frictionless MT-YBCO bearings, which allow about 43000 rpm have been developed in co-operation between IPHT Jena and IFW (Dresden, Germany). The technical parameters of devices depend crucially on the microstructure of the MT-YBCO material, e.g., the size and orientation of quasi-single crystalline grains, the presence of strong pinning centers (Y211 precipitations, stacking faults and dislocation networks), the absence of macro- and microcracks, and a high materials density. The density of MT-MT-MeBCO is approximately 85-90% theoretical one. The structure of melt-textured ceramics usually involve some pores, macro- and microcracks. Macrocracks appear at the cooling stage after the melting process. For example, in MT-YBCO because of the difference in the coefficients of thermal expansion of Y123 and Y211 phases. Microckracks form during the process of material saturation with oxygen because of the variations in volumes, for example, during the transformation of tetragonal Y123 phase into the orthorhombic one (G. Fuchs et al., 1997). The preliminary estimations show that the attained level of critical current density (j_c) in magnetic field for MT-NdBCO may be higher than that for MT-YBCO. The structure of MT-MeBCO (Me=Y, Nd) consist of one or a few quasi-single-crystalline grains of the superconductive phase (Y123 or Nd123) well oriented relative to each other. In the superconducting grains, finely-dispersed grains of non-superconductive phase Y_2BaCuO_5 (Y211) or $Nd_4Ba_2Cu_2O_{10}$ (Nd422) are present. It is well known that the more dispersed and homogeneous arrangement of Y211 or Nd422 grains promote the higher values of critical current density of the material. May be the defects that form around such non-superconductive grains can serve as pinning centers and induce an increase in critical current density, as compare with single-phased materials. In MT-NdBCO another mechanism of pinning may also exist. Nd can substitute for Ba in the Nd123 structure and thus, form the local «supersaturated» with respect to oxygen (more than 7 atoms per one unit cell) nonsuperconductive regions that might work as additional pinning centers.

The main idea of the present study was to try to improve properties (density, microhardness, fracture toughness, etc.) of MT-MeBCO (Me=Y, Nd) parts of comparatively large dimensions (up to 30 mm in diameter and 20 mm in height, such sizes allow one to manufacture, for example, constituent rotors of HTSC electromotors) using high pressure-high temperature (HP-HT) treatment. The suggested method attracts our interest due to the possibility shown earlier (T.A. Prikhna, 1997a,b,c) to save oxygen in the Me123 (Me=Y, Nd, Gd, Sm, Eu) structure and thus superconductive properties during heating up to 900-1100 °C. The results presented here deal with a short-time high pressure (up to 5 GPa) - high temperature (800-1100 °C) influence on MT-MeBCO (Me=Y, Nd) structure and properties.

3 Experimental

The high pressure-high temperature conditions have been created in the two types of high pressure apparatuses (HPA). In a recessed-anvil type HPA (Fig. 1) during the pressure treatment temperature measurements using a thermocouple have been performed. The design of high-pressure apparatus allows the creation of quasi-hydrostatic pressure, which means that the stress in the vertical direction is by 30 % higher then that in the radial directions. In a cube (Fig. 2) with the in-situ temperature control and the possibility of three-axial compression of the samples being treated.

Pressing in the recessed-anvil-type HPA were performed using 2500-ton presses and of cube-type HPA using a 14000 ton press. After pressure application the samples have been heated by the direct current flow through the apparatuses and graphite heater.

The accuracy of pressure and temperature measurements were $\pm$ 0.15 GPa and $\pm$20 °C, respectively. To protect samples from reaction with a graphite heater, during the treatment they were in contact with preannealed (at high temperature) and pre-compacted monoclinic zirconia powder. It has been shown previously (T.A. Prikhna et al., 1995) that under certain

HP-HT conditions, zirconia may prevent from oxygen losses from Me123 (Me=Y, Nd) and even be an oxygen supplier.

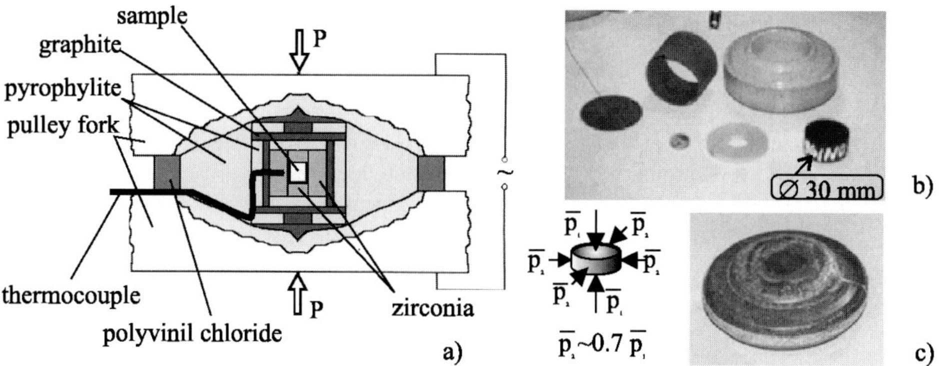

Figure 1. Recessed -anvil type HPA: (a) scheme; (b) elements of the HPA-cell; (c) the HPA-cell after pressing.

As starting samples, we use MT- $YBa_2Cu_3O_{7-\delta}$ samples (MT-YBCO) 30 mm in diameter and 17-20 mm in height that have been produced by purposeful crystallization on seeds from Y_2O_3 ($Y_{1.5}Ba_2Cu_3O_{7-\delta}$) with addition of a small amount of Pt and Ce oxides in order to decrease the size of Y211 inclusions and to attain a more uniform their distribution over the Y123 matrix. According to hydrostatic weighing, the material density was about 5.7 g/cm^3.

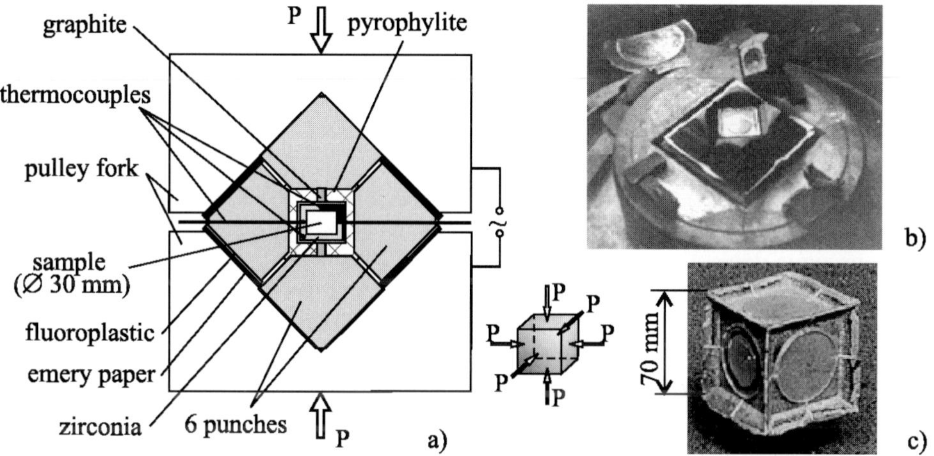

Figure 2. Cubic type HPA: (a) scheme; (b) the view of HPA and (c) view of HPA-cell after pressing.

We also use two types of melt-textured $NdBa_2Cu_3O_{7-\delta}$ samples with a density of 5.4 -5.5 g/cm^3 prepared by melt processing in a reduced oxygen partial pressure and then exposed to oxidation in flowing O_2 from (1) Nd123, Nd_2BaO_4 and Pt (of d = 22 mm, h = 10 mm) and (2) Nd123 and Nd_2O_3 (of d=30 h=17-20 mm).

The sample structure and phase composition were examined on a X-ray diffractometer DRON-2 and an automatic diffractometer ADP-1. The oxygen content of $YBa_2Cu_3O_{7-\delta}$- phase was determined by the linear equation of regression binding c and δ.

impossible to estimate oxygen content of the structure by determining the c-parameter because it is affected by both oxygen content and cation substitutions.

The structure of materials before and after thermobaric treatment was studied using polarized optical, scanning (Superprobe 733, JEOL) and transmission electron (at 200 kV) microscopes. The j_c from magnetization hysteresis loops was measured by an Oxford Instruments 3001 vibrating sample magnetometer using Bean's model. The samples density (γ) was determined by hydrostatic weighing.

Microhardness was measured employing a Matsuzawa Mod. MXT-70 microhardness tester and Berkovich Nano-indenter II. The fracture toughness was estimated from the length of the radial cracks emanating from the corners of an indent (K. Tanaka, 1987).

4 Results and Discussion

The experiments on HP-HT treatment of MT-MeBCO samples in the HPA of different types (recessed-anvil and cube) show the advantages of cube-type apparatuses for treatment of large samples. In the recessed-anvil-type HPA, thermocouples break frequently due to the flowing of the material of a gasket and the pressure medium. Besides, the volume of high pressure (HP) cell of the recessed-anvil type apparatus is too small to accommodate a sufficient amount of the pressure- medium for creation isostatic conditions of treatment of large samples and this leads to a crack formation and often to the failure of samples. The apparatuses of cube-type are of interest because of the possibility to use the HP cells of a larger volume, to realize three-axial compression and to introduce three thermocouples simultaneously, which work very stable and allow more careful measurements of the temperature field distribution.

The data given in Table 1 show the increase in MT-MeBCO (Me=Y, Nd) density and improvement of mechanical characteristics as a result of HP-HT treatment. Figure 3 presents the load-displacement curves of MT-NdBCO samples before and after HP-HT treatment, obtained with Nano Indenter-II under the 50 mN-load. Some steps (marked by arrows) are observed on the curve corresponding to the starting sample. Such steps may appear due to the closing of adjacent-to-indentor pores under indenter induced high pressure. The change of unloading curve slope of starting sample at the end of indentation cycle points to the opening of lateral crack (the chip near the indentation mark). After HP-HT treatment the indentation load-displacement curve traditional for brittle materials is observed.

Table 1. Characteristics of MT-MBCO before and after HP-HT treatment

Material (treatment conditions)	Density, g/cm^3		Vickers microhardness, GPa		Fracture Toughness, MNm$^{-3/2}$	
			(Indentation Load, N)			
	before	after	before	after	before	after
MT-YBCO (2 GPa, 900 °C, 15min)	5.7	6.3	3.4	5.3	0.71-1.46	no cracks
			(4.91)		(4.91)	
MT-NdBCO (5 GPa, 850 °C, 15 min)	5.5	6.6	4.2-5.2	5.1-5.8	-	1.19
			(1.96)		(1.96)	

In the range of 2-5 GPa (after 800-1100 °C for 10-15 min and 800-950 °C for 30 min) for MT-YBCO there was practically no variation in unit cell parameters of Y123 phase and hence in the oxygen content. The j_c (at 77K in zero field) of MT-YBCO after 2 GPa, at 800 for 30 min and at 900-950 °C for 15 min in the *ab*-plane of Y123 remained unchanged (10^4 A/cm^2) and increased in the direction of the *c*-axis (from 3×10^3 to 7×10^3 A/ cm^2). The increase in critical current density may be due to the increase of material density as well as of dislocation density (from 10^8 cm^{-2} up to 10^{12} cm^{-2}, as TEM study shows). It was not possible to conduct the quantitative microprobe analysis of MT-MeBCO (Me=Y, Nd) correctly because of very fine distribution of Y211 and Nd422 phases in the Me123 structures. SEM studies of the relative amount of different elements in MT-YBCO along the scanning line shows practically no variations in the distribution of these elements and sizes of Y211 inclusions before and after HP-HT treatment under optimal conditions (Fig. 4). The deviation from above mentioned HP-HT conditions induces the increase and recrystallization of Y211 and thus lowering the j_c.

For MT-NdBCO after 5 GPa, 800-1100 °C for 15 min the *a, b, c* of Nd123 were also practically unchanged. Up to now we failed to reduce the treatment pressure for MT-NdBCO below 5 GPa because of Nd123 decomposition. But in contrast with MT-YBCO, the variation in size of 422 inclusions was not so dramatic as of Y211 in MT-YBCO. The highest j_c for MT-NdBCO (1.2×10^4 A/cm^2) was attained after 5 GPa, 850 °C for 15 min. The MT-NdBCO with Pt additions turns out to be more stable under HP-HT.

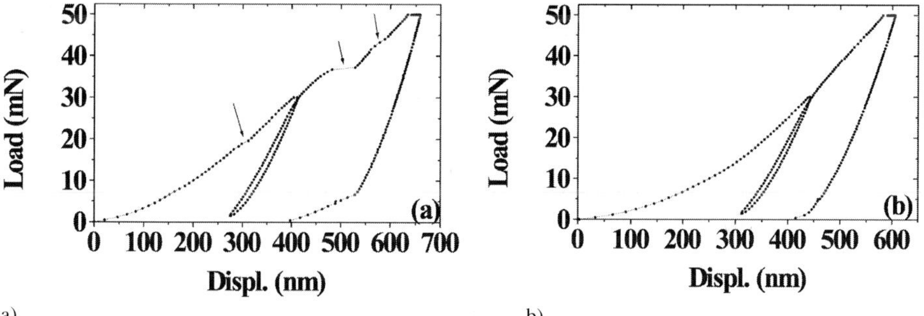

a) b)

Figure 3. Load-displacement curve of MT-NdBCO: (a) before and (b) after 5 GPa, 850 °C, 15 min.

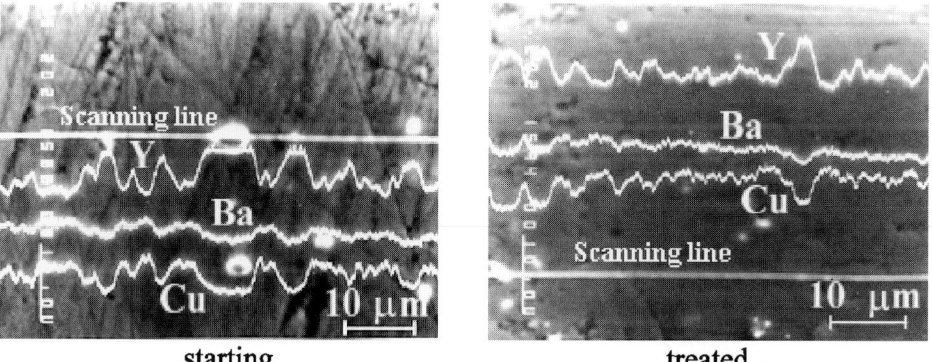

Figure 4. The relative amount of Y, Ba and Cu in the MT-YBCO structure along the line.

Thus, we may conclude that HP-HT treatment allows us for a short time to increase the density and mechanical properties of MT-MeBCO (M=Y, Nd) and even reduce the anisotropy of j_c. For HP-HT treatment of large parts of MT-MeBCO, the cube-type HPA is preferable.

5 References

5.1 Journals,

a) with Volume number

1. T. A. Prikhna, Fizika i Tekhnika Vysokikh Davlenii, 1997a, 7, 2, 109-116
2. T. A. Prikhna, J. Superhard Materials, 1997b, 4, 15-23
3. T. A. Prikhna, J. Superhard Materials. 1997c, 2, 22-27.
4. T. A. Prikhna, V. S. Melnikov, V. V. Kovylyaev and V. E. Moshchil, J. Mat. Sci. 1995, 30, 3662-3667.
5. Tanaka, J. Mater Sci. 1987, 22, 1501.

5.2 Books

b)with editor:

6. G. Fuchs, S. Gruss, G. Krabbes et al., in Proc. 10th ISS'97, October 27-30, 1997, Gifu «Advances in Superconductivity X» (Ed.: Osamura-Hirabayashi), 1997, 2, 639-644.

Materials Aspects of Solid Oxide Fuel Cells (SOFC)

Ludger Blum, Wolfgang Drenckhahn, Hermann Schichl
Siemens AG, Erlangen, Germany

1 Summary

Fuel cells are considered as a significant innovation for energy generation of the future since they are both environmentally friendly and can convert natural gas to electric current with high degrees of efficiency. Its introduction is especially suited for the decentralized provision of electricity and heat. The SOFC-technology is still at the beginning of its development, although a 100 kW-plant was already demonstrated using Westinghouse's tubular concept. The alternate to this, the planar SOFC is still in the development stages of units up to 10 kW. The introduction of high-temperature fuel cells will follow conventional combined heat and power plants (CHP) in competition. To be successful, competitive life-cycle costs are required. For this purpose a lot of innovation is necessary especially in the field of material development (low cost and high power density) and corresponding low cost manufacturing techniques.

2 Introduction

During the last 10 years decentralized power generation plants, which bring the consumer both electricity and heat to the region of demand, became an interesting alternative for local and industrial consumers of energy. The liberalization of the electricity market will probably lead to a further increase in the growth rate, at least in a midterm scenario.

The Solid Oxide Fuel Cell, SOFC, is being developed world wide in two different concepts. The development started with the tubular concept which was, in particular, pursued by Westinghouse and also by Mitsubishi Heavy Industry, and was demonstrated with power of up to 100 kW. This technology is based on the extrusion of ceramic tubes, attached sintering, and coating. The development based on the planar concept started in the eighties. This concept is being pursued in Europe e.g. by Sulzer and several research institutes and in the past by Siemens and Dornier. After having taken over the Westinghouse activities Siemens decided to stop the planar SOFC development and to concentrate on tubular, because of its further advanced development status. These activities are now performed by Siemens Westinghouse Power Corporation, SWPC.

3 Technology of the Planar SOFC

The solid oxide fuel cell uses, as the name suggests, a ceramic or more precisely an yttria stabilized zirconia as oxygen-ion-transporting electrolyte. The electrolytes are produced with a thickness of approximately 150 μm and a surface area of 10 x 10 cm². The ion-conductivity of the electrolyte can be detected at temperatures above approximately 600°C. The electrolyte is coated on the fuel-side by an anode made of nickel/ziconia-cermet, and on the air-side by a cathode made of lanthanum-strontium-manganese. The electrode-coated electrolytes are, according to their function, fuel cells in a narrower sense, and are often called Membrane Electrode Assembly, MEA .

3.1 Cells (MEA)

Having started with stoichiometric LSM, in a second step cells with a cathode of non-stoichiometric LSM: $La_{0,75}Sr_{0,2}MnO_3$ (uLSM) the so-called mono-layer-cathode were used. With this cell type a current density of 500 mA/cm² was attained at 950°C using air at the oxygen side and a mixture of 50% H_2 and 50% H_2O at the fuel side. As pointed out in Figure 1, in this case the contribution of the cathode to the cell resistance is 60%.

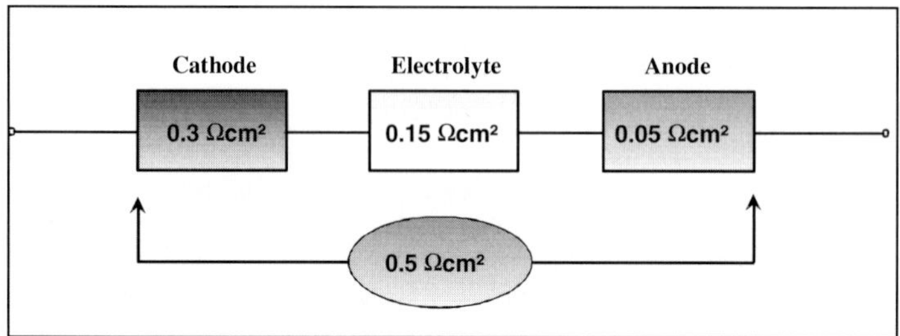

Figure 1. Resistance of the components of planar cells

It was therefore the first target to reduce the cathode polarization. Because this is mainly caused by diffusion processes an increase of the number of three phase boundaries could solve this problem. The experimental way was to prepare a screen printing paste out of a mixture of 50% uLSM and 50% yttria doped zirconia (8YSZ) using special dispersing agents. This in order to get fine branches of zirconia within the uLSM after firing with its source at the surface of the electrolyte substrate. Since this kind of cathode is targeting in a high lateral resistivity a further coating is necessary acting as a current collector layer. With such kind of double layer cathode consisting of a working and a current collector layer the polarization resistance could be decreased by a factor of three.

Now the contribution of the electrolyte to the ohmic resistance of the total cell is about 50%. Therefore the further work was concentrated on the improvement of the electrolyte. Commonly yttria - stabilized zirconia (YSZ) is used as electrolyte. The composition with 8 mol% Y_2O_3 (8YSZ) has traditionally dominated because the ionic conductivity exhibit a maximum at that yttria content. However experiments showed that the ionic conductivity of 8YSZ electrolytes decreases significantly (about 40%) by annealing at 1000°C, in contrast to the conductivity of the 10YSZ which keeps nearly constant. So after 500 hours annealing, the ionic conductivity of the 10YSZ electrolyte was higher than that of the 8YSZ. The theory that the lattice distortion of yttria doped zirconia - the ionic radius of Y^{3+} (0,92) is much larger than that of Zr^{4+} (0,79) - could be a limitation for the ionic conductivity, was leading to experiments with scandia doping instead of yttria, because the corresponding ionic radii are nearly equal (Sc^{3+} = 0,80).

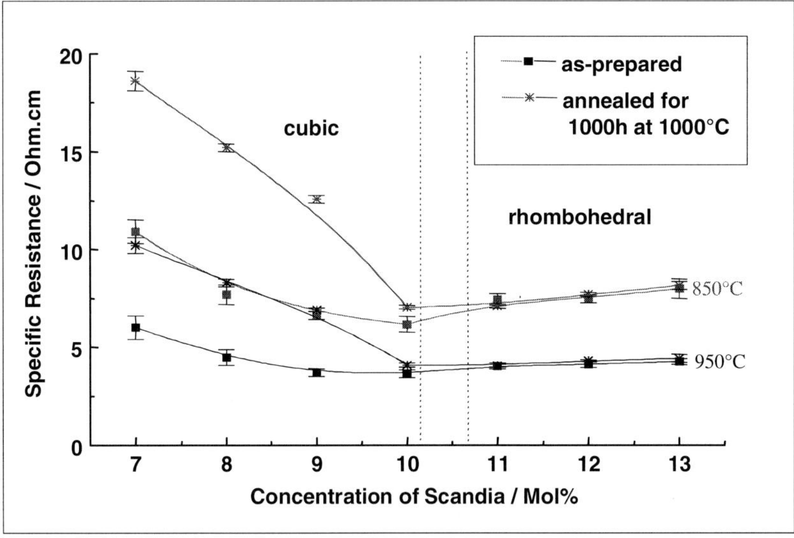

Figure 2. Resistance of scandia doped zirconia

The powder for the scandia doped zirconia (ScSZ) electrolytes have been produced by a Sol/Gel route using scandium chloride and zirconium n-propoxide as starting materials. As shown in Figure 2 the ionic resistance of scandia doped zirconia is excellent in the doping range between 8 and 12 mol% and is about half of 8YSZ. The ScSZ electrolytes with more than 10 mol% Sc_2O_3, which do not show a decrease in conductivity by annealing, undergo a cubic $\leftrightarrow$ rhombohedral martensitic phase transformation at a temperature of about 500°C, which is coupled with a volume change of less than 0.15%. Such a low volume change is not a problem for SOFC applications. As an optimum electrolytes with a Sc_2O_3 content of 10.5% was chosen. Using this material and a double layer cathode a current density at 850°C and 0.7 V of 700 mA/cm² was reached, using air and a mixture of hydrogen and water vapor (1:1), which means about 0.5 W/cm².

3.2 Interconnector

In the Siemens-concept small cells of 5 x 5 or 10 x 10 cm² are being joined together on big metallic bipolar plates in arrangements of 4 x 4 or 3 x 3 cells. The interconnector plates have been made out of a special alloy (so-called $CrFe5Y_2O_31$), which has a very well adapted thermal expansion coefficient, slightly above the electrolyte to get compression stress on the ceramic during cooling down and an excellent electrical conductivity of 10^6 S/m and a thermal conductivity of 50 W/mK, both at 1000°C. This makes it possible to build up stacks with big plates, which tolerate temperature differences between gas inlet and outlet of at least 150°C. Other developers are using doped lanthanum-chromite as interconnector material, which is more sensitive to thermal gradients and cannot be manufactured in large dimensions but has a very good chemical stability. The $CrFe5Y_2O_31$ plate material was therefore annealed for 10,000 hours in air and different fuel gases like CO, hydrogen and water vapor. The tests showed very thin corrosion layers with good adhesion to the base material. This ensures that an operation of at least 40,000 hours is possible.

In order to reduce costs and waste material a method for recycling the bipolar plates was developed and checked in several stack tests.

At operating temperatures above 800°C chromium-oxide evaporates out of the surface of the bipolar plate material. The chromium condenses at the three-phase boundary of electrolyte and cathode and blocks the active surface which causes an increased degradation. To avoid this mechanism, different protective layers were tested. Best results were achieved using a lanthanum-chromite layer applied by vacuum plasma spraying. Diffusion experiments and modelling showed that a 10 μm thick layer is sufficient for more than 100,000 hours of operation.

To get good contact between chromite and cathode a 100 μm thick layer out of lanthanum - strontium -manganese -cobaltite was applied by wet powder spraying. This layer guaranties a contact resistance below 20 $\mu\Omega.cm^2$.

On the anode side the contact between electrode and bipolar plate was realized by a nickel mesh, which is spot-welded to the metal plate. However, measurements of the contact resistance between nickel mesh and interconnector plate showed a rapid increase of the resistance, strongly dependent on temperature, caused by a reaction zone formed between nickel and chromium under presence of water vapor. To avoid this reaction different protective layers deposited on the bipolar plate were investigated. The most stable behavior was found using a nickel coating, which promises a lifetime of 40,000 hours staying below 10 $m\Omega.cm^2$, which is half of the target resistance.

3.3 Sealing

To get a gas tight separation between anode and cathode side the electrolyte is sealed to the bipolar plate via a glass sealing, which has to fulfil several requirements, as chemical stability in humid hydrogen, CO and in air up to 1000°C, deformability during the sealing process, high viscosity during operation at 800 to 900°C. A glass material, commercial available as thin foil, mixed with a certain amount of MgO, was found to fulfil all the requirements sufficiently. In addition it was possible to apply the glass by a screen printing process, which clearly reduces the manufacturing costs.

4 Technology of the tubular SOFC

First investigations on SOFC have been started at Westinghouse in the mid sixties. In the late seventies a stack technology based on the so-called thin film tubular cells was invented.

Basically similar materials as in the planar design are used. This counts especially for electrolyte and anode.

The actual concept is an air electrode (cathode) supported tubular design configured as a single cell per tube with an axial interconnection. The tube is nominally 22 mm in diameter with 1500 mm active length and one closed end, manufactured by extrusion and sintering. The total length is 1.81 m. The cathode material is also doped lanthanum - manganese. Caused by these dopants the thermal expansion of the tube matches very well to the electrolyte (see Figure 3), which is very important because of the length of the cell.

The $LaCrO_3$ interconnector is deposited in the axial direction by atmospheric plasma spraying. The deposition of a gas tight uniform electrolyte layer is done by an electrochemical vapor deposition (EVD) process. At temperatures near 1200°C and vacuum pressures near

1 mm Hg anhydrous mixtures of yttrium trichloride and zirconium tetrachloride are sublimed and passed over the exterior of the air electrode. Oxygen is provided to the tube interior. Because of the special properties of this process a uniform film of nominally 40 μm in thickness is gained. The metallic connection between the cells is made through the use of nickel felts. This electrical integration is made possible by nickel electroplating of the interconnection to provide the metallic surface for nickel felt bonding. The fuel electrode is deposited through a slurry dip process. The deposited nickel - zirconia cermet is sintered in a dual atmosphere with air in the cell interior while a reducing atmosphere is provided on the anode to avoid nickel oxidation. The final cell active area is 834 cm².

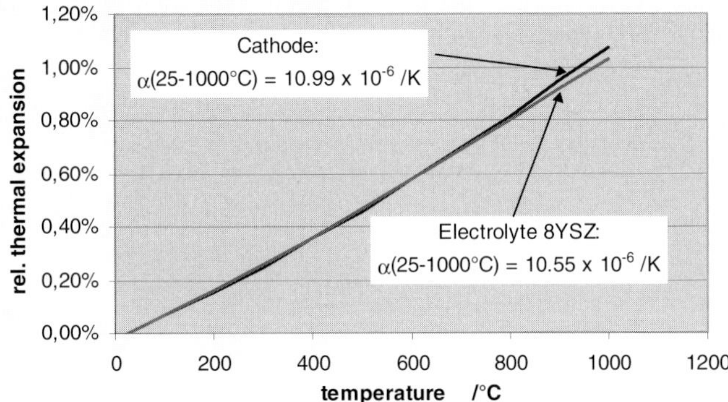

Figure 3. Thermal expansion of cathode and electrolyte

For operation air is supplied to the tube interior and fuel to the tube exterior. At atmospheric pressure, a uniform temperature of 1000°C, 85% fuel utilization and 25% air utilization, a single tubular SOFC will generate a maximum power of about 210 W. Operation at an elevated pressure of at least 3 bar will increase maximum output by about 20 percent.

In order to generate large power output the cells are arranged to bundles, each consisting of 3 by 8 cells and the bundles to bundle rows. Between each bundle row is placed an in-stack reformer. The reformer is radiantly heated by the adjacent rows of cells. For a 100 kW module 1152 of such cells are assembled within a container together with the in-stack reformer boards. The ambient air is compressed to the necessary process pressure by a blower and preheated to about 600°C by an exhaust gas heated recuperator. Natural gas is compressed and desulfurized before entering a pre-reformer for the higher hydrocarbons. Normally 85% of the fuel is electrochemically oxidized. The rest is burned nearby the module in a combustion zone. The exhaust exits the module at a temperature between 700 and 850°C.

Based on this concept an efficiency of 46% for the 100 kW-plant at full load was already demonstrated in the EDB/ELSAM plant in the Netherlands.

A pressurized plant in combination with a micro-turbine shall be put into operation at the end of 1999. This plant of 220 kW power output will already achieve an electrical efficiency of more than 55%.

5 Outlook

Today we find ourselves at the stage where the SOFC-technology is, firstly, being tested for demonstration at a larger scale. It is yet to be proven that the high expectations of the public for such technology can be fulfilled. Here, the tubular concept currently has an advantage over the planar concept because of its further advanced development stage.

Assuming that the technical problems, that still exist today, are solved, reduction of the present high costs to competitive levels through material improvements and simpler and new production methods remains a substantial task for the developer. At the end of the day, the purchasing decisions of the customers are influenced superficially by the investment costs, but ultimately by the life cycle costs. Only at a competitive price and additional non-monetary advantages, like the environment, for example, can a market-share be acquired.

Synthesis and Characterization of New Ceramic Anode Materials for SOFCs

Frank Tietz, Wolfgang Jungen and Günter Pudmich
Institute for Materials and Processing in Energy Systems (IWV-1), Research Center Jülich, D-52425 Jülich, Germany

1 Introduction

The future of SOFCs depends strongly on the exploitation of cheap fuels. Natural gas consisting mostly of methane would be a perfect solution. Unfortunately the use of humidified natural gas causes carbon deposition on the anode and poisoning by sulfur, finally resulting in a breakdown of cell performance [1]. The currently used material for anodes in solid oxide fuel cells is a Ni/YSZ cermet which is optimized for the oxidation of hydrogen. The nickel metal in the cermet tends to agglomerate after prolonged operation time leading to reduced three-phase-boundary and increasing resistance. Reforming and cleaning of methane to obtain purified hydrogen increases investment costs because of the requirement of additional components in the power plant.

Therefore, new materials with appropriate properties are a possible solution to the mentioned problems. The materials to be used have to fulfill several requirements, particularly high electrical conductivity, an appropriate catalytic activity and good stability under reducing conditions. Compatibility with the thermal expansion coefficients (TEC) of other cell components and of course no chemical reaction with the electrolyte or the interconnect, even at high temperatures and after long operation time is a necessity.

Ceramic oxides show several advantages over metals. Some work has already been performed in this field by various groups [2-5]. A search for such new anode materials was carried out in the field of perovskites, investigations concentrated on conductivity and expansion behavior, thus catalytic properties and long-term stability need further investigation.

2 Experimental

Various perovskite materials mostly based on lanthanum chromite $La_{1-x}A_xCr_{1-y}B_yO_3$ (with A = Ca, Sr; B = Ti, Fe, V, Nb) were synthesized using the Pechini method [6]. Table 1 gives an overview of selected compositions and their abbreviations. Nitrates, carbonates and oxides of the metals were complexed in an aqueous solution with citric acid and then heated with ethylene glycol. After polyesterification and removal of the excess liquid, a resin was formed, which was heated up to 300 °C to decompose the organic constituents, then ground in a mortar and finally calcined at 900 °C to produce the powder. The obtained powders were characterized in several ways, the particle size distribution measured by a centrifugal particle size analyzer; thermogravimetric measurements and differential thermal analysis (TG/DTA) were carried out with ground powder before calcination up to 1400 °C.

Table 1: Name and chemical composition of investigated powders

Name	Composition	crystalline phases after sintering at 1400 °C
LSCT-7382	$La_{0.7}Sr_{0.3}Cr_{0.8}Ti_{0.2}O_3$	perovskite [(a)]
LCCT-7382	$La_{0.7}Ca_{0.3}Cr_{0.8}Ti_{0.2}O_3$	perovskite [(b,d)]
LCCT-7328	$La_{0.7}Ca_{0.3}Cr_{0.2}Ti_{0.8}O_3$	perovskite [(a)]
LCCT-7355	$La_{0.7}Ca_{0.3}Cr_{0.5}Ti_{0.5}O_3$	perovskite [(a)]
LCCF-7382	$La_{0.7}Ca_{0.3}Cr_{0.8}Fe_{0.2}O_3$	perovskite [(a)]
LCCN-7382	$La_{0.7}Ca_{0.3}Cr_{0.8}Nb_{0.2}O_3$	perovskite
LCCV-7382	$La_{0.7}Ca_{0.3}Cr_{0.8}V_{0.2}O_3$	perovskite [(c,d)]
LCT-73X	$La_{0.7}Ca_{0.3}TiO_3$	perovskite
STN-628	$Sr_{0.6}Ti_{0.2}Nb_{0.8}O_3$	-
SLT-64X	$Sr_{0.6}La_{0.4}TiO_3$	-

(a) $LaMnO_{3.15}$(b) $La_4Ca_2Ti_5O_{18}$, $La_4CaTi_5O_{17}$ (c) $La_{0.67}Ca_2(VO_4)_2$(d) $LaCrO_3$

For dilatometric and conductivity measurements the calcined powders were pressed uniaxially at 125 MPa to sticks with about $2 \times 5 \times 40$ mm and sintered in air at 1400 °C for 10 hours. Conductivity data were collected from 100 to 900 °C in air and argon with 4% hydrogen. X-ray diffraction was carried out both with calcined powder and powder obtained after sintering and grinding.

3 Results and Discussion

3.1 Characterization of powders

The particle size distributions of the obtained powders showed a mean diameter in the range of 0.3 to 3 μm (Figure 1). The titanium-rich powders revealed a very broad particle size distribution with a mean diameter of 1 - 1.5 μm, whereas the chromium-rich samples gave a narrow distribution and a mean diameter of about 0.4 μm.

X-ray diffraction was used to investigate both calcined and sintered powders to check the formed crystalline phases (Table 1). Some of the calcined powders showed remaining small amounts of oxides, which usually disappeared after sintering. Calcined powders showed relative broad peaks, which became much sharper and more intense in the case of sintered material.

Thermogravimetric measurements indicated that the remaining organics of the resins completely disappeared at temperatures between 400 and 600 °C. DTA measurements showed for all compositions no phase transitions up to a temperature of 1400 °C.

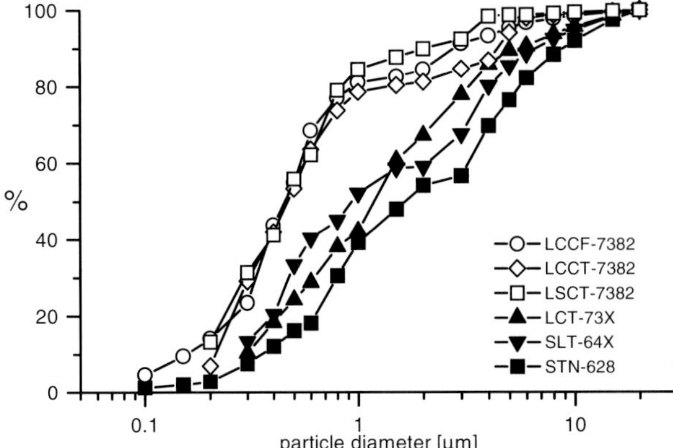

Figure 1. Cumulative particle size distributions of powders

3.2 Conductivity

Measurements were taken through an four point DC conductivity measurement both in air and argon with 4% hydrogen. Usually the conductivity dropped considerably in air about one to five orders of magnitude, depending on the composition. Only for chromium-rich samples the conductivity in air was higher than in hydrogen. At a temperature of 900 °C values up to 10^2 S/cm were obtained under reducing conditions, see Figure 2.

In the case of p-type conductors conductivity decreases in reducing atmosphere because of the decrease of charge carriers (holes) in combination with an increase of oxygen vacancies [7,8]. In the case of n-type conductors charge carriers (electrons) are formed during reduction.

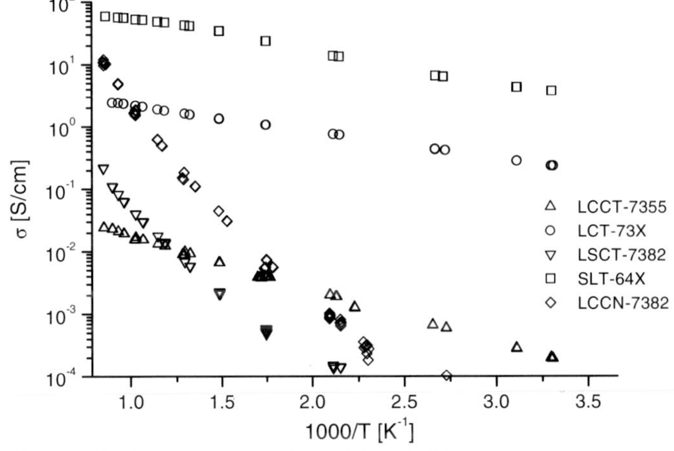

Figure 2. Conductivity of selected materials in $Ar/4\%H_2$.

3.3 Expansion due to temperature and atmosphere

Dilatometric measurements were performed only on some selected specimens. Figure 3 shows the expansion behavior of LCCT-7355 in dependence on temperature and atmosphere. Sintered sticks were heated in air with a constant rate of 3 K/min up to 1200 or 1000°C to get the thermal expansion coefficient (TEC) of the materials, then cooled down to 1000 or 800°C, respectively, with the same rate and held at this temperature for several hours to see any time-dependent changes. Then the atmosphere was switched from air to Ar/4 %H_2 and back to air after several hours. In both cases the TEC was derived from the heating segment and was in the order of 1×10^{-5} K^{-1}. The expansion and contraction of LCCT-7355 due to the atmospheric changes is very small and can only be seen in the enlarged part of Figure 3.

In the case of LCCF-7382 the behavior was completely different and a huge expansion of the material is observed, which decreases slowly at 1000°C with prolonged time. The explanation for the expansion is the reduction of Cr^{4+} and Fe^{3+}/Fe^{4+} and the formation of oxygen vacancies [8]. At 800°C in reducing atmosphere there is no further length change after the initial expansion for at least several hours. If the atmosphere is switched back to air, the expansion remains to some extend. Under the applied conditions iron can be reduced and therefore its ionic radius expands. Such a behavior can lead to stresses and mechanical failure of the fuel cell during the initial reduction step for operation.

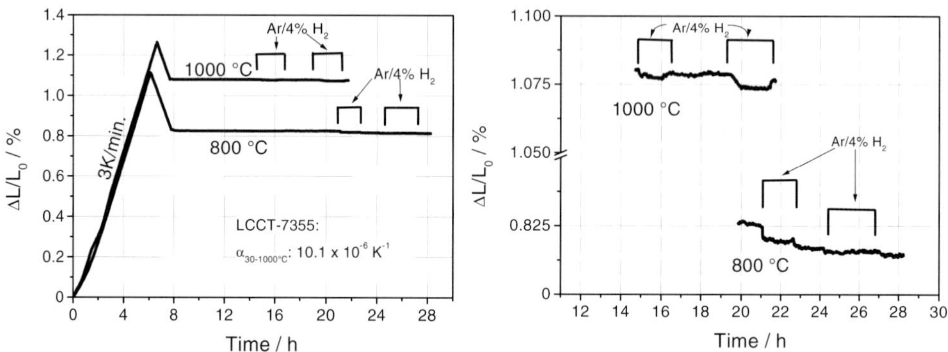

Figure 3. Expansion behavior of LCCT-7355 in different atmospheres. *Left:* Overview of experiment *Right:* enlarged scaling to show the small changes due to the alteration of oxidizing and reducing atmospheres, which cannot be seen on top

4 Summary

Synthesis of perovskites via the Pechini method is a prosperous and comfortable way of obtaining powders of high purity and small particle diameter. Nearly all materials could be synthesized single-phased. Most of the perovskites fulfill some of the requirements of a ceramic SOFC anode. High conductivity was obtained with titanium-rich materials, in the case of chromium-rich samples the conductivity dropped under reducing conditions. The thermal expansion coefficients of the materials are in the order of 11×10^{-6} K^{-1} and fit to the TEC of other used cell components. Samples containing iron showed relative large changes in the expansion coefficient when switching the atmosphere. This could lead to stresses and cracks between fabrication and operation of an SOFC, therefore iron-rich materials are not useable. Long-term stability and catalytic properties have still to be investigated.

5 Acknowledgements

The authors gratefully acknowledge the financial support of the European Commission as this work was carried out as part of the TMR-network programme "Synthesis, preparation and fabrication of alternative anodes for direct methane oxidation in SOFC" (contract-no. FMRX-CT97-0130).

6 References

[1] P. R. Slater and J. T. S. Irvine, in Proc. 3rd European SOFC Forum (Nantes, France, June 2-5 1998), Ed. P. Stevens, European Fuel Cell Forum 1998, pp.417.
[2] D. P. Fagg, S. M. Fray and J. T. S. Irvine, Solid State Ionics 1994, 72, 235-239.
[3] R. Moos and K. H. Härdtl, J. Appl. Phys. 1995, 80[1], 393-400.
[4] J. T. S. Irvine, D. P. Fagg, J. Labrincha and F. M. B. Marques, Catalysis Today 1997, 38, 467-472.
[5] P. Vernoux, J. Guindet, E. Gehain and M. Kleitz, in Solid Oxide Fuel Cells V (Aachen, Germany, June 2-5 1997), Eds. U. Stimming, S. C. Singhal, H. Tagawa and W. Lehnert, The Electrochemical Society Proceedings Series PV 97-40 (1997) 219.
[6] Pechini, M. P. US Patent 3 330 697, 1967.
[7] I. Yasuda and M. Hishinuma, Solid State Ionics 1995, 80, 141.
[8] P. H. Larsen, P. V. Hendriksen and M. Mogensen, J. Thermal Analysis 1997, 49, 1263.

Development of Thin-Film SOFC for Reduced Operating Temperature

Günter Schiller, Rudolf Henne, Michael Lang, Simone Schaper
Deutsches Zentrum für Luft- und Raumfahrt (DLR), Stuttgart, Germany

1 Abstract

A novel planar concept for solid oxide fuel cells (SOFC) based on the plasma spray technology enables the fabrication of thin-film cells thus allowing SOFC operation at reduced temperatures below 800 °C. In this spray concept the membrane-electrode assembly (MEA) with thin electrolyte and electrode layers each in the range of 30-50 μm in thickness is built up in only one process step by depositing the different layers consecutively onto a porous metallic substrate. In the present paper the principle of the spray concept and the features of the applied vacuum plasma spray process are reported. The state of development of plasma sprayed thin-film cells by means of the electrochemical performance in the range of 750 - 900 °C is presented. The fabrication method has the potential to be transferred into a future mass production process for a cost-effective SOFC production.

2 Introduction

The reduction of the operating temperature of solid oxide fuel cells to below 800 °C is an important objective in current SOFC development worldwide in order to significantly reduce production costs and to improve long-term stability. A fundamental precondition for SOFC operation at an intermediate temperature of 650 - 800 °C is given by a sufficiently high ionic conductivity and minimized ohmic losses of the electrolyte. This can principally be realized in two ways: either through the use of alternative electrolyte materials to the conventionally used yttria-stabilized zirconia (YSZ) which exhibit enhanced ionic conductivity or through novel thin-film concepts when using zirconia as the electrolyte material.

The vacuum plasma spray technique (VPS) which has been further developed at the DLR Stuttgart to meet the specific requirements of SOFC fabrication [1] has the potential to not only fabricate thin electrolytes but also the entire membrane-electrode assembly (MEA) in one consecutive spray process. Based on this technology thin-film cells of a metallic substrate supported planar SOFC concept have been developed [2]. The utilization of plasma torches with specially developed Laval-like nozzles enables the consecutive deposition of thin and dense electrolyte layers as well as of controlled porous electrodes. The state of development of such entirely plasma sprayed thin-film cells is presented.

3 Cell Design

At present, the membrane-electrode assembly is conventionally fabricated by time-consuming multistep processes based on different sintering and printing techniques. But the plasma spray process represents an attractive alternative production technique due to its high flexibility and its ability to process ceramic and metallic materials very rapidly with only one method. Hence, the multilayer MEA can be fabricated continuously in one consecutive spray process without any interruption and cooling down. The principle of the DLR spray concept for a metallic substrate supported thin-film cell design is shown schematically in Fig. 1. In this planar SOFC design the electrolyte layer needs no longer to be the mechanically supporting component thus enabling a significantly reduced electrolyte thickness compared to conventional self-supporting cells. The spray process requires a substrate to be coated for which an open porous metallic structure such as a porous plate or a felt of approximately 1 mm in thickness is used. Onto this substrate which serves as a fuel gas distributor, the anode, the electrolyte and the cathode layer, each of 30 - 50 μm in thickness, are consecutively deposited by the VPS process in only one process step. In order to provide a low ohmic contact to the bipolar plate a porous and ductile contact layer is also needed.

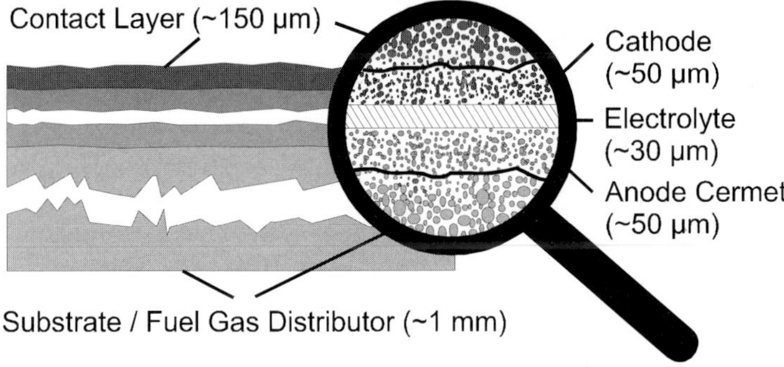

Figure 1. Principle of the VPS-adapted cell design

4 Cell Fabrication

The VPS technique and the equipment used at the DLR Stuttgart have been described in detail elsewhere [3]. Hence, only some special features of the DLR installation that are essential for the manufacture of SOFC components are emphasized. Novel plasma torches with Laval-like nozzle contours for supersonic plasma jet velocities of up to 2000 - 3000 m/s have been developed resulting in enhanced spray particle velocities of up to 800 - 900 m/s and improved spray conditions in order to achieve thin but dense electrolyte layers. The internal powder injection by several integrated powder injection ports at different positions along the nozzle allows the spraying of very different materials simultaneously, thus enabling complex coatings such as graded cermet layers exhibiting a desired material and porosity profile. Controlled porosity electrode layers can be obtained by carefully adjusting the torch nozzle, the powder injection (internally or externally), the grain size fraction of the spray powder and the spray parameters. More details on the plasma spray process applied for the fabrication of SOFC components are given in [4].

The metallographic cross-section of an entirely plasma sprayed thin-film cell is shown in Fig. 2. The three-layer compound consists of a porous zirconia/Ni cermet anode, a dense zirconia electrolyte and a porous LSM/zirconia (LSM: lanthanum strontium manganite) mixed cathode which were consecutively deposited.

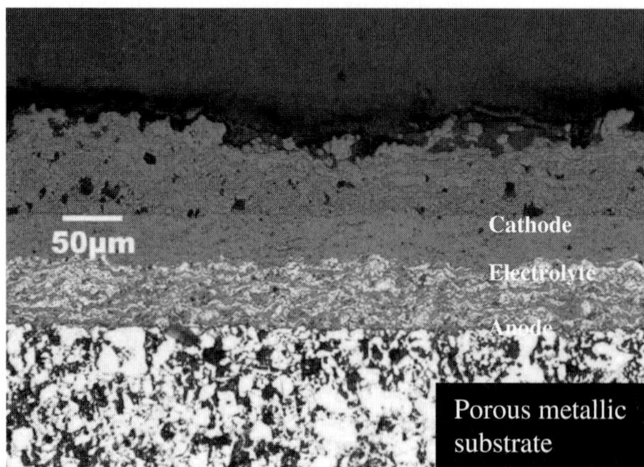

Figure 2. Metallographic cross-section of an entirely plasma sprayed thin-film cell

Dense and gastight electrolyte layers of high melting zirconia, even with a thickness in the range of only 30 -50 μm, are obtained by taking advantage of the DLR high-velocity torches with controlled expansion of the plasma providing enhanced kinetic energy and improved melting conditions for the spray particles. For an optimum cermet anode layer the zirconia and NiO powders were injected separately into the plasma which means the internal injection of zirconia and the external injection of NiO. At the SOFC operating temperature the NiO is then completely reduced by hydrogen to pure Ni. A mixed or graded microstructure for the cathode was also prepared by the internal injection of zirconia and the external injection of LSM. The cells prepared for the electrochemical characterization are circular with an area of 10 cm^2 but the fabrication process can be scaled up to produce cells with an area of up to 1000 cm^2.

5 Cell Performance

In order to evaluate the electrochemical performance of entirely plasma sprayed cells I-V characteristics and impedance spectroscopy measurements were performed in the temperature range 750 - 900 °C. The I-V characteristics of a plasma sprayed cell using scandia-stabilized zirconia (ScSZ) for the electrolyte and as an addition to the anode and the mixed cathode are shown in Figs. 3 and 4. At 900 °C power densities of 1280 mW/cm^2 at a voltage of 0.7 V for H_2 and O_2 as the operating gases and 770 mW/cm^2 for operation with H_2 and air were measured (Fig. 3). The reduction of the power density with decreasing operating temperature can be seen from Fig. 4 for H_2 and air as operating gases for the temperature range 750 - 900 °C.

At 750 °C and 800 °C reasonably high power densities of 290 mW/cm^2 and 430 mW/cm^2 respectively were achieved.

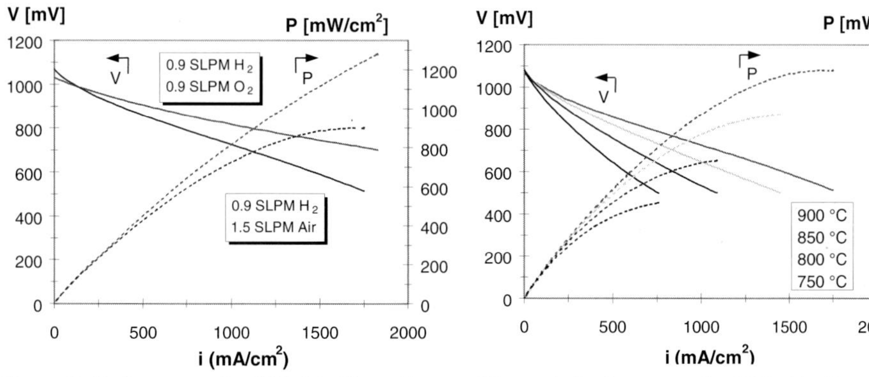

Figure 3. Performance of a VPS thin-film cell at 900 °C and H_2/O_2 and H_2/air as the operating gases

Figure 4. Performance of a VPS thin-film cell with H_2/air as the operating gases as a function of temperature

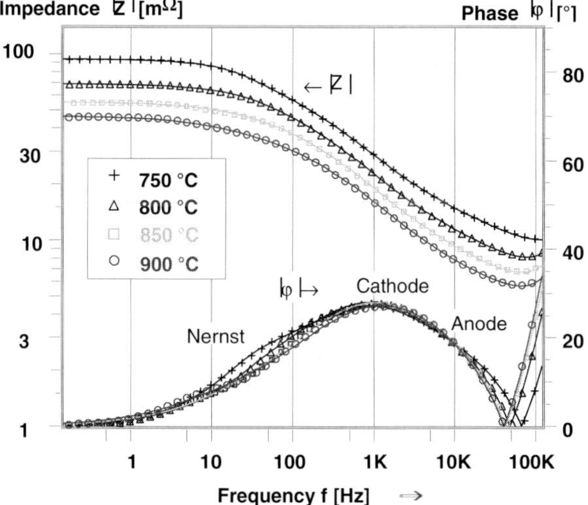

Figure 5. Impedance spectra (Bode plot) of a VPS thin-film cell with H_2/air as the operating gases at 200 mA/cm^2 as a function of temperature

Impedance spectroscopy measurements (Fig. 5) performed at a current density of 200 mA/cm^2 revealed very low ohmic area specific resistances of 0.04 - 0.08 Ωcm^2 resulting in also low total area specific resistances between 0.42 Ωcm^2 at 900 °C and 0.85 Ωcm^2 at 750 °C.

6 Conclusion

The concept of a metallic substrate supported thin-film SOFC and the fabrication of the cells by a vacuum plasma spray process has been demonstrated. The electrochemical characterization revealed good cell performance with power densities of 300 - 400 mW/cm^2 for operation with H_2/air at reduced operating temperatures of 750 - 800 °C. The further development work will focus on further improving the layers' microstructures and thus the electrochemical performance and on scaling up the cells to a size of 200 x 200 mm^2. Stacks with several cells of this size will be assembled and investigated in short-term and long-term operation in order to obtain stack performance and durability results. The VPS technology has a high potential for a continuous mass production process by using several plasma torches within a multi-chamber installation.

7 References

1. G. Schiller, R. Henne, M. Lang, Proc. Third European Solid Oxide Fuel Cell Forum 1996, 1, 123-132
2. G. Schiller, M. Fischer, R. Henne, M. Lang, Proc. Fuel Cell Seminar 1998, 515-518
3. G. Schiller, R. Henne, V. Borck, J. Therm. Spray Techn. 1995, 4 (2), 185-194
4. G. Schiller, M. Müller, R. Ruckdäschel, R. Henne, M. Lang in Thermal Spray: A United Forum for Scientific and Technological Advances (Ed. C. C. Berndt), ASM International, Materials Park, OH, USA, 1997, 27-34

Reactivity and Interdiffusion of Alternative SOFC Cathodes with Yttria Stabilized Zirconia, Gadolinia Doped Ceria and Doped Lanthanum Gallate Solid Electrolytes

G. Ch. Kostogloudis, G. Tsiniarakis, F. Riza and Ch. Ftikos
National Technical University of Athens, Greece

1 Abstract

The chemical compatibility between the cathode composition $Pr_{0.8}Sr_{0.2}Co_{0.2}Fe_{0.8}O_{3-\delta}$ and the electrolyte compositions yttria stabilized zirconia (YSZ), $Ce_{0.8}Gd_{0.2}O_{1.9}$ (CGO) and $La_{0.8}Sr_{0.2}Ga_{0.9}Mg_{0.1}O_{3-\delta}$ (LSGM) was investigated. Also, the influence of the substitution of Al for Fe on the reactivity of the cathode with YSZ was examined. All oxides were single-phase materials except for LSGM, which contained two additional phases, namely $LaSrGa_3O_7$ and $LaSrGaO_4$. Two types of experiments were performed: (a) reactivity experiments by XRD in cathode/electrolyte powder mixtures and (b) diffusion experiments by SEM/EDX analysis in cathode/electrolyte double-layer pellets. $Pr_2Zr_2O_7$, $SrZrO_3$ and $CoFe_2O_4$ were formed by the interaction of the cathode materials with YSZ. Substitution by Al at the B-site of the perovskite cathode led to a decrease of its reactivity with YSZ. No reaction products were formed for powder mixtures of $Pr_{0.8}Sr_{0.2}Co_{0.2}Fe_{0.8}O_{3-\delta}$ and CGO or LSGM electrolytes. High Co and Fe diffusion into LSGM was identified. Pr, La and Ga show a smaller tendency for diffusion. The diffusion of transition metal cations into LSGM electrolyte caused the destabilisation and disappearance of the second phases in the interdiffusion zone.

2 Introduction

The solid oxide fuel cell (SOFC) is considered as one of the most promising power generation systems, due to its high efficiency, very low pollutant emissions, low noise, potential for cogeneration and modular construction to suit load [1]. Its high temperature operation (800-1000°C), however, creates problems associated with the lack of economic balance-of-plant equipment or reliable interconnect materials [2]. This has intensified the need for operating the cell at intermediate temperatures (around 700°C). The yttria-stabilized zirconia (YSZ) solid electrolyte, which is the state-of-the-art electrolyte at high temperatures, can be used at 700°C, provided that it is manufactured as supported thick film. Other electrolytes, with higher ionic conductivity at intermediate temperatures, have also been considered, the most promising being gadolinia doped ceria and doped lanthanum gallate. The cathode material which is usually used for intermediate temperature SOFCs is based on the $La_{1-x}Sr_xCo_{1-y}Fe_yO_{3-\delta}$ perovskite oxide system, which exhibits mixed conductivity. However, it has been reported [3] that the use of Pr instead of La decreases cathode overpotentials and enables higher catalytic activities

An issue of significant importance for the successful operation of the cell is the chemical compatibility between the cathode and the electrolyte. It should be ensured that no reaction products are formed in the cathode/electrolyte interface, since this could prove detrimental for the efficiency of the cell. Furthermore, cation interdiffusion, if occurs, should not be pronounced, as it may negatively alter the properties of the materials involved.

In the present study, the chemical compatibility between the cathode compositions $Pr_{0.8}Sr_{0.2}Co_{0.2}Fe_{0.8}O_{3-\delta}$, $Pr_{0.9}Sr_{0.1}Co_{0.2}Fe_{0.8-y}Al_yO_{3-\delta}$ (y=0, 0.2, 0.4) and $Pr_{1-x}Sr_xCo_{0.2}Fe_{0.6}Al_{0.2}O_{3-\delta}$, (x=0.1, 0.2, 0.3) and the electrolyte compositions YSZ, $Ce_{0.8}Gd_{0.2}O_{1.9}$ (CGO) and $La_{0.8}Sr_{0.2}Ga_{0.9}Mg_{0.1}O_{3-\delta}$ (LSGM) was investigated by XRD and SEM/EDX.

3 Experimental

The powders of the cathode oxides were prepared by the amorphous citrate process and calcined at 1100°C for 15 h. The electrolytes used were commercial YSZ (Alfa) and LSGM (Praxair) powders, as well as CGO powder prepared in house by spray drying and calcined at 900°C for 5 h. The formation of single-phase oxides was identified by XRD in all cases, except for LSGM electrolyte, in which two additional phases were detected, namely, $LaSrGa_3O_7$ and $LaSrGaO_4$, with the former appeared in higher proportion.

For the chemical compatibility studies two types of experiments were performed: (a) reactivity experiments by XRD in cathode/electrolyte powder mixtures and (b) diffusion experiments by Scanning Electron Microscopy/Energy Dispersive X-ray (SEM/EDX) analysis in cathode/electrolyte double-layer pellets. The samples for the reactivity tests were prepared by thoroughly mixing equimolar amounts of cathode and electrolyte powders in an agate mortar, using acetone. The mixtures were compacted (200 MPa) into pellets, which were annealed in air at 1300°C for 100 h or at 1100°C for 120 h, followed by XRD to identify the formation of any reaction products. In some cases the samples were mixed with Si-powder, which served as an internal standard for d-value calibration, in order to identify a possible shift in the peaks of the electrolyte.

The double-layer cathode/electrolyte pellets for the diffusion experiments were prepared as described in [4]. These pellets were annealed in air at 1300°C for 300 h, in order to accelerate the effect of the diffusion process, and to identify which elements tend to diffuse and to what relative extend. The pellets were subsequently mounted in epoxy resin, cut perpendicularly, so that the interface was revealed, and polished finely. The cross sections of the cathode/electrolyte interfaces were coated with a thin film of carbon to provide electrical conductivity and studied by SEM/EDX (JEOL JSM-T300). Quantitative EDX measurements were performed at points across the interface.

4 Results and Discussion

4.1 Powder X-Ray Diffraction

Figure 1 shows the XRD patterns of YSZ, $Pr_{0.8}Sr_{0.2}Co_{0.2}Fe_{0.8}O_{3-\delta}$, and their mixture after annealing at 1300°C for 100 h. The main reaction product formed was $Pr_2Zr_2O_7$. $CoFe_2O_4$ and $SrZrO_3$ were also formed in small quantities. The main peak of YSZ was increased, probably due to diffusion of Sr^{2+} and/or Pr^{3+} cations from the cathode to the electrolyte. For CGO and LSGM electrolytes and the above cathode material, no reaction products could be detected by XRD after annealing. For LSGM electrolyte (Figure 2), the $LaSrGa_3O_7$ and $LaSrGaO_4$ second phases, which pre-existed in the electrolyte, were destabilised by the interaction with the cathode oxide, as confirmed by SEM/EDX measurements, discussed in the next section. The main peak of CGO and LSGM was found to exhibit a small decrease, which denotes that the diffusion of Co^{3+} and Fe^{3+} cations prevails.

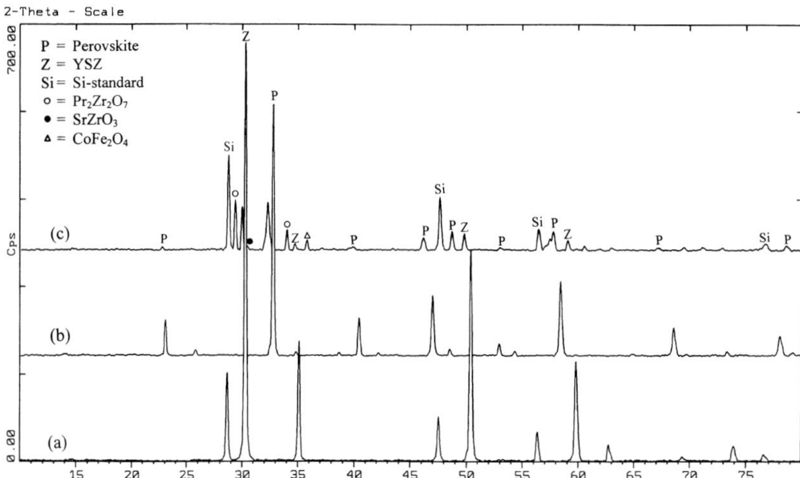

Figure 1. X-ray diffraction patterns of (a) YSZ electrolyte, (b) $Pr_{0.8}Sr_{0.2}Co_{0.2}Fe_{0.8}O_{3-\delta}$ cathode and (c) equimolar $Pr_{0.8}Sr_{0.2}Co_{0.2}Fe_{0.8}O_{3-\delta}$/YSZ mixture after annealing in air at 1300°C for 100 h.

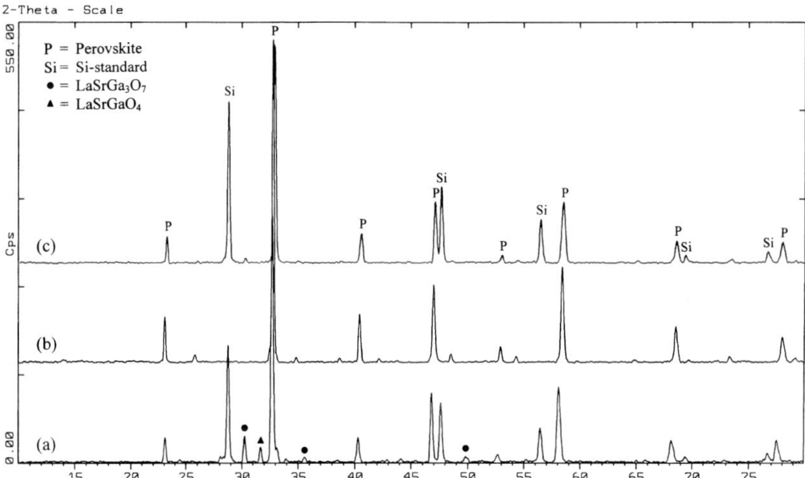

Figure 2. X-ray diffraction patterns of (a) LSGM electrolyte, (b) $Pr_{0.8}Sr_{0.2}Co_{0.2}Fe_{0.8}O_{3-\delta}$ cathode and (c) equimolar $Pr_{0.8}Sr_{0.2}Co_{0.2}Fe_{0.8}O_{3-\delta}$/LSGM mixture after annealing in air at 1300°C for 100 h.

Figure 3 shows the XRD patterns of the powder mixtures of $Pr_{0.9}Sr_{0.1}Co_{0.2}Fe_{0.8-y}Al_yO_{3-\delta}$ (y=0, 0.2, 0.4) cathodes and YSZ electrolyte, after annealing at 1100°C for 120 h. The XRD patterns of the powder mixtures of $Pr_{1-x}Sr_xCo_{0.2}Fe_{0.6}Al_{0.2}O_{3-\delta}$ (x=0.1, 0.2, 0.3) and YSZ, annealed at the same conditions, are shown in Figure 4. For both series, the main reaction products were $Pr_2Zr_2O_7$ and $SrZrO_3$, while $CoFe_2O_4$ was also formed in small amounts. The results of the XRD analysis of these mixtures are shown in Table 1. The given values are $I_{product}/I_{total}$ (%), and represent a semi-quantitative estimate of the amount of each product formed. For $Pr_{0.9}Sr_{0.1}Co_{0.2}Fe_{0.8-y}Al_yO_{3-\delta}$, the amount of $Pr_2Zr_2O_7$ decreases and that of $SrZrO_3$ does not change significantly, when the Al content increases. The increase of Sr content in $Pr_{1-x}Sr_xCo_{0.2}Fe_{0.6}Al_{0.2}O_{3-\delta}$ brings about a decrease of the amount of $Pr_2Zr_2O_7$ and an increase of the amount of $SrZrO_3$, in agreement with previous studies [5].

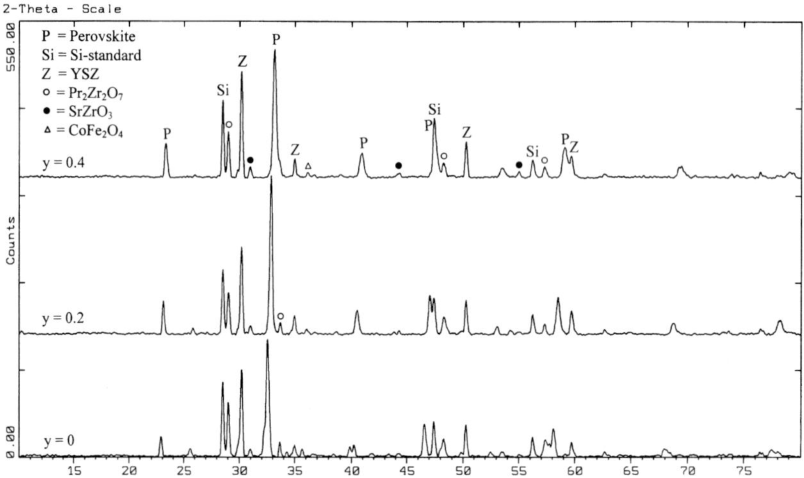

Figure 3. X-ray diffraction patterns of equimolar $Pr_{0.9}Sr_{0.1}Co_{0.2}Fe_{0.8-y}Al_yO_{3-\delta}$/YSZ powder mixtures after annealing in air at 1100°C for 120 h.

Table 1. Results of semi-quantitative XRD analysis of equimolar perovskite/YSZ powder mixtures after annealing at 1100°C for 120h. The given values represent the ratio of product intensity to total intensity (sum of intensities of all phases present), $I_{product}/I_{total}$ (%), estimated by the areas of their major peaks in the XRD patterns

Perovskite composition	$Pr_2Zr_2O_7$	$SrZrO_3$
$Pr_{0.9}Sr_{0.1}Co_{0.2}Fe_{0.8-y}Al_yO_{3-\delta}$		
y=0	15.7	2.6
y=0.2	13.3	3.4
y=0.4	12.3	3.2
$Pr_{1-x}Sr_xCo_{0.2}Fe_{0.6}Al_{0.2}O_{3-\delta}$		
x=0.1	13.3	3.4
x=0.2	9.7	3.3
x=0.3	6.6	9.7

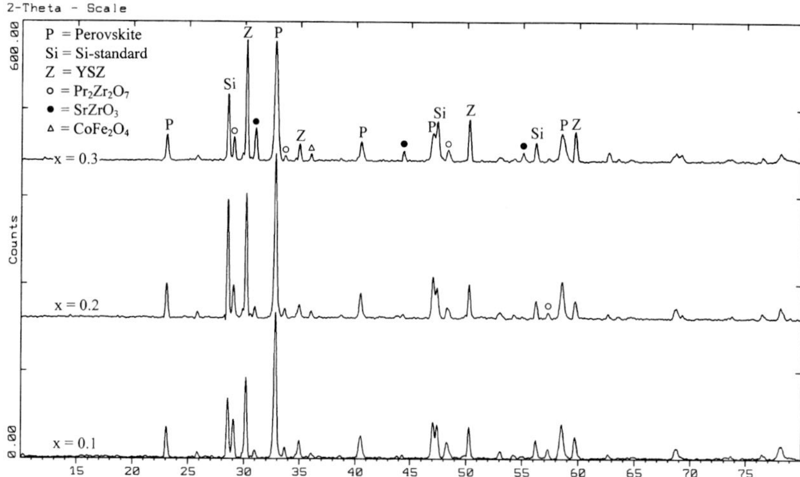

Figure 4. X-ray diffraction patterns of equimolar $Pr_{1-x}Sr_xCo_{0.2}Fe_{0.6}Al_{0.2}O_{3-\delta}$/YSZ powder mixtures after annealing in air at 1100°C for 120 h.

4.2 SEM/EDX analysis

A SEM micrograph of the cross section of $Pr_{0.8}Sr_{0.2}Co_{0.2}Fe_{0.8}O_{3-\delta}$/LSGM interface is shown in Figure 5a. The results of EDX quantitative measurements at points across this interface, expressed as g-at of metal per mol of oxide, are shown below the SEM micrograph (Figure 5b). Care was taken during point measurements in the electrolyte, that the points do not include pores or spots of the second phase. The width of the SEM micrograph, and the diffusion profile diagram below it, is the same, in order to aid the comparison between them. Regions I, II, III and IV, determined by EDX measurements, are indicated in the micrograph.

By examining Figure 5, the following conclusions can be drawn: High Co and Fe diffusion into the electrolyte was identified. Pr, La and Ga show a smaller tendency for diffusion. The perovskite parent phase of the LSGM electrolyte cannot tolerate the 20 mol% Sr content of the nominal composition ($La_{0.8}Sr_{0.2}Ga_{0.9}Mg_{0.1}O_{3-\delta}$), and half of it is depleted to form the second phase $LaSrGa_3O_7$, while the perovskite phase that remains has a composition containing 10 mol% Sr, which seemed to be stable (Region IV). The incorporation, through the diffusion process, of transition metal cations in the lattice of the perovskite parent phase of the electrolyte has as a result its stabilisation and the enhancement of its ability to accept more Sr concentration. Thus, the perovskite structure of the electrolyte can in this case tolerate the amount of 20 mol% Sr. This leads to the destabilisation of the Sr-rich $LaSrGa_3O_7$ second phase, and to its gradual disappearance in the interdiffusion zone (Region III).

It has been shown [6] that the substitution of up to 10 mol% Co or Fe for Ga is beneficial, as it increases the ionic conductivity of LSGM. Higher concentrations, though, give rise to an increase of the electronic p-type conductivity, which is disadvantageous, as it will cause the short circuit of the cell. It is, therefore, important to clarify up to what distance can Co and Fe diffuse inside LSGM, and which maximum concentrations they can reach in its lattice. Thus, further annealing experiments are needed for much longer periods and at lower temperatures to simulate actual cell operation.

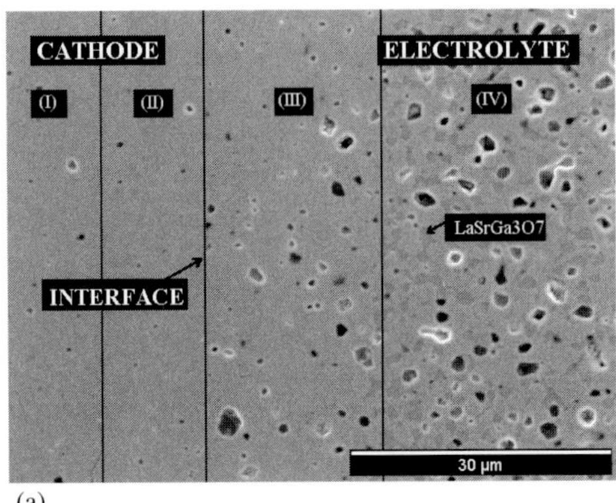

(a)

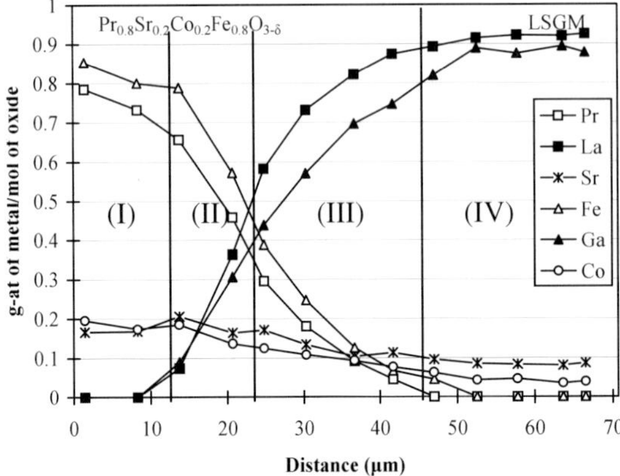

(b)

Figure 5. (a) SEM micrograph (×750) of the cross section of $Pr_{0.8}Sr_{0.2}Co_{0.2}Fe_{0.8}O_{3-\delta}$/LSGM interface after annealing at 1300°C for 300 h, and (b) results of the EDX measurements at points across the interface. Region (I): Bulk of the cathode. Regions (II), (III): Interdiffusion zone on the cathode and electrolyte side, respectively. Region (IV): Bulk of the electrolyte.

5 References

1. N. Q. Minh, J. Am. Ceram. Soc. 1993, 76, 563.
2. B. C. H. Steele in Proc. 14th Risø Intern. Symp. on Materials Science: High Temperature Electrochemical Behaviour of Fast Ion and Mixed Conductors (Ed.: F.W. Poulsen, J.J. Bentzen, T. Jacobsen, E. Skou and M.J.L. Østergård), Risø National Laboratory, Roskilde, Denmark, 1993, p. 423.
3. T. Ishihara, T. Kudo, H. Matsuda and Y. Takita, J. Electrochem. Soc. 142, 1995, 1519.
4. G. Stochniol, PhD Thesis, Forschungszentrum Jülich, 1996.
5. G.Ch. Kostogloudis, G. Tsiniarakis and Ch. Ftikos, Solid State Ionics, to be published.
6. N. Trofimenko and H. Ullmann, Solid State Ionics 1999, 118, 215.

Development of New Bi-Polar Plates Based on Electrically Conductive Filled Polymers for PEMFC

F. JOUSSE, J.F. SALAS, F. GIROUD
C.E.A. / Le Ripault, Monts, France
B. ICARD, J.Y. LAURENT, P. SERRE COMBE
C.E.A., Grenoble, France

1 Introduction

In Polymer Electrolyte Membrane Fuel Cell technology, the bi-polar plates are dedicated to:

- the current collection,
- the separation and distribution of gas (hydrogen and oxygen) at the cathode and the anode.

To achieve these functions, bi-polar plate materials must satisfy the following properties:

- high conductivity (higher than 10 S/cm),
- high chemical resistance to acid and water,
- very low permeability to hydrogen (permeability < Pe^{H2}_{Nafion} (20°C) = 7.10^{-17} m²/Pa/s).

Traditionally bi-polar plates have been designed with stainless steel or graphite. However, the cost of these plates are incompatible to transport applications, principally because of the gas channel machining step. Recently, we have noticed the work of T.M. Besmann [1] on the manufacturing of bi-polar plates based on carbon fibres and phenolic resin, processed by pyrolisis and densification on surface by a Chemical Vapour Infiltration process. However, this kind of process seems too expensive and complex for the needs of the road electric transportation industry.

Organic composites based on conductive chemical resistant fillers and processed by molding could be an alternative solution. Bi-polar plates requirements can be achieved by controlling and optimising experimental parameters such as the nature and morphology of fillers, the resin characteristics, and the process conditions. To avoid corrosion of the composite material, and then, the contamination of the cell, we have selected non metallic fillers, based on graphite or carbon black.

2 Results and Discussion:

2.1 Conductive filled composite properties

Organic conposites with high levels of conductivity require filling contents above the percolation threshold. For such composites, the level of conductivity is controlled by the nature of the electrical resistance between connected conductive particles [2, 3].

Firstly, the resistance between connected conductive particles depends on the surface area of the contact. This resistance is called constriction resistance and is defined as: $R_{cr} = \rho_i / d$, where ρ_i is the resistivity of the filler and d the diameter of the contact area. Thus, the contact surface depends on the geometry and on the viscoelastic properties of the particles (i.e. their ability to deform themselves).

The main contact resistance contribution is the tunnelling resistance [4]. This resistance is due to the formation, during the dispersion process, of an insulative coating of polymer at the surface of the conductive particles. This layer may be composed of binder additives adsorbed at the surface of the filler, or principally of a thin film of matrix resin. In the case of tunnelling resistance, the conduction mechanism is non ohmic, but performed by electronic jumps between insulated particles. The tunnelling resistance is defined by the relation: $R_t = \rho_t / a$, where ρ is the tunnelling resistivity and a the contact surface area of the particles.

The macroscopic resistance is the sum of these two contributions. Generally, the tunnelling resistance controls the macroscopic conductivity of heterogeneous middles. For example, the tunnelling resistance increases from 10^{-8} to 10^{3} Ω/cm^2 [3] when the thickness of the insulative film at the surface of the conductive particles grows from 0.5 to 12 nm. This thickness of a few nanometers corresponds to a polymeric layer adsorbed at the surface of fillers during the dispersion process.

Generally, despite filling contents much higher than the percolation threshold, the conductivity levels remain low for composites processed by extrusion and injection molding. These techniques induce high shearing stresses in the molten resin during the dispersion step, and may separate connected particles with a thin insulative coat of polymer. Moreover, during the injection molding step, capillary and thermal effects induce the formation of an insulative polymeric skin on the composite surface. Thus, as shown in the table 1, the surface conductivity of commercial composites are much weaker than volume conductivity.

Table 1: Electrical conductivity of extruded and injected composites devoted to electromagnetic shielding (Plastics Digest 1997 published by Data Business Publishing).

References	Composition	Volume conductivity (S/cm)	Surface conductivity (S/cm)
PRE ELEC TP4459	PP/Carbon black (70/30)	0.1	0.01
EMI2583	ABS-PC/graphite (70/30)	1	0.01

To improve the contact surface area between conductive particles, and then the macroscopic conductivity of composites, we have realised specific material structure by optimising the morphology of fillers and the mixing and molding processes. The SEM picture in Figure 1 shows the structure obtained in a compound based on polyamide 6 filled with carbon black.

The segregation of carbon black particules from thermoplastic resin induces a continuous conductive network (in white in the picture). The conductivity levels obtained with this technique are much higher than for directly injected composites (CEA material characteristics - table 2).

Moreover, conductive networks may be obtained with very high filling contents. In this case, the composite continuous bulk is conductive and composed with graphite. The thermoplastic binder improves the mechanical properties and the permeability of the composite (material CEA 2 and 3 - table 2). For best results, the morphology of graphite powder must be adapted.

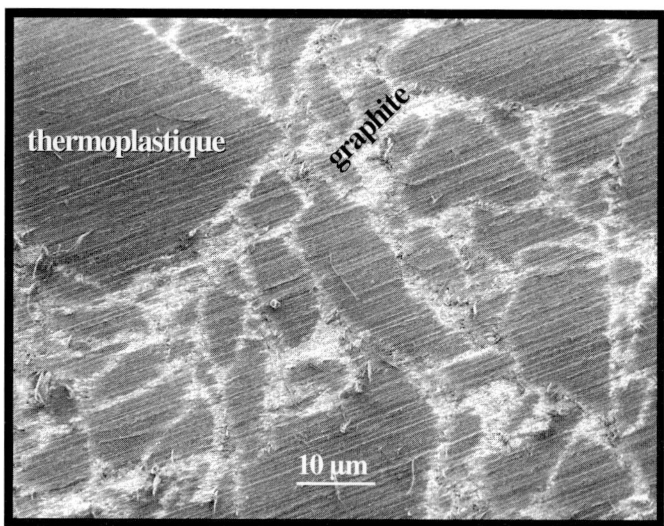

Figure 1: SEM image of the structure of a CEA material based on carbon black dispersed in Nylon 6 matrix.

Table 2: Electric conductivity of filled materials, performed in the CEA' laboratory.

Reference	Composition	Volume conductivity (S/cm)	Surface conductivity (S/cm)
CEA 1	Nylon / Carbon black (80/20)	20	30
CEA 2	Epoxy / graphite 1 (30/70)	10	>100
CEA 3	Epoxy / graphite 2 (30/70)	30	100

2.2 Electrochemical tests

Endurance tests on composite bi-polar plates in an experimental stack have demonstrated a good chemical resistance of the plates in the PEMFC environment, and a good level of conductivity.

Two plates, based on epoxy resin filled with graphite, G1 (σ_{vol} = 20 S/cm, σ_{surf} = 100 S/cm) and G2 (σ_{vol} = 10 S/cm, σ_{surf} = 150 S/cm) have been tested. The electrochemical results of these molded plates are compared to a reference machined carbon plate CGWX provided by SGL Carbon.

The 50 cm² active PEMFC stack is composed with two electrodes composed of a carbon felt filled with Pt (20% Pt/C) and a membrane of Nafion® 115 (see figure 2).

Test conditions:

Firstly, the electrolyte is humidified during 8 hours, the cell working with a 30A current provided by a 20A electronic charge. The cell temperature is set at 70°C and the (hydrogen, oxygen) dry gas pressure at 4 bars (absolute). The gas flow rates are adjusted at 4 nl/h for the cathode and 1.3 nl/h for the anode.

Three combinations have been tested:

- n°1: PM1=PM2=graphite CGWX
- n°2: PM1=CGWX / PM2=G1
- n°3: PM1=G1 / PM2=G2.

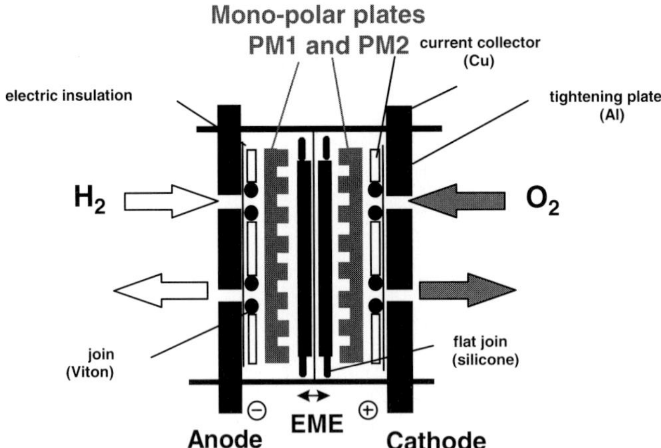

Figure 2 : Description of the electochemical cell.

Results:

The system n°3 has been tested during 400 hours with the experimental conditions described previously. The behaviour of the stack voltage is shown in figure 3.

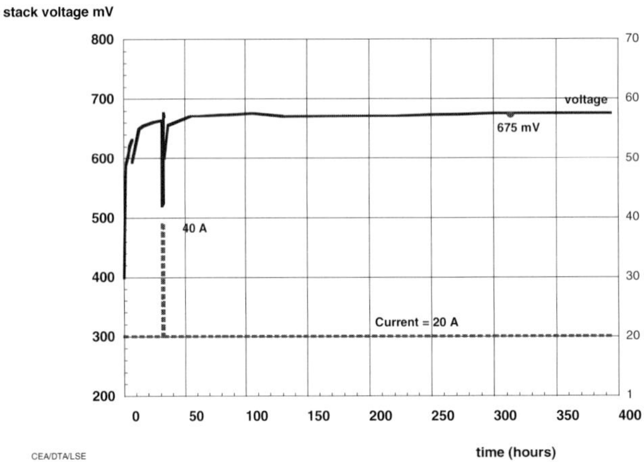

Figure 3. : Voltage (mV) versus time for system n°3 at I = 20A

This experiment demonstrates a good stability of the stack voltage during 400 hours for an 20A current. No contamination caused by the degradation of components of the stack, and particularly of the composite bi-polar plates, has been detected during this endurance test.

The U=f(I) characteristics of the three systems have been studied in similar conditions [6]. The results are described in figure 4.

From 0.1 to 1.4 A/cm², the U=f(I) characteristics are linear, with different slopes for each system. The slope increase between system 1 (PM1=PM2=graphite CGWX) and system 3 containing two composite bi-polar plates, is due to an increase of the ohmic loss of the stack. This result is confirmed by the measures of the stack resistances (table 3).

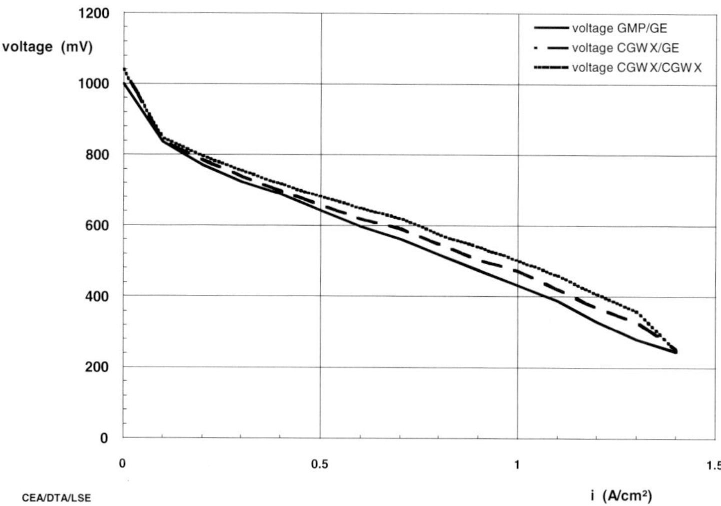

Figure 4: U=f(I) of systems 1,2 and 3.

Table 3: Ohmic losses measured with different materials for bi-polar plates.

System	Stack resistance (mΩ)	Ohmic loss (mV) at 1 A/cm²
n°1	4.6	230
n°2	5.3	265
n°3	5.7	285

The curves U=f(I) allow us to determine the voltage losses of stacks using composite bi-polar plates, in comparaison to the reference stack n°1. Thus, for a current of 1 A/cm², the stack N°2 with one composite bi-polar plate, loses 35mV compared to the reference stack N°1, and the stack N°3 with two composite bi-polar plates, loses 55 mV. These losses remain weak, and are insignificant for low current densities.

3 Conclusion

The interest of bi-polar plates based on organic composite, filled with non metallic graphite particles, is demonstrated. Experimental parameters such as filler morphology, filling contents, mixing and molding processes, allow to perform composites with bulk and surface conductivities higher than 10 S/cm.

Thanks to a judicious choice of the chemical nature of the polymeric resin and of the conductive fillers, the first 400 hours endurance experiments have demonstrated the chemical stability and the low ohmic losses of these molded composite plates, compared with traditional expensive machined graphite plates.

4 References:

1. T.M. BESMANN, PEM Fuel Cell created from light electrode material, High Tech Material Alert, vol. 15, n°10 - October 1998
2. MORI and al., Conductivity of copper with and without surface treatment, Powder Technology, 59 (1989) 191-197
3. F.GIROUD, Thèse de doctorat «Conception de matériaux composites à charges fibrillaires dispersées - Etudes de leurs propriétés radioélectriques ». Université Claude Bernard Lyon I
4. G.R. RUSHAU and al., Resistivities of conductive composites. Journal of Applied Physics, 73 3 (1992) 953-959
5. H. TANG and al., Studies on electrical conductivity of carbon black filled polymers, Journal of Applied Polymer Science, 59 (1996) 383-387
6. L. GERBAUX, Thèse de doctorat « Modélisation d'un pile à combustible hydrogène-air et validation expérimentale » Institut National Polytechnique de Grenoble

Properties of Nickel Mesh as a Methane Steam Reforming Catalyst and its Application in SOFCs

U. Flesch, R. Dahl, R. Peters, D. Stöver
Institute for Materials and Processing in Energy Systems (IWV), Research Centre Jülich (Forschungszentrum Jülich), D-52425 Jülich, Germany

1 Introduction

An essential prerequisite for high current densities in operating the planar solid oxide fuel cell (SOFC) is a good electrical contact between the electrodes and the interconnect (1.4742 steel). For this purpose, a wire mesh of pure nickel is used on the anode side. At the centre of this investigation is the question of the activity for methane steam reforming of the nickel mesh as a contact layer in comparison with the substrate anode (Ni/8YSZ).

At the Research Centre Jülich (FZJ), a planar electrode-supported design (Fig. 1) has been developed with the aim of realizing low electrical resistance by means of a thin electrolyte film and mechanical stability by a thick substrate anode. To this end, the electrolyte thickness was reduced from 150 – 200 μm to 15 – 20 μm and, at the same time, the anode thicknesses were increased from 50 – 100 μm to 1000 – 1500 μm. The components of the SOFC are a) 8YSZ (8 mole % Y_2O_3-stabilized ZrO_2) for the electrolyte, b) Ni/8YSZ cermet with 40 vol.% nickel for the anode, c) $La_{0.65}Sr_{0.3}MnO_3$ (LSM) for the cathode, and d) 1.4742 steel for the interconnect. Contact is made on the anode side with a fine-meshed nickel mesh and on the cathode side with $LaCoO_3$ paste. The fuel gas for the SOFC (methane and steam) is first catalytically reformed to H_2 and CO/CO_2 ($CH_4 + 2\ H_2O \Leftrightarrow 4\ H_2 + CO_2$ and $CH_4 + H_2O \Leftrightarrow 3\ H_2 + CO$) at the nickel surface of the anode substrate and then oxidized to H_2O and CO_2 at the electrolyte/anode boundary layer. An important function of intensive endothermic methane steam reforming is stack cooling. However, the high catalytic activity of nickel and the substrate thickness of 0.5 – 1.5 mm lead to high methane conversion rates near the gas inlet. These in turn cause large temperature gradients [2, 3] producing stresses which may lead to cracks in the stack. The nickel mesh will modify the methane steam reforming activity of the substrate in such a way that the temperature gradients are even further increased in the stack.

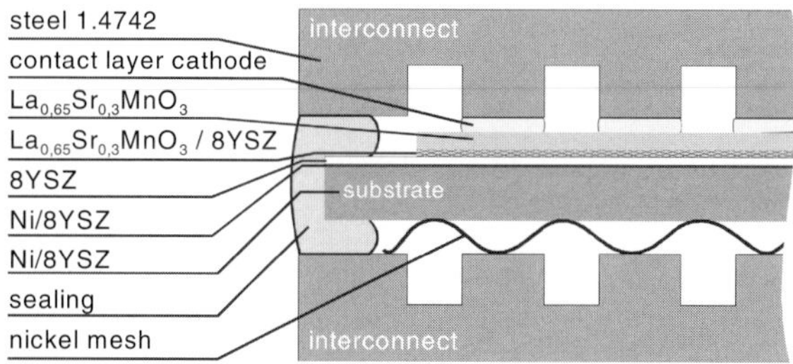

Figure 1. Schematic SOFC stack cross-section under operation conditions.

2 Experimental

The nickel mesh used (w_{Ni} = 99.8 wt %) has a mesh width of 0.20 mm and a wire size of 0.125 mm and is in direct contact with the interconnect (1.4742 steel) and the NiO/8YSZ substrate anode material and with Ni/8YSZ after stack reduction (Fig. 1). The nickel mesh is pretreated with the same process parameters as in stack assembly (sealing).

Table 1. Process parameter of the examined nickel meshes.

sample	oxidation [in air]	reduction [5 vol.% H_2 in N_2]	methane steam reforming
A	-----	-----	-----
B	2 K / min to 1173 K 3 h at 1173 K 1 K / min to 298 K	-----	-----
C	2 K / min to 1173 K 3 h at 1173 K 1 K / min to 298 K	2 K / min to 873 K 3 h at 873 K 1 K / min to 1223 K 1 h at 1223 K 1 K / min at 1188 K	first at 1188 K, than between (CH_4 : H_2O, 1:3)
D	-----	2 K / min to 873 K 3 h at 873 K 1 K / min to 1223 K 1 h at 1223 K 1 K / min to 1188 K	at 1188 K (CH_4 : H_2O, 1:3)
E	0.3 K / min to 333 K 24h at 333 K 2 K / min to 1088 K 3 h at 1088 K 2 K / min to 1173 K 3 h at 1173 K 1 K / min to 298 K	2 K / min to 873 K 3 h at 873 K 1 K / min to 1223 K 1 h at 1223 K 1 K / min to 1188 K	stack test with H_2 + H_2O at ~ 1173 K autopsy investigation of real stack under operation conditions for comparison with samples A-D

For this purpose, the nickel mesh is kept at 1173 K for three hours in air. Subsequent stack reduction in N_2 / 5 vol. % H_2 precedes methane steam reforming. For comparison, the nickel mesh was also directly used for methane steam reforming in the as-received (untreated) form. The exact process parameters can be seen in Table 1. The nickel mesh samples are mounted in a tube furnace for methane steam reforming (11 mm diameter). The catalysis apparatus can be charged with different gases (O_2, H_2, N_2, H_2O, CH_4), subsequent gas analysis is carried out with a "BINOS" IR adsorption detector (NGA 2000 from Rosemount) for CO, CO_2, CH_4 and a gas chromatograph with heat conductivity detector (Hewlett Packard) for hydrogen. Reforming takes place directly at the nickel contact at temperatures of 873 K < T < 1273 K. In order to evaluate the catalytic activity, the methane conversion ξ_{CH4} is then calculated at the gas outlet from the gas composition from the carbon balance according to $\xi(CH_4) = [n(CO) + n(CO_2)] / [n(CH_4) + n(CO) + n(CO_2)] * 100\ \%$ (with n = amount of substance per period of time).

3 Results and Discussion

The catalysis of an untreated mesh did not initially show any conversion of methane in steam reforming at 1188 K. After approx. six hours, non-steady-state operating conditions occurred and conversion up to 35 % reached. An activation with hydrogen, i.e. a supply of up to 100 vol. % H_2, did not influence this behaviour. Steady-state conversion was then finally reached by increasing the operating temperature to 1223 K. The subsequent cooling temperature program provided the results between 1153 K and 1223 K as shown in Fig. 3-D1. The analogous heating program showed different results (Fig. 2-D2). SEM photographs showed roughening and pore formation on the nickel mesh surface due to methane steam reforming (Fig. 3-D). Since this process proceeds in a time-dependent manner causing the catalytic activity to increase, the difference between curves D1 and D2 (Fig. 2) is plausible. Moreover, it can be seen that below a temperature of 1173 K methane steam reforming discontinues and only starts again at this temperature.

If the nickel mesh is oxidized analogously to the sealing process, an approx. 10 μm NiO layer (Fig. 3-B) is formed (for temperature profile see Table 1), which is not mechanically as stable as metallic nickel. Cracks in the NiO layer can be seen in Fig.3-B, which were caused by mechanical deformation of the wires or by the forces in ceramometallic grinding. Subsequent reduction of the NiO layer leads to formation of pores (Fig. 3-C) due to the NiO $\Rightarrow$ Ni volume reduction and the formation of water. These pores with their very large surface and roughness greatly increase the catalytic function of the nickel mesh. It can be seen in Fig. 2-C1 and 2-C2 that methane conversion has doubled in comparison to the untreated mesh. Both cooling and heating program curves are congruent, which is indicative of a constant catalyst activity.

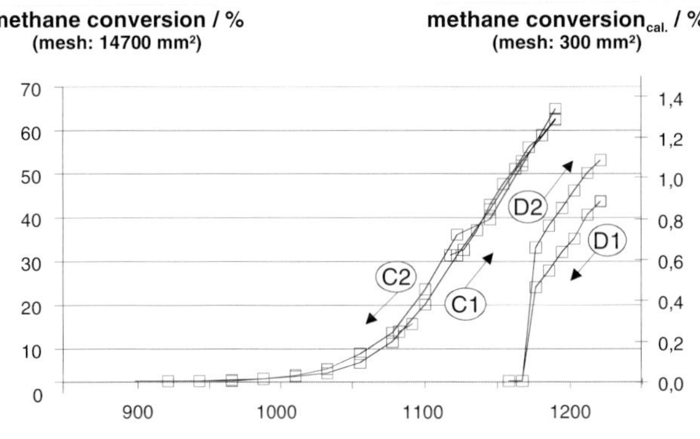

Figure 2. Results of methane steam reforming / methane conversion (31.85 L/h; CH_4:H_2O=1:3; 1.5 bar) of a nickel mesh with different pre-treatments: D1+D2 (untreated nickel mesh) and C1+C2 (oxidized and reduced nickel mesh).

Comparable standard cermets (30 x 10 mm^2) show a conversion of approx. 16 % at 1188 K. Comparing mesh and cermet, it can be seen that the nickel mesh provides a significant contribution to total reforming in the upper temperature range, especially above 1073 K. Considering that the catalytic activity of the substrate is to be lowered to approx. 1/20 [1] and taking into account that the activity of the nickel mesh has not yet been measured, it can be concluded that the nickel mesh alone is sufficient at a temperature of 1123 K to supply enough hydrogen for operating the fuel cell. The nickel in the substrate is only needed for methane steam reforming at temperatures < 1123 K.

4 Summary

The comparison measurements of the nickel meshes and the anode substrates (Ni/8YSZ cermet) have shown that a nickel mesh for methane steam reforming exhibits a very high catalytic activity after a pretreatment as applied in oxidative sealing (3 h in air at 1173 K). The surface of the nickel mesh between anode and interconnect is first oxidized to a depth of 5 - 10 µm due to oxidation in stack assembly and then completely reduced again in the preheating phase of stack operation. This produces a porous nickel surface which is catalytically highly active and considerably increases the catalytic activity of the cermet. The contribution of the nickel mesh to the total reaction is ~ 7.5 % at 1188 K, ~ 5.0 % at 1123 K, ~ 2.5 % at 1073 K and still ~ 1 % at 1023 K (anode substrate = 100%). This shows that, depending on the operating temperature, the nickel mesh has a considerable share in methane steam reforming, especially considering the 20-fold overactivity of the substrate anode. Including this aspect in an assessment, it may be concluded that above an operating temperature of 1123 K reforming by the nickel mesh alone is sufficient to supply the fuel cell with enough hydrogen.

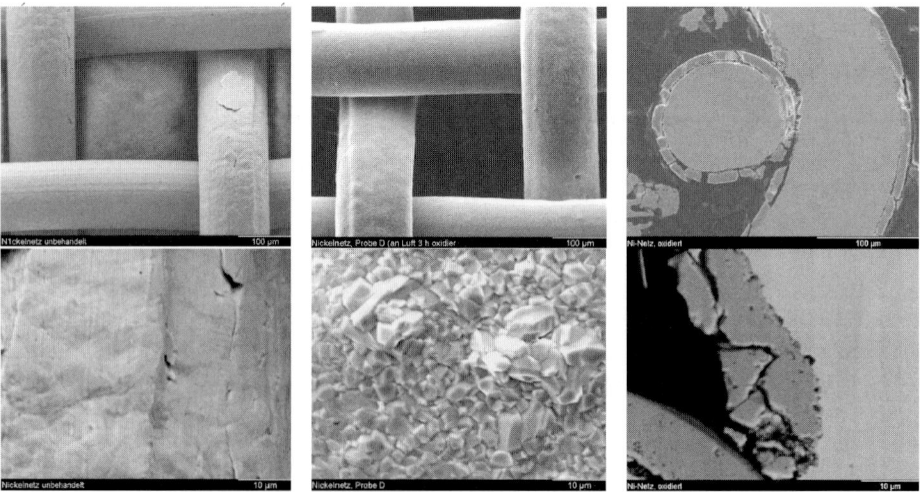

A) Untreated condition B) Oxidation

Figure 3. A),B) Comparison of the surface structures of nickel meshes untreated as supplied (A), after oxidation in air (sealing simulation) (B), after oxidation, reduction and methane steam reforming

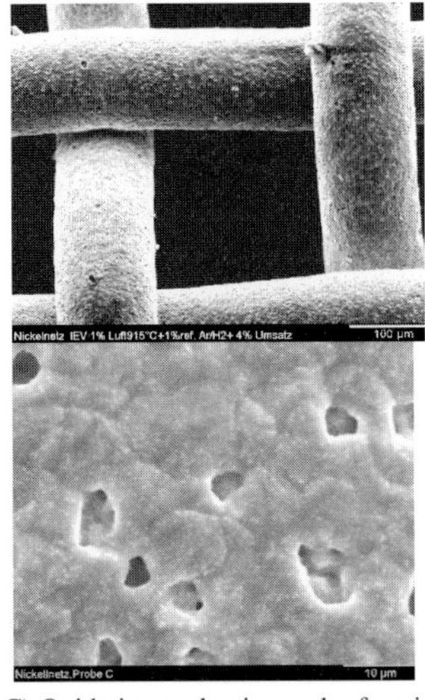

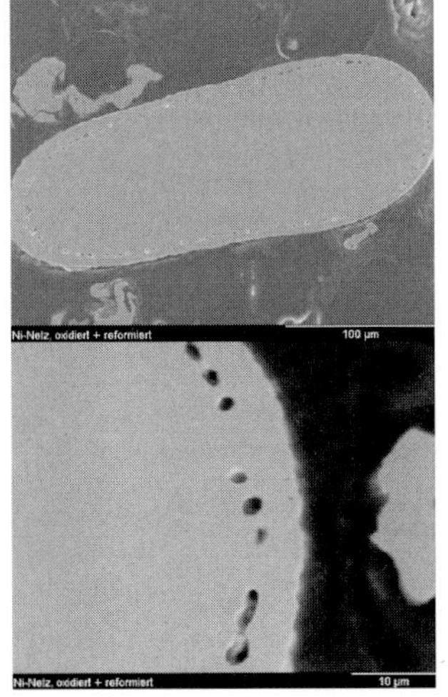

C) Oxidation, reduction and reforming

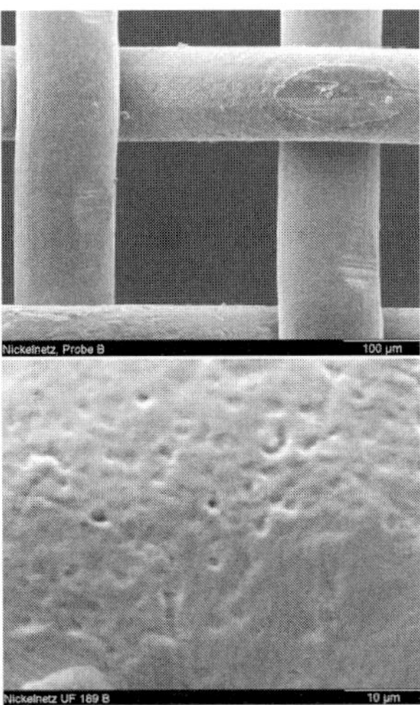

D) Reduction + reforming

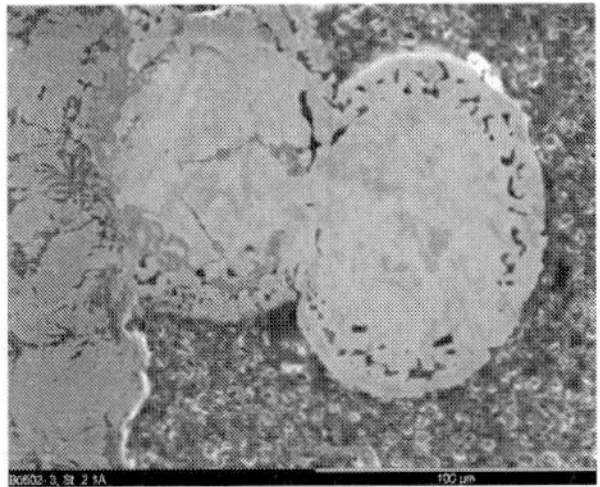

E) Cross section of a nickel wire (part of the mesh) after stack operation [7]

Figure 3. C–E) Comparison of the surface structures of nickel meshes untreated as supplied (C) without pretreatment (D) in comparison to a nickel mesh under real stack operation conditions (E).

A surface change (roughening and pore formation) can also be caused by methane steam reforming itself without a prior oxidation process. This process then seems to proceed much more slowly, but can lead to an equally high reforming activity after long operating periods as in the case of oxidative sealing. These studies will be continued.

5 Acknowledgements

The authors would like to thank all members of the SOFC team for their technical support and helpful discussions, especially W. Beckers (IWV3) and W. Fischer (IWV2), K. Bruch (IWV2) and P. Batfalsky (ZAT).

6 References

[1] U. Flesch, J. Meusinger, A. Naoumidis, D. Stöver in 5th International symposium on Functionally Graded Materials (FGM ´98) (Ed.: W.A. Kaysser), Trans Tech Publications Ltd, Uetikon-Zuerich, Switzerland, p. 788.

[2] E. Achenbach, E. Riensche, J. Power Sources 1994, 52, Iss 2, p. 283.

[3] J. Meusinger, E. Riensche, U. Stimming, J. Power Sources 1998, 71, Iss 1-2, p.315.

[4] H.P. Buchkremer, U. Diekmann, L.G.J. de Haart, H. Kabs, U. Stimming, D. Stöver, in 5th International Symposium on Solid Oxide Fuel Cells (SOFC-V) (Ed.: U. Stimming, S.C. Singhal, H. Tagawa, W. Lehnert), The Electrochemical Society Proceedings Series, Pennington, NJ, USA, 1997, p. 160.

[5] D. Simwonis, F.J. Dias, A. Naoumidis, D. Stöver, in 5th European Conference on Advanced Materials and Processes and Applications (EUROMAT 97), (Ed.: L.A.J.L Sarton, H.B. Zeedijk), Netherlands Society for Materials Science, Maastricht, Netherlands, 1997, Vol. 2, p. 375.

[6] D. Simwonis, A. Naoumidis, F.J. Dias, J. Linke, A. Moropoulou, J. Mat. Res. 1997, 12, Iss 6, p. 1508.

[7] F. Tietz, D. Simwonis, P. Batfalsky, U. Diekmann, D. Stöver in 12th IEA Workshop "Materials and Mechanisms", Annex VII - January 1999, (Ed.: K. Nisancioglu), Wadahl, Norway 1999, in press.

Preparation and Characterization of Pr-Sr-Co-Fe-Al-O Based Perovskite Oxides for SOFC Cathodes

F. Riza, G. Ch. Kostogloudis and Ch. Ftikos
National Technical University of Athens, Greece

1 Abstract

The influence of up to 40 mol% Al substitution for Fe on the properties of the perovskite oxides $(Pr_{1-x}Sr_x)(Co_{0.2}Fe_{0.8})O_{3-\delta}$ (x=0.1, 0.3) was studied. The oxides were prepared by citrate synthesis and characterized by X-ray diffraction, dilatometry and 4-point DC. The above oxides had an orthorhombic perovskite structure, with the exception of $(Pr_{0.7}Sr_{0.3})(Co_{0.2}Fe_{0.4}Al_{0.4})O_{3-\delta}$, which had the cubic structure. The lattice parameters were determined from the room temperature XRD data, using the LSUCR computer program. In general the unit cell volume decreased with increasing Al content, due to the larger size of Al^{3+} cation compared to that of Fe^{3+}. The linear thermal expansion coefficient (TEC) at 700°C decreased as Al content increased. The electrical conductivity decreased with increasing Al content. From the above oxides, $(Pr_{0.7}Sr_{0.3})(Co_{0.2}Fe_{0.6}Al_{0.2})O_{3-\delta}$, with proper tailoring of composition, seems to be the most suitable material for use as SOFC cathode. Its TEC value was 14.5×10^{-6} cm (cm °C)$^{-1}$, while its electrical conductivity had a value of 80 S cm^{-1} at 700 °C.

2 Introduction

A Solid Oxide Fuel Cell (SOFC) is an energy conversion device that produces electricity, by electrochemically combining fuel and oxidant gases and consists of two electrodes separated by the electrolyte [1]. However, operation at high temperatures (~1000°C) and high cost of precious metals (Pt), currently used for the electrodes of the anode and cathode, constitute a major problem for their wider utilization. The development of alternative components for SOFCs can lead to the substantial reduction of the operation temperature of the cell, with the consequent reduction of the total operation costs. Intermediate temperature SOFCs are expected to operate in the temperature range of 600-800°C. The problem of materials selection for intermediate temperature SOFCs concerning the electrolyte has almost been resolved; therefore the research interest is focused on problems associated with the stability of the anode and the selection of appropriate ceramic electrocatalysts for the cathode.

Perovskite oxides of the type ABO_3 with the general composition $La_{1-x}Sr_xCo_{1-y}Fe_yO_{3-\delta}$ are considered very promising cathode materials for intermediate temperature operation, due to their good electrocatalytic properties. However, problems regarding their thermal expansion (high TEC values) and chemical compatibility with the electrolyte have to be solved, while maintaining the required electrical properties. Preliminary results have shown that improvements may arise in both thermal expansion and chemical compatibility issues, by the incorporation of other rare earth cations instead of La^{3+}. The use of Pr^{+3} is reported to decrease cathodic overpotentials and thus enables SOFC operation at lower temperature [2]. Moreover, aluminum substitution for iron is believed to reduce the perovskite's TEC value and its reactivity with the electrolyte [3].

In this study the following systems were prepared and characterized:
$Pr_{0.9}Sr_{0.1}Co_{0.2}Fe_{1-y}Al_yO_{3-\delta}$, y=0, 0.2, 0.4 and $Pr_{0.7}Sr_{0.3}Co_{0.2}Fe_{1-y}Al_yO_{3-\delta}$, y=0, 0.2, 0.4.

3 Experimental

The powders of the cathode oxides were prepared by the amorphous citrate process, calcined in air at 1100°C for 15 h and milled for 48 h with acetone in a satellite-type milling apparatus. The products were dried and pressed (200 MPa) into pellets and sintered in air for 15 h at 1300°C ($Pr_{0.9}Sr_{0.1}Co_{0.2}Fe_{1-y}Al_yO_{3-\delta}$, y=0, 0.2, 0.4 and $Pr_{0.7}Sr_{0.3}Co_{0.2}FeO_{3-\delta}$) or at 1200°C ($Pr_{0.7}Sr_{0.3}Co_{0.2}Fe_{1-y}Al_yO_{3-\delta}$, y=0.2, 0.4). Small bars were cut from the pellets in order to conduct the required experiments.

The lattice parameters were calculated by the room temperature X-ray powder diffraction (XRD) data, using the Least Squares Unit Cell Refinement (LSUCR) program. A Siemens diffractometer was used (CuK_a radiation) operated at 40kV and 30mA. Theoretical densities were calculated using the above mentioned lattice parametres.

Electrical conductivity measurements were conducted with the 4-point DC method on the sintered rods. Four platinum wire contacts were made and the sample was fired at 500°C for 1 h. The measurements were performed upon heating in the temperature range 20-1000°C (heating rate: 5°C/min).

TEC values derived from dilatometric measurements in a quartz dilatometer. Thermal expansion was measured upon cooling (cooling rate: 5°C/min) in the temperature range 100-1000°C.

4 Results and Discussion

4.1 Structural study

The perovskites crystal structure, the lattice parameters and the theoretical densities were determined by the XRD data. The oxides had the orthorhombic structure, with the exception of $(Pr_{0.7}Sr_{0.3})(Co_{0.2}Fe_{0.4}Al_{0.4})O_{3-\delta}$, which had the cubic structure. The cell parameters are given in Table 1.

Theoretical densities were calculated using the following equation [4]:

$$d_{th} = z\frac{M}{V\times 0.6023} \quad (1)$$

where M is the mass in atomic mass units of the atomic ensemble, constituting one unit of the chemical formula, z is the number of such chemical units in one unit cell of the crystal and V (in $Å^3$) is the volume of the crystalline unit cell using the lattice parameters as derived from the X-ray diffraction. For the orthorhombic structure z=4, $V=abc$ and for the cubic structure z=1 and $V=a^3$. After sintering, all densities were above 93% of their theoretical values.

Table 1. Lattice parameters after sintering in air for 15h, as determined from the XRD data

Sample composition	*a* (Å)	*b*(Å)	*c*(Å)	*a′*(Å)	*V*(Å^3)
$Pr_{0.9}Sr_{0.1}Co_{0.2}Fe_{0.8}O_{3-\delta}$	5.4698	5.4939	7.7407	3.8743	232.600
$Pr_{0.9}Sr_{0.1}Co_{0.2}Fe_{0.6}Al_{0.2}O_{3-\delta}$	5.4394	5.3126	7.6941	3.8164	222.339
$Pr_{0.9}Sr_{0.1}Co_{0.2}Fe_{0.4}Al_{0.4}O_{3-\delta}$	5.4086	5.3972	7.6318	3.8189	222.782
$Pr_{0.7}Sr_{0.3}Co_{0.2}Fe_{0.8}O_{3-\delta}$	5.4505	5.4633	7.7280	3.8603	230.100
$Pr_{0.7}Sr_{0.3}Co_{0.2}Fe_{0.6}Al_{0.2}O_{3-\delta}$	5.4681	5.4382	7.6903	3.8523	228.680
$Pr_{0.7}Sr_{0.3}Co_{0.2}Fe_{0.4}Al_{0.4}O_{3-\delta}$	3.8099	3.8099	3.8099	3.8099	55.302

In order to illustrate the effect of the aluminum substitution on the lattice parameters, the pseudo-cubic lattice constant (a') was calculated, where a' is defined as the cube root of the unit cell volume per ABO_3 unit:

$$a' = (V/z)^{1/3} \tag{2}$$

where z=4 for the orthorhombic structure and z=1 for the cubic structure. For the system $Pr_{0.9}Sr_{0.1}Co_{0.2}Fe_{1-y}Al_yO_{3-\delta}$ a' decreases and then slightly increases for increasing aluminum content, while for the system $Pr_{0.7}Sr_{0.3}Co_{0.2}Fe_{1-y}Al_yO_{3-\delta}$ a' decreases. This is due to the smaller ionic radius of Al^{+3}.

4.2 Thermal expansion

Linear thermal expansion curves for $Pr_{0.9}Sr_{0.1}Co_{0.2}Fe_{1-y}Al_yO_{3-\delta}$ obtained upon cooling from 1000 to 100°C in air, are shown in Figure 1. The length increase is linear but at high temperatures the curves tend to become steeper. This may be attributed to the loss of lattice oxygen at high temperatures and the formation of oxygen vacancies [5]. Moreover, the reduction of B-cations causes a decrease in the B-O bond strength according to Pauling's second rule [6]; therefore the cell size increases.

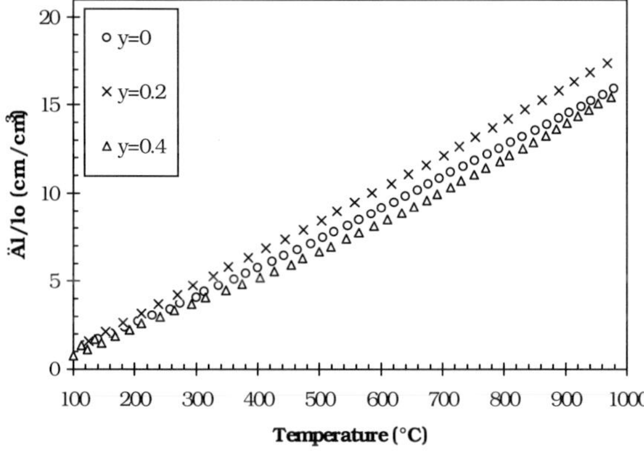

Figure 1. Linear thermal expansion curves for $Pr_{0.9}Sr_{0.1}Co_{0.2}Fe_{1-y}Al_yO_{3-\delta}$ in air

Values of the linear thermal expansion coefficient (TEC) at 700^{0}C are shown on Table 2. TEC values were rather high, but some perovskites had coefficients within the acceptable limit of ±20% of the TEC of the gadolinia doped ceria electrolyte for example (12.6 10^{-6} cm (cm°C)$^{-1}$).

Table 2. TEC values at temperature range 100-700°C

$Pr_{0.9}Sr_{0.1}Co_{0.2}Fe_{1-y}Al_yO_{3-\delta}$		$Pr_{0.7}Sr_{0.3}Co_{0.2}Fe_{1-y}Al_yO_{3-\delta}$	
y	TEC[10^{-6} cm (cm°C)$^{-1}$]	y	TEC[10^{-6}cm (cm°C)$^{-1}$]
0	13.8	0	14.9
0.2	15.2	0.2	14.5
0.4	13.4	0.4	16.3

4.3 Electrical conductivity

Electrical conductivity measured by the 4-point DC method is the total conductivity and it includes both, the electronic and the ionic term. However ionic conductivity in these oxides is about two orders of magnitude lower than the electronic term and it can be assumed that the measured values refer to the electronic conductivity alone [7].

The logarithm of conductivity (*log* σ) in air vs. reciprocal temperature is shown on the Figures 2 and 3. Conductivity decreased as aluminum content increased in accordance with other studies [3]. The temperature dependence of conductivity (equation 3) indicates a semiconducting behaviour and can be explained by the small polaron hopping mechanism [9].

$$\sigma = \frac{A}{T}\exp(-E_a / kT) \qquad (3)$$

where E_a, is the activation energy for hopping and A the pre-exponential factor. The charge carriers (holes) are localized but they possess the possibility to jump between adjacent sites. This is possible, provided sufficient energy is acquired by the hole for the transfer [6].

However, for most samples, conductivity decreased above a certain temperature, indicating a semimetallic behaviour.

The system $Pr_{0.7}Sr_{0.3}Co_{0.2}Fe_{1-y}Al_yO_{3-\delta}$ exhibited significantly higher conductivity values, due to higher Sr content [4]. The values at 700^{0}C for y=0 and y=0.2 are 260 and 80 S cm^{-1} respectively.

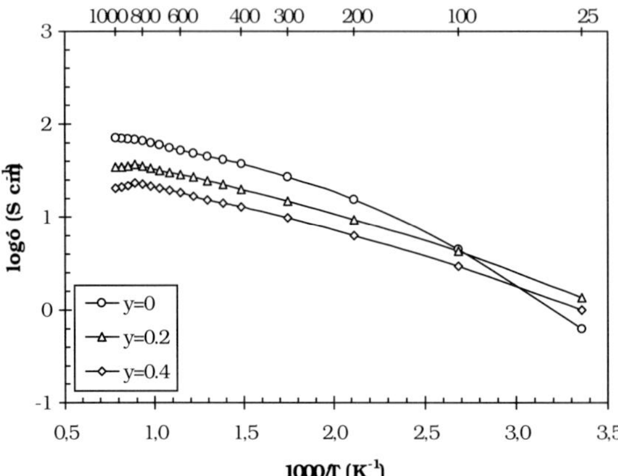

Figure 2. Log σ of $Pr_{0.9}Sr_{0.1}Co_{0.2}Fe_{1-y}Al_yO_{3-\delta}$ in air vs. reciprocal temperature

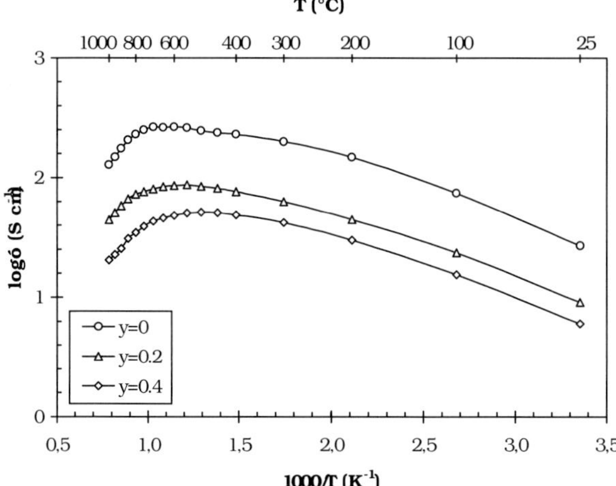

Figure 3. Log σ of $Pr_{0.7}Sr_{0.3}Co_{0.2}Fe_{1-y}Al_yO_{3-\delta}$ in air vs. reciprocal temperature

5 Conclusions

The sample with the composition $(Pr_{0.7}Sr_{0.3})(Co_{0.2}Fe_{0.4}Al_{0.4})O_{3-\delta}$ had the cubic structure, which is the ideal perovskite structure. All the other oxides had a distorted orthorhombic structure. Increase of Al content resulted generally in reduced values of the pseudo-cubic lattice constant, a'.

Some TEC values are within the acceptable limits of ±20% of the TEC of currently used electrolytes, making them suitable for use in SOFCs.

The oxides showed a semiconducting behaviour up to a certain temperature and then a semimettallic behaviour above this temperature. Electrical conductivity decreased with increasing Al content. Compositions with higher Sr content exhibited much higher conductivity values.

From the studied oxides, considering the electrical conductivity (80 S cm^{-1}) and the TEC (14.5×10^{-6} cm (cm °C)$^{-1}$), $(Pr_{0.7}Sr_{0.3})(Co_{0.2}Fe_{0.6}Al_{0.2})O_{3-\delta}$, with proper tailoring of composition, seems to be the most appropriate material for use as a SOFC cathode.

6 References

1. N. Q. Minh, J. Am. Ceram. Soc. 1993, 76, 563-88.
2. T. Ishihara, T. Kudo, H. Matsuda and Y. Takita, J. Electrochem. Soc. 142, 1995, 1519.
3. Holc J., Kuščer D., Hrovat M., Bernik S., Kolar D., Solid State Ionics 95 (1997) 259-268.
4. Raman, S. and Katz, J. L., In Handbook of X-Rays, ed. E. F. Kaelbe. Mc Graw Hill, New York, 1967, pp.29.1-29.2.
5. G. Ch. Kostogloudis, N. Vasilakos and Ch. Ftikos, J. Eur. Ceram. Soc. 17 (1997) 1513.
6. Pauling, L. The nature of the Chemical Bond, 3rd edn. Cornell University Press, Ithaca, N. Y. 1960.
7. S. Carter, A. Selcuk, R. J. Chater, J. Kajda, J. A. Kilner, B. C. H. Steele, Solid State Ionics 53-56 (1992) 597.
8. P. A. Cox, The electronic Structure and Chemistry of Solids, Oxford University Press, Oxford, 1987, p. 179.
9. G. Ch. Kostogloudis, Ch. Ftikos, Solid State Ionics 109 (1998) 43.

Interpretation of the Ageing Effect in Zirconia-based Electrolytes

Christian Haering *#, Hermann Schichl*, Andreas Roosen#
University of Erlangen-Nuremberg, Department of Materials Science,Glass and Ceramics, 91058 Erlangen, Germany
* Siemens AG, KWU BSST, Freyeslebenstrasse 1, D-91058 Erlangen, Germany

1 Introduction

Ageing at temperatures around 1000 °C reduces the electrical conductivity of 8 mol% yttria-stabilised-zirconia (8YSZ), known as a good performing material for Solid Oxide Fuel Cells (SOFC) [1-7]. This effect disappears by increasing the yttria content, but the conductivity decreases with increasing yttria content [3-7]. To improve the electrical conductivity of the planar SOFC electrolytes, the zirconia-scandia system (ScSZ) has to be considered, since it exhibits the highest ionic conductivity of the zirconia solid solutions [2,4,8-10]. A similar degradation has been observed in this system at a scandia content of 7.8 mol% too. With 11 mol% scandia, this degradation was not observed [9-12].

Many interpretations of the conductivity degradation have been given yet, but they explain isolated effects and give unsatisfactory explanations for the many different observations made. In this work, it was tried to characterise the ageing behaviour and find out an explanation for the observations made in the YSZ and ScSZ systems, which could be in agreement with the effects cited in literature.

For that purpose, the electrical conductivity and the crystal structure of specimens with 8 mol% and 10 mol% yttria were determined before and after annealing at 1000°C. Scandia-stabilised zirconia electrolytes (ScSZ) were investigated too, with an scandia content varying from 7 mol% to 13 mol%. Like for YSZ electrolytes, the electrical conductivity was measured and crystal structure was determined before and after annealing. Mechanical spectroscopy was performed on this system. Then, the electrical data and the structure analyses of both systems are correlated to each other and discussed. Finally, according to the literature and to the presented results, an interpretation is given for the ageing effect for all the zirconia-based electrolytes.

2 Experimental Procedure

8YSZ (zirconia doped with 8 mol% yttria) and 10YSZ specimen were synthesised using commercial powders like Tosoh TZ-8Y and TZ-10Y. Specimens were isostatically pressed and sintered at 1500°C for 4 h.

ScSZ electrolytes stabilised with 7 to 13 mol% Sc_2O_3 were synthesised via a Sol-Gel route, with scandium chloride (Varioline, Russia), and zirconium n-propoxide solution in ethanol (Heraeus, Germany) as starting materials. The specimens were sintered at 1400°C for 10 h. All the specimens had densities > 97 % TD.

The electrical measurements and the analyses were performed on the sintered specimens before and after annealing for 1000 h at 1000°C.

To determine the electrical conductivity, four probe d.c. conductivity measurements were performed on both the YSZ and the ScSZ specimens with the dimensions 2x5x50 mm^3, at 950°C, 850°C and 750°C.

The polished cross section of the specimens were observed under Scanning Electron Microscopy. High-resolution X-ray diffraction, neutron diffraction and STEM were performed on yttria-doped zirconia electrolytes. The crystal structure of the ScSZ specimen was determined by X-ray diffraction and mechanical spectroscopy was also performed.

3 Results

Figure 1 shows the specific conductivity of both the YSZ and the ScSZ specimens measured at 850°C, before and after annealing. It can be seen clearly that the ScSZ specimens exhibit a much higher conductivity than the YSZ ones. In the ScSZ system, the conductivity maximum is reached with specimens of 10 mol% scandia, before and after the annealing at 1000°C for 1000 h, whereas in the YSZ system the maximum is reached with the specimens of 8 mol% yttria before annealing and of 9 mol% after annealing. A 50 % conductivity degradation occurs in aged specimens with less than 9 mol% yttria or 10 mol% scandia. The conductivity of the 9YSZ and the 10ScSZ specimens shows a smaller decrease of only 10 %. The conductivity of the specimens with higher dopant content was stable after annealing.

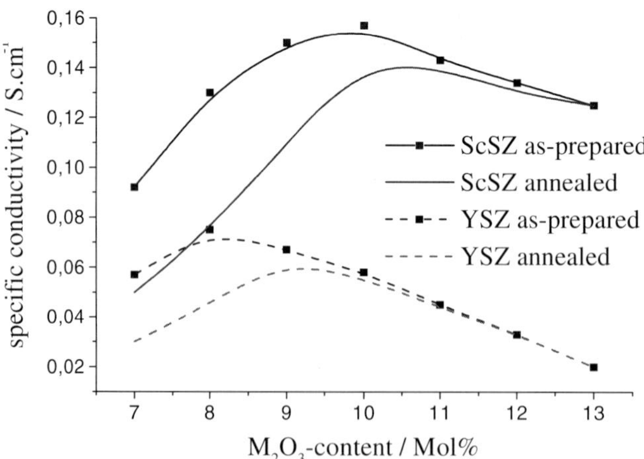

Figure 1. Conductivity before and after annealing of the YSZ and the ScSZ specimen

Figure 2 shows the X-ray diffraction patterns of the as-prepared and the annealed 8YSZ. For a better comparison, the X-ray diffraction lines of the annealed specimen were shifted by $2\theta = 1°$ from their original positions. The observed phase is in both cases the cubic phase, there is no evidence of a tetragonal one. A decrease in the peak intensities, accompanied with an enlargement of the peak areas is observed in the annealed one. The X-ray diffraction pattern of the 10YSZ specimen shows no difference before and after annealing.

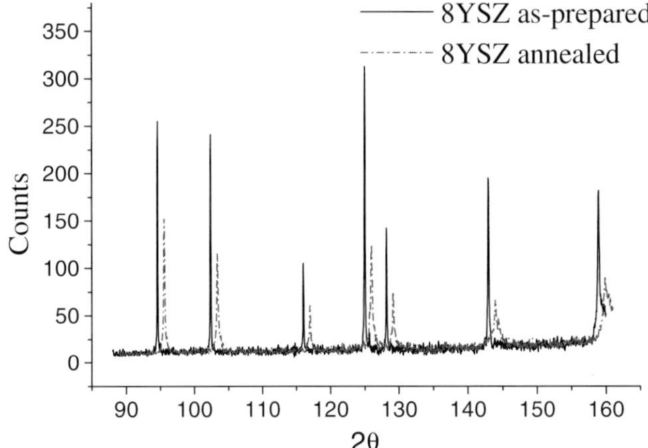

Figure 2. X-ray diffraction patterns of the 8YSZ specimen

The STEM-diffraction patterns of annealed 8YSZ and 10YSZ specimens are given on Figure 3. New reflexes appear in the pattern of the aged 8YSZ specimen, whereas the 10YSZ one shows no change of its pattern after ageing. The appearance of new reflexes are caused by a symmetry loss. The phase is still cubic; it does not have the $Fm\bar{3}m$-symmetry any more, but a $P2_13$ one.

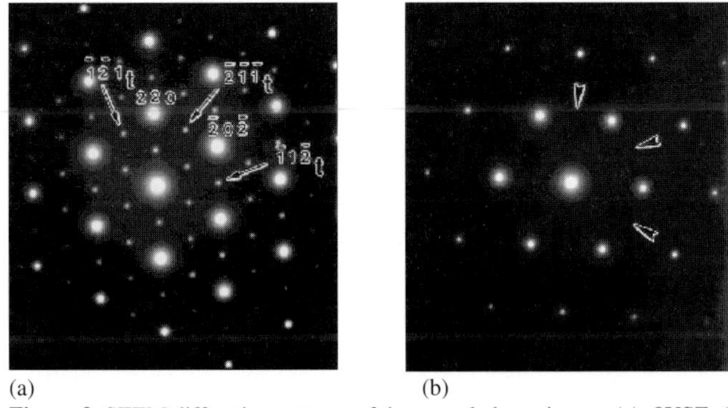

(a) (b)

Figure 3. STEM diffraction patterns of the annealed specimens: (a): 8YSZ; (b): 10YSZ

X-ray analyses of the ScSZ specimens showed a rhombohedral crystal structure with a scandia content > 10 mol%, and only observed at temperatures below 500°C (cubic at higher temperatures). The other specimens (< 10 mol% scandia content) were cubic.

After ageing of the ScSZ specimens at 1000°C for 1000 h, the only crystallographic change observed is a partial phase transformation of the 10ScSZ specimens. Part of the cubic phase transformed into the rhombohedral one. The phase of the other specimens kept unchanged after annealing. There was no phase transformation observed, even in the specimens with a low scandia content, which have a high conductivity degradation.

The internal friction is a suitable and reliable spectroscopy to define defect associates or dipoles, present in the specimens. Figure 4 shows the internal friction spectroscopy of the 9ScSZ specimen before and after annealing. The peak at 820 K (~ 550°C) in the as-prepared specimen is characteristic for a phase transition [13], which means that no reorientation of defect associates is observed, i.e. no associates are present. On the other hand, in the annealed specimen two reflexes appear, which are indications for a reorientation of associates like $[V_O^{\cdot\cdot} - Y_{Zr}']^{\cdot}$ or $[Y_{Zr}' - V_O^{\cdot\cdot} - Y_{Zr}']^{x}$ [13].

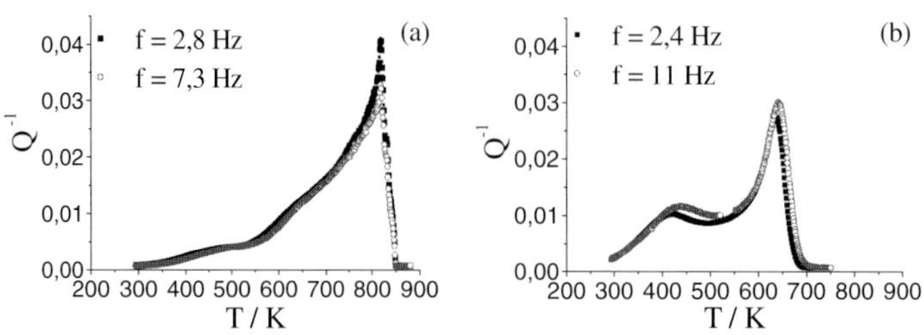

Figure 4. Internal friction of 9ScSZ specimens; (a): as-prepared; (b): annealed

4 Discussion

A symmetry loss appears in the annealed 8YSZ but it still has the cubic symmetry. On the opposite, no change was found on the diffraction pattern of the 10YSZ specimens. X-ray analyses of ScSZ specimens show no change in the structure after annealing, except 10ScSZ.

One reason given for the conductivity degradation of the zirconia based electrolyte is the phase transition or more generally the phase separation [1-3,6,7,9-12], with appearance of a dopant poor region and a dopant-rich one. This case is shown in Figure 5, and has to be discussed. Considering the phase diagram of the doped zirconia, the degradation takes place when the system is metastable, i.e. at low dopant content. At the concentration of the conductivity maximum the degradation could be due to a phase separation leading to the formation of two domains, one domain with a lower dopant content, and therefore a lower conductivity, and the other one with a higher content but having also a lower conductivity. In such a case the resulting conductivity should decrease. Considering another concentration, a phase separation would lead to the formation of two domains like in the precedent case, with the difference that one of the phases has a higher conductivity than the one of the starting composition. This would lead to an approximately constant overall conductivity. However this phenomenon can not take place because it is in contradiction with the results from Figure 1, where 7-, 8- and 9ScSZ specimens show a high conductivity loss.

In particular systems, where the two phases are built perpendicularly to the applied tension, the lower conductivity dominates (the higher resistance), and therefore the resulting conductivity decreases during annealing. If the higher conductivity dominates (the phases are parallel to the applied tension) an increase of the resulting conductivity would take place. Both cases are in contradiction to the experimental results. This explains why a phase separation can not be responsible for a conductivity degradation.

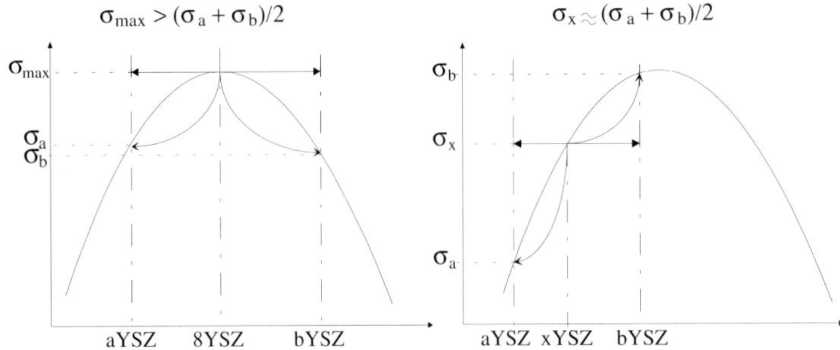

Figure 5. Conductivity as a function of the dopant content; left: content for the maximum conductivity; right: other content

Defect associates like $[V_O^{\cdot\cdot} - Y_{Zr}']^{\cdot}$ and $[Y'_{Zr} - V_O^{\cdot\cdot} - Y'_{Zr}]^x$ are well described by Frey et al. [14], who analysed the calcia doped zirconia with neutron diffraction (Figure 6).

Nakamura et al. [15] explain that, with a lower dopant content than at the conductivity maximum, only one kind of defect is present in the electrolytes, $[V_O^{\cdot\cdot} - Y'_{Zr}]^{\cdot}$. At higher dopant content than at the maximum, the conductivity decreases with increasing dopant content because the vacancies are trapped by two dopant cations (blocking effect) to form $[Y'_{Zr}\text{-}V_O^{\cdot\cdot}\text{-}Y'_{Zr}]^x$.

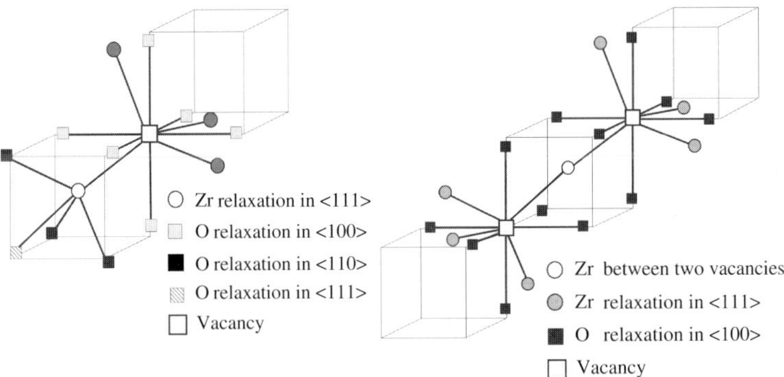

Figure 6. Single defect (left) and paired defect (right) present in the doped zirconia;

It seems that the $[Y'_{Zr} - V_O^{\cdot\cdot} - Y'_{Zr}]^x$ associates are stable whereas the like $[V_O^{\cdot\cdot} - Y_{Zr}']^{\cdot}$ ones are metastable. Indeed only the zirconia-based electrolytes doped with less than 10 mol% show a degradation during annealing. This conductivity loss can be caused by the transformation of the $[V_O^{\cdot\cdot} - Y_{Zr}']^{\cdot}$ metastable associates into the $[Y'_{Zr} - V_O^{\cdot\cdot} - Y'_{Zr}]^x$ stable ones. As a result strong lattice distorsions appear around the vacancies and their mobility is reduced. On the other hand, at higher dopant content the blocking effect stabilises the electrolytes, and no change in conductivity is observed during annealing.

The observation made on 8YSZ and 10YSZ are not in contradiction with this interpretation, since the defect structures and the lattice distorsions cause in first step a symmetry loss. When the phase of the doped zirconia is metastable (< 10 mol% M_2O_3) then the lattice distorsions in combination with the defect associates can lead to a local phase transition. This change in crystal structure can be observed or not, depending on the size and the amount of those microdomains.

5 Conclusion

Annealing the zirconia-based electrolytes causes a degradation of their conductivity when they are doped with less than 10 mol% M_2O_3. A phase transition can occur, but this is not the cause of the degradation.

In the specimens doped with more than 10 mol%, the conductivity decreases with increasing dopant content because the vacancies are trapped, i.e. they form stable associates with the dopant cations. By annealing this defect structure does not change. Therefore the conductivity of those samples shows no degradation during annealing.

On the other hand, in the specimens doped with less than 10 mol%, local lattice distorsions around the vacancies are formed during annealing, which lead to the formation of defect associates and, as a consequence, to the conductivity degradation. The aliovalent cations get nearer to the vacancies (lattice distorsions in the <111> direction) and build dipoles with them. As a result the vacancies are trapped and their mobility decreases.

This interpretation of the conductivity degradation is valid for all the zirconia based electrolytes. But the degradation is caused by defect associates which are not specifically defined for each dopant-zirconia system.

6 Acknowledgements

We would like to thank:

- Dr. Weller, Max-Planck Institute for Metal Research in Stuttgart, for the internal friction spectroscopy.
- Dr. Schnöller, Siemens AG in Munich, for the syntheses of the scandia doped zirconia specimens.
- Prof. Zimmermann, Institute for Crystallography of the University of Erlangen-Nuremberg, for the phase characterisations and interpretations.
- Dr. Bestgen, Hoechst AG in Frankfurt, for the STEM-analyses.

7 References

1 L. H. Schoenlein, L. W. Hobbs, A. H. Heuer, „Precipitation and Ordering in Calcia- and Yttria-Stabilized Zirconia“, J. Appl. Cryst. **13** (1980) 375-379.

2 A. N. Vlasov, M. V. Perfiliev, „Ageing of ZrO_2-Based Solid Electrolytes“, Solid State Ionics **25** (1987) 245-253.

3 R. Gibson, G. P. Dransfield, J. T. S. Irvine, „Influence of Yttria Concentration upon Electrical Properties and Susceptibility to Ageing of Yttria-stabilised Zirconias“, J. Europ. Ceram. Soc. **18** (1998) 661-667.

4 W. Drenckhahn, H. Schichl, R. Männer, M. Schnöller, C. Haering and A. Roosen, „Alterung von Komponenten der SOFC, Langzeitverhalten von Funktionskeramiken“, P. Otschik (ed.), Verlag Werkstoff-Informationsgesell., Frankfurt am Main (1997) 203-214.

5 H. Raeder, T. Norby, P. A. Osborg, „Ageing of Yttria-Stabilised Zirconia Electrolytes at 1000°C“, Ceram. Trans. **51** (1995) 719-730.

6 T. Yamada, T. Fukui, Y. Kodera and H. Matsubara, „Degradation in Ionic Conductivity of Yttria Full Stabilized Zirconia", Ceram. Trans. **71** (1996) 453-461.

7 J. Luo, R. Stevens, „The Effect of High Temperature Ageing on Ionic Conductivity of Y_2O_3-ZrO_2 Cubic Ceramics", Proceedings 3th European Solid Oxide Fuel Cell Forum, 2-5 June 1998 in Nantes (F), Ed. P. Stevens, (1998) 171-178.

8 D. W. Strickler and W. G. Carlson, „Electrical Conductivity in the ZrO_2-Rich Region of Several M_2O_3-ZrO_2 Systems", J. Am. Ceram. Soc. **48** (1965) 286-289.

9 S. P. S. Badwal, „Zirconia-Based Solid Electrolytes: Microstructure, Stability and Ionic Conductivity", Solid State Ionics **52** (1992) 23-32.

10 Y. Mizutani, M. Tamura, M. Kawai, „Development of High-Performance Electrolyte in SOFC", Solid State Ionics **72** (1994) 271-275.

11 S.P.S. Badwal, J. Drennan, „Microstructure / Conductivity Relationship in the Scandia-Zirconia System", Solid State Ionics **53-56** (1992) 769-776.

12 Y. Murakami, I. Nagano and H. Yamamoto, „Phase Equilibria and Phase Change During Ageing in the ZrO_2-Sc_2O_3 System", J. Mat. Sci. Lett. **16** (1997) 1686-1688.

13 M. Weller, A. Lakki, „Defects in Cubic Zirconia Studied by Mechanical Loss Spectroscopy", Ber. Bunsenges. Phys. Chem. **101** (1997) 1297-1302.

14 Th. Proffen, R. B. Neder, F. Frey, „Neutron and X-ray Diffuse Scattering of Calcium-Stabilized Zirconia", Acta Cryst. B **52** (1996) 59-65.

15 A. Nakamura, J. B. Wagner, Jr., „Defect Structure, Ionic Conductivity, and Diffusion in Yttria-Stabilized Zirconia and Related Oxide Electrolytes with the Fluorite Structure", J. Electrochem. Soc. **133** [8] (1986) 1542-1548.

Determination of the Oxygen Ion Diffusion Coefficients in $SrMnO_{3-y}$ Mixed Conductor by the Reoxidation Kinetics

Ralf Kriegel / Christian Kaps, Bauhaus-Universität Weimar, Professur Bauchemie

1. Introduction

During the past several years the concepts of oxygen permeation with mixed conducting ceramic membranes are of growing interest. A mixed conducting material, which has both high oxygen ionic and electronic conductivity, is able to separate oxygen from an oxygen containing gas (mostly air) efficiently without an external electrical circuit. There are several applications where such materials are of interest, including membrane reactors, membranes, gas sensors and solid oxide fuel cells.

The whole permeation flux depends on the electronic conductivity, the surface exchange kinetics and the oxygen ion diffusion. Usually the permeation flux is limited by the oxygen ion diffusion, for thin membranes by the surface exchange kinetics [1]. Whereas the exchange kinetics could be optimized by surface enlargement (porous support) [2] and by catalysts deposited on the membrane surface [3], but the ion diffusion depends mainly on the material composition. Therefore, an efficient material screening requires a simple, fast determination of the oxygen diffusion coefficient. The diffusion measurement setup used in the present work is based on a special diffusion model called "infinite sheet" [4].

Mixed conductors are mostly consist of compounds of the general composition ABO_{3-y} crystallizing in the perovskite structure. Although the original perovskite is build up by a cubic dense package of oxygen and A ions, many non-cubic compounds are also ranked with the perovskite family. The influence of the crystal structure on the oxygen diffusion rate was rarely described, however cubic compounds should be better suited for oxygen permeation [5]. Usually, the finding out of the optimum of the phase content plays a subordinated part in the screening process. This may result in a misjudgment of the material potential.

An example is $SrMnO_{3-y}$, which does not matter for oxygen separation applications. The system $SrMnO_{3-y}$ was extensively investigated by Negas and Roth [6],various modifications were obtained. Chemically reduced $SrMnO_{3-y}$ is formed at high temperatures and can be obtained as a metastable modification by cooling in an oxygen-free atmosphere [7]. The reoxidation of the reduced samples in an oxygen-containing gas mixture forms metastable β-$SrMnO_3$ below 840 °C, otherwise α-$SrMnO_3$ [7] is formed. The high content of oxygen vacancies in $SrMnO_{2.57}$ and the fast reoxidation of dense samples of β-$SrMnO_3$ refer to a remarkable oxygen diffusion rate. Because the hexagonal α-$SrMnO_{3-y}$ is the stable modification in air, the oxygen diffusion of the cubic high temperature modification β-$SrMnO_{3-y}$ was not described up to now.

2. Experimental

Samples of $SrMnO_{3-y}$ were prepared by conventional ceramic technology. Dense bodies of the composition $SrMnO_{2.57}$ with >96 % of the x-ray density were obtained. Samples with higher oxygen stoichiometry were synthesized by partly reoxidation in a closed reaction chamber with a limited oxygen amount. The oxygen content and the phase content of the samples were determined by chemical analysis and x-ray diffraction. Samples with $y > 0.4$ consist of a single orthorhombic phase (superlattice of perovskite structure), the sample reoxidation forms pure cubic β-$SrMnO_3$ below 840 °C. The sample preparation and characterization was described earlier in detail [7].

The measurement of the oxygen ion diffusion was carried out by the characterization of the isothermic reoxidation kinetics of chemically reduced, dense samples. It has been assumed that the reoxidation is controlled by the oxygen diffusion. The diffusion could be supposed as one-dimensional, because the ratio of diameter to thickness was higher than 10 : 1 for each sample. The isothermic reoxidation was measured with two different experimental setups designated as pressure experiment and as gravimetric experiment.

In the first one, the oxygen pressure was measured during the reoxidation of $SrMnO_{3-y}$ in a closed reaction chamber of an automated sorptometer (Porous Materials, Inc.). The chemically reduced sample (cylinders of $SrMnO_{2.57}$) was placed in a platinum crucible which was in contact to the control thermocouple of an electric oven. The reaction chamber was heated up to the working temperature under vacuum, the connection to the vacuum pump was closed and a known amount of oxygen was introduced into the reaction chamber. The time-dependent oxygen consumption of the sample (oxygen pressure) was recorded by the control computer. The pressure was recalculated into the molar amount of oxygen with the known volume of the measurement system.

The second setup was based on a thermal balance (TGA 92, Setaram). Because the high diameter of the samples (18 mm), the thermal balance was used without a conventional sample holder. The specimen was placed in a platinum net nearly to the control thermocouple. Because each sample should be used for several diffusion measurements, the reduction was carried out within the thermal balance in a flux of purified argon at 1550 °C. After cooling, the reduced sample was heated up to the working temperature under vacuum, then the vacuum valve was closed and the carrier gas valve was opened. The sample was reoxidized in a flux of purified air, the growing sample mass was recorded by the control computer.

3. Results

Figure 1 shows the oxygen pressure in a closed reaction chamber at 220 °C, which contains chemically reduced samples of $SrMnO_{3-y}$ with an oxygen content of $3\text{-}y = 2.57$ and 2.75, respectively. The present oxygen gas generates an oxygen pressure at the beginning, which is constant for a few minutes. After a time delay, the oxygen pressure decreases with growing diffusion time to a constant final value. The end of the oxygen consumption of the sample is indicated by the linear variation of pressure with time. The variation of the oxygen content of the samples was calculated using the measured volume of the equipment. The sample mass gain after the reoxidation was used to determine the oxygen stoichiometry independently. No significant differences were observed. The same measurements were carried out on samples with different start stoichiometry (2.57 - 2.93). The same functional behavior was observed for all samples.

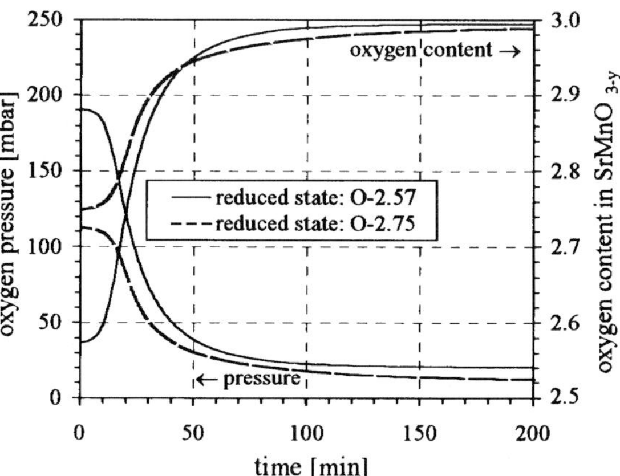

Figure 1. Oxygen pressure and oxygen content in $SrMnO_{3-y}$ during the reoxidation of $SrMnO_{2.57}$ and $SrMnO_{2.75}$ at 220 °C

The sample mass during the reoxidation in a thermal balance is shown in Figure 2. The reoxidation of $SrMnO_{2.54}$ was carried out in purified air at 300 and 500 °C, respectively. The first entry of oxygen into the evacuated thermal balance generates a small peak in the sample mass, because the air flows down to the sample. The following mass gain is caused by the reoxidation of the reduced sample. At the beginning, the reoxidation runs faster than the oxygen entry into the sample chamber. Therefore, the oxygen (air) flow controls the mass gain, and the reoxidation proceeds linearly with time. If the reoxidation becomes slower than the oxygen flux, the measured mass gain gets closer to a constant value. Only the non-linear mass gain could be evaluated to calculate the diffusion coefficient. The variation of the oxygen stoichiometry was not shown, because the sample mass is direct proportional to the oxygen content. If the variation of lattice constants with the oxygen content is neglected, the diffusion rate can be calculated by the sample mass variation. The maximum mass gain is a little bit smaller at 500 °C compared to 300 °C, because the maximum oxygen content of $SrMnO_{3-y}$ decreases with growing temperature.

The described measurements were evaluated with a special diffusion model called "infinite sheet", schematically drawn in Figure 3. A species with a start concentration c_0'' ($c_0'' \neq c_0'$) migrates in a plane or sheet with finite thickness h and infinite lateral expansion (y, z >> h). Therefore, the diffusion flux should

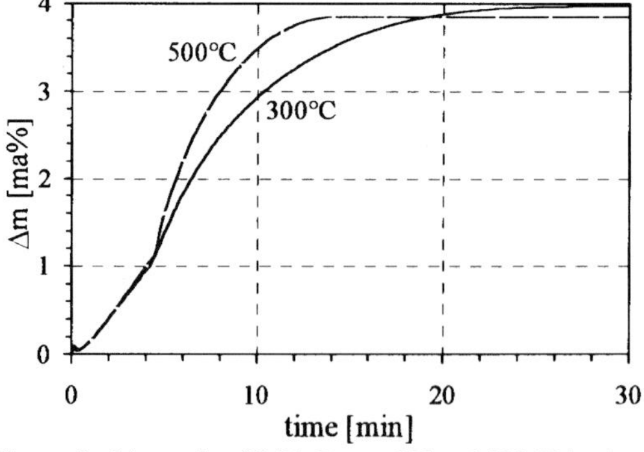

Figure 2. Mass gain of $SrMnO_{2.54}$ at 300 and 500 °C in air

be limited on one dimension. In this special case, the analytical solution of Fick's second law for a constant diffusion coefficient is based on the evaluation of infinite series, which was described in detail by Jost [4]. The mathematical expressions are simplified by the introduction of the relative concentration c_r, which is evaluated from the averaged concentration c_m (integrated over the whole sample thickness) and from the start and the end concentrations (c_a, c_e). The normalization to the concentration range is expressed by the following equation.

$$c_r(t) = \frac{c_m - c_e}{c_a - c_e} = \frac{8}{\pi^2} \sum_{\nu=0}^{\infty} \frac{1}{(2\nu+1)^2} \exp\left(-\left(\frac{(2\nu+1)\pi}{h}\right)^2 D\,t\right) \tag{1}$$

The series evaluation could be cancelled past several summations, because the result gets close to a constant value for finite times ($\Sigma \rightarrow$ const.). In this case the following simple expression results.

$$c_r(t) \approx \frac{8}{\pi^2} \exp\left(-\frac{\pi^2 D t}{h^2}\right). \tag{2}$$

This relation could be transformed into a linear equation using the natural logarithm.

$$\ln(c_r) = \ln\left(\frac{8}{\pi^2}\right) - \left(\frac{\pi^2 D}{h^2}\right) \cdot t \tag{3}$$

Therefore, a linear decrease of $\ln(c_r)$ with the time is expected for a constant diffusion coefficient. The diffusion coefficient D could be calculated with the sheet thickness h and with the gradient b of the normalized curve.

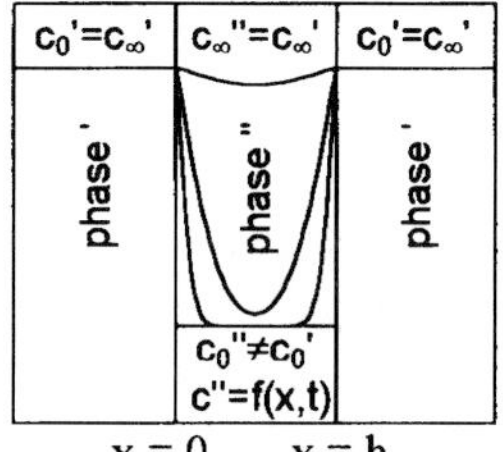

Figure 3. Diffusion model „infinite sheet"

$$D = \frac{b\,h^2}{\pi^2} \tag{4}$$

Figure 4 shows selected measurements on $SrMnO_{3-y}$ already shown in Figure 1 and Figure 2, which were transformed corresponding to the equation (3). A pressure experiment with a start pressure of 190 mbar on $SrMnO_{2.57}$ was selected, because this value was nearly equal to the oxygen partial pressure in the thermal balance (air). The oxygen diffusion coefficients were calculated with equation (4) for each temperature by the gradient of the linear part of the normalized curve. With the usually used Arrhenius equation in the logarithmic form

$$\ln(D) = \ln(D_0) - \frac{E_A}{kT}$$

we obtained an activation energy $E_A = 0{,}35 \pm 0{,}02$ eV and a pre-exponential factor $D_0 = (7{,}2 \pm 0{,}4) \cdot 10^{-3}$ cm^2/s (linear regression, correlation coefficient 0.998). The measurement results are summarized in table 1. In figure 4 the normalized relative concentration $\ln(c_r)$ decreases linearly in with the diffusion time, which indicates a constant diffusion coefficient. This was the prerequisite for the use

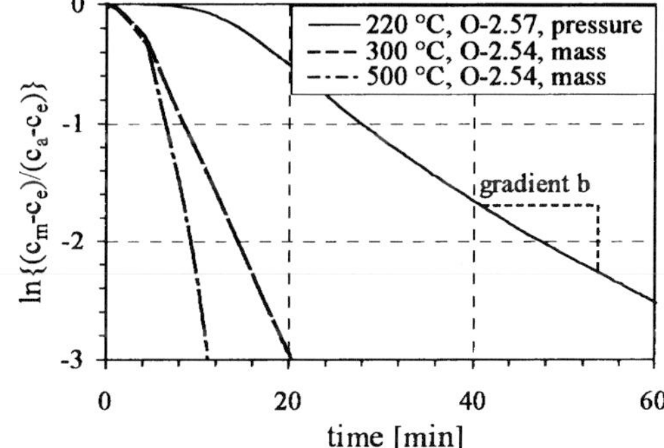

Figure 4. Normalized reoxidation of $SrMnO_{3-y}$ resulting from two experimental setups at different temperatures

Table 1 Selected results of reoxidation measurements

used experimental setup	pressure	gravimetry	gravimetry
start composition	$SrMnO_{2.57}$	$SrMnO_{2.54}$	$SrMnO_{2.54}$
temperature [°C]	220	300	500
D [cm²/s]	$1.9 \cdot 10^{-6}$	$5.3 \cdot 10^{-6}$	$3.7 \cdot 10^{-5}$

of the special diffusion model "infinite sheet".

In contrast to the previous described experiments, all normalized measurements under variable oxygen pressure show distinct deviations from a linear behavior for high oxygen contents and low oxygen pressure. Gravimetric measurements, which run at constant pressure, does not show such effects. Therefore we assume, that the oxygen diffusion coefficient should not depends on the oxygen concentration in $SrMnO_{3-y}$.. However, the decreasing oxygen pressure during the reoxidation reduces the oxygen surface concentration $c_{surface}$ as well. The boundary conditions of the used special solution of the second Fick's law are not kept in this situation, and no analytical solution exists to approximate the influence of a surface concentration decreasing with growing diffusion time.

Numerical calculation methods promise a higher versatility for the solution of ordinary differential equations. A windows® software [8] was used for the one-dimensional diffusion simulation at different conditions. The calculation is based on the approximation of the deviations in Fick's second law by finite differences. The diffusion dimension x is subdivided in layers. With the numerical approximated deviations, the partial differential equation could be transformed into a system of ordinary differential equations. The time-dependent variation of the concentration is now calculated for each layer using a finite time step Δt. The software uses a four-step method (Runge-Kutta) because it's numerical error is relatively low (error ~ Δt^4).

Figure 5 shows the oxygen stoichiometry of $SrMnO_{2.70}$ during a reoxidation recorded by the variation of the oxygen pressure (start: 132.9 mbar, end: 16.2 mbar). The normalized curve $\ln(c_r)$ deviates distinctly from a linear behavior. The influence of a decreasing surface concentration $c_{Surface}$ was simulated with the as-described software. The environmental concentration was set to an arbitrary start value (100 %), which was reduced during the simulation by the subtraction of the cal-

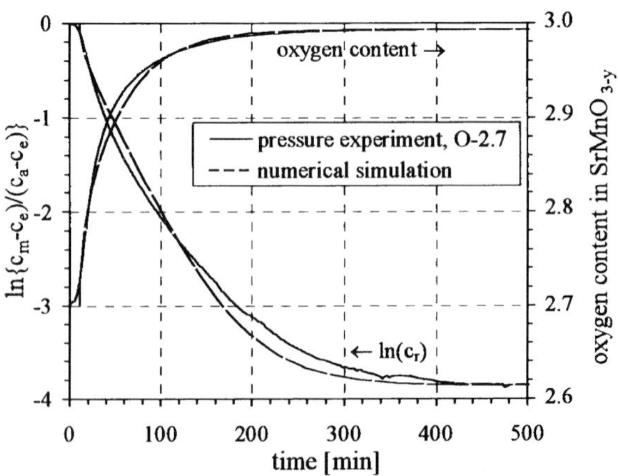

Figure 5. Normalized reoxidation of $SrMnO_{2.70}$ (pressure experiment) and numerical simulation ($D = 5 \cdot 10^{-7}$ cm²/s; time delay 11 min) with decreasing surface concentration

culated oxygen flux consumed by the sample. Best results (as shown in Figure 5) were obtained by a decrease of the surface concentration from 100 % to 70 %. Because the relation between the oxygen pressure and the effective surface concentration was unknown, no direct relationship to the oxygen pressure was obtained.

4. Summary

It was shown in this work, that metastable cubic $SrMnO_{3-y}$ has a high potential for oxygen permeation applications because it's high oxygen diffusion coefficient ($D = 2.9 \cdot 10^{-4}$ cm^2/s, extrapolated for 1000 °C). The phase transition prevents the use of the material at temperatures above 840 °C. Therefore, controlled substitution on the A or B site of the Formula ABO_3 should be done, also with the aim of the improvement of the whole permeation process.

Usually, an experimental setup for the measurement of oxygen diffusion or permeation requires the exchange of sealing and following the total exclusion of leaks, frequently a time-consuming search. These (and further) disadvantages are surmounted by the as-described experimental setups are based on the diffusion model "infinite sheet". The normalization makes possible to detect error sources (boundary conditions, non-constant diffusion coefficient etc.).

A wide variety of mixed conductors should be measurable with the described method, if a small reoxidation effect is detectable. For this reason, we was encouraged to develop a simple measurement equipment for the oxygen diffusion measurement. This equipment is based on a volume measurement at nearly constant pressure and first results will be published shortly.

5. References

[1] H. J. M. Bouwmester, H. Kruidhof, A. J. Burggraaf: Sol. St. Ionics 1994, 72, 185 - 194

[2] R. M. Thorogood, R. Srinivasan, T. F. Yee, M. P. Drake: US 5,240,480, 1992

[3] M. F. Carolan, P. N. Dyer: EP 0 663 232 A2, 1995

[4] W. Jost, Diffusion, Darmstadt 1957, p. 52

[5] T. L. Cable, T. J. Mazanec: EP 0 705 790 A1, 1996

[6] T. Negas and R. S. Roth, J. Sol. State Chem. 1970, 1, 409 - 418

[7] R. Kriegel, N. Preuß, Thermochimica Acta 285 (1996), 91 - 98

[8] R. Kriegel: "Win_Diff - a diffusion simulation programm", Weimar 1999

Mechanical Properties of Doped $LaCrO_3$ After Exposure to Reducing Atmospheres

F. Meschke and R.W. Steinbrech
Institut für Werkstoffe und Verfahren der Energietechnik Forschungszentrum Jülich GmbH, 52425 Jülich

1 Abstract

The mechanical properties of two doped lanthanum chromite perovskites were evaluated with respect to strength, fracture toughness curve and elastic modulus. With decreasing oxygen partial pressure during annealing at 1000°C the room temperature mechanical properties improved significantly. Calculations show that the flaw size is not affected by the annealing.

2 Introduction

Doped lanthanum chromite has gained considerably attraction in the last decade due to good electrical conductivity and a high temperature stability. In particular, advantage of these properties is taken in both high temperature heating elements /1/ and in solid oxide fuel cells (SOFCs) /2/. For example, lanthanum chromite became important as interconnect material for planar SOFC stacks since the thermal expansion coefficient can be adjusted with appropriate dopants close to that of the commonly used zirconia electrolyte membrane /2/. In the interconnect application the $LaCrO_3$ material also must separate fuel from air at the anode and cathode side, respectively. Thus the $LaCrO_3$ faces simultaneously reducing and oxidizing atmosphere at opposite surfaces, i.e. low and high oxygen partial pressure. At oxygen partial pressures $p(O_2) < 10^{-12}$ bar, oxygen-vacancies related swelling of doped lanthanum chromite occurs /3,4/, which can causes bending of the interconnect plate. Compositional changes of the dopants allow to suppress to a certain extent the swelling /5/. However, comparably little attention has been directed to date towards the influence of the oxygen vacancy concentration on the mechanical properties of the $LaCrO_3$-based interconnect perovskites, i.e. the impact on strength, toughness and elastic modulus.

The mechanical results reported in literature are still contradictory. Paulik et al. /6/ claimed that both strength and toughness of Sr- and Ca-doped $LaCrO_3$ are smaller after annealing in reducing atmosphere, whereas Milliken et al. /7/ reported a strength increase of 50 % and Montross et al. /8/ observed no effect. Depending on composition, the elastic modulus results are found to decrease or increase with decreasing $p(O_2)$ /9/. In the present work two materials with a distinct difference in swelling behavior are examined. The changes in mechanical properties as a function of oxygen partial pressure are described.

3 Experimental

3.1 Materials

Two types of doped $LaCrO_3$ have been examined, one being doped with 20% Sr and 3% V (Risö National Laboratory, DK), the other having a proprietary dopant composition based on substitution with Al and Mg (Dornier, D). The materials are referred to as LSCV and LCMA, respectively. Both materials exhibit swelling at 1000°C under low oxygen partial pressure. The swelling becomes noticeable below $p(O_2) = 10^{-12}$ bar and rises up to 0.06 % and 0.2 % at 10^{-18} bar for LCMA and LSCV, respectively. The LSCV microstructure consisted of 3 – 5 μm large matrix grains (Fig. 1a) with second phases at triple points. LCMA exhibited grain sizes is the same range, however, significantly less second phase pockets appeared (Fig. 1b).

Bend bars (3x4x45 mm^3, LSCV) or discs (ϕ = 22 mm, t = 0.5 mm, LMCA) were cut from the as-supplied perovskite materials. In total 100 specimens were prepared with specimen surfaces polished to a 1 μm finish. Ten samples of each material were annealed at 1000°C under various oxygen partial pressures. The $p(O_2)$ ranged from 10^{-20} bar (10^{-18} bar represents the fuel gas atmosphere in SOFCs) to 0.21 bar (air). Different pressures were adjusted by wetting $Ar/4\%H_2$ gas with H_2O. The exposure time was chosen at 100 h to ensure proper, through-specimen thickness, reduction. In order to freeze the reduction level achieved at high temperature, the specimens were rapidly cooled to room temperature, but quenching was not so severe to introduce thermal shock cracks. From ten annealed specimens five samples were notched for fracture toughness measurements according to the SEVNB method. The toughness was also determined as a function of crack extension (K_R-curve) by using a long distance microscope for crack length measurements.

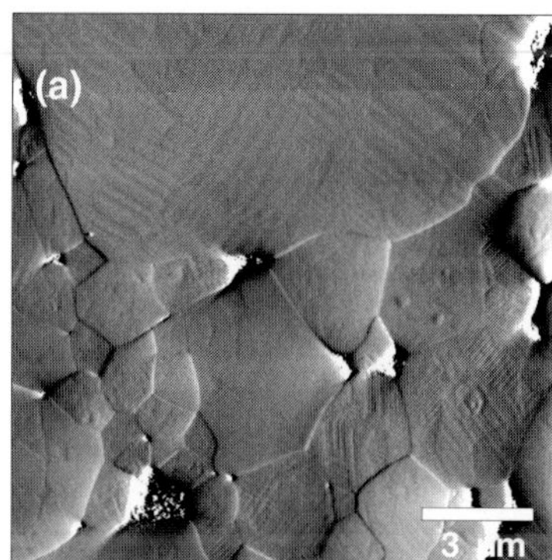

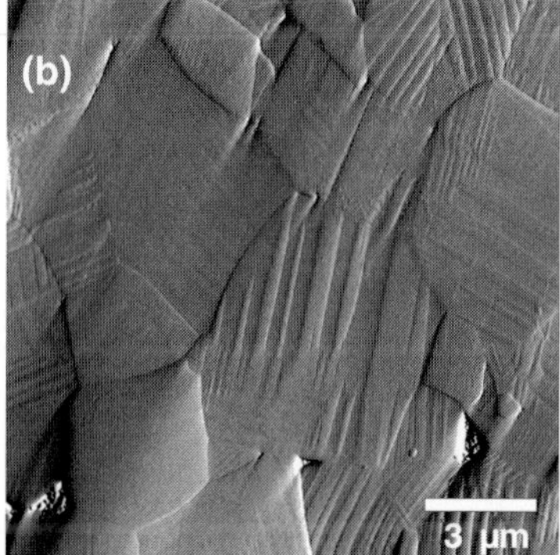

Figure 1. AFM micrographs of (a) LSCV and (b) LCMA.

4 Results and Discussions

4.1 Strength

The room temperature measurements of LSCV and LCMA show that the fracture stress of both perovskites depends on the oxygen partial pressure during the proceeding annealing (Fig. 2a). The strength clearly increases with decreasing $p(O_2)$. After exposure to 10^{-20} bar strength is 40 % higher compared to the air value. Since all samples were polished identically and different annealing atmospheres did not alter the grain size, the effect can not be attributed to a decrease in flaw size. Therefore, primarily an increase in toughness is considered to be responsable for the strength increase.

Comparing the strength of the two materials, that of LCMA turned out to be higher than that of LSCV. This is indicative for a smaller defect size of LCMA. However, it has to be emphasized that two different methods were used to determine fracture stress. Thus one reason for the lower strength of LSCV might also be the larger tested volume.

4.2 Toughness

Both perovskites show rising K_R-curve behavior (Fig. 2b). After annealing in air, the K_R-curve of the LSCV material starts with an initial value of 1.1 MPa√m and increases to 1,4 MPa√m after crack extension of 500 μm. A plateau value of 1.5 MPa√m is observed after 1 mm. LCMA exhibits significantly higher toughness values. The respective K_R-curve starts with an initial value of 1.4 MPa√m and increases to 2.1 MPa√m after a crack extension of 500 μm. Before reaching the plateau value, crack growth typically became unstable and the SEVNB-bend bars fractured catastrophically.

With decreasing $p(O_2)$ the fracture toughness curves of both materials are to higher values without significant change in slope. The initial value of LSCV increases from 1.1 MPa√m after annealing in air to 1.5 MPa√m after annealing at $p(O_2) = 10^{-15}$ bar. Similarly, the initial toughness value of the LCMA material changes from 1.4 to 2.0 MPa√m. Annealing at $p(O_2) = 10^{-17}$ bar causes only a small additional increase to a total value of 2.3 MPa√m. Lowering the $p(O_2)$ to 10^{-20} bar had no further influence on the initial toughness, but K_R-curves could not be measured since the specimens fractured after crack extension in the order of some 10 microns. The obtained maximum starting toughness of 2.3 MPa√m corresponds to an increase of about 60 %.

Rising K_R-curve behavior has also been observed recently in ferroelectric perovskites like $BaTiO_3$ /10/. In these materials stress-induced ferroelastic domain switching and crack face interaction are considered as main toughening mechanisms. It is not unlikely, but needs further experimental evidence, that the same mechanisms are opperative in the electrically conductive $LaCrO_3$ perovskites. Note that domain structures are visible in the micrographs of Fig. 1.

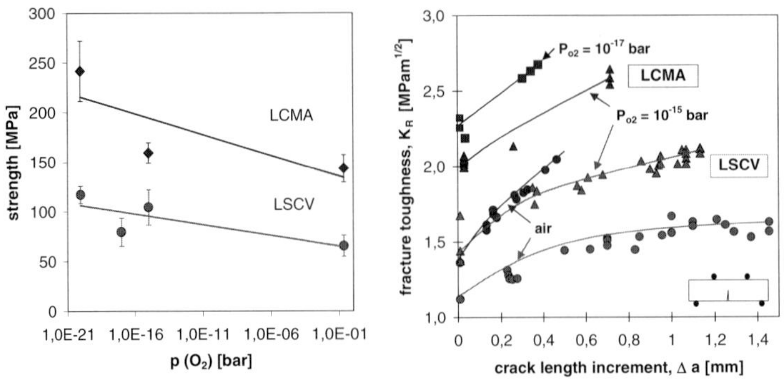

Figure 2. Fracture behavior of two multiple-doped $LaCrO_3$ perovskites as a function of oxygen partial pressure during annealing at 1000°C. a) fracture stress b) fracture toughness curve (K_R-curve)

4.3 Pseudo-Plasticity

The stress-strain curves as derived from the central deflection during bending clearly show a pseudo-plastic deformation behavior of the $LaCrO_3$ perovskites (Fig. 3). The degree of plastic deformation depends on $p(O_2)$ during annealing. In general those specimens exhibit highest pseudo-plasticity which were annealed in air. It is suggested that ferroelastic domain switching is the origin of the pseudo-plasticity. Very similar behavior has been observed in perovskites like $BaTiO_3$ and PZT under both bending and compression loading /11/ and domain switching has been verfied as the underlying mechanism.

The onset of non-linear deformation shifts to lower stress with increasing $p(O_2)$. It is believed that the increasing amount of oxygen vacancies inhibits increasingly the motion of domain walls during the switching process.

4.4 Elastic Modulus

The stress-strain curves of both perovskites also reveal an increase in elastic modulus with decreasing $p(O_2)$. The elastic modulus of LCMA nearly triples when $p(O_2)$ is decreased from air to 10^{-20} bar. LSCV shows a similar behavior. The reason for the increasing stiffness is still unknown. Discussion in literature /9/ currently focuses on oxygen-vacancy related differences in crystallographic bonding properties.

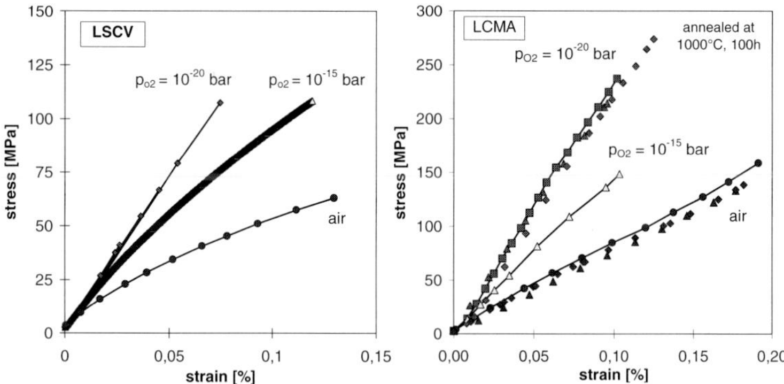

Figure 3: Stress-strain curves of LSCV (a) and LCMA (b) after annealing under different oxygen partial pressures.

4.5 Interrelations

Applying fracture mechanics and assuming that strength correlates with the starting value of the K_R-curves, $K_{R,0}$, (note that the measured K_R-curves are comparably shallow) the failure controlling flaw size, a, can be determined from the Griffith's instability criterion

$$a = \frac{K_{R,0}^2}{Y^2 \sigma_f^2} \qquad \text{(Eq. 1)}$$

Using the fracture stress data, σ_f, from Fig. 2 and the geometry factor $Y = 1.3$ (assuming semicircular surface flaws), constant flaw sizes of $a_{LCMA} \approx 60\ \mu m$ and $a_{LSCV} \approx 80\ \mu m$ were calculated according to Eq.1. The independence of flaw size from $p(O_2)$ proves that annealing treatment under various reducing atmospheres indeed does not change the failure relevant defect size.

5 Conclusion

It has been demonstrated that the level of oxygen partial pressure during annealing sensitively controls the mechanical properties of electrically conductive doped lanthanum chromites. Strength, fracture toughness and the elastic modulus increase with diminishing oxigen partial pressure, i.e. increasing amount of oxygen vacancies. Bending models based on asymmetric swelling of SOFC interconnect components have to be reviewed by also incorporating the observed change of elastic modulus.

6 Acknowledgements

The authors thank K. Hilpert and F. Boroomand for providing the specimen and for many fruitful discussions. The financial support by the Deutsche Forschungsgemeinschaft DFG (Ni 178-17-1) is gratefully acknowledged.

7 References

1. A.J. Moulsen and J.M. Herbert, Electroceramics, Chapman and Hall, London, 1990, p.122.
2. N.Q. Minh, J. Am. Ceram. Soc., 1993, 76 [3], 563-88.
3. P.V. Hendriksen, J.D. Carter, and M. Mogensen, Proceedings of SOFC-IV, The Electrochemical Society, Pennington, 1995, 944-951.
4. T.R. Armstrong, J.W. Stevenson, L.R. Pederson, and P.E. Raney, J. Electrochem. Soc., 1996, 143 [9], 2919-2925.
5. P.H. Larsen, P.V. Hendriksen, and M. Mogensen, Proceedings of the 3rd European SOFC Forum, Oberrohrdorf, CH, 1998, pp. 181.
6. S.W. Paulik, S. Baskaran, and T.R. Armstrong, J. Mater. Sci., 1998, 33, 2397-2404.
7. C. Milliken, S. Elangovan and A. Khandar, Proceedings of the 3rd International Symposium on SOFC, 1993, The Electrochemical Society, Pennington, p.335.
8. Ch. Montross, H. Yokokawa, and M. Dokiya, Extended Abstracts of the 60th Meeting of the Electrochemical Society of Japan, 1993, p.263.
9. Ch. Montross, J. Europ. Ceram. Soc., 1997, 18, 353-358.
10. F. Meschke, A. Kolleck, and G.A. Schneider, *J. Eur. Ceram. Soc.*, 1997, 17 [8] 1143-49.
11. A.B. Schäufele, and K.H. Härdtl, J. Am. Ceram. Soc., 1996, 79 [10] 2637-40.

Performance of Different Glas-Ceramic Sealants for a Planar SOFC Concept

T. Schwickert, U. Diekmann
Forschungszentrum Jülich
P. Geasee, R. Conradt
RWTH Aachen,

1 Introduction: Planar Solid Oxide Fuel Cells (SOFCs)

Planar solid oxide fuel cells are a promising approach for cost- effective electrochemical energy conversion. They are especially suitable for stationary production of electrical power in small decentralized units. This means the electrical power is generated where it is needed. In this application, the planar SOFC has the following main advantages:

- high efficiency (> 60% electrical)
- low emissions
- fossil fuels and hydrogen applicable

2 Requirements for Sealants

Planar concepts require sealants with long-term stability at high operation temperatures of 700 - 900 °C. Basic requirements for these operation conditions are:

- high thermal expansion coefficient (TEC)
- high electrical resistance
- thermochemical compatibility with the fuel cell materials.

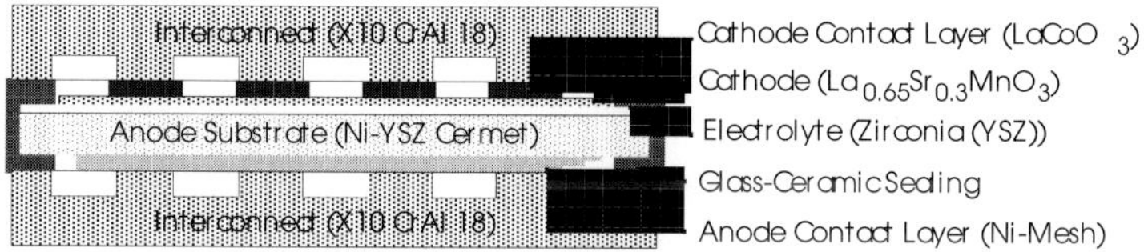

Figure 1. Geometry of the sealing in the SOFC stack.

2 Glass Preparation

Different alkali-free glasses of the alkaline earth-alumosilicate type were melted in batches of 200 g in a Pt crucible in an induction furnace at 1480 °C for 2 hours. During soaking at 1480 °C, the glass was stirred to enhance homogeneity. After soaking, the glass was fritted in ice water.

The glasses were milled in acetone adding 0.1% PVB and using agate balls. Final particle size was < 63 μm.

3 Characterization and Results

3.1 Differential scanning calorimetry (DSC)

Glass was analyzed by DSC at a heating rate of 20 K/min from room temperature to 1200 °C. It was found that the barium- silicate-rich glass 1 shows a stronger crystallization tendency than glass 2, glass 3 and glass 4.

3.2 Dilatometry

Dilatometer samples from all glasses were pressed at 20 MPa and sintered at 900 °C for one hour in air. The dilatometric measurements were performed at a heating rate of 3 K/min in air.

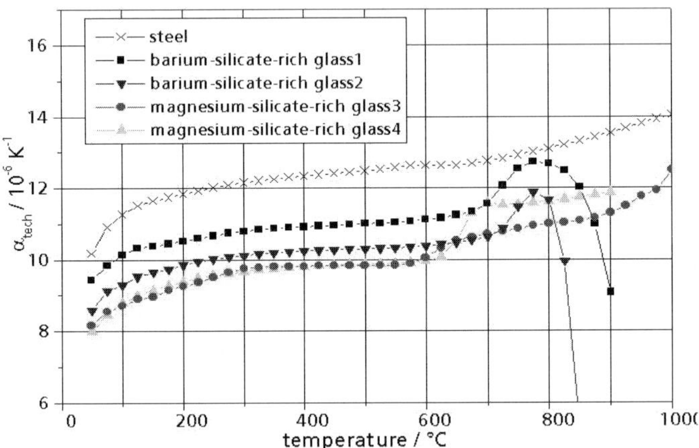

Figure 2. Dilatometric results

Besides the dilatometric results of the four glasses, figure 3 shows the thermal expansion coefficient for the steel X10 CrAl 18, which is used as an interconnect in the SOFC stacks. At about 800 °C glass 1 and glass 2 start softening (dilatometric softening point). The thermal expansion coefficients of glass 3 and glass 4 increase with temperature in two steps: the first one at about 300 °C, the second one between 600 °C and 700 °C. The TEC of all glasses is lower than that of the steel. After further crystallization for 10 hours at 900°C, the barium-silicate-rich glasses have a significantly higher TEC than before (see glass 1 in figure 2).

3.3 Sintering behavior

The flow and shrink behavior during sintering of the glasses was characterized by a modified dilatometer. A sample for this dilatometer consists of a layer of glass about 400 μm thick sandwiched between two steel plates. During the dilatometer test, this specimen is pressed with a pressure of 39 kPa. The measurements were performed at a heating rate of 2 K/min in air.

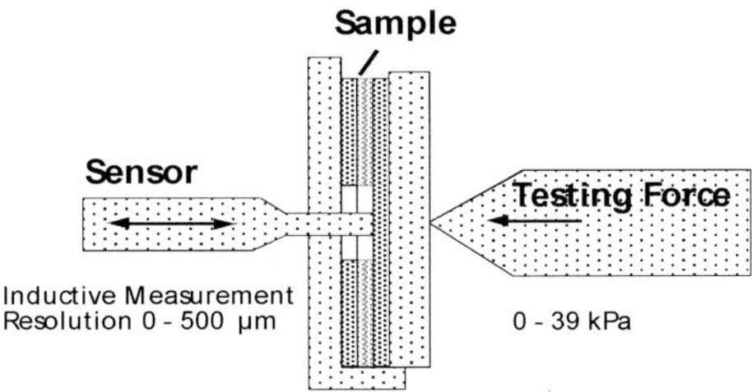

Figure 3. Modified dilatometer for sintering tests

All glasses start shrinking at about 800 °C. The magnesium-silicate-rich glasses 3 and 4 shrink in two steps, the second step taking place at about 900 °C.

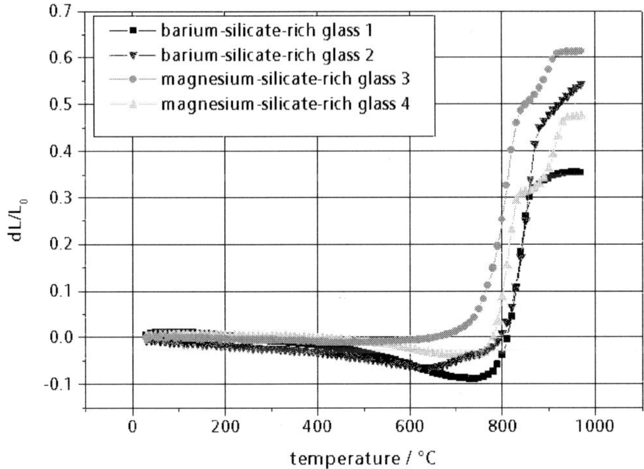

Figure 4. Flow behavior during sintering

3.4 Joining tests

Various joining tests with steel and ceramic samples of the SOFC were performed simulating the different joints in the stack. After crystallization, magnesium-silicate-rich glass ceramics showed interfaces adhering more tightly to the steel surface than the investigated barium-silicate-rich glass ceramics, despite the TEC mismatch of the former. None of the glasses investigated had any difficulty in adhering to the zirconia and the Ni-zirconia cermet

4 Conclusions

The adhesion properties of the magnesium-silicate-rich glass ceramics are superior, whereas the barium-silicate-rich glass ceramics are, because of their well-adjusted TEC, superior in thermal cycling. An optimum still has to be found.

5 References

1. H. Dörsing, R. Conradt and U. Diekmann, DVS-Berichte, Dt. Verl. für Schweisstechnik, Düsseldorf, 1998, Vol. 192, 323-324
2. P. Batfalsky, U. Diekmann, J. Godziemba-Maliszewski, T. Koppitz in DVS-Berichte, Dt. Verl. für Schweisstechnik, Düsseldorf, 1997, Vol. 184, 72 – 76
3. Y.S. Touloukian, Thermophysical Properties of High Temperature Solid Materials, Vol. 4, Macmillan, New York 1967
4. Engineering Materials Handbook, Vol. 4, USA; 1991

Solar Control Coatings for Windows

Helen Rose Wilson

Affiliation: Interpane E & BmbH, Lauenförde, Germany, Contact Address: Fraunhofer Institute for Solar Energy Systems, Oltmannsstr. 5, 79100 Freiburg, Germany

1 Abstract

With increasingly large areas of glazing being a feature of modern architecture, the question of solar gain control is being addressed with considerable intensity. In predominantly hot climates, the aim is generally to admit sufficient light for visual comfort but to exclude excess solar radiation which would otherwise cause overheating. At medium to high latitudes, the requirements on solar gain vary throughout the year. During winter, all solar radiation is welcome as a source of space heating, and the low U values of modern glazing allow large window areas to be used as apertures for solar energy. However, the same features, large areas and low U values, mean that there is the danger of overheating in summer, so that the requirements then are similar to those of low-latitude locations. As a response to this situation, different types of solar control coatings have been developed for windows. Some rely on the spectral selectivity of thin film systems alone, others also include the property of angular selectivity. A third class of coatings, encompassing thermotropic and thermochromic materials, responds to temperature change with variation of transmittance and reflectance properties. A fourth class, including electrochromic and gaschromic systems, allows "active" switching in response to an external parameter. This paper will present the underlying physical principles, the optical properties and the effect on windows containing materials from the first three classes; representatives of the fourth class will be presented in another paper at this conference[1].

2 Introduction

"Solar control" as a term applied to architectural glazing refers to deliberately limiting the transmittance of at least part of the incident solar radiation in order to avoid overheating and/or glare. In order to achieve this aim at a given location, not only the total intensity of the solar radiation but also its spectral, angular and temporal distribution must be taken into account. Some options for solar control which may be appropriate for other regions of the world will be less suitable in the European context considered here.

Figure 1(a) illustrates the spectral distribution for solar radiation, as incident after transmission through 1.5 thicknesses of atmospheric air mass (AM1.5) on a terrestrial surface tilted at 37 ° [2]. The global radiation spectrum is shown, i.e. the sum of the radiation component transmitted directly from the sun and the diffuse component after scattering from the sky. This spectrum is one which has been chosen internationally [3] as a standard to weight characteristic glazing spectra to obtain integrated values for solar transmittance, reflectance or absorptance, T_{sol}, R_{sol} and A_{sol}, respectively. Other solar spectra such as AM1.0 global [4] are also in use for this purpose. No such "standard" can be fully representative of the solar spectral variation found for different atmospheric path lengths, meteorological conditions and receiving surface orientation. Similarly, among the range of spectral sensitivity curves describing the response of the human eye under different lighting conditions, the photopic sensitivity spectrum V(λ) [5] weighted with the AM1.5 global solar spectrum, also shown in fig. 1(a), has been chosen to obtain visible properties, e.g. T_{vis}, R_{vis}, A_{vis}. The spectra in fig. 1(b), the Planck blackbody radiation spectra for temperatures of 283 K, 300 K and 323 K define the range for which the emittance ε of window glazing surfaces are of interest.

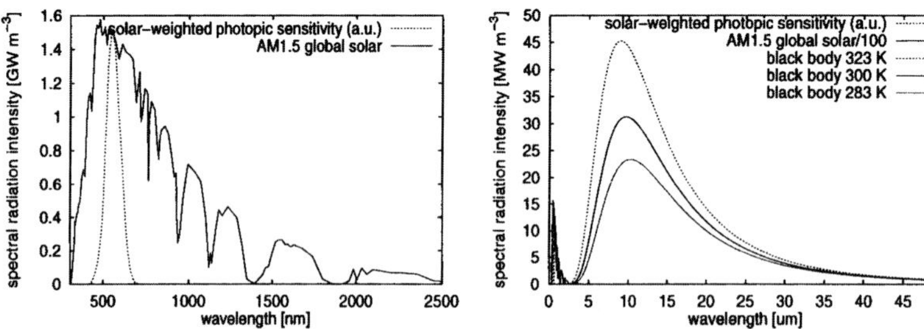

Figure 1. (a) AM1.5 solar radiation spectrum and the solar-weighted photopic sensitivity spectrum V(λ); (b) the spectra of (a) and the Planck blackbody radiation spectra for temperatures of 283 K, 300 K and 323 K. (The scaled spectra of (a) are also included for reference.)

For any latitude (ϕ), the sun's altitude at solar noon will vary throughout the year from (90 - ϕ) - 23.5° to (90 -ϕ) + 23.5° [6]. The diurnal variation in azimuth between sunrise and sunset is superimposed on the annual cycle.

For completeness, the general trend for temporal variation of solar radiation intensity is noted: the amplitude for the daily cycle is greatest near the equator (reaching a maximum of ~1000Wm^{-2}), whereas the amplitude for the annual cycle is greatest at high latitudes, resulting in strong temperature differences between the seasons.

In addition to the solar and visible properties mentioned above, two further glazing characteristics should be introduced: the U value, the heat transfer coefficient describing heat transport due to conduction, convection and radiation through 1 m^2 glazing area per Kelvin temperature difference across the glazing, and the total solar energy transmittance TSET (also known as g value of solar heat gain coefficient SHGC), the sum of T_{sol} and internal heat gains resulting primarily from reradiation of solar energy absorbed in the glazing unit.

3 Solar Control Coatings - Materials and Optical Properties

3.1 Spectrally selective solar control coatings

The oldest and possibly still most widely known form of solar control glazing is body-tinted glass containing high concentrations of iron and other transition metal oxides, which absorb in the visible and NIR solar spectral range. However, the broad aborption band brings a major disadvantage for solar control; even for a low T_{sol} value, achieved with a high oxide concentration, the TSET value is high due to the high emissivity of the glass surface, which is similar to that of common soda-lime float glass ($\varepsilon_n = 0.89$) [7].

A more recent and sophisticated approach is to apply a spectrally selective coating on the glass surface which guarantees high NIR and IR reflectance, and thus low values for T_{sol} and ε, while still retaining sufficiently high T_{vis} values for the intended application. The most effective solution is derived from so-called "low-e" coatings, which are in widespread use for thermally insulating glazing [8, 9]. They are based on a single layer of noble metal, which has high IR reflectance and thus low emittance due to its high conductivity resulting from the intrinsically high bulk electron density ~6×10^{22} cm^{-3}. One of the challenges in manufacturing such films is to achieve bulk metallic properties in a thin film. A variety of vacuum deposition methods is used, but the most widespread is planar magnetron sputtering. The tendency of very thin sputtered coatings to form porous structures as a result of nucleation and island growth is counteracted by precise control of sputtering parameters and additional measures such as ion-assisted sputtering or the inclusion of blocker or adhesion layers such as NiCr. Silver is preferred to gold or copper, as their absorption in the visible range causes colour distortion in transmission and reflection. In low-e coatings for thermal insulating glazing, T_{vis} and T_{sol} are kept relatively high by combining the thinnest conductive silver layers with anti-reflective coatings of dielectrics characterised by high refractive indices (Bi_2O_3, In_2O_3, SnO_2, TiO_2, ZnO, ZnS). For spectrally selective solar control coatings based on this principle, two thin silver layers are embedded between the same choice of dielectrics. This approach is preferred to using a single thicker silver layer to reduce T_{sol} and thus the TSET value, firstly because the total silver thickness and thus material costs are lower, and secondly because the spectral selectivity can be more finely tuned. Figure 2 shows the transmittance and reflectance spectra for single coated panes used in commercially available glazing: a "classic" low-e coating on glass intended for thermal insulation ($U = 1.1$ $Wm^{-2}K^{-1}$, TSET = 0.58), and two solar control coatings, optimised for a high ratio of T_{vis}:TSET value ($U = 1.1$ $Wm^{-2}K^{-1}$, TSET = 0.34) and for a low TSET value ($U = 1.1$ $Wm^{-2}K^{-1}$, TSET = 0.23) respectively. The stated U and TSET values refers to a sealed insulating glazing unit (IGU) consisting of an uncoated and a coated pane, with the coating on a surface adjacent to the 16 mm Ar-filled cavity. In general, a solar control coating will be positioned on the outer pane, whereas a low-e coating for thermal insulation will be on the inner pane. The sealed double glazing configuration improves the thermal insulating properties significantly compared to those of the single coated glazing. If necessary, they could be further enhanced with a "classic" low-e coating on the inner pane, which also further reduces the T_{sol} value. Of course, this option is also open for all the forms of solar control glazing discussed in this paper.

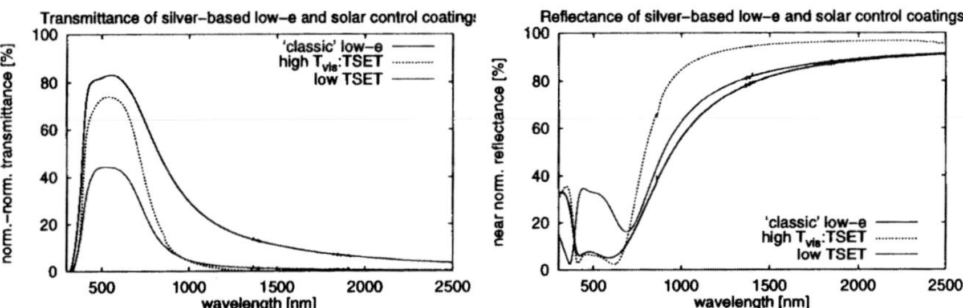

Figure 2. (a) The normal-normal transmittance spectra of a low-e coating for thermal insulation and solar control coatings optimised for spectral selectivity and low TSET value respectively; (b) the corresponding near-normal reflectance spectra.

Selective solar control coatings based on doped oxide semiconductors are reviewed in [8]. Oxides of zinc, cadmium, indium, tin, thallium, lead and alloys of these have a sufficiently wide bandgap for interband absorption in the solar spectral range to be low, so that T_{vis} can be high. By doping with foreign elements (SnO_2:F, SnO_2:Sb, In_2O_3:Sn, ZnO:Al) or creating a moderate oxygen deficiency, the electron density can be raised to a maximum of about 10^{21} cm^{-3}. However, this electron density is more than an order of magnitude lower than for the silver films, so that the electrical conductivity and IR reflectance are too low to achieve the low values of ε desirable for solar control purposes ($\varepsilon_n = 0.15$ for the coating in fig. 3). In addition, as can be seen for the pyrolitically deposited SnO_2:F film in fig. 3, the onset of IR reflectivity only occurs at $\lambda > 1.5$ µm, so that T_{vis}:T_{sol} selectivity is limited.

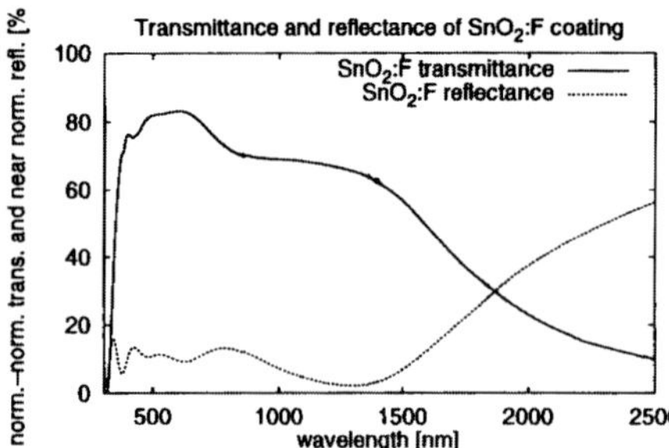

Figure 3. Normal-normal transmittance and near-normal reflectance spectra of an In_2O_3:Sn coating.

Nitrides of chrome, titanium, iron, zirconium and hafnium have also been investigated for solar control purposes [10, 11]. As with the doped semiconductor films, the emittance is higher (0.14 - 0.4 for TiN_x) than for silver-based films. The interest in this development seems to be primarily motivated by applications requiring hard coatings for external exposure, or if deliberately coloured glazing is to be used which is at least superior in its emittance properties to body-tinted glass.

3.2 Angle-selective solar control coatings

Earlier work on angle-selective thin films on glass concentrated on structures which achieved angular selectivity in visible transmittance as a measure to reduce glare from sunlight or sky-scattered light, while retaining visual contact in the usual near-horizontal viewing directions [12]. More recently, as reported in a comprehensive review on angle-selective glazing [13], a double-layer coating has been made by depositing the same type of film onto ITO-coated glass, with the specific aim of improving solar control properties by reducing the values of ε and T_{sol} without significantly affecting T_{vis}.

These angle-selective thin-film coatings are based on composite media with an inclined columnar structure achieved by vacuum deposition techniques ensuring oblique incidence of the depositing species. Good column formation depends strongly on the choice of deposition angle, deposition rate, degree of collimation of depositing species and the choice of materials. The first films studied consisted of metal columns of diameter typically less than 40 nm, separated by voids. The refractory metals Cr and Ta proved to be particularly suitable, as their high melting point and thus low mobility after deposition enhanced column formation. However, these films all suffered the disadvantage of mechanical weakness, being soft and not strongly bonded to the substrate.

Excellent adherence and hardness were obtained by using oxides (Al_2O_3, SiO_2, TiO_2, Ta_2O_5, and WO_3) rather than air as the embedding matrix. Still more importantly, they allowed better tailoring of the optical properties by widening the choice and distribution of the metal within the film. Whereas the oxide is usually transparent, the metal should be absorptive across the whole solar spectral range, as the angular selectivity is determined by anisotropic absorbing properties of the composite film. Metals such as Al and Ag, with low melting points, proved to be favourable, as their mobility after deposition enhances column formation due to preferential metal-metal bonding in the spaces between the dielectric columns. Compositional analysis with Rutherford back scattering identified a metal content ranging from between 3 - 5 % for some Ag cermets to 20 - 30 %, depending on the deposition method - filtered cathodic arc, reactive sputtering or electron-beam deposition of the oxide, with co-deposition of the metal by thermal or electron-beam evaporation.

The optical properties are most commonly characterised by transmittance spectra for the plane of incidence which includes both the column tilt direction and the deposition direction. This is usually the plane where angular selectivity is most pronounced; there is no angular selectivity for planes normal to this plane. Figure 4(a) illustrates the angle-dependent transmittance spectra in this plane for a 100 nm thick Al/TiO_3 film on a bare glass substrate [14]. It represents the best spectral selectivity (T_{vis}/T_{sol}) at high negative incidence angles reported for a single-layer film to date. In the plane represented, the transmittance spectra for p-polarised radiation are asymmetric for angles relative to the surface normal, whereas they are symmetric for s-polarisation i.e. equal for positive and negative angles of the same absolute value. (In this and following figures, negative angles of incidence correspond to an upward viewing direction through vertically mounted glazing.)

Better solar control is achieved with a double-layer coating, an Al/SiO_2 film on an ITO-coated glass substrate, as characterised by the spectra in fig. 4(b). The T_{vis}/T_{sol} selectivity is not as large as for the purely spectrally selective coatings of fig. 2, but, depending on the latitude, this may not be the decisive criterion. The angle-dependent curves for T_{vis} and T_{sol} in fig. 5 show that T_{sol} is < 20 % for incidence angles < -60 °, corresponding to summer zenith angles at latitudes lower than 50 °. Although the T_{vis} values in the same range of incidence angles are not much higher than T_{sol}, they would still usually be high enough to guarantee sufficient daylighting from direct solar radiation. At the lower light levels associated with overcast skies, the higher T_{vis} values at incidence angles > -60° become relevant for the diffuse sky radiation.

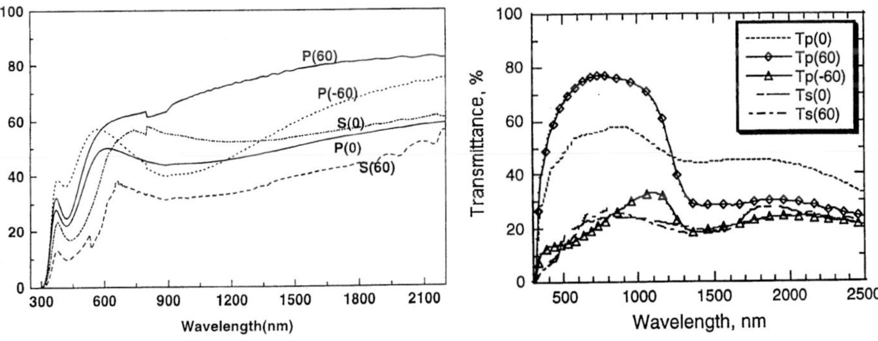

Figure 4. Transmittance spectra at selected angles of incidence and s- and p-polarised light of (a) an Ag/TiO_2 film of thickness 100 nm [14], (b) an Al/SiO_2 film on In_2O_3:Sn [13].

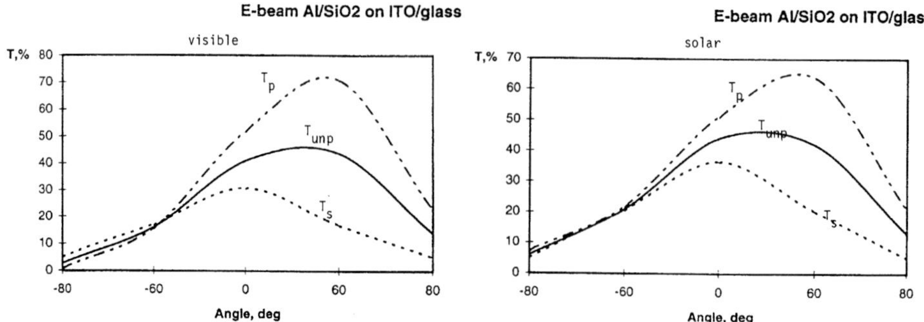

Figure 5. Visible and solar transmittance as a function of angle of incidence for the coating of fig. 4(b) [13].

Clearly the additional degree of freedom introduced by fabricating a coating with biaxial anisotropy allows differentiation between the aspects of preventing overheating and/or glare and guaranteeing sufficient daylight. However, it also increases the complexity of determining the correct configuration for a given location and of manufacturing, which may explain why angle-selective coatings are not yet produced on an industrial scale.

3.3 Thermochromic solar control coatings

An alternative approach to preventing overheating is to respond to the seasonal temperature variation with coatings with transmittance which automatically decreases when the coating temperature increases. Both thermotropic coatings, which will be presented in the next section, and inorganic thermochromic thin films possess this property.

Many transition metal compounds display thermochromism, a phase change from a semiconductor or dielectric state to a metallic state on temperature increase, accompanied by increased IR reflectivity, reduced emissivity and lower T_{sol} values. Among these, doped vanadium oxide was early identified as the most suitable candidate for energy control in buildings [15]. The transition temperature of 68 °C for bulk undoped VO_2 can be further reduced by alloying with metals such as W, Mo, Ta, Nb or Li, replacing O with F or promoting monocrystalline rather than polycrystalline thin film growth [16, 17]. For solar control purposes, the most succesful approach has involved co-doping with tungsten, achieved by dual-cathode sputtering or sputtering from an alloyed target. The transition temperature is depressed by 23 K/at. % W, as shown in fig. 6 [18]. The transition temperature can be reduced continuously to about 280 K, which is lower than the value needed for switching in an architectural glazing application. At the same time, interesting optical properties can be achieved, as illustrated in fig. 7 for an 80 nm film of $V_{0.986}W_{0.014}O_2$, with a transition temperature of 37°C [19]. While T_{vis} remains about 30 %, T_{sol} is reduced on switching from 33 % to 23 %, so that there is significant spectral selectivity for the higher-temperature state. From the transmittance and reflectance values at 2500 nm, it can be inferred that ε switches from less than 0.40 at high temperatures to less than 0.20 at low temperatures, so that it is always significantly lower than for uncoated glass, even if not equal to the values for low-e coatings based on noble metals.

Recent investigations have examined the effect of co-doping VO_2 with tungsten and fluorine simultaneously by reactive sputtering with CHF_3 from an alloyed target [20]. Increasing each dopant concentration was found to depress the transition temperature independently and additively. Similarly to doping with W or F alone, the decrease in transition temperature was accompanied by an increasingly metallic character for the low temperature state, so that for 0.7 % W doping, no switching occurred for F concentrations exceeding about 2 %.

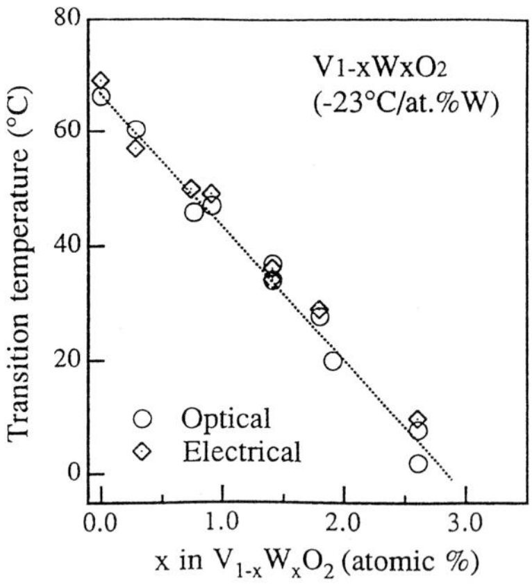

Figure 6. The optical and electrical transition temperature vs x in $V_{1-x}W_xO_2$ films [18].

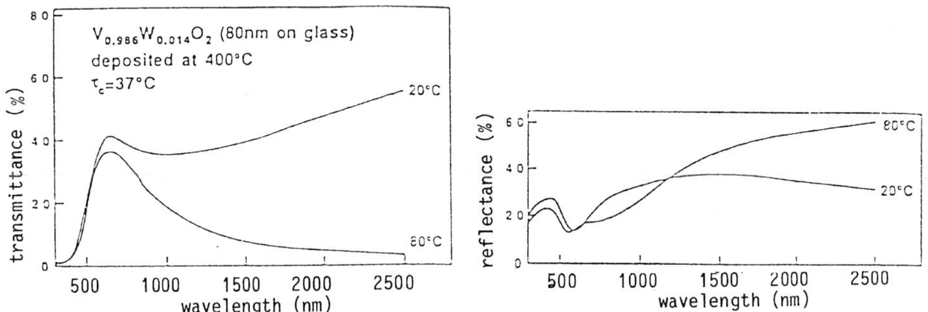

Figure 7. Normal-normal transmittance and near-normal reflectance spectra for an 80 nm film of $V_{0.986}W_{0.014}O_2$ [19].

By comparing figures 7 and 8, both for coatings of 80 nm thickness, it can be seen that fluorination acts to raise both T_{vis} and T_{sol} for films switching at similar temperatures. This would make it more suitable for combination with a second, low-e coated pane for use in climates with warm summers but cool winters. Growth of the thermochromic film was improved by depositing it onto an intermediate rutile (TiO_2) film.

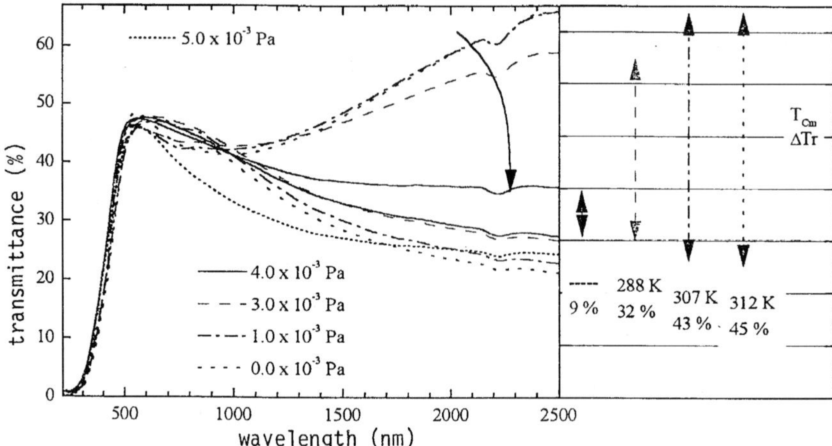

Figure 8. Normal-normal transmittance spectra of the double layer TiO_2-$V_{0.993}W_{0.007}O_{2-y}F_y$ for different values of y. The corresponding transition temperatures and the switching range at λ = 2500 nm are indicated in the box to the right [20].

3.4 Thermotropic solar control coatings

Whereas the previous three classes of coatings employed inorganic thin films, thermotropic solar control coatings all include polymers in their composition. They consist of two main components which are segregated at higher temperatures into domains of differing refractive index. As the domain dimensions are comparable to those in the solar spectral range, the radiation is scattered. If the proportion of back-scattering and amount of multiple scattering is high enough, much of the incident radiation is diffusely reflected, providing solar control. At lower temperatures, the composite material is optically homogeneous and transparent.

As these coatings are based on dielectric materials, their emittance values will be high, similar to that of uncoated glass. This means that combination with a low-e coated pane in an insulating glass unit will improve the thermal insulation to a greater extent than for the coating classes previously discussed. In addition, as the switching effect of such radiation-scattering coatings is usually greatest in the visible spectral range, the spectral selectivity of the low-e coating would effectively enhance the dynamic range of T_{sol} for the whole unit by filtering out much of the NIR component.

Although based on the same physical principles, the results reported in a previous review on this subject [21] no longer represent the current status, as the performance of some materials reported there has been improved, and investigation of some new types of materials has started recently.

The historic Christiansen effect, in which the temperature-dependent light scattering property results from the differing temperature dependence of the refractive index for two interdispersed but permanently segregated materials, has been reexplored for solar control purposes [22].

Figure 9 shows the temperature dependence of selected organic monomers which can be used for this purpose. The aliphatic compound characterised there, with its strongly temperature-dependent refractive index, was dissolved together with polyester in an aromatic solvent, applied to the substrate and dried under conditions which promoted growth of aliphatic crystals of appropriate dimensions in the polyester matrix. The normal-hemispherical transmittance spectra for the 80 μm thick film above and below the transition temperature of 35 °C are shown in fig. 10. For solar control purposes, the value of $T_{sol} = 0.55$ in the blocking state is still too high, but if the application technology allows it, could be reduced by increasing the film thickness. More effective switching is reported in another paper by the same group at this conference [23].

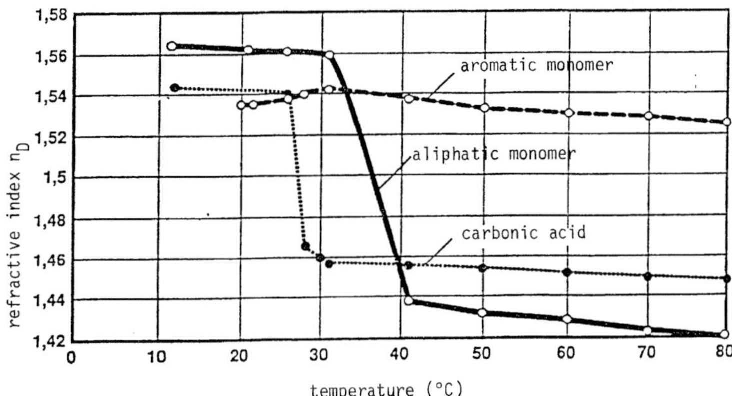

Figure 9. Temperature dependence of the refractive index of selected organic monomers (adapted from [22]).

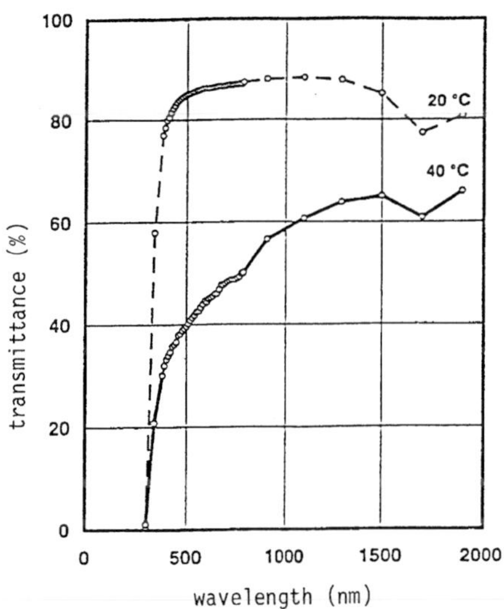

Figure 10. Normal-hemispherical transmittance spectra for an 80 μm thick thermotropic coating based on the Christiansen effect, below and above the transition temperature of 35 °C [22].

Whereas the Christiansen effect applies to permanently segregated mixtures, the following thermotropic materials consist of partly miscible components, with their miscibility decreasing with increasing temperature over the range of interest. A liquid crystal polymer gel is an example of such a material being investigated for thermotropic windows [24]. Domains of a liquid crystal within a polymer gel network undergo temperature-dependent transitions between an isotropic and an optically anisotropic phase. Cross-linked polyvinyl alcohol is favoured as the matrix, while polyethylene glycols, polyvinyl pyrrolidone or ethoxylised polydimethyl siloxane are the preferred liquid crystal candidates. Water is present in both components. The results published in [24] on temperature-dependent "transparency" do not yet provide a basis for assessing this material's effectivity in solar control. More recent results will be presented in another paper at this conference [25].

In polymer blends developed as solar control coatings, polystyrene-co-HEMA, with a refractive index of 1.58, forms the polymer network for the more mobile component, polypropylene oxide with a refractive index of 1.45 [26]. Distinct domains consisting predominantly of polypropylene oxide separate from the PS-HEMA network when the temperature is raised, and cause light to be scattered. The coatings are adhesive, so would normally be used as a laminating layer between two panes of glass. The temperature-dependent normal-hemispherical transmittance spectra for an 0.8 mm layer between low-iron glass slides, measured at Fraunhofer Institute for Solar Energy Systems in Freiburg, are shown in fig. 11(a), with the corresponding values for T_{vis} and T_{sol} being shown in fig. 11(b). Measurements and Monte Carlo simulation of the multiple scattering processes [27, 28] indicate that the values for the higher-temperature states of the polymer blend between two panes of 4 mm float glass are similar to those shown. The measurement error for small scattering samples, which leads to apparent transmittance values which are too low, is similar in magnitude to the additional absorption in two panes of float glass. A prototype glazed unit with an area of 0.46 m^2 has been fabricated with this polymer blend.

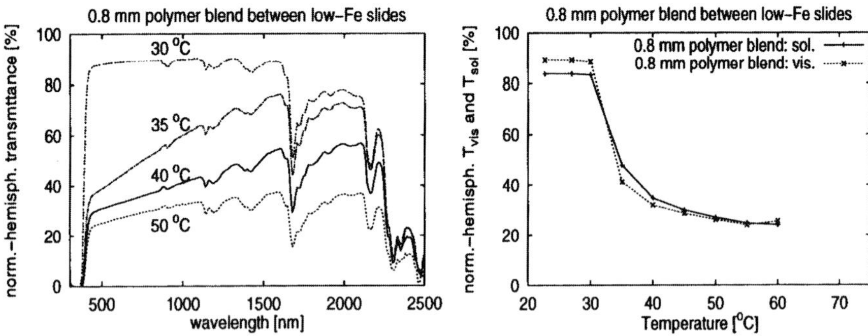

Figure 11. (a) Temperature dependence of normal-hemispherical transmittance spectra for an 0.8 mm thick layer of polymer blend between two low-Fe glass substrates. (b) Temperature dependence of T_{vis} and T_{sol} for the polymer blend of (a).

The most promising results published so far for thermotropic solar control glazing have been achieved with a hydrogel consisting primarily of water, with the remaining additives - a water-soluble polymer with hydrophobic groups, an amphipathic molecule and sodium chloride - being responsible for the thermotropic effect [29]. At low temperatures, the hydrophilic groups are hydrated, so that the water-soluble polymer and the amphipathic molecules are dissolved homogeneously, forming a transparent, isotropic aqueous solution. At higher temperatures, hydrophobic bonding between the polymer (a cellulose derivative) and the amphipathic molecule causes them to aggregate and results in gel formation. Light scattering occurs due to the difference in refractive indices between the aggregates and the surrounding free water. The degree of aggregation is controlled by the concentration of sodium chloride, which promotes the separation of free water.

Table 1. Comparison of the optical properties of the reviewed solar control coatings.

Coating	Glass substrate	Variable	Comment	Solar spectrum	T_{vis} [%]	T_{sol} [%]	$R_{vis, coating}$ [%]	$R_{sol, coating}$ [%]	ε_n
MO_x-Ag-MO_x	4 mm soda-lime	-	"classic" low-e	AM1.5 global	82	53	6	32	0.04
MO_x-Ag-MO_x-Ag-MO_x	6 mm soda-lime		high T_{vis}: TSET	AM1.5 global	73	37	5	42	0.03
MO_x-Ag-MO_x-Ag-MO_x	6 mm soda-lime		low TSET	AM1.5 global	44	23	29	42	0.04
SnO_2:F	4 mm soda-lime		"hard coating"	AM1.5 global	83	71	11	11	0.15
In_2O_3:Sn-Al/SiO_2	low-Fe	+ 60 ° unpol.	angle-selective	AM1.5 global	43	41	11 [31]	16 [31]	not spec.
"	"	0 ° unpol	"	"	41	44	1 [31]	3 [31]	<0.4
"	"	-60 ° unpol.	"	"	11	17	11 [31]	16 [31]	not spec.
$V_{0.986}W_{0.014}O_2$	Pyrex ™	T=25° C	thermo-chromic	AM2 direct	33	33	15	24	<0.2 (est.).
"	"	T=80° C	T_{trans} =37°C	"	30	23	13	22	<0.4 (est.)
aliphatic crystals in polyester	poly-carbonate	T=20°C	thermo-tropic	not spec.	86	84	not spec.	not spec.	>0.8
"	"	T=40°C	T_{trans} =35°C	"	43	55	"	"	"
PPO in PS-co-HEMA	2x low-Fe	T<30°C	thermo-tropic	AM1.5 global	89	83	not spec.	not spec.	0.89
"	"	T=40°C	"	"	32	35	not spec.	not spec.	"
"	"	T=50°C	"	"	26	27	not spec.	not spec.	"
1 mm hydrogel	2x 3mm soda-lime	T<28°C	thermo-tropic	not spec.	>80 (est.)	>75 (est.)	<10 (est.)	<10 (est.)	0.89
"	"	T≅38°C	"	"	<10	<10	>40	>35	"

Switching from the transparent to the saturated scattering state occurs typically over a narrow temperature range of less than 10 K [30]. The transmittance spectra in fig. 12(a) illustrate the clear and saturated scattering states for different thicknesses of a hydrogel layer between two 3 mm panes of float glass, in which the onset of scattering occurs at 28 °C. The thickest layer, with 1 mm, reduces T_{sol} to less than 10%. Combination with the reflectance spectra of fig. 12(b) reveals that solar absorptance probably still exceeds 30 %, but that this is primarily due to absorption in the float glass. The absorption due to the water bands does not affect T_{sol}, as atmospheric water has already removed these radiation bands from the terrestrial solar spectrum. Prototype glazing with an area of 1 m^2 has been prepared using this hydrogel.

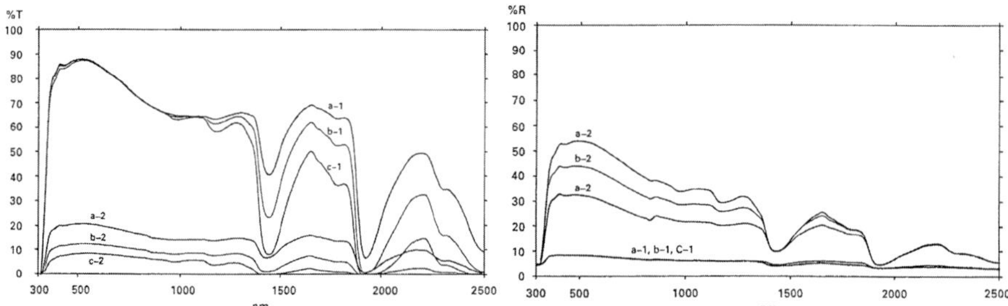

Figure 12. (a) Normal-hemispherical transmittance spectra for three layer thicknesses of a hydrogel between two 3 mm float glass substrates, at temperatures below and above the onset of switching at 28 °C. (b) The corresponding near normal-hemispherical reflectance spectra. Thicknesses: a - 0.25 mm, b - 0.5 mm, c - 1.0 mm [29].

4 Conclusion

Four distinct types of coatings have been presented, which can be used to enhance architectural glazing to achieve solar control. The selective nature of the properties summarised in table 1, whether spectral, angular or temperature-dependent, means that complete characterisation is inevitably complex. It also means that their solar control performance is strongly influenced both by the local climatic conditions and the building type in which they are installed. As yet, only spectrally selective solar control glazing has been produced commercially, but it will be interesting to observe which of the other types enters the market within the next ten years and contributes to more energy-efficient buildings.

5 Acknowledgements

I would like to thank my colleagues Markus Arntzen, Rolf Blessing, Lothar Herlitze and Manuela Scholz-Rehling from Interpane E & BmbH, Ekkehard Jahns from BASF AG, Alexandra Raicu from Sto AG, Geoff Smith from the University of Technology in Sydney, Kazuki Yoshimura from NIRIN in Nagoya and Volker Wittwer and Peter Nitz from the Fraunhofer Institute for Solar Energy Systems for their support in preparing this paper. This work has been partly funded by the German Ministry for Education, Science, Research and Technology (BMBF).

6 References

1. A. Georg, W. Graf, V. Wittwer. Conf. Proc. Euromat99, 27-30/9/1999, Munich, Germany.
2. R. Hulstrom, R. Bird, C. Riordan. Solar Cells 1985, 15, 365 - 391.

3. ISO 9050 Glass in building - Determination of light transmittance, solar direct transmittance, total solar energy transmittance, ultraviolet transmittance and related glazing factors 1997.
4. EN 410 Glass in building - Determination of luminous and solar characteristics of glazing 1998.
5. G. Wyszecki, W.S. Stiles, Color Science, 2nd ed., Wiley, New York, USA, 1982, p. 256.
6. J.A. Duffie, W.A. Beckman, Solar Engineering of Thermal Processes, 1st ed., Wiley, New York, USA, 1980, p. 13.
7. EN 673 Glass in building - Determination of thermal transmittance (U value) - Calculation method 1997.
8. C.G. Granqvist in Materials Science for Solar Energy Conversion Systems (Ed. C.G. Granqvist, Pergamon Press, Oxford, Great Britain, 1991, Chapter 5.
9. K. Häuser, B. Kramer, R.W. Schmid, R. Walk, Gestalten mit Glas, (Ed. INTERPANE GLAS INDUSTRIE AG), Beverungen, Germany, 1997.
10. G. Bräuer, Surf. Coat. Technol. 1999, 112, 358 - 365.
11. G.B. Smith, P.D. Swift, A. Bendavid, Appl. Phys. Lett. 1999, 75, 630 - 632.
12. G.W. Mbise, D. Le Bellac, G.A. Niklasson, C.G. Granqvist, SPIE 1994, 2255, 182 - 192.
13. G.B. Smith, S. Dligatch, R. Sullivan, M.G. Hutchins, Solar Energy, 1998, 62, 229 - 244
14. F. Jahan, G.B. Smith, Thin Solid Films 1998, 333, 185 - 190.
15. G.V. Jorgenson, J.C. Lee, Sol. En. Mat. 1986, 14, 205 - 214.
16. G.V. Jorgenson, J.C. Lee in Large-Area Chromogenics: Materials and Devices for Transmittance Control (Ed. C.M. Lampert, C.G. Granqvist), SPIE IS 4, 1990, in Part 4.
17. P. Jin, K. Yoshimura, S. Tanemura, J. Vac. Sci. Technol. A 1997 15 1113 - 1117.
18. P. Jin, K. Yoshimura, S. Iwama, S. Tanemura, SPIE 1996, 2531, 51 - 59.
19. P. Jin, T. Miki, K. Yoshimura, M. Tazawa, Y. Tai, S. Tanemura, Conf. Proc. 1st Int'l Conf. New Energy Systems and Conversions, Yokohoma, Japan, 1993, 301 - 304.
20. T.J. Christmann, Herstellung und Charakterisierung transparenter Metalloxide: VO_2 und ZnO, Doctoral Thesis, 1998, Justus-Liebig-Universität, Gießen, Germany.
21. H.R. Wilson, SPIE 1994, 2255, 214 - 225.
22. K. Gertis, H. Gödeke, Conf. Proc. Bauphysik-Kongreß, Berlin, Germany, 1997, 137 - 142.
23. H. Gödeke, Conf. Proc. Euromat99, 27-30/9/1999, Munich, Germany.
24. A. Seeboth, H.-R. Holzbauer, Int. Z. Bauinstandsetzen 1998, 4, 507 - 520.
25. H.-R. Holzbauer, T. Fischer, J. Kriwanek, A. Seeboth, Conf. Proc. Euromat99, 27-30/9/1999, Munich, Germany.
26. W. Eck, H.R. Wilson, H.-J. Cantow, Adv. Mater. 1995, 7, 800 - 803.
27. A. Raicu, H.R. Wilson, E. Jahns, P. Nitz, M. Rommel, K. Wendker, Conf. Proc. Eurosun '98, Portoroz, Slovenia, 1998, IV.1.14.1 - 7.
28. P. Nitz, J. Ferber, R. Stangl, H.R. Wilson, V. Wittwer, Sol. En. Mat. Solar Cells 1998, 54, 297 - 307.
29. H. Watanabe, Sol. En. Mat. Solar Cells 1998, 54, 203 - 211.
30. H. Watanabe, SPIE 1996, 2531, 42 - 50.
31. G.B. Smith, private communication.

The Gasochromism of Sputtered WO_3 Films

A. Georg , W. Graf, V. Wittwer
Fraunhofer Institute for Solar Energy Systems, Oltmannsstr. 5, 79100 Freiburg, Germany

1 Abstract

Sputtered films of WO_3 were coated with platinum and colored by exposure to diluted hydrogen gas. Bleaching was achieved with diluted oxygen. Because of its simple layer structure, this so-called gasochromic device is suitable for large-area window applications.

Sputtered WO_3 has a columnar structure with pores, which reach to the substrate.

During the coloration, H_2 is dissociated on the platinum, transferred into a pore of the WO_3 and creates water and an oxygen vacancy. The O vacancy diffuses into the interior of the WO_3 and the water slowly leaves the film. The bleaching in argon or vacuum is the reverse reaction. The bleaching by oxygen is the dissociation of the O_2 on the platinum, transfer into a pore and recombination with the oxygen vacancy.

Chemisorbed water is essential for the diffusion of the hydrogen along the pore surface. Films with a small amount of water show a slow coloration due to a slow diffusion.

The main aging process in systems using the air for bleaching was detected to be the blocking of the reactive WO_3 surface by adsorption of an impurity in the air.

2 Introduction

Gasochromic films consist of an electrochromic layer such as WO_3 and a very thin coating of a catalyst (fig.1). Hydrogen gas is dissociated on the catalyst into H atoms, which color the electrochromic film. Diluted hydrogen is used, such that the H_2 concentration is below the combustion limit. The coloring mechanism is similar to the electrochromic reaction.

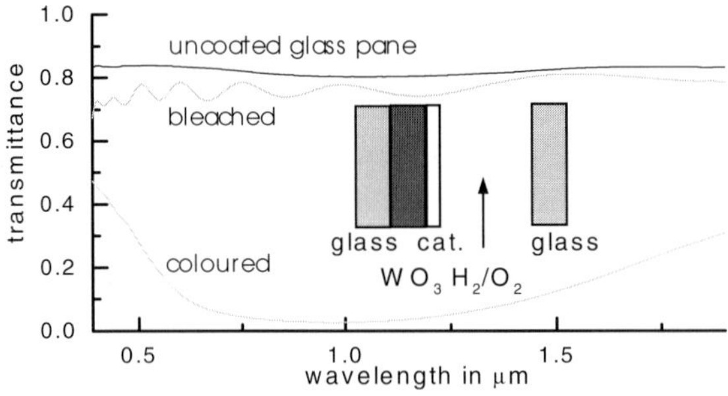

Figure 1. Typical spectrum of the transmittance and layer structure of a gasochromic device

Both approaches are being investigated for application in windows with variable transmittance. Gasochromic windows differ from electrochromic windows in employing a simpler inexpensive coating configuration, which results in a higher solar transmittance. On the other hand, electrochromic windows are simpler with regard to the system technology. With a double glazing unit in the gasochromic configuration, switching times below 10s and a switching range in the solar transmittance of 72 - 5% may be achieved. To demonstrate the long term stability, a window with an area of 1.1 m*0.6 m was switched 20,000 times over two years without its performance changing significantly. System aspects including simulations of the building energy savings were published recently [1].

3 Film Preparation

Four different types A, B, C and D of WO_3 films were produced by sputtering and evaporation. IR spectroscopy revealed a decreasing amount of W=O (970cm^{-1}), W-O-H (1420cm^{-1}), H_2O (1620cm^{-1}) and O-H (2850cm^{-1}, 3010cm^{-1}, 3040cm^{-1}) bonds in the order A, B, C and D (fig. 2). XRD identified an amorphous structure for all four production methods [2].

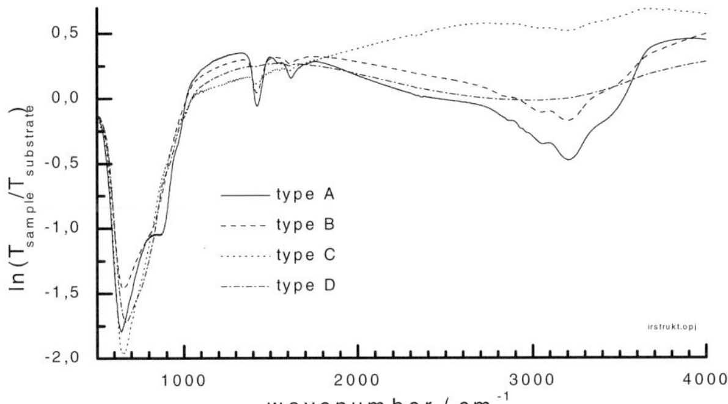

Figure 2. IR transmittance of types A, B, C,and D deposited on Si in the range of the W-O (500..1000cm^{-1}), W-O-H (1420cm^{-1}), H_2O (1620cm^{-1}) and O-H (2850cm^{-1}, 3010cm^{-1}, 3040cm^{-1}) bonds.

The thickness of the WO_3 layers was kept in the range of 200nm up to 1000nm.

Very thin films of Platinum were sputtered onto the WO_3 layers. The thickness was well below the percolation threshold, as could be concluded by determining the dependence of the electrical conductivity on the Pt film thickness. Thus the Pt layer consists of isolated islands. The very thin film of Pt does not disturb the optical measurements.

From Auger Electron Spectroscopy, we know that the platinum, which was sputtered onto the WO_3 films, remains on the surface and does not penetrate into the pores of the WO_3.

4 The Porous Structure of WO_3

The porous structure of the WO_3 films was demonstrated by a simple experiment: the platinum was deposited directly onto the glass substrate and then the WO_3 layer was added.

These films colored and bleached as quickly as the regular films with the Pt on top of the WO_3. However, a catalyst must be present to attain a reasonable response time with the four types of WO_3 investigated. For a WO_3 film without any Pt, very slow coloration was observed (10^{-9} times the coloration velocity of the WO_3 films with Pt!), which corresponds to a switching time of about 7 years. Thus we can conclude that the films contain pores which reach down to the substrate and are large enough to allow rapid diffusion of H_2 and O_2.

A columnar structure with columns of a size of about 150nm of similarly prepared films was demonstrated in [3]. However, the resolution was too bad to allow the detection of structures in the range of 10nm. TEM measurements, which were done on film type B with a thickness of about 15nm showed islands with a typical distance of about 10..20nm. These may be the nuclei for the columns. Columns with a size of 10..30nm are also reported in refs. [4] and [5] for evaporated and in [6] for sputtered WO_3 films.

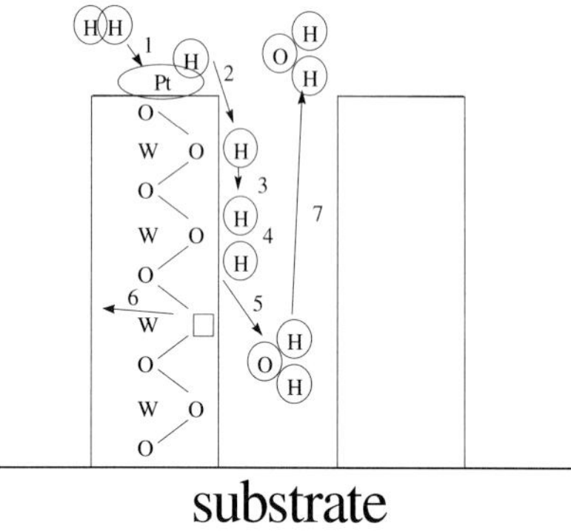

Figure 3. Mechanism of coloration by H_2.1: adsorption and dissociation of the H_2 on the Pt, 2: transfer of the H from the Pt onto the WO_3 surface 3: diffusion of H along the interior WO_3 pore surface, 4: formation of an intermediate state, 5: formation of an oxygen vacancy and H_2O, 6: diffusion of the oxygen vacancy, 7: escape of the H_2O.

5 The Mechanism of the Coloration

The mechanism of gasochromic coloration of porous WO_3 coated with a platinum catalyst is shown in fig. 3 [7]. The WO_3 consists of columns, which enclose pores reaching down to the substrate. The H_2 is adsorbed and dissociated on the Pt (step 1), and transfers onto the WO_3 surface (step 2). There it adsorbs onto and diffuses along the interior pore WO_3 surface (step 3), forms an intermediate state (step 4), and then finally an oxygen vacancy and H_2O (step 5). The oxygen vacancy, the so-called "color center", diffuses into the interior of the column (step 6), and some of the H_2O leaves the film (step 7). Steps 4, 5 and 6 can be treated as a single process with one time constant [7]. Fig. 3 is only schematic. Actually, water is very important for the transport process of the H along the surface. Formation of H_3O^+ and e^- from the adsorbed H_2O and H is probable [8]. The importance of adsorbed water for rapid proton diffusion is shown in [9] and [10]. Bleaching in Ar or vacuum is the reverse process, whereas during bleaching in O_2, the oxygen is dissociated on the Pt surface, diffuses into the WO_3 pores, and recombines with the oxygen vacancies, which diffuse from the interior to the surface of the column.

6 The Optical Response

The kinetics of coloring and bleaching were studied at room temperature by suddenly increasing the hydrogen or oxygen gas concentration and recording the optical response of the device. Before each coloring or bleaching process, the sample chamber was flushed with argon. The optical density, i.e. the decadic logarithm of the transmittance T, was measured in situ with a LED (wavelength 650 nm) and a Si photodiode. The velocity of coloration and bleaching is defined as the derivative of the optical density with respect to time at the beginning of the coloration or bleaching process.

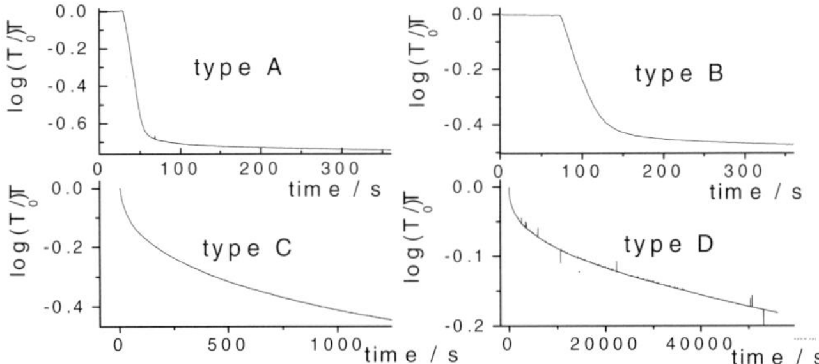

Figure 4. Optical response of the four types A, B, C and D. The coloration of type C was measured at 60°C to accelerate the coloration. For types A and B, the H_2 flushing was started at 30 and 70s, respectively.

By using a set of flow controllers and mixing H_2 or O_2 with dry Ar and Ar with 100% relative humidity, it was possible to generate various gas concentrations and humidities.

From electrochemical experiments, we know that the optical density is proportional to the concentration of the color centers, and the coloration efficiency, i.e., the ratio of the optical density to the inserted charge per area was evaluated to be 50 cm^2/C at 650 nm for all four types.

The optical response shows an exponential behavior for the sample types A and B (fig. 4), which is typical for a surface reaction [7], whereas the square root responses of types C and D are due to diffusion.

Variation of the film thickness of type D did not change the optical response. Therefore, it is the diffusion along the pore surface, which determines the kinetics. Slow diffusion of the color centers into the columns should lead to a linear increase of the coloration rate with the film thickness, which is not observed.

Drying of the WO_3 films of type A and B in vacuum at 100°C or 200°C decreases the water content of the films, as could be detected by IR spectroscopy. The optical responses of such dried samples show a square root.

The coloration velocity of fresh films of type B is independent on the WO_3 thickness. However, for films which were exposed to unfiltered air in a closed box for several days or for several hours in a cell with a steady stream of air, the coloring in H_2 and bleaching in Ar as well as in O_2 is reduced drastically [11]. Furthermore, then the coloration velocity depends linearly on film thickness. This is caused by an adsorption of an impurity of the air, which adsorbs on the WO_3 surface, and reduces the reactive surface [12]. A poisoning of the catalyst, i.e. a reduction of the hydrogen dissociation or spill-over was not observed. Thus, for fresh films, the reaction on the Pt is slow compared to the reaction on the WO_3, and for films aged in air the reaction on the WO_3 surface is slow compared to the Pt. In both cases, the escape of the water (step 7) is even slower than these steps.

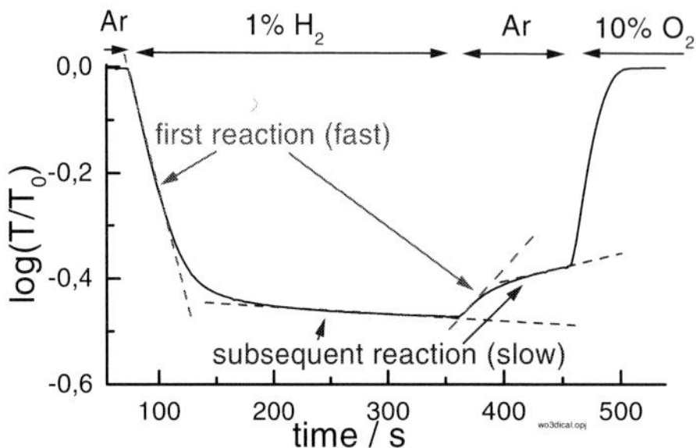

Figure 5. Coloring in 1% H_2 and bleaching in Ar and in 10%O_2 (50% rel.hum.) of a film of type B.

Fig. 5 displays a typical optical response of a film of type B, which was poisoned by air. Its shape remains the same for films of type A. The first process, which defines the coloration velocity, is determined by the combined time constant of steps 4, 5 and 6. Then follows a slow steady increase of the optical density. This second process is due to the escape of the water (step 7). Similarly, for the bleaching in argon, there is a first time constant, which is the reverse reaction of steps 1-6, and a second process due to the condensation of water on the film out of the atmosphere.

7 References

[1] A. Georg, W. Graf, D. Schweiger, V. Wittwer, P. Nitz, H.R. Wilson, Solar Energy 1998, **62**, 215-228

[2] A. Georg, W. Graf, V. Wittwer, Solar Energy Materials and Solar Cells 1998, **51,** 353-370

[3] D. Schweiger, A. Georg, W. Graf, V. Wittwer, Solar Energy Materials and Solar Cells 1998

[4] A. Agrawal, M. Habibi, Thin Solid Films 1989, **169,** 257-270

[5] F.G.K. Baucke, K. Bange, T. Gambke, Displays 1988, **9**, 179-187

[6] A.P. Giri, R. Messier, Mater. Res. Soc. Symp. Proc. 1984, **24**, 221-227

[7] A. Georg, W. Graf, R. Neumann, V. Wittwer, accepted by Solid State Ionics (1999)

[8] M. Stolze, L.-K. Thomas, Verhandl. DPG (VI) 1999, **34**, 510

[9] T.C. Arnoldussen, J. Electrochem.Soc. 1981, **128**, 117-123

[10] J. Vondrák, J. Bludská, Solid State Ionics 1994, **68**, 317

[11] A. Georg, W. Graf, R. Neumann, V. Wittwer, proceedings of the 2nd EuroSun conference 1998 in Portoroz, Slovenia.

[12] A. Georg, W. Graf, R. Neumann, V. Wittwer, submitted to Solar Energy Materials and Solar Cells

Thermoelectric Performance and Long Term Stability of Plasma Spray Formed Iron Disilicide

K. Schackenberg, E. Müller, W.A. Kaysser
German Aerospace Center (DLR), Institute of Materials Research, D-51170 Cologne, Germany.
J. Schilz
EG&G Heimann Optoelectronics, D-65020 Wiesbaden, Germany.

1 Introduction

Iron disilicide is considered to be very promising as an economic material for thermoelectric (TE) low power generation and thermal sensor applications. It offers attractive solutions for supplying electronic control and display elements in self-standing systems, preferentially, when a heat flux over a temperature difference of a few hundred Kelvin is available. In our study, thermal spray forming turned out to be a promising way for mass production of thermoelectric devices [1]. The method is capable of consolidating large amounts of material, with simultaneous lateral patterning and composition control. Due to their large technological relevance, shrouded plasma and vacuum plasma spray formed samples of p- and n-type $FeSi_2$ (doped with Al or Co) have been surveyed. We investigated their thermoelectric transport properties as depending on their distinct microstructure and phase composition. A comparison to hot pressing as standard consolidation technique is drawn.

A technological challenge is to keep the exact stoichiometry during the consolidation process in order to obtain high thermoelectric performance. The influence of oxygen incorporation, a deviation from the $FeSi_2$ stoichiometry and the local distribution of the doping elements will be discussed. Special respect is attributed to diffusion and rearrangement processes during an extended high temperature treatment.

2 Consolidation of $FeSi_2$ from Gas Atomized Powders

2.1 Characterization of the Powders

The starting material for the consolidation of $FeSi_2$ bulk material consists of gas atomized powders with the nominal stoichiometry $FeAl_{0.06}Si_{1.95}$ for p-type and $Fe_{0.95}Co_{0.05}Si_2$ for n-type semiconducting iron disilicide, respectively. The powder particles are almost spherically shaped having a diameter $\leq$ 53 µm. Because of the rapid cooling during the powder processing, the high temperature phase composition consisting of α-Fe_2Si_5 and ε-FeSi is quenched [2]. In electron probe microanalysis (EPMA) investigations the monosilicide has been found to form precipitates of µm size. A chemical analysis of the powders revealed a silicon loss during the gas atomizing process. Due to the initially over-stoichiometric content of silicon and aluminum in the p-type material the stoichiometry shift is smaller here, than in case of n-type $FeSi_2$:Co.

2.2 Hot Pressing and Plasma Spraying

The consolidation of standard specimens was carried out in a hot uniaxial press (HUP) at 900 °C under vacuum (< 10^{-5} mbar) for 20 min at 30 MPa. From a broad variety of thermal spray processes, plasma spraying under reduced atmosphere (200 mbar, vacuum plasma spraying, VPS) and shrouded plasma spraying (with inert gas shielding, SPS), have been investigated. During the spraying process, the powder is fed into a hot plasma jet, where it is heated and accelerated. In that way, the powder particles are sprayed onto a cooled substrate; there they solidify, forming a bulk material with a flake like structure. The thickness of sprayed $FeSi_2$ layers ranges from below 1 up to 10 mm.

2.3 Oxygen Incorporation

In order to prevent oxidation, the plasma spraying process was carried out either under low pressure (VPS) or with an inert gas shielding (SPS). Nevertheless, an oxygen content of 0.2 wt% has been found by chemical analysis of HUP and SPS specimens. For the VPS material the oxygen content was significantly higher (0.6 wt%). The oxygen is incorporated as SiO_2 due to its high formation enthalpy. In case of aluminum doping, the formation of Al_2O_3 is even more probable. In any case, both of the above-mentioned oxidation mechanisms will lead to an under stoichiometric composition and finally result in additional ε-phase precipitates. This will be discussed in the following section.

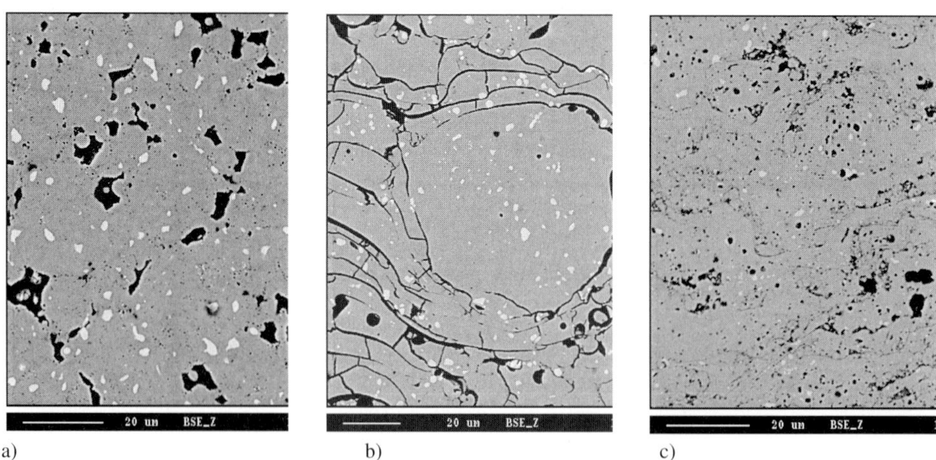

Figure 1. Microstructure of iron disilicide from different consolidation techniques after 50 h annealing at 800 °C: a) HUP-$FeSi_2$:Co, b) SPS-$FeSi_2$:Co, c) VPS-$FeSi_2$:Al.

3 Microstructure

3.1 Comparison of Hot Pressed and Plasma Spray Formed Material

Immediately after consolidation the specimens from all consolidation processes contain a phase mixture of α-Fe_2Si_5, ε-FeSi, and semiconducting β-$FeSi_2$. Only the latter is desired for TE applications. The content of high temperature α-phase was at maximum for the SPS material. This is due to the rapid cooling of relatively small specimen during the spraying process, which is accelerated additionally by the inert gas shroud. On the other hand, for the larger hot pressed specimens the cooling down to room temperature takes about 2 h, which is long enough to partially transform α-Fe_2Si_5 + ε-FeSi to the low temperature β-phase. After a heat treatment of 1 h at 800 °C the high temperature phase was exhausted in all specimens. According to the lack of silicon in the gas atomized powder, there was still some residual ε-FeSi found by X-ray diffraction (XRD).

The microstructure of hot pressed and plasma spray formed iron disilicide looks much different. The hot pressed material shows a few clearly separated pores. They are arranged like chains around areas of the original powder particles' dimension (**Figure 1**a). The SPS $FeSi_2$ shows a distinctly coarser microstructure with macroscopic separations between almost undisturbed areas – the so called splats (**Figure 1**b). For the VPS material, separate splats can be distinguished as well, but the separations between neighboring splats are not as clear as for the SPS material (**Figure 1**c).

The matrix of all samples consisted of β-$FeSi_2$. The content of cobalt in the n-type material was almost as high as expected from the powder's composition. On the other hand, the matrix of all p-type specimen contained much less aluminum than the powder. The structure of all Co-doped samples and of VPS $FeSi_2$:Al contained some inclusions which have been identified as ε-phase by EPMA. This formation of ε-precipitates was already expected from the stoichiometric shift.

Contrarily, there were no monosilicide inclusions found in HUP and SPS $FeSi_2$:Al. Instead, we identified precipitates with a much higher aluminum content than in the matrix. Obviously, the solubility of aluminum in β-$FeSi_2$ is limited in a way promoting the formation of Al rich precipitates. From [3] we suppose, that these inclusions contain a ternary phase $Fe_{32}Al_{38}Si_{30}$ (called τ_8) besides ε-FeSi and β-$FeSi_2$. Instead, VPS-$FeSi_2$:Al contains enough oxygen incorporated during the consolidation process to oxidize more than half of the Al doping. The remaining amount of freely available Al doping can be solved completely in the β-matrix of the VPS material.

3.2 Aging Behavior under Extended Heat Treatment

The heat treatment was carried out at 800 °C under vacuum for an overall annealing time of 1000 h. The extend of microstructure disorder unexpectedly did not decrease significantly during the annealing process.

Furthermore, we observed a decrease of the Co concentration within the iron disilicide matrix of the plasma spray formed samples. At the same time, we perceived a growing volume portion of ε-inclusions. Due to the higher solubility and, apparently, a higher chemical potential of Co doping in ε-FeSi than in β-$FeSi_2$, the growth of monosilicide precipitates causes the above-mentioned loss of doping within the matrix. For VPS-$FeSi_2$:Co we found up to 20% of iron in the monosilicide replaced by cobalt after 1000 h annealing, whereas in the matrix there were only about 5% of Fe substituted by Co.

For the $FeSi_2$:Al, the reduction of the doping concentration was even stronger. After 1000 h, the Al content corresponded very well to the value estimated from the amount of Al_2O_3 formed by reaction with the oxygen incorporated during consolidation. Obviously, the complete oxidation of the aluminum doping did not take place during the consolidation process but mainly during the extended heat treatment.

The aging behavior of both Al and Co doped $FeSi_2$ can be explained by the following mechanism: Preferentially at the surface of the powder particles, the formation of SiO_2 during the consolidation process causes a thin Si depletion layer. While annealing the corresponding Fe excess is balanced by diffusion processes. At the phase boundary between β-matrix and ε-inclusions the Fe excess is compensated. Thus, some β-$FeSi_2$ transforms to ε-FeSi, there, finally leading to the growth of ε-precipitates. At the same time, Si and Al diffuse away from these inclusions. The aluminum is moved to the SiO_2. Because Al_2O_3 is much more stable than silicon dioxide, the redox reaction $4\ Al + 3\ SiO_2 \rightarrow 3\ Si + 2\ Al_2O_3 + 622$ kJ/mol takes place, finally resulting in a continuous diminution of freely available aluminum. According to this aging mechanism, an enrichment of Al and O has been established at the splat boundaries of SPS- and VPS-$FeSi_2$:Al after 100 h annealing at 800 °C (**Figure 2**).

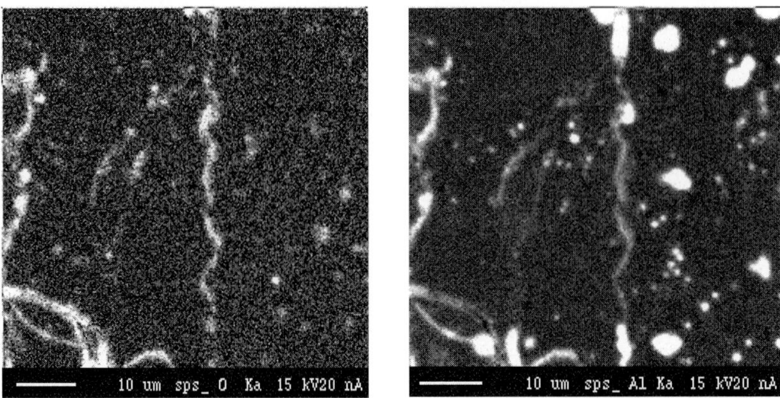

Figure 2. EPMA mappings of SPS-$FeSi_2$:Al after 100 h annealing at 800 °C showing an enrichment of O (left) and Al (right) at the splat boundaries and τ_8-precipitates (only Al enriched).

4 Thermoelectric Performance

For the evaluation of the thermoelectric performance the thermoelectric figure of merit is defined as $Z = S^2\ \sigma / \kappa$ [4]. Here S is the Seebeck coefficient, σ and κ represent the electrical and thermal conductivity, respectively. Z has to be maximized for thermoelectric converter applications.

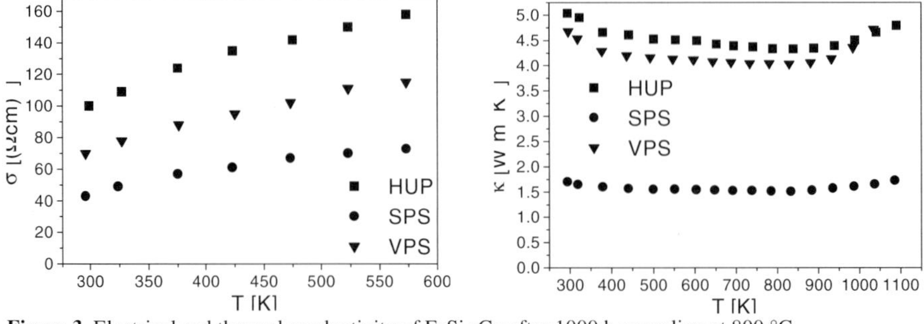

Figure 3. Electrical and thermal conductivity of $FeSi_2$:Co after 1000 h annealing at 800 °C.

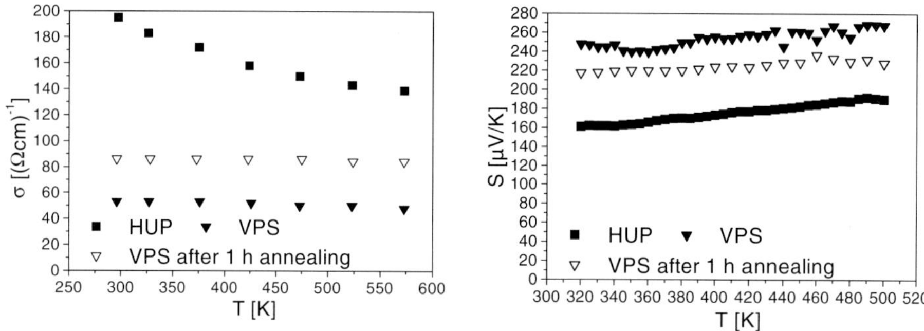

Figure 4. Electrical conductivity and Seebeck coefficient of $FeSi_2$:Al after 1000 h at 800 °C.

5 The Influence of Microstructure Disorder

The distinct microstructure of the plasma spray formed iron disilicide correspondingly effects a decrease of the electrical and thermal transport properties. Thus both σ and κ are reduced compared to the hot pressed standard material. For the SPS formed $FeSi_2$:Co the diminution of the electrical conductivity is significantly lower than this of the thermal conductivity (**Figure 3**). Due to the Seebeck coefficient being almost independent from the consolidation technique, the figure of merit of the Co-doped SPS material has been improved by about 30%. Obviously, the thermal contact between neighboring splats is worse than the electrical. The increased σ/κ was almost unaffected by annealing.

For the VPS-$FeSi_2$:Co the σ/κ ratio is reduced compared to the hot pressed reference. We suppose the existence of oxide layers between the splats providing a better thermal than electrical contact between neighboring splats.

5.1 The Influence of Oxygen Incorporation

Especially for the VPS-$FeSi_2$:Al the doping concentration within the $FeSi_2$-matrix was remarkably reduced during the long term annealing. Accordingly, the charge carrier density [5] and the electrical conductivity were reduced at the same time, whereas the Seebeck coefficient was significantly increased (**Figure 4**). Because the thermal conductivity was almost unchanged by the heat treatment, the decrease of the electrical conductivity finally resulted in a degradation of the thermoelectric figure of merit. Because of the higher Seebeck coefficient, we nevertheless found a figure of merit Z slightly above the hot pressed standard's value.

6 Conclusion

The formation of iron disilicide by plasma spraying provides bulk material with a thermoelectric performance that is comparable or even better than that of hot pressed standard $FeSi_2$. The oxygen incorporation during consolidation leads to the formation of ε-precipitates with an enrichment of Co doping and thus effects a reduction of the Co concentration within the material's matrix during long term annealing. In VPS-$FeSi_2$:Al an oxidation of the doping has been observed, finally leading to a reduction of the charge carrier density and the electrical conductivity.

We are greatly indebted to Dr. G. Schiller, DLR Stuttgart, and Dr. R. Mathesius, RWTH Aachen, for the fabrication of the plasma spray formed specimens.

7 References

1. J. Schilz, M. Riffel, R. Mathesius, G. Schiller, R. Henne, R.W. Smith, Proc. XV Int. Conf. on Thermoelectrics ICT96, IEEE, Piscataway, NJ, USA, 1996, 71-74.
2. J.-P. Piton, M.-F. Fay, C. R. Acad. Sc. Paris 1968, 266, 514-516.
3. G. Ghosh in Ternary Alloys, Vol. 5 Al-Cu-S to Al-Gd-Sn (Ed.: G. Petzow, G. Effenberg), VCH, Weinheim, 1992, 394-438.
4. H.J. Goldsmid in CRC Handbook of thermoelectricity (Ed.: D.M. Rowe), CRC Press, Boca Raton, 1995, Chapter 3.
5. K. Schackenberg, E. Müller, J. Schilz, H. Ernst, W.A. Kaysser, Proc. XVII Int. Conf. on Thermoelectrics ICT98, IEEE, Piscataway, NJ, USA, 1998, 422-425.

Optimization of Spectrally-Selective Coatings for Solar Absorbers

Zorica Crnjak Orel, Marta Klanjšek Gunde
National Institute of Chemistry, Ljubljana, Slovenia

1 Introduction

The inexpensive selective surfaces for solar absorbers were prepared by application of black paint on the high-reflective substrate. The layers have to be transparent in the infrared in order to support the low thermal emittance of the substrate. For this purpose, the optical properties of coatings have to be optimized to minimize the expense of the final product. The selectivity was attained by the mutual effect of a highly absorptive black paint layer and low emitting (i.e., infrared reflecting) metal substrate. Optimized paint coatings are not thicker than a few micrometers and exhibit high opacity, leading to energy-efficient selective coatings for solar collector applications. The painted samples are characterized by high absorption, finite sample thickness, nonideal support material, and smooth front surface. These properties distinguish our samples from those of other studies in this field. To design a functional pigmented layer, the optical properties of all constituents have to be known separately. Due to this reason the diffuse reflectance of black thickness-sensitive spectrally selective (TSSS) paints was analyzed. For theoretical consideration of paint layers, the simple Kubelka-Munk (KM) theory was used. It is the almost universally applied theoretical approach within the color using industry (1). It relates diffuse reflectance of a pigmented layer to two phenomenological coefficients, absorption (K) and scattering (S), thickness of the layer, and reflectance of the substrate. The optical properties of layer material are involved in both coefficients. This enables optimal thickness calculation (2), i.e. the theoretical prediction of the best thickness value that will give the highest solar absorptance and simultaneously, the lowest thermal emittance of the respective paint. The KM coefficients depend also upon addition of fumed silica (dispersive agent). Applying KM theory, the degree of pigment dispersion was quantified (3). This approach was an original contribution that modify the widespread used flocculation gradient technique to be useful also for black-pigmented paints what was not possible before. Applying the KM theory, the solar absorptance gradient was predicted in advance (4). By KM theory, also the optimal pigment-to-volume concentration ratio (PVC) of paints was determined.

In this article, our many-years research in this field is discussed and some new results are presented. Applying the KM theory together with some exact optical theories, the behavior of solar energy flux inside the coating system was analyzed. These results were the great help in defining the optimal PVC from the optical point of view: preparation of paint from higher PVC does not necessarily show the best covering efficiency in the UV-VIS spectral region. Low PVC is beneficial in transparency needed in the IR spectral region and for better adhesion properties. With suitably low PVC the price of solar collector is also reduced.

2 Experimental with Calculation

The paint was prepared from 50 % solution of phenylmethyl-polysiloxane in mixture of organic solvent with an addition of pigment PK 3060 (Ferro, Holland), fumed silica (Aerosil 380 - Degussa), silane coupling agent, and dispersive agent. The primary particles are almost spherical with average size of 0.4 μm.

Paints with different PVC were prepared (PVC = 0.16, 0.18, 0.20, 0.24, and 0.28) from pigment concentrate with PVC = 0.32 by adding the resin. They were applied over Al foil by draw-bar coater. Each paint was applied at eight different thicknesses.

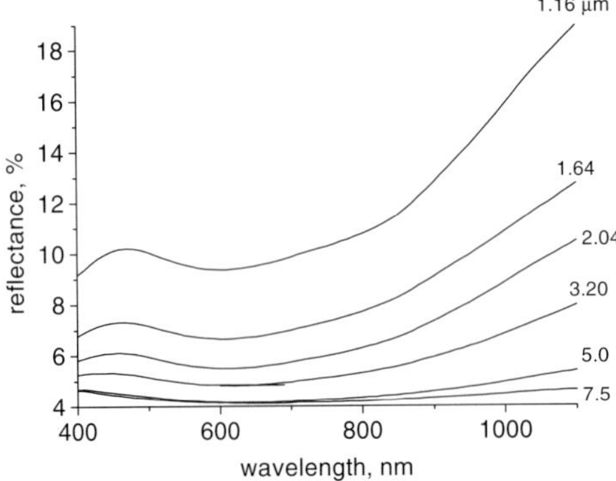

Figure 1. Diffuse reflectance spectra of paint with PVC=0.18 at different coating thickness.

Diffuse reflectance spectra of all prepared samples, as well as of the bare chromated aluminum substrate, were recorded in the 400-1100 nm spectral region. The measurements were performed on the ACS Spectra-sensor II (Datacolor International) with barium sulphate coated integrating sphere in the specular included mode. For paint with PVC=0.18 they are represented in Figure 1.

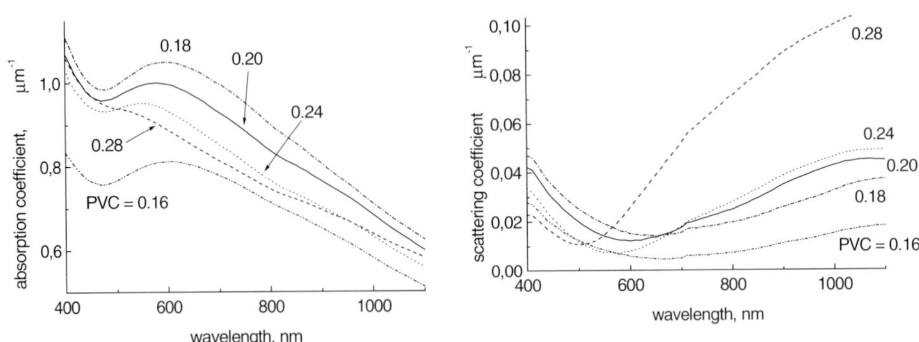

Figure 2. Spectral dependence of Kubelka-Munk absorption and scattering coefficients for paints with different PVC.

The spectral dependence of the KM absorption (*K*) and scattering (*S*) coefficients was determined from measured *R(d)* by the curve fitting procedure. The calculation was performed on a personal computer using standard interactive software package for scientific and engineering numeric computation. By repeating the calculation for each wavelength in the considered spectral region, K and S spectra were obtained. They are shown in Figure 2. Applying these coefficients, cross-sections for absorption and scattering were calculated and the efficiencies for both processes (Figure 3). In all calculations, spherical shape of pigment particles was assumed.

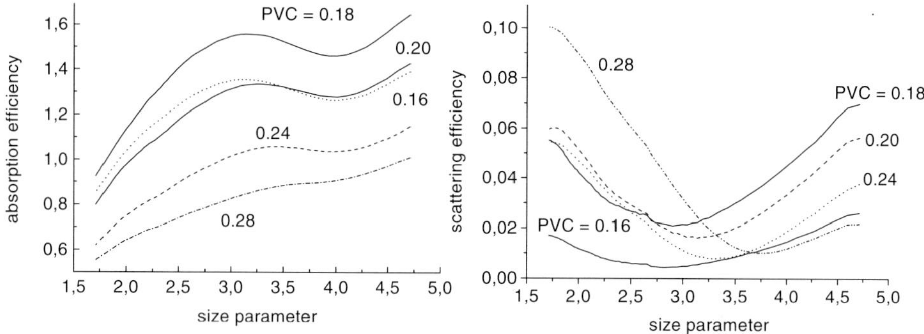

Figure 3. Absorption and scattering efficiencies for paints with different PVC.

3 Discussion

The KM coefficients provide a quantitative measure for absorption and scattering abilities of the entire coating system. The resin and all additives in our paints are optically homogeneous and colorless all over the considered spectral region. Therefore we suppose that the calculated K and S spectra predict the optical performance of coatings only in terms of pigment particles involved.

When PVC increases, it influences directly on K and S values that are expected to be higher due to the bigger amount of pigment particles where absorption and scattering can take place. However, the region where K and S increase with PVC is limited. For our samples, absorption reaches maximum at PVC = 0.18 all over the considered spectral region. In scattering, such behavior was obtained only for $\lambda < 480$ nm (see Fig. 2). Deviation from linearity of the S(PVC) curve at low PVC and also it's diminishing after the maximum value are considered as the effect of particle crowding (packing or shadowing effect) (5). The minimum in S is a critical point and the corresponding PVC value is called the critical pigment-to-volume concentration ratio (CPVC). At this pigment loading, pigment particles are not able to achieve full contact to the binder system. Increase of S-values beyond CPVC is considered due to porosity within the film, i.e. the effect of pigment-air interfaces (5). It is not economical to produce coatings at PVC beyond the maximum S. At present no approach is available to calculate this value before preparing and measuring samples. Contrary, the pigment's CPVC can be calculated from pigment's oil absorption data (5). For our pigment, CPVC = 0.35. The pigment concentrate used to prepare our paints had lower PVC (0.32). The considered paints were prepared from $0.16 < \text{PVC} < 0.28$. Therefore we may suppose, that the obtained deviation from increase of K and S upon PVC is caused by distribution of pigment particles inside polymer matrix.

The efficiencies of absorption and scattering were calculated by dividing corresponding cross section with particle cross-sectional area (Fig. 3). The efficiency of absorption is the highest at PVC = 0.18 and then diminishes. In the scattering efficiency curves the interference effect is much more pronounced that it is in the absorption. This effect was interpreted as interference between the incident and the forward-scattered light beam (6). Most likely the *S* curves are located at the first minimum of the interference structure. According to theoretical calculations (7) the minimum shifts to larger size parameter when elongation of the spheroid increases. Our curves obey properties of these calculations if the pigment particles form spheroidal groups with average orientation perpendicular to the coating surface and if these groups become longer when PVC increases.

This way of particle gathering helps to understand also the diminish of absorption efficiency when PVC increases beyond 0.18. The black pigment is one of highly absorbing materials. In an elongated group, the beam of light may cross only first few particles. When such groups are longer (at higher PVC), and when their average orientation is perpendicular to the coating surface, the number of particles crossed by light will be smaller. The absorption efficiency of coatings may therefore diminish although PVC increases. Such behavior is illustrated in Fig. 4.

This phenomenon explains why practically the same solar absorptances were obtained for all samples with PVC > 0.18 (8). Extinction of solar radiation inside coatings do not increases with larger amount of imbedded pigment particles. This helps to define the most efficient PVC.

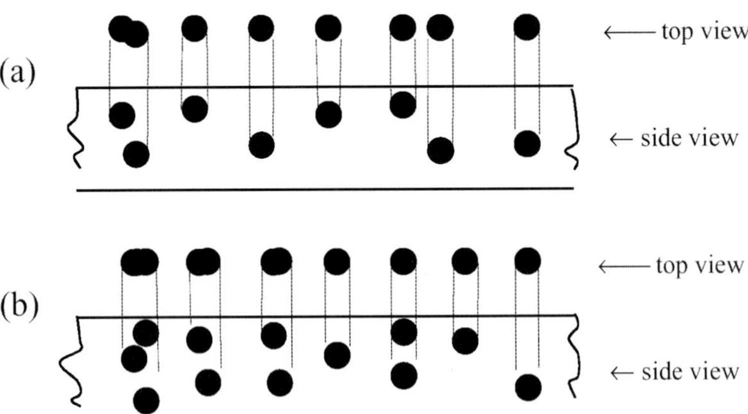

Figure 4. Two-dimensional sketch of pigment particle distribution within dry coatings when PVC increases. Increase of PVC from (a) to (b) corresponds to PVC=0.18 and PVC=0.28.

4 Acknowledgement

This work was supported by the Ministry of Science and Technology of Republic Slovenia (Contracts J2-1468 and P1-0512).

5 References

1. D.C. Rich, J. Coatings Technol. 1995, 67, 53-60.
2. M. Klanjšek Gunde, J. Kožar Logar, Z. Crnjak Orel, B. Orel, Thin Solid Films 1996, 277, 185-191.
3. M. Klanjšek Gunde, Z. Crnjak Orel, J. Kožar Logar, B. Orel, Appl. Spectrosc. 1995, 49, 1756-1761.
4. Z. Crnjak Orel, M. Klanjšek Gunde, B. Orel, Prog. Org. Coat. 1997, 30, 59-66.
5. F.B. Stieg, Jr., in Pigment Handbook (Ed.: T.C. Patton), Wiley, New York, 1973, 203-207.
6. C.F. Bohren, D.R. Huffman, Absorption and Scattering of Light by Small Particles, Wiley, New York, 1983.
7. S. Asano, Appl. Opt. 1979, 18, 712-723.
8. Z. Crnjak Orel, Sol. Energy Mat. Sol. Cells 1999, 57, 290-301.

Nanocrystalline Sol-Gel Nb_2O_5 Films for Photovoltaic Energy Conversion

Y. P. Guo, M. A. Aegerter[1]
Institut für Neue Materialien - INM, Saarbrücken (Germany)

1 Introduction

Niobium pentoxide (Nb_2O_5) is finding more and more new applications in the areas of reversible cathodes, electrochromic devices, solar cell, catalyst, ferroelectric ceramics, and optical coatings [1-8]. For many of these applications, the microstructure of niobium oxide plays an important role. Sol-gel chemistry provides a convenient approach for the preparation of wide-band-gap semiconductors with distinct advantages over other semiconductor processing routes. Surface and bulk properties can be tailored under ambient conditions.

In this study we will focus on how the microstructure of the Nb_2O_5 electrode affects the photoelectrochemical properties of nanocrystalline dye-sensitized solar cells by modifying morphology of sol-gel Nb_2O_5 coating by adding polymeric ligand and carbon soot to the sol.

2 Experimental Details

Preparation of the nanocrystalline film electrodes have been described in our previous works[4]. The solution was made by a partial hydrolysis of $NbCl_5$ in a solution of water and ethanol. Nanocrystalline Nb_2O_5 sol-gel films were deposited on conducting glass substrate (FTO, 10 Ω) and 0.4 mm AF45® (DESAG, Germany) glass substrate by spin coating (1000 rpm), respectively. After deposition, each layer was dried at room temperature and then heat treated at 400°C for 10 min. The deposition process was repeated to increase the thickness of the coating. The last coating was sintered for 30 min in air at different temperatures from 500°C to 600°C.

X-ray diffraction patterns of the films were measured with a Siemens Diffractometer D 500. The topography of the films was observed by scanning white light interferometry using a new view 100 system (Zygo® USA) and their surface area and pore size were determined using a Micromeritics ASAP 2400 N_2 adsorption-desorption equipment.

A single compartment three electrode cell (volume 30 ml) used for the photoelectrochemical measurement and the solar cells were fabricated in glove box under nitrogen atmosphere, as already reported[4]. The amount of adsorbed dye was determined by desorbing the dye from the Nb_2O_5 surface into a solution of 0.1M NaOH in water and ethanol (50/50 by volume) and measuring its absorption spectrum with a CARY 5E spectrophotometer. Electrochemical impedance spectroscopy were measured using IM5d system (Zahner® Germany). The impedance data was obtained in the frequency range of

[1]Corresponding author. Tel: +0049-681-9300317, Fax: +0049-681-9300249, e-mail: aegerter@inm-gmbh.de

0.01 Hz-100k Hz. The photovoltaic properties of the solar cells were measured with an EG&G 273A potentiostat under irradiation with a 450 W Xenon lamp. The power of the light was measured with a Radialux global sensor (Heraeus Instruments) in the spectral range 300 - 800 nm.

3 Results and Discussion

The X-ray diffraction pattern Nb_2O_5 sol-gel film prepared with Nb_2O_5 sol and Nb_2O_5+PRG+C sol and sintered at different temperatures for 30 min is shown in Figure 1. The crystallization is observed at 400°C. The crystallites are partially oriented along the (001) direction. The structure of the sample made with Nb_2O_5 sol consists of different phases, one of them being the hexagonal phase (JCPDS No: 28-317). Many peaks observed at low sintering temperatures get smaller or disappear at 600°C. At 600°C only this phase is observed (Fig. 1 left). The crystallite size increases from 32 nm to 41 nm ((001) peak) and from 13 nm to 19 nm ((100) peak) for samples heat treated at 400°C and 600°C, respectively. The structure of the sample made with Nb_2O_5+PEG+C sol consists of only hexagonal phase (Fig. 1 right). The crystallite size decreases from 37 nm to 35 nm ((001) peak) and from 22 nm to 15 nm ((100) peak) for samples heat treated at 400°C and 600°C, respectively.

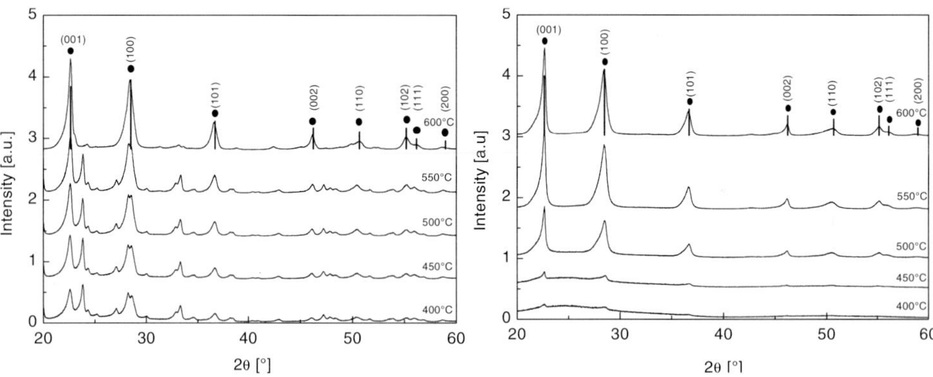

Figure 1. X-ray pattern of the Nb_2O_5 films prepared with different sols sintered at different temperature from 400°C to 600°C for 30min. Left: Nb_2O_5 sol, right: Nb_2O_5+PEG+C sol. The position of X-ray peaks of JCPDS 28-317 is also shown.

BET surface area of the coatings prepared with three different Nb_2O_5 sols are given in Table 1. An increase of the BET surface area from 26 to 38.4 and to 44.1 m^2/g was obtained by adding PEG and PEG+C to the sols, respectively. The average particle size decreases from 46 to 31 and to 27 nm. The BJH adsorption and desorption average pore diameters are only slightly altered. The films still possess pores with an ink-bottle shape and narrow neck [9].

Table 1 Average pore diameter, surface area and average particle size of films prepared with Nb_2O_5, Nb_2O_5+PEG and Nb_2O_5+PEG+C sols

Sol	BJH adsorption average pore diameter (nm)	BJH desorption average pore diameter (nm)	Surface area (m^2/g)	d_{BET}* (nm)
Nb_2O_5	9	5	26	46
Nb_2O_5+PEG	9.2	6.3	38.4	31
Nb_2O_5+PEG+C	9	6.3	44.1	27
TiO_2powder (P25 Degussa)	16	17	53	30

*Average diameter back-calculated from the surface area, assuming that the particles are spherical and using a Nb_2O_5 density of 5 g/cm^2 [10].

Table 2 shows the effect of the different preparation sols of the Nb_2O_5 films on the morphology of the electrodes. The amount of dye adsorbed on the film surface increases from 3.3×10^{-8} to 8.8×10^{-8} mol/cm^2(10μm) and the roughness factor from 200 to 530 for film made by using the Nb_2O_5+PEG+C sol. The surface area calculated from adsorbed amount of dye is smaller than that of the BET surface area value. This difference is attributed to some small pores where the dye molecules can not access.

Table 2 Comparison of density, porosity, adsorbed amount of dye, surface area and roughness factor of films prepared with Nb_2O_5 and Nb_2O_5+PEG+C sols

Sol	Density (g/cm^2)	Porosity (%)	Adsorbed amount of dye (mol/cm^2(10μm)	Surface area*	Roughness factor*
Nb_2O_5	2.74	45.2	3.3	8.7	200
Nb_2O_5+PEG+C	2.70	46	8.8	16.6	530

*Assuming that each dye molecule occupies an area of 1 nm^2 [11].

Figure 2 shows 2D (left) and 3D (right) scanning white light interferometry topography of Nb_2O_5 films prepared with different sols sintered at 520°C for 30 min. They have a porous structure. The roughness of sample prepared with Nb_2O_5+PEG+C sol increases substantially. This result agrees with the BET surface area measurement and the amount of adsorbed dye measurement. The surface of the Nb_2O_5 agglomerates formed by the highly packed Nb_2O_5 nanoparticles is also observed.

The complex impedance plane diagrams of nanocrystalline dye-sensitized Nb_2O_5 phtoelectrodes prepared with Nb_2O_5 sol and Nb_2O_5+PEG+C sol under illumination and at open circuit condition is shown in Figure 3, which consists of one semicircles in the high frequency region and another semicircle and a straight line in the low frequency region. The extrapolated curves intersect the real axis at 8.6, 25.1 and 38.8 Ω for the photoelectrode prepared with Nb_2O_5 sol and at 8.6, 21.1 and 32 Ω for the photoelectrode prepared with Nb_2O_5+PEG+C, respectively. The complex impedance spectrum can be analyzed in terms of an equivalent circuit model, which can be assumed to be made up of three elements in series: one resistive and two RC elements. The resistive element, which is independent of the morphology of the electrode, is characteristic of the bulk of electrolyte. The RC element at higher frequencies, which is morphology-dependent, is characteristic of mass transfer in porous electrodes. The RC element at low frequencies also depends on the morphology. This is assigned to chemical reaction, ionic diffusion, space charge in niobia film and charge transfer at the substrate/Nb_2O_5 interface.

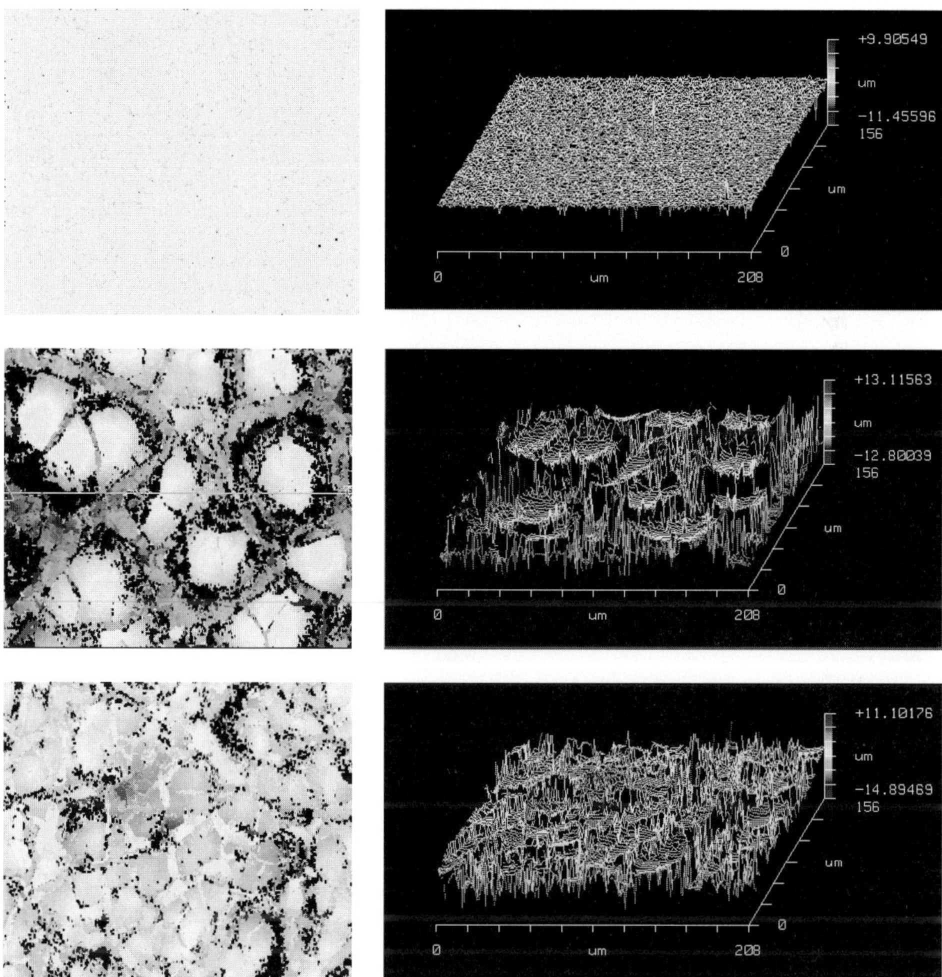

Figure 2. Scanning white light interferometry topography of the Nb_2O_5 films prepared with different sols sintered at 520°C for 30min. Left: two dimension, right: three dimension. Top: Nb_2O_5 sol, middle: Nb_2O_5+PEG, bottom: Nb_2O_5+PEG+C sol.

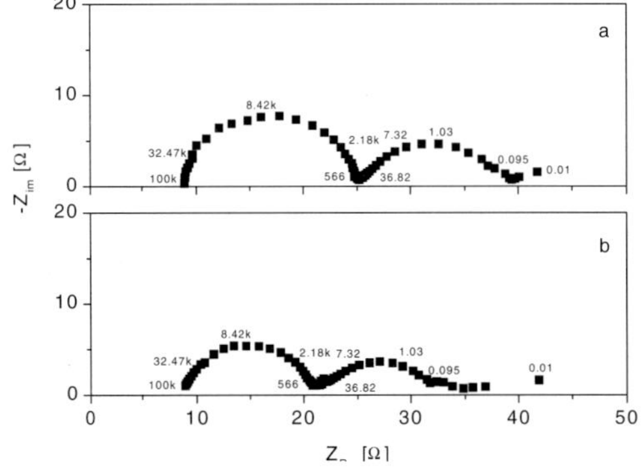

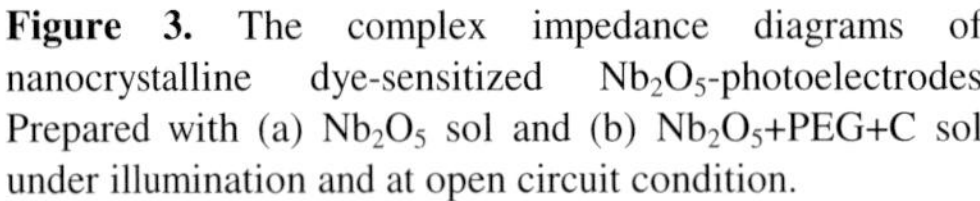

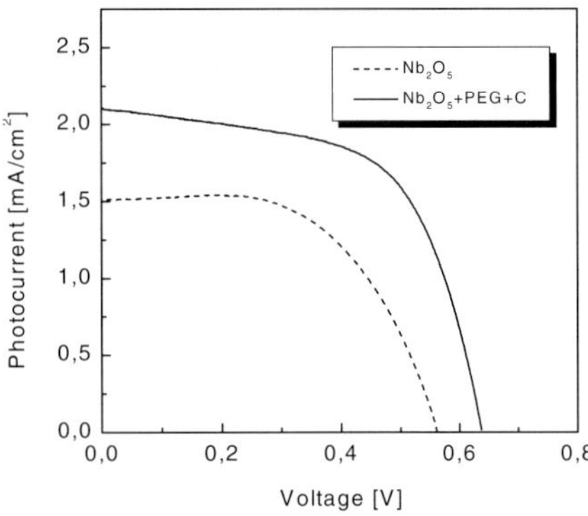

Figure 3. The complex impedance diagrams of nanocrystalline dye-sensitized Nb_2O_5-photoelectrodes Prepared with (a) Nb_2O_5 sol and (b) Nb_2O_5+PEG+C sol under illumination and at open circuit condition.

Figure 4. Photocurrent-voltage characteristics of Ru(ii)-sensitized sol-gel Nb2O5 film solar cells made with films prepared with Nb2O5 sol and Nb2O5+PEG+C sol under 100 W/m2 light irradiation.

Table 3. Photovoltaic parameters of 0.2 cm^2 Ru(II)-sensitized sol-gel Nb_2O_5 film solar cells made of the films prepared with Nb_2O_5 sol and Nb_2O_5+PEG+C sol under several light intensities.

Light intensity (W/m^2)	V_{oc} (V)	i_{sc} (mA/cm^2)	FF	η (%)
Electrode prepared with Nb_2O_5 sol				
100	0.57	1.66	0.55	4.9
250	0.59	3.36	0.47	3.7
430	0.60	4.58	0.45	2.9
900	0.61	7.0	0.44	2.2
Electrode prepared with Nb_2O_5+PEG+C sol				
100	0.64	2.11	0.60	8.1
250	0.66	4.50	0.56	6.3
450	0.66	5.70	0.59	5.1
950	0.67	6.94	0.62	3.0

The photocurrent-voltage characteristics of 0.2 cm^2 Ru(II)-sensitized sol-gel Nb_2O_5 film solar cells is shown in Figure 5 and their photovoltaic parameters under several light intensities are presented in Table 3. The photocurrent , photovoltage and fill factor of cells made with films prepared with Nb_2O_5+PEG+C sol are improved. The solar cell efficiency under 100 W/m^2 light irradiation increases from 4.9 to 8.1%, a value close to that obtained for TiO_2 solar cell. Nevertheless, the cells efficiency still decrease down to ca. 3.0% under 950 W/m^2 illumination. This is due to the fact that the core diameters of the neck (Table 1) were not improved and that the pores have still an ink-bottle shape which inhibit the regeneration of the photoelectrochemical process at high illumination level.

4 Conclusions

Thick Nb_2O_5 films have been prepared by sol-gel process. The morphology and structure of the films can be altered by adding polyethylene glycol and carbon soot to the sol. The material start to crystallize at 400°C and the structure consists of single hexagonal phase. An increase of BET surface area and roughness factor leads to increase the amount of dye adsorbed on the film surface resulting enhancement in the light harvesting efficiency. The photovoltaic characteristics of the solar cell were improved. The impedance measurement has shown that the films prepared with the modified sol are preferable photoelectrochemical reaction to that prepared with Nb_2O_5 sol. The films still possess pores with an ink-bottle shape and narrow neck, which causes a decrease of conversion efficiency of the solar cell with the increase of the illumination intensity.

5 Acknowledgements

Research financed by the „Bundesministerium für Bildung und Forschung (BMBF)“ under contract 2A67/03N9040 and the state of Saarland (Germany). We also thank Mr. Z.-F. Liu for measurement of scanning white light interferometry topography.

6 References

[1] N Kumagai, K. Tanno, T. Nakajima, N. Watanabe, Electrochem. Acta, 1983, 28, 17.
[2] M. Schmitt, S. Heusing, M. A. Aegerter, A. Pawlicka and C. Avellaneda, Sol. Energy Mater., Sol. Cells, 1998, 54, 9.
[3] B. O´Regan, M. Grätzel, Nature, 1991, 353, 737.
[4] Y. P. Guo, M. A. Aegerter, Thin Solid Films, 1999, 351, 290.
[5] K. Sayama, H. Sugihara, H. Arakawa, Chem. Mater., 1998, 10, 3825.
[6] Jih-Mirn Jehng, I. E. Wachs, Catalysis Today, 1990, 8, 37.
[7] Y. Xi, H. Mckinstry, L. E. Cross, J. Am. Ceram. Soc., 1983, 66, 637.
[8] S. Parraud, L. G. Hubert-Pfalzgrat, H. G. Floch, J. Am. Ceram. Soc., 1992, 75, 2289.
[9] S. J. Gregg, K. S. W. Sing, Adsorption, Surface Area and Porosity, 2nd ed., Academic Press, London 1982.
[10] L. K. Frevel, H. W. Rinn, Anal. Chem. 1955, 27 1329.
[11] M. K. Nazeeruddin, P. Liska, J. Moser, N. Vlachopoulos, M. Grätzel, Helv. Chem. Acta, 1990, 73, 1788.

New Solvent-Based and Water-Borne Paints for Solar Collectors

Zorica Crnjak Orel[1], Alojzij Lenček[2], Marta Klanjšek Gunde[1]
[1]National Institute of Chemistry, Ljubljana, Slovenia
[2]Helios, Domžale, Slovenia

1 Introduction

The requirement for high performance, low-cost and consumer - oriented spectrally selective coating has been the main goal in our research for the past ten years. Our investigations have been directed towards the preparation and characterization of spectrally selective coatings. Selective absorbers were obtained by the optimization of paint thickness on highly reflective metal surface in order to achieve maximal absorption in solar spectrum and the transparency in the infrared region (1-3). Several such tandems have been produced with solar absorbance ($a_s \sim 0.9$) and thermal emittance ($e_T \sim 0.2$) mainly by high-tech coil-coating technique (4-6). The curing process for this application technique was performed at high temperature (300°C). In many cases this method could not be used because of special requirements of users for an air-dried paint that cured at room temperature.

In this paper we present some new results for preparation two different types of spectrally selective paints. Solarect Z-LT was prepared due to the demands for preparation of air drying coatings. Solarect Z-WB was a new water-based paint where the use of volatile organic solvents was restricted. In both cases the silicon resins were used as a matrix component. The properties of coatings are a complex combination of properties of the individual components and interfaces between them and polymer and substrate. Different variations for the preparation of final coating over three substrates (aluminium, cooper and stainless steel) were applied in order to obtain maximum solar absorptance and minimum thermal emittance. The durability of coatings has been studied by exposing samples to elevated temperature.

2 Experimental

All paints were prepared with standard industrial or laboratory equipment (high-speed dispersers, pearl mill) using raw materials that are currently available on the world market (polysiloxane and epoxy/silicone resins as binders; black, high temperature stable spinel type pigments (P3060, Ferro Co., and rheological additives). The primary pigment particle size was 0.400 μm of black pigment that allows deposition of the paint in very thin layer 1-2 μm.

Paints were applied on Al, Cu and stainless steel foils by spray deposition technique. In order to achieve different weights of paint per m^2, spraying time was changed. By draw-bar coater Solarect Z-LT paint was also applied.

All paints dry in only few minutes. Solarect Z-LT started to cure at room temperature due to addition of some curing agents. For Solarect Z-WB high curing temperature (200-250°C) is required to attain adequate thermal, weathering and mechanical resistance of the coating.

When paint is applied on copper foil, the curing temperature should not exceed 200°C to prevent the formation of copper oxides layer. For Al foil curing can be performed at temperature above 220°C.

FT-IR spectra were measured on Perkin Elmer FTIR System 2000 equipped with near-normal (6°) reflectance cell.

Thermal emittances (e_T) were measured by emissiometer (AE DAS model) using standardized reference coatings. Solar absorbance was measured on an Alphameter (D&S model).

3 Results and Discussion

Solarect Z-LT air dry paint that cured at room temperature was applied by spray deposition techniques on Cu, stainless steel, and Al substrates (Table 1, 2, 3) and by draw-bar coater on Cu substrate (Table 4 and 5).

Spectral selectivity (a_s and e_T values) of coating prepared from Solarect Z-LT depends strongly on deposition technique, thermal emittances of the substrate used and applied thickness of the paint. Better values were obtained for aluminium and copper substrate (Table 1, 2). When the thickness was higher from 3.5 g/m^2 the selectivity value change rigorously (e_T > 0.4 and a_s ~ 0.92). To improve adhesion properties of Solarect Z-LT paint on stainless steel, the substrate was previously treated with adhesion promoter. Unfortunately it influences on e_T value. After the application of adhesion promoter (at thickness of 0.43 g/m^2 was e_T = 0.13) due to decreased reflectivity of substrate (Table 3). Results presented in Table 4 and 5 reveal that when Solarect Z-LT paint was applied on Cu substrate (thermal emittance e_T = 0.046) at the thickness of about 2 μm it exhibited e_T = 0.15 and a_s = 0.89. At approximately same thickness of paint layer with Cu foil possessing higher emittance value (e_T = 0.09), the emittance of paint coatings was e_T = 0.195 and a_s = 0.89. It means that thickness of about 2 μm of paint increased final emittance by about 0.10 units.

Legalization related to environmental and safely issues have affected all areas of the industry (7). The main pressure on surface coating manufacturers is to develop coatings that are less harmful to the environment. The paint industry started looking for cheap more readily available solvent. Water is convenient for all these demands. All these reasons were the main cause why we started with preparation of water-based paint. Our research was concentrated to find the appropriate resin that has to fulfill some requirements closely connected with desired final application of coatings. For these purposes we tested IR transparences of some silicones resins. As example, for very thin film of resin on Al substrate e_T for thickness 0.72 g/m^2 was only 0.13. Spectrally selective paint was prepared from Silicophen P 40/W. The preparation of pigment dispersion can be understood as three-step process (8). First, particle surfaces must be wetted and absorbed air must be replaced with molecules of liquid medium (dispersing); second, the agglomerate particles must be broken up; and finally the stabilization against flocculation must be performed. The main problem in our preparation of paint was dispersion of pigment. Different additives (Schwego watt EFKA 4550 and Tego dispers) were tested in order to find the proper one (Tego disperse 720W). After the preparation of stabile dispersion, paints with different pigment to volume concentration ratio (PVC) were obtained with addition of water based silicon resin (1,2).

Four different paints at PVC 32, 24, 21 and 18 were prepared. Unfortunately the adhesion for all prepared paint at desired thickness was poor. To improve adhesion (9) properties paint was applied on previously treated Al substrate with silane what helps the paint adhere to the substrate. First the optical properties were tested as is presented in Table 6 and 7. Obtained results show that pretreatment with silane (at thickness 0.8-0.9 g/m^2) changes the final optical properties of selective coatings and influences on emittance for paint at thickness about 2 g/m^2. However, to obtain environmentally acceptable coatings, lower performance standards will have to be accepted.

Table 1. Thermal emittance (e_T) and solar absorbance (a_s) as a function of thickness (g/m^2) for Solarect Z-LT on aluminium substrate (spraying deposition).

Sample	paint thickness g/m^2	e_T	a_s
Al1	1.37	0.128	0.790
Al2	1.89	0.185	0.87
Al3	2.8	0.327	0.91

Table 2. Thermal emittance (e_T) and solar absorbance (a_s) as a function of thickness (g/m^2) on copper substrate (spraying deposition).

Sample	paint thickness g/m^2	e_T	a_s
Cu1	2.08	0.20	0.90
Cu4	3.7	0.332	0.93
Cu5	3.96	0.427	0.91

Table 3. Thermal emittance (e_T) and solar absorbance (a_s) as a function of thickness(g/m^2) for Solarect Z-LT on stainless steel (stainless steel previously coated with primer) (spray deposition)

Sample	adhesion promoter	paint thickness g/m^2	e_T	a_s
r2	0.41	1.45	0.32	0.88
R3	0.42	2.64	0.40	0.94
R4	0.43	3.75	0.55	0.93

Table 4. Thermal emittance (e_T) and solar absorbance (a_s) as a function of thickness (g/m^2) for Solarect Z-LT for copper substrate (draw-bar coater) on Cu foil with $e_T = 0.09$.

Sample	paint thickness g/m^2	e_T	a_s
sp0	2.06	0.19	0.90
sp1	2.65	0.29	0.91
sp2	3.22	0.41	0.93

Table 5. Thermal emittance (e_T) and solar absorbance (a_s) as a function of thickness (g/m^2) for Solarect Z-LT for copper substrate (draw-bar coater) on Cu foil with e_T = 0.046.

Sample	paint thickness g/m^2	e_T	a_s
F1* (sp0)	2.12	0.15	0.89
F2* (sp1)	2.3	0.26	0.91
F3*	2.92	0. 39	0.93

The durability of any organic coating can be defined as its resistance to undesirable changes caused by the natural environment to which it is exposed during the service life (10). The rate at which polymer degradation occurs will vary widely, depending upon temperature, exposure site, substrate, etc. all of which makes the evaluation of coating durability difficult. Accelerated life testing offers the possibility to forecast service life and proper material selection.

Thermal stability of all paint coatings was studied at 200°C for a period up to 21 days. Changing of normalized peak intensities for Solarect Z-LT paints on two different substrates (Al and Cu) are presented in Fig. 1. In Solarect Z-LT paint the peak intensities of the characteristic bands at about 1036 cm^{-1} (Si-O-Si vibration) and 1509 cm^{-1} (aromatic C-C stretching) change upon heat-treatment. Bigger changes were noted at the beginning (up to 76 hours) due to the unfinished curing process in the paints. On both substrates stability tests of Solarect Z-LT paint at 200°C show some degradation or the evaporation of epoxy resin owing to the diminishing of peak intensities at 1509 cm^{-1} (only presented on Cu substrate) in Fig. 1. The presence of Si-O-Si vibration bands and good adhesion of paint layer prove satisfactory stability of Solarect Z-LT paint on Al-substrate.

To improve adhesion properties on Cu substrate and to prove its thermal stability, paint was applied on pure and precoated Cu-substrate. The shape of the Si-O-Si stretching band is also influenced by variation in the bonding character, density, and probably with the change of the porosity of the film. Paint coating stability is closely connected also with the substrate stability. Uncoated cooper substrate after the heat treatment at 200°C for only 42 hours shows the formation of Cu_2O and CuO, and also the reflectivity value in all IR spectral region was drastically reduced. It shows that paint can be used also for protection of Cu substrate.

Stability test of Solarect Z-WB paint was performed on Al and previously pretreated Al substrate with silane (Fig. 2). In the beginning (up to 36 hours) some changes were observed for band at 1036 cm^{-1} (Si-O-Si) vibration and bigger for band at 1260 cm^{-1} (C-O Aril-ether) that can be explained with same unfinished curing process. Nevertheless, both coating shows that (up to 506 hours) good thermal stability is obtained and proves their use for a spectrally selective surface for solar collectors.

Table 6. Thermal emittance (e_T) and solar absorbance (a_s) as a function of thickness (g/m^2) for Solarect Z-LT on aluminium substrate.

Sample	paint thickness g/m^2	e_T	a_s
V1	1.98	0.20	0.89
V2	3.07	0.36	0.94
v4	4.94	0.47	0.95
v5	5.02	0.47	0.94

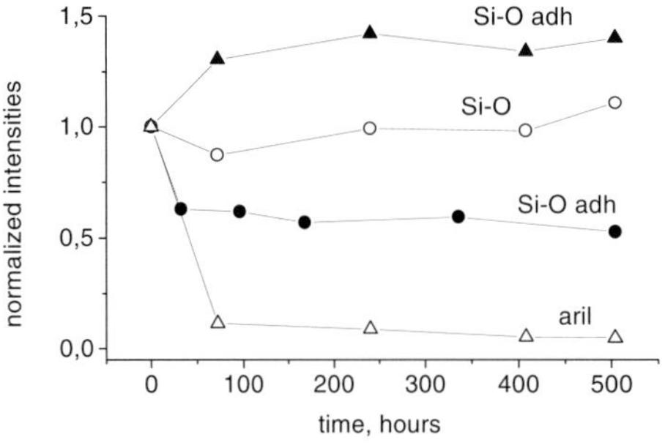

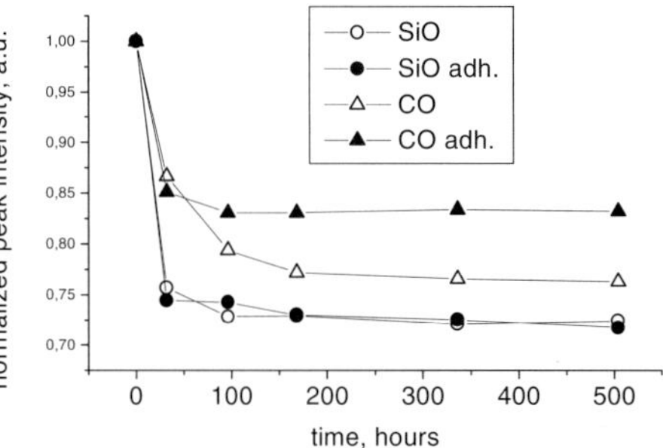

Figure 1. Normalized peak intensities of air drying paint vs. duration of heat treatment (200°C) on Al (●) and Cu (▲ Δo) substrate. Si-O stands for Si-O-Si band (1036 cm^{-1}); Aril stands for aromatic stretching ν_{C-C} at 1507 cm^{-1}.

Figure 2. Normalized peak intensity of water based paint vs. duration of heat treatment (200°C) on Al plate (thickness of the paint ~2 g/m^2). ● ▲ Al plate precoated. SiO stands for Si-O-Si band (1036 cm^{-1}), CO stands for Aril ether (ν_{CO} at 1260 cm^{-1}).

Table 7. Thermal emittance (e_T) and solar absorbance (a_s) as a function of thickness (g/m^2) for Solarect Z-LT on Al substrate previously coated with adhesion promoter.

Sample	adhesion promoter	paint thickness g/m^2	e_T	a_s
v7	0.82	2.0	0.30	0.90
v8	0.82	2.74	0.38	0.92
v9	0.89	4.17	0.49	0.94

4 Acknowledgement

This work was supported by the Ministry of Science and Technology of Republic Slovenia (Contract No. L2-0703).

5 References

1. Z. Crnjak Orel, Sol. Energy Mat. Sol. Cells 1999, 57, 291-301.
2. Z. Crnjak Orel in XIV. SPIE proc., Vol. 2531, Washington, 1995, 296-307.
3. M. Klanjšek Gunde, J. Kožar Logar, Z. Crnjak Orel, B. Orel, Thin Solid Films 1996, 277, 185-191.
4. Z. Crnjak Orel, M. Klanjšek Gunde, B. Orel, Prog. Org. Coat. 1997, 30, 59-66.
5. Z. Crnjak Orel, N. Leskovšek, B. Orel, M.G. Hutchins, Sol. Energy Mat. Sol. Cells 1996, 40, 197-204.
6. M. Klanjšek Gunde, J. Kožar Logar, Z. Crnjak Orel, B. Orel, Appl. Spectrosc. 1995, 49, 623-629.
7. V.D. McGinniss, Progress in Org. Coat 1996, 27, 153-161
8. R. Craft, Modern Paint and Coating 1991, 38-43.
9. P. Bugnon, Progress in Org. Coat. 1996, 29, 39-43.
10. B.W. Johnson, R. McIntyre, Progress in Org. Coat. 1996, 27, 95-106.

Control of P-Type Conductivity in CdTe by $CdCl_2$ Vapor Phase

J. Hiie, V. Valdna
Tallinn Technical University, Tallinn

1 Introduction

CdS/CdTe solar cells are among the devices for low cost and efficient photovoltaic modules [1]. $CdCl_2$ treatment of CdTe films in CdS/CdTe/glass solar cells is necessary in order to remove the defects in the CdTe film by recrystallization and possibly to dope it with V_{Cd}–Cl complexes which behave as p-type dopants [2].There has been several attempts to control the recrystallization and doping processes in CdS / CdTe thin films by heat treatment with $CdCl_2$ vapor phase [3] and good results has been received.

This work is an attempt to investigate the equilibrium between CdTe crystals and $CdCl_2$ vapor phase to clear up between the doping and recrystallizing activities of $CdCl_2$ in CdTe. The bulk crystals were used to eliminate the recrystallization effects of $CdCl_2$ from Cl doping activity

2 Experimental Procedure

Undoped, unoriented semiinsulating (10^7 - $10^8\Omega$cm) single-crystal CdTe (5N purity) cutted - unpolished wafers with cleavage edges, with thickness of 0.4 mm and masses about 100 mg were used. 5N purity $CdCl_2$ from the Aldrich Chem. Co, Inc. was used.

The anneals were carried on in the evacuated sealed off two-zone quartz ampoules near to the minimum-total-pressure condition at 863 K for CdTe. The mixture of $CdCl_2$ and CdTe was added to the lower temperature zone in amount and composition sufficient for formation of saturated vapor, liquid and solid phases between the room temperature and 848 K. Duration of the anneals varied from 10 minutes to 50h. The Lindberg / Blue M three.zone tube furnace was used.

Conductivity type was determined by measuring the polarity of the thermovoltage. The dark resistance R_D of the crystal wafers was measured by calibrated teraommeter E6-3 between two indium plastic contacts assureing their ohmic behaviour. The light resistance R_L was measured using a 100 W incandescent lamp. Photoluminescence spectra of CdTe crystals were measured in the Janis CCS – 150 He cryostat at 8 K. The 441,6 nm line of Omnichrome He-Cd laser with CC8 and 1M $CuSO_4$ solution filter was used as the excitation source. The PL signal was dispersed using the monochromator SPM-2. A chopped signal from the InGaAs photodetector was amplified,recorded and corrected on a computer. So the arbitrary units on the all figures of the photoluminescence spectra have the same corrected value.

3 Theoretical Background

In the course of the first experiments we met with the fenomenon of $CdCl_2$ condensation from the cooler part to the hot end of the ampoule onto CdTe crystals. Solubility of CdTe in the liquid $CdCl_2$ is high [4] and one should to consider that in the hot zone the vapor pressure of $CdCl_2$ (P''_{CdCl2}) on CdTe crystals must be less than the P'_{CdCl2} on the liquid phase of $(CdTe - CdCl_2)_{(l)}$ in equilibrium with solid $CdTe_{(s)}$ at the hot zone temperature:

$$P''_{CdCl2} < P'_{CdCl2} \quad (1)$$

In the opposite case $CdCl_2$ will be condensated from the cool zone onto the crystals. The inequality condition (1) is regulated by the temperature difference between the hot and cool zones. But when the difference of the temperatures is too small, then one should add enough CdTe to $CdCl_2$ to saturate the $CdCl_2$ phase and to guarantee the steady presence of solid $CdTe_{(s)}$ in the cooler zone in contact with $CdCl_2$. So the reproducible value of cadmium chloride partial pressure P''_{CdCl2} over the entire volume of the ampoule can be assured.

Such high activity of cadmium halides has not been considered in the halogen diffusion investigations [ ] where in the isothermal conditions $CdCl_2$ was undesirably condensated on the CdTe crystals despite of the membranes used. From the other side the high activity of $CdCl_2$ has been used for even recrysallization of CdTe thin films through the gas phase in the solar cells technology [ ]. To control the process the future studies of halogenide partial pressures will be needed.

4 Results and Discussion

Noticeable doping effect began at T_{CdCl2}=470 K with increasing of photosensitivity and resistance of crystals up to 530 K followed by the region of low photosensitivity and "semiinsulating" medium resistance of 10^5—10^6 Ω. The photosensitivity and resistance of chlorine-doped samples increased with the time of $CdCl_2$ annealing after their overheating in vacuum.

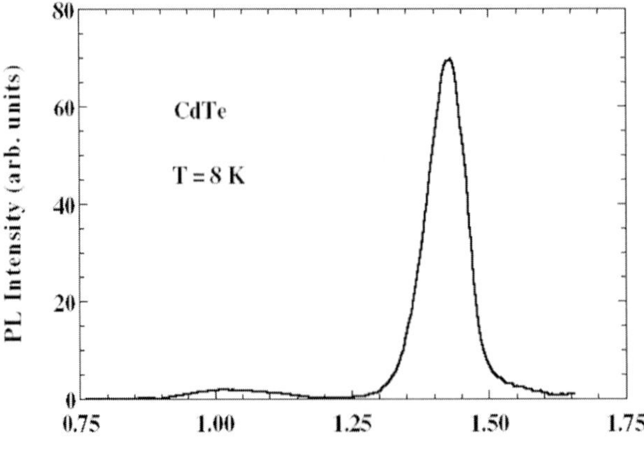

Figure 1. PL spectrum of CdTe crystal vacuum annealed at 863 K / Room T for 240(min), sample 1F (Table 1.)

Table 1. Resistance values of CdTe crystals versus $CdCl_2$-zone temperature (T_{CdCl2}) and annealing time[(a)]

No	Sample No	Cool zone T_{CdCl2},K	Annealing time,(min)	R_D, Ω	K_F
1	1F[(b)]	Room	240	$1{,}5x10^7$ -- $4{,}4x10^7$	1,4 - 2,7
2	1E[(b)]	455	120	$2{,}4x10^6$ -- $9{,}2x10^6$	1,1 - 1,6
3	1E	455	120	$7{,}5x10^6$ -- $8{,}1x10^6$	1,6 - 1,8
4	G	530	130	$1{,}2x10^7$ -- $5{,}5x10^8$	13,8 - 383,3
5	1D	557	570	$3{,}1x10^4$ -- $1{,}0x10^5$	1,02- 1,03
6	3G	733	300	$2{,}9x10^4$ -- $6{,}9x10^4$	1,0
7	1H	787	130	$6{,}3x10^4$ -- $1{,}3x10^5$	1,1 - 1,2
8	2H	848	100	$3{,}1x10^5$ -- $7{,}7x10^5$	1,2 - 1,6
9	2F	848	480	$1{,}8x10^5$ -- $2{,}8x10^5$	1,02- 1,2
10	2G	848	1200	$2{,}2x10^5$ -- $3{,}4x10^5$	1,6 - 2,5

(a) R_D is dark resistance; K_F is dark to light resistance ratio (min and max values of the 4 samples); T_{CdTe} = 863 K is the hot-zone temperature for all annealings
(b) without $CdCl_2$ in the zone with lower temperature

High dark resistivity and photoconductivity can be achieved by vacuum annealing, i. e. by evaporation of chlorine. The same result can be obtained by vapor phase treatment with $CdCl_2$.

Table 2. Resistance values of CdTe:Cl samples successively annealed in $CdCl_2$ and tellurium vapors and in vacuum[(a)]

No	Sample No	T_{CdCl2}, K	T_{Te2}, K	Annealing time, (min)	R_D, Ω	K_F
1	1E[(c)]	455	--	120	$2{,}4x10^6$ -- $9{,}2x10^6$	1,1 – 1,6
2	4I	--	848	300	$6{,}0x10^3$ -- $7{,}6x10^3$	1,00
3	1G	530	--	130	$1{,}2x10^7$ -- $5{,}5x10^8$	13,8 – 383,3
4	4L	--	848	300	$4{,}2x10^3$ -- $1{,}2x10^4$	1,0 – 1,03
5	2H	848	--	100	$3{,}1x10^5$ -- $7{,}7x10^5$	1,2 – 1,6
6	3B	--	848	300	$8{,}2x10^3$ -- $2{,}2x10^4$	1,00
7	2G	848	--	1200	$2{,}2x10^5$ -- $3{,}4x10^5$	1,6 – 2,5
8	3A	--	848	300	$7{,}0x10^3$ -- $9{,}4x10^3$	1,01 - 1,03
9	2H	848	--	100	$3{,}1x10^5$ -- $7{,}7x10^5$	1,2 - 1,6
10	4B	--	--	420 vac[(b)]	$4{,}6x10^5$ -- $1{,}2x10^7$	1,07 - 1,5
11	2G	848	--	1200	$2{,}2x10^5$ -- $3{,}4x10^5$	1,6 - 2,5
12	4A	--	--	420 vac[(b)]	$6{,}2x10^7$ – $5{,}9x10^9$	3,9 – 281

(a) every pair of rows corresponds to the successive serial annealings. R_D is dark resistance; K_F is dark to light resistance ratio (min and max values of 4 samples);
T_{CdTe} = 863 K is the hot-zone temperature for all annealings
(b) vacuum annealing under 10^{-3} (Torr), hot zone at 863 K, cooler zone at room temperature
(c) without $CdCl_2$ in the zone with lower temperature

Working with 5N CdTe, unintentionally doped with donor impurities (like halogens, Al, Ga, In), the main role of $CdCl_2$ will be as the rcrystallization and sintering agent – generation of liquid flux.

Results of our investigation are in accordance with the conception that the acceptor defect responsible for the effective p-type doping, and appearing with $CdCl_2$ treatment, may be identified with a singly ionised cadmium vacancy – halogen complex $(V_{Cd}Cl_{Te})^{-}$, which is reported to have ionization energy for hole generation 1.20 eV above the valence band edge in CdTe [ ].

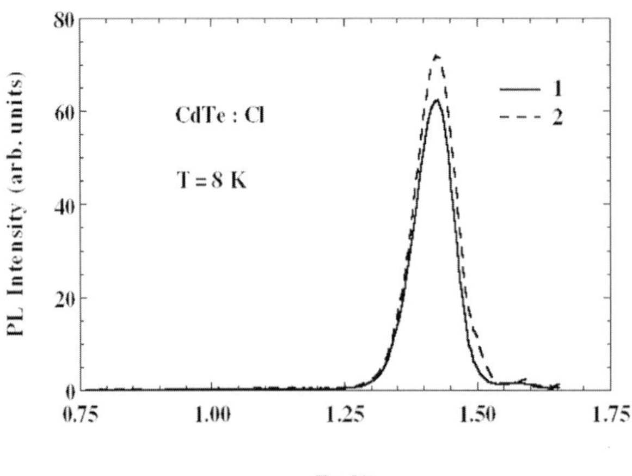

Figure 2. PL spectra of CdTe crystals, $CdCl_2$ - annealed at 863 K / 848 K for 100(min) - 1 and 1200(min) – 2 samples 2H and 2G (Table 1.)

In comparision with the work of B. McCandless et al [ ], our results are similar, despite of the oxygen addition to $CdCl_2$ vapor, used in [ ]. The most favorable region of $CdCl_2$ vapor activity between 570 – 740 K. (Table 1) is advantageous for CdS / CdTe thin films technology and needs more detailed investigation.The possible explanation of oxygen necessity in the many CdS / CdTe thin films technologies is the low melting point of cadmium and tellurium oxides eutectics with $CdCl_2$ [ ]. Rate of the recrystallization in the presence of the liquid phase exceeds the velocity of the solid state process by several orders of magnitude [ ]. But the presence of oxygen compounds in principal does not promote to the higher efficiency and stability of devices. This impels to find alternative fluxes with lower melting point suitable for soda-lime glass substrates.

Thickness (400 μm) of used crystals was the reason that we were not able to get high p-type conductivity as the result of annealings in comparision with thin films [ ]. Similarity of the photoluminescence spectra of the annealed crystals (Fig. 1 – 5) and thin CdTe films [ ], show the identity in the mechanisms of the treatment processes in the bulk crystals and thin films.

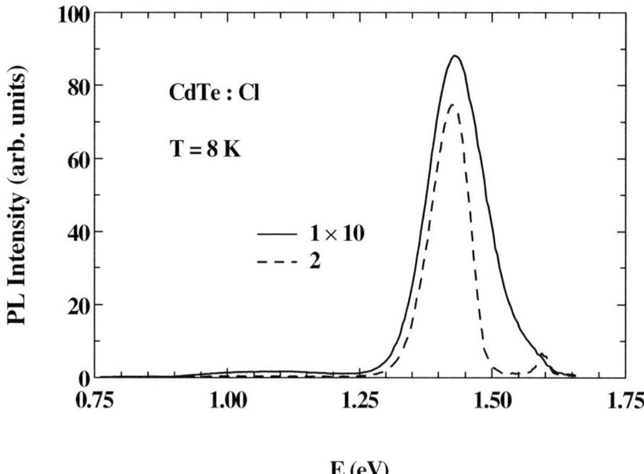

Figure 3. PL spectra of CdTe crystals, $CdCl_2$ - annealed at 863 K / 848 K for 100(min) - 1, 1200(min) - 2 and followed by vacuum annealing at 863 K / Room T for 425(min), samples 2H – 4B and 2G – 4A (Table 2.)

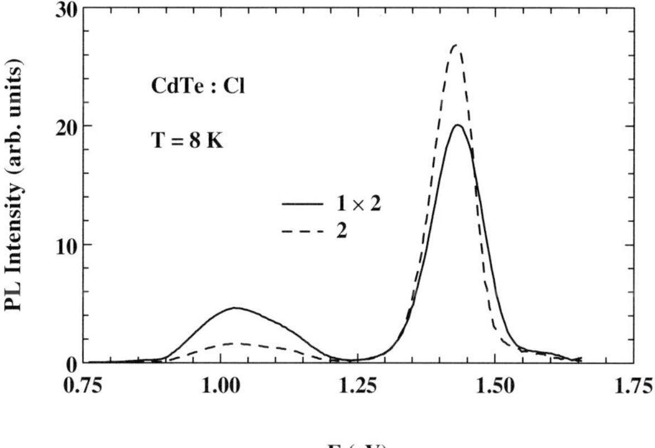

Figure 4. PL spectra of CdTe crystals, $CdCl_2$ - annealed at 863 K / 848 K for 100(min) - 1, 1200(min) - 2 and followed by Te_2 - annealing at 863 K / 848 K for 300(min), samples 2H – 3B and 2G – 3A (Table 2.)

5 Conclusions

It is concluded that $CdCl_2$ is an active p-type dopant in CdTe. It is possible to control the doping process in a two-zone apparatus by the temperature of the equilibrium liquid-solid phase [($CdCl_2$-CdTe)l-(CdTe)s] in the cooler zone.

The thermal annealing under $CdCl_2$ vapor pressure is a reliable technological method to control the p-type conductivity and photoelectrical characteristics of CdTe.

6 Acknowledgements

Financial support by the Estonian Science Foundation under the Grant # G3397 is gratefully acknowledged. Special thanks to the Bruker AXS, Inc., who provided part of the chemicals and equipment. Authors are grateful to dr J. Krustok for help in the measurements of PL spectra.

7 References

1. D. Bonnet, P. Meyers, J. Mater. Res. 1998, 13, 2740 – 2753.
2. N. Romeo, A. Bosio, R. Tedeschi and V. Canevari, Proc. 2nd WCPEC, 1998, pp. 446 – 447.
3. B.E. McCandless, H. Hichri, G. Hanket, R.W. Birkmire, Proc. 25th PVSC, 1996, pp. 781 – 784.
4. V.N. Tomashik, V.I. Griziff, Phase Diagrams of II-VI Semiconducting Compounds, Naukova Dumka, Kiev, 1982, pp. 87,88 (in Russian).
5. D. Shaw, E. Watson, J. Phys. C: Solid State Phys. 17 (1984) 4945.
6. V. Valdna, in: Polycrystalline Semiconductors V – Bulk Materials, Thin Films, and Devices, J.H. Werner, H.P. Strunk, H.W. Schock (Eds)., in Series "Solid State Phenomena", SciTech Publ., Uettikon am See, Switzerland, 1999, to be published.
7. J. Hiie, M. Altosaar and E. Mellikov, in: Polycrystalline Semiconductors V – Bulk Materials, Thin Films, and Devices, J.H. Werner, H.P. Strunk, H.W. Schock (Eds)., in Series "Solid State Phenomena", SciTech Publ., Uettikon am See, Switzerland, 1999, to be published.

Investment Casting of Intermetallic NiAl-Ta-Cr Alloys for Low Pollution Power Conversion

F. Scheppe, P.R. Sahm
Giesserei-Institut, RWTH Aachen
M. Palm, G. Sauthoff
Max-Planck-Institut für Eisenforschung, Düsseldorf
W. Hermann
Siemens AG KWU, Mülheim
J. Preuhs
TFB Feingusswerk Bochum GmbH

1 Introduction

Intermetallic compounds, such as NiAl, offer new opportunities for developing low density, high strength structural alloys which could be used at temperatures higher than currently possible with conventional titanium- and nickel-base alloys [1]. The strong bonding between aluminum and nickel, which persists at high temperatures, can provide high strength at elevated temperatures such that the specific strength of intermetallics could be competitive with superalloys and ceramics. NiAl-Ta-Cr alloys offer four key advantages:

- High melting point of above 1460 °C which is about 100 ° to 250 °C higher than those of the superalloys;
- thermal conductivity is about four times that of nickel-base superalloys;
- density of 6.2 - 6,35 g/cm^3 is approximately 75% the density of state of the art superalloys;
- good oxidation resistance.

However, the high strength is usually associated with poor ductility at room temperature. With respect to ductility, intermetallics lie between metals and ceramics.

NiAl-Ta-Cr alloys are subject of an ongoing development of new high-temperature materials for application in gas turbines. Simple component geometries were assembled in production size clusters for investment castings (e.g. heat shields, turbine blades).

2 Experiments

High temperature properties, thermal expansion and chemical behavior of the ceramic mold system has to be compatible with the properties of the cast alloy [2], [3]. To process NiAl alloys, the components of the mold are very close to their application limits. A special ceramic mold was used to prevent cracking and excessive mold-metal reactions. The mold, must be very strong at the casting temperature, but also weak enough at low temperatures to avoid cracking of the low-plasticity alloy due to differential contraction during cooling. For NiAl alloys only SiO_2 free ceramic molds were used. The molds were manufactured in the standard shell mold process, using a steam autoclave for dewaxing. All experiments were carry out in a controlled atmosphere furnace, Fig. 1. The ductile-to-brittle transition requires a rather small solidification rate. As the result of these condition, coarse grains will be arise. Various casting conditions (solidification rate, mold temperatures) in combination with two different ceramic molds were employed to investigate their effect on the grain structure. During the first test series, the effect of various cast conditions (solidification rate, mold temperature) on the grain structure was investigate, **Tab. 1**. The second test series was used to investigate the effectiveness of a grain refining ceramic.

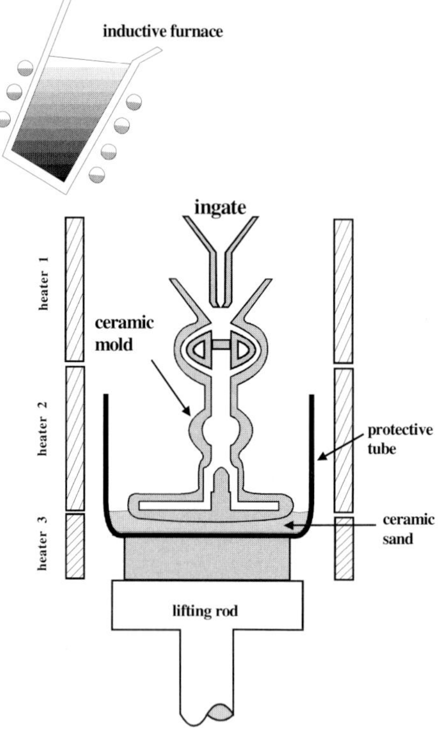

Fig. 1: All experiments were carry out in a controlled atmosphere furnace. For the casting process, the mold is preheated. The transition from ductile to brittle behavior at about 800 °C, requires a rather small solidification rate.

For the experiments a NiAl-Ta-Cr alloy was used. The alloy is characterized as a precipitation hardening, laves-phase reinforced alloy (FG 75) [4], [5]. It contains 45at.% Ni, 45 at.% Al, 7,5 at.% Cr and 2,5 at.% Ta.

Meanwhile for this purpose the numerical simulation of moldfilling and is a standard tool. However, the numerical simulation has still to be adapted as well to the special properties of the investment casting process as to the special properties of the NiAl-Ta-Cr-Alloys. Moldfilling and solidification were simulated with the program MAGMAsoft.

Tab. 1: During the first test series, the effect of various casting conditions (solidification rate, mold temperature) on the grain structure was investigate. The cast temperature was 1700 °C.

experiment	mold temperature [°C]	Solidification rate between 1100°C and 600°C [K/min]
1	1200	2,5
2	1200	5,0
3	1200	7,5
4	1250	2,5
5	1250	5,0
6	1250	7,5
7	1300	2,5
8	1300	5,0
9	1300	7,5

3 Results

With the used SiO_2-free ceramic mold it was possible to produce heat shields without any cracks. **Fig. 2** shows a crack free heat shield.

The samples to determine the grain size were take from definite positions in the heat shield. These method guaranteed, that different morphologies in the cast unit were include. For each analyze 15 micrographs were use.

The trend of grain refining as a result of increasing the solidification rate and decreasing the mold temperature shows Fig. 3. Fig. 4 shows the result of the second test series. The grain refining effect is plain to see. The comparison between the first and second test series demonstrate the effect of the used grain refining ceramic, see also Fig. 5.

The simulation of the solidification confirm the controlled solidification from the corners to the middle of the heat shield, Fig. 6. Further results of the numerical Simulation with magmasoft will be present on the “Modeling of Casting, Welding and Advanced Solidification Process IX” from 20. till 25. August 2000 in Aachen, Germany.

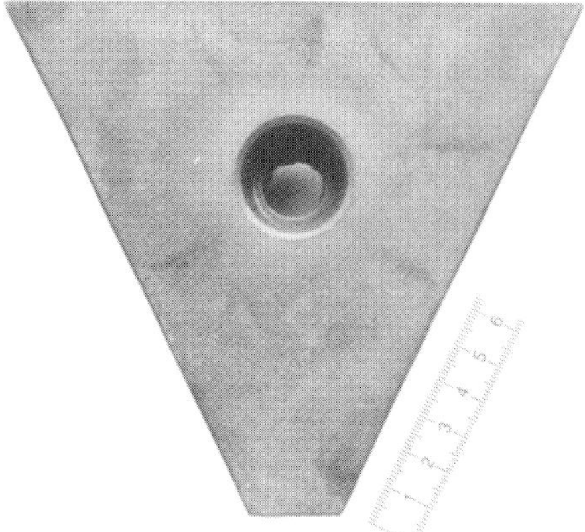

Fig. 2: This crack-free heat shied was produced with a SiO_2-free ceramic mold (T_{cast}= 1700°C, T_{mold} = 1300°C).

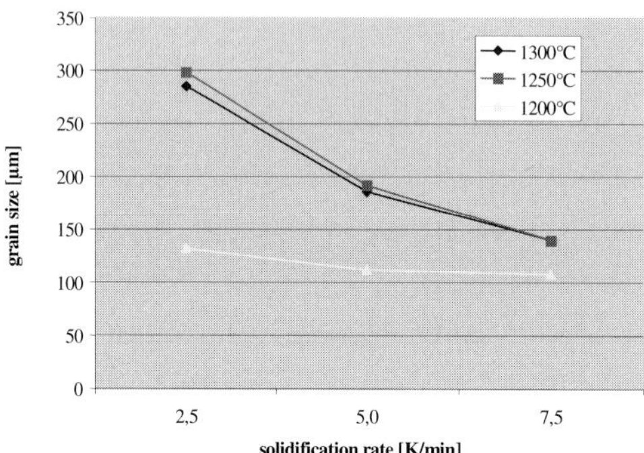

Fig. 3: The solidification rate was adjust between 1100 °C and 600 °C. The grain refining as the result of various cast condition is plain to see.

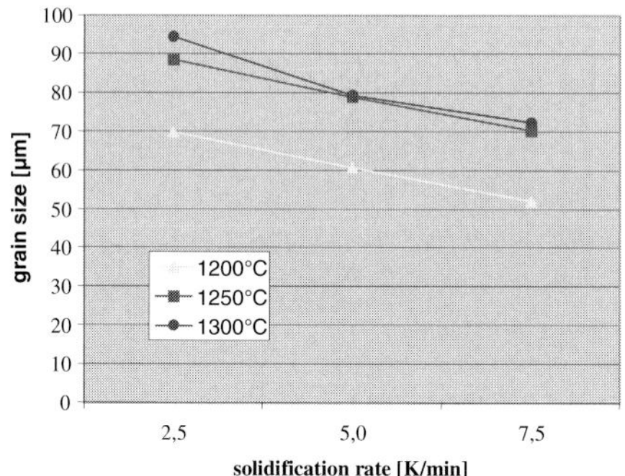

Fig. 4: The solidification rate was adjust between 1100 °C and 600 °C. The grain refining as the result of various cast condition is plain to see. But the average grain size of the second test series is smaller than the average grain size of the first test series. The experiments shows, that both possibilities are complement one another.

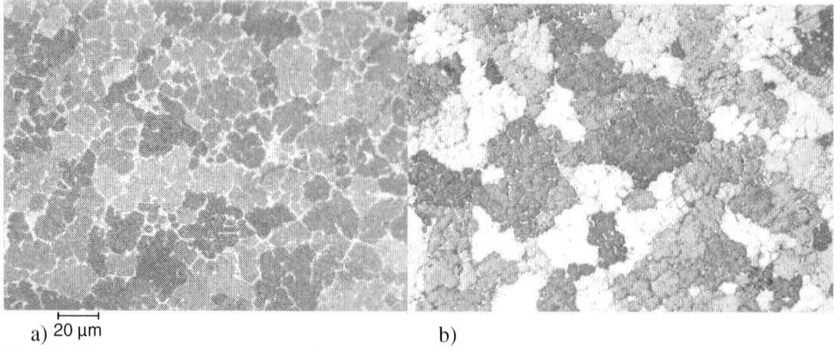

a) b)

Fig. 5: The casting with the grain refining ceramic (a) shows smaller grain size than the casting with the normal SiO_2-free ceramic mold. The cast parameters for both castings are the same (mold temperature = 1300 °C, solidification rate = 2,5 K/min).

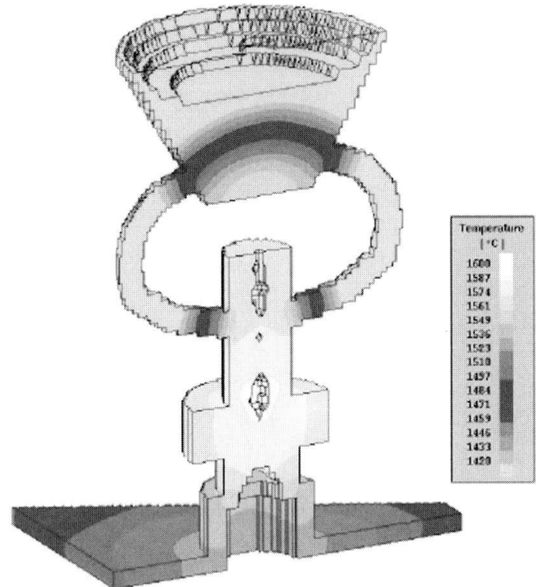

Fig 6: The simulation of the solidification confirm the controlled solidification from the corners to the middle of the heat shield.

4 Conclusion

This research showed the possibility to use a SiO_2-free mold in the standard shell mold process. The cast experiments made the requirement clear to develop a weakening of shell mold for the production of crack-free cast parts containing undercuts (e.g. heat shields, turbine blade).

The result of the visual inspection of the cast parts showed little or no reaction between the melt and the shell mold during the solidification. The research also showed the influence of the mold temperature of the manufacture of crack-free heat shields. To produce crack-free heat shields it is necessary to minimize the cooling rate. But all the experiments demonstrate, that the proportion between the cast parameters and the production of crack-free parts depends on the geometry (e.g. undercuts).

The grain refining can be successful with the used NiAl-Ta-Cr-alloy as a result of the variation of the mold temperature and solidification rate. Another successful way for a grain refining of this NiAl-alloy was the use of a grain refining ceramic mold. The results of the first test series show, that a relative solidification rate and a low mold temperature can aim a good grain refining. The results of the second test series make clear, that both possibilities of grain refining complement on another.

In further work, the properties of the cast parts have to be determined. In the next step, the lab scale results have to transferred to industrial scale (e.g. production in clusters).

5 Acknowledgment

The authors would like to thank BMBF for the financial support.

6 Reference

1. Darolia, R.: NiAl Alloys for High Temperature Structural Application, JOM 43(3), 1991, S. 44-49
2. Goldman, E.H.: Single Crystal Processing of Intermetallics for Stuctural Applications, High Temperature Ordered Intermetallic Alloys V, eds. Baker L., Darolia R., Whittenberger J.H.,1993
3. Oti, J.A., Yu, K.O.: Production Processing of Investment Cast Complex Shaped NiAl Single Crystal Airfoils, Proc. Int. Symp. On Stuctural Intermetallics, TMS, Warrendale, 1993, S. 505-512
4. Zeumer, B.A.: Zur Entwicklung Laves-Phasen-verstärkter NiAl-Basislegierungen für Anwendungen bei hohen Temperaturen, VDI-Verlag, Düsseldorf, 1994
5. Hermann, W., et al.: Intermetallische NiAl-Komponenten zur umweltfreundlichen Energiewandlung, Tagungsband Werkstoffwoche 98, Band III, München, 1998

Improvement of the Efficiency of Silicon Solar Cells by Electrochemical Passivation of High Leakage Current Areas in the pn-Junction

M. H. Al-Rifai, J. Carstensen and H. Föll

Faculty of Engineering, Christian-Albrechts University of Kiel, Kaiserstr. 2, D-24143 Kiel, Germany E-mail: mhr@techfak.uni-kiel.de, http://www.techfak.uni-kiel.de/matwis/amat/, Tel.+49 431 77572 505, Fax +49 431 77572 503

1 Abstract

The efficiency of silicon solar cells can be strongly degraded by localized defects. Characterizing silicon solar cells locally by a new method (CELLO), problematic areas are detected and subsequently disconnected from the cell by cutting through grid and p-n-junction around the defect. Applying anodic bias to the solar cells in contact to a KOH electrolyte allows to completely passivate the surface states introduced by this procedure. Although the total area of the solar cell is decreased by this procedure, the efficiency of the complete solar cell can be increased.

2 Introduction

All multi crystalline solar cells have areas, where the diffusion length of the minority carrier is high and the leakage current is low (good areas) but they have also bad areas, e.g. with small diffusion length, large leakage currents or small shunt resistance, that decrease the over all efficiency [1]. Since improving the efficiency of Si solar cells is a very important task for the large-scale use of solar energy, several methods for obtaining local information about those local defects have been developed, the most prominent one being the "Laser beam induced current" (LBIC) technique [1,2]. LBIC, however, only probes the local short-circuit current which is not sufficient to identify all "bad" areas or to show improvements of e.g. the open circuit voltage or the series resistance after passivation. For identifying "bad" areas of all kind and for monitoring the success of passivation treatments we therefore used the newly developed "CELLO" (for "CELl LOcal) technique [7] which allows to measure all relevant solar cell parameters locally (including the open circuit voltage U_{OC}, the series and shunt resistance (R_{ser} or R_{sh}) or even the complete local IV-characteristics).

3 Experimental

"Bad" areas of various kinds were identified using the CELLO technique. The solar cell was then coated with a (positive) photo resist which in turn was structured to provide a closed curve with an opening of about 1 mm around the bad area by a simple lithographic procedure. The resulting open areas were used to dissolve the grid. After removal of the resist, the solar cell was coated with a relatively thick (several 10 μm) spray-on polymer mask and finally, for the sake of simplicity, a scratch through the polymer layer was made with a diamond scribe (about 150 μm wide and reaching about 10-50 μm into the Si) following the contours of the lithographically defined contour. The remaining grid thus was totally covered by the polymer. This is important because in the following passivation procedure, anodization of the grid must be avoided.

IV-characteristics were measured throughout the procedure [1,3] to monitor the over-all changes in the solar cell performance. All measurements were taken at 25°C. Generally, the leakage current and the series resistance increases strongly because of the scratch.

For the electrochemical passivation of the scratch a large variety of electrolytes and etching parameters was tried and optimized. For the purpose of the experiments described in this paper, a solution of 10% KOH [4,5,6] proved to be the best choice. A constant anodic current density of about 300 mA/cm^2 was applied to the solar cell backside using a PC controlled current supply (see Figure 3).

The potential was measured as a function of the time and all parameters were computer controlled via a *IEEE* interface.

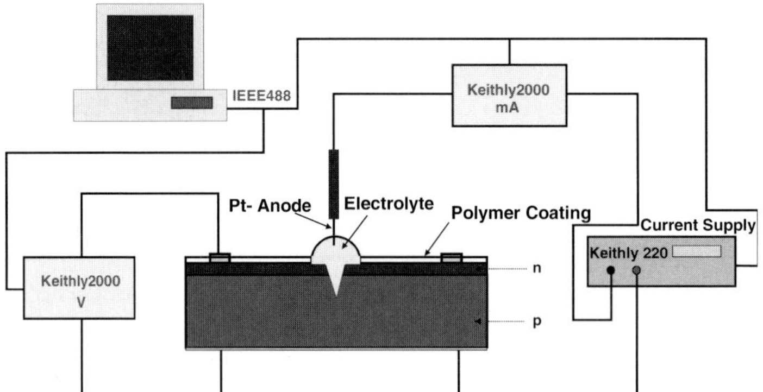

Figure 3 Passivation procedure of the scratch using KOH as electrolyte.

If all goes well, the "bad" area is completely separated from the remaining solar cell without any negative influences of the separation procedure. To establish the claim, the remaining solar cell must now be characterized again and, if so desired, the separated "bad" area may also be measured.

4 Results

Figure 4 shows CELLO maps of a 4.9 x 4.9 cm^2 multi-crystalline silicon solar cell before (a,b) and after separation and passivation of the bad area (c,d). The short circuit current I_{SC} map of the as-received solar cell (a) shows a pronounced bad area in the middle of the right half of the cell which is also seen in the serial resistance R_{ser} map (b). The latter also shows a bad area in the upper part of the right hand side and increased resistance around the edges.

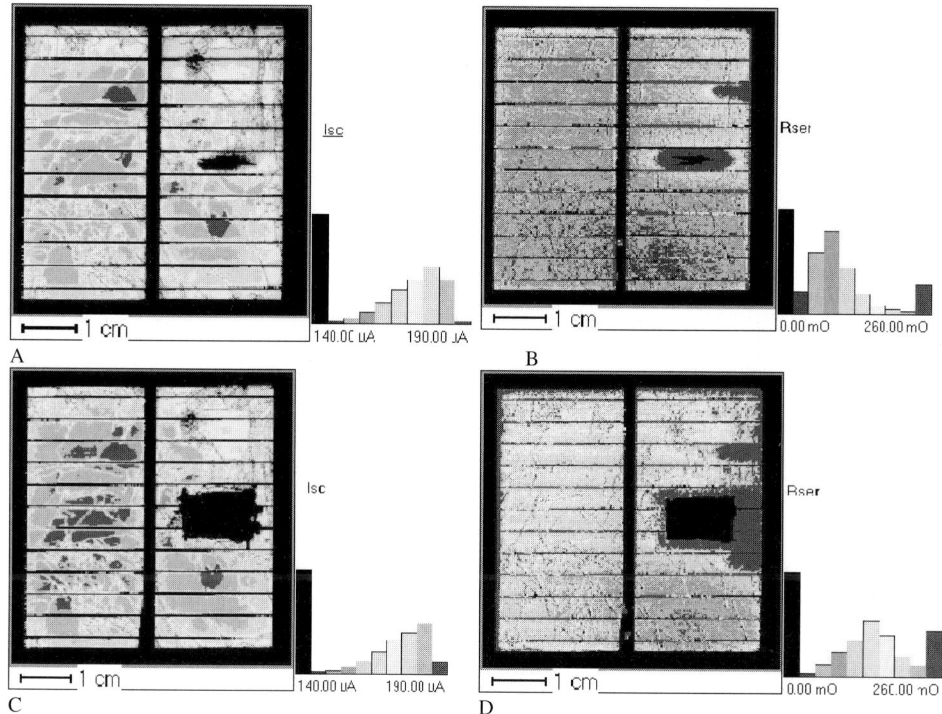

Figure 4 I_{SC} (on the left) and R_{ser} (on the right) map of a multi-crystalline silicon solar cell as measured with the CELLO technique before (a,b) and after the (c,d) the isolation and passivation of the defect. The boundaries of the separated area do not show increased recombination in the I_{SC} map.

The area which had been chosen to isolate the "bad" area seen in the I_{SC} map was a rectangle of 2.29 cm^2 completely enclosing the bad part. As seen in Fig. 4c,d, the procedure succeeded in complete separation of the bad area (monitored totally black, i.e. giving no signal) without introducing reduced diffusion lengths at the boundaries from the cell walls (Figure 4(c,d)). The increase of the series resistance around the separated area in the R_{ser} map, though completely natural even for totally passivated scratches, does depend on the quality of the passivation process and can be used (with some experience and against a reference) as a qualitative measure of the degree of passivation. A better measure, however, are quantitative IV-measurements which for this solar cell are shown in Fig. 5.

Fig. 5 shows a remarkable improvement in the characteristics of the treated solar cell. It remains to be seen if this translates into an improvement of the over-all characteristics of the treated solar cell (which now has lost about 10% of its area) in comparison to the as received cell. In other words, do the improved characteristics overcompensate the loss of the bad area which, though "bad" still contributed somewhat to the over-all performance?

Figure 6 shows the IV-characteristics under illumination in a direct comparison. Predictably, the short-circuit current I_{SC} is somewhat reduced, but the fill factor and the maximal power point is much improved (before the passivation $FF_1 = 0.38$, $P_{m1} = 0.1714$ Watt and after the passivation $FF_2 = 0.55$, $P_{m2} = 0.2326$ Watt), the total efficiency η of the now smaller cell thus was improved by 35% (50% if referenced to the reduced size). This is a direct demonstration that "bad" areas may not only contribute to the performance of a solar cell but may consume a sizeable part of the power generated in the good parts.

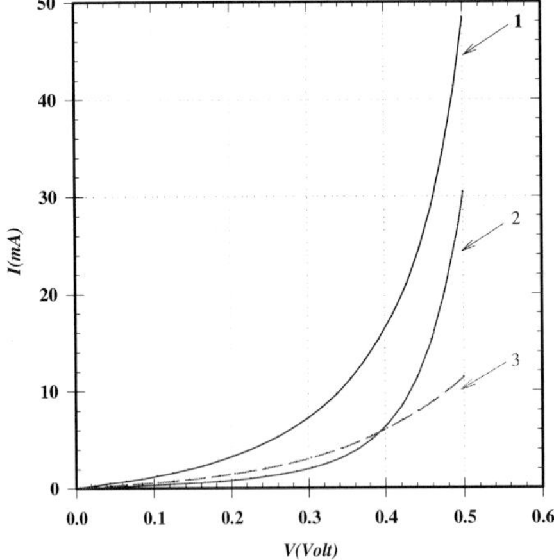

Figure 5 IV-characteristic (without illumination) of the solar cell. The numbers denote: 1) as received, 2) after separation and passivation, 3) characteristics of the separated bad area.

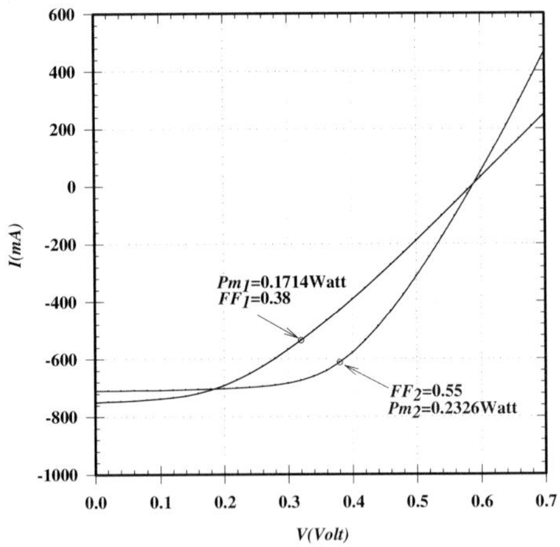

Figure 6 The IV-characteristics of the solar cell under illumination before (FF_1) and after (FF_2) the passivation.

5 Discussion

The results demonstrate that the separation method employed can be used to isolate specific areas on solar cells for either improving the over-all efficiency or studying bad areas in detail. The CELLO maps after the passivation (Figure 4 c,d) show that the isolated area is electronically no longer existent. The I_{SC} map does not show any decrease in lifetime at the edges of the scratch, indicating good passivation of scratching induced defects. The R_{ser} map of course shows an increased series resistance around the isolated area because parts of the grid are now missing. However, the over-all effects in this extreme case of reducing the area by about 10% and cutting through the grid are still very positive. This shows that bad areas are not only bad "providers" of current but act as very efficient internal current sinks consuming a sizeable part of the current generated elsewhere.

6 Conclusion

It is demonstrated that it is possible to improve the over-all efficiency η of multi crystalline solar cells by neutralizing bad areas. The method employed achieved the separation of the bad areas by scratching and subsequent passivation of the scratches. Substantial improvements are possible as shown by IV-characteristics and local measurements of solar cell parameters employing the CELLO method. The method in its present form is not applicable to large scale production but may be modified to accommodate specific tasks in a fixed technological environment.

7 References and footnotes

1. Goetzberger, J. Knobloch, B. Voss, Crystalline Silicon Solar Cells, WILEY, England, 1998, p.201-225.
2. G. Agostineli, W. Zaaiman, C. Helmke, 2nd World Conference of Photovoltaic Solar Energy Conversion, Vienna, 1998, 184-187.
3. S. M. Sze, Physics of Semiconductor Devices, 2nd Edition, WILEY, NEW YORK, 1981, p. 805-811.
4. O. J. Glembocki, R. E. Stahlbush, J. Electrochem. Soc., 1984, 132, 145-151.
5. E. D. Paltik, O. J. Glembocki, I. Heard, J. Electrochem. Soc., 1987, 134, 404-409.
6. L. Chen, M. Chen, C. Lien , C. Wan, J. Electrochem. Soc. 1995,142, 170-175.
7. To be published; see also the homepage of the authors.

III Polymers

Contemporary Applications of Rubberformed Advanced Thermoplastic Composites

D.R. Manten

Delft Thermoplastic Composites, Rotterdamseweg 35, 2289 AC Rijswijk, The Netherlands

A. Beukers

Delft University of Technology, Faculty of Aerospace Engineering Kluyverweg 1, 2629 HS Delft, The Netherlands

1 Abstract

Advanced thermoplastic composites (ATC) have been around now for almost two decades. It was not until recently however that their specific processing methods were developed far enough to make their use more widespread. One manufacturing technique has been very important in this: rubberforming. Rubberforming combines the advantages of matched metal-die forming and conventional rubberpressing of sheet metal. Typical rubberforming tooling consist of a rigid metal mould, a matching flexible rubber contra-mold and sometimes a blankholder. With this process more and more high-end composite products are produced on a successful commercial basis. Typical series range from 100 to 5000 identical parts. Examples of these products are aircraft structural parts like ribs, aircraft interior parts, EMC-shielding laptop casings, housings for electronics, springs for bicycle suspension, and more.

2 Introduction

Thermoplastic resins in composite materials were introduced by the US military when developing the B2-bomber, early 1980's. The main reason behind the switch from thermosets to thermoplastics, was the need for a composite material that could be processed easier, quicker and, most important, with more control over quality. Additional benefits were to be found in improved material characteristics as fatigue insensitivity, chemical resistance and an elevated service temperature. Materials at which development was focussed in that time were high performance thermoplastics like PEEK and PEI. Later, materials were also developed with less expensive engineering thermoplastics like polyamides and polypropylene, focussed more on high volume markets as the automotive industry.

Thermoplastics loose molecular coherence above T_G (glass transition temperature) or T_M (melting temperature) for respectively amorphous and semi-crystalline thermoplastics. At the processing temperature the thermoplastic softens to a low viscosity at which the ATC material is easily formable. After the material is draped quickly over or into a cold mould, the fabric reinforced thermoplastic hardens and retains the shape and texture of the mould. The processing temperature determines the time and energy needed for a production cycle, the lower the processing temperature the shorter the cycle time.

Since the processing of thermoplastics is based on phase transitions instead of chemical reactions, application of thermoplastic composites offers an immense reduction in manufacturing time, compared with the in general time consuming processes of thermoset composites. Product quality can be reproduced accurately by controlling the three main process parameters time, temperature and pressure. Fiber placement can be predicted (with programs like Drape®), controlled and reproduced as the draped fabric geometry is determined solely by the geometry of the product surface and the fabric parameters (see figure 1).

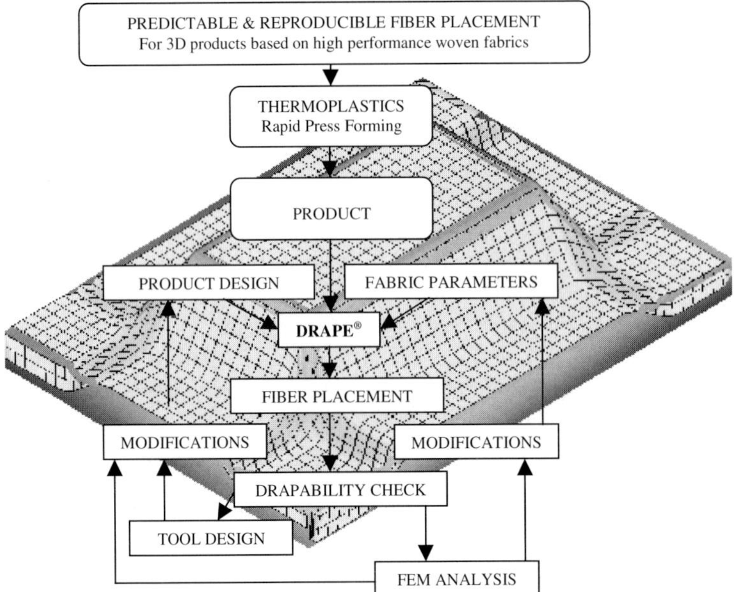

Figure 1. Predictable and reproducible fiber placement

Many kinds of advanced thermoplastic composites are available today. Fibers like carbon, Kevlar® and glass are combined with many kinds of amorphous and semi-crystalline thermoplastics like PEI, PEEK, PPS, PBT, PA and PP. The selection of the thermoplastic matrix is based on different criteria: cost, specific weight, chemical resistance, fire-smoke-toxicity–criteria, glass/melt temperature, processing temperature (cycle time), service temperature or water absorption.

3 Rubberforming

Rubberforming combines matched metal die forming and rubberpressing of sheet metal as used in the aerospace industry. In rubberforming the tooling consists of one rigid mold with the product definition and a flexible counterpart that is made from an elastomeric material like polyurethane or silicone rubber. The elastomeric tool can be both positive or negative, depending on which side the product needs exact definition. The flexible mould forms the ATC laminate in or over the mould and applies the consolidation pressure.

The rubberforming cycle starts with the heating of a flat pre-consolidated ATC sheet. The heating can be done in several ways of which conduction heating in a hot platen press and radiation heating with infrared panels are the two most commonly applied techniques. The heating with infrared panels has the advantage of being quick and without surface contact and is most commonly used. When the processing temperature is reached, the thermoplastic matrix is soft and has a low viscosity. The hot laminate is quickly transported from the heater to the press and is placed between the two molds. In a few seconds then, the flexible stamp forms the product in or over the rigid mould. If the product is deep and large shearing angles of the fabric are required, a blankholder can be used to introduce tensile forces at the edges of the laminate to prevent wrinkling and buckling of the fibers. As the ATC laminate touches the mold it starts cooling down and reconsolidates under an increased surface pressure (typically 10 - 50 bar) until the thermoplastic matrix regains a sufficient solid state. Then the product can be removed and be trimmed and machined as required. Figure 2 shows the set-up for the rubberforming production of a halve of a pressure vessel. Due to the flexibility of the rubber contra-mold a quasi-hydrostatic pressure distribution is obtained, necessary for the reconsolidation of the ATC material. The use of rubber stamps has more advantages like the short lead time of tool manufacturing, the less strict tolerances on matched tooling and the possibility to form small undercuts. The relatively low cost of a mold set makes it possible to produce high-end composite parts cost effectively from series starting at 100 parts a year. Rubberformed ATC is typically used in load carrying structures and in protective shells. Examples are ribs, brackets and stringers in aircraft, high speed fans, helmets, computer casings, and all kinds of other protective covers. Other properties of ATC like the high failure strains and excellent fatigue properties have found their way into typical application like springs. In the following a few applications are highlighted.

Figure 2. Set-up for the rubberforming production of Glass/PP pressure vessels

4 Contemporary Applications

4.1 Casing for military laptops

Computers used by the military are subject to strict requirements for the operational reliability of these systems and just as important, the safekeeping of the data they process. Computers can not be affected in their performance by electromagnetic interference nor may it be possible to eavesdrop on sensitive information. Besides intrinsically safe architecture of the electronics itself, a casing is needed that shields off the computer completely from electromagnetic waves coming in or getting out. Carbon reinforced polyetherimide, with its good conductive properties, gives an ideal material for a casing that has to be strong, lightweight and impact resistant at the side too. As a matrix the chemically stable polyetherimide was chosen to fulfil mechanical requirements under extreme circumstances. The thermoplastic composite design consists of a total of 4 shells that enclose motherboard, keypad and screen. The use of thermoplastic composites offers besides a 70% reduction in weight a substantial cost-saving of more than 60 percent over the traditional steel casings. The main cost saving factor lies in the speed of processing: The use of the rubberform process enables to form a shell every 60 seconds, making it possible to be competitive with the previously used (machined) stainless while the raw material is 15 times more expensive.

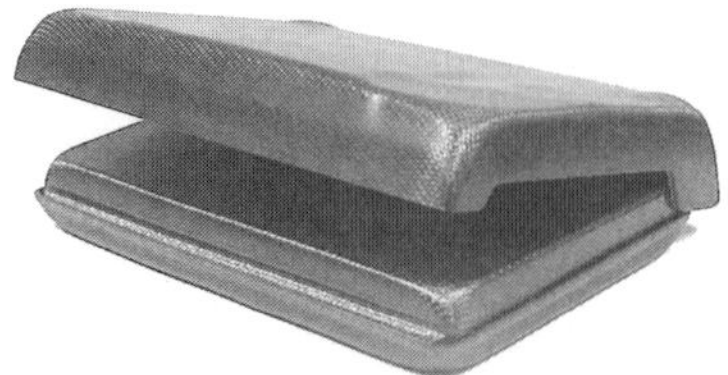

Figure 3. Carbon-polyetherimide laptop casing

4.2 Torsion link®

The function of a torsion-link® (figure 4) is to transfer torsion-loads between two axis while offering flexibility in axial direction. The torsion-link® is a typical application of ATC that would be impossible to realize from metal or even from thermoset composites. The anisotropic nature of fabric reinforced composite makes it possible to tune the ratio between bending stiffness and torsional stiffness. This makes it possible that a material can be flexible in one direction while rigid in the other. Because of the compact geometry of the torsion-link® high strains develop at the maximum travel of the suspension. ATC material can resist these strains without fatigue failure or plastic deformation which would normally occur with thermoset or metal. A thick laminate of carbon fabric with a Polyamide-12 matrix was selected as it was the only material that could fulfil these requirements. Torsion-links can be found in high speed machinery but also on bicycles with a mono-fork.

4.3 Smoke detector cover

In 1996 an aircraft crashed in The Everglades because of an undetected fire in the cargo hold. As a result of the accident a retrofit of smoke and fire detectors was ordered for the cargo holds of more than 3000 aircraft. Smoke detectors are fitted in the ceiling of the cargo hold and they are protected with a cover. This cover also serves as the bracket to which the smoke detector is mounted. With thermoset lay-up techniques, a maximum productivity of 6-8 parts per person per day could be achieved. With more than 8.000 parts to be produced in a year this was a major bottleneck in production. An alternative version from glass fabric with a polyetherimide matrix was developed for the rubberform process and cycle times were reduced down to 3 minutes. Besides saving time on making the part itself, trimming speed could be increased due to the increased toughness of the material. Next to the reduced manufacturing cost the thermoplastic smoke detector cover also outperforms the thermoset version in fire-smoke-toxicity (FST) and mechanical properties.

Figure 4. Carbon-polyamide torsion link®

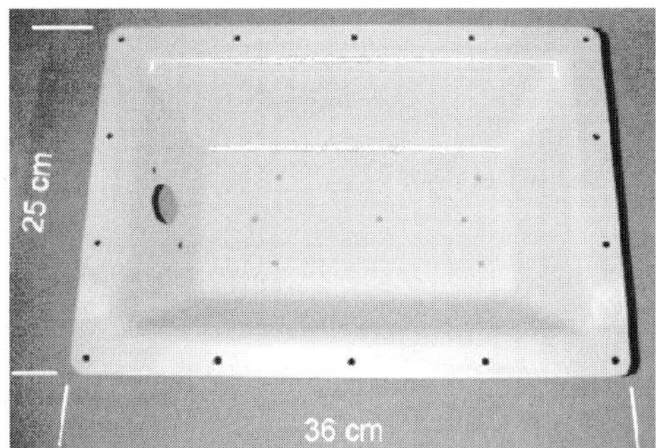

Figure 5. Glass-PEI smoke detector cover

5 Conclusion

Rubberforming of continuous fiber reinforced thermoplastic composites has reached the level of maturity that is necessary for successful commercial application. It has proven to be a cost effective manufacturing technique for small and medium series of high-end applications in thermoplastic composite materials.

6 References

1. L.M.J. Robroek, The development of rubber forming as a rapid thermoforming technique for continuous fiber reinforced thermoplastic composites, Delft University Press, Delft, 1994

Ultrafast Crosslinking of Styrene-Butadiene Rubbers

Christian Decker, Nguyen Thi Viet Trieu
Département de Photochimie (CNRS) – Mulhouse (France)

1 Introduction

Thermoplastic elastomers, like styrene-butadiene rubbers (SBS), are usually made of triblock copolymers which form segregated glassy and elastomeric domains [1]. These polymeric materials exhibit a high elasticity and resiliency, but a poor resistance to organic solvents. Moreover, they start to flow upon heating above 80°C, the polystyrene glass transition temperature, when the physical network loses its cohesion. An effective way to make these polymers more resistant to solvents and elevated temperatures, as required for hot-melt adhesive applications, is to create covalent bonds between the elastomeric chains.

Such a chemical network can be readily formed by photoinitiated crosslinking through the vinyl and butene double bonds located on the polybutadiene chain [2,3]. Upon UV-curing, the Shear Adhesion Failure Temperature (SAFT) was found to rise from 80°C to over 160°C [4,5], while the polymer remained tacky and became completely insoluble [3]. The main advantage of photoinitiation is that crosslinking is achieved within seconds at ambient temperature, selectively in the illuminated areas, by a solvent-free process. This technology is therefore of great interest for adhesives and sealants applications, as well as for the manufacture of flexographic printing plates. We report here how the curing performance of photosensitive SBS-based resins can be substantially improved by using a trifunctional thiol as crosslinking agent.

2 Experimental Part

Two types of polystyrene-*block*-polybutadiene-*block*-polystyrene (SBS) from SHELL were used in this study. They differ only by the content of pendent vinyl double bonds : 8% in SBS-8 and 59% in SBS-59 of the total unsaturation content. An acylphosphine oxide (Lucirin TPO from BASF) was selected as photoinitiator. A trifunctional thiol, trimethylolpropane mercapto-propionate [TRIS] from Evans Chemetics, was used as crosslinker. In a typical experiment, a 20 µm thick film was cast on a KBr crystal and exposed briefly to the radiation of a medium pressure mercury lamp at a light intensity of 0.6 W cm^{-2} in the UV-range.

The cure kinetics was studied by FTIR spectroscopy, by following the disappearance of the vinyl and butene-2 double bonds at 910 cm^{-1} and 965 cm^{-1}, respectively. The amount of insoluble polymer formed after UV-irradiation was determined by gravimetry, using toluene as solvent. The hardness of the sample, before and after UV-exposure, was determined by monitoring the damping time of the oscillations of a pendulum (Persoz hardness).

3 Photocrosslinking of SBS Rubber

When the styrene-butadiene sample SBS-8 was exposed to UV radiation in the presence of a photoinitiator ([Lucirin TPO] = 3 wt%), a few vinyl double bonds were found to react rapidly, while the butene-2 unsaturation content remained essentially unchanged. Figure 1 shows the decrease of the two types of double bonds, as measured by infrared spectroscopy, upon UV-exposure for up to 2 seconds in the presence of air, at ambient temperature. After 0.5 s and a 10% vinyl loss, the reaction is slowing down because of the fast consumption of the photoinitiator. That crosslinking has occurred was demonstrated by the nearly complete insolubilization of the irradiated sample, as well as by the increase of the polymer hardness (Figures 2 and 3). Rising the vinyl content of the polybutadiene chain up to 59% was found to increase only slightly the rate of gelation (Figure 2). An even less pronounced effect was observed for the increase of the polymer hardness upon UV-exposure (Figure 3). These results suggest that, in addition to the intermolecular reaction which leads to the formation of the polymer network, an intramolecular polymerization is taking place in SBS-59 between neighboring vinyl groups located on the same polybutadiene chain.

The difference in behavior between the two SBS samples is clearly apparent in Figure 4 where we have plotted the gel fraction as a function of the amount of vinyl groups which have polymerized. In SBS-8, total insolubilization was achieved at a vinyl concentration of 0.14 mol kg^{-1}, which corresponds to 17 double bonds polymerized per polymer chain. In SBS-59, these numbers rise to 0.52 mol kg^{-1} and 62, respectively, which means that an increase of the SBS vinyl content will mainly favor the intramolecular polymerization, without much effect on the crosslinking process. Further studies were therefore performed on the low vinyl content sample (SBS-8), which is very similar to typical KRATON thermoplastic elastomers.

After UV-curing, the Shear Adhesion Failure Temperature was found to rise from 80 to over 160°C, while the adhesive peel strength remained high, provided that a plasticizer was introduced in the resin formulation. For hot-melt adhesive applications, it is important that the polymer material retains its elastomeric character and tackiness after crosslinking, i.e. no hardening upon UV exposure. To fulfil this requirement, the photoinitiator concentration was reduced to 1 wt% and the exposure time to 1 s. Under those conditions, the crosslinking polymerization proceeded less extensively and insolubilization was far from being complete (gel fraction of 60%). To make the resin more reactive, small amounts of a crosslinking agent, a trifunctional thiol, had to be introduced in the UV-curable formulation.

4 Photocrosslinking of Thiol-SBS Resins

The photoinduced reaction of a thiol with an olefinic double bond was used to generate crosslinks between polybutadiene chains. The polymerization proceeds by a step growth addition mechanism which is propagated by a chain transfer reaction involving the thiyl radical $RS^{\bullet}$ [6]. With a trifunctional thiol (TRIS), the overall curing process is represented schematically in Figure 5. By adding as little as 1% by weight of TRIS to the SBS-8 + Lucirin TPO (1 wt%) formulation, crosslinking was found to occur 10 times as fast as in the absence of thiol. Figure 6 shows some typical curves of insolubilization and hardening upon UV exposure.

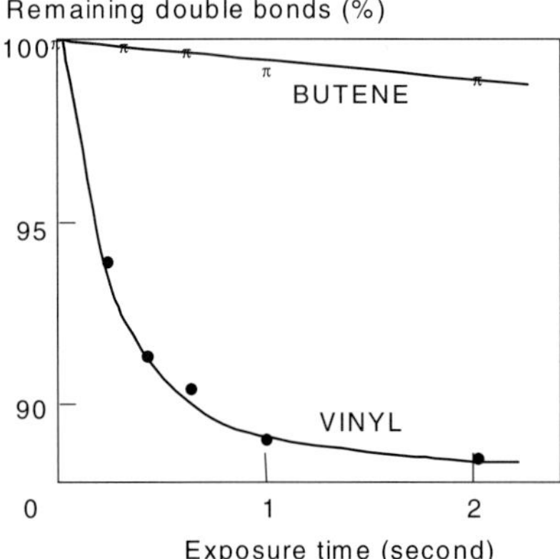

Fig. 1 : Loss of the polybutadiene double bonds upon UV irradiation of SBS-8. [Lucirin TPO] = 3 wt%

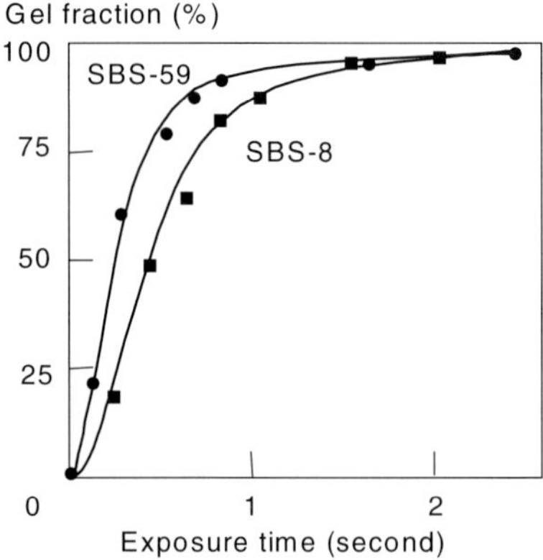

Fig.2 : Influence of the vinyl content on the insolubilization of a SBS rubber upon UV-irradiation. [Lucirin TPO] = 3 wt%

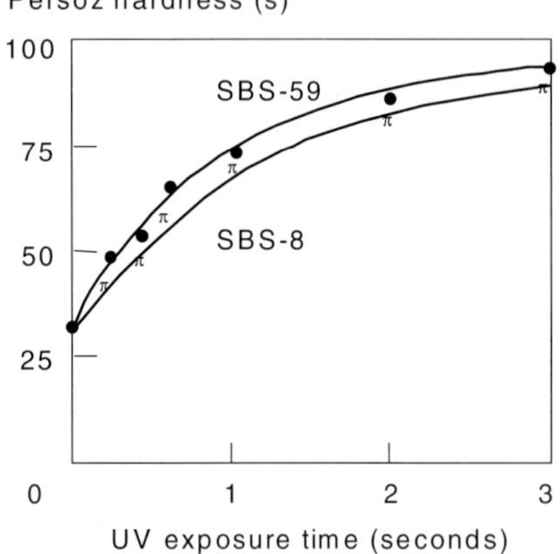

Fig. 3: Influence of the vinyl content on the hardening of a SBS rubber upon UV-irradiation. [Lucirin TPO] = 3 wt%

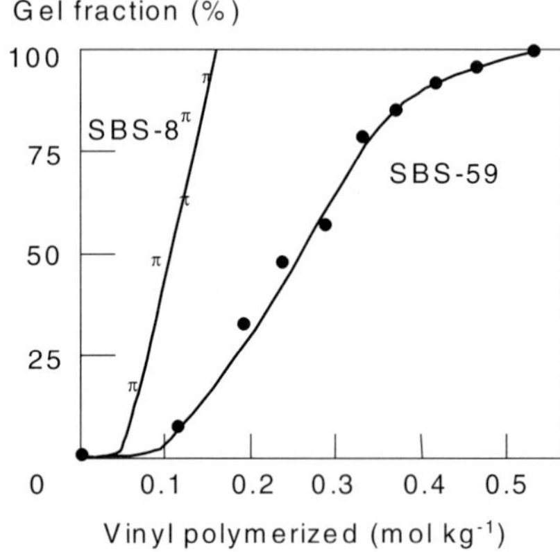

Fig.4 : Dependence of the gel fraction of a photocrosslinked SBS rubber on the degree of vinyl conversion. [Lucirin TPO] = 3 wt%

Insolubilization was already achieved within 0.3 s, once essentially all the thiol had reacted. From the quantity of reactive groups consumed, the number of branch points needed to achieve insolubilization was calculated to be 15 attachments *per* macromolecule, i.e., a value similar to that found in neat SBS. After a 0.3 s UV exposure the hardness increase was limited and the cured polymer retained its elastomeric character. Here again the addition of plasticizers (partly hydrogenated hydrocarbons) was needed to achieve the high peel strengh requested for adhesive applications.

Eventhough thiyl radicals were found to be 10 times as reactive toward vinyl than toward butene double bonds [7], the thiol-butene copolymerization is still accounting for nearly half of the curing process, because of the large number of butene-2 double bonds in SBS-8 (92% of the polybutadiene unsaturation). On the contrary, homopolymerization of the few vinyl groups (like in neat SBS) provides only a minor contribution to the overall crosslinking process, because the propagating alkyl radical reacts preferentially with the thiol group than with the vinyl double bond [7]. Figure 7 illustrates these results by showing the relative importance of the three reactions in UV-curing of the SBS-8/TRIS system. For the SBS sample with a high vinyl content (SBS-59), these proportions are changed, as expected, the contribution of the butene copolymerization becoming negligible, while vinyl-based reactions are gaining in importance (Fig.7).

For some applications of these photocured thermoplastic elastomers, like flexographic printing plates, glass laminates or composite materials, a few millimeter thick samples need to be produced. This can still be achieved by a photochemical reaction by increasing the penetration of UV light into the sample through a decrease of the photoinitiator concentration. The resulting drop in the initiation rate leads of course to a slower cure than in the previous experiments. But, owing to the high reactivity of the thiol-ene system, one can still achieve a fast deep through-cure of 2 mm thick samples within a few seconds when the photoinitiator concentration was lowered down to 0.1 wt%. The fast photobleaching of Lucirin TPO allows UV radiation to penetrate deeper into the sample, thus promoting a frontal polymerization. Figure 8 shows the variation with the photoinitiator concentration of the vinyl double bond consumption, gel fraction and pendulum hadness in a SBS-8/TRIS sample UV irradiated for 0.5 s. A Lucirin TPO concentration of 0.1 wt% appears to be enough to obtain a crosslinked elastomer showing the required performance with respect to photoreactivity, viscoelastic properties and solvent resistance.

Besides being used as hot-melt adhesives, these photocured thermoplastic elastomers are expected to find new applications in various industrial sectors. They proved to be well-suited for the manufacture of safety glasses because of their outstanding adhesion on glass and their excellent impact resistance. Flexible composite materials have been readily obtained upon UV-irradiation, after introducing various translucid mineral fillers (silica, carbonates) into the resin formulation. Other specific applications take advantage of the fact that crosslinking occurs exclusively in the illuminated areas. Flexographic printing plates can thus be made of SBS-based photocurable resins. High resolution images can also be generated onto photosensitive substrate by means of a focused laser beam (Ar^+ or Kr^+), operated in the UV range on a continuous wave mode.

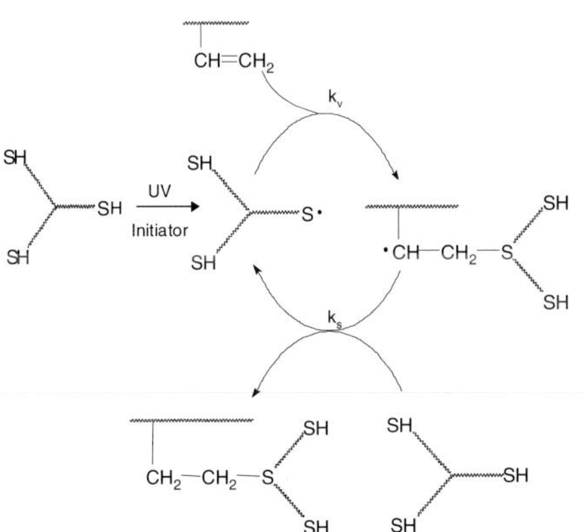

Fig. 5 : Photoinitiated crosslinking polymerization of the thiol-polyene system

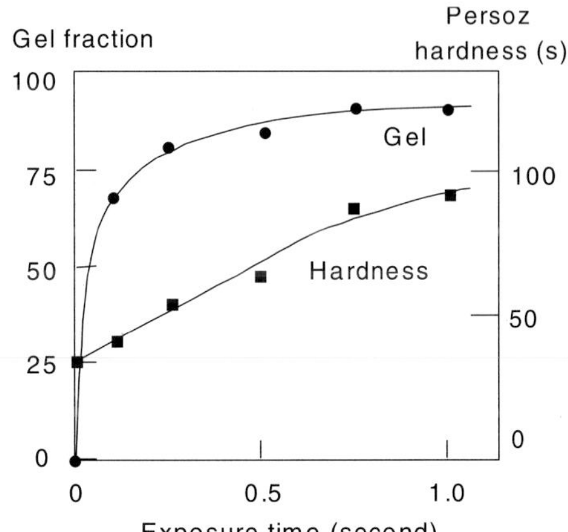

Fig.6 : Photocrosslinking kinetics of the SBS-8/TRIS resin. [TRIS] = 1 wt% ; [Lucirin TPO] = 1 wt%

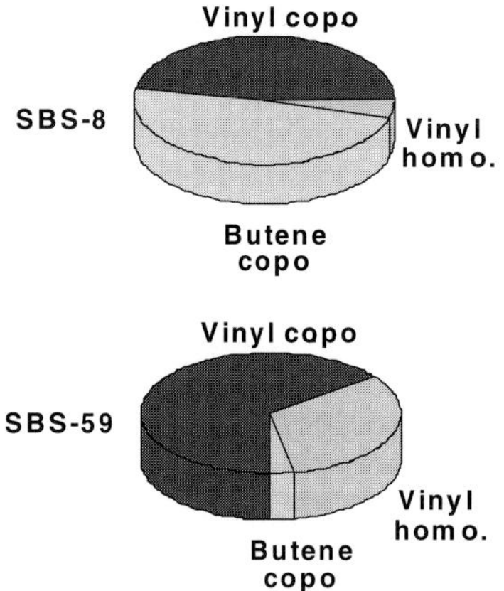

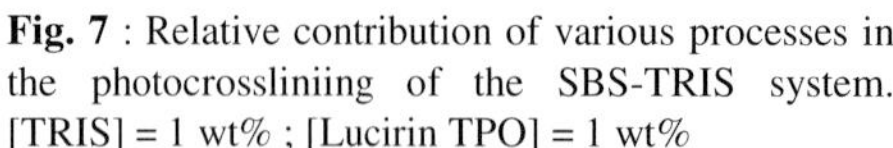
Fig. 7 : Relative contribution of various processes in the photocrossliniing of the SBS-TRIS system. [TRIS] = 1 wt% ; [Lucirin TPO] = 1 wt%

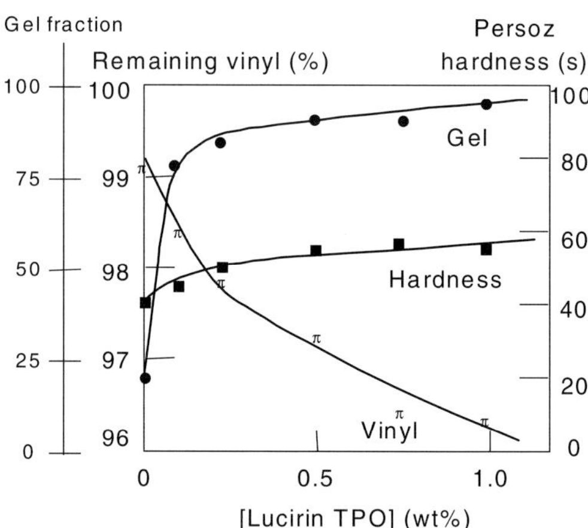

Fig.8 : Influence of the photoinitiator concentration on the photocrosslinking of SBS-8. [TRIS] = 1 wt% ; UV exposure = 0.5 s

5 Conclusion

Styrene-butadiene thermoplastic elastomers are readily crosslinked at ambient temperature by a short UV exposure in the presence of a radical-type photoinitiator. The polymerization of the pendent vinyl double bonds leads within seconds to the formation of a tridimensionnal polymer network which shows a great resistance to organic solvents and heat. By reinforcing the physical network, the covalent bonds created in the elastomeric mid-phase impart cohesion to the polymer material at elevated temperatures, as required in hot-melt adhesive applications. The addition of small amounts (1 wt%) of a trifunctional thiol was shown to increase drastically the reaction rate, to yield an insoluble material within a fraction of a second. The radical-induced thiol-ene polymerization proceeds by a step growth addition mechanism involving the thiyl radical which reacts with both the vinyl and, to a lesser extent, the butene double bonds of the polybutadiene moiety. The photoinitiator concentration can be lowered down to 0.1 wt%, thus allowing relatively thick samples to be cured, because of the increased penetration of UV radiation and the fast photobleaching of the initiator.

The photocrosslinking technology offers a number of advantages, such as on-line processing, high-speed curing in air, solvent-free formulations, low-energy consumption and selective cure in the illuminated areas. The main applications of photocurable thermoplastic elastomers are expected to be found in industrial sectors where cure speed, physico-chemical properties and spatial control are critical issues, such as for the manufacture of hot-melt adhesives and sealants, safety glasses and flexible printing plates and composite materials.

6 Acknowledgements

The authors wish to thank SHELL RESEARCH (Louvain la Neuve, Belgium) for a research grant.

7 References

1. L.H. Sperling and V. Mishra, in *Polymeric Materials Encyclopedia*, Vol.5, J.C. Salomone (ed.) CRC Press 1996, p.3292
2. H.F. Huber in *Radiation Curing in Polymer Science and Technology*, Vol 4, J.P. Fouassier and J.F. Rabek (eds), Elsevier Applied Science, London 1993, p.51
3. C. Decker and T. Nguyen Thi Viet, *Macromol.Chem.Phys.*200, 358 (1999)
4. M. Dupont and N. De Keyser, *Proc.RadTech Europe, Conf. Maestricht* 1995, p.174
5. M. Dupont and C. Mayenez, *Kleben Dichten Adhäsion*, 43 (3), 22 (1999)
6. C.R. Morgan, F. Magnolta and A.D. Ketley, *J.Polym.Sci., Polym.Chem.Ed.* 15, 627 (1977)
7. C. Decker and T. Nguyen Thi Viet, *Macromol.Chem.Phys.* (in press)

LCP-Polysulfone Multiblock Copolymers: Combination of High Performance Polymers

Doris Pospiech[1)], Liane Häußler[1)], Kathrin Eckstein[1)], Hartmut Komber[1)], Dieter Voigt[1)], Andreas Janke[1)], Antje Gottwald[1)], Dieter Jehnichen[1)], Hans R. Kricheldorf[2)]

[1)] Institute of Polymer Research, Dresden, Germany

[2)] University of Hamburg, Department of Technical and Macromolecular Chemistry, Hamburg, Germany

1 Introduction

The coupling of segments of different polymers in block copolymers has been proven to generate compounds with property profiles that show a combination of properties of the parent homopolymers. The property combination is caused by the formation of microphase-separated structures by the incompatible polymeric segments, allowing to develop the properties of each polymer involved [1]. Thus, materials with tailor-made properties for very specific applications can be developed [2]. The combination of high performance polymers in block copolymers should therefore allow to obtain polymers with superior properties that could be useful in microelectronics. Our recent work on multiblock copolymers with a general structure shown in Figure 1 lead to the conclusion that the connection of polysulfone (PSU) and liquid crystalline polyester (PES)- or poly(ester imide) (PEI) segments indeed resulted in a combination of, e.g., high thermostability of PSU on the one hand and high tensile strength and E-modulus of LCP on the other hand [3-5]. The investigation of PSU-PEI multiblock copolymers gave first hints that the development of synergistic effects is due to both, the occurrence of liquid crystallinity as well as occurrence of microphase separation in the block copolymers [6,7].

Figure 1. Schematic structure of LCP-PSU multiblock copolymers (MBCP)

In this paper, we would like to discuss the influence of the chemical structure of different poly(ester imide) segments as well as their molecular weight on the resulting macroscopic properties of multiblock copolymers with polysulfone. The chemical structure of the LCP segments was varied from semiaromatic [poly(ethylene terephthalate)-co-poly(ester imide)], as shown in Formula 1, to fully aromatic poly(ester imides) with two different substituents, shown in Formula 2.

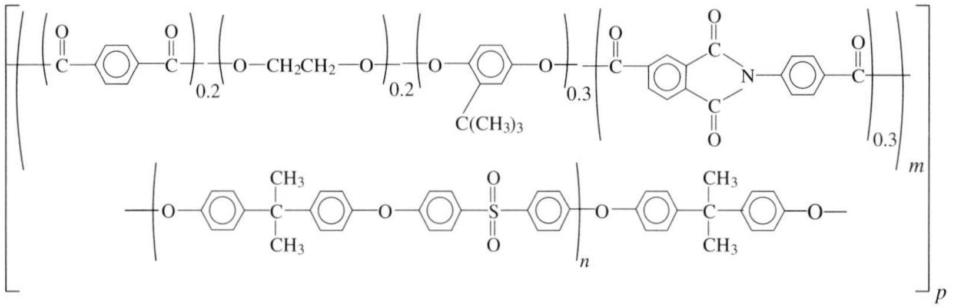

Formula 1. Chemical structure of PSU-(PET/PEI) multiblock copolymers (MBCP **1**)

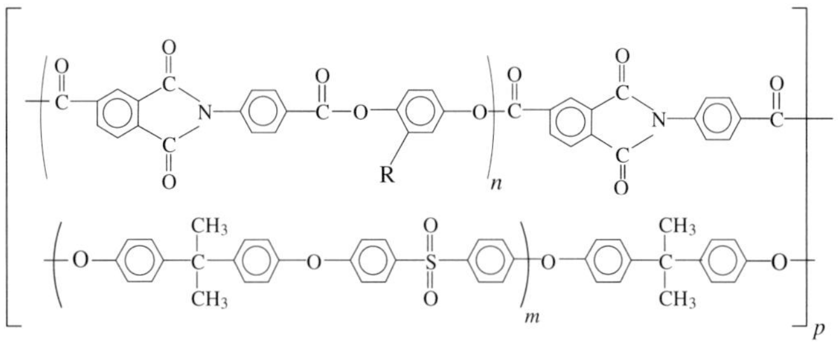

Formula 2. Chemical structure of PSU-PEI multiblock copolymers
(MBCP 2: R = $-C(CH_3)_3$; MBCP 3: R = Ph)

2 Experimental

A melt transesterification polycondensation procedure as described in [3] and [7] and in [6] for MBCP **1** was employed for MBCP synthesis. MBCP **2** and **3** were prepared by copolycondensation of acetoxy-terminated polysulfone oligomers, t-butyl-hydroquinone diacetate (HQDA), (MBCP **2**), or phenyl-hydroquinone diacetate (MBCP **3**), respectively, with N-(4-carboxyphenyl)trimellitimide (TMI). The poly(ester imide) blocks were formed *in situ* during polycondensation. Their molecular weight was controlled by using different comonomer ratios. Conversion into block copolymers was proven by ^{13}C NMR by disappearance of end-groups (-$\underline{C}$OOH (t-butyl-HQDA: at 175.17/175.33 ppm; Ph-HQDA: at 174.75/174.90 ppm; CH_3-$\underline{CO}$O-PSU: at 169.71 ppm; $\underline{C}H_3$-COO-PSU at 22.10 ppm) and formation of signals for ester groups linking PSU and TMI units (MBCP **2**: δ = 150.30/166.20; MBCP **3**: δ = 149.86/150.25/165.5 ppm)

The samples were characterized by solution viscosity (trifluoroacetic acid/ $CHCl_3$ 1/1 vol/vol; 20 °C), GPC (Knauer; eluent: cresol/$CHCl_3$, 1/3 vol/vol; PL Minimix-C separation column), ^{13}C NMR (Bruker; trifluoroacetic acid/$CDCl_3$), DSC (Perkin Elmer DSC 7; heating and cooling rate: 20 K/min), DMA (Eplexor 150 N; Gabo Qualimeter; heating rate: 2 K/min, 10 Hz) and TGA (TGA 7, Perkin Elmer; heating rate: 10 K/min) and dynamic mechanical analysis (ARES, Rheometrics, cone-plate geometry, inert gas).

3 Results and Discussion

The molecular weights of both segments coupled in the multiblock copolymer were varied systematically to obtain MBCP with different composition. Polysulfone oligomers having 2300, 4900, 7200 and 14600 g/mol were linked to PEI with molecular weights ranging from 700 to 7200 g/mol. The chemical composition of the series of MBCP **2** and **3** containing PSU 7200 g/mol blocks and their characterization are given in Table 1.

Table 1. Chemical composition of the synthesized PSU-PEI multiblock copolymers (selected samples)

MBCP	PSU segment M_n [1)] (g/mol)	Ratio TMI[2)]/ HQDA	PEI segm. M_n [3)] (g/mol)	MBCP Solution viscosity (dl/g)	MBCP $M_{w,GPC}$ [4)] (g/mol)	MBCP M_w/M_n	MBCP Melting range [5)] (°C)
2a	2300	1/0	280	0.67	72700	4.30	265 i
2b	7200	1/0	280	0.45	43400	3.29	240 i
2c	7200	2/1	700	0.51	52600	4.11	250 i
2d	7200	5/4	2000	0.83	71500	5.22	290-350 LC
2e	7200	10/9	4100	1.17	-	-	340 LC
2f	7200	16/15	6900	insoluble	-	-	330 LC
3a	7200	2/1	740	0.56	61300	4.48	240 i
3b	7200	5/4	2100	part. sol.[6)]	-	-	300 i
3c	7200	10/9	4400	part. sol [6)]	-	-	330 LC
3d	7200	16/15	7200	insoluble	-	-	290 LC
PEI (t-b)	-	1/1		1.77	-	-	345 LC
PEI (Ph)	-	1/1		1.58	-	-	270 LC

1) determined by end group titration and ^{1}H NMR spectroscopy as reported in [4]
2) HQDA: substituted hydroquinone diacetate (R as given in Formula 2)
§) calculated from the ratio of TMI/substituted hydroquinone diacetate
4) determined by GPC
5) determined by an hot-stage polarizing microscope (LC: liquid crystalline melt; i: isotropic melt)
6) only partially soluble in TFA/$CHCl_3$ and pentafluorophenol/$CHCl_3$

Table 1 shows that the phase behavior in the melt (LC or isotropic melt) can actually be controlled by the block copolymer composition, thus allowing to obtain a melting behavior as desired.

In each block copolymer series, the onset of phase separation, which depends on the block molecular weights, was determined by DSC. Non-phase-separated samples showed a single T_g whereas the occurence of two glass transitions indicated phase separation. The T_g behavior of PET-PEI-PSU and PEI-PSU block copolymers is illustrated for series with constant LCP segment molecular weight (Figure 1).

The T_g's of the PET/PEI-PSU MBCP are lower than those of the fully aromatic PEI-PSU MBCP series, independently of the substituent used. This is caused by the semiaromatic character of PET/PEI. However, a significant T_g increase compared to the pure PET/PEI polyester has to be noticed. PEI-PSU MBCP **2** have even higher T_g values. Here, phase separation results in the occurence of two T_g, one of them higher than the original PEI-T_g, one of them lower. This behavior is typical for phase-separated samples.

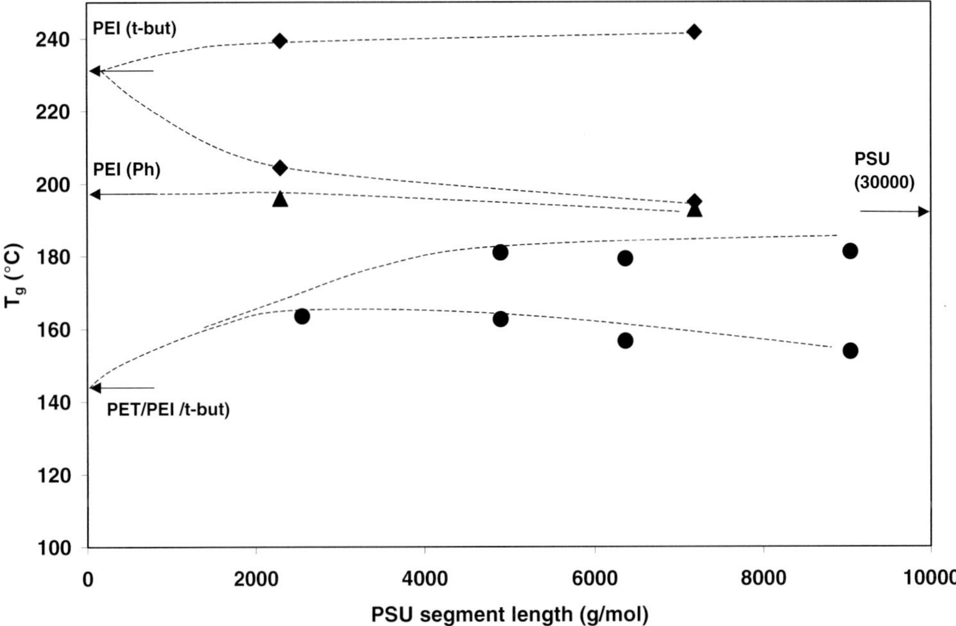

Figure 1. Glass transition behavior of PET/PEI (7450 g/mol)-PSU MBCP 1, PEI (t-b) (4100 g/mol)-PSU MBCP 2 and PEI (Ph) (4400 g/mol)-PSU MBCP 3 with constant poly(esterimide) segment and PSU segments with different molecular weight.

The relaxation behavior of the MBCP is influenced by the chemical structure of the LC segments. Figure 2 shows the storage modulus curves vs temperature of PSU MBCP with different LC segments having almost comparable segment molecular weights. The initial E-modulus depends significantly on the chemical structure of the LC segment. MBCP with semiaromatic LC segments (PET/HBA- and PET/PEI) have the lowest E-modulus. Intermediate values are shown by MBCP with fully aromatic PEI segments having phenyl substituents, whereas MBCP with t-butyl-substituted PEI possess the highest value, i.e., increasing chain stiffness of the LC segment raises the storage modulus. Additionally, the modulus can be tailored by the segment molecular weights. Longer PEI segments result in further modulus increase, which was detected for MBCP **1** [6] and MBCP **2** and **3**, as well.

Figure 3 shows for PEI(t-butyl)-PSU MBCP that the complex melt viscosity (measured at 380 °C) of the block copolymers can also be controlled by PEI and PSU segment molecular weights. The high melt viscosity of the fully aromatic PEI can effectively be dropped by MBCP formation with PSU, thus increasing the processability of PEI which can originally barely be handled in the melt [8].

The thermostability of the MBCP is high and exceeds the stability of the parent homopolymers. No decomposition can be observed until 380 °C in all MBCP.

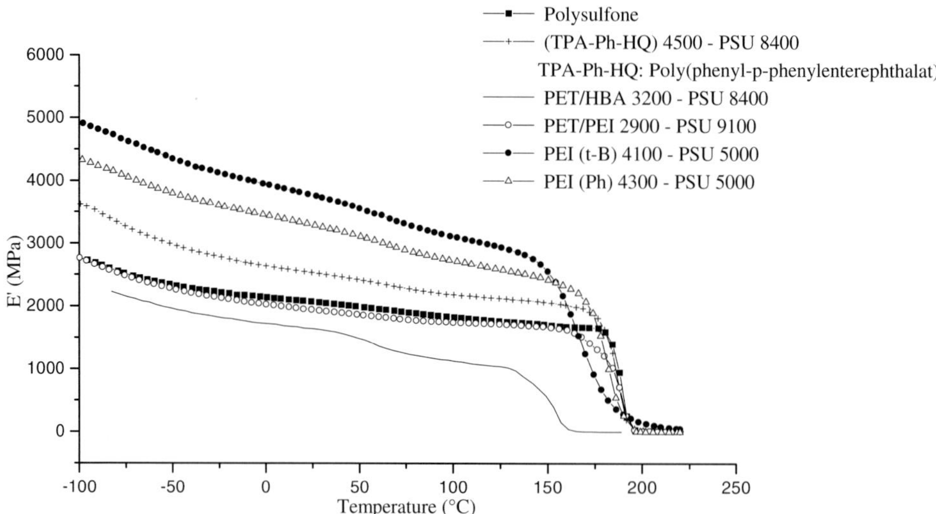

Figure 2. Storage modulus of PSU-LCP multiblock copolymers with different types of LC segments (the numbers after each segment give the segment molecular weights in g/mol)

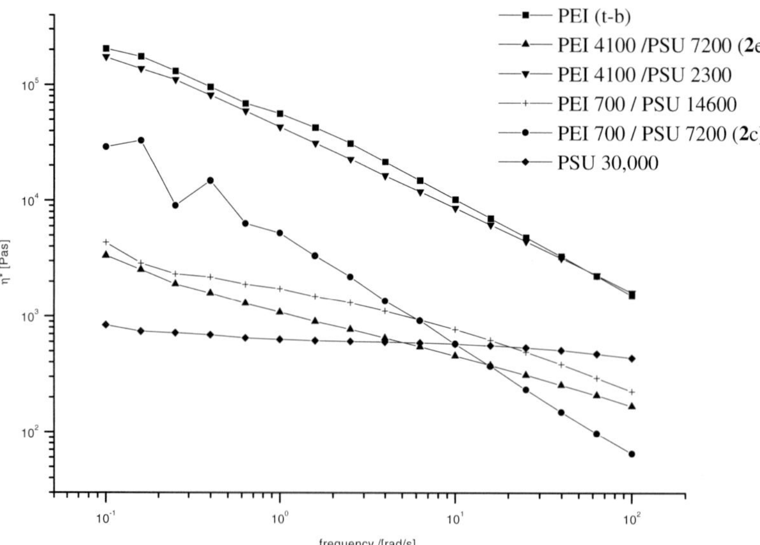

Figure 3. Complex melt viscosity as function of measuring frequency of PEI(t-butyl)-PSU MBCP **2** with different segment molecular weights of PSU segments, at 380 °C.

3 Summary

The investigations showed that the properties of PSU-LCP multiblock copolymers (e.g., phase behavior, glass transition temperature, thermostability, melt viscosity and solubility) can be effectively controlled by the chemistry and molecular weight of the blocks linked together in the block copolymer. Block copolymer formation actually allows to combine advantageous properties of polysulfone (high thermostability) and liquid crystalline polyesters (high E-modulus) within one material. It also prevents macrophase separation, as observed in corresponding blends of the materials. Some drawbacks of the original materials, such as high melt viscosity and weak solubility in the case of fully aromatic poly(ester imides), can be altered, resulting in improved processability. All multiblock copolymers synthesized showed a very high thermostability and might therefore be used as new high performance engineering materials.

4 Acknowledgements

We would like to thank Mrs. Monika Dittrich for preparation of melt pressed samples, Mrs. Petra Treppe for her efforts with GPC analysis and Mrs. Sabine Görner for performing dynamic mechanical analysis of the block copolymers.

5 References

1. M. Walther, H. Finkelmann, Prog. Polym. Sci. 1996, 21, 951-979.
2. R. Mülhaupt, U. Buchholz, J. Rösch, N. Steinhauser, Angew. Makromol. Chem. 1994, 223, 47-60.
3. D. Pospiech, L. Häußler, B. Voit, F. Böhme, H. R. Kricheldorf in Solvent-Free Polymerizations and Processes: Minimization of Conventional Organic Solvents (M. Hunt, T. Long, Eds.), *ACS Symp. Series No.* **713**, American Chemical Society, Washington, DC, USA, 1998, Chapter 2.
4. D. Pospiech, L. Häußler, D. Voigt, D. Jehnichen, A. Janke, A. Baier, K. Eckstein, F. Böhme, J. Appl. Polym. Sci. 1996, 62, 11,1819-1833.
5. D. Pospiech, L. Häußler, E. Meyer, A. Janke , R. Vogel, J. Appl. Polym. Sci. 1997, 64, 619-630.
6. D. Pospiech, L. Häußler, K. Eckstein, H. Komber, D. Voigt, D. Jehnichen, E. Meyer, A. Janke, H. R. Kricheldorf, Designed Monomers and Polymers 1998, 1, 2, 187-206.
7. D. Pospiech, K. Eckstein, H. Komber, D. Voigt, F. Böhme, H. R. Kricheldorf, Polym. Prepr., Am. Chem. Soc. 1997, 38, 2, 398-99.
8. H. R. Kricheldorf, A. Domschke, G. Schwarz, Macromolecules 1991, 24, 1011-1016.

Exploring Mechanical Design in Biodegradable Polymers

Prabhu Kandachar, Joke Reijnhoudt, Rolf Koster & Ineke Goedhart
Faculty of Design, Engineering and Production, Delft University of Technology, The Netherlands

1 Abstract

Comparative investigations have been carried out to examine the mechanical properties of injection molded test specimens in biodegradable polymers and conventional plastics. A simple product has also been injection molded in biodegradable polymers. This product has been experimentally tested for its load carrying capacity. A finite element model has been made to predict its mechanical response.

2 Introduction

During the last decades, both the public and the policymakers have shown increased awareness as well as concern about decreasing material resources to match with increasing population, economy and the corresponding increase in consumption. Plastics, based on one of the non-renewable material resources (crude oil), have shown to be capable of meeting a large variety of product requirements, both for consumer and for industrial applications (ref. 1). The new family of biodegradable polymers, based on renewable material resources, could challenge the supremacy of conventional plastics, as feedstock for some specific consumer applications.

Biodegradable polymers, however, are still in a development stage. Although they promise biodegradability and compostability in professional composting units, problems for application on a large scale, like sensitivity to moisture, price economics, definitions, standards, legislation, etc., have not been resolved yet (ref. 2). The optimism of early 70's of the biodegradable polymer industry has made place for realism in the recent years(ref. 3). Although biodegradable polymers are still being developed, designing products in these new materials has to be understood as well to facilitate market introduction. This paper is a contribution towards that goal.

3 Experiments

Two materials have been used for this investigation: polyhydroxybutyrate (PHB), a maize-based material supplied by Biomer (ref. 4), and polyester-amide, a synthetic biodegradable plastic supplied by Bayer (ref. 5), with mechanical properties similar to thermoplastic elastomers. The PHB material did not have hydroxyvalerate added and therefore represented the most brittle version available, with mechanical properties resembling polystyrene.

Tensile specimens (ISO 3167 A) and tensile-impact specimens (ISO 8256-3) were injection molded. Additionally, several dishes (for demonstration of feasibility) were injection molded.

Figure 1 shows all three test specimens. Between production and testing the test specimens were conditioned for at least 14 days at 20°C and 60% relative humidity. Tensile testing was done at constant clamp separation speeds based on ISO 527 recommendations for the tensile specimens: 50 mm/min for polyester-amide (ductile) and 5 mm/min for PHB (brittle). Initial modulus of elasticity, maximum stress, strain at maximum stress, stress at failure and strain at failure were determined from stress-strain diagrams obtained during tensile testing. During testing, the specimen was held in a closed chamber maintained at 60% relative humidity and 20°C.

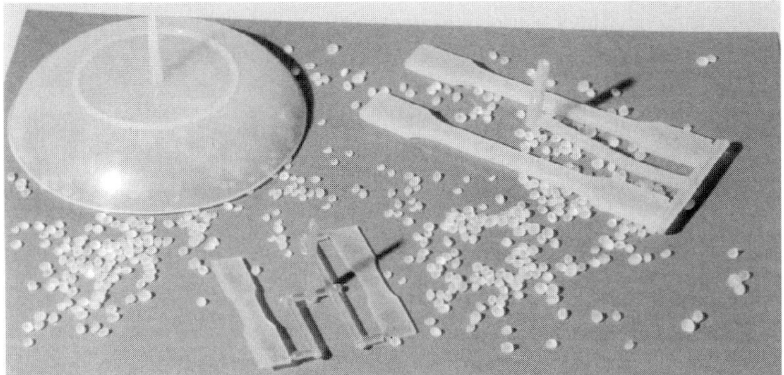

Figure 1. The three test specimens. Left: dish (diameter 155 mm, wall thicknessess 2.5 to 3.5 mm). Right: tensile test specimen. Front: tensile test-impact specimen, with one of the specimens with a double gate.

4 Results and Discussion

The test results for the tensile specimens are best for comparison with standardized test results and hence these will be used in this paper. These results for both PHB and polyester-amide are summarized in **Table 1** while Figure 2 shows typical stress-strain diagrams of these materials, along with those of some conventional plastics. The test results on tensile-impact specimens are discussed in ref 6.

Table 1. Tensile test results

Property	PHB	polyester-amide
Initial modulus of elasticity, N/mm^2	1695	447
Engineering strain at maximum stress, %	5	72
Maximum engineering stress, N/mm^2	21	16

Strain values of PHB are comparable to those of PMMA and SAN, and PHB can be considered brittle. Strengthwise, PHB has lower values than these materials and can be considered equivalent to HDPE. Polyester-amide is comparable to LDPE both as regards tensile strength and elongation at break. The calculated initial modulus of PHB shows little scatter with standard deviations no more 1.5% of the average values. There was a large variation in the strain at break, however, with strain values between 3% and 7%. All tensile specimens in polyester-amide elongated more than 100 %. Although polyester-amide does seem to exhibit reasonably clear yielding, a further examination revealed that small cracks developed in the specimen, coalescing together to form cavities.

5 Modelling

The modelling procedure consisted of the following steps:

- numerical modelling of the tensile specimen
- verifying the results of the model with actual test data
- extending the model to the dish (demonstrator product)
- comparing the results of the dish model with experimental data of the dish

The purpose of this exercise was to verify whether the numerical approach will provide sufficiently accurate results when compared to experimental data such that the numerical approach can be extended to other (and more complex) geometries.

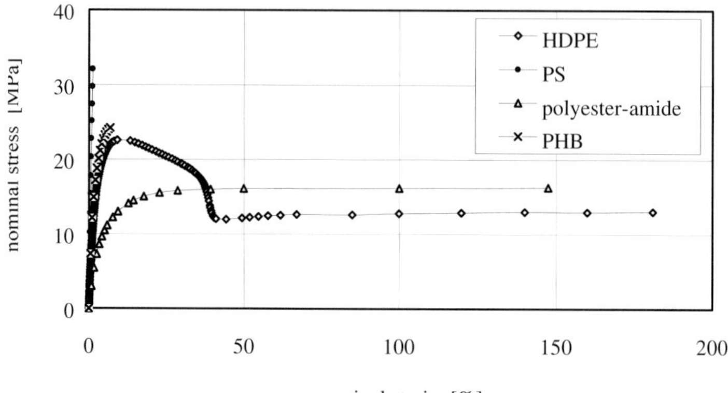

Figure 2. Typical stress-strain diagrams of PHB, polyester-amide, PS and HDPE

The tensile test specimen is geometrically symmetric, and therefore only a quarter of the specimen was modelled and copied to make a model of the full sized specimen. The model was built up of 280 (8-node brick) elements, 28x5x2, with three degrees of freedom at every node. MARC Finite Element Analysis (FEA) package was used for analysis purposes. Typical experimental test data for both materials were implemented. From these data true stress and strain values, assuming constant volume, were determined and used for the material model. Measurements encompassed both the elastic and the plastic regions and the material was modelled according to isotropic elastic-plastic behavior.

Similar to actual testing, the FEA calculations were performed with a constant displacement rate at the end of the specimen, where it was clamped. The calculations on the PHB specimen resulted in a maximum true stress of 26.4 N/mm^2. For comparison, the calculated true stress in the specimen after a 7 mm extension during the experiment was 25.9 N/mm^2 . Thus a difference of only 2% was observed. A comparison of the results from the experiments with those from FEA calculations on PHB reveals that the model closely describes the experimental results, a maximum difference of 4% was achieved (Figure 3). Although limited experimental data were used to generate this model the model of the PHB tensile test specimen approached quite closely the experimental data up to an extension of approximately 7 mm. A similar approach on polyester-amide resulted in a maximum true stress of 39.9 N/mm^2 in the model, while from the experiments the maximum true stress was calculated at 39.1 N/mm^2. Again the numerical model was able to predict the measured data with a 4 % difference in the applied force. The results obtained with this simplified material model gave us confidence to use this model to study the dish, which was subjected to small deformations only.

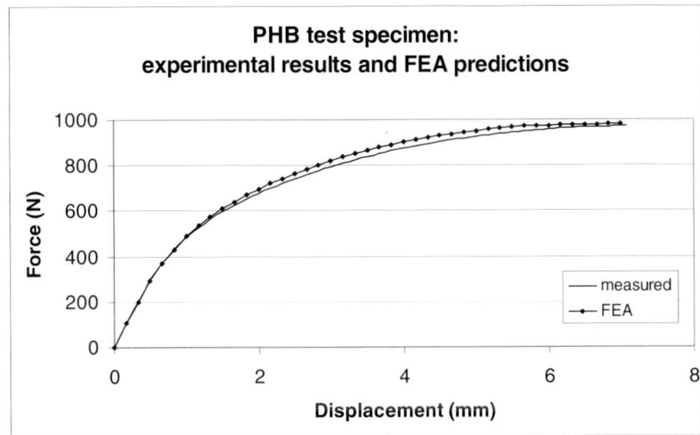

Figure 3. Output from experiments on PHB compared with numerical modelling.

The circular symmetry of the dish allowed modelling only a cross section from the periphery to the center. The axisymmetric shell element had two nodes. The dish model was built up of 34 elements with varying thickness. The node in the center of the dish was fixed in the y-direction, while the last node on the periphery was fixed in the x- and y-directions. The node just off-center is subjected to a prescribed displacement in several small steps up to 14 mm maximum displacement in the x-direction. The maximum equivalent stress calculated for an applied displacement of 14 mm just at the center of the PHB dish was 25.8 N/mm^2. Similar calculations on polyester-amide dish revealed a maximum equivalent stress of 9.7 N/mm^2. To verify whether this model can correctly predict the mechanical behavior of the dish, the following experiments were conducted:

- placing the dish upside down (with the gate side facing up) on the testing plateau
- imposing a displacement of 14 mm (and measuring the required compressive force) at the center of the dish

The deformation rate was chosen such that the calculated maximum strain rate in the dish was within the same range as during testing of the tensile specimens to minimize rate effects. Figure 4 shows the results of the experiments as well as of numerical modelling of PHB dish. For displacements up to approximately 4 mm, the three PHB dishes showed similar behavior, with a slight divergence for displacements up to 7 mm. On increasing the displacement the dishes started cracking. For displacements up to 7 mm the numerical model of PHB also predicted the applied force very well. For higher displacements the FEA results diverged from measurement data. During testing, the polyester-amide dishes regained their original shape after removing the compressive load, indicating that the deformation under load was largely elastic. The model predicted the experimental results reasonably accurate. Note that in none of these material models the viscoelasticity of the material was taken into account. Despite the simplifications made it was possible to model the dish reasonably accurately. Therefore the model can be used in further studies on other geometries in PHB or polyester-amide.

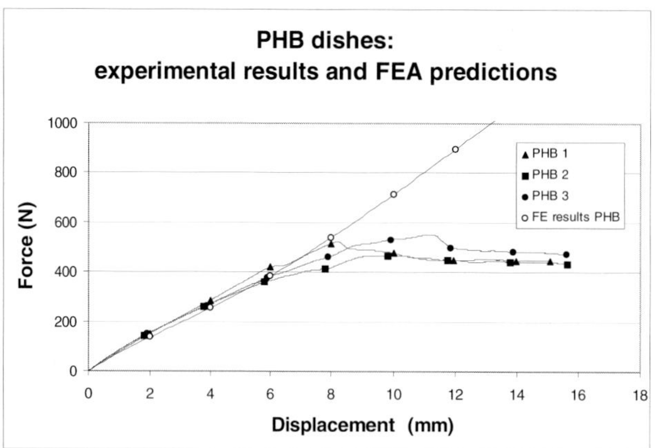

Figure 4. Experimental data and results of numerical modelling on PHB dishes

5 Concluding Remarks

By carrying out tensile tests on biodegradable polymers and subsequently using the measured material properties in a numerical model, it is possible to describe the mechanical behavior of relatively simple products made of new biodegradable materials. Further research based on this exploratory work is in progress.

6 References

1. J. van Liempd, Proceedings of the Wageningen Symposium on Renewable Bioproducts, June 1997, published by the European Commission, p.19-25
2. Warmington, European Plastics News, May 1999, p. 22.
3. Editorial, European Plastics News, May 1999, p. 5
4. Biomer biodegradable polymers – Processing recommendations.
5. Bayer LP BAK 402 Data Sheet, Version 1.14.12.98
6. Prabhu Kandachar, Joke Reijnhoudt, Rolf Koster & Ineke Goedhart, Injection Molding in Biodegradable Plastics; paper submitted to EUROMAT 1999

7 Acknowledgments

The authors are grateful to Dr U. Hänggi of Biomer, Germany and to Mr. Drießen and Dr Schlitte of Bayer, Germany for providing materials, data and processing guidelines. Mr. Carel Hermans provided valuable experimental support at our laboratory. We had fruitful discussions with Mr. Johan Beijer on the experimental approach and interpretation of the results.

Hyperbranched Polyesters and Poly(ether amide)s – Synthesis, Modification, Melt Rheology and Application in Blends

Thomas Huber, Petra Pötschke, Dirk Schmaljohann, Brigitte Voit
Institut für Polymerforschung Dresden e.V.

Introduction

Dendritic polymers feature a large number of functional groups, low solution and melt viscosities and a globular, three-dimensional structure[1]. Due to these properties, dendrimers and hyperbranched polymers are discussed as suitable new functional materials for a variety of applications[1]. Whereas dendrimers might be used in medical application e.g. as a drug-carrier or in gene delivery systems due to their well-defined structure and molar mass, the non-perfect analogues to the dendrimers, the hyperbranched polymers, can be used in less sensitive fields. Most commonly discussed applications are those as melt modifiers or blend components. But the large number of functional groups accessible for further modification reactions can also be used to change the polarity or to introduce special (reactive) groups e.g. for crosslinking reactions in coating systems[2].

Dendrimers and hyperbranched polymers have been used as blend material following different aims. Kim et al.[3] blended hyperbranched polyphenylene with polystyrene. The resulting blends exhibited improved thermal stability and a reduced melt viscosity at high temperatures and high shear rates. Blends of hyperbranched polyesters[4], aryl ester dendrimers[5] and PAMAM dendrimers[6] with different linear polymers such as polyesters[4,5], polyamides[4], polycarbonate[4], and poly(vinylchloride)[6] and poly(vinyl acetate)[6] have been studied with regard to compatibility and change in mechanical properties. Khadir et al.[7,8] prepared an arborescent polystyrene and blended it with linear polystyrene and polymethylmethacrylate, respectively, resulting in materials with very low apparent viscosity. The use of dendrimers as carrier molecules for additives in blends has been reported by de Brabander-van den Berg et al.[9].

We will focus on the synthetic approaches to hyperbranched polyesters and poly(ether amide)s, their application in different blend systems and the modification of these structure to adjust compatibility and polarity.

Results and Discussion

Synthesis and modification of polyesters and poly(ether amide)s

Hyperbranched aromatic polyesters are readily prepared using melt polycondensation techniques and activated monomers based on 3,5-dihydroxybenzoic acid as established by Kricheldorf[10] Fréchet[11] and Voit[12]. **P1-OH** is based on bis-3,5-(trimethylsiloxy)benzoylchloride **1** which leads to hyperbranched aromatic polyesters with trimethylsiloxy end groups in high yields and high molar masses. The silylgroups are hydrolyzed during workup to leave reactive phenol groups (Scheme 1). 3,5-(Bisacetoxy)benzoic acid **2** can be converted in a melt polycondensation at 200 °C into an aromatic hyperbranched polyester with acetoxy end groups (**P2-OAc**)[12] (Scheme 1). Both polymers can be obtained in molar masses above M_n = 50,000 g/mol, with broad molar mass distributions, high solubility in organic solvents e.g. THF and a degree of branching of 50%.

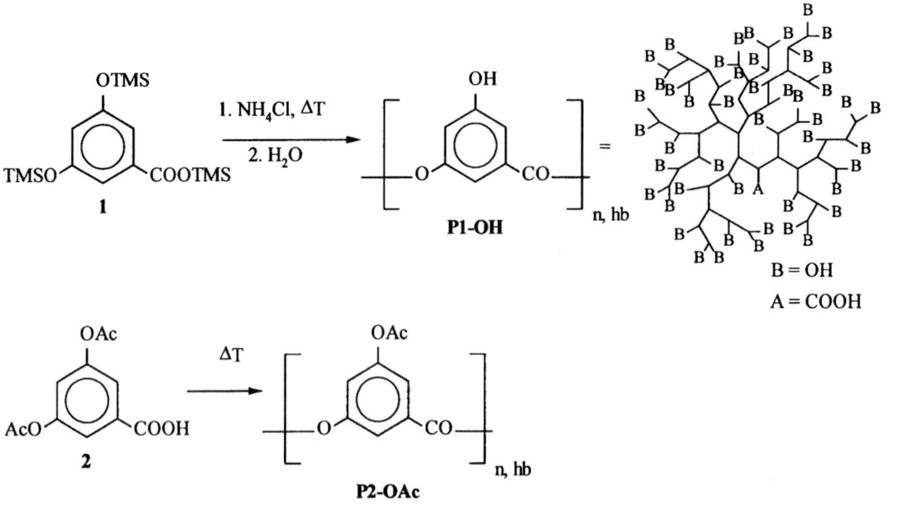

OTMS; TMSO; COOTMS; **1**; 1. NH_4Cl, ΔT; 2. H_2O; OH; O; CO; n, hb; **P1-OH**; B = OH; A = COOH

OAc; AcO; COOH; **2**; ΔT; OAc; O; CO; n, hb; **P2-OAc**

Scheme 1

Starting from the hyperbranched polyester with OH end groups, a variety of different products can be obtained by polymer analogous reaction with aliphatic acid chlorides[13]. Thus, aliphatic chain ends of different length and in different ratios compared to OH functions can be introduced which allows to adjust the solubility and the polarity of the polyesters. When dodecyloyl chloride or hexadecyloyl chloride are used for quantitative modification of all OH groups, the resulting hyperbranched polymers react like amphiphilic molecules with solubility as well as in hexane as in THF. This behavior was studied in detail for the C12-modified product **P1-OCOC11** (Scheme 2)[13].

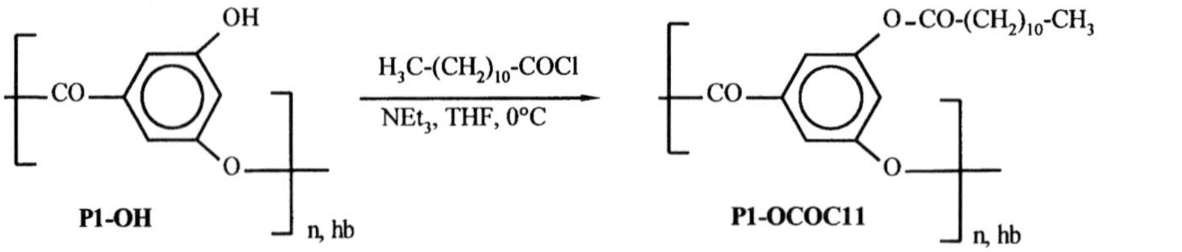

OH; CO; O; **P1-OH**; n, hb; H_3C-$(CH_2)_{10}$-COCl; NEt_3, THF, 0°C; O-CO-$(CH_2)_{10}$-CH_3; CO; O; **P1-OCOC11**; n, hb

Scheme 2

Whereas several hyperbranched polyesters with aromatic, aliphatic or aromatic-aliphatic backbones are now well known, the synthesis of well defined hyperbranched polyamides is still a challenge.

Kim[14], Jikei[15] and Russo[16] described the synthesis of fully aromatic structures via classical polycondensation techniques. We were able to prepared aliphatic-aromatic hyperbranched poly(ether amide)s via a nucleophilic ring opening reaction of oxazolines (Scheme 3), a reaction which was used previously for the synthesis of linear polymers[17]. For the synthesis of hyperbranched polymers, 2-(3,5-bishydroxyphenyl)-1,3-oxazoline **3** has been synthesized as AB_2 monomer and converted at temperatures between 180 and 220 °C in bulk or solution into the hyperbranched poly(ether amide) **P3-OH**[18].

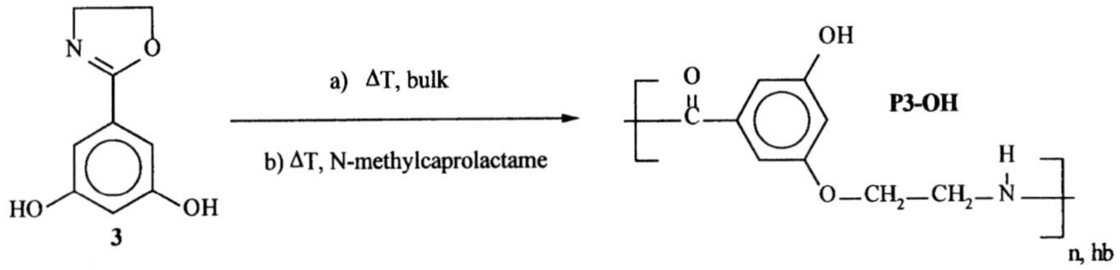

a) ΔT, bulk

b) ΔT, N-methylcaprolactame

3

P3-OH

Scheme 3

The aliphatic-aromatic hyperbranched polyamide is fully soluble in polar organic solvents e.g. DMSO or DMAc with low solution viscosities. It can be obtained in high molar masses ($M_n > 40{,}000$ g/mol determined via GPC, linear calibration), and exhibits a degree of branching of 50% which allows the conclusion that a random reaction took place with equal reactivity of both phenol groups in **3**. The degree of branching could be determined via quantitative high resolution NMR experiments after assignment of dendritic, terminal and linear units in 2D experiments. All units could be clearly assigned and no side reaction of the oxazoline group could be detected via NMR spectroscopy.

Melt rheology of hyperbranched polymers

Hyperbranched polymers are well known to exhibit low solution viscosities compared to analog linear polymers. However, much less is known on their behavior in the melt[19,20]. We performed melt viscosity measurements using a using a Rheometrics ARES with plate-plate-geometry in oscillation mode under nitrogen atmosphere[13,21,22]. The plate diameter was 25 mm and the gap ranged from 1.2 to 1.95 mm. A frequency range between 0.1 and 100 rad/s and a strain within the linear viscoelastic range were used. The samples were investigated as extruded strands or after extensive drying at elevated temperatures and in vacuum.

Fig. 1 shows the complex viscosity plots versus the frequency for different samples measured all app. 60 °C above their glass transition temperature to be able to compare the data. It can be seen that unmodified polyesters and poly(ether amide)s with polar OH end groups (M_n app. 50,000 g/mol) exhibit a rather high complex viscosity at low frequencies with no leveling out but a strong decrease of η^* towards higher frequencies. Therefore, it is not possible to determine zero shear viscosity data. In contrast to this, the alkyl modified sample **P1-OCOC11 of** similar molar mass shows much lower melt viscosity data and a pronounced plateau area. This allows the conclusion that the interaction of the polar end groups determine strongly the rheological behavior of hyperbranched polymers. This is further supported by studies on linear model polymers with similar repeat units but without functional groups. A linear model poly(ether amide), which could be obtained only at low molar masses, also exhibits a plateau region (Fig.1, e)). In addition, the poly(ether amide) sample used for the measurement c) showed an further increase in molar mass upon heating at 230 °C which led to a strong increase in viscosity in repeated experiments (compare Fig. 1 c) and d).

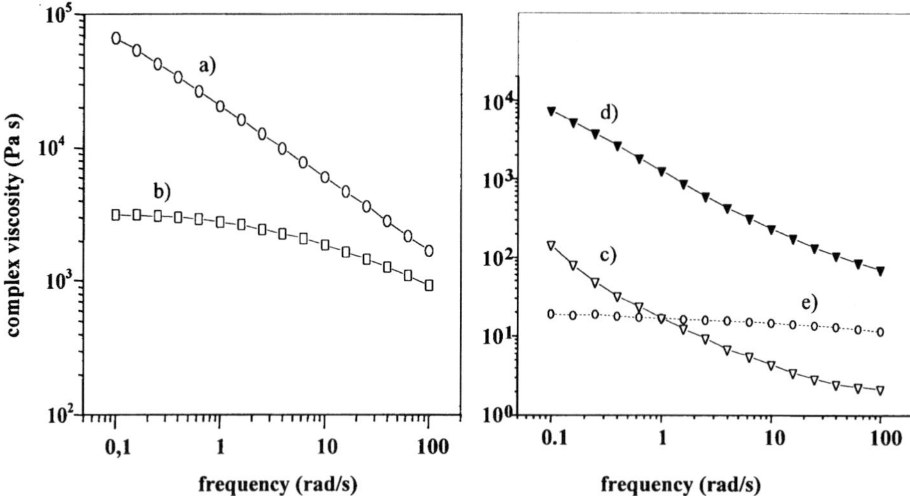

Figure 1. Melt viscosity of a) **P1-OH** at 280 °C, b) **P1-OCOC11** at 110 °C, c) **P3-OH** at 230 °C, 1st run, d) P3-OH at 230 °C, 6th run after about 1.5h, e), linear model poly(ether amide) (M_n only 1800 g/mol) at 155 °C (all measurements app. 60 °C above T_g)

Hyperbranched polyesters in blends

The hyperbranched polyesters with different end groups have been studied in blends with linear polymers with the focus on determining the compatibility and the effect of the hyperbranched material on the mechanical properties of the blends[4,19].

Melt and solution blends of **P1-OH** and **P2-OAc** with different commercially available linear polyesters and polyamides as well as with polycarbonate have been prepared and studied via differential scanning calorimetry[4]. **P1-OH** was fully compatible with most of the studied polyesters and polyamides (e.g. Nylon-6, Trogamide-T, MXD-6 polyamide, poly(butylene adipate), copolyesters from terephthalic acid, isophthalic acid, ethyleneglycol, and 1,4-bishydroxymethylcyclohexane or bisphenol A) as determined by a single Tg observed in 50/50 mixtures of the two blend components. **P2-OAc** exhibited a lower compatibility with linear polyesters and polyamides but was still fully compatible in melt blends with poly(butylene adipate) and some of the copolyesters. The high number of polar end groups seems to enhance the compatibility. No full compatibility was achieved with bisphenol A polycarbonate (Lexan 145), but electron microscopy revealed at least a partial compatibility and comparable strong interactions between the polycarbonate and the polyesters. Mechanical tests on these samples with 25 wt% of **P1-OH** or **P2-OAc** showed that hyperbranched molecules act mainly as reactive fillers with increase in tensile modules but loss of toughness of the blend material.

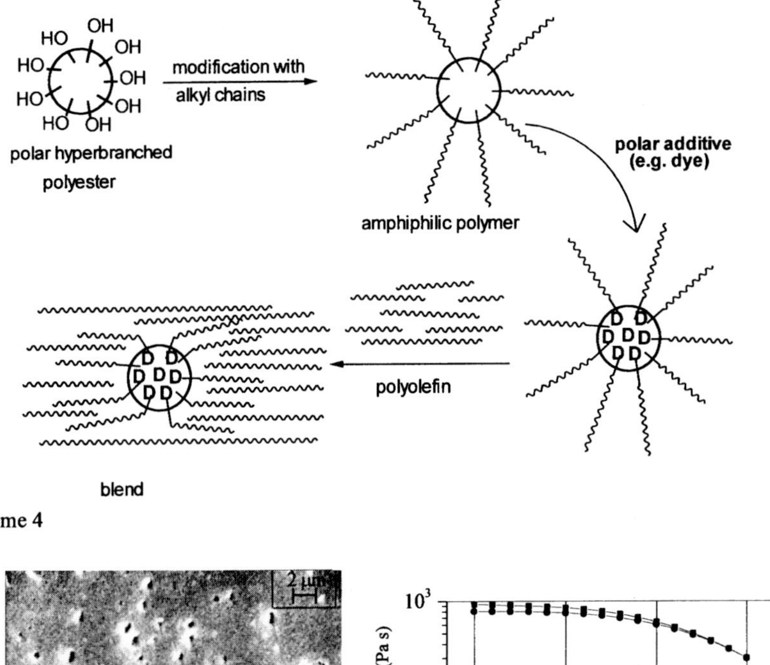

Scheme 4

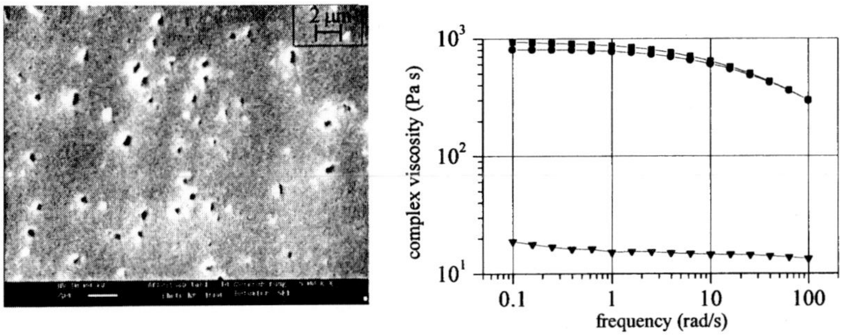

Figure 2. Electron microscope picture (left) of a blend of 5 wt% **P1-OCOC11** and *i*-PP and complex viscosity (right) of *i*-PP (■), *i*-PP + 5wt% **P1-OCPC11** (●) and **P1-OCOC11** (▼)

A strong incompatibility was found between polyolefins and hyperbranched aromatic polyesters. However, by modifying the polyesters with long alkyl chains, the interactions between the hyperbranched material and the polyolefin (polyethylene and polypropylene) improved (Scheme 4)[19]. Since the modified polyesters exhibit amphiphilic properties like micelles with a polar core and a less polar outer shell, it was also possible to incorporate polar additives (e.g. an organic blue dye) into the polar core simply by coprecipitation and separation of non-incorporated dye by washing. The polyolefin and the modified, partially dye loaded hyperbranched polyesters (5 wt% and less) were mixed in the melt, extruded and analyzed via electron microscopy, rheology and dynamic mechanical measurements.

Even though the system polyolefin/hyperbranched polyester was still phase separated (Fig.2, left), the domains of the modified were very small (below 0.3 μm) and the organic dye could be distributed very evenly which was not possible without use of the hyperbranched polymer. In addition, the mechanical properties were still determined by the polyolefin matrix with only a small decrease in melt viscosity (Fig.2, right).

References

1) D.A. Tomalia, H.D. Durst, *Top. Curr. Chem.* 1993, **165**, 194 ; J.M.J. Fréchet, „Synthesis and Properties of Dendrimers and Hyperbranched Polymers" in: *Comprehensive Polymer Science*, Second Suppl., S.L. Aggarwal, S. Russo, Eds., Elsevier Science Ltd., Oxford 1996, p. 71-133; G. Newkome, C.N. Moorefield, F. Vögtle, „Dendrimers - Concept and Perspectives", VCH, Weinheim 1996; B. Voit, *Acta Polymer.* 1995, **46**, 87; B.I. Voit, S.R. Turner, „Hyperbranched Polyesters" in: *Polymeric Materials Encyclopedia*, Vol. 5, J.C. Salamone, Ed., CRC Press, Boca Raton 1996, p. 3177 ff., and references therein

2) M. Johansson, E. Malmström, A. Hult, *J. Polym. Sci. Part A: Polym. Chem.* 1993, **31**, 619.

3) Y.H. Kim, O.W. Webster, *Macromolecules* 1992, **25**, 5561.

4) D.J. Massa, K.A. Shriner, S.R. Turner, B.I. Voit, *Macromolecules* 1995, **28**, 3214.

5) P.L. Carr, G.R. Davies, W.J. Feast, N.M. Stainton, *Polymer* 1996, **37**, 2395.

6) C.M. Nunez, A.L. Andrady, R.K. Guo, J.N. Baskir, D.R. Morgan, *J. Polym. Sci. A, Polym. Chem.* 1998, **36**, 2111.

7) A. Khadir, M. Gauthier, *Polym. Mater. Sci. Eng.* **1997**, *77*, 174.

8) A. Khadir, M. Gauthier, *Ann. Tech. Cont. - Soc. Plast. Eng.* **1997**, *55*, 3732.

9) E.M.M. de Brabander-van den Berg, P.E. Froehling, S. Stevelmans, J.C.M. van Hest, Int. Pat. WO97/1998, assigned to DSM Research, 1996; *Chem. Abstr.* **1997**, *127*: 82286.

10) H.R. Kricheldorf, Q.-Z. Zang, G. Schwarz, *Polymer* 1982, **23**, 1821.

11) C.J. Hawker, R. Lee, J.M.J. Fréchet, *J. Am. Chem. Soc.* 1991, **113**, 4583.

12) S.R. Turner, B.I. Voit, T.H. Mourey, *Macromolecules* 1993, **26**, 4617.

13) D. Schmaljohann, J.A. Loontjens, L. Häusler, P. Pötschke, B.I. Voit, *Macromol. Chem. Phys.*, accepted.

14) Y. H. Kim, *J. Am. Chem. Soc.* 1992, **114**, 4947.

15) G. Yang, M. Jikei, M. Kakimoto, *Macromolecules* 1998, **31**, 5964.

16) S. Russo, A.Boulares, *Macromol. Symp* 1998. **128**, 13

17) C. Wörner, P. Müller, R. Mülhaupt, *Polym. Bull. (Berlin)* 1995, **34**, 301.

18) T. Huber, F. Böhme, H. Komber, J. Kronek, J. Luston, D. Voigt, B. Voit, *Macromol. Chem. Phys.* 1999, **200**, 126.

19) Hawker, C.J. Farrington, P.J., Mackay, M.E., Wooley, K.L., Fréchet, J.M.J., *J. Am. Chem. Soc.* **1995**, *117*, 4409

20) Uppuluri, S.; Keinath, S.E.; Tomalia, D.A.; Dvornic, P.R.; *Macromolecules* , **1998**, *31*, 4498.

21) D. Schmaljohann,; Häusler, L.; Pötschke, P.; Voit, B.I. P.E. Froehling, B. Mostert, J.A. Loontjens, *Macromolecules* 1999, **32**, internet prepublished

22) T. Huber, P. Pötschke, B. Voit, unpublished results

Electrosynthesis and Studies on Polypyrrole Composites

Tran Trung[a], Nguyen Van Tri[b],
[a] Dept. of Electrochemistry, Faculty of Chemisrty, [b] Dept. of Magnetic Resonance, Hanoi University of Techonology, No. 1 DaiCoViet Road, Hanoi, Vietnam

1 Introduction

It has previously been known that polypyrrole in both its neutral-insulating state and oxidized conducting state exhibits a strong ESR signal [1,2]. The first, it is due to the existence of the defect paramagnetic centers as neutral π-radicals with energy below the Fermi level. And later, there exist charge carriers as positive paramagnetic polarons. A presence of anions and doping impurities incorporated within polypyrrole matrix could strongly change physical and electrochemical properties. Recently, polypyrrole composites with oxides of transition metals as well as their lithiated spinels have remarkably attended [3,4]. These impurities built the electronic structure can be viewed as consisting of orbitals of sp^2 hybridized carbon atoms and Pz^2 of heteroatoms and d-orbitals of the impurity counterions localized near by polymer chains. Consequently, the influence of the heteroatoms on the defect paramagnetic centers and positive paramagnetic polarons become stronger (Fig. 6). From useful information obtained during ESR studies on the polypyrrole composites, we can clearly understand the promising electrochemical properties of these composites.

2 Experiment

The detail description of the supporting electrolyte and experiments was shown in elsewhere [5]. β-MnO_2 and $LiMn_2O_4$ powder was made [7]. Data of XRD spectrum of the $LiMn_2O_4$ powder was shown in Fig. 1. The polypyrrole composites denoted PPy/ClO_4, PPy/ClO_4.β-MnO_2 and PPy/ClO_4.$LiMn_2O_4$ were electrosynthesized at a constant potential of 0.65V/SCE in the electrolyte with 50mM pyrrole monomers (solution 1), solution 1 containing 25mg/l β-MnO_2 and solution 1 containing 25mg/l $LiMn_2O_4$, respectively. PPy^o denotes polypyrrole electrosynthesized in aqueous solution only containing 50mM pyrrole monomers. Teflon sealed plate platinum working electrode (area = 1 cm^2) was controlled with PAR model 362 for the electrosynthesis. ESR measurements carried out using Electrnenrezonenz - Spektrometer ESR220.

3 Results and Discussion

3.1 ESR spectrum of PPy°

Data of ESR spectrum of PPy° shown in Fig. 2 indicate that there exist two types of resonance lines. The first with the line width $\Delta H_1 = 1.06$ G is very more narrow than the later with the line width $\Delta H_2 = 144$G. The first is attributed to radical nitrine $>N^\bullet$, later to metastable radical carbine $\equiv C^\bullet$. The mean distance between two nitrine radicals is about 20.6Å, and about 4Å between two carbine radicals. The metastable state even causes the saturated ESR phenomenon (Fig. 2). It may be a triplet state in polymer chains with dicarbine radicals that are like the one observed on several other polymers [6], there many dicarbines whose distances are around 4Å. But a type of dicarbine shown in Fig.4 could be most stable. Spin packet of these dicarbines is firstly saturated as we measured. Schematic interpretation of ESR obtained signal of saturated spin packet was shown in Fig. 3. The very intensive one spin packet saturated ESR line suggests that there are different populous packets of carbine radicals. This phenomenon also signifies that some carbine radicals appear together in the different position as well as crystal fields on the polypyrrole chain segment.

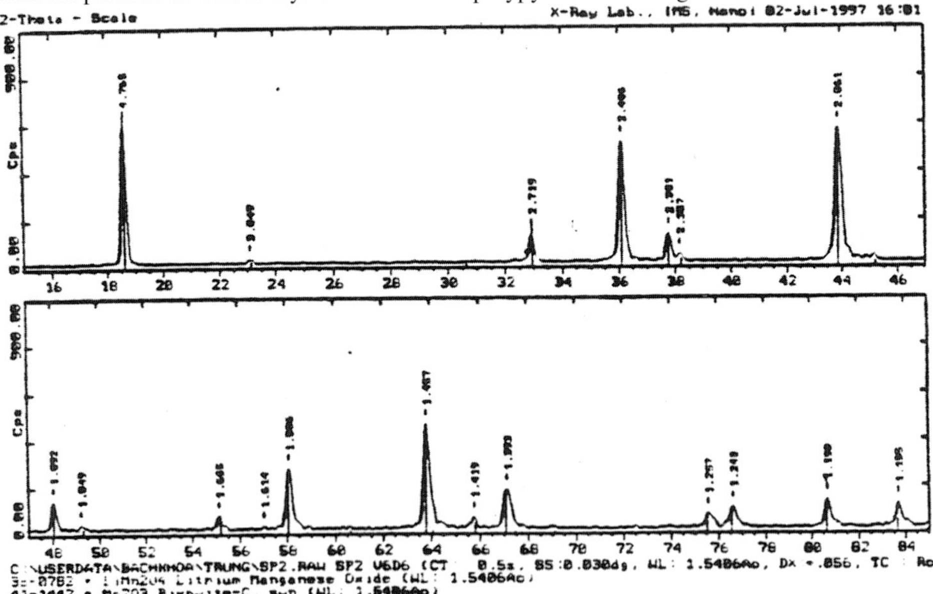

Fig.1: X-ray diffractogram of $LiMn_2O_4$ synthesized by [7] and analyzed by using D5000.

3.2 ESR spectra of PPy/ClO_4 and $PPy/ClO_4.\beta\text{-}MnO_2$

The ESR data measured on PPy/ClO_4 and $PPy/ClO_4.\beta\text{-}MnO_2$ in the range of temperature between 83K and 353K were shown in Fig. 6 and Fig. 7, respectively. The sample of aqueous solution only containing $LiClO_4$ 0.1M didn't exhibit resonance signals. The ESR spectra of PPy/ClO_4 at different temperatures showed that dicarbines existed within sample of PPy° (with spin number of $1.5\text{x}10^{21}$) still appear in the sample of PPy/ClO_4, but their spin concentration already decreased equal to about 10^{20} spin/gram. It may be due to that anions ClO_4^- are localized next to dicarbines. And another effect is more important that the part of the dicarbines, due to the presence of doping anion ClO_4^-, can transmit to excited states (singlet states with opposite spins), and other part can separate into two respective carbines and to take shape of odd electrons. The values of the ESR parameters estimated from ESR measurements of PPy/ClO_4 were lead into table 2. The experimental spectrum given in Fig. 5 showed that there exists the hyperfine structure of a five-line parallel spectrum. And the hyperfine splitting is about 7.1 Gauss. The middle line was covered with a line of carbine radicals. Based on the constancy of the hyperfine splitting and the 1:2 ratio of two left lines intensities, we can guess that the remarkable hyperfine interaction between a radical electron spin (S=1/2) and two nitrogen atom nuclear spins (I=1) has been occurred in PPy/ClO_4β-MnO_2 sample.

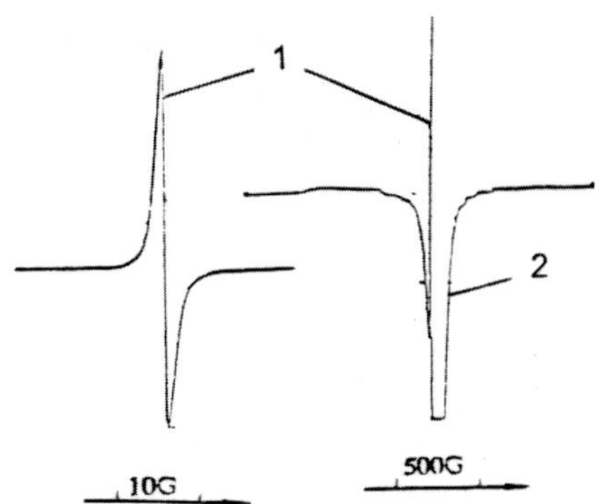

Fig. 2: ESR spectrum of PPy° at 300K

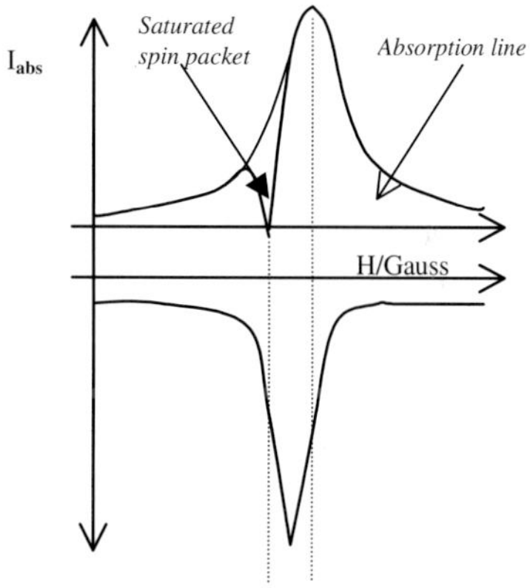

Fig. 3: Graphic interpretation for lineshape of ESR signal shown in Fig. 2.

Table 1. ESR parameters estimated from ESR data of Fig.2

ESR parameters of carbine			ESR parameters of nitrine		
Spin number [spin/gram]	g-Factor	Line width [G]	Spin number [spin/gram]	g-Factor	Line width [G]
$1.5x10^{21}$	> 2.000	144	$5.36x10^{18}$	2.0034	1.06

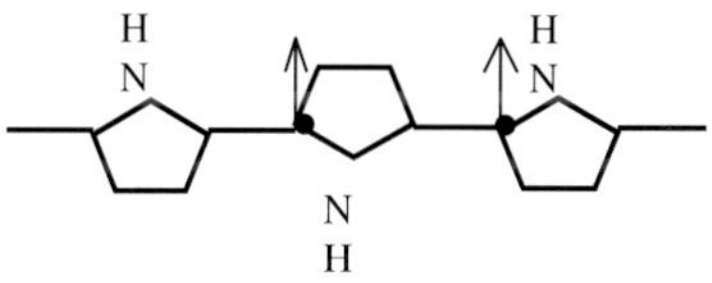

Fig. 4: Schematic representation of a metastable dicarbine.

The origin of the hyperfine structure can be explained in Fig. 8. This structure suggests that here the unpaired electrons were dislocalized to rather large area on the polypyrrole chain. Especially, it is very notable that the measured spin number in the PPy/ClO_4.β-MnO_2 composite sample changes very distinctly vs. the sample temperature in both sides with increasing and decreasing in temperature (Fig.7). The unusual phenomena can be only explained by the coexistence of the different spin complexes, i.e. a spin center here is not a single ion (or radical) but a spin exchange coupling combination from two or more ions (radicals).

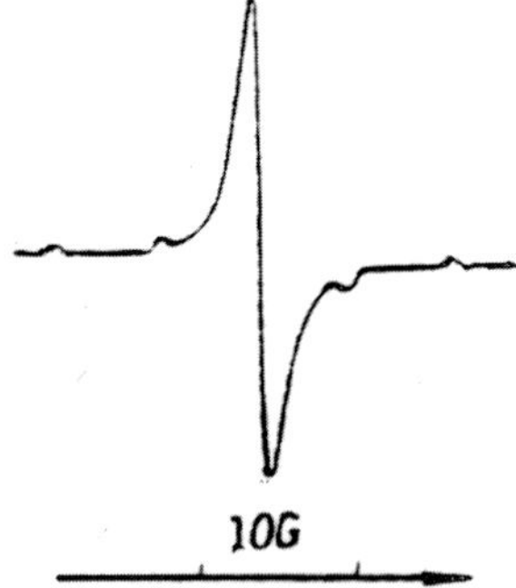

Fig. 5: ESR spectrum of PPy/ClO$_4$.β-MnO$_2$ at 83K with spacing 7.1 Gauss.

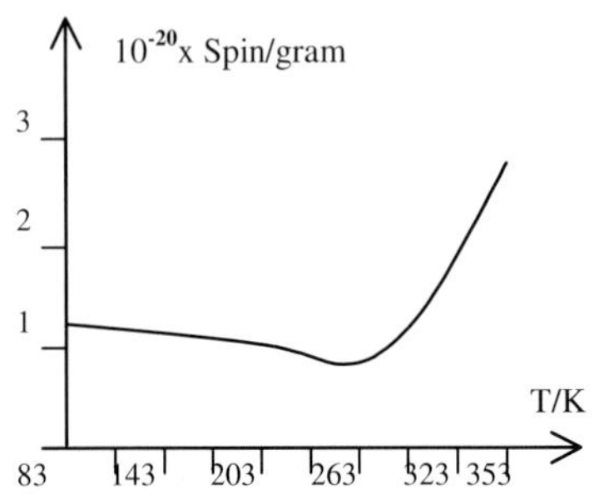

Fig. 6: Dependence of the spin number of PPy/ClO$_4$ on temperature

Table 2: ESR parameters estimated from ESR measurements on PPy/ClO$_4$

T/K	ΔH/G	Spin/gram	g
83	0.90	$1.134x10^{20}$	2.0028
113	0.84	$1.046x10^{20}$	2.0032
143	0.84	$1.090x10^{20}$	2.0031
173	0.80	$1.041x10^{20}$	2.0032
203	0.78	$1.019x10^{20}$	2.0033
233	0.70	$0.855x10^{20}$	2.0032
263	0.80	$1.034x10^{20}$	2.0034
293	0.74	$1.165x10^{20}$	2.0034
323	0.73	$2.297x10^{20}$	2.0034
353	0.72	$3.042x10^{20}$	2.0034

Table 3: ESR parameters estimated from ESR measurements on PPy/ClO$_4$.β-MnO$_2$

T/K	ΔH/G	Spin/gram	g
83	2.04	$7.11x10^{18}$	2.0018
113	2.08	$6.49x10^{18}$	2.0019
143	2.38	$6.50x10^{18}$	2.0021
173	2.54	$5.13x10^{18}$	2.0022
203	3.12	$2.00x10^{18}$	2.0022
233	3.51	$3.02x10^{18}$	2.0024
263	3.72	$7.32x10^{18}$	2.0024
293	3.38	$8.96x10^{18}$	2.0024
323	3.10	$10.23x10^{18}$	2.0024
353	2.62	$11.43x10^{18}$	2.0029

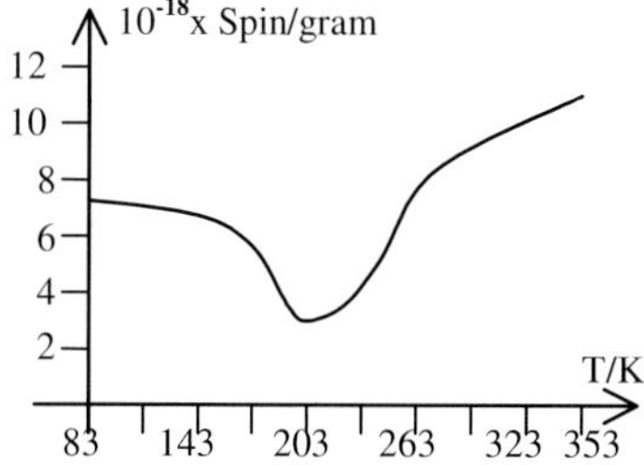

Fig. 7: Dependence of the spin number of $PPy/ClO_4.\beta\text{-}MnO_2$ on temperature

Data lead in table 3 show that in comparing with the spin concentration in the PPy^o sample ($\sim 10^{21}$) or PPy/ClO_4 ($\sim 10^{20}$), the spin concentration intensively decreases in $PPy/ClO_4.\beta\text{-}MnO_2$. It may be due to that the presence of $\beta\text{-}MnO_2$ impurity makes neighboring radicals up into the couples of two radicals with opposite spins. And the electron hopping between these two radicals were taken place by bridging impurity.

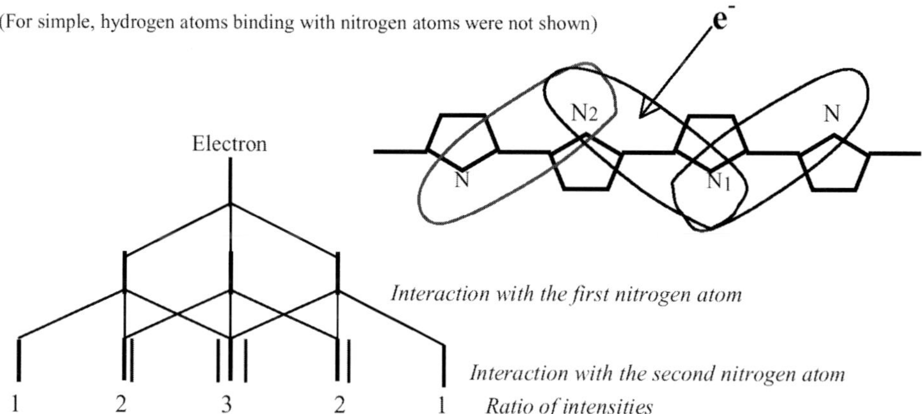

Fig. 8: Schematic interpretation of the formation of the hyperfine structure shown in Fig.5.

Another possibility is that the super exchange interaction of two radicals were carried out by bridging nitrogen atom and exhibited with the hyperfine structure (Fig.5). The first case is relevant to two radical localized on the two neighboring polymer chains. The later is relevant to two neighboring radicals localized on a polymer chain. Intrinsically, the interaction didn't destroy radicals but a part of radicals disappear apparently. These interactions make the wave functions of radicals overlapping together.

3.3 ESR spectrum of $PPy/ClO_4.LiMn_2O_4$

The ESR spectrum of the $PPy/ClO_4.LiMn_2O_4$ sample shown in Fig. 9 show a very outstanding appearance of the dislocalized state Mn^{+6} with concentration of $4.50\text{x}10^{20}$ spin/gram which is several hundreds times larger than the one ($1.25\text{x}10^{18}$ Spin/gram) in precursor lithiated spinel $LiMn_2O_4$. And the precursor states Mn^{+2} and Mn^{+3} also with a concentration of about $3.80\text{x}10^{20}$ spin/gram are almost not more measured when the $LiMn_2O_4$ has been doped in polypyrrole. In The other hand, the observed hyperfine structure with the Mn nuclear spin and the measured very markedly becoming

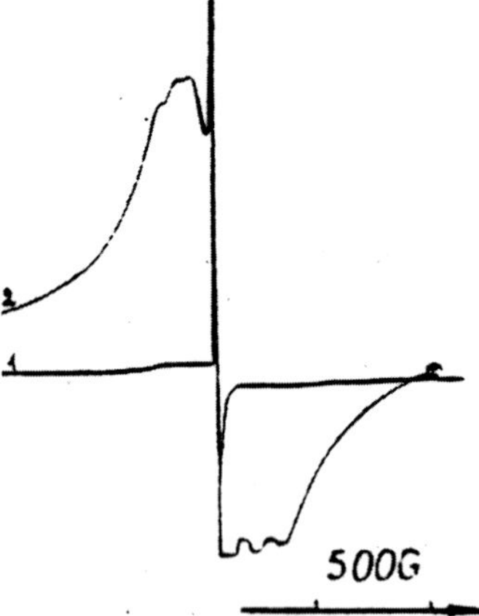

Fig. 9: ESR spectrum of $PPy/ClO_4.LiMn_2O_4$ at 83K

g-factor of 2.0076 of the radicals show a strong overlapping of the pπ-orbitals of the radicals on polypyrrole chains and the d-orbitals of Mn^{+6}. These both phenomena suggest that the Mn^{+6} structure was formed during electropolymerization and localized near by radicals, that creates an electron super exchange bridge between the respective polypyrrole chains. In our previous report [7], the electrochemical impedance studies on PPy/ClO_4, $PPy/ClO_4.\beta\text{-}MnO_2$ and $PPy/ClO_4.LiMn_2O_4$ films also showed the existence of the two depressed visual semicircles at high frequencies of the Nyquist plots of PPy/ClO_4 and $PPy/ClO_4.\beta\text{-}MnO_2$ films, meanwhile there exists only one charge transfer semicircle in the electrochemical impedance spectrum of the $PPy/ClO_4.LiMn_2O_4$ film. It means that the $PPy/ClO_4.LiMn_2O_4$ film acts in electrochemistry as a system with one electron exchange. So that, according to a presence of the dopant $LiMn_2O_4$, charge carriers, polarons and bipolarons were generated with very high concentrations and simultaneously combined to form electron exchange coupling complexes.

4 Conclusion

ESR studies on these polypyrrole coposites indicated the coexistence of two types of radicals as carbine and nitrine in the polypyrrole matrix. Especially, the coupling of two carbine radicals formed a specially and metastable state that is the origin of the ESR saturated phenomenon as observed. ESR studies also supported direct evidences to prove the coexistence of domains of couple with parallel spins and domains of couples with opposite spins. And a presence of impurities as $\beta\text{-}MnO_2$, especially, $LiMn_2O_4$ markedly increases the interaction between polypyrrole chains as well as appear the remarkable hyperfine interaction between a radical electron spin and nuclear spins of two nitrogen atoms and the observed hyperfine structure with the Mn nuclear spin. The achieved results give possibility of leading to an initial conclusion that the composite structure in the doped polypyrrole compounds can promote the unpaired electrons to a specially excited state (shall become conducting state), in which their wave function overlapping is extended to the entire polypyrrole chain and between chains. A special excited state may be like a new polaron lattice with a singlet state and triplet state [8].

5 References

1. Scott J.C., Pfluger P., Krounbi M.T., Street G.B., Phys. Rev. B 28 (1983) 2140.
2. Hari S. Nalwa, Phys. Rev. B 39 (1989) 5964.
3. Susumu Kuwabata, Akira Kishimoto, Takayoshi Tanaka, and Hiroshi Yoneyama, J. Electrochem. Soc., 141 (1994) 10.
4. Ali H. Gemeay, Hiroshi Nishiyama, Susumu Kawabata, and Hiroshi Yoneyama, J. Electrochem. Soc., 142 (1995) 4190.
5. Tran Trung, Truong Ngoc Lien, Nguyen Thanh Thuyet, Nguyen Van Tri, J. Chemistry, Vol. 37, No. 1 (1999) 95, National Centre for Science and Technology of Vietnam.
6. Electron Spin Resonance Ed. John E. Wert and Jaims R. Bolton New York, 1972, p.265.
7. Tran Trung, Truong Ngoc Lien, Nguyen Thanh Thuyet, Nguyen Van Tri, J. Chemistry, Vol. 35, No. 3 (1997) 90, National Centre for Science and Technology of Vietnam.
8. Shi-Jie Xie, Liang-mo Mei and D.L. Lin, Phys. Rev. B, 50 (1994) 13364.

Build-up of Polyelectrolyte Multilayer Assemblies: A Useful Tool for Controlled Modification of Microfiltration and Pervaporation Membranes

Jochen Meier-Haack, Theresia Rieser, Wolfgang Lenk, Siegfried Berwald, Dieter Lehmann,
Institute of Polymer Research; Dresden

1 Introduction

Surface modification of membranes has become a versatile tool to design membranes with tailor-made filtration properties [1 - 3]. For example, surface modification such as permanent hydrophilization and/or permanent introduction of charges to membrane surfaces can help to reduce irreversible fouling in certain applications. Furthermore, ultra- and microfiltration (MF) membranes have been converted into other types of membranes such as pervaporation (PV) and gas separation membranes by the surface modification technique [4]. Few papers have dealt with the formation of polyelectrolyte (PEL) complex layers by consecutive alternating polyelectrolyte adsorption on modified membrane surfaces [5]. This method, which was developed by Decher et al. [6], opens up the opportunity to control the surface and filtration properties by the number of adsorbed polyelectrolyte layers and the choice of polyelectrolytes used.

In this work two examples are given to demonstrate the effectiveness of modification with polyelectrolyte complex layers on the separation and filtration properties, and on the fouling behavior of polyamide-PV-membranes and polypropylene-MF-membranes.

2 Experimental

The effect of polyelectrolyte complex layers on the separation and filtration properties and on the fouling behavior of surface and bulk modified membranes were studied using two different separation processes, namely pervaporation and microfiltration. Therefore two different membrane types were employed. The microfiltration tests were conducted using a commercially available polypropylene membrane (Celgard 2400, Celgard GmbH, Germany). The pervaporation experiments were performed on a home-made polyamide-6 pervaporation membrane. The properties of materials used throughout this work, and the key experimental conditions are summarized in table 1.

Table 1. List of the used materials and the key experimental conditions

	microfiltration	Pervaporation
Membrane	Polypropylene (PP) Celgard 2400	mod. Polyamide-6 (PA-6) [7]
Polycation	Poly(diallyldimethylammoni-umchloride) (PDADMAC) M_W 99000 (FhG Teltow)	Poly(ethyleneimine) (PEI) M_W 750000 (Aldrich)
Polyanion	Poly(acrylic acid) (PAAc) M_W 90000 (Polyscience)	Poly(acrylic acid) M_W 90000 (Polyscience)
Concentration of PEL-solutions	20mM (ref. to monomer units)	1 wt.-% or 20 mM (ref. to monomer units)
Test solutions	Human-Serum-Albumin (HSA) 0.1 wt.-%	Water/2-Propanol 5 - 100 wt.-% Water
Filtration apparatus and test conditions	stirred dead-end filtration cell Millipore Type 8050 25°C, 300 kPa	Laboratory test cell P28 (CMCelfa); 50°C; 20 mbar (down stream)

2.1 Surface modification of polypropylene membranes

The polypropylene membranes were treated first with a CO_2 plasma (experimental conditions: 70 W; 3 min; 10 sccm CO_2; 10 Pa). The plasma treatment creates peroxide species on the membrane surface which were used as initiators for the subsequent grafting of acrylic acid. Nitrogen was bubbled through the aqueous monomer solution to remove oxygen before the grafting reaction was started. The monomer concentration was 3 wt.-%. The grafting reaction was carried out at 70°C for 30 - 120 min. The samples were treated with hot water for 24 hours to remove residual monomers and non-grafted homopolymer. The graft yield was between 0.1 and 5 wt.-%, depending on the reaction time.

2.2 Preparation of modified polyamide support

Modified polyamides for the preparation of pervaporation membranes were obtained by reacting polyamide-6 with poly(α-methylstyrene-alt-ε-caproic acid maleimide) (M_n 30.000 g/mol) in the melt phase. A polyamide having a comb-like structure with polyamide-6 side chains was obtained. The calculated molecular weight of the polyamide chains was approximately 4520 g/mol (DP $\approx$ 40). The carboxyl end group concentration was about 203 μmol/g.

2.3 Preparation of the polyamide pervaporation membranes

The phase-inversion-process [8] was used to prepare asymmetric membrane from the melt modified polyamide. The membrane, which consists of a thin dense top layer and a highly porous support layer, was used as a substrate for the polyelectrolyte complex layers. The polyamide membrane was formed by casting a polymer solution containing 15 wt.-% polyamide in formic acid/water (85/15 wt./wt.) on a glass plate. The film thickness was set to 200 μm. After an appropriate time (10 min), during which the solvent was allowed to partially evaporate, the polymer film was immersed in cold water (4°C) for precipitation. The resulting asymmetric membrane was placed in water for several days to leach out residual formic acid, and was finally dried at room temperature.

2.4 Surface modification of the PA- and PP-membranes by polyelectrolyte adsorption

The formation of the polyelectrolyte complex layers was carried out by alternative dipping of the modified membrane into aqueous solutions of oppositely charged polyelectrolytes (Fig 1). The adsorption time was 30 min for the MF-membranes and 60 min for the PV-membranes. After each adsorption step the membrane was rinsed extensively with pure water to remove weakly bound polyions. The poly(acrylic acid) grafted microfiltration membranes were modified with two additional polyelectrolyte layers consisting of poly(diallyldimethylammoniumchloride) (PDADMAC, polycation) and poly(acrylic acid) (PAAc, polyanion). The outermost surface was therefore negatively charged. The modification of the polyamide membrane was done with poly(ethyleneimine) (PEI, polycation) and poly(acrylic acid). The total number of dipping cycles varied between 10 and 12.

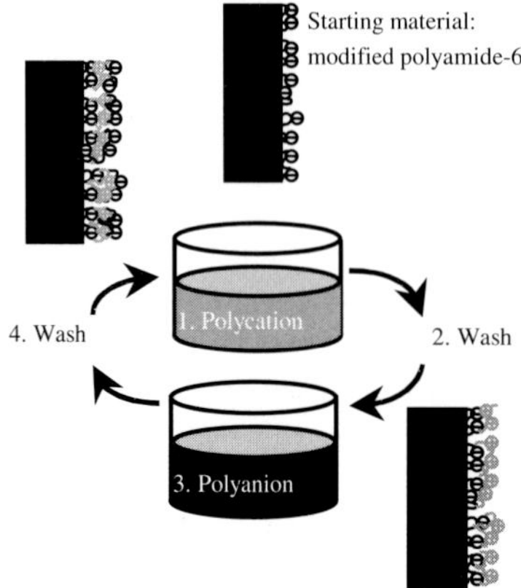

Figure 1: Schematic drawing of the polyelectrolyte deposition process on modified polyamide-6 membranes

2.5 Characterization of the modified membranes

The human serum albumin (HSA) concentration was determined by UV/VIS spectroscopy (CADAS 100, Dr. Lange, Germany) using the typical protein absorption at 287 nm.

The feed and permeate compositions were determined by density measurements using a DMA 58 density meter (A. Paar KG, Austria). The separation factor α was calculated using equation (2) where c is the concentration of water and alcohol in the feed and the permeate, respectively as symbolized by the subscripts W, A, F, P.

$$\alpha = \frac{c_{W,P} \cdot c_{A,F}}{c_{W,F} \cdot c_{A,P}} \qquad (2)$$

The permeate flux was measured gravimetrically in pervaporation experiments and volumetrically in microfiltration experiments.

3 Results and Discussion

3.1 Modification of polyamide pervaporation membranes

The bulk and surface modified polyamide membranes were characterized by streaming-potential measurements at different pH values. The polyelectrolytes were adsorbed from aqueous solutions containing 1 wt.-% of the respective polyion. Figure 2 shows typical ζ-potential vs. pH plots for an unmodified membrane, a bulk modified membrane, and modified membranes with a positive surface charge and with a negative surface charge after the adsorption of poly(ethyleneimine) and poly(acrylic acid), respectively.

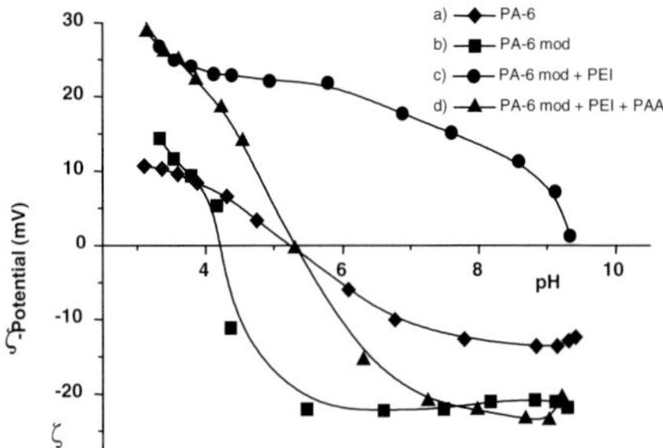

Figure 2: -potential vs. pH plots of a) modified polyamide; b) sample a + PEI; c) sample b + PAAc; c) sample a + PDADMAC; d) sample c + PSS

The unmodified polyamide membrane has an isoelectric point (IEP) at the weakly acidic pH of 5.4. The ζ-potential curve of the bulk modified sample shows a definite plateau above pH 6 and a much more negative ζ-potential than the starting material. This is attributed to the larger number of carboxyl end groups in the bulk modified polyamide and the resultant higher hydrophilicity. Furthermore, a shift of the IEP to a lower pH value was observed (pH 5.0), which is consistent with an increased number of carboxyl groups at the membrane surface. The adsorption of poly(ethyleneimine) led to a reverse of the surface charge. This is diplayed by a definite shift of the isoelectric point from the acidic to the basic pH range (pH 9.4) and a sign-reversal of the ζ-potential over the pH-range investigated. Upon further reverse of the surface charge achieved by the adsorption of poly(acrylic acid), the IEP shifts again to the acidic pH range but the IEP of the bulk modified polyamide was not reached. This observation is explained by the fact that a large number of carboxyl groups of the adsorbed poly(acrylic acid) is fixed by complexation with basic amino groups on poly(ethyleneimine).

The influence of polyelectrolyte multilayer complexes on the separation properties of the polyamide pervaporation membranes is shown in Fig. 3. This plot shows the results obtained for membranes which were modified with aqueous solutions containing equimolar amounts of the polyion (20 mmol/l). The unmodified polyamide membrane has low selectivity but high permeability for a water/2-propanol-mixture containing 30 wt.-% water. The consecutive adsorption of oppositely charged polyelectrolytes on the membrane surface led to a step-wise improvement in selectivity. A significant rise in selectivity was observed after 4 to 5 dipping cycles. The increased selectivity is accompanied by a decrease in permeability as it is often observed in membrane performance. The effect of the polyelectrolyte complex layers on the separation properties can be explained an enhanced hydrophilicity which favors the uptake of water by the adsorbed polyelectrolyte multilayer. Furthermore, defects such as pinholes or micropores in the substrate are covered with the polyelectrolytes.

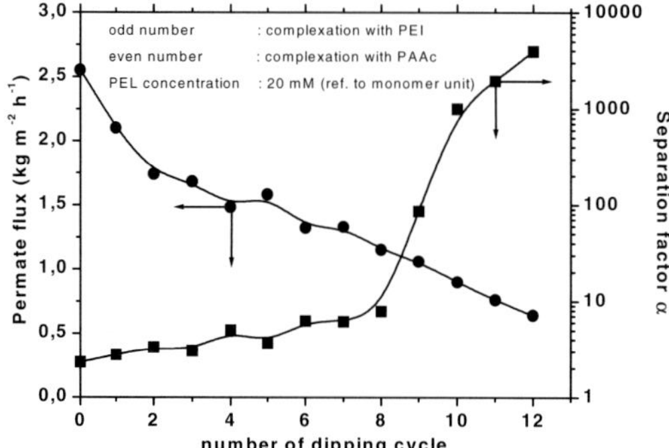

Figure 3. Pervaporation separation properties of polyelectrolyte composite membranes

3.2 Modification of PP-MF-membranes

To modify the hydrophobic polypropylene membranes with polyelectrolyte complex multilayer, poly(acrylic acid) was grafted on the plasma treated membranes as a first polyelectrolyte layer. Further polyelectrolyte adsorption was performed on the PAAc-grafted polypropylene membranes.

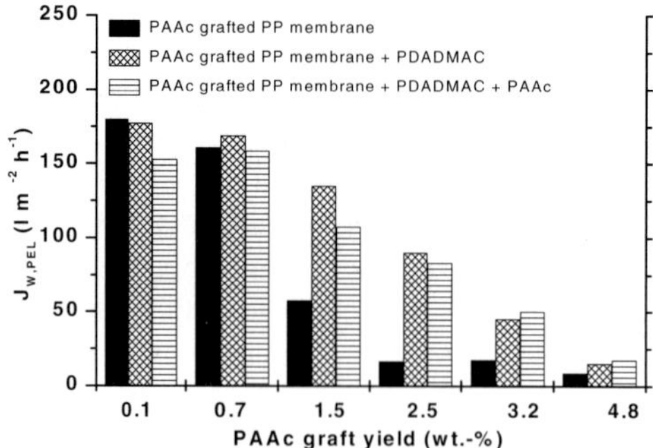

Figure 4. Influence of polyelectrolyte multilayer assemblies on the filtration properties of MF-membranes

The influence of the adsorbed polyelectrolyte layer on the filtrate flux is visualized in Fig 4. Water flux decreases with increasing graft yield due to a reduction in the pore diameter. Interestingly, a flux increase was observed when a second polyelectrolyte layer (PDADMAC) was adsorbed onto the grafted poly(acrylic acid). This effect was most pronounced in the PAAc graft yield range between 1.5 and 3.3 wt.-%. It is likely that the adsorption of the polycation induces a conformational change of the grafted poly(acrylic acid) chains. In the uncomplexed state, the poly(acrylic acid) chains are thought to be in a strechted conformation to minimize the repulsive force between equally charged carboxyl groups. As the charge of the grafted poly(acrylic acid) is partially compensated by complexation with the polycation, the repulsive force is weakened and the conformation becomes more compact. This conformational change results in enlarged pore diameters and therefore the observed flux enhancement. Similar observations have been reported recently by Osada et al. [10].

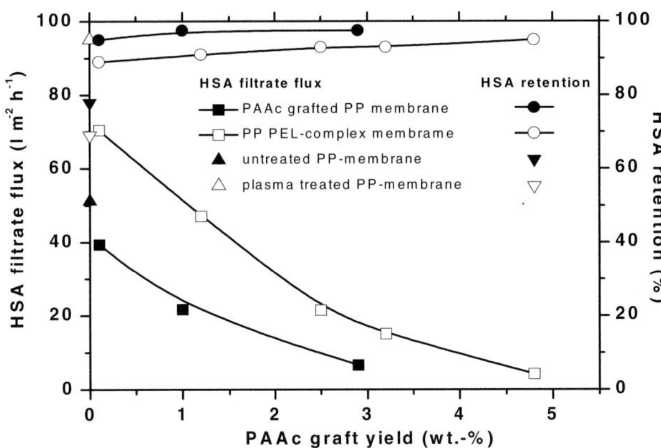

Figure 5. Influence of polyelectrolyte multilayer assemblies on HSA retention and filtrate flux

The same flux enhancement after the formation of polyelectrolyte complex layers was observed during protein filtration of human serum albumin (HSA, 0.1 wt.-%, pH 6.4). Figure 5 shows that the filtrate flux through a PP-3-layer membrane is approx. double that of a just grafted membrane. Furthermore, for low PAAc graft yields the protein filtrate flux was higher than for unmodified membranes. Protein retention by the PEL complex membranes was found to be in the range of 90%, which differs slightly from protein retention by the just grafted membranes, although a higher filtrate flux was observed for the former membranes. To explain these results not only the conformational changes upon complexation have to be taken into account but also topographical effects and changed interactions between the membrane surface and feed components (i.e. the PEL complex layers and the HSA) must be considered.

4 Summary

The consecutive alternating polyelectrolyte adsorption was found to be a versatile tool for the preparation of membranes with controlled surface charge properties and hydrophilicity as demonstrated by ζ-potential measurements.

Dense membranes, used in pervaporation, for example, showed an improved water selectivity after the formation of a PEL complex layer system. The permeate flux decreased with increasing selectivity and increasing layer number. It is possible to control the properties of pervaporation membranes by the number of adsorption steps.

The adsorption of only two polyelectrolyte layers onto a PAAc-grafted microfiltration membrane surface resulted in a reduced protein adsorption in the case of repulsive electrostatic interactions between the membrane surface and the feed components (e.g. proteins). Higher protein retention was simultaneously observed. From these results a reduced fouling of PEL complex layer membranes was deduced when the membrane surface and the feed components are equally charged. These results could easily be achieved with membranes having a low degree of grafting (< 1 wt.-% PAAc).

5 References

1. Ulbricht, M.; Belfort, G. J. Membr. Sci., 1996, 111, 193 - 215
2. Stengaard, F. F. Desalination, 1988, 70, 207 - 224
3. Wolff, J.; Steinhauser, H.; Ellinghorst, G. J. Membr. Sci., 1988, 36, 207 - 214
4. Hirotsu, T.; Isayama, M. J. Membr. Sci., 1989, 45, 137 - 154
5. Rieser, T.; Lunkwitz, K.; Berwald, S.; Meier-Haack, J.; Kessler, B.; Simon, F. PMSE, 1997, 77, 351 - 352
6. Decher, G.; Hong, J.-D.; Schmitt, J. Thin Solid Films, 1992, 210/211, 831 - 835
7. Meier-Haack, J.; Lehmann, D.; Berwald, S.; Lenk, W.; Hansen, P.; Berwald, F.; Rieser, T.; Lunkwitz, K. in Polymerwerkstoffe ´98, Tagungsband-1 Vorträge, Merseburg, (Eds. H.-J. Radusch, J. Vogel), 1998, 327 - 334
8. Loeb, S.; Sourirajan, S. Adv. Chem. Ser., 1962, 38, 117 - 132
9. Müller, M.; Rieser, T.; Lunkwitz, K.; Berwald, S.; Meier-Haack, J; Jehnichen, D. Macromol. Rapid Commun., 1998, 19, 333 - 336
10. Osada, Y.; Honda, K.; Otha, M. J. Membr. Sci., 1986, 27, 327 - 338

ORMOCER®s for Optical Interconnection Technology

R. Buestrich[1], F. Kahlenberg[1], M. Popall[1], P. Dannberg[2], R. Müller- Fiedler[3], O. Rösch[3]

[1] Fraunhofer Institut für Silicatforschung, Würzburg,

[2] Fraunhofer Institut Angewandte Optik und Feinmechanik, Jena,

[3] Robert Bosch GmbH, FV/FLD, Stuttgart

1 Abstract

New inorganic-organic hybrid polymers (ORMOCER®s) for integrated optical and opto-electronic devices were synthesized by sol-gel processing of functionalized alkoxysilanes. Process parameters (catalyst, temperature etc.) were optimized to achieve highly reproducible low cost materials which are photopatternable even in higher layer thickness (presently 100 μm within one step).

The resulting materials have low optical losses at the most important wavelengths for telecommunication in the NIR range (0.2 dB/cm at 1310 nm , 0.5 dB/cm at 1550 nm) and a variety of additional advantageous properties for optical interconnection technology and production of opto-electronic devices: good wetting and adhesion on various substrates (e.g. glass, silicon and several polymers), low processing temperatures (postbake at 160 °C) and high thermal stability (decomposition at 270 °C) compared to alternative opto-polymers for NIR applications. A further advantage is a tunable refractive index, which can be achieved by mixing different resins.

2 Introduction

ORMOCER®s are inorganic–organic hybrid polymers which offer great potential for use as optical and dielectrical materials. Although the materials mentioned in this article also have good dielectric properties ($\varepsilon = 3.3$, at 10 kHz), we just focus here on ORMOCER®-optimization for optical applications in the NIR region, especially at wavelengths 1310 nm and 1550 nm. The article is divided into two parts: first, the chemical development of the materials is discussed. Afterwards a very detailed application example is given: the fabrication of a digital optical switch by direct UV-patterning.

The chemical synthesis of ORMOCER®s is a two-step process in which initial inorganic polycondensation of the silane molecules via sol-gel reaction is followed by crosslinking of the organic side chains resulting in three-dimensional duroplastic materials [1].

Starting point are organoalkoxysilanes which have become commercially available in a large variety. Dependent on the number of sol-gel reactive substituents, mono-, di-, tri-, and tetrafunctional precursors (described as M, D, T, and Q units) can be distinguished. The precursors are linked up during sol-gel reaction by establishing the inorganic Si-O-Si network [2]. For example, if a difunctional silane is cocondensed with a trifunctional silane the resulting resin is expected to consist of oligomers which have chains or rings and sphericle elements. Application of the resin by UV- direct patterning is done as follows: the viscous ORMOCER® resin is mixed with solvent and UV-initiator, spun on a wafer and patterned by UV-irradiation, which results in radical methacrylic polymerization.

3 General Aspects

Main objectives of material development for integrated devices are low optical absorption in the NIR region (especially at 1310 and 1550 nm) and additional UV-patternability. As discussed elsewhere [3, 4] low absorption can be achieved with low Si-OH and low C-H aliphatic content in the ORMOCER®. These considerations make us believe that a combination of diphenylsilanes, which promise good optical properties, with 3-methacryloxypropyltrimethoxysilane (MEMO) for UV-patternability of the resulting material, provides an auspicious precursor system for the desired material. A simplification of the sol-gel processing can be obtained by focussing on alkoxylation which is best performed by adding no water to the silanes [5].

We react diphenysilanediol (P2) with MEMO at higher temperature with a suited catalyst according to the following reaction scheme:

$$RSi(OMe)_3 + R'_2Si(OH)_2 \xrightarrow{[cat.]} \text{polysiloxane} + 2\ MeOH \qquad \text{equation 1}$$

R = 3-methacryloxypropyl
R' = phenyl

An outstanding characteristic of the resulting polysiloxane network is a remarkably low content of silanol groups. Alkoxylation is favoured by the sterical demand of the bulky phenyl groups on the diol component which prevent self-condensation (see control experiment 1). The use of higher temperatures (60-100°C) in combination with a suited catalyst results in short reaction times (kinetical control) favouring alkoxylation.

For comparison, starting with a mixture of diphenyldimethoxysilane and MEMO and adding water and catalyst does not result in low Si-OH siloxanes (see control experiments 4,5). This reaction, as well as reaction 3, seems to proceed over hydrolyzed species, of which the silanol groups afterwards cannot completely react in the conden-sation steps due to sterical reasons.

Control experiments (same catalyst concentration, same temperature, time etc.):

1. P2 + P2 (solvent for homogeneity); result: much slower reaction rate than eq. 1
2. P2 + MEMO (3:2 molar ratio); result: no further decrease of remaining Si-OMe in the polysiloxane compared to the 1:1 product (sterically hindered)
3. P2 + MEMO + water; result: no low Si-OH product
4. diphenyldimethoxysilane + water + MEMO; result: no low Si-OH product
5. [(diphenyldimethoxysilane + water) pre-hydrolysis] + MEMO; result: no low Si-OH product

The control experiments show that the chosen reaction conditions are the most suitable to get reproducible, homogeneous low Si-OH containing ORMOCER® resins which promise good optical properties. Index-tuning can be achieved by varying the composition of the educts in the ORMOCER® system. Substitution of MEMO to a small extent by tri-fluoropropyltrimethoxysilane results in a resin with lower refractive index.

4 Results

Diphenylsilanediol (ABCR), MEMO (Hüls) and the catalyst are mixed and brought to reaction at elevated temperature. Afterwards volatile products (here methanol) are evaporated resulting in the viscous, storage stable ORMOCER® resin.

The weight loss of the reaction mixture during evaporation of volatile reaction products corresponds exactly to the calculated weight loss in equation 1. The amount of water in the evaporated methanol was determined with Karl-Fischer-titration to be 0.08 wt%. The water content of the resulting resins is 0.01 wt%. Reproducibility from batch to batch is quite good in terms of refractive index $n = 1.5382 \pm 0.0005$ and viscosity $\eta = 5.4 \pm 0.2$ Pas. IR spectra of the resin show no Si-OH peak in the range 3600-3200 cm^{-1} whereas a peak detected at 2840 cm^{-1} is due to Si-OMe vibration.

The ^{29}Si NMR spectrum of the resin was obtained on a Bruker DPX 400 NMR spectrometer from a 30 vol% resin in $CDCl_3$. A proton decoupling pulse program with a 30 s delay was used. As relaxation reagent, $Cr(acac)_3$ was added in a concentration of 30 mg/l to make sure that the applied 30 s delay was sufficient for the relaxation of all silicon nuclei. All chemical shifts refer to TMS as the internal standard.

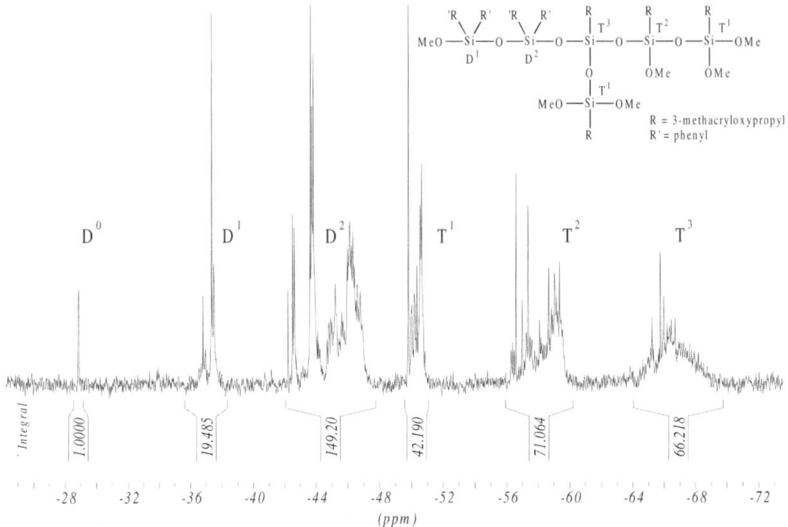

Figure 1: ^{29}Si NMR spectrum of the described P2-MEMO resin with peak assignment.

The comparison between $\sum D^i = 169$ and $\sum T^i = 179$ (theoretical value = 1:1) indicates that the integration is feasible, i.e. the ^{29}Si spin-lattice relaxation time of D and T is shorter than the delay time used. Traces of diphenyldimethoxysilane (= D^0) can be removed by heating the resin in vacuum. Up to now only an interpretation of the first coordination-sphere of the different Si-species (D, T) can be given. The assignment of the D- and T shifts is reasonable compared to literature values [6, 7].

Thermal properties of the cured resin were checked with DTA-TG in the range of 25 °C to 400 °C under air at a heating rate of 5 °C/min. No weight loss occurred until degradation started at 270 °C. DSC measurements in the range of -100 °C to 200 °C at a heating rate of 5 °C/min exhibited no glass transition temperature.

FT-IR/DRIFT measurements were carried out to estimate the stability of the cured resin towards humidity. A powdered sample was stored at 85 °C and 85% relative humidity for 7 days. No decrease of the Si-OMe peak integral was detected, proving the stability of the polymerized material towards hydrolysis.

5 Discussion

IR and NMR spectra as well as weight loss and water-content measurements are taken into account for the mechanistic interpretation. Although the complex sol-gel process was simplified in order to focus on alkoxylation, the elimination of side reactions was not quantitatively successful. Even with the reaction conditions discussed above there are no pure $-(D-T)_n-$ polymers available. Although nearly no water in the evaporated methanol could be detected, an *in situ* water production must be stated to explain the final distribution between D and T species in the resin (Figure 1). Assuming that the water-producing P2-P2 condensation is suppressed by kinetics (see control exp. 1) and the attack of P2 on T^2 is sterically hindered (see control exp. 2), the following reaction scheme (Figure 2) shows a possible side-reaction responsible for the production of T^3-species. As stated in the literature esterification reactions are much faster than condensation reactions [8].

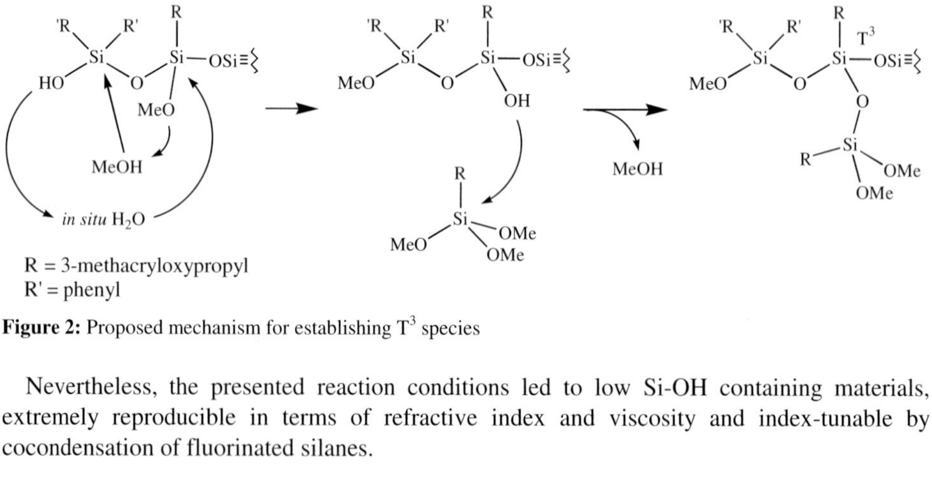

'R R' R Si Si—OSi≡ HO O MeO MeOH *in situ* H_2O → 'R R' R Si Si—OSi≡ MeO O OH R Si MeO OMe OMe → MeOH 'R R' R T^3 Si Si—OSi≡ MeO O O R Si OMe OMe

R = 3-methacryloxypropyl
R' = phenyl

Figure 2: Proposed mechanism for establishing T^3 species

Nevertheless, the presented reaction conditions led to low Si-OH containing materials, extremely reproducible in terms of refractive index and viscosity and index-tunable by cocondensation of fluorinated silanes.

6 Application – ORMOCERs for Integrated Optics

Essential criteria for the decision on the waveguide fabrication technology is the compatibility to microelectronics technologies. Direct UV-patterning (Figure 3) fulfils these demands. In this technology, a silicon substrate is covered with an ORMOCER layer of precise thickness by spin-coating. After polymerization of this lower buffer layer, a second slightly fluorinated ORMOCER layer (core) with a thickness of 6 μm is spin-coated. The definition of the waveguide structures is realized by UV-polymerization through a mask resulting in waveguide channels. The non-polymerized core material is dissolved similar to an development process in photolithography. In the next step, an upper cladding layer is spin-coated and polymerized. Finally, the heating electrodes are sputtered on top. One of the advantages of the UV-patterning technology is the low roughness of the structures. This was proven by SEM and by Atomic Force Microscopy. A rms-value of 20 nm was determined across the top face and the sidewalls of the waveguide structure.

Embedded channel waveguides, 7.1 cm long, with a cross section of 6x6 μm^2 (Figure 4) have been fabricated and characterized. The perfect planarization of the ORMOCER leads to a smooth surface of the cladding layer.

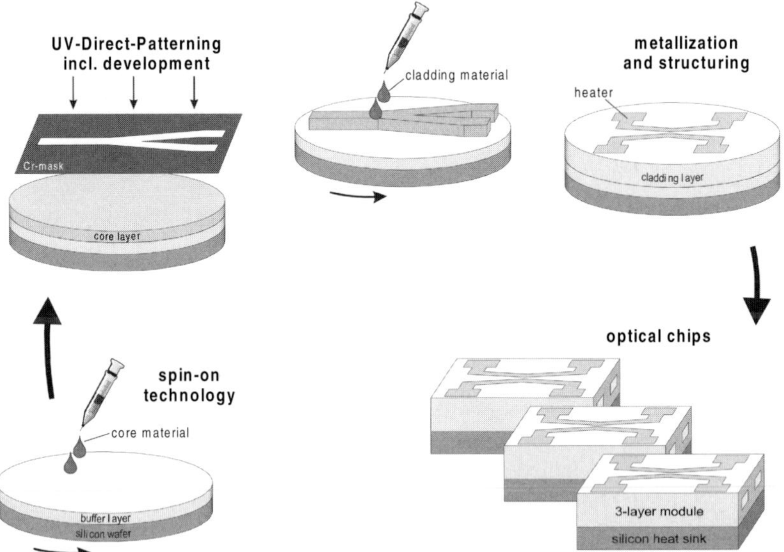

Figure 3: Direct-UV-Patterning for Integrated Optics

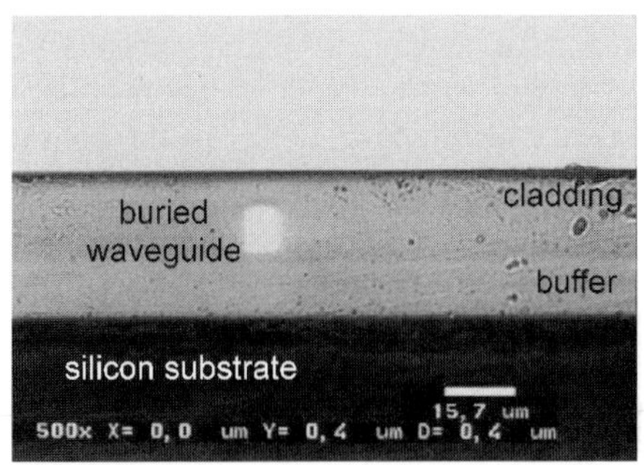

Figure 4: embedded strip waveguide consisting of ORMOCER

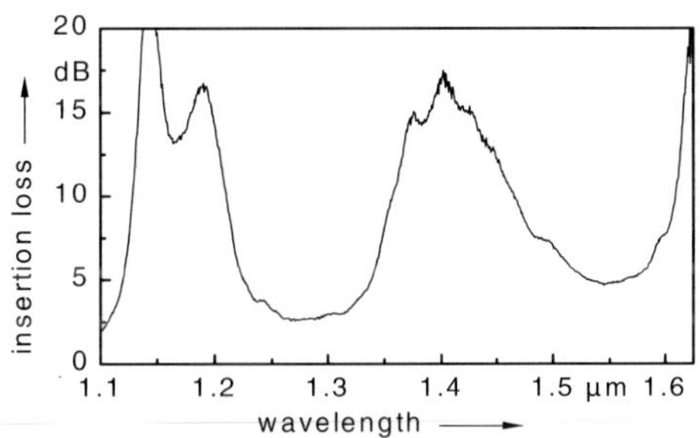

Figure 5: Spectral attenuation of 7.1 cm strip waveguide incl. fiber-chip-coupling

The insertion loss including two fiber coupling interfaces was 2.4 dB at $\lambda = 1310$ nm and 4.75 dB at $\lambda = 1550$ nm. The spectral attenuation is shown in Figure 5. More detailed information about loss mechanisms was obtained by using the cut-back method. The insertion losses were measured for different waveguide lengths and the intrinsic losses were determined to be 0.25 dB/cm at $\lambda = 1310$ nm and 0.51 dB/cm at $\lambda = 1550$ nm. In the 7.1 cm long waveguide we achieved a polarization dependent loss of max. ± 0.3 dB in the optical windows. Therefore the length dependence is below ± 0.05 dB/cm for our integrated optical compounds.

The metal electrodes of the *digital optical switch (DOS)* are defined by wet etching using a photoresist mask. For the characterization, the samples are held by vacuum and the electrodes are contacted by probe tips. The input and output fibers have to be controlled precisely by piezo driven manipulators. By applying a heating power of 60 mW we achieved an insertion loss including the fiber-chip-coupling of 0.7 dB @ 1.32 μm and 1.5 dB @ 1.55 μm and an isolation ratio of 30 dB and 29 dB, respectively. The speed of the switches is characterized with a digital oscilloscope, which shows simultaneously the steplike changing electrical heating current and the optical response. It takes 3.5 ms to switch into the ON state. According to the thick buffer layer it takes nearly two times longer to switch off the branch. Nevertheless, both switching times are well below the required time delay for optical crossconnects of about 10 ms.

7 Conclusion

The synthesized inorganic-organic hybrid polymers are low cost materials, photopatternable even for higher layer thickness (presently 100 μm within one step without cracking or delamination). Optical losses of 0.2 dB/cm at 1310 nm and 0.5 dB/cm at 1550 nm (strip-waveguide) are excellent values for most of board-size applications in telecommunication.

Further advantages for use in interconnection technology are good wetting and adhesion on various substrates, low processing temperatures (postbake at 160 °C), high thermal stability (decomposition at 270 °C) and a tunable, reproducible refractive index, which can be achieved by mixing different resins.

Integrated optical devices, especially digital thermo-optic switches were fabricated from the material (negative resist behavior) resulting in smooth side-walls with extremely low additional scattering losses (< 0.1 dB/cm) [3].

So we can present a new hybrid-polymer material for Integrated Optics. Direct UV-patterning provides low cost fabrication compared to dry etching methods because of a reduced number of fabrication steps in combination with high reproducibility of the optical components.

8 Acknowledgement

The authors would like to gratefully acknowledge partly financial support by the German Federal Ministry BMBF in the framework of the Photonik II program (01 BP 464/4).

9 Literature

1. M. Popall, A. Dabek, M.E. Robertson, G. Gustafsson, O-J. Hagel, B. Olsowski, R. Buestrich, L. Cergel, M. Lebby, P. Kiely, J. Joly, D. Lambert, M. Schaub and H. Reichl, Proc. EC/TC, Seattle, WA, 1998, 1018-1025.
2. R. J. P. Corriu and D. Leclercq, 1996, 35, 1420-1436.
3. O. S. Rösch, W. Bernhard, R. Müller-Fiedler, R. Buestrich, C. Roscher, M. Popall, P. Dannberg, G. Pompe, and A. Neyer, Proc. POF (POF), Berlin, pp 332-338, 1998
4. O. S. Rösch, W. Bernhard, R. Müller-Fiedler, P. Dannberg, A. Bräuer, R. Buestrich, and M. Popall, SPIE, 1999
5. E. Wipfelder, K. Höhn, Die Angewandte Makromolek. Chemie 1994, 218, 111-126
6. Brunet, J. of Non-Cryst. Solids 1998, 231, 58-77
7. D. Hoebbel, T. Reinert and H. Schmidt, J. of Sol-Gel Science and Techn. 1996, 7, 217-224
8. S. Rankin, C.W. Macoosko and A. V. McCormick, Chem. Mater. 1998, 44, 1143-1156

Carbon Fibers Reinforced Ceramics Obtained with Polymer Precursors with Silicon

Carmen Novac, Mihaela Carmen Bunescu, Eugeniu Vasile
SC METAV SA – Bucharest, ROMANIA, P.O.Box 18/3

1 Abstract

The processes during the pirolytic transformation of the preceramic polymers to CMC-s (ceramic matrix composites) with Si_3N_4/SiC/residual C or SiC/SiO_2/residual C are presented. The thermal decomposition (pirolysis) up to 1100^0C in Ar/N atmosphere produces an amorphous ceramic with a good resistance against oxidation. Thus, the transformation of the organo-metallic polymers into an amorphous matrix is influenced by the atmosphere of the reaction (Ar, N). The pirolytic transformation is accompanied by the loss of volatile, resulting high-density differences between the precursors (~ 1 g/cm^3) and the ceramic residuum (~ 2.5 – 3.2 g/cm^3). The residual porosity and the contraction (~ 70 - 80%) are very high. Four types of preceramic polymers with Si have been used. The pirolysis at 1200, 1400 and 1600^0C produced the experimental CMC-s. The microstructure analysis has evidenced the SiO_2 and SiC phases. With the same precursors but in N atmosphere and pirolysis between 1200-1400^0C CMC-s with Si_3N_4/SiC/residual C matrix have been obtained.

2 Introduction

The thermal decomposition of the polymeric precursors up to 1100^0C in Ar/N produces an "amorphous ceramic" Si/C/N. In order to obtain preceramic polymers with mix coordination of the tetragonal Si atoms (e.g. polysylazan, polycarbosylan, polysyloxan, polyborosyloxan and polycarbosyloxan) cloro-organo-sylicil compounds should be used. If monomers with variable molar function are used the structure of the polymers is modified from linear to highly branched.

The pirolysis reaction has 3 important steps:

- the transformation organometalic–inorganic (500-800^0C), producing an amorphous hydrogenated residuum with tethraedric structure;
- the nucleation of the crystalline precipitates, like SiC/Si_3N_4 or SiO_2, and a secondary carbonic phase (1200-1400^0C);
- the growth of the crystals/grains by transformation of the residual amorphous phase and decreasing the oxygen content due to the evaporation of the gaseous compounds (SiO and CO).

Crystalline CMC-s can be produced if reactive filler materials are used in the immersion solution. The usual dimension of the reactive filler materials is 1-10 μm and they can be partially substituted by inert filler powders (e.g. SiC, Si_3N_4, Al_2O_3). These materials reduce the contraction and porosity generated by the transformation polymer – ceramic compound.

An usual heating route starts with a fast increasing of temperature up to 400^0C (heating speed 5^0C/min), maintaining for 1-2 hours, followed by a second increase with 1-2^0C/min, up to the final temperature (1200-1500^0C). The reaction between the filler material with the atmosphere of N_2 the Si oxynitride is produced.

During the pirolysis in reactive atmosphere (N_2, NH_3, BN_3) the filler material can react producing nitrides or borides that increase the weight. Thus, the type of the polymer allows the control of the obtained material regarding the porosity, contraction and characteristics.

The pirolysis of the preceramic precursors with low free C produces evaporation of CO (g) and SiO (g) at temperatures higher than 1400^0C, due to the reducing reaction of SiO_2. For a high content of free C the decomposing products are: SiC(s) + CO (g). This is the situation for the ceramic matrix composites reinforced with carbon fibbers, presented in this paper.

The ceramic composites obtained from polysyloxan precursor can be used in applications at temperatures up to 1400^0C. For higher temperatures CMC-s with Si-C-N type matrix with low oxygen content are recommended. In the usual conditions the polysyloxan precursors are easier to handle and process than the polymers with low oxygen content. The advantage of the silicon system is the stability against air and the high variety of reticular reactions already studied.

3 Experimental and Results

Pirolysis experiments of many-layer preformers in Ar/N have been performed. The preformers were obtained from carbon fibbers immersed in preceramic polymers. The sequential process of immersion and pirolysis consisted of:

- producing of many-layer preformers;
- heat treatment of the "green" many-layer preformers at 1000^0C in reducing atmosphere, in order to produce a uniform and protective coating of pirolytic graphite;
- immersion in organo-metallic precursor;
- pirolysis in Ar at 1400-1600^0C;
- Filler materials (TaC, TiC and graphite) have been added to the "green" preformers in order to increase the mechanical resistance. In addition, the reactive materials, like Ta, Ti, Cr, etc., are beneficial for the pirolysis reaction of the ceramic polymers.

The preceramic precursors used were:

- trimetoxymethilsylan (TMM)
- trietoxyvynilsylan (TEV)
- triclorooctadecylsylan (TCOD)

TMM and TEV have been used both as hydrolyzed and not hydrolyzed, and TCOD has been used as not hydrolyzed. For every compounds two types of hydrolysis have been applied. The following differences between hydrolyzed and not-hydrolyzed compounds can be observed:

- hydrolysis produces a decrease of the SiO_2 content due to the production of sylicic acids as sols;
- the viscosity of the hydrolyzed solution increases in time, due to the poly-condensation reactions in the hydrolyzed solution; the polymerization process is so advanced, leading to a decrease of the flowing characteristics of the hydrolyzed solution; thus, it becomes useless in order to obtain the ceramic matrix by immersion; the alteration process of the

hydrolyzed solution can be appreciated by the viscosity of the stored solutions; this can be 5-8 times greater than the viscosity of the fresh solution;

- pH decreases due to the HCl additions that acts as cathalisator for the hydrolysis reaction.

The experiments also included hydrolysis and testing of TMM samples from MERCK.

Table 1 shows the values for density and viscosity of TMM and TEV, with and without hydrolysis (two types).

Table 1. Density and viscosity of hydrolyzed and not-hydrolyzed TMM and TEV

No.	Type of compound	Characteristic	Value		
			not-hydrolyzed	hydrolyzed 1	hydrolyzed 2
1	TMM	η[cP]	6.74	10.38	9.44
		ρ[g/cm^3]	0.954	0.9072	0.9075
2	TEV	η[cP]	6.47	10.385	11.983
		ρ[g/cm^3]	0.9101	0.9496	0.9449

Four tests of pyrolysis in Ar and one test of pyrolysis in N have been performed. For the first test a TMM precursor with ~ 40% SiO_2 has been used. The powder obtained after the heat treatment of the gel resulted from TMM was supplementary heat-treated at 1400^0C. Microstructure evaluation by scanning electron microscopy (Philips SEM-515) and X-ray diffraction (diffractometer type HZG-4A, CuKα, λ = 1.5465Å, Ni filter) has been performed.

For the tests 1, 3 and 4 a heat treatment at 1600^0C of the gel obtained from TMM, TEV and TCOD has been applied.

During the pyrolysis of the polysylozans the formation of the intermediary compounds can initiate some complex chemical reactions that depend on the structure of the starting precursor and on the pyrolysis conditions.

The X-ray diffraction patterns of the samples pyrolysed in Ar/1600^0C show an amorphous structure, the same as for the pyrolysis at 1000^0C, but with different chemical structure. The following transformations happen during the pyrolysis.

The mechanism of the pyrolysis has the following steps (Fig.1):

1. the organo-metallic transformation in the temperature range 550-800^0C; a solid, amorphous and hydrogenated product, with thetraedric structure Si(C_xO_y) (x and y depend on the initial components of the polymer) results;
2. the nucleation of the crystalline precipitation, like SiC or SiO_2, and of a carbonic phase, in the temperature range 1200-1600^0C;
3. the growth of the grains by consumption of the residual morphous phases and decrease of the oxygen content, due probably of the evaporation of SiO and CO.

Complex thermal analyses both for the preceramic precursors with Si and for the many-layer preformers of carbon fibers have been applied.

The pyrolysis of the organo-metallic precursors to SiC is supported by mechanisms with free radicals. The radicals are formed on the Si atoms by braking the Si-H bonding in the first step at 500-700^0C. In the second step of the pyrolysis, at 750-1550^0C, the C-H bonding is broken and the radicals are formed on the C atoms.

The microstructure of the resulted CMCs from the sequential immersion/pyrolysis process have been analyzed. Thus, the densification process and the continuous crack formation have been observed.

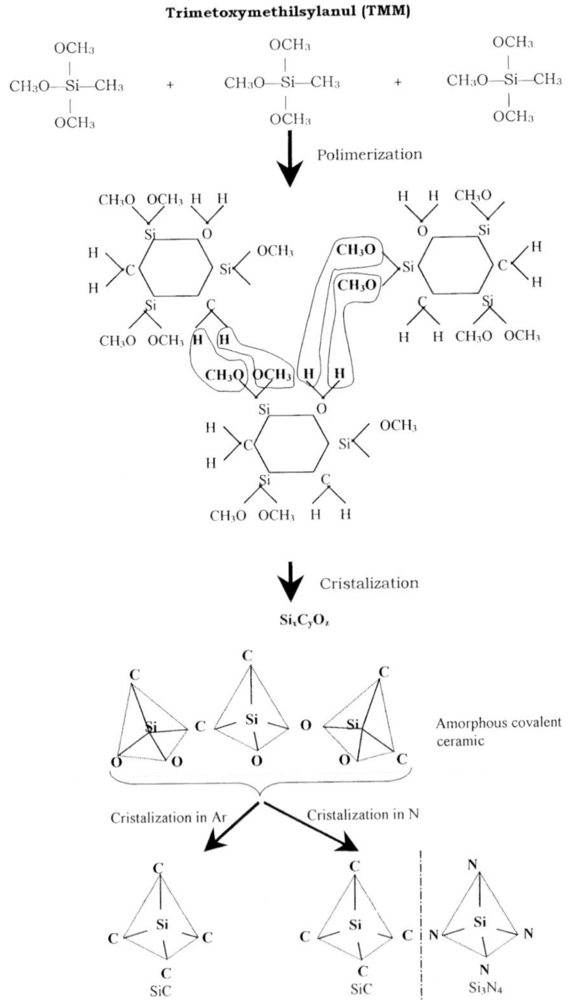

Fig.1. Structural transformations during the pyrolysis of the preceramic polymer

Samples 1, 2 and 3 have been examined by transmission electron microscopy (TEM) and X-ray diffraction (XRD). The TEM examination was performed on replicas, due to the heterogeneous composition of the samples that did not allow preparing thin foils. Although, in a few attempts, the bright field image in Fig.2 shows parallel stripes along a carbon fiber on a thin foil transparent to electrons.

This distance of 3.46Å corresponds to the interplanar distance of the (002) planes of the crystalline packing zones in the carbon fiber. It is obvious that the (002) planes are perpendicular on the fiber, with a distribution of packing around this direction, underlined by the elongation of the spots as a half-moon.

The same samples have been analyzed by scanning electron microscopy (SEM). Thus, the image in Fig.3 shows fragments of carbon fibers broken during the sample preparation, and the ceramic matrix formed by the pyrolysis at 1500^0C in Ar. The morphology of the matrix shows the beginning of a densification process.

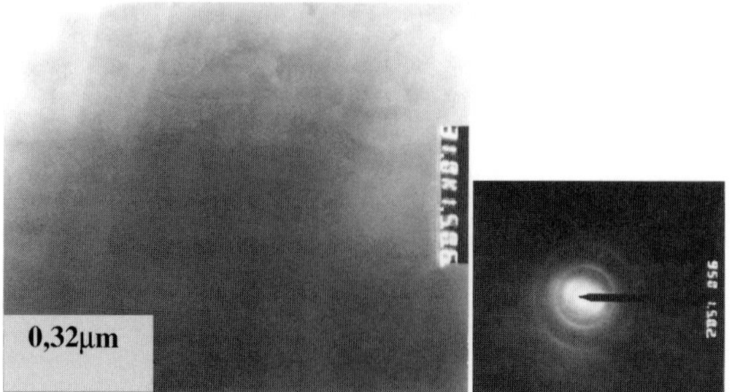

Fig.2. TEM bright field image and selected area electron diffraction (SAED) pattern – sample 3

According to the literature [1,2], SiC can be formed at 1500^0C by the C (pyrolitic and from the fibers), and the Si supplied by the organo-metallic precursor. The pyrolitic graphite has a higher reactivity for this reaction. Thus, in the area where a uniform and homogeneous pyrolitic graphite layer has been formed and an optimum impregnation of the organo-metallic precursor has been produced there is a good opportunity for SiC formation. In Fig.4 SiO_2 melted coating around the carbon fibers is observed. The adherence of this coating is limited by the pyrolytic graphite layer uniformly distributed along the fiber.

Fig.3. Microstructure of a polished sample (x1250)

Fig.4. Agglomeration of components of the ceramic matrix (x625)

The melted zones produced during the heat treatment at 1500^0C have an uniform distribution in the many-layer preformer showing a good impregnation of the preformer, heat treated at 1000^0C in reducing atmosphere.

4 Conclusions

The CMCs with carbon fibers have been obtained using multilayer carbon fiber preforms and preceramic precursors at temperatures of between 1200-1600°C. The presence of SiO_2 and SiC/Si_3N_4 in the ceramic matrix was identified by SEM and XRD.

SiO_2 is the major phase in the ceramic matrix (like crystobalit) along with a low proportion of SiC formed at the interface of the carbon fiber coated with pyrolitic graphite by reaction with the silicon from the preceramic precursor used.

5 References

[1] G.ZIEGLER, The Ceramic Society of Japan, 448, (1991)
[2] P.GREIL, J.of Am.Ceram.Soc., 78,[4],835,(1995);

Strength and Stability of Rubber O-Seals

Brice Cassenti and Alexander Staroselsky
United Technologies Research Center, East Hartford, CT 06108, USA

1 Introduction

Polymer O-rings are widely used as seal elements in engineering practice in general and in airplane engines in particular. Now that twin engine airplanes have been approved for flights between Europe, or Asia, and the Americas, failure of these seals will result in engine shutdown and flights continuing on the one remaining engine. During installation O-ring seals are usually stretched and twisted. Twist can result in elastic instability while the shape of the deformed ring does not remain round. If the O-ring is set on a rigid cylinder, its shape is defined and additional stresses, sometimes very localized, arise. The deformed state of this elastic material causes residual stresses that in turn lead to cracking of the O-ring. Additional thermal stresses appear due to heating, because the thermal expansion coefficient for rubber-like materials is negative and the heating of the installed ring causes additional stretching. Hence, improper installation can lead to premature seal, and subsequently engine, failure.

The problem of estimating the stress in stretched and twisted rubber o-rings is considered here in the context large-strain deformations. Material parameters are chosen to fit the constitutive law from experimental uniaxial stress-strain data. We have analyzed two related problems, namely (*i*) homogeneous stretching followed by axi-symmetric twisting of the ring and (*ii*) the full three-dimensional twisting-torsion problem for o-rings. Under certain physically based assumptions the first problem allows a closed form solution. The second problem is solved numerically using the finite element technique.

As one may see from a dimensionless analysis, the maximum stresses acting in the isothermal axisymmetric deformed ring are functions of the material properties and the ratio between the ring thickness (minor radius a_0) and the major ring radius R_0. The objective of this paper is to develop a model and simulation tool for the prediction of seal elastic behavior and to understand the effects of numerical and physical instabilities.

2 Modification of Constitutive Models of Rubber Elasticity

Rubber-like materials possess relatively low shear moduli while supporting extremely large elastic strains. This makes standard small strain elasticity inappropriate for analyzing components such as o-rings. Even in small strain cases the linear elasticity equations need to be modified. Using indicial notation the equations for incompressible materials can be summarized as (Treloar, 1975):

$$s_{ji,j} + p_{,i} = 0\,, \tag{1}$$

In eq. (1), $\mathbf{s}$ is the deviatoric Piola-Kirchhoff stress, and p is the pressure. One may see that the system (1) can be generally resolved within an additive parameter p. In other words, when compared to classical elasticity the pressure is now an independent variable, and there is a constraint on the displacement gradient from the fact that there is no volume change. In ordinary polymers the strains can become extremely large requiring a more general formulation than equations (1). In this case the deformation gradient, $\mathbf{F}$, can be decomposed into a rotation, $\mathbf{R}$, and a right stretch tensor, $\mathbf{U}$. The stretch tensor can be additionally decomposed into the product of a dilation, $\underline{\Delta}$, and a deviatoric stretch, $\underline{\Lambda}$, or

$$F_{ij} = R_{ik}U_{ki} = R_{ik}\Delta_{kl}\Lambda_{lj} . \tag{2}$$

The dilation is a diagonal tensor given by

$$\Delta_{ij} = J^{1/3}\delta_{ij} \tag{3}$$

where $J = \det\{\mathbf{F}\}$ and $\det\{\underline{\Lambda}\} = \det\{\mathbf{R}\} \equiv 1$.

Kinematics and equilibrium are not sufficient to calculate the stress in the rubber-like materials; we also need a mechanically motivated constitutive relationship for nonlinear elasticity. A good constitutive model should represent the three-dimensional nature of stress-stretch behavior using a minimal number of parameters to represent physically the deformation process. The eight chain model of (Aruda and Boyce 1993) accurately captures the cooperative nature of network deformation while requiring only two material parameters, an initial shear modulus G and a limiting chain extensibility where the parameter N is the number of chemical crosslinks per length. This parameter is related to the locking stretch as follows: $\lambda_L = \sqrt{N}$. We modify this model by incorporating the term reflecting the volume changes and the energy function (i.e., the Hemholtz free energy), W has been taken in the form:

$$W = G\left(\frac{1}{2}(I_1 - 3) + \frac{1}{20N}\left(I_1^2 - 9\right) + \frac{11}{1050\,N^2}\left(I_1^3 - 27\right) + \ldots\right) + \frac{1}{2}\cdot K\cdot(J-1)^2 \tag{4}$$

where $I_1 = \lambda_1^2 + \lambda_2^2 + \lambda_3^2$ is the first invariant of the diagonal stretch tensor and K is a bulk modulus. In this modified model eigenvalues of only deviatoric stretch tensor $\underline{\Lambda}$ denoted through $\{\lambda_i\}$, $i = 1, 3$ and volume change determine the elastic energy.

In order to find the true stress components one should find a derivative of (4) with respect to "total" stretch ratios subject to the constraint imposed by finite volume change J. It can be shown for the case of one-term Aruda-Boyce model (or in other words Mooney-type model) that the principal components of the stress (S_1, S_2, S_3), that are conjugate to the principal stretch ratios $(J^{1/3}\lambda_1, J^{1/3}\lambda_2, J^{1/3}\lambda_3)$, are given by

$$S_i = (G\lambda_i - P/\lambda_i)/J^{1/3} \quad i = 1,2,3 \tag{5}$$

where

$$P = KJ(1-J) + \frac{1}{3}G(\lambda_1^2 + \lambda_2^2 + \lambda_3^2) . \tag{6}$$

Material models, as in equation (5), allow the analyst to investigate arbitrary stress states, including materials with an infinite bulk modulus. This allows a direct evaluation of the effects of a finite bulk modulus.

We calibrated our constitutive model by comparing the model predictions and test results for simple tension up to 200% deformation. Parameters with values of $G = 160$ psi; N = 18 are ensure curve-fitting errors of less than 5% for the o-ring material we considered (see Fig. 1).

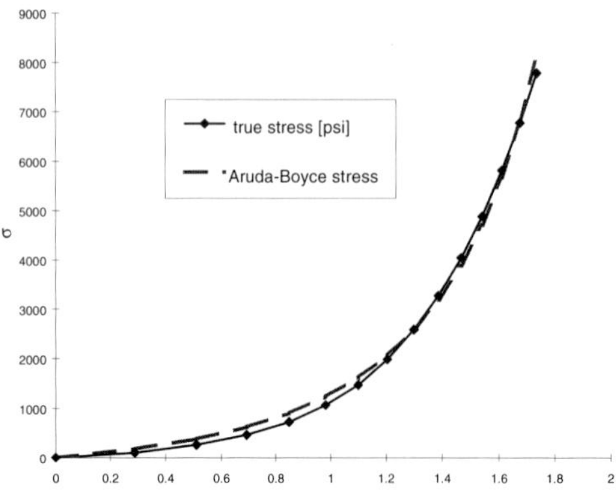

Figure 1. Experimental and predicted true stress-strain curves for tension

3 Model Verification and Results of Simulations

O-rings provide simple yet practical examples for use in simulations. The loads to be considered will be either compression or twisting, both of which can develop instabilities. For example, if an o-ring is twisted about the major circumference so that the outer major circumference is on the inside, the o-ring will be in a state of unstable neutral equilibrium. Similarly, if a uniform hydrostatic pressure is applied the volume change will be small, but, because of the small shear modulus, non-uniformity in the loads or the geometry can result in significant o-ring deformations. Both the twisting and pressure instabilities are geometric, and it can be readily shown that the constitutive model of Eq. (5) has no material instabilities.

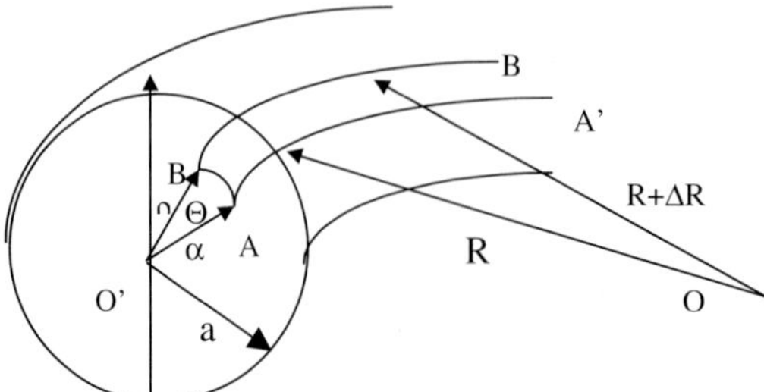

Figure 2. Cross-section and part of the twisted O-ring.

In addition to material and geometric instabilities there exists the possibility of numerical instabilities. This greatly complicates interpreting results that fail to converge. Model problems are needed in order to gain experience in determining the source the instabilities. For the model calibration we obtain an analytical solution for the axisymmetric o-ring torsion problem and compare it with the numerical predictions and show that both results are accurate.

To obtain the exact solution, first of all, we start with the kinematics (i.e. the motion of points in the material). The basic undisturbed geometry consists of an o-ring with major radius R_0 and minor radius a_0. For the case of axisymmetric torsion, we assume that during the deformation each cross-section (circle with the radius ρ) remains in its original plane. We also assume that due to symmetry of the problem the principal strains are in hoop, radial, and circumferential directions. From Fig. 2, the sign and magnitude of this deformation depends on the position of point A (or angle α) and on the position of point B (or angle $\alpha+\theta$), as well as on radius R_0. Under these assumptions the stretch ratio $\mu_1 = J^{1/3}\lambda_1$ in the hoop direction is given by

$$\mu_1 = [1 - \frac{\rho}{R_0}\cos(\alpha+\theta)]/[1 - \frac{\rho}{R_0}\cos\alpha] \tag{7}$$

We have considered two model cases of o-ring axysimmetric torsion. First, a test case used the deformation illustrated in Fig. 2 by specifying a rotational displacement. This model allows the volume change, which increases in the "tension" zone and decreases under compression. From pure geometrical considerations, it is easy to show that the volume ratio is given by $J = \mu_1$.

$$\frac{S_1}{G} = \frac{2}{3}\left(1 - \frac{1}{J^2}\right) - K\frac{J-1}{J^{1/3}} \tag{8}$$

Note that the remaining two principal stretch ratios are one (i.e., $\mu_2 \equiv J^{1/3}\lambda_2 = \mu_3 \equiv J^{1/3}\lambda_3 = 1$). The graphs of stress calculations versus rotation angle for two surface points of the o-ring are shown in Fig. 3 and 4.

The second considered case is based on the additional assumption of incompressibility of rubber-like material. It immediately leads to $\mu_i = \lambda_i$, for all principal directions or in other words to $J = \lambda_1 \lambda_2 \lambda_3 \equiv 1$. The hoop stretch ratio equals λ from (7). Since we have assumed a uniaxial stress state, the remaining two principal stretch ratios can now be found as $1/\sqrt{\lambda}$. The uniaxial stress can now be found by substituting for the stretch ratios in the expression for elastic energy (material constitutive model) and the differentiating with respect to the stretch as follows:

$$\frac{S_1}{G} = \left(\lambda^2 - \frac{1}{\lambda}\right) + \frac{1}{5N}\left(\lambda^2 + \frac{2}{\lambda}\right)\left(\lambda^2 - \frac{1}{\lambda}\right) + \ldots \tag{9}$$

The graphs of stress calculations based on these model assumptions are also presented in Figs. 3 and 4 together with Finite element method (MARC) results. This comparison between numerical and analytical solutions has been used to verify the finite element model predictions.

The MARC finite element code was used to perform numerical investigations. Subroutines were written to incorporate the Aruda-Boyce model and a finite bulk modulus as in eq. (4). A test of the model for uniaxial loading was performed and the results compare exactly as shown in Figs. 3 and 4. The hoop stress S_1 from eqs. (8,9) was compared to the second Piola-Kirchkoff stress predictions from the MARC simulations. The comparison is good as shown in Figs. 3-4 where the shear modulus is 160 psi and the Poisson ratio is 0.4. These results clearly indicate that material volume changes cause additional stresses, mostly associated with an additional hydrostatic pressure. As one may see from these figures, kinematic boundary conditions imposed the in finite element code MARC correspond to the first type of the axysimmetric problems.

Several full 3-D cases were run using the MARC finite element code including a test case for the axisymmetric deformation case described above. The most interesting from the experimental point of view case is twisting of one side of the o-ring while the opposite part (180 degrees apart) is to be held. One may see, that the ring has the new equilibrium state after

180 degrees rotation and this equilibrium shape is not round anymore and has so-called “chair”-type shape. A comparison between FEM predictions and actual o-ring shape is shown in Figure5. If the ring is set on rigid cylinder, its shape is defined and additional stresses, very localized, arise. The model predicts the deformation extremely well.

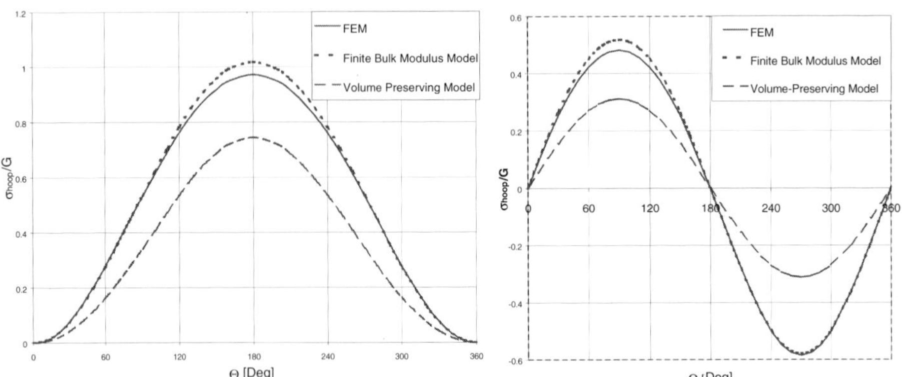

Figure 3. Normalized True Stress vs. Angle for Inside-Out Rotation $\alpha=0$

Figure 4. Normalized True Stress vs. Angle for Rotation of the point $\alpha=90$

Several full 3-D cases were run using the MARC finite element code including a test case for the axisymmetric deformation case described above. The most interesting from the experimental point of view case is twisting of one side of the o-ring while the opposite part (180 degrees apart) is to be held. One may see, that the ring has the new equilibrium state after 180 degrees rotation and this equilibrium shape is not round anymore and has so-called “chair”-type shape. A comparison between FEM predictions and actual o-ring shape is shown in Figure5. If the ring is set on rigid cylinder, its shape is defined and additional stresses, very localized, arise. The model predicts the deformation extremely well. However, besides the real “physical” instability, specific for o-ring deformation, the numerical instability, which can be dependent on the material model, effects computational results. Generalized Ogden and Mooney models were also investigated. If the bulk modulus was set to infinity, numerical instabilities occurred more frequently, especially when more approximate loads or the boundary conditions were input.

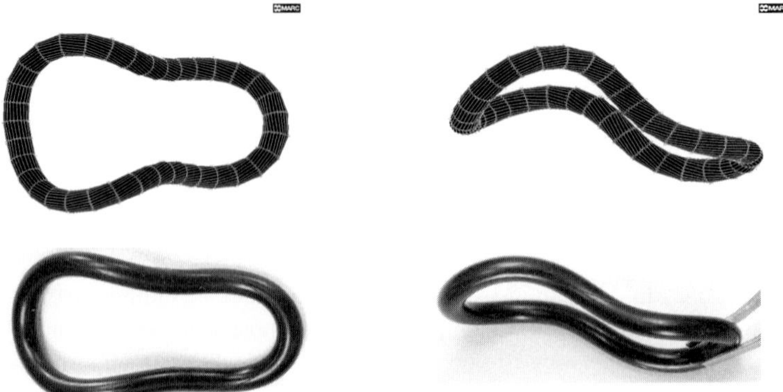

Figure 5. O -ring Twist Deformation: FEM predictions vs. Nature

4 Conclusions

We have evaluated the role of the both types of deformation: axisymmetric torsion and twist in the formation of stress-strain state in the ring. This is related to the manufacturing process when the ring may be installed inside out by device or by twisting. Twist causes elastic instability and the shape of the deformed ring is not round anymore. Complete stress states may be predicted using the modified material model described in the paper. Numerical results show that pure o-ring twist causes localized hoop stresses up to $0.4G$. Heating causes rubber contraction, therefore, during engine operation, additional tension stresses appear in the O-ring seal.

The simulations using various material models showed that a model that allows volume changes is numerically more stable. A model that fails to converge leaves the analyst with little in the way of information on corrections. It appears that most of the 'failure to converge' results were due to numerical instabilities. The models did though predict instability just beyond the inside-out location. The modified Aruda-Boyce material model is more accurate than the generalized Mooney or Ogden models, and more accurately predicted the instability.

5 References

1. Treloar L., The Physics of Rubber Elasticity, Clarendon Press, Oxford, 1975, p. 310.
2. Aruda E. M., and Boyce M. C., "A Three-Dimensional Constitutive Model for the Large Stretch Behavior of Rubber Elastic Materials," J. Mech. Phys. Solids, 1993, **41**, 2, pp.389-412.

Competing Reactions during Block Copolymer Synthesis by Melt Transesterification

A. Gottwald, D. Pospiech, D. Jehnichen, P. Pötschke, A. Janke
Institute of Polymer Research Dresden, Hohe Str. 6, 01069, Dresden, Germany

1 Introduction

Block copolymers (BCP) are modern materials that combine properties of the parent polymers and compatibilize immiscible blends. Economically as well as ecologically new methods of synthesis could support their further propagation. Such a method is the melt transesterification polycondensation, successfully used by Pospiech et al. [1] to synthesize BCP of polysulfone (PSU, Fig.1a) and the liquid crystalline (LC) poly-(ethylene terephthalate-*co*-ester imide) (PET/PEI, Fig.1c). Applying the same method to form BCP of PSU and LC poly-(ethylene terephthalate-*co*-oxybenzate) (PET/HBA, Fig.1b) [2] led not only to the desired BCP but also to a byproduct. It was characterized by NMR and GPC to be chain-extended PET/HBA, i.e. high molecular [3]. Obviously, not only BCP formation but also a side reaction took place.

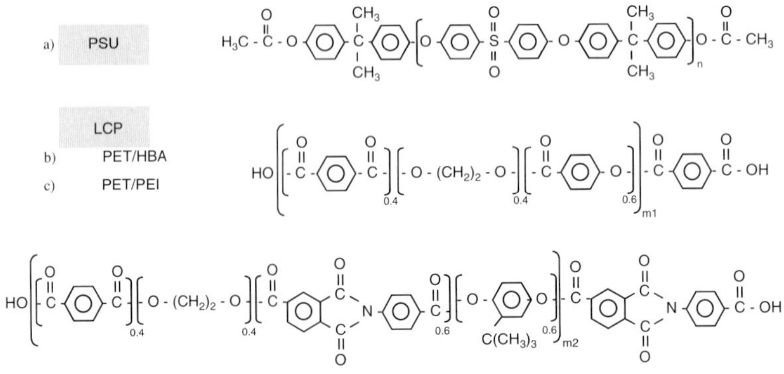

Figure 1. Components of the multiblock copolymers, a) PSU, b) PET/HBA, c) PET/PEI

Both systems seem to be quite similar (Tab. 1). The BCP is formed by transesterification of acetoxy-terminated PSU and carboxy-terminated LCP (Fig. 2a) In spite of the vacuum, the formed acetic acid can initiate side reactions in the LCP phase (Fig. 2b), also transesterifications. Because the BCP is formed in the phase boundary, the main reaction will be favored at the expense of the side reaction by increasing the area of phase boundary and the final influence on the quality of the product will be the quality of mixing.

It should be mentioned that PET/HBA is not a completely random copolyester. In contrast to PET/PEI, it tends to form HBA blocks and separates into a PET-rich and an HBA-rich phase [4]. This fact will be discussed shortly in chapter 3.

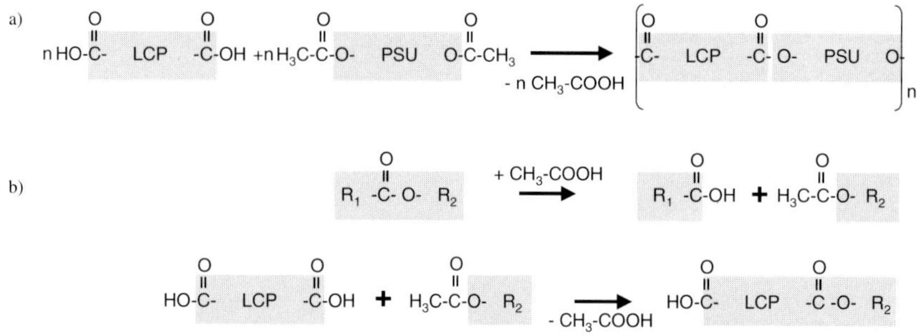

a) $n\,HO\text{-}\overset{O}{\overset{\|}{C}}\text{-}$ LCP $\text{-}\overset{O}{\overset{\|}{C}}\text{-}OH + n\,H_3C\text{-}\overset{O}{\overset{\|}{C}}\text{-}O\text{-}$ PSU $O\text{-}\overset{O}{\overset{\|}{C}}\text{-}CH_3 \xrightarrow[-\,n\,CH_3\text{-}COOH]{} \left[\text{-}\overset{O}{\overset{\|}{C}}\text{-}\right.$ LCP $\text{-}\overset{O}{\overset{\|}{C}}\text{-}O\text{-}$ PSU $\left.O\text{-}\right]_n$

b) $R_1\,\text{-}\overset{O}{\overset{\|}{C}}\text{-}O\text{-}\,R_2 \xrightarrow{+\,CH_3\text{-}COOH} R_1\,\text{-}\overset{O}{\overset{\|}{C}}\text{-}OH + H_3C\text{-}\overset{O}{\overset{\|}{C}}\text{-}O\text{-}\,R_2$

$HO\text{-}\overset{O}{\overset{\|}{C}}\text{-}$ LCP $\text{-}\overset{O}{\overset{\|}{C}}\text{-}OH + H_3C\text{-}\overset{O}{\overset{\|}{C}}\text{-}O\text{-}\,R_2 \xrightarrow[-\,CH_3\text{-}COOH]{} HO\text{-}\overset{O}{\overset{\|}{C}}\text{-}$ LCP $\text{-}\overset{O}{\overset{\|}{C}}\text{-}O\text{-}\,R_2$

Figure 2. Transesterifications, a) BCP formation (main reaction), b) side reaction

Table 1. Properties of the components and the reaction products

Components	PET/PEI	PSU_1	PSU_2	PET/HBA
molecular weight [g/mol]	7500	9500	7400	7700
glass temperature(s) [°C]	141	175	158	T_{g1}: 55, T_{g2}: 75
melting point/range [°C]	210 - 230	-	-	190
phase behavior	LC (nematic)	amorphous		LC (nematic)
$T_{isotrop.}$ [°C]	275 - 355	-		disintegr. (350)
solubility in chloroform	non	yes		partially

Products	PSU + PET/PEI	PSU + PET/HBA
products	BCP	BCP + PET/HBA
phase separation	microscopic	macroscopic
number of glass transitions	2	2
solubility in chloroform	yes	partially

2 Experimental

Using comparable reaction conditions (e.g. temperature, rotational speed, scale), the mixing quality depends on, e.g., the interaction parameter, the differences in melt viscosity of the starting oligomers and block molecular weights. The three compositions PSU (9500 g/mol) - PET/PEI (7500 g/mol), PSU (9500 g/mol) - PET/HBA (7700 g/mol) and PSU (7400 g/mol) - PET/HBA (7700 g/mol) were investigated.

Interaction parameter χ

The interaction parameter χ between two polymers can be calculated from their solubility parameters δ_p according to eq. 1. That were determined experimentally by measuring the solution viscosity and using the incremental method of Small. (p - polymer, s - solvent)

$$\chi = (\delta_{p1}-\delta_{p2})^2 * V_{ref} / RT \qquad \text{(eq. 1)}$$

(V_{ref} = 100 cm³/mol, arbitrary reference volume, R - gas constant, T - temperature)

The higher the solvent efficiency is the more the polymer coils are expanded, and hence, the viscosity is increased. The quality of the solvent is described by the difference of the solubility parameters of polymer and solvent. Thus, the solubility behavior of the polymer should be expressed by the solubility parameter of the best solvent. [5]

Solvents with different δ_s were provided by mixtures of chloroform ($\delta_{s,1} = 18.9_0$ MPa$^{0.5}$) and hexafluoro-2-propanol ($\delta_{s,2} = 17.9_8$ MPa$^{0.5}$). The δ_s of the mixture consists of the $\delta_{s,i}$ of their components as in eq. 2. ($x_{s,i}$ - volume ratio of i)

$$\delta_s = x_{s,1} * \delta_{s,1} + x_{s,2} * \delta_{s,2} \qquad \text{(eq. 2)}$$

The viscosity of the polymer solutions (5 mg/ml) and the solvents were measured with an Ubbelohde-Micro-Viscometer (capillary 53610-I) at 20°C. δ_p is defined by the position of the maximum in the $\eta_{rel}(\delta_s)$-plot; the relative viscosity η_{rel} is defined as $\eta_{solution}/\eta_{solvent}$. In contrast to the height of the maximum, its position does not depend significantly on molecular weight and concentration.

The second method for determining δ_p, used here, based on Small's formulae eq. 3 [6].

$$\delta_p = \Sigma F_i \,/\, V_{0,m} \qquad \text{(eq. 3)}$$

$V_{0,m}$ is the molar monomer volume. It can be calculated by $V_{0,m} = \Sigma V_i$. The V_i and F_i represent the contributions of the constituting atomic groups to $V_{0,m}$ and to energetic interactions, respectively. Several authors have presented tables for F_i with sometimes considerable differing values (e.g. [7], [8]). In this paper, we used van Krevelen's values for F_i as well as for V_i. [8]

Melt viscosity

The melt viscosity η was measured with the ARES rheometer (Rheometrics) using plate-plate geometry (∅ = 25 mm, gap 1-2 mm) at 280°C (according to the reaction temperature), in a frequency range from 0.1 to 100 rad/s and with a strain of 3-5%.

Microscopy

Samples were planed with a microtom (Leica Ultracut S) at room temperature. The phase separation was visualized by SEM (LEO 435 VP) and by AFM (NanoScope III, Digital Instruments).

3 Results and Discussion

The solubility parameters were determined from Fig. 3a as $\delta_{PSU} = 18.5_5$ MPa$^{0.5}$ and $\delta_{PET/HBA} = 18.2_0$ MPa$^{0.5}$. The resulting χ is 0.01. From the incremental method follows: $\delta_{PSU} = 21.7_9$ MPa$^{0.5}$, $\delta_{PET/HBA} = 20.9_2$ MPa$^{0.5}$ and hence, $\chi_{PSU\text{-}PET/HBA} = 0.03$. Although, the δ-values, obtained by both methods, differ somewhat, the χ-values are comparable. Using the same F_i-table for PET/PEI as for PET/HBA the solubility parameter $\delta_{PET/PEI}$ is 25.4_1 and consequently, $\chi_{PSU\text{-}PET/PEI} = 0.5$.

The interaction parameter imply that PSU would be better miscible with PET/HBA than with PET/PEI. This result is in contrast to the observation of the byproduct in the system PSU-PET/HBA, whereas the system PSU-PET/PEI is macroscopically homogeneous. Two explanations for this are possible: 1st both methods are suggested only for amorphous polymers. 2nd The calculations of δ and the resulting χ depend strongly on the chosen F_i (e.g. [6] or [7]), deviations up to one decade were found. So, the interaction parameter, obtained here, reflect - in the best case - tendencies.

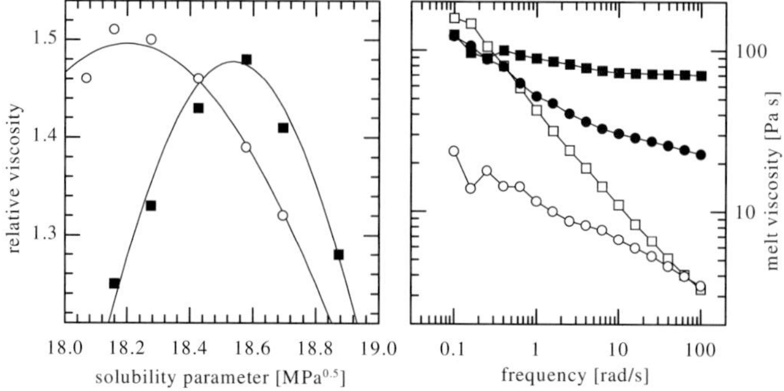

Figure 3. a) relative viscosity η_{rel} of solutions of PSU (30000 g/mol) (■) and PET/HBA (13500 g/mol) (○) vs. δ_s; b) melt viscosity of the oligomers: PSU (9500 g/mol, -■-, 7600 g/mol, -□-), PET/HBA (7700 g/mol, -○-) and PET/PEI (7500 g/mol, -●-) vs. frequency at 280°C

Considering the results from the melt viscosity, depicted in Fig. 3b, the systems, investigated here, are more complex than supposed. Particularly, both PSU oligomers (7600 and 9500 g/mol, respectively) are hardly comparable. At first, we will discuss the higher molecular PSU. It could explain the differences in the phase separation behavior by means of the viscosity ratio λ. This parameter strongly affects the mixing behavior in blends. If its value exceeds four, the low viscous matrix is unable to transmit shearing forces to the disperse phase. An effective dispergation is avoided. Although, there are many differences between a blend and a reaction melt, this value may give a first indication of the mechanical mixing behavior.

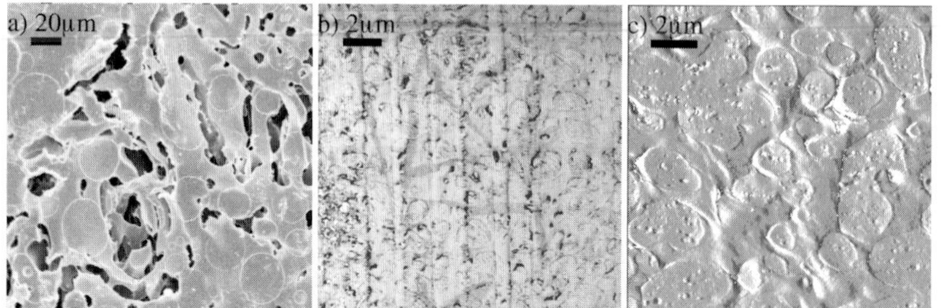

Figure 4. a) PSU (9500 g/mol) - PET/HBA (7700 g/mol) (SEM, sample etched with $CHCl_3$), b) PSU (9500 g/mol) - PET/PEI (7500 g/mol) (AFM, phase mode) c) PSU (7500 g/mol) - PET/HBA (7700 g/mol) (AFM, amplitude mode)

In the case of PSU (9500 g/mol) - PET/HBA λ is larger than seven over the whole frequency range. This may cause the unsuitable mechanical mixing and, consequently, the macrophase separation, shown in Fig. 4a. In the system PSU (9500 g/mol) - PET/PEI λ is about 3 at high frequencies and nearly one at lower. Fig. 4b illustrates a clearly diminished macrophase separation. Regarding the viscosity of the shorter PSU and the resulting morphology of PSU (7400 g/mol) - PET/HBA, it becomes quite clear that the λ-parameter alone does not describe the mixing behavior completely. Although the PSU viscosity decreases rapidly with frequency and is equal to the viscosity of PET/HBA at 100 s^{-1}, the macrophase separation is not significantly suppressed (Fig. 4c). This could be explained by differences between a non-reactive blend and a reaction melt. On the other hand, as mentioned in the introduction, PET/HBA is phase separated itself. It consists of an isotropic PET-rich and an anisotropic (LC) HBA-rich phase. It is already known from microscopical investigations that PSU does not mix with both. Further studies, considering χ and λ between PSU and the PET-rich phase and the HBA-rich phase separately, may help to understand this complex three-phase-system.

4 Conclusions and Outlook

Two quite similar systems, PSU-PET/PEI and PSU-PET/HBA, used for the BCP synthesis, were investigated. The main aspect of the investigation was the dependence of mixing behavior on the interaction parameter χ, the melt viscosity ratio λ as well as the block molecular weights. An influence of χ could not be unambiguously proven. It could be shown that λ has an important influence on the mixing behavior not only in the corresponding blend (as already known) but also in reaction melts. Especially, the high λ-value for PSU (9500 g/mol) - PET/HBA prevents a suitable mixing, whereas the smaller value for PSU (9500 g/mol) - PET/PEI allows a better mixing. However, the macrophase separation in PSU (7400 g/mol) - PET/HBA in spite of the low λ is a clear sign for differences between a non-reactive blend and a reaction melt and requires further investigations.

5 Acknowledgements

The authors would like to thank Mrs. K. Eckstein for help with viscosity measurements, Mrs. M. Franke for microtoming the samples, Prof. K.-F. Arndt and Dr. B. Morgenstern (TU Dresden) for helpful discussions about χ and the German Science Foundation for financial support.

6 References

[1] D. Pospiech, L. Häußler, K. Eckstein, H. Komber, D. Voigt, D. Jehnichen, E. Meyer, A. Janke, H. Kriecheldorf, Designed Monomers and Polymers 1998, 1, 187-206
[2] D. Pospiech, L. Häußler, H. Komber, D. Voigt, D. Jehnichen, A. Janke, A. Baier, K. Eckstein, F. Böhme, J. Appl. Polym. Sci., 1996, 62, 1819-1833
[3] A. Gottwald, D. Pospiech, D. Jehnichen, J. Macromol. Sci. - Physics (in press)
[4] D. Scheller, J. Kressler, H.W. Kammer, F. Böhme, D. Voigt, D. Leistner, M. Rätzsch, Polymer Bull., 1989, 21, 585-591
[5] E. Schröder, G. Müller, K.-F. Arndt, Polymercharacterization, Hanser, München, 1989
[6] P.A. Small, J. Appl. Chem., 1953, 3, 71
[7] K.L. Hoy, J. Paint Technology, 1970, 42, 76
[8] D.W. van Krevelen, Properties of Polymers, Elsevier, Amsterdam, Oxford, New York, Tokyo, 1990, p. 87, p. 195

Composite Membranes for Pervaporation Separation Processes Based on Polyelectrolyte Multilayer Assemblies

Jochen Meier-Haack, Theresia Rieser, Wolfgang Lenk, Siegfried Berwald, Dieter Lehmann
Institute of Polymer Research; Dresden

1 Introduction

Pervaporation is nowadays recognized as an effective and energy saving membrane based technique for separating azeotropic, close-boiling, or aqueous organic mixtures. Especially the dewatering of low molecular weight alcohols such as ethanol has attracted much attention and great efforts have been made to develop effective membranes [1]. For the dehydration by pervaporation, hydrophilic membrane materials are favorable to the permselective separation of water. Additionally, thin separating layers are recommended, because the transport through the membranes is diffusion controlled [2]. Both requirements can be combined in so-called composite membranes consisting of a highly permeable support layer coated with a thin and highly selective separation layer.

A promising tool to prepare thin coatings on solid surfaces is given with the layer-by-layer deposition of oppositely charged polyelectrolytes, which has been studied extensively in recent years [3, 4]. First attempts to prepare composite membranes for pervaporation and gas separation processes by the alternating adsorption of polyanions and polycations on porous or pore-free substrates have been reported recently [5 - 7]. The substrates used in these studies were polypropylene (porous), polydimethylsiloxane [5], poly(4-methyl-1-pentene) [6], and polyacrylonitrile (porous) [7]. In some cases, the surfaces were pretreated by a wet-chemical technique [6] or by plasma [7] to ensure an improved adhesion of the polyelectrolyte multilayer film on the initially hydrophobic supports. However, up to 60 double layers consisting of poly(allylamin hydrochloride) and poly(sodium styrene sulfonate) had to be deposited on the support to obtain membranes with sufficient properties.

Recently, the use of melt-modified polyamides for pervaporation membranes [8] and the influence of polyelectrolyte multilayer systems adsorbed on such membranes was reported together with preliminary results of the dewatering of 2-propanol by pervaporation [9]. Here we present a study of 2-propanol-water pervaporation through polyamide/polyelectrolyte multilayer composite membranes. The effect of different polyelectrolyte pairs on the pervaporation performance was investigated. Poly(acrylic acid) and poly(styrene sulfonic acid) were used as polyanionic components and poly(ethyleneimine) and poly(diallyldimethylammoniumchloride) were used as polycations

2 Experimental

2.1 Materials

Polyamide-6 (Miramid SH3, M_n 22.600 g/mol) was purchased from Leuna-Werke AG (Germany). Poly(α-methylstyrene-*alt*-ε-caproic acid maleimide) (M_n 30.000 g/mol) was provided by D. Lehmann.. Formic acid (85 wt.-%) and 2-propanol both of analytical grade were obtained from Merck (Germany). Poly(acrylic acid) (M_w 90.000 g/mol) and poly(styrene sulfonic acid) (M_w 70.000 g/mol) were from Polyscience (Germany). Poly(diallyldimethyl-ammoniumchloride) (M_w 99.000 g/mol) was purchased from FhG Teltow (Germany), and branched poly(ethyleneimine) (M_w 750.000 g/mol) was obtained from Aldrich (Germany). Milli-Q-water (resistance $\geq$ 17 MΩ/cm) was used as solvent and for purification. All materials were used without further purification

2.2 Methods

2.2.1 Modification of polyamides

Modified polyamide-6 for the preparation of the membrane supports was obtained from reacting polyamide-6 with poly(α-methylstyrene-*alt*-ε-caproic acid maleimide) in the melt phase according to the procedure reported earlier [8]. The resultant comb-like copolymer had polyamide side chains with a molecular weight per polyamide chain of approx. 4520 g/mol (DP $\approx$ 40).

2.2.2 Preparation of supporting materials

The phase-inversion-process was used to prepare asymmetric membranes from the melt-modified polyamide [10]. A polymer solution containing 15 wt.-% modified polyamide in formic acid/water (85/15 wt/wt) was casted on a glass plate. The film thickness was set to approx. 200 µm. After an appropriate time (10 min), during which the solvent was allowed to partially evaporate, the polymer film was immersed into cold water (4°C) for precipitation. The resulting asymmetric membrane, which consists of a thin dense layer and a highly porous support was placed in water for several days to leach out residual formic acid.

2.2.3 Surface modification of polyamide membranes by polyelectrolyte adsorption

The formation of the polyelectrolyte multilayer assemblies was carried out by alternative dipping of the modified membranes into aqueous solutions of oppositely charged poly-electrolytes. The adsorption time was 30 min. After each adsorption step the membranes were rinsed extensively with pure water to remove weakly bound polyions. The modifications were done with the polycations poly(ethyleneimine) (PEI), and poly(diallyldimethylammoniumchloride) (PDADMAC) and with the polyanions poly(acrylic acid) (PAAc), and poly(styrene sulfonic acid) (PSS). The concentration of each polyelectrolyte was 2×10^{-2} monomole/l (monomole = mole of monomer unit). The number of dipping cycles varied between 0 and 10.

2.2.4 Characterization of modified membranes and set-up pervaporation experiments

The successful adsorption of polyelectrolytes on the membrane surfaces was monitored by streaming potential (ζ-potential) measurements at different pH values using an Electrokinetic Analyzer (EKA, A. Paar KG, Austria).

For pervaporation experiments membranes with an effective area of 30 cm^2 were placed in a thermostated stainless steel cell (PM 28, CM-Celfa AG, Switzerland). The operating temperature was maintained at 50°C by a thermostat. Binary water/2-propanol mixtures containing 5 to 95 wt.-% water were used as feed. The downstream pressure was 20 hPa and was controlled by a vacuum constanter (CVC2, VaccuBrand, Germany). The permeate was collected in cold traps kept at -20°C. The permeation rates were determined gravimetrically. The composition of feed and permeate mixtures was obtained from density measurements using a DMA 58 digital density meter (A. Paar KG, Austria). Calibration was done with water/2-propanol mixtures of known compositions. The separation factor α was calculated using equation (1) where c is the concentration of water and alcohol in the feed and the permeate, respectively as symbolized by the subscripts W, A, F, and P.

$$\alpha = \frac{c_{W,P} \cdot c_{A,F}}{c_{W,F} \cdot c_{A,P}} \tag{1}$$

3 Results and Discussion

An asymmetric membrane, prepared by the well-known phase-inversion-process from modified polyamide-6 was used as a support for the deposition of the polyelectrolytes. The modified polyamide-6 was obtained from the reaction of a linear polyamide-6 with α-methylstyrene-*alt*-ε-caproic acid maleimide in the melt phase [8]. This modified polyamide has a comb-like structure with a carboxyl group at each chain end. The calculated end group concentration was approx. 203 µmol/g. This end group concentration ensures a good adhesion between the substrate and the first polyelectrolyte layer without further surface modification steps. The deposition of the polyelectrolytes on the membrane surface was carried out according to the procedure described by Decher et al. [3] and which is outlined schematically in Fig. 1. The initially anionic support is dipped into an aqueous solution a polycation (PEI or PDADMAC). This membrane equipped with a cationic surface is carefully rinsed with water and subsequently dipped into an aqueous solution containing a polyanion (PAAc or PSS) and was rinsed with water afterwards. This dipping cycle was repeated up to 10 times. The properties of membranes with different numbers of adsorbed polyelectrolyte layers were determined with a feed mixture containing 30 wt.-% water and 70 wt.-% 2-propanol.

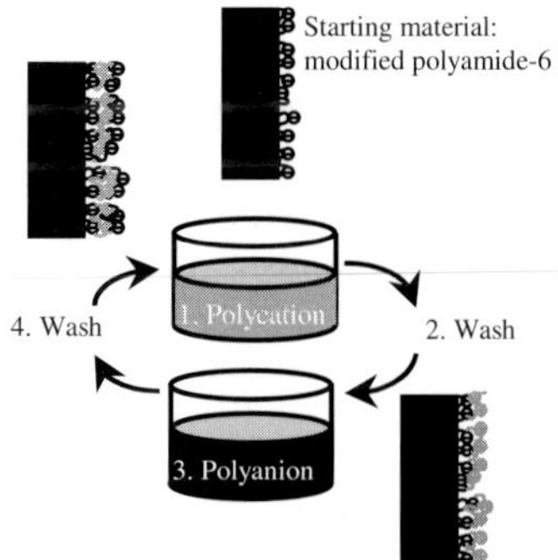

Figure 1. Schematic drawing of the polyelectrolyte deposition process on modified polyamide-6 membranes

The determination of the ζ-potential as a function of the pH is a frequently used method for the characterization of polymer surfaces. This simple method was applied for monitoring the successful deposition of the polyelectrolytes on the membrane surface. Figure 2 shows typical ζ-potential vs. pH plots for positively and negatively charged surfaces as present after the adsorption of polycations (e.g. PEI, PDADMAC) and polyanions (e.g. PAAc, PSS). The adsorption of the polycations PEI or PDADMAC on the initially anionic polyamide surface (Fig. 2a) led to a reverse of the surface charge, which is indicated by the positive ζ-potential over a wide pH range investigated and by a shift of the isoelectric point (IEP; ζ-potential = 0) to the basic pH range. Upon further reverse of the surface charge achieved by the adsorption of the polyanions PAAc and PSS, respectively, the IEP shifts again to the acidic pH range and a sign-reversal of the ζ-potential over a wide pH range is observed.

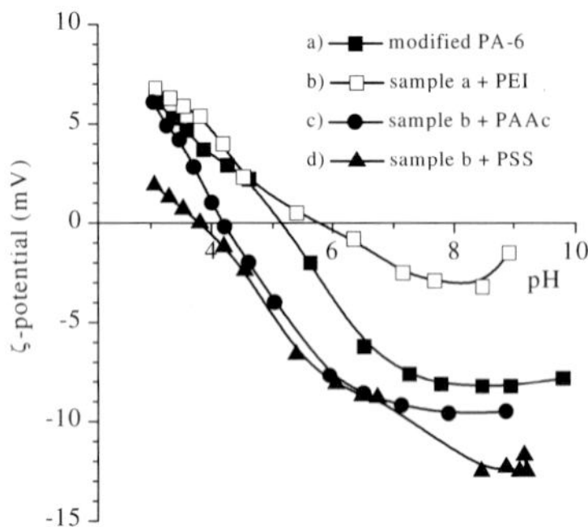

Figure 2. ζ-potential vs. pH-plots of a) modified polyamide-6; b) sample + PEI; c) sample b) + PAAc; d) sample b) + PSS

Pervaporation experiments were carried out with membranes having different numbers of adsorbed polyelectrolyte layers and different polyion pairs used for the built-up of the various multilayer systems. In figure 3 the selectivities and permeabilities of pervaporation membranes which were coated with various polyelectrolyte pairs versus the number of dipping cycles are displayed. The unmodified polyamide-6 membrane had low selectivity but high permeability for the test mixture consisting of 30 wt.-% water and 70 wt.-% 2-propanol. The consecutive alternating adsorption of oppositely charged polyelectrolytes led to a step-wise improvement in selectivity. The increased selectivity is accompanied by a decrease in permeability as it is often observed in membrane performance. However, strong differences in selectivity were observed for the various polyelectrolyte multilayer systems investigated, although each polyion could be adsorbed successfully on the membrane surfaces as it was confirmed by ζ-potential measurements. For example, high permeabilities and low selectivities were observed when strong or permanently charged polyelectrolytes such as PSS or PDADMAC were used for the built-up of the polyelectrolyte multilayer assemblies.The selectivities of these membranes differs only slightly from that of the unmodified sample. On the other hand membranes with high selectivities were obtained from multilayer systems consisting of poly(acrylic acid) and poly(ethyleneimine).

It is proposed that the polymer chains of polyelectrolytes, that contain permanent charges such as quaternary poly(ammonium salts) or polyacid salts as well as strong polyacids (e.g. PSS), which are completely dissociated in water, adopt a stretched rod-like conformation.

This conformation will help to minimize the repulsive electrostatic interaction between the equally charged functional groups in the polymer backbone. For that reason, these polyions can only adsorb in relatively thin layers on oppositely charged surfaces. A thickness of about 2 nm for one double layer consisting of poly(sodium styrene sulfonate) and poly(allylamine hydrochloride) has been reported in the literature [11]. On the other hand weak polyelectrolytes which have a lower degree of dissociation such as poly(acrylic acid) or poly(ethyleneimine) may adopt a more coiled conformation. Since an increase in layer thickness was observed when inter- and intramolecular electrostatic interactions were reduced by the addition of low molecular weight electrolytes to the aqueous polyelectrolyte solutions, it is probable that poly(acrylic acid) or poly(ethyleneimine) adsorb also in thicker layers which finally lead to membranes with improved separation properties with a relatively low number of polyelectrolyte layers.

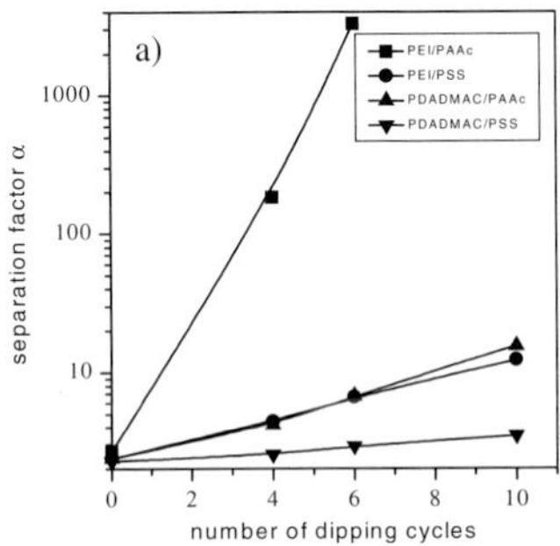

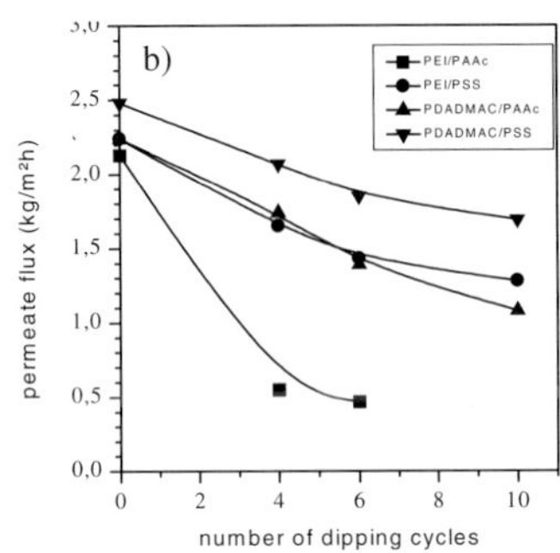

Figure 3. a) selectivity; b) permeability of polyamide membranes modified with different polyelectrolyte pairs vs. the number of dipping cycles.

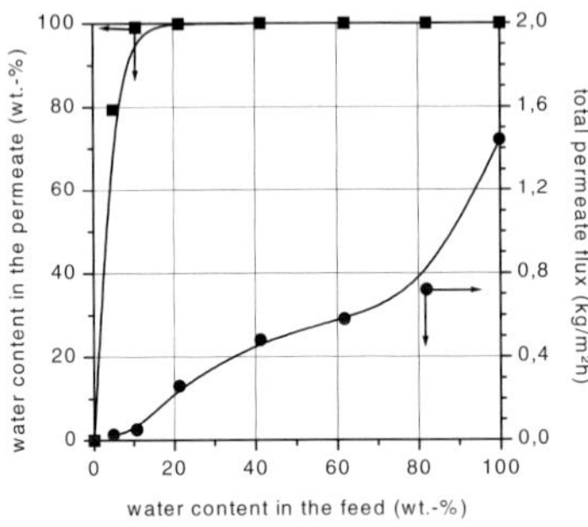

Figure 4. Effect of feed composition on permeate flux and permeate composition (sample: modified polyamide-6 membrane after 7 dipping cycles using poly(acrylic acid) and poly(ethyleneimine)

For one composite membrane the influence of the feed composition on the membrane properties permeate flux and selectivity was studied using water/2-propanol feed mixtures with water contents in the range from 5 to 100 wt.-%. The polyelectrolyte complex layer was built-up by seven dipping cycles using poly(ethyleneimine) as polycation and poly(acrylic acid) as polyanion. The results are displayed in figure 4. This membrane showed a high water selectivity over the whole range of feed composition. For feed mixtures containing more than 5 wt.-% water, the water content in the permeate was higher than 99 wt.-%. Furthermore, no significant influence of the feed mixture on the permeate composition was detectable.

Contrary to these findings, a strong effect of the feed composition on the permeate flux was found. The permeate flux decreased strongly with decreasing water concentrations in the feed. It can be inferred that interactions between the polyelectrolyte coating and the water lead to a swelling of the separating layer which allows the water molecules to permeate quickly through the membrane. The degree of swelling is decreased on reducing the water content in the feed. Therefore the water uptake of the separating layer is hindered and the permeation through the polyelectrolyte multilayer system decreases with decreasing water content in the feed.

4 References

[1] J. Néel in Pervaporation Membrane Separation Processes (Ed.: R. Y. M. Huang), Elsevier Science Publisher, Amsterdam, 1991, p. 42
[2] Y. Cen, R. N. Lichtenthaler in Membrane Separations Technology Principles and Applications (Eds.: R. D. Noble, S. A. Stern), Elsevier Science Publisher, Amsterdam, 1995, p. 89
[3] G. Decher, J.-D. Hong, J. Schmitt, Thin Solid Films 1992, 210 /211, 831 - 835
[4] G. Decher, Science 1997, 227, 1232 - 1237
[5] P. Stroeve, V. Vasquez, M. A. N. Coelho, J. F. Rabolt, Thin Solid Films 1996, 284/285, 708 - 712
[6] J.-M. Leväsalmi, T. J. McCarthy, Macromolecules 1997, 30, 1752 - 1757
[7] L. Krasemann, B. Tieke, J. Membr. Sci 1998, 150, 23 - 30
[8] J. Meier-Haack, D. Lehmann, S. Berwald, W. Lenk, P. Hansen, T. Rieser, K. Lunkwitz in Werkstoffwoche ´98 Band VIII Symposium 10 Polymere (Eds.: W. Michaeli, R. Mülhaupt, M. Möller, H. Riedel), Wiley-VCH, Weinheim, 1999, p. 229 - 234
[9] J. Meier-Haack, T. Rieser, W. Lenk, D. Lehmann, S. Berwald, S. Schwarz, Chemie Ingenieur Technik 1999, 71, 839 - 844
[10] S. Loeb, S. Sourirajan, Adv. Chem. Ser. 1962, 38, 117 - 132
[11] G. Decher, J. Schmitt, Prog. Colloid Polym. Sci. 1992, 89, 160
[12] G. B. Sukhorukov, J: Schmitt, G. Decher, Ber. Bunsenges. Phys. Chem. 1996, 6, 948 - 953

Injection Molding in Biodegradable Plastics

Prabhu Kandachar, Joke Reijnhoudt, Rolf Koster & Ineke Goedhart
Faculty of Design, Engineering and Production, Delft University of Technology, The Netherlands

1 Abstract

An exploratory investigation was conducted to establish the importance of injection molding conditions for mechanical behavior of two biodegradable thermoplastics. Tensile test results are reported from specimens with 2 mm and 4 mm wall thicknesses, as well as a 2 mm thick specimen with a cold flow weld line. One of the challenges in injection molding with these materials was a limitation to the process window because of narrow margins for melt plasticizing temperatures, pressures and screw rotation rates. The results show that mold filling pressure and melt front speed have a strong effect on tensile behavior, paricularly on strain at yield or break.

2 Introduction

Biodegradable plastics are still in a development stage. The raw material cost is comparatively high because they are only supplied in limited quantities by a few manufacturers. Designing in biodegradable plastics is difficult because insufficient material data is available for most of these materials. Biodegradable plastics have mainly been applied for foils and a few other packaging products. Although mechanical properties of some biodegradable plastics appear promising, molded products for load-bearing applications have hardly been manufactured.

The work reported in this paper is part of a project to explore the consequences of different geometries and injection molding settings for biodegradable thermoplastics products. Tensile test results are presented for two types of specimens representing several geometrical characteristics including a flow weld line. The project is intended to contribute to predictibility of mechanical behavior of injection molded products of biodegradable plastics to support product development.

3 Experimental

Two materials have been used: polyester-amide, a synthetic biodegradable plastic supplied by Bayer, with mechanical properties similar to thermoplastic elastomers, and polyhydroxybutyrate (PHB), a maize-based material supplied by Biomer. The PHB material did not have hydroxyvalerate added and therefore represented the most brittle version available, with mechanical properties resembling polystyrene.

Tensile specimens (ISO 3167A) and tensile-impact specimens (ISO 8256-3) have been injection molded. The specimens are shown in **Figure 1**, together with an injection molded saucer which was used in another investigation. The mold for the tensile-impact specimens has a single-gated and a double-gated cavity. The double-gated specimen has a cold flow weld line in the center of the test region, at the end of the two melt flow paths. The tensile-impact specimens have 2 mm wall thickness and flow lengths 80 mm (single-gated) or 40 mm (double-gated) inside the mold cavities. The tensile specimens have 4 mm wall thickness and 175 mm flow length inside the mold cavities. The gates are narrow for the tensile-impact specimens and generous for the tensile specimens.

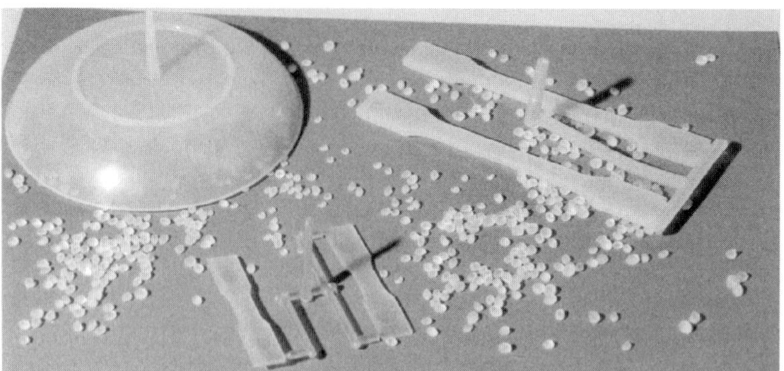

Figure 1. Injection Molded Test Specimens; Right: Tensile Specimens, Front: Tensile-Impact Specimens, Left: Saucer (diameter 155 mm, wall thicknessess 2.5 to 3.5 mm)

One of the challenges in molding both materials was the stickiness of the melts. This caused the plasticizing times at low pressures or low screw rotation rates to be rather long for molds with larger shot sizes such as the tensile specimens and the saucer. However high plasticizing pressures or screw rotation rates had to be avoided as well, to prevent thermal degradation. For the same reason, temperature control of the melt had to be tight just as for conventional heat-sensitive thermoplastics. On the other hand, too low melt temperatures made it difficult or impossible to properly fill the cavities of the tensile-impact specimens with their restrictive gates. For some of the double-gated PHB specimens the weld line was found to be away from the center at locations ranging from a few tenths of millimeters from the center to all the way near one of the gates. Restrictive gates, intended to induce shear-thinning in conventional thermoplastics, are not recommended by the manufacturers of most biodegradable plastics because of the risk of shear-induced overheating. Both materials are not particularly shear-thinning according to their suppliers. It should be noted that we may not yet have found the very best molding conditions for these new materials: one of our near-future plans is a further investigation of molding conditions.

An important molding variable for PHB is the cool rate (controlled by the mold coolant temperature), which determines the time the melt is within the crystallization temperature range. Other molding variables investigated for both materials were mold filling pressure and melt front speed during filling. These parameters influence shear heating of melt in the gate and orientation of macromolecules during melt flow.

Specimens were only tested when they met basic shop-floor quality criteria: mold cavities completely filled and surface quality (sink marks) at least satisfactory. Double-gated specimens were only tested if the weld line was within 1 mm from the center. Fortunately it was quite easy to obtain surfaces with hardly noticeable sink marks for both materials, even for the tensile specimens with 4 mm wall thicknesses: molding conditions were not critical in achieving this remarkable result.

Tensile testing was done at constant clamp separation speeds based on ISO 527 recommendations for the tensile specimens: 50 mm/min for polyester-amide (ductile) and 5 mm/min for PHB (brittle). The clamp separation speeds for the tensile-impact specimens were 5 mm/min (polyester-amide) or 0.5 mm/min (PHB).

4 Results and Discussion

Most PHB specimens fractured at the maximum nominal (engineering) stress. The polyester-amide specimens fractured at nominal strains well above the yield strain; for some of these specimens we found strain hardening prior to fracture. In this paper the nominal stresses and strains at break (PHB) or at yield (polyester-amide) are reported since they indicate an upper limit for practical applications. A full analysis of the tensile curves and the phenomena occurring during testing will be published in the near future.

Average values of tensile stress and strain at yield (polyester-amide) or at break (PHB) of all specimens injection molded with reference conditions are listed in Table 1. Each value is the average of 20 individual tests for the PHB tensile specimens and 10 individual tests for all other values.

Table 1. Tensile Test Results - Reference Specimens

	maximum tensile stress [MPa]	strain at maximum tensile stress [-]
polyester-amide:		
tensile specimens	18.0	1.71
tensile-impact specimens	17.3	1.39
tensile-impact specimens with weld line	14.3	0.70
PHB:		
tensile specimens	21.7	0.036
tensile-impact specimens	21.0	0.037
tensile-impact specimens with weld line	19.1	0.028

Table 2 shows the effects of injection molding conditions for polyester-amide specimens and Table 3 for PHB specimens. The tensile stresses and strains are displayed as a percentage of the reference values shown in Table 1. Each value in Table 2 and Table 3 is the average of 10 individual tests for the tensile specimens and 5 individual tests for the tensile-impact specimens. The standard deviations for the test results on polyester-amide specimens range from 0.5% to 15% of the average for the yield stresses and 3% to 25% of the average for the yield strains. For the PHB specimens the standard deviations are 6% to 9% of the average for the yield stresses and 5% to 25% of the average for the yield strains. The standard deviations for polyester-amide are high because strain hardening influenced the apparent yield point in the stress-strain curves for some specimens. For brittle thermoplastics such as PHB these standard deviations are normal.

Considering the standard deviations, it can be seen in **Table 2** that some molding variables have an important effect on yield strain of polyester-amide specimens. Increasing fill pressure causes lower yield strains: this is common for some semicrystalline plastics because higher pressure increases orientation in the melt in the flow direction, which is also the tensile direction, promoting crystallization transverse to this direction. For tensile specimens the yield strain is lowest at the intermediate of three ram speeds. Increasing ram speed delays freezing of the gate, allowing more hot melt to enter the mold cavity. This results in higher frozen-in stresses caused by higher temperature gradients. However for an even higher ram speed (reference condition) the addition of more hot melt lowers the cooling rate and thus increases crystallization.

Table 2. Effects of Molding Parameters on Tensile Properties - Polyester-Amide

	tensile stress at yield	tensile strain at yield
tensile specimens:		
higher injection nozzle temp (180°C → 195°C)	93 % of reference	102 % of reference
higher fill pressure (+25%)	96 % of reference	68 % of reference
lower plasticizing pressure (-40%) and higher fill pressure (+130%)	92 % of reference	44 % of reference
lower injection ram speed (-62%)	96 % of reference	29 % of reference
lower injection ram speed (-87%)	101 % of reference	102 % of reference
tensile-impact specimens:		
lower plasticizing pressure (-33%)	100 % of reference	94 % of reference
lower plasticizing pressure (-33%) and higher injection ram speed (+100%)	93 % of reference	83 % of reference
lower injection ram speed (-67%)	102 % of reference	95 % of reference
tensile-impact specimens with weld:		
lower plasticizing pressure (-33%)	98 % of reference	64 % of reference
lower plasticizing pressure (-33%) and higher injection ram speed (+100%)	94 % of reference	55 % of reference
lower injection ram speed (-67%)	95 % of reference	46 % of reference

The opposite dependence of yield strain on ram speed for polyester-amide is seen in **Table 2** for the thin-walled tensile-impact specimens with restrictive gates: the effect of higher frozen-in stresses caused by delayed freezing of the gate is noticed when ram speed is increased from the lowest to the intermediate value. At further increase of ram speed the importance of orientation in the melt in the flow direction and hence crystallization transverse to this direction appears to dominate. **Table 2** also shows that a lower injection pressure decreases relative weld line strength and a lower plasticizing pressure even more. This indicates the importance of pressure on the two opposing melt fronts when they meet to form the cold weld line.

For the PHB tensile specimens it is seen in Table 3 that increasing the cooling rate by decreasing the mold coolant temperature from 65°C to 25°C results in a more ductile material (higher tensile strain at break). This indicates a lower degree of crystallinity. For the tensile-impact specimens, both with and without weld line, a decrease of the coolant temperature from 75°C to 60°C has no noticeable effect on tensile properties. However, earlier ejection from the mold (shorter cool time) results in higher ductility of specimens without weld line, indicating a lower degree of crystallinity. Apparently a higher cooling rate is obtained by earlier exposure to ambient temperature. Earlier ejection decreases cycle time but some caution is necessary since it may also increase post-molding warpage in practice.

Table 3. Effects of Molding Parameters on Tensile Properties - PHB

	tensile stress at break	tensile strain at break
tensile specimens:		
lower mold coolant temperature (65°C → 25°C)	100 % of reference	118 % of reference
tensile-impact specimens:		
lower mold coolant temperature (75°C → 60°C)	97 % of reference	98 % of reference
lower mold coolant temperature (75°C → 60°C) and shorter cool time (-50%)	102 % of reference	114 % of reference
tensile-impact specimens with weld:		
lower mold coolant temperature (75°C → 60°C)	98 % of reference	94 % of reference
lower mold coolant temperature (75°C → 60°C) and shorter cool time (-50%)	96 % of reference	96 % of reference

5 Concluding Remarks

Although there are limitations to the process window because of narrow margins for plasticizing parameters, other molding variables such as mold filling pressure and melt front speed have a strong effect on tensile behavior, particularly on strain at yield or at break. Additional efforts have been planned to find optimum molding conditions for these and other biodegradable plastics. We aim to gain better understanding of the phenomena observed by experiments such as tensile tests at different strain rates and tensile-impact and flexural impact tests to study mechanical behavior in more detail, annealing to observe whether additional crystallization can be induced, and testing after longer storage times to check for possible degradation.

6 Acknowledgments

The authors are grateful to Dr U. Hänggi of Biomer, Germany and to Mr. Drießen and Dr Schlitte of Bayer, Germany for providing materials, data and processing guidelines. Mr. Fred den Elzen and Mr. Carel Hermans provided valuable experimental support at our laboratory.

Fabrication of SiC/SiC Composites by 2-step Reaction-Bonding Method

Eiji Tani, Kazuhisa Shobu, Kazushi Kishi, Seiki Umebayashi, and Shigeru Hanzawa*
(Kyushu National Industrial Research Institute, Tosu, Japan,
* NGK Insulators, Ltd., Nagoya, Japan)

1 Introduction

Reaction-bonding method is a low cost fabrication process for ceramic matrix composites because of a short reaction time and no shrinkage during sintering process.Liquid silicon infiltration into porous carbon matrix was generally used for silicon carbide matrix composites [1,2].However, it is difficult to control the reaction between fiber and molten silicon and the fiber needs expensive coating of BN [2].We had reported that unidirectional carbon fiber reinforced SiC composites could be fabricated at 1420°C by reaction-bonding with carbon from phenolic resin and silicon powder as matrix precursor, and there was neither residual silicon in the products nor reaction between carbon fiber and silicon [3].However, this reaction caused ~38% volume reduction, so it is difficult to get dense composites.

The reaction zone of molten Si on glassy carbon is much smaller than that on graphite [4].Therefore, the dense amorphous carbon layer from phenolic resin could protect fiber from the attack of molten Si.The silicon for the liquid infiltration method comes from outside of the preform, then the reaction between amorphous carbon in the preform and molten Si increases the volume of ~ 56%.Therefore, this preform needs appropriate porosity to complete this reaction.Porous SiC matrix is easily obtained by our reaction-bonding method.So we combined these reaction-bonding method and liquid silicon infiltration method to obtain dense SiC/SiC composites.

The purpose of the present study was to investigate a new procedure for SiC/SiC composites by this 2-step reaction bonding method without BN coating, and the effect of the amorphous carbon coating on SiC fabric on the protection from molten Si.

2 Experimental

The SiC (Si-Zr-C-O) fiber of 8-harness satin woven fabric was ZMI (Ube Industries Ltd., Japan).The bulk density and average diameter of the fiber were 2.48 Mg/m^3 and about 11 μm.The average tensile strength, elastic modulus and elongation at failure of the fiber were 3.0 GPa, 200 GPa and 1.7 %, respectively.After the sizing of SiC fabric was removed through washing with hot water at 80°C, the fabric was dipped in the phenolic resin (Novolac type; Hitachi Chemical Co., Tokyo, Japan) dissolved in ethanol, and dried at 70°C for 12 h.Some SiC fabric dipped in phenolic resin or carbon fiber paper (E-204; Kureha Co., Ltd., Japan) were dipped in the mixed slurry composed of silicon powder (<74 μm in size; Sumitomo Sitix Co., Amagasaki, Japan), phenolic resin (the molar ratio of the composition of the slurry was

Si/C=0.8 for SiC fabric, 1.5 for carbon fiber paper), and ethanol.The prepregs were cut to size, stacked in a steel die, treated under a pressure of about 25 MPa at 135°C for 20 min.The SiC fabric dipped in phenolic resin and the mixed slurry (Si/C=0.8) were stacked together (Sample; 0.8S).However, two sheets of the SiC fabric dipped in phenolic resin and two sheets of carbon fiber paper dipped in the mixed slurry of Si/C=1.5 composition were stacked alternately and the carbon fiber paper located outside of the preform (Sample; 1.5SC).

The samples were produced by pyrolyzing the die-pressed sample at 1000°C for 1 h in Ar and impregnating it with phenolic resin dissolved in ethanol, then repeating this pyrolysis/impregnation process, and finally reaction-sintering and liquid silicon infiltrating the compound at 1450°C for 1 h in vacuum simultaneously.

Flexural strength was measured on as-fired samples that were ~ 1.8 mm thick, 12 mm wide and 45 mm long.Flexural strength measurements were performed at room temperature using a three-point loading device with a span of 30 mm and a crosshead speed of 0.5 mm/min.The elastic modulus of the composites was calculated using beam theory.

3 Results and Discussion

The fracture behavior of the 0.8S-composites as a function of weight ratio of phenolic resin to ethanol (Phe/Et) for the first dipping of SiC fabric was shown in Fig. 1.The fracture behavior of the composites with Phe/Et= 0 (without phenolic resin dipping) showed a brittle manner.However, the nonlinear fracture behavior of the composites after a linear stress-stress response increased with Phe/Et ratio.This behavior clearly means that the carbon coating from phenolic resin could protect SiC fabric from the attack of molten Si.The bulk density of the composites with Phe/Et= 0 was as high as 2.69 Mg/m^3, however, that of the other composites was 2.1 ~ 2.2 Mg/m^3.These results also supported that carbon coating from phenolic resin could protect SiC fabric.Because the formation of SiC (ρ= ~ 3.2 Mg/m^3) with the reaction between Si (ρ= ~ 2.33 Mg/m^3) and amorphous carbon (ρ= ~ 1.5 Mg/m^3) increases the bulk density of the composites.The open porosity of the composites with Phe/Et=0 was 0.5 %, however, that of the composites with Phe/Et= 20 and 40 wt% was as high as 11.6 and 7.1 %.And the appearance of these composites also showed that the liquid Si infiltration seemed to be inhomogeneous. These results probably means that the thickness of the porous reaction-bonded SiC layer, the channel for silicon infiltration, might to be too narrow.

Table 1.Some properties of the 0.8S-composites.

Phe/Et (wt%)	Bulk density (Mg/m^3)	Open porosity (%)	Flexural strength (MPa)	Elastic modulus (GPa)
0	2.69	0.5	185	215
20	2.14	11.6	84	81
40	2.22	7.1	118	145
60	2.25	2.1	131	113
80	2.14	1.4	106	88

Carbon fiber paper dipped in the mixed slurry of Si/C=1.5 was used to control the thickness of porous reaction-bonded SiC layer.The fracture behavior of the 1.5SC composites was shown in Fig. 2-A.The fracture behavior of this sample also showed nonlinear behavior after a linear stress-stress response.Open porosity and bulk density of this sample were 0.5 % and 2.25 Mg/m^3.Flexural strength and elastic modulus of this sample were 150 MPa and 140 GPa.The liquid Si infiltration seemed to be homogeneous from the appearance of this sample.These results imply that the liquid silicon infiltration could be controlled with the thickness of porous reaction-bonded SiC layer.

The fracture behaviors of 1.5SC composites oxidized at 1000°C for 1h in air were also shown similar behavior of the sample before oxidation (Fig 2- B,C).One sample (in Fig. 2-C) was cut at both ends before oxidation for neglecting Si or SiC coating at the outside edge.The flexural strength and elastic modulus of the oxidized 1.5SC sample were 125, 112 MPa and 118, 107 GPa for as-infiltrated and cut both ends samples, respectively.A little degradation of the flexural strength and elastic modulus of the oxidized 1.5SC samples was observed.However, 110 MPa in flexural strength and 110 GPa in elastic modulus were enough high values.Weight loss of the oxidized 1.5SC sample was 1.3 and 2.8 % for as-infiltrated and cut both ends samples.Weight loss and the increase of open porosity (~ 12%; for as-infiltrated and cut both ends samples) of the oxidized 1.5SC samples seemed to result from the oxidation of carbon coating on SiC fabric.

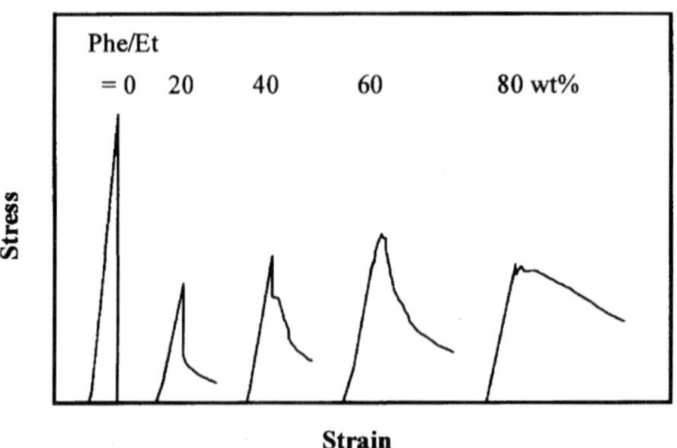

Figure 1. The fracture behavior of the 0.8S-composites as a function of weight ratio of phenolic resin to ethanol (Phe/Et) for the first dipping of SiC fabric.

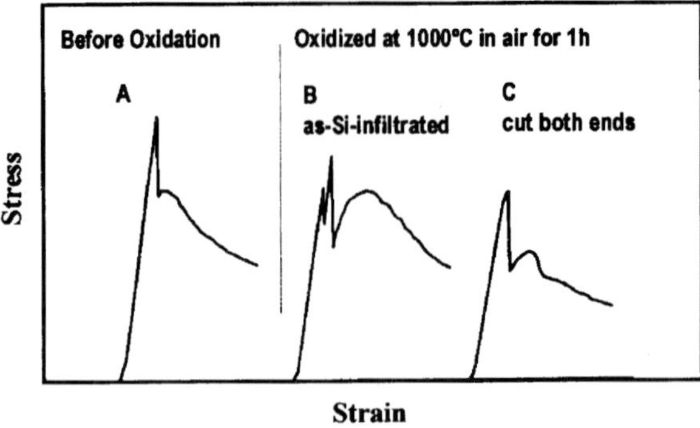

Figure 2. The fracture behavior of the 1.5SC-composites (A, B: as-infiltrated sample, C: the sample was cut at both ends before oxidation).

4 Conclusions

8-harness satin woven SiC fiber fabric (Tyranno, ZMI) reinforced SiC composites were fabricated by 2-step reaction-bonding method.The bulk density and open porosity of the composites were ~2.2 Mg/m^3 and < 3%, respectively.The fracture behavior was linear elastic up to the first blip at ~140 MPa, and showed continuous stress decreased.The carbon coating from phenolic resin could protect SiC fabric from the attack of molten Si.The flexural strength and elastic modulus of this composites were ~140 MPa and ~150GPa, respectively.After oxidation at 1000°C for 1h in air, the weight loss of the specimen was < 3 %, and only a little degradation of mechanical properties at room temperature after oxidation was observed.

5 References

1. K.Luthra, R.Singh, M.Brun, High Temperature Ceramic Matrix Composites, (Ed. R.Naslain, J.Lamon, D. Doumeingts) Woodhead Publishing, Abington Cambridge, **1993**, p. 429.
2. T. Kameda, A. Sayano, N. Amiji, H. Ichikawa, H. Hamada, A. Fujita, T.Uozumi, Ceram. Eng. Sci. Proc., **1997**, 18 [3] , p.419.
3. E.Tani, K.Shobu, T.Watanabe, J. Ceram. Soc. Japan, **1992**, 100, 596-98.
4. T.J.Whalen, A.T. Anderson, J.Am.Ceram.Soc., **1975**, 58, 396-99.

CONFORMATION REGROUPINGS IN DEFORMATION PROCESS AT VARIOUS TEMPERATURES OF THE ORIENTED FLEXIBLE-CHAIN LINEAR CRYSTALLINE POLYMER

Ulmas Gafurov

Institute of Nuclear Physics, Ulugbek,Tashkent 702132, Uzbekistan
Tel: 99871-260-6108, Fax: 99871-264-2590, e-mail: ulmas@suninp.tashkent.su

Abstract: The conformation regroupings of macromolecular chains at various stretching deformation temperatures of linear high oriented polyethylene samples was considered.Measurement of the rotary isomers relative concentration was carried out by IR Spectroscopy method.

The conformation regroupings happen in the highoriented polymer after definite value of elastic stretching of the sample irrespective of deformation temperature for creating the steric conditions of cooperative transition of the coiled-isomers in the straighten trans-isomeric conformation state. The polymer stretching in elevated temperature conditions happen with considerable increase of the elastic deformation value and of the conformation regroupings of amorphous molecular sections. Slippage of macromolecular chain with already straightening amorphous section is reason that deformation at more high temperature happens with more slow decreasing of the coiled isomer concentration.

The contribution in amorphous region elongation of the chain stretching in polymer crystallite was defined using of the Frenkel-Kontorova's crowdion model.

Keywords: polyethylene, rotary isomers, deformation, molecular regroupings.

1. Introdution

According to conformational structure of macromolecules of flexible chain polymers [1-3] it should to be expected that stretched deformation of the flexible chain crystallinne polymers will be accompanied by redistrubution contents of different rotary isomers and by conformational strethening macromolecar chain. Orientational drawing of the linear polyetylene is accompanied by decreasing of the coiled isomers contents and increasing of the straightening trans-isomer sec-tion of molecular chain mainly in an amorphous region of polymer [4]. By the IR Spectroscopy method on example of linear polyethylene it was shown by us also the conformation mechanism of elastic deformation of oriented flexible chain crystalline polymer [5]. The molecular reorga-nizations in the polymer deformation depends on ordering and the conformation state of the amorphous parts of the passage macromolecules. Using different conformation models of kink-isomers of the amorphous sections we have calculated the number of rotary-isomeric transforma-tion corresponding to given lengthening. It has been shown that stretching is linearly dependent on decreasing of number of coiled isomers.

In this paper we consider the character of the conformation reorganizations at various deformati-on temperatures and the molecular regroupings at facilitating conditions of the passage macro-molecules slippage in deformation process of the linear polyetylene. It is interesting for farthest understanding of molecular mechanics of the flexible-chain polymers deformation and to establishing connection between structural parametrs and the molecular ones in deformation processes of yhe flexible-chain crystalline polymers.

The experimental methodics were described in [4,5]. Fibers and films o f highoriented linear polyetylene samples were used.

2. Results and Discussion

The IR spectrum in region of displaing of wagging vibrational modes of CH_2 grouppings in molecular sections ith different coiled isomeric states and its change in stretching deformation of the highoriented linear polyetylene at temperature 80 C is shown in fig. 1.

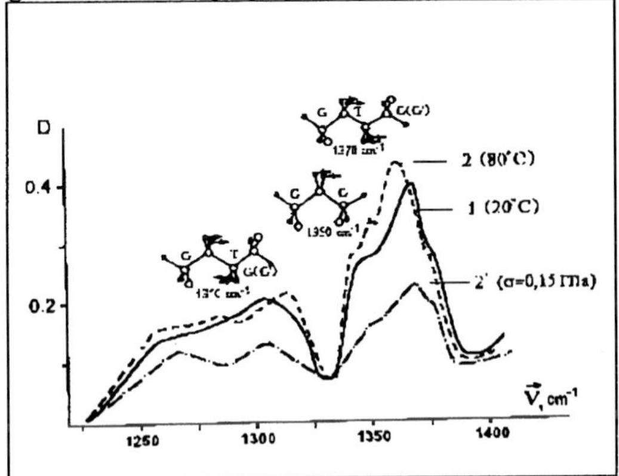

Fig. 1. IR spectrum of highoriented linear polyethylene in absorption region 1200-1400 cm^{-1} [1]

1 --- unloaded initial sample;
2 -- at temperature 80 C;
2' -- deformed sample at 80 C.
D - the optical density.

Above on fig is presented molecular section in different coiled rotary isomeric states.

Change of relative contents of coiled rotary isomers with the polymer stretching is shown in fig.2 and fig.3.

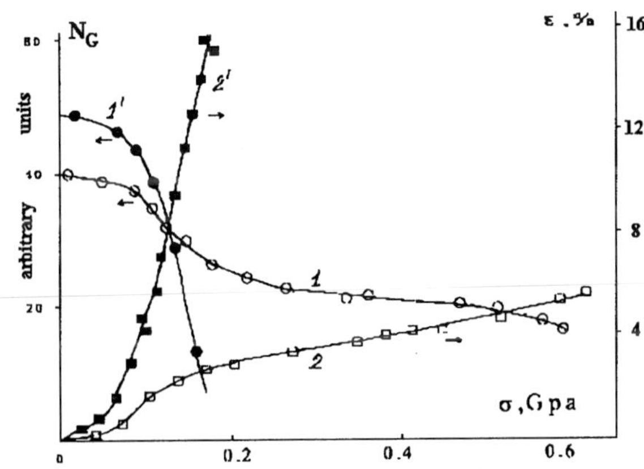

Fig. 2 Change of IR bands intensity proporcially to concentration of the coiled

rotation isomers (1,1') and elastic deformation value (2,2') vs stretching load in various deformation temperature conditions:

1,2 - 20 C, 1',2' - 80 C.

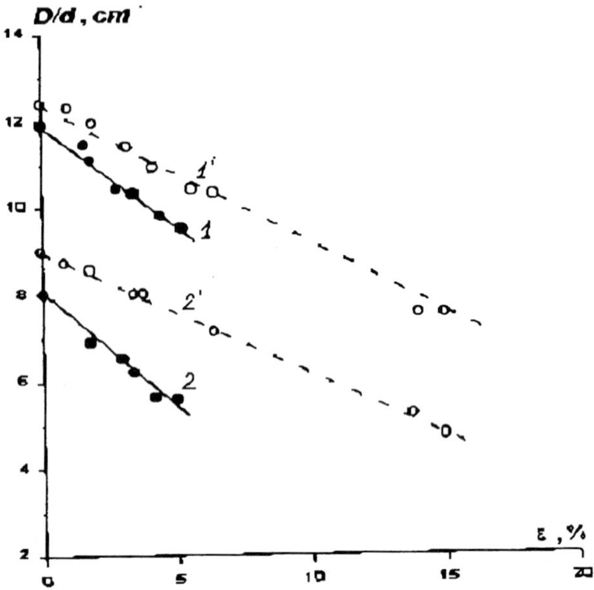

Fig. 3. Change of the absorption band intensities at 1370 cm^{-1} (1,1') and at 1350 cm^{-1} (correspondly contents of molecular sections in the coiled conformation states) elastic deformation value.
The deformation temperature: 1,2 - 20 C, 1',2' - 80 C.

The data of the measurements shows that in oriented linear polyethylene conformation reorganization happens with raise of stretching load value after definite value of elastic stretching of sample irrespective of temperature of deformation. This preliminary lengthening creates the steric conditions for cooperative transition of the kink-isomers in the straightening trans-isomeric states. Because of steric conditions (a chain is fixed inside crystallits) and discreet character for rotary-isomeric transformations some preliminary elastic lengthening of passing sections is demanded in order to realize their of conformation reogroupings. As a results of this deformation such highoriented polymer sample acquires the accelerated character. Limits of loads of deforma-tion within which conformation reogroupings takes place characterize dispersions on lengths and conformation structure of passing amorphous sections. The more homogeneous and ordering is molecular structure of a polymer the more accelerating character has conformation reconstructi-ons of the passing chains and correspondingly deformation in tension of oriented crystalline polymer.

Afterwards the straightening of relatively short passage amorphous parts of macromolecules for further conformation reorganizations more long amorphous sections with kink-isomers additional lengthening of the amorphous regions is necessary. The lengthening is reached

by increase of stretching load or with slippage and rupture of short straight stressed macromolecular sections.
Under constant external load further conformation reconstructions could take place only if rupture or slippage of straightened passing molecular sections occurs.

With the weakening of the intermolecular interactions and increase of molecular mobility with temperature raise, the polymer chains slippage is facilitated. The slippage of macromolecules with stretching amorphous section promotes further reorganizations of more long passage amorphous sections with increase of value of elastic deformation of polymer sample.

Slippage of macromolecular chain with already straightening amorphous section is reason that deformation at more high temperature happens with more slow decreasing of coiled isomers concentration.
We notice, that especially at relatively large loads the contribution in amorphous region elonga-tion the chain stretching in polymer crystallite follows to take account of. This contribution (L_{cr} to tension of macromolecules for semiinfinite polymer crystallite has been calculated with use of the Frenkel-Kontorova's crowding model [7 -10].
According to these:

$$\Delta L_{cr} = \frac{a}{\pi} \arcsin \frac{f_a}{\sqrt{2k\Delta U_0}}$$

where **a** - is period of crystal lattice, **k** - is elastic constant of macromolecular link, **(U_0** - height of the intermolecular potential barrier, $\mathbf{f_a}$ - local load on an amorphous molecular section of passage macromolecar chain.

3. Conclusion

The measurements show that the conformation reogroupings in highoriented polymere happen after definite value of elastic stretching of sample irrespective of deformation temperature. This preliminary lengthening creates the steric conditions for cooperative transition of the kink-isomers in the straighten trans-isomeric states. As a results of this deformation such highoriented polymer sample acquires the accelerated character.

The polymer stretching at elevated temperatures happens with considerable increases of the conformation reogroupings of the molecular sections.
Slippage and ruptures of passage macromolecules lead to redistribution of local loads,conformati-on reogroupings and lengthening the amorphous sections of other passing chains. Slippage of macromolecular chain with already straigh-tened amorphous section is reason that deformation at more high temperature happens with more slow decreasing of coiled isomer concentration.

The contribution in amorphous region elongation the chain tension in polymer crystallite to take account in polymer deformation at lower temperature (at rela-tively large local loads on molecular chain).

References

1. M.V. Volkenshteyn, Configuration statistic of polymer chains, AS USSR, Moskow, 1959, in Russian.
2. T.M. Birshteyn, O.B.Pticin, Conformation of macromolecules, Nauka, Moskow, 1964.
3. P.J. Flory, Statistical mechanivs of chain molecukes, Interscience Publishers , a division of John Wiley & Sons, New York-London-Sydney-Toronto, 1969.
4. U.G.Gafurov,Visokomolekulyarnie ssoedineniya, 1972, A14, 873.
5. U.G.Gafurov, I.I.Novak, Mechanika polimerov, 1973, 4, 584.
6. R.J.Snyder, J Chem Phys, 1967), 47, 1316.
7. A.D.Chevichelov, Visokomolekulyarnie ssoedineniya, 1966, A8, 49.
8. H.H. Kausch, D.Langbein, J. Polym. Sci.(Polym. Phys. Ed.), 1979, 11, 1201.
9. H.H. Kausch and C.J.G.Plummer, Polymer, 1994, 35, 18, 3848.
10.U.G.Gafurov, Uzbek. fizicheskiy jurnal, 1994, 1.

PropaFoam® and FMP® New Engineered PP Foams for Automotive Applications

Jan Zwaan
Sentinel Polyolefins LLC - US

1 Introduction

1.1 Sentinel Polyolefin Foam Range

Sentinel offers the largest and widest range of extruded polyolefin foams versus any competitor worldwide. Using various dedicated manufacturing processes, the foams are available in PE and PP based formulations, both physically and chemically blown. In addition, technology has allowed Sentinel to provide crosslinked foams using a variety of polymers including PEs, EVAs, PPs, EPDM-type rubbers and other specialized metallocene based elastomeric compounds.

The product range is available in continuous rolls, sheets and planks to fit the needs of the customer. Compared to the competition, this is the broadest product breadth available and provides the customer a single source for foam products for all types of applications.

This paper will present Sentinel's family of two different polypropylene foam product lines. Each of the product lines – PropaFoam and FMP – possess various features and provide differing benefits to any end application.

1.2 PropaFoam®

Physically blown non-crosslinked PP foam made from special High Melt Strength (HMS) Polypropylene blended with Ethylene-Propylene Impact Copolymers. Available in continuous rolls and planks in densities ranging from 25 kg/m3 (1.7 pcf) to 150 (9 pcf).

1.3 FMP®

Flexible Metallocene-modified Polypropylene Foam is produced in the proprietary silane-grafted crosslinking process using a conventional polypropylene blended with elastomeric metallocene. These crosslinked products are available in continuous rolls possessing densities between 32 kg/m3 (2 pcf) through 128 kg/m3 (8 pcf).

2 Foams

Thermoplastic foams are currently used in the automotive industry. All of the competitive materials utilize basic polymers in their manufacture. The most common polymers used for foaming are Polystyrene (PS), Polyvinyl Chloride (PVC) and Polyethylene (PE). Recently, Polypropylene (PP) has been requested as a polymer for foam due to its thermal stability properties. All of the above polymers produced in conventional production methods will generate foams with good properties, but they will also possess some drawbacks and shortcomings.

Sentinel Polyolefins, with its unique grafting technology and processing, can gain the positive attributes from all polymers used. For example, FMP combines the benefits of PP and metallocene catalyst resins. The PP provides the thermal stability requirements while the metallocene provides a foam that has increased softness, greater tensile strength and excellent thermoforming capabilities versus conventional PP products. This material combines the thermal stability advantages of PP with the fabrication advantages of LDPE foam.

Polypropylene as a polymer contributes a wide temperature range, higher dimensional stability, good thermoformability characteristics and minimal environmental concern compared with other polymers.

2.1 Thermoformable Polyolefin Foams

The main requirement for the ability to thermoform a cellular product is usually created during the crosslinking phase of the production process. Conventional crosslinking methods consist of two means – physical or chemical. Physical means would require irradiation i.e. the bombardment of an e-beam on the extruded sheet. Chemical crosslinking would utilize peroxide or other crosslinking agent in the process. Sentinel Polyolefins is unique, as it has commercialized the crosslinking of foams through silane grafting utilizing metallocene catalyzed polyethylene.

Without crosslinking the polymer chains of a product, the polyolefin foam would not be thermoformable. Through the introduction, development and successful use of a new generation of foams based on High Melt Strength Polypropylene, Sentinel Polyolefins can provide a PP foam called PropaFoam that possesses the ability to thermoform without the need of crosslinking.

There are two foam alternatives as presented here today are:

1. *Silane grafted* crosslinked polypropylene foams using a conventional *homo-PP* blended with elastomeric metallocene.
2. *Non-crosslinked* polypropylene foams based on the utilization of *high melt strength PP* resins.

2.2 Comparison of Polyolefin Foams

Sentinel Polyolefins LLC possesses a family of polyolefin foam products each utilizing polyolefin polymers in their manufacture. Although there is confusion about the difference of these product lines and achieving a proper material selection, various properties can be tailored to design foam "fit for purpose".

If we compare the two basic product offerings on five key criteria, you see the difference in properties between the materials.

The *dimensional structure* is principally different between non-crosslinked and crosslinked foam. Non-crosslinked foam will have a two-dimensional string of linear polymers so the extrusion direction will always be easy to determine. These structures create differences in the mechanical properties in the machine direction compared to the cross machine direction. Crosslinked foams with their three-dimensional network and infrastructures have similar properties in both directions. In addition, their strength properties are improved versus non-crosslinked foam at similar densities. Crosslinked foams are aesthetically more pleasing due to the fine cell structure and smoother skin.

Cell structure will also vary between crosslinked and non-crosslinked foams. The non-crosslinked products possess larger, courser and more irregular cells, while crosslinked foams exhibit a finer, more uniform cell structure.

On high *temperature resistance*, Sentinel's products based on PP content have a temperature resistance up to 140° C (280 F). The conventional crosslinked products utilizing LDPE will possess resistance up to 93° C (200 F).

An important aspect is *recycling*. Sentinel's non-crosslinked PP foam is a thermoplastic and their crosslinked foams, both PP and polyolefin based, are partial thermosets, partial thermoplastics. Various recycling alternatives do exist for both of these products.One option is to granulate any waste and use in the extrusion process again. The second option is to granulate any waste, densify and use these by-products for other end applications.

Thermoformability is possible with both Sentinel's PropaFoam and FMP. The HMS based PP foams, due to the nature of the polymer, have a smaller operating window to achieve and complete this molding. Crosslinked foams are easy to fabricate and provide a cost-effective means to generating a molded end product. The preferred method to thermoforming crosslinked foams is to pre-heat the blank before submitting to the mold.

2.3 Polypropylene Foam

PP foam in itself is not a novelty. Polypropylene has been used in various industries for some time. Foamed PP film can be extruded with chemical foaming agents in densities from 500-700 kg/m^3 and a thickness range from 0.2 to 1.5 mm. The production takes place on conventional cast film lines. Special production technologies and/or special raw materials can only obtain lower densities.

For foamed PP in densities below 200 kg/m3, a High Melt Strength (HMS) PP is required for the non-crosslinked PP foam process. For the crosslinked foam process, the PP homopolymer must be blended with PE's to allow for the processing of these polymers.

PP foam sheet is available in densities from 200-500 kg/m^3 in thickness' ranging from 1 to 3 mm. The physical foaming process can make these products on direct gassing lines. In order to achieve these densities and a good quality product, Sentinel formulates a blend of elastomeric metallocene resins with the conventional PP.

The two PP-based product lines presented in this paper are all low in density and share:

- An excellent price/performance ratio
- A tremendous flexibility in product formulations to meet performance parameters
- Excellent mechanical physical properties
- Ease of fabrication and conversion
- Rigidity at lowest foamed density possible
- Ability to compete in a wider spectrum against other materials which was not possible before.

2.4 Extruded PP Foam Range

PropaFoam® is a product line based on a high percentage of High Melt Strength PP blended with some ethylene-propylene impact modifiers.

In a physical foaming process, using hydrocarbon gasses, rolls or planks are extruded in densities between 25 and 150 kg/m^3. This process allows Sentinel to tailor the required dimensions and the formulation to meet the requirements.

FMP, Flexible Metallocene-modified Polypropylene, consists of a blend of elastomeric metallocene and conventional polypropylene.FMP offers superior linear shrinkage properties while maintaining the feel of a conventional TPE at low foamed densities. The foam exhibits the following properties:

- Shrinkage of 2.5% at 93 C for 50 hours
- Compression Deflection at 25% of 3 – 6 psi or 20 – 41 kPa
- Enhanced Compression Molding and Vacuum Forming
- Enhanced Lamination to Various Substrates

2.5 Features & Benefits

Tables 1, The main features of these foams provide our customers several benefits.

Feature	Benefit
Low density	Weight reduction
	Cost reduction
Non-crosslinked	Recycle post process scrap
	Recycle used product
Crosslinked	Thermoformability
	Superior physical properties at low foamed densities
Closed cell	Good temperature insulation
	Good sound absorption
	Low water absorption
Polypropylene	Good chemical resistance
	Excellent water barrier
Foam	Increase stiffness at low weight
	Heat and sound insulation
Wide service temperature	Good cold impact
	High temperature resistance
Thermoformable	Convert on vacuum forming lines
	Convert via compression molding

2.6 Converting

All our foams are semi-finished products. The next step in the chain is further processing into parts. The crosslinked and non-crosslinked PP foams can be further fabricated in a range of processes depending on the requirements outlined by the end application.

Thermoforming was already discussed: the PP foam sheet can be converted on the same lines, with basically the same tools - except for calculating less shrinkage - and will be converted at minimum the same output. Simple to medium shapes can be converted directly, for more complex shapes plus-assist systems are used. The PP foam planks, which are used as inner cores in a sandwich construction, are made with the pressmolding technique. A special feature is that in the mold comparable materials can be laminated together, skipping the pre-laminating step.

Furthermore the PP foams can be converted with the traditional mechanical converting means or laminated to PO based or other substrates using various bonding techniques.

3 Automotive Market

The main motivation for the development of the PP foam range is the emerging automotive market where plastic consumption is growing to over 100 kg per car. This percentage has increased steadily over the past years and is expected to continue. The use of PP as the preferred polymer is also driving this demand. In addition, the usage of foams over films is growing as the benefits provided by foams are increasing.

Below are a few trends what we, as a material supplier in the automotive industry, experience daily. In parts design, the following issues are driving the further use of PP in auto parts:

- Desire to construct an all PP part
- Demand for a high performance car which last 10 years
- Low material emissions to reduce window fogging
- Improved performance without cost increase
- Introduction of energy and sound management systems
- Added luxury to achieve the 'soft touch' feel
- Weight reduction

In material selection:

- Plastics are replacing metals
- Foams are replacing films
- More emphasis on environmental selection criteria

In the supply chain:

- More of a systems approach; system integrator, component producer, semi-
- finished material manufacturer and raw material producer working closely
- together
- More responsibility down the chain to the suppliers

Overall, we see higher performances requested combined with a market demand for lower cost per unit.

Sentinel has developed various automotive applications together with Tier One Suppliers for PP foams in Molded Interior Trim and Technical Components such as door water shields, acoustic absorbers, sun visors, headliners, package trays, TPO lining for various interior trim panels, boot- and trunk lining, spare wheel covers and sandwich cores.

For these existing applications, the new foams give our automotive customers several additional benefits over materials used before.

4 Automotive Application Examples

4.1 PropaFoam® in Door Water Shields

A well known part: a plastic water- and dust-barrier between the metal frame and the glass window. This part is also used increasingly to reduce noise within the car body. The primary requirement is cost reduction. Furthermore, there are several requirements for increased material performance and improved processing.

Traditionally, various films such as PP, PVC and sometimes ABS were used. Crosslinked polyethylene foams were also utilized as a substrate layer. Comparing these traditional alternatives with the novelty non-crosslinked PP foam on relevant criteria, we can conclude the following.

The use of PropaFoam PP roll stock reduces the system cost compared to crosslinked PE foams. Although all materials give the required moisture barrier; acoustics and sound insulation are improved by introducing foams.

PropaFoam has an increased tear resistance and a high temperature resistance, compared to the PE foam. In fabrication, thermoforming for all products is acceptable, but die cutting and the bond of the necessary adhesives are better using PropaFoam. In addition, recycling the PP foam is favorable.

The introduction of PropaFoam as an alternative gained this product line a substantial share of this market. Only two years after its introduction, this material can be found in several high volume production models and will be introduced in several more shortly.

4.2 PropaFoam® as GMT Sandwich Core

Another example is PropaFoam PP plank used as in sandwich core in combination with Glass Matt Thermoplasts (GMT). Although this is still experimental, testing is in process for the use of these materials in load bearing systems as load floors and high impact absorbing systems such as engine covers. Preliminary feedback is very promising.

The PP plank fulfills the requirement for a PP monomaterial and reduces the weight of a part over a third compared to a full GMT part. Processed in a compression molding process, the PP plank gives an excellent bond with the GMT – PP, adding to the flexural strength.

We see good possibilities for these non-crosslinked PP foam plank to replace planks of various other plastics as well as PP Honeycomb materials.

4.3 FMP® Soft Touch in Interior Trim

This application requires a cellular material to provide maximum flexibility with minimum linear shrinkage. Furthermore, ease of converting in laminating to a TPO skin and thermoforming the skin/foam laminate in parts as door panels or instrument panels is important.

The current trend is to the need and use for an all-PP composite. This part, such as a dashboard, would be constructed from Crosslinked PP foams topped with a TPO skin. This TPO skin needs to be backed by a crosslinked PP foam to provide improved strength properties, increased thermal stability and the 'soft touch' feel.

FMP, Flexible Metallocene Modified Polypropylene, is a silane-grafted crosslinked PP foam that can provide the required soft touch and heat properties as required. In addition, the thickness of the TPO may be reduced without deteriorating the grain during the forming process. In addition, FMP can be laminated directly after extruding the TPO skin in a one-step process.

Sentinel Polyolefins' FMP has successfully replaced conventional elastomers, PVC blends and elastomeric urethanes as the soft touch substrate in this application. FMP meets, and even exceeds, all of the requirements for this part and provides a foam material with a lower overall cost. An additional benefit is the reduction of window fogging imparted by the material emissions. FMP utilizes no plasticizers in its production.

Major suppliers in Europe and North America are currently evaluating and prototyping this new material in all PP interior trim development.

5 Conclusion

These three examples are only a few of the applications currently using our technology. We hope we demonstrated to you that Sentinel Polyolefins' PropaFoam® and FMP® are new alternatives in the automotive foam market.

Intelligent use of these novelty foams can reduce the overall system cost and further improve the performance of these parts. However, these materials require a joint development within the whole chain including supplier, converter, and end customer.

Table 2, Sentinel's Product Line Offerings

Supplier	Physical Blown PE Plank	Physical Blown PP Plank	XL PE Plank	Physical Blown PE Sheet	Physical Blown PP Sheet	XL PE Sheet	XL PP Sheet	Metallocene Elastomers	Rubber Foams
Sentinel Polyolefins LLC	•	•	•	•	•	•	•	•	•
Alveo			•			•	•		
Dow Chemical	•			•					
Fagerdala				•	•				
HT Troplast		•				•	•		
Sealed Air Corp.				•					
Toray Plastics						•	•		
Zotefoams			•						

Micro-CT for Polymers and Composite Materials

Alexander Sassov and Erik Buelens
SkyScan, Aartswlaar, Belgium

1 Introduction

Non-destructive information from the internal structure of polymer and composite materials in natural or adjustable conditions can be obtained by transmission X-ray microscopy. Combination of X-ray microscopy technique with tomographical reconstruction allows getting three-dimensional information about the internal microstructure nondestructively [1]. The spatial resolution and detectability in the micron range of sizes can be improved by using of phase contrast.

2 The Equipment

A desktop X-ray microscope-microtomograph "SkyScan-1072" has been used for nondestructive 3-dimensional reconstruction of the internal microstructure of polymers and composites. Block-diagram of equipment in shown on Fig.1.

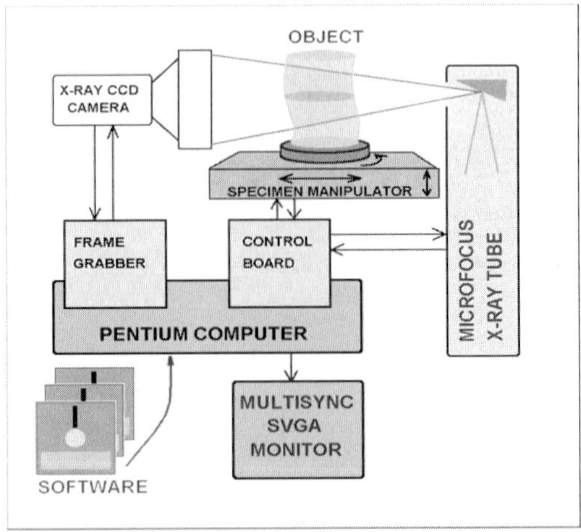

Figure 1. Block-diagram of the X-ray microtomograph

It consist of a microfocus sealed X-ray source 20-80kV / 100uA with <8um spot size, a precision object manipulator for object rotation and movement inside the conical x-ray beam, an X-ray CCD-camera with field of view 25x25mm and an internal Dual Pentium III computer for system control and 3-dimensional reconstruction. System operated under

Windows NT with software as a multiprocessor 32-bit application. Software package include possibilities for X-ray radiography, stereomicroscopy, three-dimensional microtomographical reconstruction, image correction, analysis and realistic 3D-visualization. The X-ray magnification is continuously variable in the range between 10 and 120. Reconstruction time is near 20 sec per cross section. A single cross section as well as full 3D internal structure can be reconstructed after one complete object scan. During investigation the specimen displaced in natural conditions with no need for any preparation or coating. Powerful image processing and analysis possibilities integrated into the software package allows calculating the numerical characteristics of object's internal microstructure non-destructively. Visualization procedures contain of cross section by cross section displaying and realistic visualization of complete 3D-object onto the screen with possibilities for rotation, movements, zoom, recalculation of new cross section in any alternative direction and flight around or through the object's microstructure.

3 Phase Contrast

Conventional X-ray radiography and tomography are based on X-ray attenuation inside an object. However, for light materials and hard X-rays, the attenuation can be small that it puts strong limits on the attainable resolution. Fortunately the attenuation is not only one physical phenomenon for X-ray transmission imaging. Another kind of information can be obtained by phase contrast. The phase contrast, rather than attenuation contrast, since the phase change, due to changes in index of refraction, can be up to 1000 times larger than the change in amplitude for visualization of light objects. In the same time the phase contrast have another geometrical behavior which allows using this kind of information for improvement in spatial resolution. However, phase contrast techniques require the disposal of monochromatic X-ray sources, such as synchrotrons, combined with special optics, such as double crystal monochromators and interferometers [2].

Recently it has been shown [3] that one can also obtain phase contrast by using a polychromatic X-ray source provided the source size and detector resolution are small enough to maintain sufficient spatial coherence. This opens perspectives for using phase contrast information in a microfocus tomographic system. Because the phase and absorption contrast presents in microfocus X-ray systems simultaneously, the optimization of geometry can create the way for performance improvements by right combination of both of them.

Absorption contrast in any real system contains some unsharpeness as a result of finite spot size in the source. By another hand the phase contrast can be described as a small-angle reflection from the object surface which improve the sharpness of the image. Specific x-ray geometry has been found in which the blur produced by spot size in the source compensated by the phase contrast signal. It allows inside the system with 8 microns spot in the source visualize the objects with several microns in size. Improvement in visualization and reconstruction for the objects of 2-5 microns in size very important for the investigation of plastic foams and fiber-foam composite with the wall thickness in this size range.

4 Application Examples

Polymers and polymer-based composite materials are the important application areas for microtomography. This technique allows reconstructing the internal structure of the objects nondestructively and investigates reaction to external influences. The system allows to obtain transmission image through the composite materials and to reconstruct nondestructively any cross section or complete internal 3D-structure. The sample can be loaded in situ to investigate the reaction of matrix and enforcing structures to external strain. Absorption and transmission by liquids can be shown directly in three dimensions. Small detail up to 3um in size can be visualized inside the specimen by phase contrast enhancement. Fig.2-3 shows possibilities of microtomography reconstruction for foam/fibers plastic composite material. One of transmission x-ray image (microradiography) through the object shown in the Fig.2a. From the hundreds of transmission images obtained during object rotation an internal object microstructure can be reconstructed cross section by cross section. One of reconstructed cross-sections through the central part of specimen has shown in Fig.2b.

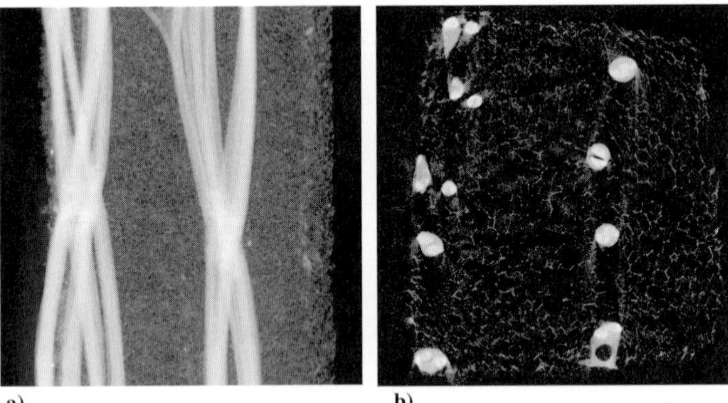

a) **b)**

Figure 2. Microtomographical reconstruction of the foam/fibers plastic composite: a – transmission X-ray image (radiography), b –reconstructed cross-section.

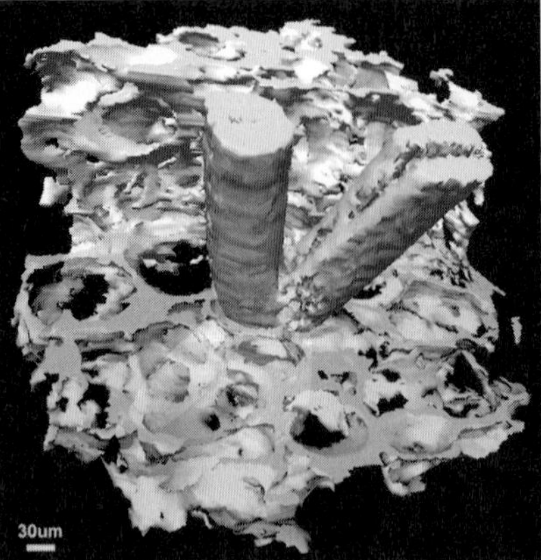

Figure 3. Full 3D-reconstruction from one joint-point of fibers in composite

After full reconstruction the set of reconstructed cross sections can be converted to the realistic 3-dimensional model of the internal microstructure. Full 3D-visualisation for one of internal joint point of fibers by this way is shown in Fig.3.

Another example of microtomographical reconstruction is taken from the epoxy-based composite. A shadow x-ray image (microradiography) of the sample with multilayer fiber structure in epoxy is shown in Fig.4a. Fig.4b shows the results of microtomographical reconstruction in two different orientations. Position and orientation of fibers in different layers can be easily obtained. Non-destructive reconstruction by microtomographical methods allows also inspecting for different kinds of defects in the fibers packing and internal voids.

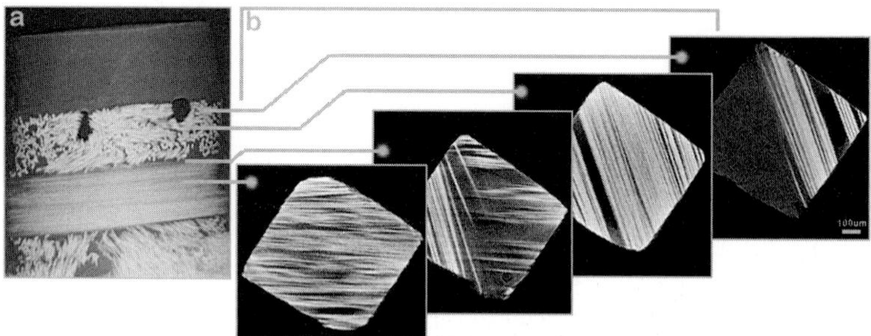

Figure 4. X-ray shadow image (a) and reconstructed cross sections in different orientations (b) from the epoxy-based composite

5 Conclusions

Desk top microtomography system can be applied for nondestructive investigation of internal microstructure of polymeric materials and composites. Phase contrast improves resolution for details in the micron range. Non-destructive reconstruction can be done without any preparation, in normal environmental conditions as well as in interaction with liquids or under mechanical tests.

6 References

1. A.Sasov, D.Van Dyck. Microscopy and Analysis (European edition), March 1998, pp. 21-23.
2. V.N Ingal, E.A.Beliaevskaya, (1995), J. Phys. D: Applied Physics 28, 2314.
3. S.W.Wilkins, T.E.Gureyev, D.Gao, A.Pogamy and A.W.Stevenson. Nature, 1996, 384, 335.

IV Sensors and Actuators

Modeling of piezoelectric actuators

Thorsten Steinkopff, Andreas Wolff, Siemens AG, Corporate Technology, Munich

1 Introduction

Piezoceramic actuators are established in many fields of applications due to their fast electromechanical response and compact size. In the last few years, multilayer actuators made by cofiring technique gain in importance because of their advantages in performance like fast switching time, high blocking force, and low driving voltage. These features are opening a wide field of industrial applications for example
- in the automotive area for fast injection valves (fig. 1) and
- in textile industry for automated weaver´s looms (so-called Jacquard technology, fig. 2).

Fig. 1: Multilayer piezoceramics stack actuators for fast injection valves in common rail systems.

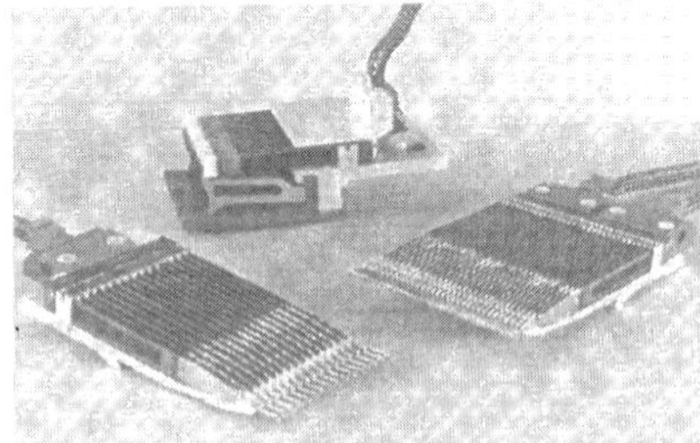

Fig. 2: Weaver´s loom control modules for textile machines built up with piezoceramic trimorph bending elements.

Besides the performance the long-term reliability of piezoelectric components represents a key feature for the broadening of application. For example, tension stresses and dynamic driving conditions can cause crack initiation and crack growth in the ceramic component which may dramatically decrease the mean time to failure. Because of the nonlinear and strongly coupled electromechanical behavior of the ferroelectric material which is based on domain wall motions the quantitative prediction of failure mechanisms proves to be difficult. New terms like fracture electromechanics have been introduced.
In the present paper the question should be answered: How can modeling and simulation contribute to a better performance and reliability of piezoelectric components?
First a short review of materials behavior of piezoelectrics will be given. The understanding of deformation and polarization mechanisms is necessary for formulating the nonlinear constitutive equations. Phenomenological and micromechanical approaches will be discussed. From the comparison with experimental results we have learned that the grain-to-grain interaction must be taken into account.
In the second part examples of performance and loading analyses of piezoelectric components will be presented. Stack and bending actuators are suitable to discuss typical deflection profiles and distributions of electrical field strength and mechanical stresses.
Finite element methods are proven and powerful tools also for the analysis of piezoelectric field problems and can contribute to improve both the performance and the reliability.
For the analysis of large signal applications the occurrence of ferroelastic depolarization must be taken into account. For this reason the use of nonlinear FE methods is unavoidable.
In technical applications the piezoelectric components interact with other electrical and mechanical components. In order to optimize such systems dynamic models of piezoelectric components must be provided.
In the third part a short introduction into the most widely used simulation tool SIMULINK will be given. Linear and nonlinear models of the stack actuator will be presented. Special attention will be paid to the hystereses properties.

2 Modeling of materials behavior

2.1 Materials behavior of piezoceramics

Piezoelectricity in polycrystalline materials like PZT ceramics is a complex phenomenon of both linear and ferroic electromechanical coupling which is based on the single crystal piezoeffect and on domain switching, respectively. Domain switching causes additional contributions to both strain S and polarization P and affects the nonlinear strain versus electric field behavior known as butterfly shaped curve (fig. 3 /1/).

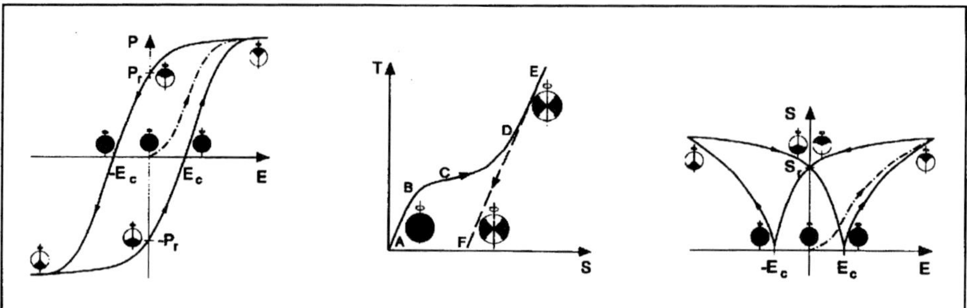

Fig. 3: Polarization and deformation behavior of piezoelectric ceramics /1/.

The domain processes can be induced by mechanical and/or electrical loading T and E respectively. In analogy to the definition of the coercive field strength E_c in the case of ferroelectricity, a coercive stress T_c is defined for the ferroelastic behavior. This coercive stress depends on the electrical field /1/, and vice versa, the coercive field strength is a function of the applied mechanical stress.

In reality domain switching means domain wall motion: Favorably oriented domains grow at the expense of unfavorably oriented domains. Thus changes in the so-called polarization texture are caused which are shown in fig. 3 by means of the circular symbols.

This also yields to changes of the macroscopic linear piezo moduli among others.

The total response of piezoelectric material is traditionally expressed by means of small and large signal coefficients. It seems to be clear that domain switching essentially contributes to the large signal coefficients. But even the small signal coefficients are influenced by domain wall motions.

Our special interest is directed to the nonlinear deformation behavior. In fig. 4 experimental results for both the uniaxial loading and the four-point bending /2/ are shown. The asymmetry between tension and compression

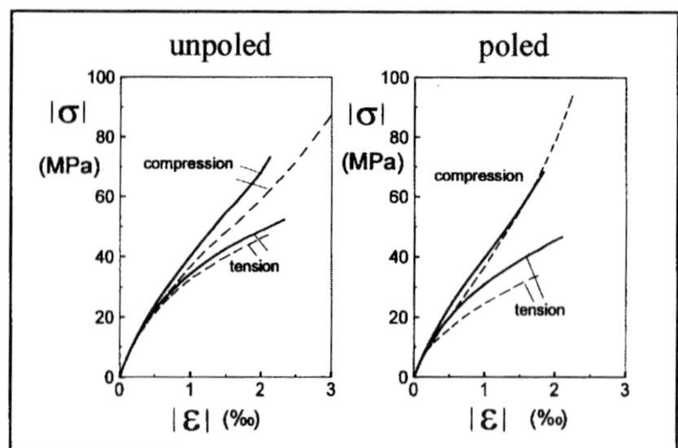

Fig. 4. Experimental results for both uniaxial loading (dashed lines) and four-point bending (solid lines) /2/.

can be made clear by means of different numbers of domains which can switch into plane and out of plane respectively.

2.2 Phenomenological and micromechanical modeling

There are two principal ways to describe the material behavior: Phenomenological and micromechanical approaches. Phenomenological models confine to reflect the observed behavior. They don´t attempt to explain its physical origin. This type of materials law is favored about all because of its easy handling. But for more complex loading situations like stress triaxility and for nonlinear electromechanical coupling plausible assumptions are necessary.
In micromechanical models a set of physical equations is solved on a small material volume from which the macroscopic behavior is calculated by averaging. The materials behavior is described by means of physically meaningful variables. The main advantage of these models lies in their predictive capability.
For instance, Kamlah and Jiang /3/ use two microstructural parameters to describe the polarization texture for the one-dimensional case. The parameters correspond to the volume fractions of domains with polarization axis within the so-called 45° cones. The different effects of electrical fields and mechanical stresses give rise to distinguish between a strain texture parameter and a polarization texture parameter corresponding to fig. 5. Kamlah and Jiang have succeeded in formulating evolution equations for these internal variables. The model response exhibits rate effects.

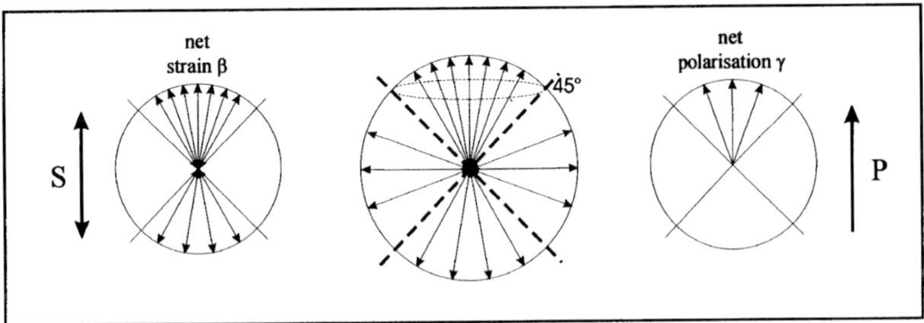

Fig. 5. Inner variables β and γ describing the net strain and net polarization state respectively /3/.

2.3 Micromechanical models for domain switching

Micromechanical models are often based on so-called switching criteria: The work done by switching is assumed to exceed a critical value. Under uncombined uniaxial loading this critical value corresponds to a critical stress or a critical electric field . As a consequence of the combined energy criterion the critical stress value is dependent on the applied electric field and vice versa.
Different authors investigated the deformation and polarization behavior of piezoceramics without considering the electrical and mechanical grain-to-grain interaction. Own analytical investigations /4/ are based on a well-defined

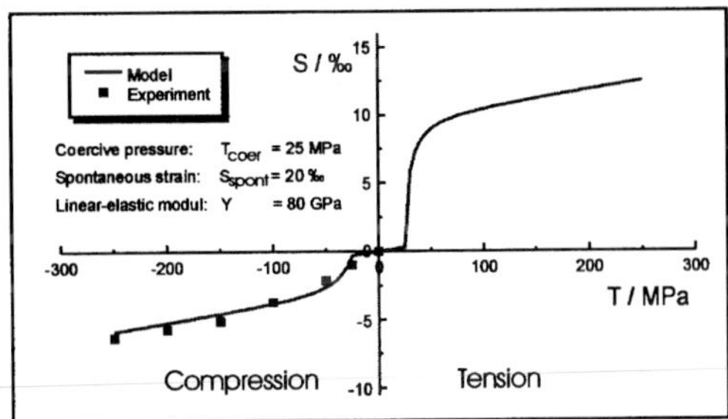

Fig. 6. Deformation behavior without grain-to-grain interaction: Analytical results of 1D modeling. Experimental results taken from /1/.

polarization texture parameter which was introduced by Michelitsch and Kreher /5/. The analytical results and other numerical results are in contradiction to experimental observations on PZT four point bending specimens which only show a low stress-strain asymmetry. Probably the asymmetry in the simulated deformation behavior (fig. 6) is caused by neglecting the grain-to-grain interaction.

Obviously, in piezoelectric ceramics the changes of both the ferroelastic strain and the ferroelectric polarization of a single domain cannot take place without clamping reactions of neighboring domains and grains. First of all, the different grain orientations are responsible for the inhomogeneous reaction of polycrystalline materials.
Hwang et al. /6/ investigated the averaged reaction of a piezoelectric ceramic by means of the so-called Eshelby formalism. That means an inclusion in a mean field is considered. The switching energy depends on both the orientation of the inclusion and the electrical and mechanical interaction with the matrix. The macroscopic behavior is calculated by averaging over all possible single crystal orientations. The numerical results (fig. 7) show a significant autocatalytic effect which is also observed in experiments. It must be noted that the best fit of the numerical results to the experimental ones only succeeded by means of two steps: 1. a significantly lower elastic module in comparison to experimental small-signal results and 2. a different weighting of electrical and mechanical contributions to the domain switching energy.

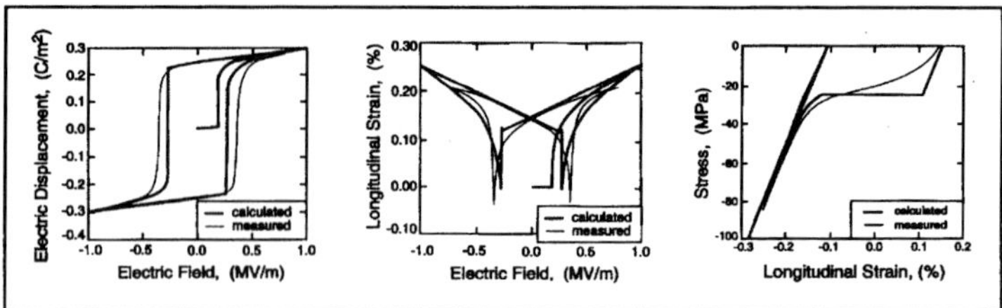

Fig. 7. Results of Eshelby formalism for piezoelectric ceramics /6/.

2.4 Switching model, Finite element results

In the 3D micromechanical model /4, 7/ used here the work done by switching is assumed to exceed a critical value:

$$\underline{T}\cdot\cdot\Delta\underline{S}^s + \vec{E}\cdot\Delta\vec{P}^s = \Delta w \leq \Delta w^c$$

which corresponds to critical stress or critical electric field under uncombined uniaxial loading. According to experimental observations /1/ the critical stress value is linearly dependent on the applied electric field.
Finite element methods are very suitable to investigate the macroscopic behavior of inhomogeneous materials.
In the present model the single crystals are represented by so-called subelements. In an untextured sample each of the randomly oriented single crystals initially consists of 6 tetragonal domains with the same volume fractions. The domain walls are not modeled. During loading the volumes of favorably oriented domains increase at the expense of the other domains of the same single crystal. In accordance to the theory of ideal plasticity the constitutive equations have been derived. That means for each loading increment the changes of spontaneous polarization ΔP^s and spontaneous strain ΔS^s are calculated.
The polarization texture is given by the whole of the volume fractions of all domains of all single crystals which change during loading. In fig. 8 the pole angle distribution for a poled sample is shown as a result of FE simulation on the basis of the domain switching model.

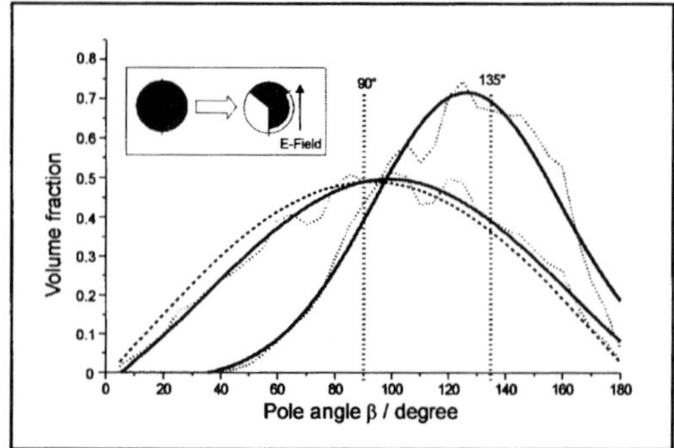

Fig. 8. Changes in the polarization texture during poling: Results of Finite element modeling.

All FE results shown here are based on calculations by means of the software package ANSYS /8/.
In the first step the uncoupled mechanical and electrical behavior has been simulated on the basis of the nonlinear 3D structural element SOLID 45 Fig. 9 shows the results for the ferroelastic behavior of an untextured sample.
Because the Finite element simulation automatically takes the grain-to-grain interaction into consideration only a small degree of nonsymmetry between tension and compression is observed which is in accordance to the experimental results /9/.
From the ferroelectric hysteresis (fig. 10) a characteristic dependence of the experimentally observed coercive field on the micromechanically defined critical field has been observed. The coercive field strength appears as a cooperative switching of 180° domains which is often called autocatalytic effect.

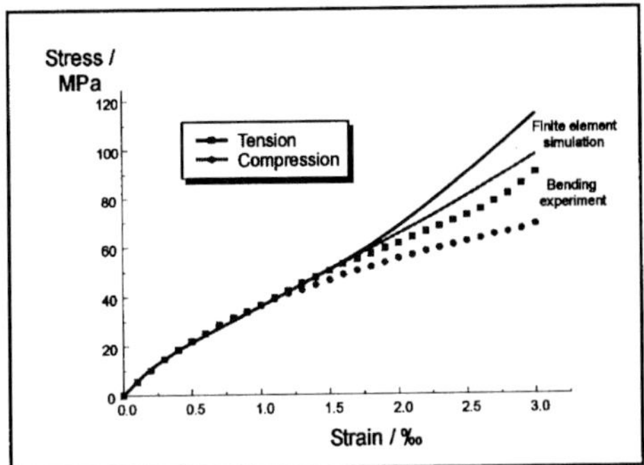

Fig. 9. Finite element simulation of the ferroelastic deformation behavior. Experimental results taken from /9/.

In the second step the full coupled behavior has been simulated on the basis of the 3D piezo element SOLID 5 which has been extended to the nonlinear option. The texture dependent piezoelectric activity has been taken into consideration.

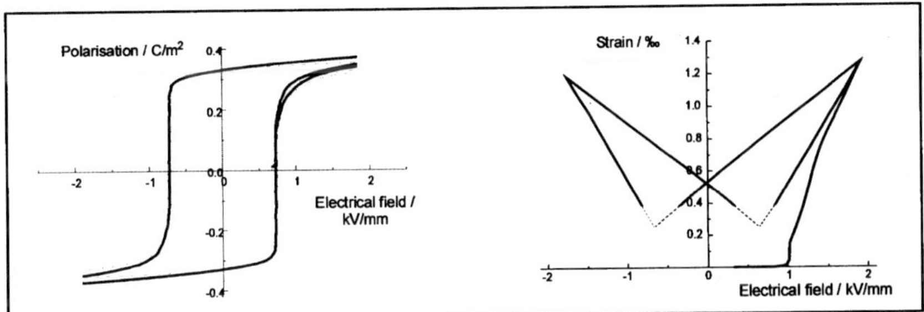

Fig. 10. Ferroelectric hysteresis and butterfly-curve as results of Finite element simulation.

The fig. 10 also shows a typical butterfly hysteresis simulated by means of the nonlinear piezoelectric element. In fact, the results of Finite element simulations are very similar to the results of the Eshelby formalism.

3 Simulation of piezoelectric components

3.1 Common aims, piezoelectric Finite element method

Simulation is very helpful for development and optimization of components with regard to function, size, shape, driving and loading conditions. In the field of piezoelectric actuators the efforts are concentrated to both large deflection and high blocking force.
For simulation of complex structures and coupled field problems the Finite element method is its own recommendation. The knowledge of internal electrical fields and stresses is indispensable for the understanding of damage phenomena and failure mechanisms. In fact, problems of reliability are gaining more and more in importance.

In most cases the linear piezoelectric FE analysis of ferroelectric components is used for both the small and the large signal simulation. There are first attempts to insert nonlinear properties of piezoceramics into FE simulations, for example for high precise positioning.
The FE solution of linear piezoelectric problems is based on the following constitutive law:

$$T = c^E \cdot S - e \cdot E$$
$$D = e \cdot S + \varepsilon^S \cdot E$$

From these equations the generalized stiffness matrix *K* is derived. In structural analyses the stiffness matrix brings the nodal point displacements u in relation to the nodal point forces F. In piezoelectric analyses these vectors are extended to the electric potentials U and free charges Q respectively.
As a consequence of the electromechanical coupling the generalized stiffness matrix *K* is an antisymmetric hypermatrix. By multiplying second equation with (-1) usual linear solver can be used for the coupled problem. Note, that 1. the so-modified matrix *K* has a semi-definite form and 2. the solution yields (-Q) indeed of Q. The latter is important in the case of coupling to so-called circuit elements.
A second feature of the matrix *K* is that stiffnesses and capacities are of very different order. For reasons of numerical precision the physical units should be chosen very carefully.

3.2 Deflection and loading analyses of piezoelectric stack actuators

One important type of piezoelectric actuators is the stack actuator because of its compact shape and high blocking force. Longitudinal strains up to 0.15 % are reached because of large-signal coefficients d_{33} of about 700 pm/V.
In fig. 11 the design of a cofired multilayer stack actuator is shown. The gray regions represent electrodes with alternate electric potential. The transparent regions represent PZT layers with alternate remanent polarization. For cofired multilayer actuators, the dependence of the magnitude of deflection on shape and size of inactive regions is of special interest.

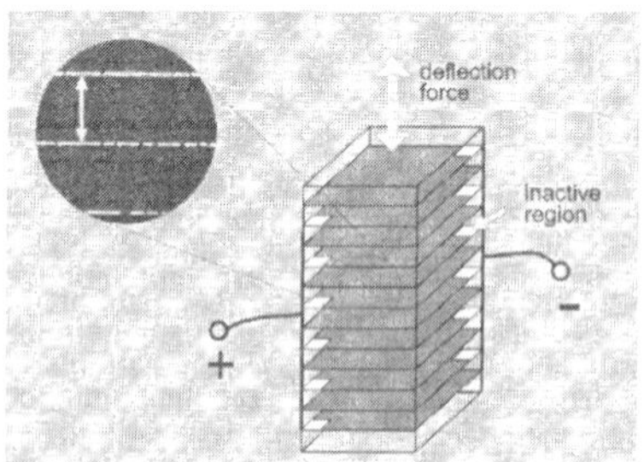

Fig. 11. Design of cofired multilayer stack actuator.

Fig. 12 shows deflection profiles of stack actuators with different designs of the inactive regions calculated by means of piezoelectric Finite element analyses. These regions of inactive piezoceramic are forced to elongate by the active regions and vice versa – the deflection of active parts is reduced by the inactive regions.
As a result high tension stresses occur within the inactive regions. The tension stress distribution (fig. 13) is dependent on actuator length whereas the maximum stress value is essentially determined by driving voltage. Because of the high tension stresses delamination cracks are initiated first of all during the poling process. These poling cracks are caused to growth at further cycling. On the other hand, cyclic operation of the actuator may cause fatigue of the external electrical connections which directly affects the components reliability. In order to find an optimal design for the actuator unit numerical simulations can be very helpful in this field, too.

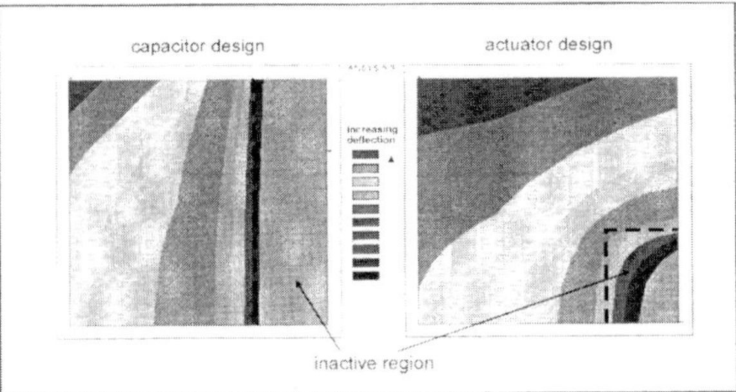

Fig. 12. Deflection profiles of stack actuators with different designs of the inactive regions.

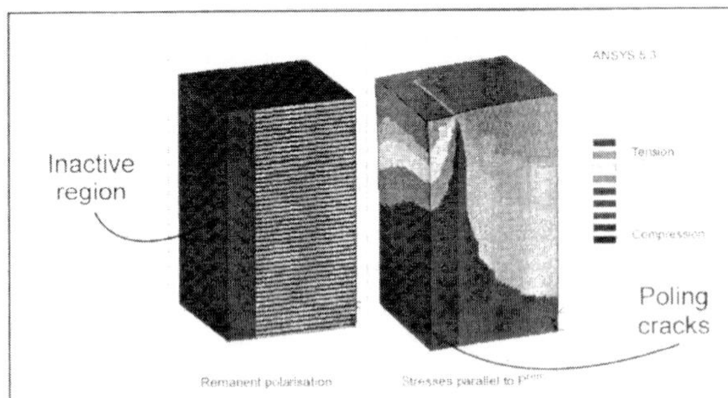

Fig. 13. Stress profile of one quarter of a stack actuator.

3.3 Deflection and loading analyses of piezoelectric bending actuators

The deflection behavior of bending actuators is very different from that of stack actuators. For bending applications the 31-component of piezoelectric matrix is used. Fig. 14 shows the geometry of a so-called trimorph element which consists of 2 PZT layers and an passive middle layer. Under electrical driving a contraction of the upper

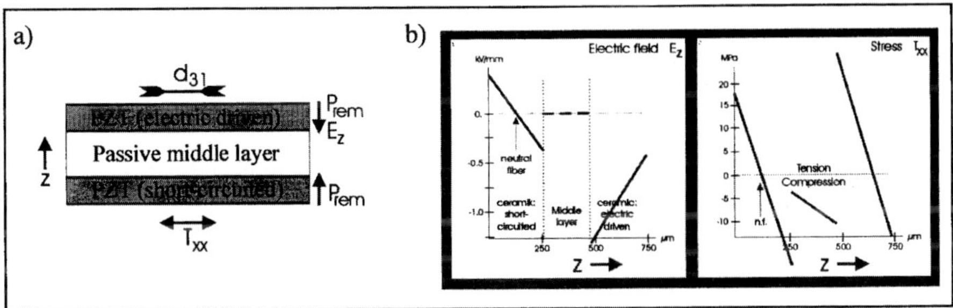

Fig. 14. a) Trimorph geometry. b) Results of linear analysis of trimorph bending actuator.

PZT layer causes the trimorph element to bend upwards. At the right hand side typical profiles of mechanical and electrical loading as results of the linear large-signal analysis are shown. Because of bending linear distributions of the electric field parallel and of the stress perpendicular to the electrodes are observed. Even the electric field strength can considerably deviate from the ratio of voltage per PZT layer thickness.

It is well known that high tension stresses like observed at the bottom of the lower PZT layer can cause ferroelastic depolarization which is of special interest in fig. 15.

The diagram at the left hand side shows experimental observations on the field dependence of the coercive pressure in a soft PZT /1/. The higher the applied field the higher the coercive compression stress in the direction of remanent

polarization. Coming back to the loading situation in the trimorph bending actuator. The tension stresses T_{xx} in the bottom layer which in praxis are much higher then the coercive stress cause ferroelastic depolarization. The electric field parallel to the remanent polarization (fig. 15b) counteracts this tendency but it seems to be to weak to maintain the polarization state.
As a result of the depolarization nonlinear electric field and stress distributions and a modified deflection behavior are observed.

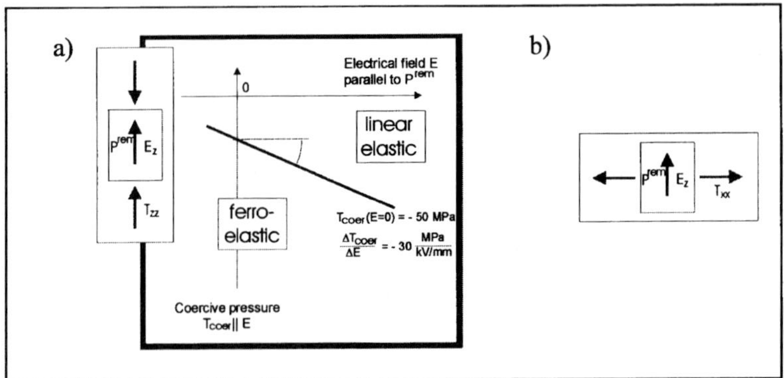

Fig. 15. a) Experimental observations on coercive stress /1/.
b) Loading situation at the bottom of the lower PZT layer in the trimorph bending actuator.

4 Dynamic analysis of piezoelectric systems

4.1 The dynamic analysis system MATLAB/SIMULINK

SIMULINK /10/ is the most widely used software package for modeling and simulating dynamical systems which are described by ordinary differential equations. These equations or systems of equations are solved by means of Finite difference methods.
The main advantages of the SIMULINK software are
- the capability for solving linear and nonlinear systems,
- the graphical user interface which is easy to handle and
- the possibility to create blocks, submodels and libraries.

SIMULINK includes a comprehensive block library of sources and sinks, linear and nonlinear components, and connectors. To solve a model different integration methods are offered. The user can see the simulation results while the simulation is running. In addition, he can change parameters and immediately see what happens. The simulation results can also be processed and visualized by MATLAB.

4.2 Linear response of a piezoelectric stack actuator

In technical systems piezoelectric components interact with other electrical and mechanical components. For instance injection systems for automotive applications essentially consist of 3 components: the stack actuator unit, the injector and the power supply corresponding to fig. 16.

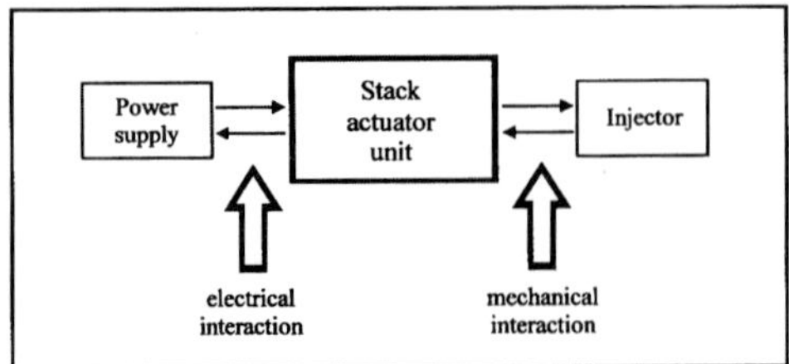

Fig. 16. Electrical and mechanical interaction between the components of an injection system.

The interaction between the electrical and mechanical components is realized by voltage U, force F, charge Q (or current I), and elongation x (or velocity v). Despite of its complexity the dynamic analysis can be successful in optimizing the system performance first of all in regard to the injection volume rates. This aim requires a realistic simulation of the nonlinear and time-dependent behavior of the piezoelectric actuator.
Fig. 17 shows the model of the stack actuator unit which consists of the piezoelectric actuator, the elastic housing with spring (stiffness k, mass m) and the electrical connection (resistance R). By means of R the open-circuit actuator unit ($R \rightarrow \infty$) and the short-circuit actuator unit ($R = 0$) can be realized.

4.3 Nonlinear response on the basis of a hysteresis model

The injection system is a typical large-signal application of piezoceramic actuators. For this reason the typical nonlinear characteristics like polarization vs. electric field and strain vs. electric field must be modeled in detail. On the basis of the linear actuator model a nonlinear model has been developed. First of all this model must meet hysteresis properties like

- path dependence because of loading history effects and
- minor loops which are enveloped by the major loop.

Remember that the hysteresis behavior of piezoceramics are caused by domain switching. As a result of this load and path-dependent process the mechanical, the dielectrical and the piezoelectrical behavior is strongly nonlinear.

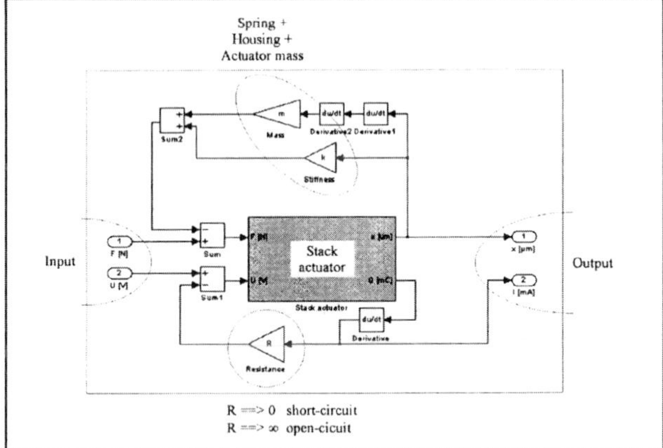

Fig. 17. SIMULINK model of the piezoelectric stack actuator unit

Actually all coefficients must be taken into the nonlinear consideration:

$$dS = s^E \cdot dT + d \cdot dE$$
$$dD = d \cdot dT + \varepsilon^T \cdot dE$$
$$s^E,\ d,\ \varepsilon^T = f_i(T, E, \dot{T}, \dot{E}, P^{rem})$$

We succeeded in finding a suitable hysteresis model which is applicable to all coefficients mentioned above. Fig. 18 shows that the modeled hysteresis fulfils the property of minor and major loops mentioned above.

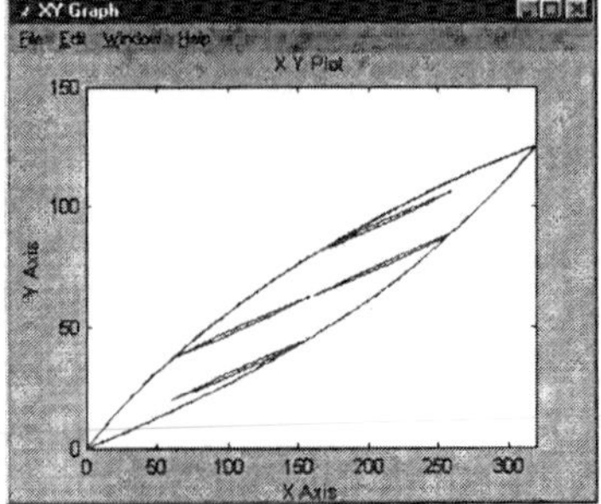

Fig. 18. Modeled hysteresis with major loop and minor loops.

Fig. 19 and 20 show the results of an electrically and mechanically driven stack actuator with hysteresis behavior. In both cases the energy vs. time plots show that the hysteresis behavior causes loss of electrical and mechanical energy respectively. In the current vs. time plots the deviation from the sinus-like behavior is significant.

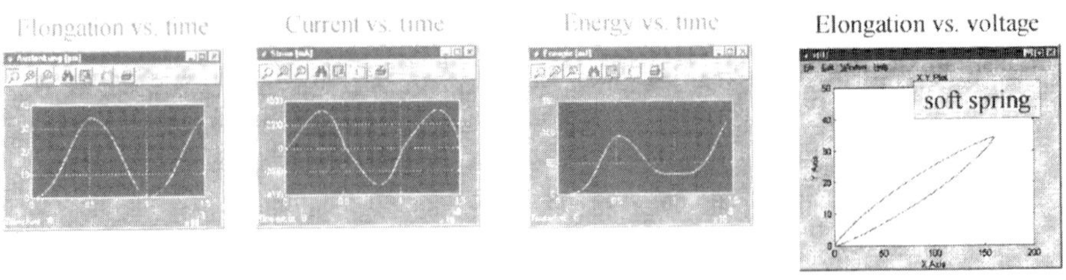

Fig. 19. SIMULINK results of a voltage-driven nonlinear stack actuator.

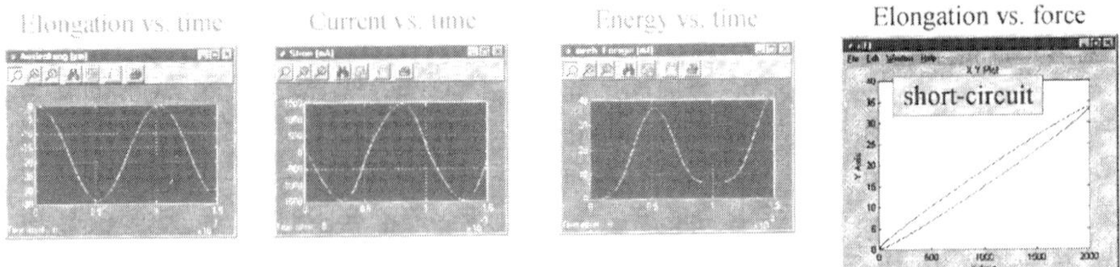

Fig. 20. SIMULINK results of a force-driven nonlinear stack actuator.

5 Outlook

During the last few years useful progress has been made in the field of modeling and simulation of piezoelectric actuators which should be demonstrated here. But there is still a lack of realistic numerical models particularly for large-signal applications under complex loading situations which must base on a comprehensive understanding of the nonlinear, coupled material behavior. Next efforts will be aimed at using the domain switching processes in order to improve and to stabilize the performance of piezoelectric components.

References

1. A.B. Schäufele, K.H. Hädtl, J. Am. Ceram. Soc. 1996, 79, 2637-2640
2. T. Fett, D. Munz, G. Thun, J. Am. Ceram. Soc. 1998, 81, 269-272
3. M. Kamlah, Q. Jiang, Forschungszentrum Karlsruhe, 1998, FZKA 6211
4. T. Steinkopff, Ferroelectrics 1999, 222, 383-387
5. T. Michelitsch, W. Kreher, Acta mater. 1998, 46, 5085-5094
6. S.C. Hwang, J.E. Huber, R. McMeeking, N.A. Fleck, J. Appl. Phys. 1998, 84, 1530-1540
7. T. Steinkopff, J. Europ. Ceram. Soc., 1999, 19, 1247-1249
8. ANSYS,Inc., Southpointe, 275 Technology Drive, Canonsburg, PA 15317
9. H. Weitzing, G.A. Schneider, Unpublished results, 1998
10. MathWorks, Inc., 24 Prime Park Way, Natick, MA 01760-1500

Crystalline Piezoelectric Materials with Special Consideration of $GaPO_4$

K. Jacobs, P. Hofmann, J. Reichow, and P. Görnert[*)]
Institut für Kristallzüchtung Berlin
[*)]Innovent e.V., Jena

1. Introduction

Piezoelectricity, or "pressure electricity", was discovered in 1880 by J. and P. Curie, studying the symmetry of crystals. Later, it was found that piezoelectricity does not only occur in single-crystalline materials, but also in polycrystalline ceramics and in both biopolymers and synthetic polymers. The origin of piezoelectricity is the separation or unequal displacement of the centres of gravity of the positive and negative charges under the influence of mechanical stress or an electric field. Prerequisite for the occurrence of piezoelectricity in crystalline materials is the lack of a symmetry center. Thus, in materials from 20 crystal classes piezoelectricity will occur. (Actually, 21 from the 32 crystal classes are non-centrosymmetric, but in the non-centrosymmetric cubic class 432 all the piezoelectric moduli vanish due to the high symmetry).

It was not until World War I that piezoelectricity found technical application. Today there is a tremendous variety of applications relying on either the direct or the converse piezoelectric effects, especially for acoustic transducers, frequency control devices, actors and sensors. In principle, for each application there is a special material best suited, because an optimum combination of piezoelectric, thermal, dielectric, mechanical and other properties has to be found. Last but not least, also technological and economic considerations play an important role. Therefore, all classes of piezoelectric materials, single crystals, polycristalline ceramics and organic polymer films, are in practical use. They are complementary, rather than competetive.

In this paper we will briefly discuss the merits of single-crystalline piezoelectric materials. Special emphasis will be paid to our own research on gallium orthophosphate, a material that has attractive properties, but is difficult to grow.

2. Common single-crystalline piezoelectric materials

Comparing a certain material in either single crystalline or polycristalline form, the magnitude of polarisation is much higher in the single crystal. Ceramics are complex solids with multidimensional variability of composition. Grain boundaries, porosity, different crystallite orientations and internal stress in the polycrystalline ceramics prevent the alignment of the polarized domains or the coalescence into one single uniform domain. Their properties depend strongly on manufacturing methods, poling procedure and temperature-stress history. For the reliable determination of fundamental piezoelectric constants of a certain material only single crystals of this material can be used. Another important point for the use of single crystals instead of possibly cheaper ceramics or polymers is the "Q factor". It covers the quality of a resonator with respect to its energy storage, transfer and loss per cycle of resonance. The Q factor of single-crystalline materials is orders of magnitude higher than that of others. Also aging and damping of surface acoustic waves is much lower in single crystals.

The piezoelectric effect is most distinct in highly polar single crystals. Single crystals of a number of compounds are well investigated and partly in technical use. As already mentioned, an optimum combination of properties is required, the piezoelectric ones being only one set among others. Piezoelectricity is an effect, composed from mechanical and electrical variables. For its satisfying description not only piezoelectric, but also elastic and dielectric constants with their orientation and temperature dependence are required [1]. In Table 1, essentially taken from [2], a number of common materials and selected properties are compiled.

Table 1: Characteristics of some "classical" piezoelectric materials

material	formula	point group	ij	d_{ij} [pC/N]
ammonium dihydrogen phosphate (ADP)	$NH_4H_2PO_4$	$\bar{4}2m$	36	48
ammonium dihydrogen phosphate (deuterated) (AD*P)	$ND_4D_2PO_4$	$\bar{4}2m$	36	75
barium titanate	$BaTiO_3$	4mm	15	587
lithium niobate	$LiNbO_3$	3m	22	21
lithium iodate	$LiIO_3$	6	33	93
lithium tantalate	$LiTaO_3$	3m	22	7.5
potassium dihydrogen phosphate (KDP)	KH_2PO_4	$\bar{4}2m$	36	21
quartz	α-SiO_2	32	11	2.3
rochelle salt	$NaKC_4H_4O_6\ 4H_2O$	222	14	1570
tourmaline	$(Na,Ca)(Mg,Fe,Al,Li)_3Al_6$-			
	$(BO_3)_3(Si_6O_{18})(OH,F)_4$	3m	15	3.7
zinc oxide	ZnO	6m	33	12

By far the most important single-crystalline material is α-quartz. It comes in quantities second only to silicon. Quartz frequency control devices utilize the piezoelectric effect in two rather different ways, either surface acoustic waves (SAW) or bulk acoustic waves (BAW). SAW devices are presently the largest use of quartz. Its piezoelectric coefficient d_{11}, relating the electric polarization in x direction to the mechanical stress in x direction, or alternatively, the strain in x direction to the component of the electric field in the same direction, is not very high, but the Q factor of quartz resonators is several hundred thousand times higher than those of most other electrical components. The usability of quartz, however, has two essential restrictions: One is the temperature range of operation, which is limited by the phase transition between α-quartz and β-quartz, occurring at about 573 °C. The other is its comparatively small electromechanical coupling coefficient k_{ij}.

Another important material, used in single-crystalline form, is lithium niobate. About 80 tons are estimated to be grown worldwide during 1999. This material shows besides piezoelectricity interesting optical properties, such as large birefringence and a distinct electro-optic effect. Both quartz and $LiNbO_3$ are grown today as single crystals with dimensions enabling the production of wafers of 100 mm diameter. Quartz is grown from aqueous solutions at temperatures around 380 °C and pressures > 1000 bar; more details will be discussed in section 3. Lithium niobate crystals are grown by the Czochralski method at the material's melting point, 1255 °C.

3. New single-crystalline piezoelectric materials

In recent years the desire or necessity has arosen to operate electronic components and devices under extreme environmental conditions, especially at high temperatures. This is useful in order to come closer to the point of use, without delay and risk of losses via transmission lines. For example, combustion processes, such as in engines, can be monitored and controlled more precisely.

New materials that potentially can overcome the limitations of quartz are under investigation for about twenty years now, but no actual breaktrough has been reached so far. The most promising candidates are langasite, $La_3Ga_5SiO_{14}$, and gallium orthophosphate, $GaPO_4$. Some of their interesting properties are compiled in Table 2.

Table 2: Properties of some emerging piezoelectric materials

	α-quartz	$AlPO_4$	$GaPO_4$	$La_3Ga_5SiO_{14}$
piezoelectric module d_{11} [pC/N]	2.3		4.5	6.2
Coupling coefficient k [%]	8.5	11.0	18.4	16
Surtension coefficient Q	$3{\cdot}10^6$	$1{\cdot}10^6$	$> 5{\cdot}10^4$	$4{\cdot}10^4$
Phase transition temperature [°C]	573	586	none	none

Also included in the table are quartz for comparison, and berlinite, $AlPO_4$. The latter material, however, suffers also from a phase transition, occurring at relative low temperatures [3] (cf. Table 2).

Langasite, the name is derived from its components, also belongs to the trigonal system, crystal class 32, with the space group P321. It has been grown and studied since about 1980, mainly by Russian scientists. Its melting point is 1470 °C. Langasite is grown by the Czochralski method. This has the advantage that high growth rates can be achieved and round wafers with geometries common for other electronic materials are easily obtained. Mass production of three-inch diameter langasite wafers has been developed in Russia. The technology has been licensed to Mitsubishi Materials, Japan. The properties of langasite are not yet fully explored. The adjustment and reproducibility of the correct stoichiometry is not easy, and there are also some uncertainties concerning the electrical resisitivity at high temperatures. Nonetheless, at present this material is the most promising one for new demanding piezoelectric applications. Important properties of langasite are also given in Table 2.

Another promising material is gallium orthophosphate, $GaPO_4$. Properties and growth of this material are studied for about 20 years, mainly by the AVL company in Austria [4, 5], by Philippot and coworkers at the University of Montpellier, France, [6, 7] and by scientists from the Institute of Crystallography of the Russian Academy of Sciences in Moscow [8, 9]. This material is, as $AlPO_4$, a III-V analogous to α-quartz. The Si atoms in $(SiO_2)_2$ are alternately substituted bei either Ga or P atoms. It has a number of properties for piezoelectric applications that

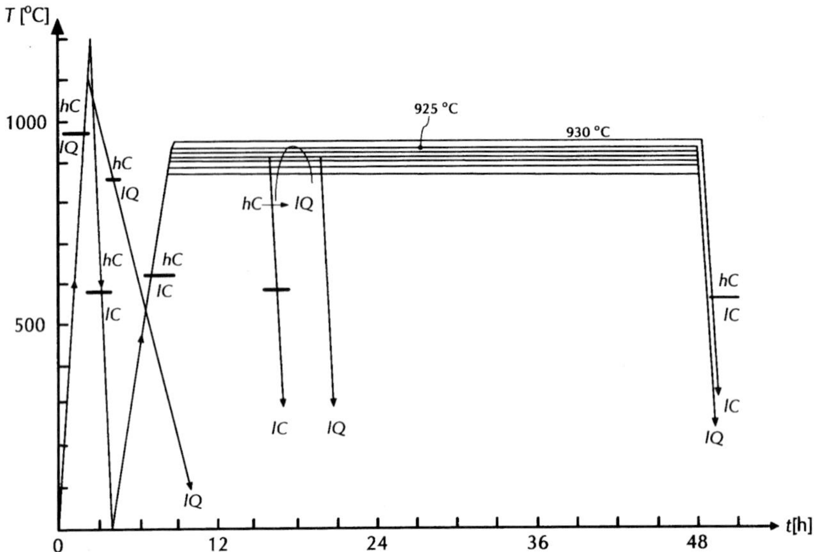

Figure1. Existence regions of different polytypes of $GaPO_4$

lQ - low-quartz analogous; *lC,hC* - low- and high-cristobalite modifications, respectively

are superior to those of quartz, as can be seen from Table 2. The growth of $GaPO_4$ crystals, however, is difficult and not satisfactorily solved yet. Melt growth of this material is impossible, due to the condensation of orthophosphate anions to polyphosphates or even glass formation. In the literature sometimes another reason is mentioned, the "irreversible phase transition" occurring at about 930 °C. This, however, is not correct. Our own investigations show that this transition between the low-quartz and the high-cristobalite modifications is reversible. Comparatively fast cooling from the high-cristobalite modification leads to the low-cristobalite analogous modification, while after slow cooling down to room temperature the low-quartz modification is found again. Figure 1 shows the phase transitions occurring in $GaPO_4$ under different thermal treatments [10].

The material is most commonly grown from saturated solutions of $GaPO_4$ in concentrated phosphoric acid. Up to temperatures of about 300 °C the solubility of $GaPO_4$ in H_3PO_4 is retrograde, i.e. it decreases with increasing temperature. Therefore, the supersaturation is established either by slow heating the nutrient or by applying a reverse temperature gradient with the hottest point at the growing interface. The advantages of H_3PO_4 in comparison with other solvents are the chemical identity of the anions, the high solubility of $GaPO_4$ and growth rates not too anisotropic. Figure 2 shows the growth rates of $GaPO_4$ crystals grown at 150 °C in two orthogonal directions as a function of the temperature difference between the bulk solution volume and at the growth interface. Opposite to most literature data [11] we observe higher growth rates in z direction, parallel to the c axis. The use of concentrated H_3PO_4 as a solvent is not very pleasant. The acid is chemically aggressive; it has a very high viscosity. The tendency to the formation of polyphosphates increases wit higher temperatures. Sometimes it is discussed that this should be advantageous or even a necessary pre-requisite for the successful growth of $GaPO_4$ crystals [8]. A serious problem is the evaporation of water from the solvent, changing the supersaturation and other properties of the solution. Stable conditions can only be obtained in closed systems, preventing the uncontrollable loss of water from the nutrient. Hydrothermal growth, i.e. the growth from hot aqueous solutions under high pressure in closed systems, has been well developed on an industrial scale for the growth of quartz crystals.

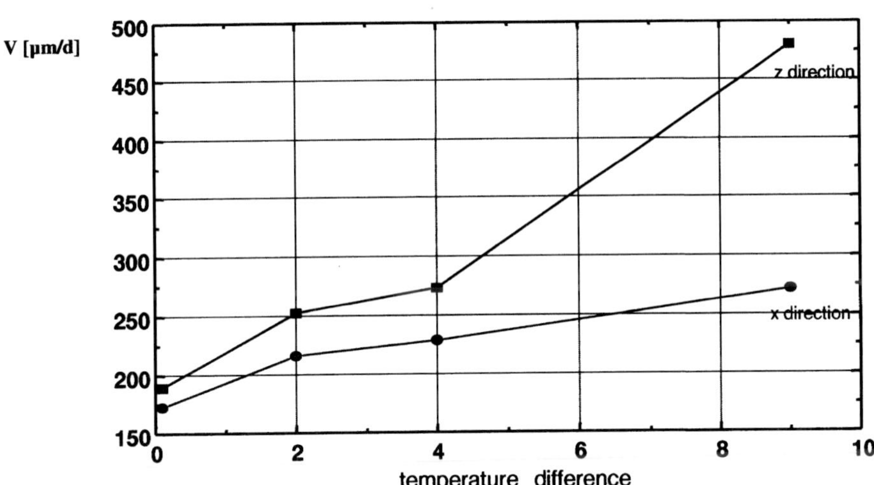

Fig. 2: Growth rates V [µm/d] of GaPO4 crystals as a function of temperature difference between bulk solution and growth interface (1.19 mol $GaPO_4$ dissolved in 85 % H_3PO_4 at 150 °C)

For this material, higher temperatures, well above the boiling point of the nutrient solution under atmospheric pressure, are required in order to get a sufficient concentration of the solute SiO_2 raw material. A reasonable solubility for the hydrothermal crystallization has been determined generally to be 2 to 5 weight-% [12]. Besides high temperatures the addition of a few percent of NaOH or Na_2CO_3 is useful, leading to a 1 M alkaline solution. The dissolving temperature of SiO_2 is about 40 °C – 50 °C above the growth temperature of 360 °C - 380 °C. Depending on the degree of fill of the autoclave, a pressure of 1.3 - 2 kbar is established in the gas room.

Also for the growth of $GaPO_4$, a closed system is absolutely necessary in order to prevent the evaporation of water. The solubility of $GaPO_4$ in concentrated (85 %) H_3PO_4, however, is about 10 weight-% at 150 °C, decreasing with temperature increase. Comparing with quartz hydrothermal growth, from the point of view of sufficient solubility there is no need for growing $GaPO_4$ crystals at higher temperatures. Nonetheless, growth at higher temperatures appears necessary for several reasons: With increasing crystallisation temperature less water is incorporated; improving the electroacoustic properties of the material. The viscosity of the solution is reduced and transport processes in the solution are faster. Also the rates of chemical reactions and surface processes at the growing interface are faster, increasing exponentially with temperature. We believe that the nutrient contains $GaPO_4$ essentially in colloid form [13]. If this holds true, crystal growth requires the decay of colloid agglomerates. This can also be expected to proceed faster at higher temperatures.

4. Summary

Besides quartz, new single crystalline piezoelectric materials are emerging. The most promising ones are langasite and gallium orthophosphate. Both of them show electromechanical properties making them potentially an attractive substitute for quartz. Czochralski-grown langasite is more advanced and could be the finally winning material. Gallium orthophosphate is handicapped because it can for principal reasons only be grown from solution.

References

[1] W. R. Cook, H. Jaffe, in Landolt-Börnstein Numerical Data and Functional Relationships in Science and Technology (Ed. K.-H. Hellwege), Springer, Berlin, Heidelberg, New York 1979, vol. 11, chapter 3

[2] S. C. Abrahams, K. Nassau in Encyclopedia of Materials Science and Engineering (Ed.: M. B. Bever) Pergamon Press, Oxford, New York, Toronto, Sydney, Frankfurt 1995, vol. 5, p. 3524 - 3528

[3] T. Rey, Z. Kristallogr. 1966, 123, 263

[4] P. Krempl, G. Schleinzer, W. Wallnöfer, Sensors and Actuators 1997, A 61, 361-363

[5] P. Krempl, F. Krispel, W. Wallnöfer, Ann. Chim. Sci. Mat. 1997, 22, pp. 623-626

[6] E. Philippot, A. Ibanez,vA. Goiffon, M. Cochez, A. Zarka, B. Capelle, J. Schwartel, J. Détaint, J. Crystal Growth 1993, 130, 195 – 208

[7] D.Palmier, A. Goiffon, B. Capelle, J. Détaint, E. Philippot, J. Crystal Growth 1996, 166, 347 – 353

[8] A. S. Shigarov, O. V. Zvereva, Yu. M. Mininzon, Kristallografiya 1994, 39, 717 -719

[9] O. V. Zvereva, L. N. Demyanets, Kristallografiya 1995, 40, 1065 - 1068

[10] K. Jacobs, P. Hofmann, D. Klimm, J. Reichow, M. Schneider; to be published in J. Solid State Chem.

[11] M. Cochez, A. Ibanez, A. Goiffon, E. Philippot, Eur. J. Solid State Inorg. Chem. 1993, 30, 509 - 519

[12] A. A. Ballman, R.A. Laudise in The Art and Science of Growing Crystals (Ed.: J. J. Gilman), Wiley, New York, London 1963, p. 231 - 251

[13] K. Jacobs, in preparation

Sol-Gel Derived TiO_2-Films for Gas Analysis in Nitriding Atmospheres

Sandra Beling/ Heinrich Klümper-Westkamp/ Franz Hoffmann/ Peter Mayr
Stiftung Institut für Werkstofftechnik, 28359 Bremen

1 Introduction

Due to the increasing automation of thermochemical heat treatment processes, online gas monitoring during nitriding and nitrocaburizing has gained much interest. At the moment zirconia sensors are used for oxygen detection and sometimes even for monitoring of the nitriding potential [1,2]. But these sensors require reference phases with constant oxygen pressure and they are rather expensive. On the other hand gas sensors based on semiconducting oxides like TiO_2 work without reference phase. With TiO_2-sensing elements the detection of oxygen in atmospheres with oxygen partial pressures as low as 10^{-18} Pa is possible [3,4]. In this work the sensing abilities of titania films derived by sol-gel techniques are investigated in nitriding atmospheres. The microstructure of the films was investigated by electron microscopy. Conductance measurements were carried out in mixtures of N_2, H_2 and NH_3 at temperatures between (500-600) °C which were realized in a small industrial nitriding furnace of 90 l volume.

2 Experimental

Titania films were prepared by sol-gel techniques. Titaniumtetraisopropoxide (TTIP, $Ti[OCH(CH_3)_2]_4$), 2-butanol ($CH_3(COH)C_2H_5$), acetic acid (HAc, CH_3COOH), acetylacetone (AcAc, $CH_3COCH_2COCH_3$), polyethylenglycol 400 (PEG) and distilled water were used as precursor. TTIP and 2-butanol were mixed in a beaker. AcAc was added as a chelating agent. The molar ratio of TTIP:AcAc was 1:1. For hydrolysis and stabilization of the sol water and HAc were added. The mixture was stirred until a clear solution was formed. PEG was used as a drying control additive. Further details of sol preparation are described somewhere else [5].

Two different types of samples called TiO_2-SA and TiO_2-SA-IDS respectively were used for conductance measurements. For preparation of TiO_2-SA samples polished sheets of alumina of (20×20) mm^2 size were coated with TiO_2 by dip coating with constant pull-out speed. After drying of the sol in air for a few minutes the samples were heat treated at 600 °C 10 min. Coating and heat treatment was repeated 5 to 6 times to obtain a film of about 1 μm thickness. Electrical contacts were realized with two AISI 304 sheets which were pressed against one surface. TiO_2-SA-IDS samples were prepared on commercial (3.5×3.5) mm^2 sized Al_2O_3-substrates [6] with Pt-interdigital electrodes of 1 μm height, 20 μm width and 30μm distance from each other and a 30 Ω Pt-heater. The substrates were coated with a drop of sol using a pasteur pipette and the film was dried in air for a few minutes. Finally the samples were heat treated at 600 °C 10 min. Coating and heat treatment was repeated 5 times.

The Microstructure of the films was investigated by field emission scanning electron microscopy (SEM) and transmission electron microscopy (TEM). The thickness d of the films was measured by ball cratering method. With mass m of the film per coated surface area A the density ρ can be determined as

$$\rho = \frac{m}{d \cdot A}. \tag{1}$$

The relative density was determined as the ratio of measured density ρ and the average density of the three existing phases of TiO_2, $\rho_t = 4{,}09\ g\ cm^{-3}$.

DC conductance measurements were carried out in a small computer driven industrial nitriding furnace with about 90 l in volume. Different mixtures of N_2, NH_3, H_2 and dry air were realized by mass flow controllers. The furnace contained a zirconia cell for online monitoring of oxygen partial pressure pO_2. For further gas monitoring a heated λ-probe and a infrared ammonia sensor were placed in the exhaust. The resistance of the TiO_2-layer was measured with a HP 34401 multimeter at temperatures between 500 °C and 580 °C in different mixtures of N_2, NH_3, H_2 and dry air.

3 Results and Discussion

Corresponding to former XRD phase analyses [5] all films which were heat treated at 600 °C 10 min consisted of anatase phase. The repeatedly coated films were 0.9 μm to 1.3 μm thick which means that the thickness of one coating was (200 –300) nm. The films were approximately 80 % dense. The surface of the TiO_2 films did not show structures in SEM photographs while grains could only be detected at fractured faces as illustrated in fig. 1. The TEM-photograph in fig. 2 shows TiO_2-grains of 20 nm diameter on average.

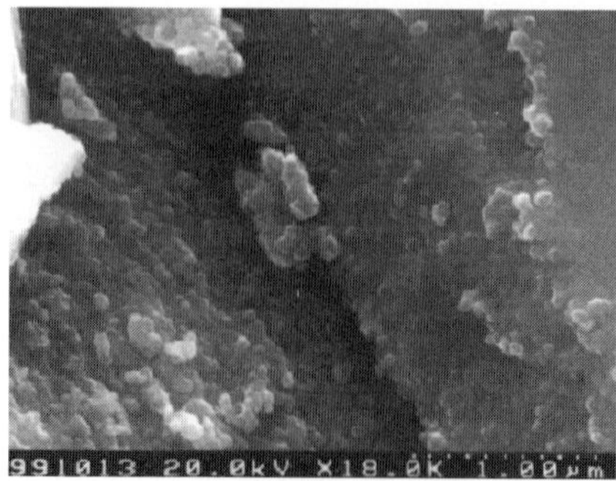

Figure 1: SEM photograph of surface of TiO_2 film on polished AISI 304 steel

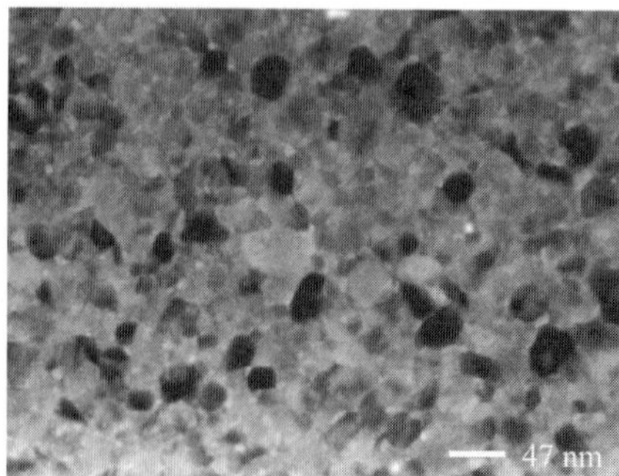

Figure 2:. TEM photograph of titania film.

Conductance measurements have been carried out in atmospheres containing different mixtures of N_2, H_2, NH_3 and dry air. All processes were carried out under constant gas flow. This was realized by controlling N_2 flow such that the sum of all gas flows was 600 lh^{-1}. At this flow the theoretical time for one volume exchange of the furnace is 9 min. In fact the exchange time should be longer due to the relatively large volume of 90 l and due to sensor housings and sample holder in the furnace. In N_2 conductance of TiO_2-SA films was

(0.1-1.0) μS at temperatures between 500 °C and 580 °C. Conductance of TiO_2-SA-IDS films was (100-600) μS under the same conditions.

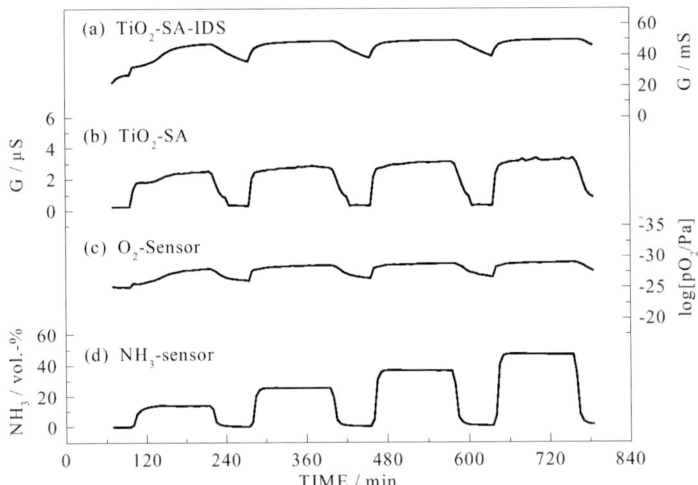

Figure 3: Conductance of TiO_2-films in nitriding furnace at 580 °C in mixtures of NH_3 and N_2

To test the reaction of TiO_2-films in different reducing atmospheres the furnace was heated to 580 °C. Then constant flows of [100, 200, 300, 400] lh^{-1} NH_3 were led through the furnace for 2 h each. Between the NH_3 flows the furnace was rinsed with N_2 for 1 h. As mentioned before entire flow was kept at 600 lh^{-1} during the complete process by controlling the N_2 flow. A corresponding process was carried out with [25, 50, 75, 100] lh^{-1} H_2. Fig. 3 (d) shows the NH_3 analysis in volume-% during such a process with different NH_3 flows. At high temperatures H_2 and NH_3 react like

$$2\,NH_3 \leftrightarrow 3\,H_2 + N_2 \quad (2)$$

$$2\,H_2 + O_2 \leftrightarrow 2\,H_2O \quad (3)$$

which means that oxygen partial pressure pO_2 and amount of reducing gases depend on each other, i. e. more NH_3 results in less O_2. Fig. 3 (c) shows the logarithm of pO_2 which was measured by the zirconia cell. pO_2 was evaluated from the equipotential voltage U_{eq} of the zirconia cell by using the Nernstian equation

$$U_{eq} = -0.0496 \cdot T \cdot \log[pO_2/pO_2^{ref}] \quad (4).$$

T is the temperature in K and $pO_2^{ref} = 2090$ Pa the oxygen partial pressure of the reference air. The conductance of both TiO_2-SA-IDS in fig. 3 (a) and TiO_2-SA film in (b) respectively corresponds to the pO_2 measurement of the zirconia cell. The TiO_2-SA film has a larger difference between the conductance at low pO_2 and high pO_2 and a faster transient.

In order to raise pO_2 at the start of NH_3 flow, the furnace was rinsed with 100 lh^{-1} air for 5 min during each rinsing with N_2 as indicated in fig. 4 (d). At the start of air rinsing, pO_2 in fig. 4 (c) raises and stays high until NH_3 flow begins. Conductance of TiO_2-SA film behaves similar as indicated in fig. 4 (b). But both TiO_2-SA and zirconia cell do not behave much different compared to the process without air rinsing shown in fig. 3. Conductance of the TiO_2-SA-IDS film, fig. 4 (a), in contrast decreases rapidly at the start of each air rinsing and raises again at the end of air rinsing.

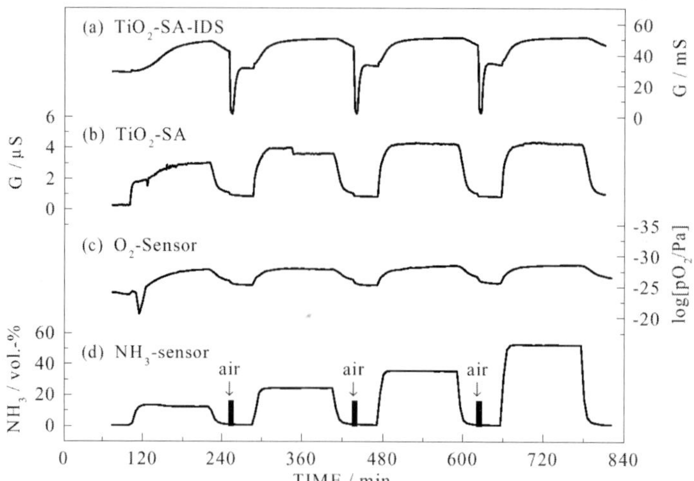

Figure 4. Conductance of TiO_2-films in nitriding furnace at 580 °C in mixtures of NH_3 and N_2, 100 lh^{-1} air for 5 min during N_2 rinsing.

The difference in the behavior of the TiO_2-SA and the TiO_2-SA-IDS film may result from different structures of the two films. TiO_2-SA is prepared by drain coating on a plain substrate which results in a plain crack-free titania film of about 1 µm thickness. If the substrate in contrast is structured, e.g. with interdigital electrodes like the TiO_2-SA-IDS, the sol cannot drain from the substrate between the electrodes which results in higher film thickness. Thus crack free films cannot be obtained during drying and heat treatment but layers of thin titania sheets are formed. Thus the TiO_2-SA-IDS film probably has a higher surface to volume ratio than the TiO_2-SA film which might be a reason for the faster reaction of the latter [3].

Fig. 5 illustrates the dependency of TiO_2-SA-IDS film resistance on pO_2. The squares are experimental data of two samples measured at 580 °C in mixtures of N_2 and either NH_3 or H_2. The resistances of sample 1 were taken from six processes equal to the one shown in fig. 3 which were carried out in a time span of one month. Sample 2 was added after four weeks, so that the last three measurements were taken at both samples. Sample 2 had been freshly prepared from a slightly different sol. Between the mentioned measurements several different processes were carried out to test the stability of the TiO_2-film. Partial oxygen pressure was measured with the ZrO_2 cell. The straight line is a least square fit with the known dependency of R upon pO_2

$$R = A \cdot \exp(E_A / kT) \cdot pO_2^{1/m} \qquad (5).$$

A is a constant, E_A is the activation energy, k is the Boltzman constant and m is determined by the conduction mechanism [7]. Fig. 5 also illustrates the reproducibility of the sample resistances. The logarithm of the resistance of sample 1 varies less than ±3 % during one month. It must be taken into account that TiO_2-films and ZrO_2 cell were placed about 20 cm distant from each other for technical reasons and the atmosphere might not have been homogeneous in the whole furnace due to sensor housings and sample holder for example. This explains variation of sample resistance at a given pO_2.

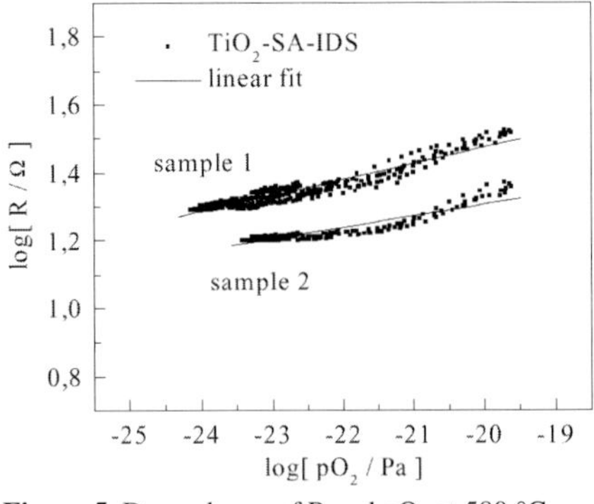

Figure 5. Dependency of R and pO_2 at 580 °C.

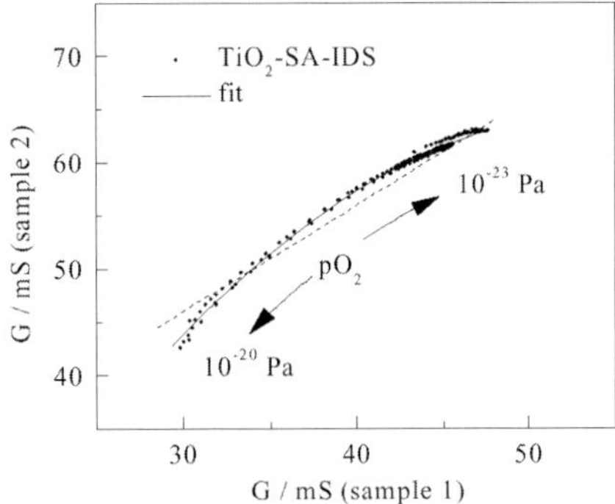

Figure 6. Comparison of two TiO_2-SA-IDS samples

The correspondence between the conductance of TiO_2-SA-IDS sample 1 and sample 2 is shown in fig. 6. The experimental data are taken from the three processes including both samples mentioned above. The TiO_2-SA-IDS samples were placed 3 cm distant from each other. The approximate range of corresponding partial oxygen pressures is indicated, too. Absolute identical sensors would correspond to a straight line with slope 1 and offset 0. The data correspond quite good to a straight line with slope 1 (dashed line in fig. 6), but conductance of sample 2 is about 16 mS higher than conductance of sample 2. Deviation from ideality might be due to different structure of the films, e.g. film thickness, grain size or porosity. The coating by dropping the sol onto the substrate does not allow exact control of film thickness. On the other hand the measured resistances are very close to the fitted line in fig. 6. The variation is less than 1 % indicating the high reproducibility of the sensor resistance.

5 Summary

TiO_2-films with an approximate density of 80 % and a grain size of (30-50) nm were prepared by sol-gel techniques. The TiO_2-films can be operated as oxygen sensors in nitriding processes. The films have a linear relation between log[R] and log[pO_2] in the examined pO_2 range, independent whether NH_3 or H_2 are present in the atmosphere. The log[R]-log[pO_2] characteristic which was tested in a 90 l nitiriding furnace has a good stability with a variation of less than 3 % in one month.

6 Acknowledgement

This work was supported by the Materials Science & Technology Cooperation (MATEC) Bremen.

7 References

[1] H. Klümper-Westkamp, F. Hoffmann, P. Mayr, et al., HTM 52, 4 (1997) 193
[2] B. Edenhofer, HTM 52, 4 (1997) 202
[3] E. M. Logothetis, W.J. Kaiser, Sens. Act. 4 (1983) 333
[4] U. Kirner, K.D. Schierbaum, W. Göpel, et al., Sens. Act. B1 (1990) 103
[5] S. Beling, A. Mehner, H. Klümer-Westkamp et al., in: 9th CIMTEC – World Forum on New materials, Symposium IX – Solid State Chemical and Biochemical Sensors, P. Vincenzini, L. Dori (Editors) 1999
[6] UST Umweltsensortechnik, Geraberg
[7] M. Li, Y. Chen, Sens. Act. B32 (1996) 83

Investigation of the kinetic of self-supported open porous oxygen sensors based on $Sr(TiFe)O_3$

Barbara Hippauf, Sylvia Schöllhammer, Albert Krügel, Ellen Ivers-Tiffée; Institut für Werkstoffe der Elektrotechnik (IWE), Karlsruhe.

1 Introduction

The development of resistive sensors for the determination of the rest oxygen content in the automobile exhaust gas is being investigated. There are used as oxygen sensors in a lean mix engine (λ>1) which is expected to reduce the fuel consumption .

Prior to using this sensor in a lean mix engine, the following conditions must be fulfilled: signal stability, signal reproducibility and response time of below 10 ms.

To avoid problems such as diffusion cations into the supporting ceramic (like Al_2O_3), a self-supporting sensor has been developed. This paper just stress on the investigation of the kinetic behaviour of the oxygen resistive sensor.

2 Production

Every sample is based on $Sr(Ti_{0,65}Fe_{0,35})O_3$, a material which the electrical conductivity depends on the oxygen partial pressure of the ambient atmosphere. A faster response of time can be obtained by increasing the surface area (or increasing the pore volume) which enhances the diffusion of oxygen. Image 2.1 shows the structure of three open porous bulks with respective calcination temperature of 1225, 1350 and 1380 °C.

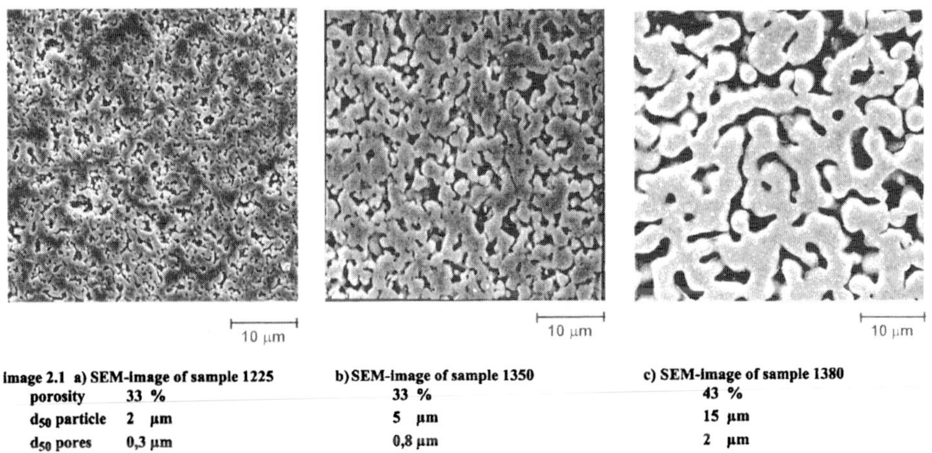

Image 2.1	a) SEM-Image of sample 1225	b) SEM-Image of sample 1350	c) SEM-Image of sample 1380
porosity	33 %	33 %	43 %
d_{50} particle	2 µm	5 µm	15 µm
d_{50} pores	0,3 µm	0,8 µm	2 µm

Every sample were sintered at 1300 °C, maintaining it for 10 minutes and brought to room temperature with 10 K per minute.

The homogeneous pore distribution of all the samples is shown.

Due to the systematic increase in the calcination temperature, different mean particle sizes where obtained. The mean particle sizes of the samples are d_{50} = 2 µm, 5 µm and 15 µm respectively. In order to obtain a faster rsponse of the sensor signal for the application in the automobile exhaust gas of internal combustion engine, it is necessary to know what kind of sensor parameters that influence the dynamic behaviour of the signal and which kinetically determining processes are possible. Then the sensor parameters can be optimized to obtain a required response of the signal.

3 Kinetic measurements

In order to obtain a certain information about the response of the sensor signal, kinetic measurements have been done by carrying out with temperature ranging from 700 °C up to 900 °C. Therefore the sensor is exposed to a pressure modulation with variable frequency ranging from 0.1 Hz to 2000 Hz. The pressure wave moves in the direction of the sensor which is four-point connected. It is feeded with a constant current I. The potential fall ΔU in consequence of the pressure modulation Δp is measured. With the fraction $I/\Delta U$, the changing in the electrical conductivity $\Delta\sigma$ of the signal is obtained. The amplitude of $\Delta\sigma$ and the phase responses have been obtained as a function of the wave frequency ω.

The length of the sensor where 7 mm and the width 6 mm. Only the thickness were varied from 0.25 to 1 mm. Image 3.1 shows the amplitude of the sensor signal $\Delta\sigma$ against the wave frequency ω dependent on the thickness of the sample.

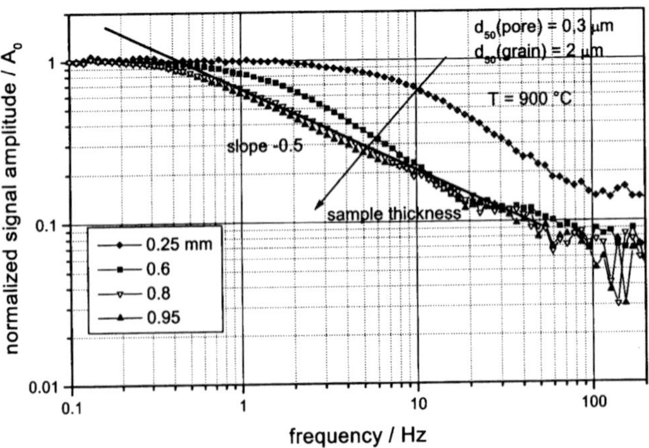

image 3.1: dependence of the signal amplitude on the thickness, sample 1225

The low pass filter character of the sensor signal can be distinguished. The thinnest sample (0.25 mm) shows a critical frequency of 10 Hz (the critical frequency is defined by a damping of 3 dB). The amplitude falls with a slope of –0.5 over the corner frequency.

Image 3.2 shows the signal amplitude against the wave frequency with the temperature as a parameter. The temperature were varied from 700 to 900 °C.

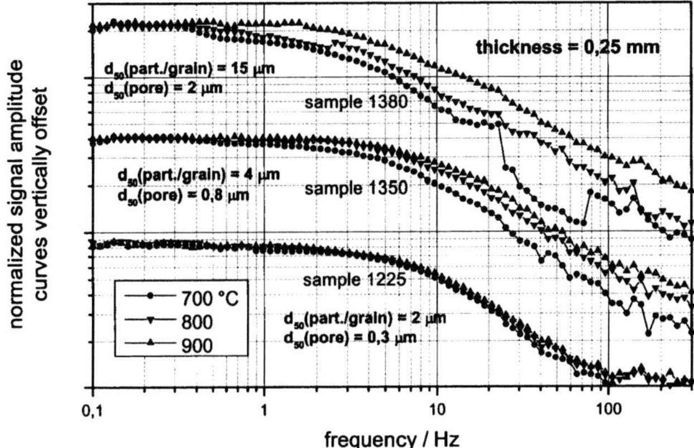

image 3.2 dependence of the signal amplitude of the three samples on the temperature

The image shows an increase of signal dependence on the temperature and a deceleration of the response of signal by increasing the grain size. A faster response is expected by increasing the mean pore size; oppositely, it suggests that the influence of the grain size is a response time limiting factor.

Based on these, the postulation of a model will be presented in order to understand the kinetics of open porous polycrystalline materials in a more detailed manner.

4 Kinetic model

There are various effects which influence the sensor kinetic. The kinetically determining processes are the sonic pressure propagation, the gas diffusion, volume diffusion and grain-boundary diffusion. The most important processes are the sonic pressure propagation and the volume diffusion.

4.1 Sonic pressure propagation

Firstly, a simplification is made: the sonic pressure wave propagates in a cylindrical open tube. Considering this, the solution of the respective equation of motion is searched. With the wave equation

$$\frac{\partial^2 p}{\partial y^2} = \frac{1}{c^2}\frac{\partial^2 p}{\partial t^2} + \frac{R}{\rho c^2}\frac{\partial p}{\partial t} \tag{4.1}$$

its solution in the case of expansion of a sonic wave in an open porous body [1]

$$p(y,t) = \hat{p} \cdot e^{i\omega t} e^{-\gamma y}. \tag{4.2}$$

is obtained, where p is the sonic pressure, y the propagation direction, c the sound velocity, t the time, ρ the density of the medium and R the aerodynamic resistance.

Equation 4.2 are composed of a time dependent term $e^{i\omega t}$ and a space dependent term $e^{-\gamma y}$. As for the following calculations, only the space dependent term is required because the amplitude is measured. The propagation constant γ in the case of porous materials becomes

$$\gamma = i \cdot \frac{\omega}{c} \cdot \sqrt{\chi} \cdot \sqrt{1 - i\frac{R \cdot P}{\omega\rho\chi}} \tag{4.3}$$

where χ is the structural factor which considers the elongation of the diffusion path through the pore labyrinth and P is the volume porosity.

Due to the highest aerodynamic resistance R of the samples (> 10^7 kg/m^{3}s), the term $\frac{R \cdot P}{\omega\rho\chi}$ becomes greater than 1 and equation 4.3 converts to

$$\gamma = \frac{1}{c}\sqrt{i\frac{R \cdot P}{\rho}} \cdot \sqrt{\omega}\,. \tag{4.4}$$

From equation 4.4 follows that the real term of γ and as a consequence of it, the damping is proportional to $\sqrt{\omega}$.

As the electrical conductivity is measured, a function $\Delta\sigma = f(\Delta p)$ is needed. $\Delta\sigma \sim \Delta p$ is a material property, so why the electrical conductivity can be written as follows:

$$\Delta\sigma \sim \hat{p} \cdot e^{-\gamma y} \tag{4.5}$$

$\Delta\sigma$ will then be proportional to the space dependent term of equation 4.2.

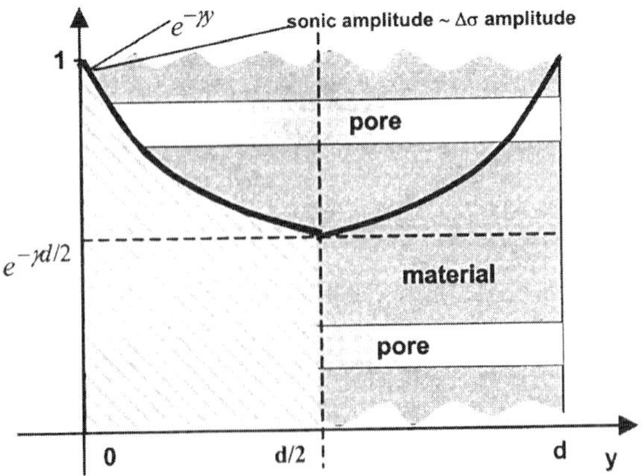

Image 4.1: dependence of $\Delta\sigma$ on the sample thickness (schematic)

If the sonic pressure propagation is considered as the only time determining process, it could be intergrated over the whole thickness d of the sensor (image 4.1). The whole changing in the electrical conductivity $\Delta\sigma_{ges}$ in the case of the changing in the sonic pressure mode Δp inside the open porous sample will then be obtained.

$$\Delta\sigma_{ges} \sim \frac{2}{d} \cdot \frac{1}{\gamma}\left(1 - e^{-\frac{\gamma d}{2}}\right), \tag{4.6}$$

If $\gamma d \gg 1$ then $\Delta\sigma_{ges} \sim 1/\gamma$ and as a consequence of it $\Delta\sigma_{ges} \sim 1/\sqrt{\omega}$, if $\gamma d \ll 1$ then $\Delta\sigma_{ges}$ is constant.

As the propagation constant γ is independent on temperature, equation 4.6 is independent on temperature too. The dependence of the sound velocity and the density of the medium on the temperature balance out.

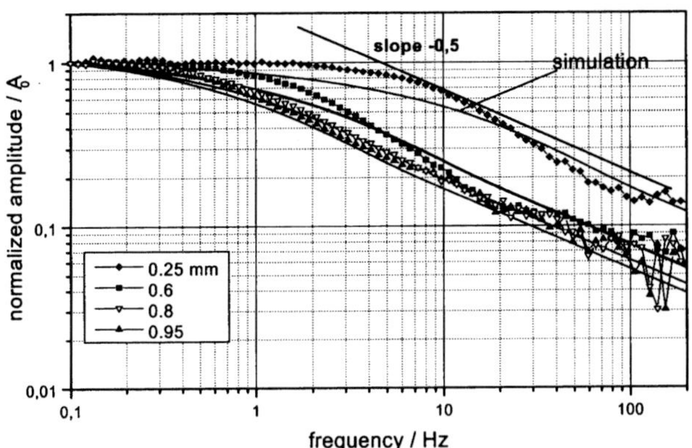

Image 4.2: comparison between experiments and simulation, sample 1225

Image 4.2 shows the comparison between the experiments and the simulation in the case of the sonic pressure propagation as the only response time determining process. The agreement is excellent.

4.2 Diffusion model

The first derivative of the oxygen partial pressure with respect to time conduces to a period vacancy concentration at the surface of the material. The Fick's second law

$$\left(\frac{\partial K}{\partial t}\right) = D \cdot \left(\frac{\partial^2 K}{\partial x^2}\right) \tag{4.7}$$

describes the propagation of the oscillation within the grain (intersolid diffusion), where K is the mean vacancy concentration and D its corresponded diffusion coefficient. The solution of the differential equation in the case of a sinusoidal wave for a finite-dimentional body with a defined thickness will be [2]

$$K(x,t) = \hat{K} \cdot \frac{\cosh\left(\frac{x(l+i)}{\delta}\right)}{\cosh\left(\frac{l(1+i)}{\delta}\right)} \cdot \sin(\omega t) \tag{4.8}$$

where l is the half grain length. The penetration depth δ is defined as $\sqrt{2D/\omega}$ and x represents the space coordinate. After the integration of equation (4.8) over the half grain length, the solution of the whole electrical changing of the conductivity $\Delta\sigma_{fest}$ inside the grain will be obtained.

$$\Delta\sigma(\omega)_{fest} \sim \frac{\tanh\left(\frac{l(1+i)}{\delta}\right)}{\frac{l(1+i)}{\delta}} \tag{4.9}$$

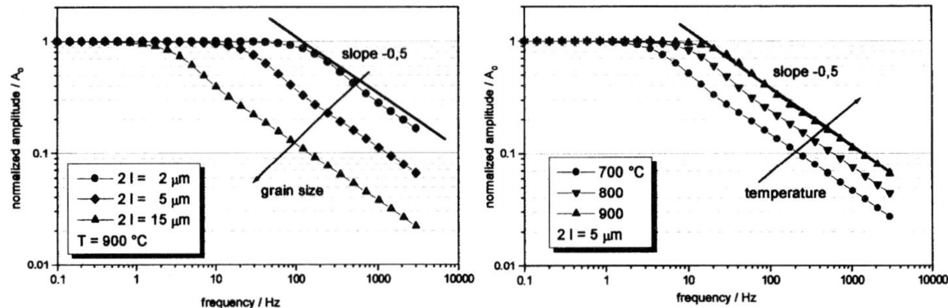

image 4.3 a) Kinetic as a function of the grain size (simulation) image 4.3 b) Kinetic as a function of the termperature in the case of the volume diffusion as kinetically determining process

The result of plotting the electrical conductivity $\Delta\sigma_{fest}$ (4.9) against the frequency ω is shown in the images 4.3 a) and 4.3 b) respectively. If the mean grain size is chosen as a parameter by fixed temperature (image 4.3 a) the response of signal becomes faster when the mean grain size decreases; whereas when the temperature is a parameter and the mean grain size becomes constant, the set of curves will become a high temperature dependence. The simulation agrees excellently with the experimental kinetic behaviour (compare with image 3.2).

5 Conclusions

Due to the open porous self-supported gas sensor, the kinetic is determined by the sonic pressure propagation, the gas diffusion inside the pores as well as by the grain-boundary diffusion and the volume diffusion. The kinetically determining processes are the sonic pressure propagation and the gas diffusion by small pores. The volume diffusion will be the determining process if the mean grain size increases.

To obtain an optimal fast response of the signal, it must be considered that the mean pore radius must be greater than 1 μm and the mean grain size lower than 2 μm.

Literature

[1] Frank, P. und Mises, R. v., Die Differential- und Integralgleichungen der Mathematik und Physik, Band II, Vieweg, Braunschweig, 1961, S. 570-576.
[2] Föllinger, O., Laplace- und Fourier-Transformation, Hüthig, Heidelberg, 1993, S. 161-181.

$FeSi_2$ for Sensor Applications – Control of Functional Properties by Composition

E. Müller, K. Schackenberg, H. Ernst, E. de Groote, and W. A. Kaysser
German Aerospace Center (DLR), Cologne, Germany

1 Introduction

Semiconducting β-iron disilicide, known as a technologically easy to handle high temperature resistant low-cost material for thermoelectric (TE) energy conversion [1], is due to its exceptional high Seebeck coefficient [2] attractive as a functional material for thermal sensors as well, to be applied significantly above room temperature [3,4].

Numerous applications for control and modeling of high temperature processes in industries, automotive and aerospace require thermal sensors able to stand temperatures of several hundred degrees C during longer periods of operation. Heat flux measurement at high temperature is required e. g. for determination of thermal loads in aerodynamic test facilities for hot gas engines such as stationary or airplane turbines.

So far, metallic sensors basing on the resistive or thermoelectric functional principles have been used for heat flux measurement. On the contrary, lowly doped semiconductors are suitable to fabricate arrays of highly responsive small-size sensors for a locally resolving two-dimensional heat flux measurement, when patterned by modern thin film techniques (evaporation, sputtering or other coating methods). Another pre-condition for operation under hot-gas conditions is chemical and thermomechanical stability of the semiconducting material and its composites with other components of the sensor structure in a design which is restricted to not more than a few hundred microns and which does not protrude neither into the constructional part nor invade into the hot gas stream.

An obstacle to smart sensors, so far, is linked to the relatively narrow temperature range of high Seebeck values. Good responsivity is obtained merely in a restricted T range, whereas in adjacent ranges strongly depends on temperature. For calibration, an accompanying T measurement is required, reducing the speed in dynamic measurement.

2 Iron disilicide as a high temperature thermoelectric material

While iron disilicide, occupying merely a restricted spectrum of application cases as a thermoelectric material, mainly due to the moderate conversion efficiency, having its advantage mostly in low material and processing price, environmental sustainability and unlimited availability, the functional properties are very suitable for high temperature heat flux measurement. Compared to optimized TE converter material, where a thermopower of about 160 to 200 μV/K is achieved, the doping content of the semiconductor has to be reduced for heat flux measurement. Two to three times higher Seebeck values

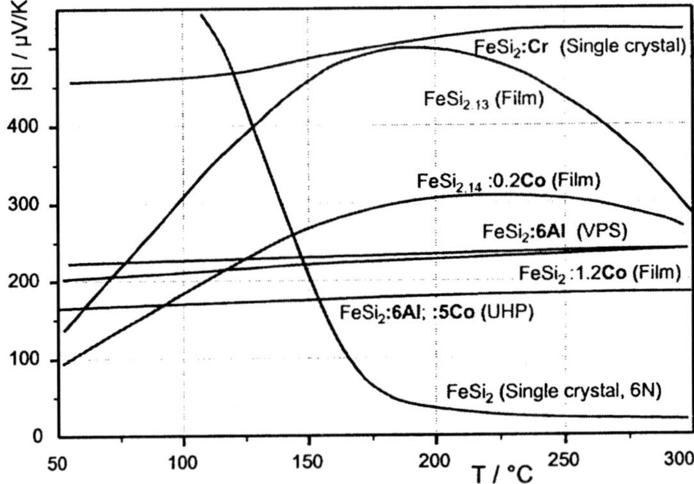

Figure 1. Seebeck coefficient of iron disilicide - Dependence of magnitude and temperature curve on doping, composition, and preparation technique
UHP – Uniaxially hot pressed; VPS – Vacuum plasma sprayed;
Film – Magnetron sputtered thin films [3,4]
Single crystal – Gas phase grown high purity single crystals [2]

have been reached in this way, with the magnitude and temperature behavior depending (beyond the type and concentration of doping) furthermore on stoichiometry and the influence of preparation technique. Fig. 1 reveals a striking difference in the temperature behavior, depending on the doping content. Whereas doping with more than about 1 at% Al or Co leads to relatively flat Seebeck curves $S(T)$ (independently whether they are made by hot pressing or plasma spray of bulk material or sputtering of thin films), lower doped samples reach much higher maximum values but exhibit bell shaped temperature curves with narrow range of highest values and steep slopes, which are not desired for sensors to be operated over a broader temperature range. Extremely high Seebeck values have been found for high purity gas phase transport grown single crystals, which are out of consideration for sensor use for practical reasons. Nevertheless, they give very remarkable proof that flat shaped curves are even possible at highest magnitude.

For heat flux meters, the thermo-power is the most significant quantity, to be increased for high sensitivity. Thermal conductivity affects the dynamic behavior and sensitivity to some minor extent.

3 Doping variation for maximization of the Seebeck coefficient

A series of five samples with stepwise changed Al doping content has been prepared by hot pressing of gas-atomized powders [5]. Their thermoelectric properties have been investigated. The Al concentration was varied from undoped, referred to as "A0", to 6 at% Al doped (pre-alloyed during gas-atomizing) – "A100". Mixtures of both these powders containing 25%, 50%, and 75% of the A100 material have been milled together in a planetary ball mill for one hour. Samples were consolidated by uniaxial hot pressing for 15 min at 940 °C under vacuum and afterwards annealed for a few hours to complete

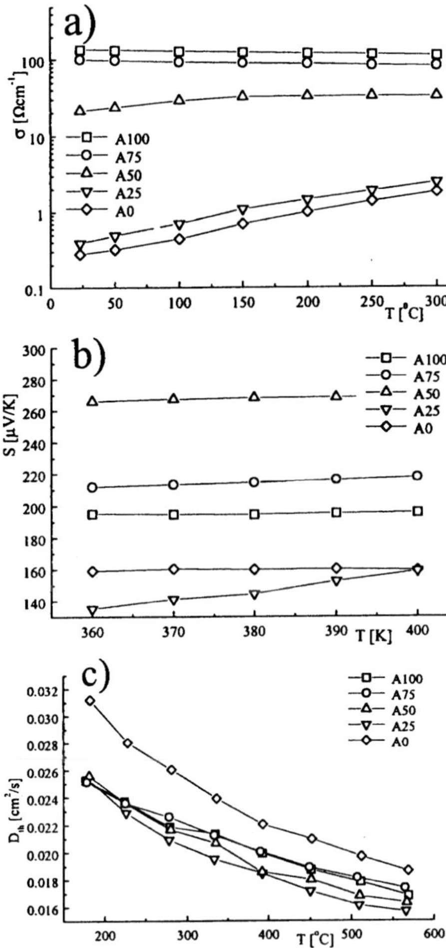

Figure 2. Electrical conductivity (*a)*), Seebeck coefficient (*b)*), and thermal diffusivity (*c)*) of $FeSi_2$:Al with changing Al content

the phase transformation into the desired semiconducting low-temperature phase (β-$FeSi_2$).

Seebeck coefficient, electrical conductivity and thermal diffusivity have been measured in dependence on temperature (Fig. 2). As expected, the electrical conductivity decreases with decreasing doping content, but stronger than according to the nominal doping concentration. This proves the occurrence of other mechanism of Al consumption, which could be linked to the oxides incorporation during processing [5].

A remarkable feature is the change of the temperature slope of the electrical conductivity. While for heavy doping ordinary band conduction can be assumed, the positive temperature slope at low doping (A0, A25) indicates carrier excitation processes as they are typical for hopping conduction. This change of conduction mechanism is emphasized by the doping dependence of the Seebeck coefficient *S*. Its increase with falling doping according to band conduction is continued down to about nominally 3 at% Al ("A50"). There, a maximum is reached, with much lower Seebeck values at lower doping. The exact maximum position on the concentration scale has to be cleared out during further investigation. In particular, not only the magnitude, but the temperature slope of *S* as well depend on the conduction mechanism (A25), hence the slope may directly be controlled in the low doping composition range. This is of importance for sensors, which responsivity shall not depend on temperature in a certain range of application.

The thermal properties are not significantly influenced by the observed change of conduction mechanism. The thermal conductivity is predominantly linked to lattice thermal conduction. However the tendency apparent from fig. 2 is another one: The thermal conductivity is significantly reduced for all samples containing Al, compared to the non-doped $FeSi_2$.

In continuation, the Al concentration value where the conduction mechanism is changing has to be identified more precise. In particular, this concentration range will offer a very interesting subject to study the influence of structural disorder on transports. Since the Seebeck coefficient is quite high there and is dependent on the balance of two significantly contributing effects, it becomes an indicator on the content of disordered regions in powdermetallurgical $FeSi_2$. By controlling preparation parameters influencing the state of disorder of the material, maximum Seebeck coefficient shall be reached and stabilized.

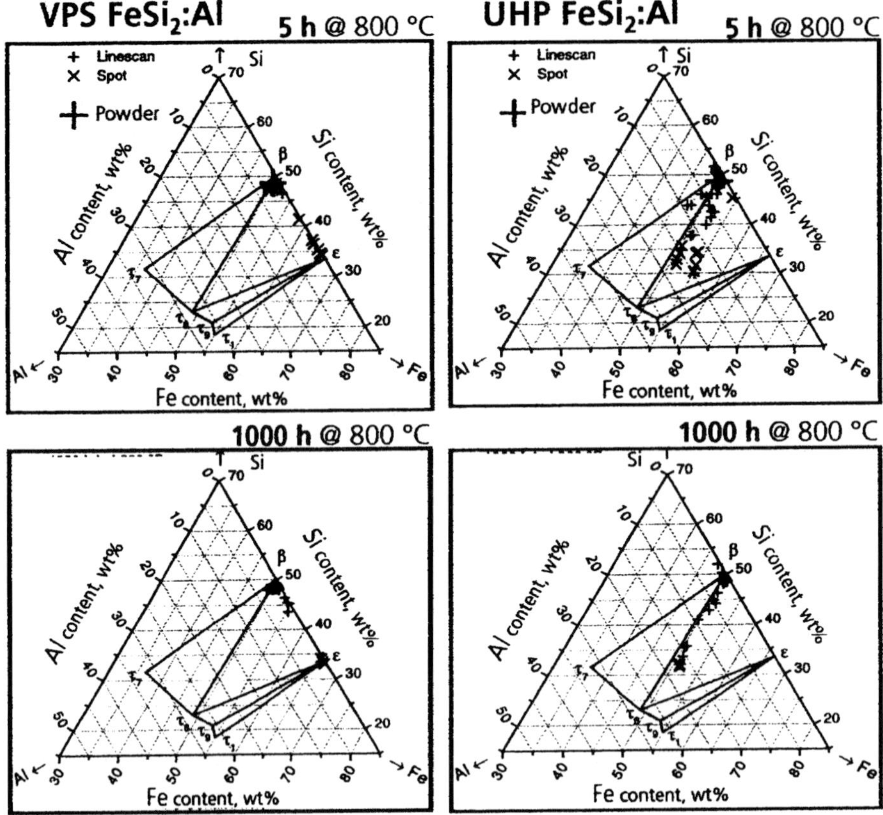

Figure 3. Diagnostics of the phase constitution of heterogeneous powdermetallurgical iron disilicide: Ternary plots of EPMA scans on polished sample surfaces (representation of the local elemental concentrations), soon after solidification (5 h annealing at 800 °C) and after longer high temperature treatment (1000 h @ 800 °C)
VPS - Vacuum plasma sprayed material, strong oxidation influence (0.7 wt% oxygen content)
UHP - Hot pressed material, low oxygen incorporation (0.2 wt%)

4 Compositional effects in $FeSi_2$:Al

The phase constitution and aging behavior of $FeSi_2$:Al during long-term annealing has been investigated in some detail for hot pressed and thermally sprayed $FeSi_2$ [6,7]. It could be shown that the incorporation of oxides during preparation, where preferentially

SiO_2 is accumulating at the particles' surface, leads to a long-term aging process visible through a continuous slow drift of the transport properties.

During annealing at 800 °C, diffusion brings Al from the particles' interior to their surface, where finally Al_2O_3 as the chemically most stable oxide is formed. This is accompanied by an immediate doping loss, by oxidic precipitation near material separations and at the particles' surface [6].

Oxidation to SiO_2 and moreover to Al_2O_3 brings along Si depletion in the semiconductor, leading to the formation of Fe rich phases (ε or τ_7, τ_8, τ_9, depending on the Al and Fe excess in the sample; see fig. 3) which are degrading the thermoelectric properties of the sample due to their metallic character. Phase analysis of annealed doped $FeSi_2$ revealed, that these phases act as drains for alloying elements: The τ_i phases of the ternary Fe-Si-Al phase diagram are containing much more Al than the pre-alloyed β-$FeSi_2$:Al, thus reducing the electrically active doping of the semiconductor.

The transport properties react in a very typical way during several hundred hours of aging [7]: For band conduction, the Seebeck coefficient increases with doping loss. The electrical conductivity increases with the healing out of material separations (a lot of them are found in thermally sprayed material) but decreases with doping loss, and with the formation of oxide layers. The thermal conductivity as well initially increases due to microstructural healing but then gradually decreases due to accumulation of fine distributed oxide precipitation during sintering.

It's of crucial importance for the feasibility of high temperature sensors from iron disilicide whether these aging processes come to a stable final state.

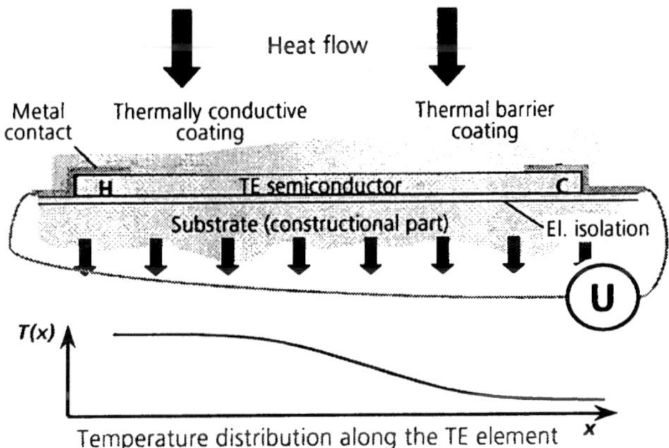

Figure 4. Schematic of a flat-shaped thermoelectric heat flux meter structure for high temperature operation, to be mounted at the surface of a constructional part in a hot gas engine

5 Design concept

High stream hot gas conditions and structures working near their strength limit imply hard restrictions to sensor design. Small size flat structures mounted to hot surfaces are desired.

A **design concept** is followed, meeting these criteria. The application of a semiconductor as an active material enables high signal and local resolution. This concept starts out of a quite simple set-up (fig. 4): A stripe-shaped layer of thermoelectric material, not necessarily thicker than several ten microns, is deposited electrically isolated but thermal well anchored to a substrate (e.g. a metallic constructional part). It is contacted electrically at both its ends which are covered with electrically isolating thin layers of different thermal properties, one of them offering a low thermal resistance to the cross-plane heat flux ("hot" side of the thermoelectric stripe), whereas the other layer shall possess an as high as possible one ("cold" end of the stripe). Conventional thermal barrier coating techniques may be applied here.

6 Conclusion

Iron disilicide has been found to be a suitable thermoelectric material for small-size heat flux meters, due to its high signal responsivity (its very high thermo-power *S*) and its good chemical high temperature stability. *S(T)* can be tuned by doping, stoichiometry, and preparation. The influence of oxidation has been proved.

Al incorporation into $FeSi_2$ leads to faster diffusion, better densification of the material and stronger phonon scattering. As a major mechanism of aging, besides microstructural healing Si and Al oxidation have been identified, leading to Fe rich phases acting as dopant drains. Thus, Fe:Si stoichiometry control is of central interest to obtain long-term stable material.

Al has been found to be interesting as an additive to powdermetallurgical iron disilicide beyond its effect as a dopant. In small amounts (1 at% and below), it can be used to reduce other oxides, which might have drastic TE effects, to an inert final state, to decrease thermal conductivity and accelerate sintering processes during preparation and aging.

We want to express our gratitude to Dr. G. Schiller (DLR Stuttgart) for fabricating thermal spray samples for performance evaluation.

7 References

[1] U. Birkholz, E. Groß, U. Stöhrer, *Polycrystalline Iron Disilicide as a Thermoelectric Generator Material*, in: D. M. Rowe (Ed.), CRC Handbook of Thermoelectrics, CRC Press, Boca Raton 1995, 287-298.

[2] G. Behr, J. Werner, G. Weise, A. Heinrich, A. Burkov, C. Gladun, *Preparation and Properties of High-Purity β-FeSi₂ Single Crystals*, phys. stat. sol. (a) **160**, 549 (1997)

[3] J. Schumann, H. Griessmann, A. Heinrich, *Doped β-FeSi₂ as Thin Film Thermoelement Sensor Material*, XVII. Int. Conf. on Thermoelectrics, Nagoya, Japan, ISBN 0-7803-4907-5, Piscataway, NY 1998, 221-225.

[4] H. Griessmann, A. Heinrich, J. Schumann, D. Elefant, W. Pitschke, and J. Thomas, *Thermoelectric Transport Properties, Structure Investigations and Application of Doped β-FeSi₂ Thin Films*, XVIII. Int. Conf. on Thermoelectrics (ICT'99), Aug. 29th to Sept. 2nd, 1999, Baltimore, MD.

[5] E. Müller, K. Schackenberg, *Funktionsgradiertes FeSi₂ für thermische Sensoren*, DGM-Symposium „Verbundwerkstoffe und Werkstoffverbunde", October 5th to 7th, 1999, Hamburg, Germany.

[6] K. Schackenberg, E. Müller, W. A. Kaysser, J. Schilz, *Thermoelectric Performance and Long Term Stability of Plasma Spray Formed Iron Disilicide*, Euromat'99, Sept. 27th to 30th 1999, Munich, Germany.

[7] K. Schackenberg, E. Müller, J. Schilz, H. Ernst, W. A. Kaysser, *Carrier Density Behavior During Aging of Doped Plasma Spray Formed Iron Disilicide*, XVII. Int. Conf. on Thermoelectrics, 25.-29. 5. 1998, Nagoya, Japan, ISBN 0-7803-4907-5, Piscataway, NY 1998, 422-425.

Microstructurable Glasses and Technologies

Alf Harnisch, Dagmar Hülsenberg
Technical University of Ilmenau

1 Abstract

Special glasses from the system Li_2O-Al_2O_3-SiO_2 are microstructurable by a UV-lithographic process. If you use the basic process it is possible to prepare two dimensional through structures in a glass wafer.

A new exposure technique allows a 3D etching process in photosensitive glass. A pattern with a defined width and mesh are made in a chromium layer on vitreous silica. By use of these mask a different UV-intensity results in the glass during the exposure. Due to the different intensities, the exposure of various depths is possible.

The advantage of this technology is the production of structures with two dimensional free design, various depths and a defined angle of walls in one process.

2 Material Basics

All used glasses were developed and produced at the Technical University of Ilmenau [1,2,3]. Glasses from the Li_2O-Al_2O_3-SiO_2 system are modified by micro additions, to achieve photosensitivity. Doping agents are the oxides of Ag, Ce, Sn and Sb. The photosensitive glasses have thermal expansion coefficients of $\alpha_{20-400\,°C} = 6 - 13{,}6 \cdot 10^{-6}\ K^{-1}$.

3 Technological Basics

The basic photo-structuring process with the partial process steps UV-lithography, thermal treatment and etching is shown in fig. 1.

The structuring of photosensitive glass is based on different etching rates of exposed (crystallized) and unexposed glass. The first step is a UV-exposure of the glass. A thin structured chromium layer on vitreous silica is used as a mask, which lays on top of the glass wafer. After the UV-exposure a thermal treatment follows. During these process the exposed areas crystallize. The crystals can be etched away with an anisotropic etching process in hydrofluoric acid.

Lines with widths of 150-200 µm and various depths can be realized without any difficulties. Depending on the glass wafer thickness, holes and grooves of 10 µm up to several millimeters width can be fabricated. The slope angle of the walls is approximately 2-6 degrees (depended on the kind of glass and the technological conditions). Glass wafers of 0.5 mm up to 2.0 mm thickness can be well handled. It is possible to produce through structures by the basic process and structures with a various depth by an etch stop process. A multi step structuring technology allows complex structured glass elements (through structures and structures with a defined depth) [4].

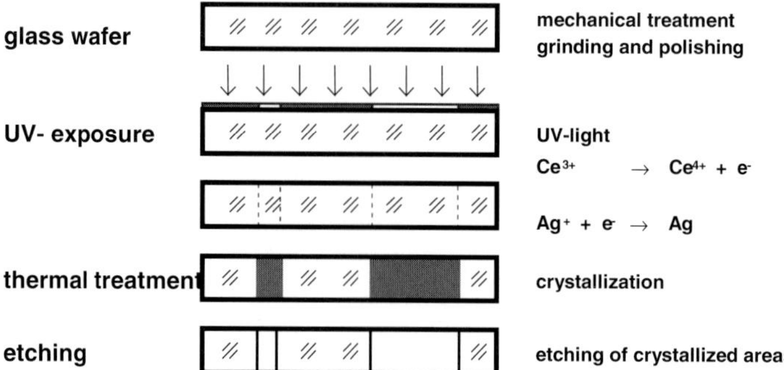

Fig. 1: Basic process of photostructuring

This technology needs only low investments. Because of this it is well suitable for small and medium companies.

4 Modified Mask Technology

Experiments have shown, that holes smaller than 10 µm can not be etched through an 1 mm thick wafer, because these holes are not completely exposed in depth. Using this effect a mask was modified with a pattern (fig. 2). Figure 3 shows the depth h of structures and speed of etching v_E depending on ratio of transmission TA and time of etching t_E.

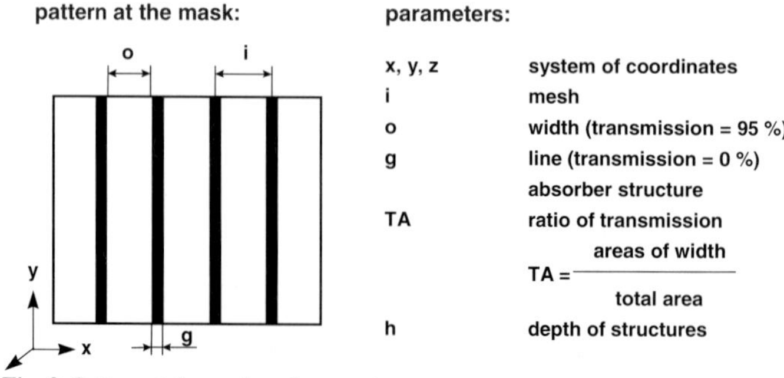

Fig. 2: Pattern at the mask and parameters

The depth of the structures approaches to a threshold value of depth for long etching times. The threshold value depends on the value of transmission and the density of exposure energy.

At the beginning of the etching process, the speed of etching is high. For long etching times the speed of etching decreases (fig. 3). Then the etching speed of the exposed and unexposed areas is nearly the same, the depth is nearly constant.

With this technology two dimensional free design, various depths and a defined angle of walls in one process are possible. Restrictions of resolution exist only in the grey-scale pattern. Steps in etched structures of a 0.5 μm step pattern are clearly recognizable (fig. 4). Using 0.1 μm steps nearly continuous forms can be made (fig. 4).

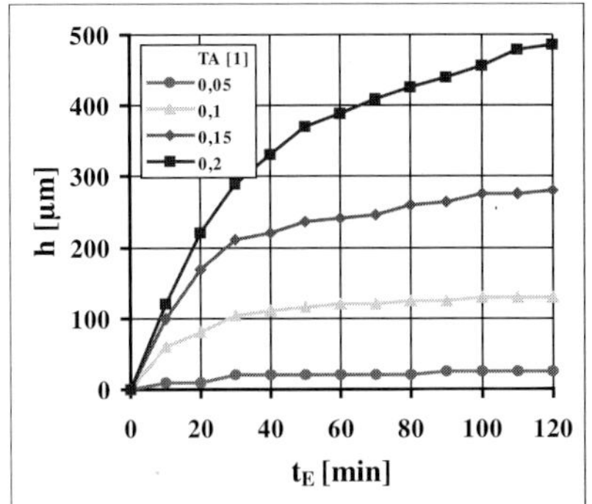

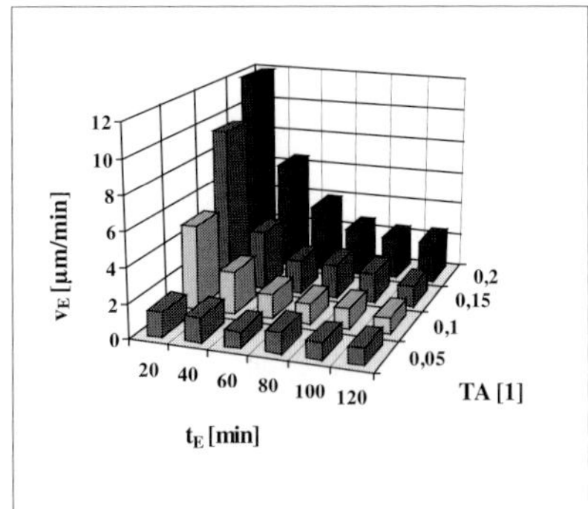

Fig. 3: Depth h and speed of etching v_E depending on ratio of transmission TA and time of etching t_E (mesh: 20 μm)

The depth of the structures approaches to a threshold value of depth for long etching times. The threshold value depends on the value of transmission and the density of exposure energy.

At the beginning of the etching process, the speed of etching is high. For long etching times the speed of etching decreases (fig. 3). Then the etching speed of the exposed and unexposed areas is nearly the same, the depth is nearly constant.

With this technology two dimensional free design, various depths and a defined angle of walls in one process are possible. Restrictions of resolution exist only in the grey-scale pattern. Steps in etched structures of a 0.5 μm step pattern are clearly recognizable (fig. 4). Using 0.1 μm steps nearly continuous forms can be made (fig. 4).

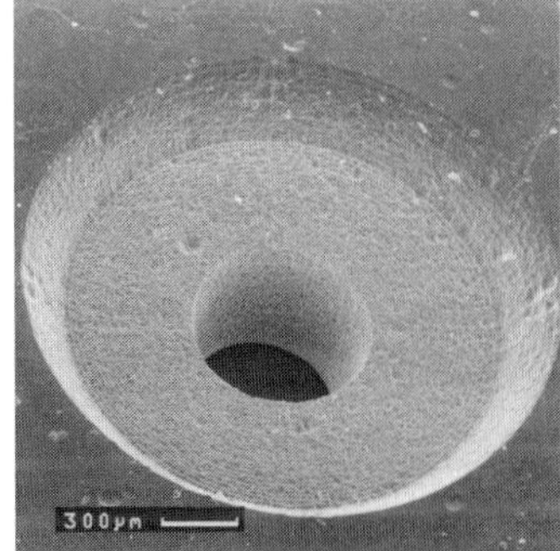

Fig. 4: Samples, structured by modified mask technology

5 Calculation of Distribution of Intensity

The Fresnel-Huygens-Approach (At all not covered points of a wave front starts a spherical elementar wave. All elementar waves interfere.) was used. The principle of calculation and the defined parameters are shown in Figure 5.

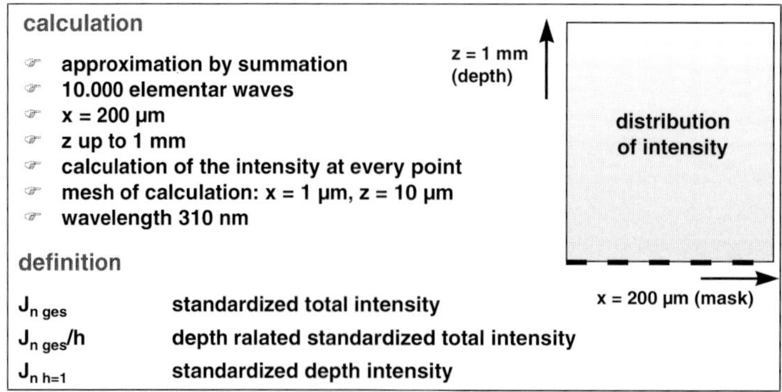

Fig. 5: Principle of calculation and defined parameters

The intensity at every point in the area (x= 200 µm and z=1 mm) were calculated. The standardized total intensity is the total amount of the calculated intensity at every point. Figure 6 shows the standardized total intensity and the depth of structures depending on the geometry of pattern.

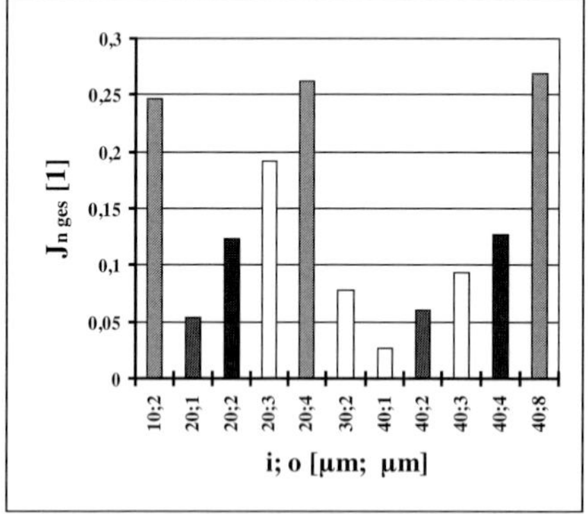

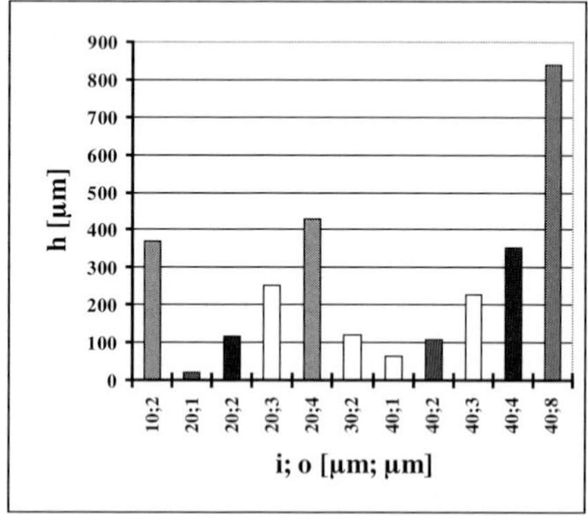

Fig. 6: Standardized total intensity (J_{nges}) and depth of structures (h) depending on mesh (i) and width (o)

Patterns with the same TA (e.g. 20;2 and 40;4) have nearly the same standardized total intensity. The depth of these structures (same TA) is not the same. The depth of the structure with the decreased width is lower than the depth of the structure with the increased width. This effect is not evidently by a transmission controlled process of structuring. Especially at small widths it is also a diffraction of the radiation during the exposure process. Figure 7 shows the distribution of intensity in the calculated area for various geometry's of pattern.

The diffraction restrict the depth of structures for small widths. The smaller width of the pattern results in a shorter depth of structures. The transition between transmission controlled structuring and diffraction controlled structuring depends on the geometry of pattern. At a mesh of 20 µm the transition is at a width of 2 - 3 µm.

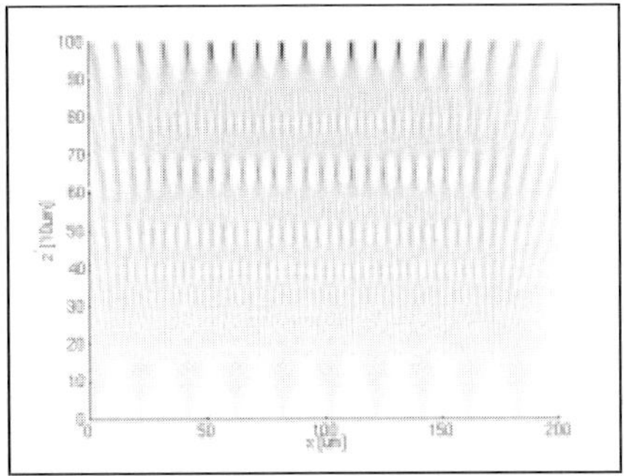

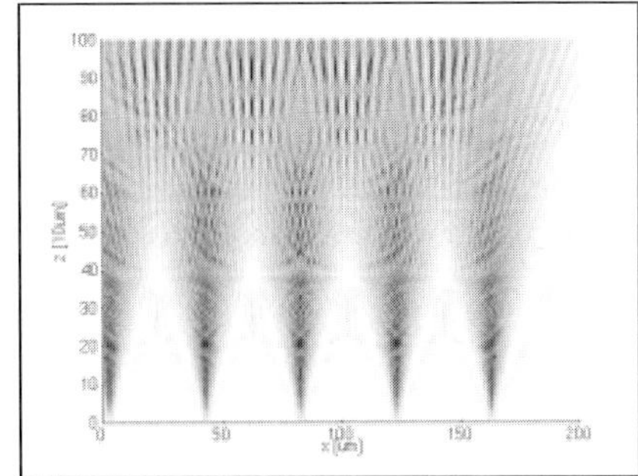

Fig. 7: Distribution of intensity (left: mesh: 20 µm, width: 2 µm; right: mesh: 40 µm, width: 4 µm)

The technique of modified masks makes it possible to produce 3dimensional free structures out of undercut. The etching process not stops at any depths, there is only a reduction of the speed of etching. The etched depth depends on the geometry of pattern and the energy of exposure. Figure 8 shows the roughness at the bottom of structures.

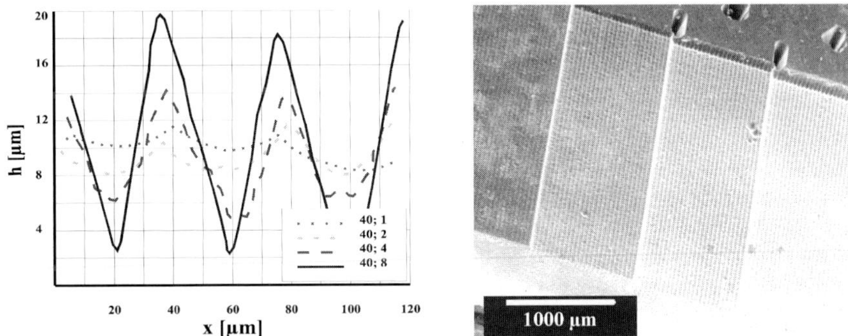

Fig. 8: Roughness at the bottom of structures. The geometry of the pattern is imaged at the bottom of structure. An increase of the width of the pattern causes a higher roughness at the bottom.

6 Conclusion

The thermal expansion coefficient of photosensitive glass can be adapted to different materials. Related to other structuring technologies (e.g. LIGA-technique), the glass structuring is inexpensive. In dependence of the necessary geometry a lot of technological variations are possible. The modified mask technology opens new possibilities of 3d-structuring of photosensitive glasses.

Applications of the structured glass elements are deformation elements and elastic glass structures [5], microactuators [6], fluid components [7] and microgripping devices [8].

7 Acknowledgement

This paper based on a project supported by BMBF (Bundesministerium für Bildung, Wissenschaft, Forschung und Technologie), number of support: 03N1013B0. The responsibility of the content have the authors.

8 References

1. A. Ehrhardt, B. Straube, D. Hülsenberg: Microstructurable glasses with high thermal expansion coefficients. MicroMat 97, Berlin 97
2. Hülsenberg, D.; Harnisch, A.; Ehrhardt, A.; Baumgart, H.; Straube , B. Mikrostrukturierbare Gläser hoher thermischer Dehnung. Werkstoffwoche 1998, Symposium 12, München 1998
3. Ehrhardt, A.: Beitrag zur Entwicklung fotostrukturierbarer Gläsern und Glaskeramiken mit unterschiedlichen thermischen Ausdehnungskoeffizienten, Ilmenau, Dissertationsschrift, 1999
4. Alf Harnisch: Beitrag zur Entwicklung von Herstellungstechnologien für komplexe Bauteile aus mikrostrukturiertem Glas. Dissertationsschrift, Technische Universität Ilmenau, 1998
5. Scherge, M.; Schäfer, J., Mollenhauer, O.: The role of water on the microtribology of MEMS. Proceedings of MST 98, Potsdam 1998
6. Feindt, K.; Harnisch, A.; Hülsenberg, D.; Kallenbach, E.: Miniaturised electromagnetic actuators in glass, 9-IPES/UME 4 Braunschweig, May 1997, Procedings Volume 2 714-717
7. Harnisch, A.; Hülsenberg, D.; Feindt, K.; Kallenbach, E.; Töpfer, H.; Mollenhauer, O.; Lösche, S.; Hartleb, M.: Anwendungen von mikrostrukturierbaren Gläsern. Werkstoffwoche 1998, München, Okt. 1998
8. Zöppig, V.: Untersuchungen zur Entwicklung von Miniaturgreifern für die industrielle Mikromontage. Dissertationsschrift, Technische Universität Ilmenau, 1998

A New Ultra-light Flexible Large Area Thin Film PSD Based on Amorphous Silicon

Elvira Fortunato, Isabel Ferreira, Franco Giuliani, Rodrigo Martins
Department of Materials Science, CENIMAT, Faculty of Sciences and Technology, New University of Lisbon and CEMOP-UNINOVA, 2825-114 Caparica, Portugal

1 Introduction

The technical and theoretical concepts of large area thin film position sensitive detectors based on the amorphous silicon technology was introduced some years ago [1, 2, 3], where most of the research done was concentrated in the development of devices deposited on rigid substrates, such as glass. Nevertheless, in applications where the device is mounted on a rotating axis of a motor or determination of the angular position of the rotor of a motor is required, it should be of great importance to have a curved sensor with a cylindrical shape. To do so, it is important to have the sensing element deposited on a flexible substrate with the required position accuracy and reliability, besides having a high enough response frequency. In this paper we present the electro-optical performances of one dimensional (1D) a-Si:H thin film position detectors deposited on polymeric substrates (Kapton – polyamide foil), compatible with all the steps of the deposition of the materials that constitute the final device as well as with the photolithographic techniques used. Apart from that, we also demonstrate that when improved interfaces are produced the back metal contact does not need to cover the entire back area of the sensor to produce the required reference equipotential.

2 Experimental Procedures

The substrates used in this work are from Kapton foil (polyamide) type HN, with 125 μm thickness. The main reason for choosing such polymer is that its thermal, mechanical, chemical and electrical properties are retained over a wide range of temperatures (-270 °C to 400 °C). Besides these properties it is transparent (slightly amber coloured) with a refractive index of 1.66 and has a total transmission of about 70%, for a wavelength (λ) of 650 nm, which allows the utilisation of diode lasers (630-670 nm) as the optical source for the position sensors developed. In table 1 we compare the most important physical properties of Kapton foil type HN, (125 μm thickness) and Corning 7059 glass.

The a-Si:H position sensors developed have a pin structure deposited onto intrinsic tin oxide (SnO_2), which had deposited previously onto the polymer substrates. The SnO_2 layers were produced by spray pyrolysis at a temperature of 325 ºC using a solution of $SnCl_4{\cdot}5H_2O$ (Riedel-de Haën, ref. 14550 with 98% purity) dissolved in isopropilic alcohol (Pronalab with 99.7% purity). The resistivity (ρ) of this layer is of the order of 10^{-2} Ωcm and presents an average transmittance (T) over the visible spectrum) of 85%. The relatively high value for ρ of the SnO_2 is necessary to avoid the existence of a quasi equipotencial layer [4, 5], which could annul the appearance of the required lateral photovoltage effect [6].

Table 1. Comparison of the main physical properties of corning glass 7059 and Kapton foil type HN (125 μm).

	Corning Glass (7059)	Kapton foil type HN
Density (gcm^{-3})	2.76	1.42
Refractive index (visible range)	1.53	1.66
Thermal expansivity ($10^{-6}K^{-1}$)	4.6	20
Roughness (μm)	0.01	0.20
Upper working temperature (°C)	639	400
Dielectric constant (1 MHz)	5.84	3.40
Volume resistivity (Ωcm)	1.3×10^{13}	1.5×10^{17}
Young modulus (Pa)	6.7×10^{10}	2.5×10^{9}
Thermal conductivity ($Wm^{-1}K^{-1}$)	0.16	1.09

The pin structures were deposited in a single chamber reactor by conventional Plasma Enhanced Chemical Vapour Deposition (PECVD) technique using a frequency discharge of 13.56 MHz. Table 2a shows the process parameters used to produce the different a-Si:H layers, including the gases and corresponding percentages, while table 2b shows the electro-optical properties obtained for the individual layers produced.

Table 2a. Deposition parameters of p, i and n type a-Si:H layers.

Deposition parameters	p layer $[B_2H_6(0.2)SiH_4(44.2)$ $CH_4(33.4)H_2(22.2)\%]$	i layer $[SiH_4(100\%)]$	n layer $[PH_3(0.7)SiH_4(15.1)H_2(2)$ $He(82.2)\%]$
Temperature (°C)	240	230	220
Pressure (mTorr)	0.2	0.5	0.2
Power (mW/cm^2)	50	100	50
Flux (sccm)	20	7	10

Table 2b. Main electrical and optical properties of the p, i and n a-Si:H layers.

Properties	p–layer	i–layer	n–layer
Dark conductivity, σ_d $(\Omega cm)^{-1}$	10^{-5}	8×10^{-11}	4.5×10^{-3}
Activation energy, ΔE (eV)	0.20	0.79	0.15
Optical gap, E_{op} (eV)	1.70	1.60	1.80
Photoconductivity, σ_{ph} $(\Omega cm)^{-1}$	---	8×10^{-6}	---

The back contacts are of Al and were deposited by electron-gun thermal evaporation. Due to the improved i/n interface the metal contact was restricted to a strip with an area of about 7 mm^2. This achievement not only proves our previous models concerning the lateral effect in pin junctions [7] being, in our opinion, a major technical breakthrough in producing large area position sensors. Now, one less production step (and mask) is needed and the yield is improved since the possibility to have micro-short circuits decreases [8], besides leading to an overall enhancement of the device response time.

A schematic cross section of the a-Si:H position sensor structure on the polymer substrate is shown in figure 1. The three top contact stripes are used separately and the objective is to perform different measurements at different contact positions, aiming to establish how far the carrier's collection could be performed with the required linearity. In this paper the results presented have been done between the bottom and one of the lateral strip contacts, 2 cm lateral away from where the top metal contact was performed. Figure 2 illustrates some of the fabricated flexible a-Si:H position sensors. To test the position sensors, the processed substrates were cut in strips 1 cm wide and 2.5 cm long.

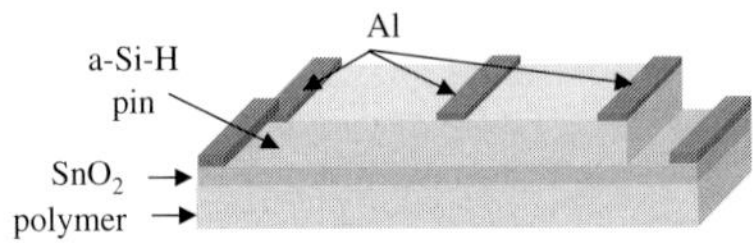

Figure 1. Schematic cross section of a-Si:H position sensor deposited on Kapton foil 125 □m thickness, showing the location of 3 metal strips .

Figure 2. Photograph showing some of the fabricated flexible a-Si:H position sensors developed within this work.

These strips were glued around a glass cylinder (see Figure 2) with a radius of curvature of 14 mm, under convex bending. All the electrical and optical measurements have been done in the bent state, to evaluate the properties of the sensors under stress conditions.

3 Results

Figure 3 shows a typical normalised spectral response of the 1D position sensor as a function of λ, measured under short circuit conditions, showing a maximum for λ=590 nm. The low efficiency for λ < 590 nm is due to the absorption of the polymer.

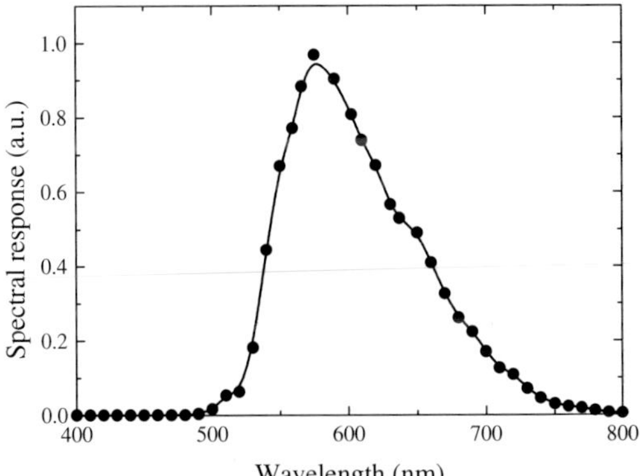

Figure 3. Normalised typical spectral response for a-Si:H position sensors deposited on Kapton foil 125 μm, with 1 cm wide and 2 cm long.

To evaluate the light intensity (Φ) dependence on the photocurrent (I_{ph}), neutral density filters under monochromatic light (λ= 600 nm) have been used to change Φ.

Figure 4 shows I_{ph} as a function of Φ, varying from 10^{-4} Wcm^{-2} to 10^{-7} Wcm^{-2}. For the range used a power law dependence, $I_{ph} = 425.6\Phi^{0.804}$ was fit to the data, even for low light intensities. The correlation coefficient of the fit was R = 0.999. This value refers to monomolecular recombination kinetics and it is usually found in intrinsic a-Si:H samples with the Fermi level near the midgap [8].

Figure 5 shows the linearity measurements performed in the 1D a-Si:H position sensor 1 cm wide and 2 cm long between the bottom and one of the lateral strip contacts. These measurements were made using a diode laser (6 mW; 670 nm) and the data were fit with a linear function y = 0.758 – 0.215x, with R = 0.996 and a standard deviation of 0.01401, confirming the feasibility of the produced sensors.

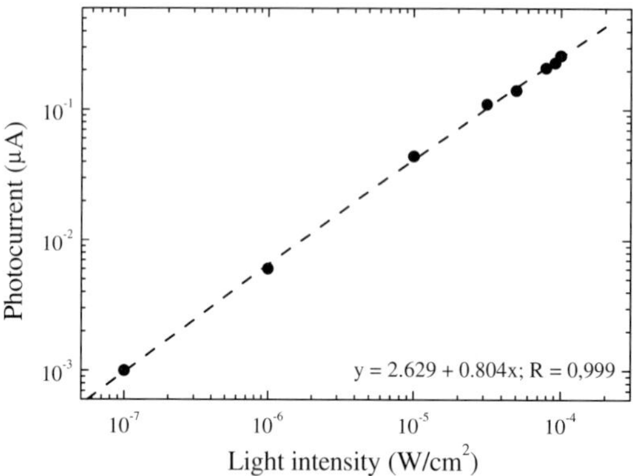

Figure 4. Dependence of I_{ph} on Φ, under 600 nm monochromatic light.

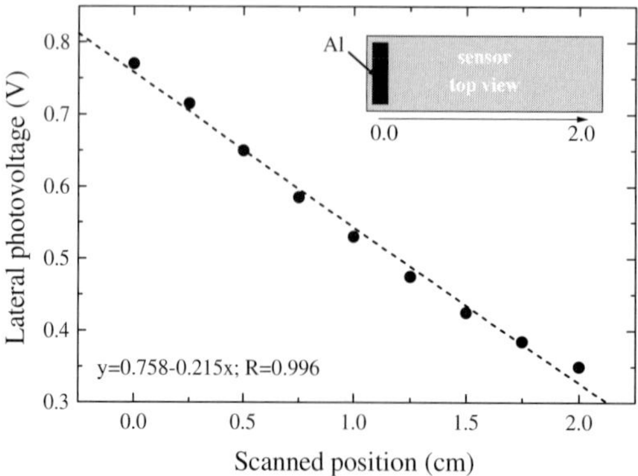

Figure 5. Nonlinearity measurements performed in a-Si:H flexible position sensor with 1 cm wide and 2 cm long. The optical source used was a diode laser with 6 mW and 670 nm.

As the metal back contact does not cover the entire area we are not using a fully equipotential and so, the reference is not zero and it is not located in the centre of the device

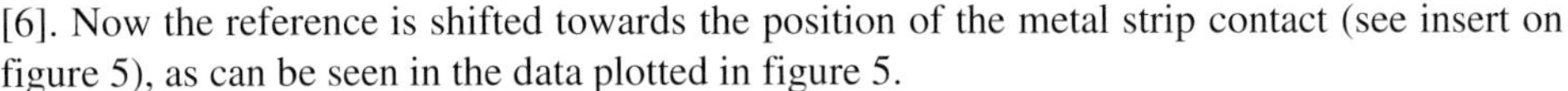

[6]. Now the reference is shifted towards the position of the metal strip contact (see insert on figure 5), as can be seen in the data plotted in figure 5.

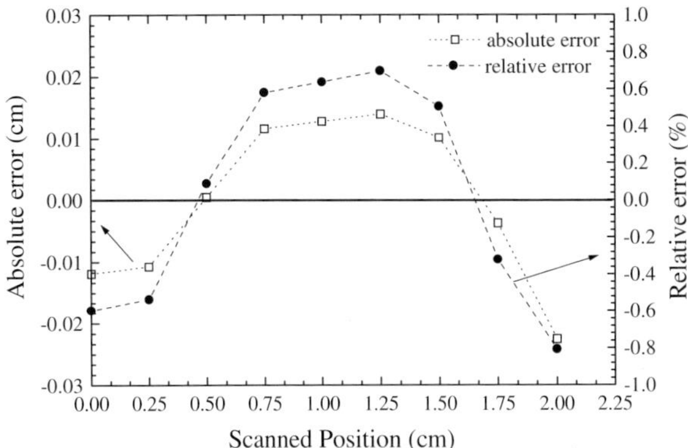

Figure 6. Absolute (open squares) and relative (full circles) errors of the linear fit as a function of the scanned position. The solid line refers to 0% of deviation of the measurement.

Figure 6 shows the absolute and the relative error of the measured scanned position concerning the straight linear line fit used. The minimum deviation (negative) is obtained for distances close to the top contact (0.0112cm and 0.59%, respectively), while the largest deviation (also negative) is obtained for the largest lateral distance from the top contact (0.026 cm and 0.72%, respectively). In the intermediate region (from about 0.50 cm to about 1.50 cm), the deviation obtained is positive, not exceeding, in absolute value, the ones recorded for the edges of the sensor. These data are comparable to those a-Si:H position sensors fabricated on glass substrates [9, 10].

4 Discussion

The spectral response data shown in figure 3 reveals a decrease in the spectral response of the sensor for $\lambda < 600$ nm, which is due to the light absorption by the polyamide film, as can be seen from the T spectra data of the Kapton film shown in figure 7. There, we also show T of the SnO_2 and of the intrinsic a-Si:H films deposited on Kapton substrate. There, we notice that the λ cut-off is located around 500 nm and for $\lambda > 590$ nm (spectral response maximum, see figure 3). The average T of the Kapton film coated with SnO_2 is about 68%, decreasing to about 54%, when coated to a-Si:H fim, which explains the spectral response data obtained in the near red region of the light spectrum for the sensor analysed. That is, a strong optical losses are detected when comparing these data with the ones obtained on position sensors grown on glass substrates [10].

The dependence of I_{ph} on Φ follows a power law of about 0.8, even for $\Phi < 0.1\ \mu W/cm^2$). This dependence is correlated with the monomolecular recombination kinetics associated with intrinsic a-Si:H samples with the Fermi level near the midgap [11]. This correlation agrees with the activation energy inferred from the dark conductivity measurements [see table 2b)].

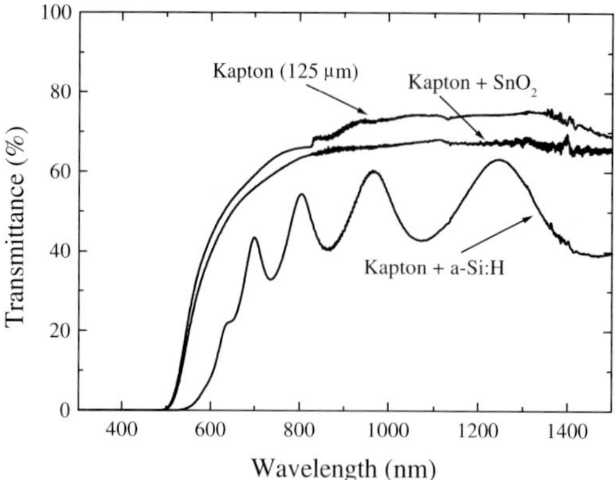

Figure 7. T spectra of: Kapton, Kapton+SnO_2 and Kapton+a-Si:H, for 300 nm< λ < 1500 nm.

From the linearity measurements we have obtained a good fit between the recorded data and the spatial position of the light spot on the device surface. The experimental data are well fitted by the equation y = 0.758 – 0.215x, with a correlation better than 0.996. The device nonlinearity (position detection error - δ defined as [12]: $\delta = 2\sigma/F$ where σ is the rms deviation from the regression line data and F the measured full scale), was below ±1%. This error is mainly ascribed with edge effects associated with the electric field distribution along the sensor, as can be seen from the data plotted in figures 5 and 6, respectively. This deviation can be attributed to limitations imposed by the fall-off parameter, the maximum distance that the carriers can be detected under linearity conditions. This parameter is given by $(\sigma_d\sigma_s 4a^2)/WW_r$ ratio, where σ_s is the conductive of the collecting layer, a is the width of the sensor, W the thickness of the i-layer and W_r is the thickness of the collecting layer, [10]. In this case, without using a metal equipotential layer, it is expected to have a fall-off parameter of about 0.3 cm^{-1}. The deviations obtained in the linearity can also be attributed to some non-uniformity's of the SnO_2 film and of the a-Si:H pin structure, close to the sensor edges.

5 Conclusions

We have successfully fabricated large area flexible and ultra-light a-Si:H position sensors using standard thin film and microelectronic technologies. The sensors present good performances concerning the position nonlinearity, ±1%, comparable to those fabricated on glass substrates. These sensors also present adequate electrical and optical properties, adequate for the new type of lightweight and unbreakable sensors, even for other optoelectronic device applications.

6 Acknowledgements

The authors thank to Ana Domingues and Ana Cabrita for some of the electrical and optical characterisation and Ana Sofia Alves for the SnO_2 deposition by spray pyrolysis. This work was supported by the Fundação para a Ciência e a Tecnologia through Pluriannual Contracts with CENIMAT and by the projects TMR ERBFMRXCT970141, PRAXIS/ 3/3.1/MMA/1788/95 and PRAXIS/P/CTM/12094/98.

7 References

[1] E. Fortunato, M. Vieira, L. Ferreira, C.N. Carvalho, G. Lavareda, R. Martins, Mat. Res, Symp. Proc. 1993, Vol. 297, p. 981-986.

[2] E. Fortunato, M. Vieira, L. Ferreira, G. Lavareda, R. Martins, J. Non-Cryst. Solids 1993, 164&165, 797-800.

[3] E. Fortunato, G. Lavareda, M. Vieira, R. Martins, Rev. Sci. Instrum. 1994, 65, 3784-3786.

[4] J.T. Wallmark, Proc. IRE 1957, 45, 474-484.

[5] E. Fortunato, R. Martins, Rev. Sci. Instrum. 1996, 67, 2702-2707.

[6] R. Martins, E. Fortunato, IEEE Transactions on Electron Devices 1996, 43 (12), 2143-2152.

[7] R. Martins, E. Fortunato, J. Appl. Phys. 1995, 75, 3481-3487.

[8] E. Fortunato, Doctoral Thesis, New University of Lisbon, 1995.

[9] E. Fortunato, R. Martins, in Hydrogenated Amorphous Silicon (Ed.: H. Neber-Aeschbache), Vol. 44-46, Part 2, Scitec Publications, Zurich, 1995, p. 883-930.

[10] R. Martins, E. Fortunato, in R. Street (Vol. Ed.), Technology and Applications of Amorphous Silicon, Vol. 37, Springer-Verlag, Berlin, 1999, (in press).

[11] H. Dersch. L. Scweitzer, J. Stuke, Phys. Rev. B 1983, 28, 4678-4683.

[12] A. Kawasaki, M. Goto, Sensors & Actuators 1990, A21-A23, 534-537.

Microstructure and Piezoelectric Properties of Lead Metaniobate $PbNb_2O_6$

S. Ray, Y. Sauermann, F. Paul, E. Günther, H.-J. Ritzhaupt-Kleissl
Forschungszentrum Karlsruhe GmbH, Institut für Materialforschung III

1 Abstract

In this contribution an overview is given of the influence of various processing parameters, sintering parameters in particular, on the development of the microstructure of piezoceramic samples made of lead metaniobate $PbNb_2O_6$. Furthermore, the influence of the microstructure on the resulting piezoelectric properties of the material will be discussed briefly.

2 Introduction

Lead metaniobate $PbNb_2O_6$ (PN) shows interesting piezoelectric properties, e.g. an extremely low mechanical Quality factor Q_m leading to a large band width as ultrasonic device[1; 2]. Lead metaniobate also exhibits a strong anisotropy of the coupling coefficients k_t and k_p: In spite of a high k_t along the polar axis the values of k_p perpendicular to it are negligible, often immeasureable[1; 2]. Furthermore, the high T_c of undoped lead metaniobate of 570° C allows applications of PN up to 250°C[1; 2].

Poling of lead metaniobate, however, is known to be difficult. Typical problems are either dielectric breakdown of the samples during poling or an inadequate piezoelectric effect after the polarization process.

Adequate polarization parameters and resulting piezoelectric properties turned out to be strongly dependent on the microstructure of the material.

3 Experimental

3.1 Specimen fabrication

Samples were prepared by two different processing routes:

At first, one batch of lead metaniobate powder was prepared by the Thermal Two Stage Process. Details of the process are described elsewhere [3; 4].

Starting materials are lead acetate ($Pb(CH_3COO)_2$) and Niobium pentaethylate ($Nb(OC_2H_5)_5$). By mixing the precursors an intimate mixture is generated. By spray drying this precursor solution a precursor powder is produced which is subsequently converted into the ceramic powder by calcination at 700 °C for one hour.

Another batch of lead metaniobate powder was prepared by a thermochemical process. Details of this process will also be published elsewhere [5]. In this process instead of the most expensive Niobium pentaethylate, a cheaper Niobium starting material is used. By mixing the starting materials the precursor powder is obtained. After the heat treatment at 1250°C for 1 hour pure stoichiometric lead metaniobate ceramic powder is obtained.

Lead metaniobate exists in a number of stable and metastable phases [1; 2], but only the metastable phase derived from the tetragonal high-temperature phase by orthorhombic distortion exhibits piezoelectric properties [1; 2].

Therefore the produced powders are heated to 1250 °C for 1 hour and quenched in ambient air in order to obtain the piezoelectric phase. Subsequently, the powders are ceramically processed and characterized with respect to phase content, particle size distribution and specific surface according to BET.

Uniaxial dry pressing with a pressure of 170 ... 300 MPa is used to prepare cylindrical pellets of the diameter 5.9 mm and a length of 10 mm. These pellets are sintered in a dilatometer (Netzsch, DIL 402 C) with different heating cycles.

Before measuring a piezoelectric effect, the statistically distributed electric domains have to be orientated by a polarization process. This is obtained by applying a strong electric field to the samples.

In table 1 the processing routes, heating schedules and polarization conditions of the samples are summarized:

Table 1: Experimental conditions

Sample	Processing route	Heating schedule	Polarization conditions
TTP 1280	Thermal Two Stage	2 K/min --- 1280 °C; 2 h	3...3,6 KV/mm; 4 h
TTP 1290/2	Thermal Two Stage	2 K/min --- 1290 °C; 2 h	3...3,6 KV/mm; 1.5 h
TTP 1290/6	Thermal Two Stage	2 K/min --- 1290 °C; 6 h	4 KV/mm; 0.5 h
TTP 1295/2	Thermal Two Stage	2 K/min --- 1295 °C; 2 h	3...3,6 KV/mm; 1.5 h
TTP 1295/4	Thermal Two Stage	2 K/min --- 1295 °C; 4 h	3...3,6 KV/mm; 1.5 ... 15 h
TC 1280	thermochemical	2 K/min --- 1280 °C; 2 h	3,6 KV/mm; 4 h
TC 1290/2	thermochemical	2 K/min --- 1290 °C; 2 h	3,6 KV/mm; 4 h
TC 1290/4	thermochemical	2 K/min --- 1290 °C; 4 h	3,6 KV/mm; 4 h
TC 1295/2	thermochemical	2 K/min --- 1295 °C; 2 h	3,6 KV/mm; 4 h
TC 1295/4	thermochemical	2 K/min --- 1295 °C; 4 h	3,6 KV/mm; 4 h

3.2 Characterization

The sintered specimens are characterized with regard to density, phase content by XRD and microstructure by scanning electron microscopy (SEM). The chemical composition is determined by X – Ray Fluorescence Analysis.

The piezoelectric properties are characterized as follows: The (low frequency) charge constant d_{33} and the voltage constant g_{33} is measured by a quasistatic method at a low cycling frequency of 120 Hz using a Berlincourt meter (CADT 3300 by Channel Products Inc., USA). The dielectric loss tan δ and the relative dielectric constant ε_r are measured with a LCR – meter (SR 720 by Stanford Research systems Inc., USA).

Subsequently, the piezoelectric properties are measured by impedance spectroscopy (Impedance Phase Analyzer HP4194 A by Hewlett Packard).

4 Results

4.1 Ceramic characterization

In table 2 the results of the ceramic characterization are summarized:

Table 2a: Powder characterization

	unit	TTP	TC, after 5h milling
Phase content before heating		Rhombohedral	Rhombohedral
Phase content after quenching		Orthorhombic	Orthorhombic
d_{50}	µm	12.84	6.5
d_{90}	µm	30.78	15.8
Specific surface (BET)	m^2/g	0.66	0.58

Table 2b: Pellet characterization

	unit	TTP 1280	TTP 1290/2	TTP 1290/6	TTP 1295/2	TTP 1295/4
Sintered density	% th.D	88.4	89.5	91.7	90.2	91.8
Mass loss	%	0.87	0.88	1.77	0.97	1.4
stoichiometry	PbO 44.57 weight-%, Nb_2O_5 54.90 weight-%					
Phase content	Orthorhombic, JCPDS 11 – 0122					
	unit	TC 1280	TC 1290/2	TC 1290/4	TC 1295/2	TC 1295/4
Sintered density	% th.D	78.4	81.1	81.1	82.8	81.8
Mass loss	%	1.13	1.39	2.40	1.16	2.60

The microstructure of the sintered, polished and thermally etched specimens are displayed in Fig. 1a and 1b:

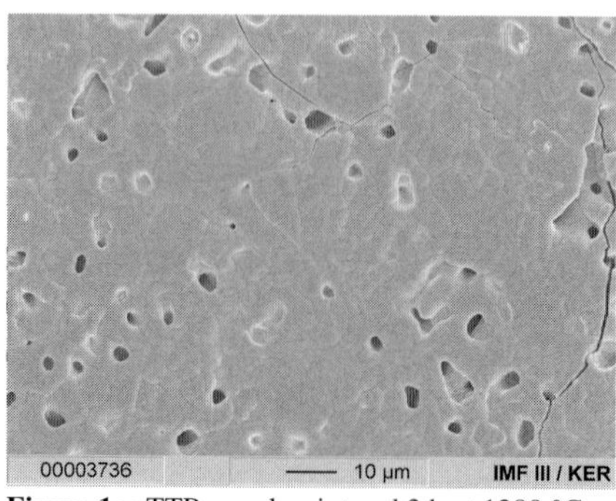

Figure 1a: TTP sample, sintered 2 h at 1280 °C,

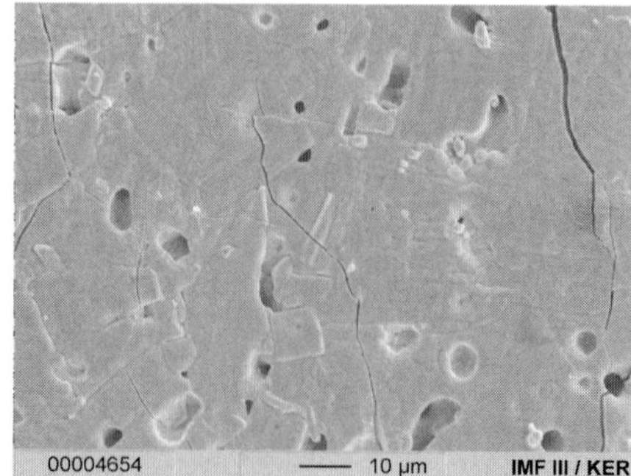

Figure 1b: TC sample, sintered 2 h at 1280 °C

4.1 Electric characterization

The piezoelectric material data are given in Tab. 3. Polarization conditions are 3.6 kV/mm for 4 hours.

Table 3: Piezoelectric material data

Quantity	symbol	unit	TTP 1280	TTP 1290/2	TTP 1295/4	TC 1280	TC 1290/2
Dielectricity const.	ε_r	-	280	390	307	160	220
Loss factor	tan δ	10^{-3}	9	14,6	1.5	1.1...9	52
Quantity	symbol	unit	TTP 1280	TTP 1290/2	TTP 1295/4	TC 1280	TC 1290/2
coupling coeff.							
Planar	k_p	-	0.113	0.261	0.128	*	*
Transversal	k_t	-	0.32	0.42	0.384	0.37	*
Charge constant	d_{33}	pC/N	57	71	58	70	59
Voltage constant	g_{33}	10^{-3}Vm/N	20	20	22	45	34
Frequency constant	N_t	MHzmm	1383	1769.7	1317	*	*
Quality factor	Q_m	-	26	17.3	-	21	*

* = Not measurable

5 Discussion

An increase in the sintering temperature leads to reduced porosity, while the grain size increases considerably, especially at the sintering temperature of 1295 °C. This is also true for the extended sintering times of 4 hours. In this case or for a sintering temperature of 1295 °C, some pores can also be found within the grains. The size of the pores is larger. Both is a sign of slight oversintering.

The piezoelectric data seem to decline gradually with increasing density or grain size, but more experimental work has to be carried out to investigate the decisive reason.

Comparing the different processing routes the samples display a completely different densification behaviour during sintering:

Considering the fact that sintering of lead metaniobate is difficult in general, the powder from the Thermal Two Stage Process ("TTP – powder") shows a relatively good densification behaviour. By increasing the sintering temperature from 1280 °C to 1295°C the density can be increased considerably. The same can be achieved by extending the sintering time from 2 hours to 4 or 6 hours. By increasing only the temperature no excessive mass loss by PbO evaporation can be noticed, whereas the mass losses at extended sintering times are considerably.

The thermochemically processed powder ("TC – powder") on the other hand shows only very reluctant densification. Increasing the temperature or extending the sintering time only leads to a minor increase in density. The mass loss is much larger than that of the TTP samples and is increased much more by extended sintering times.

Due to the higher density of the TTP sample sintered at 1280 °C for 2 hours compared to the TC sample sintered with the same heating schedule, the porosity of the latter is also much higher. Little densification has taken place at that temperature. The grain size distribution of the TC sample seems to be larger than that of the TTP sample. The same effect but even much stronger can be seen for the increased sintering temperatures and times, e.g. the samples sintered at 1295 °C for 4 hours. In this case the pore size distribution is different with much larger pores in the TC sample and so is the grain size distribution with a much larger distribution for the TC samples (3 ...50 μm) compared to the TTP samples (3 ...15 μm).

The piezoelectric material data also differ considerably between the samples from different processing routes: The TTP samples show a much higher dielectricity constant ε_r, but the voltage constant g_{33} and the transversal coupling coefficient k_t of the TC samples are considerably higher than the corresponding values of the TTP samples. An explanation can be found in the different microstructures of the TTP and the TC samples but additional experiments are necessary in order to investigate whether the piezoelectric data depend more on the different grain size distribution or on the higher porosity of the samples. Furthermore, the ceramic processing of the TC samples has to be improved in order to achieve higher sintered densities, necessary for application.

6 Summary

With the Thermal Two Stage process a well investigated process is available for the production of pure stoichiometric lead metaniobate, which shows satisfying piezoelectric material data. It is possible to obtain different microstructures depending on the heating schedule.

The novel thermochemical process offers a quick, inexpensive method for the production of pure stoichiometric lead metaniobate. In spite of the satisfying piezoelectric material data, however, the ceramic processing still has to be improved in order to obtain higher sintering densities, which are indispensable with regard to the application of the piezoceramic material as ultrasonic device.

7 References

[1] B. Jaffe, W.R. Cook, H. Jaffe: „Piezoelectric Ceramics“, pp 213 - 222 Academic Press, London 1971

[2] A.J. Moulson, J.M. Herbert: „Electroceramics“Chapman and Hall, London – New York 1990

[3] E. Günther, U. Maciejewski 8th Cimtec, Florenz, Juli 1994

[4] S. Ray, E. Günther, H.-J. Ritzhaupt-Kleissl: „Preparation and Piezoelectric Properties of Stochiometric Lead Metaniobate $PbNb_2O_6$ synthesized by a Thermal Two-Stage Process“ *to be published in: Journal of Material Science, 1999*

[5] F. Paul, Diplomarbeit Fachhochschule Isny, 1999

The "Size-effect" on the Fatigue and Fracture Properties of Thin Metallic Foils

A. Hadrboletz, G. Khatibi and B. Weiss
Institute of Materials Physics, University of Vienna , 1090 Vienna, Austria

1 Introduction

Thin foils, sheets or wires of metallic, semiconducting or non - metallic materials resemble essential parts of microelectronic and micro-electro-mechanical systems (MEMS). These parts are often subjected to mechanical and thermal fatigue causing premature failure. In numerous microdevices thin foils act as switches operating at frequencies from less than 1 Hz up to 1 MHz. Therefore the fatigue behavior of thin foils or wires attains considerable technical interest in order to assure a safe operation of MEMS. In the past mainly fatigue tests of given devices have been performed mostly for the verification of the design. As the field of MEMS matured basic material properties of thin foils will be required for suitable design and life prediction.

Various investigations revealed that fatigue data of bulk material can not be adopted to almost two dimensional structures such as thin foils or films due to the "size effect". Recently the most prominent fatigue testing and fatigue data of materials used in MEMS have been presented in [1].

This review yields that the shapes of the specimens and/or their adjustments in the experimental set-up and the experimental procedure may be very different and sometimes rather complicated. Usually stress vs. fatigue life curves are measured in bending [2]. Also an equipment for tension-tension testing of thin Cu-films has been reported in [3].

Almost no information on fatigue crack growth behavior of free-standing thin foils is available. Therefore, the objective of the present investigation was to develop a suitable testing method to investigate the crack growth behavior of thin metallic foils with thicknesses ranging from a few microns up to 200μm. Fatigue tests have been accompanied by the observation of dislocation structures due to plastic deformation in the vicinity of the crack. The size of the plastic zone in relation to the geometric dimensions have been studied for explanation of the fracture behavior. The applicability of the commonly used LEFM-based parameters is critically evaluated.

A comparison of experimentally determined data of bulk material is added to focus on the effect of dimensional constraint. Furthermore the influence of the foil thickness on fracture topography has been investigated in tension testing.

2 Experimental

The specimens investigated were recrystallized commercial rolled Cu-foils with a nominal purity of 99,95 % in thicknesses ranging from a few microns up to 200 μm. Additionally electrodeposited foils (35 and 105μm thick) of identical purity have been investigated. For comparison bulk material (commercial E-Cu) as rolled strips – each 6mm thick- was investigated. Grain sizes have been determined by the linear intercept method (twin boundaries are not considered) yielding values in the range of a few microns for electrodeposited foils and between 20 and 200μm for recrystallized foils depending on the heat treatment as presented in table 1. The mechanical properties are determined in a commercial micro-tensile-test machine (Meßphysik Fürstenfeld, Austria); the strain was measured by use of the non-contacting laser-speckle correlation technique [4]. This technique was also applied for the determination of the Young`s modulus. As commonly known values of the Young`s modulus are strongly dependent on texture. Therefore the angle between the direction of crack propagation - and the rolling-direction plays an important role. The variation of the Young`s modulus with this angle is added in Tab. 1.

Table 1. Mechanical properties of heat treated and electrodeposited foils and of bulk Cu

Thickness (μm)	Grain size (μm)	R_m (MPa)	$R_{p0,2}$ (MPa)	A (%)	E (GPa)		
					Angle between crack growth- and rolling direction		
					0 °	45 °	90 °
9)*	30	120	25	13	112	104	112
19)*	40	142	20	17	68	108	82
35)*	60	155	15	17,5	115	125	132
50)*	80	114		19	111)***	133)***	111)***
100)*	130	165	18	20	116	87	112
200)**	40						109
Electrodeposited foils as received							
35	1	260		2,6			92
105	5 - 10	255		8			102
Bulk material							
6000	20 - 30	210	14	> 50			126

)* Heat treated 900 ° C/ 2h
)** Heat treated 600 ° C/ 2h
)*** calculated from [5]

The experimental set-up for fatigue testing is shown in figure 1.The foil-specimens with dimensions of 20x20mm are glued with a commercial strain gauge adhesive to a bar shaped holder of high strength Al alloy. The specimen holder is part of a resonance - testing equipment which is excited to longitudinal push-pull vibrations with zero mean stress (i.e. R=-1) at a testing frequency of 20kHz. A standing wave is built up so that the length of the specimen-holder corresponds to half of the wavelength. The distribution of displacement and strain varies sinusoidally along the specimen-holder with the maximum of the strain in the mid-section of the holder. In this location a slot of 1,2mm width and 0,5mm depth is machined into the upper side of the holder so that the foil glued across this slot is free-standing in this area. The width of 1,2mm assures that the strain value is kept almost constant at its maximum value; depth (0,5mm) is chosen so that there is enough space for eventually penetrated adhesive and is otherwise not too large to guarantee the maintenance of a proper mode of loading. This is also the reason why an identical slot is introduced in the lower side of the holder just opposite to the upper one. Since the foil thicknesses are very small compared to the thickness of the holder (t=8mm) detrimental effects such as heating can be neglected.

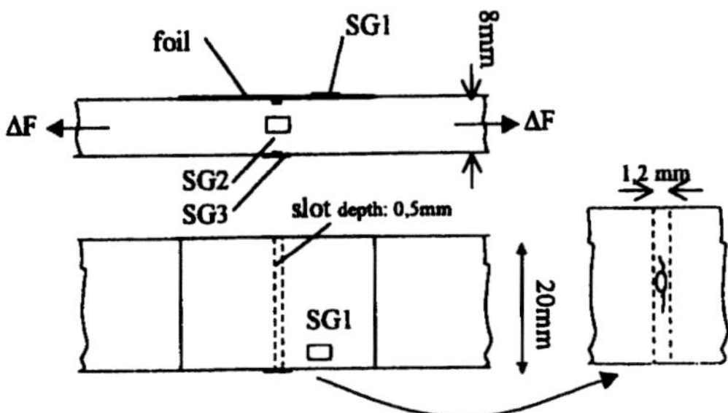

Figure 1: Experimental set-up for fatigue testing of foils, schematically

For the monitoring of the strain of the loaded foil miniature strain gauges of the type LY 11/HBM with an active length of 0,6mm have been positioned in the locations SG1, SG2 and SG3. From calibration measurements it is feasible to use each of these positions for controlling testing. Exactly identical wave-forms of the strain signals could be observed for all three strain gauges. This implies an appropriate loading of the foil without buckling. Taking into account that the free-standing zone of the foil (which is of a rectangular shape of B=20mm x b=1,2mm) is loaded along the much longer side of 20mm this behavior is not surprising since the critical force for buckling is proportional to B^3/b. For fatigue crack growth testing a lancet-shaped notch (1,5mm long, ~0,5mm wide) is introduced in the mid-section of the foil by EDM. Crack growth can be monitored by a traveling light microscope with a resolution of approximately 1µm. The strain value may be recorded by SG1, SG2 or SG3 alternatively. Crack growth was recorded continuously with a computer-aided data acquisition system. Crack growth rates were in the range of 10^{-10} and 10^{-13}m/cycle. Dependence of crack length as function of loading cycles were obtained for constant strain amplitude loading. Intermittently specimens were investigated in a scanning electron microscope using the electron-channeling contrast imaging (ECCI) technique to reveal changes in the global dislocation arrangement. ΔK-values have been calculated using the commonly known equation for through cracks according linear elastic fracture mechanics (LEFM) considerations. These have been determined using the appropriate value of the Young´s modulus, which dependency in texture has been considered. To determine ΔK_{th} the applied load was decreased stepwise in increments of 5% of the preceding strain-value until the crack growth rate was smaller than 10^{-12}m/cycle.

3 Experimental Results and Discussion

3.1 Crack growth curve

In Fig. 2 a variety of selected crack growth curves at constant strain-ranges for specimens of different thicknesses is presented. For the purpose of comparison a crack growth curve of bulk material (6mm thick) is added. For all foils up to 100μm Fig. 2 indicates that the crack propagation rate decreases with increasing number of cycles. Compared to the behavior of bulk material including a foil with a thickness of 200μm the curvature of the presented curves is opposite. In addition the crack growth curves are characterized by the occurrence of intermediate saturation values for thicknesses up to 100μm. This anomalous behavior is investigated in more details for a 100μm thick foil as an example.

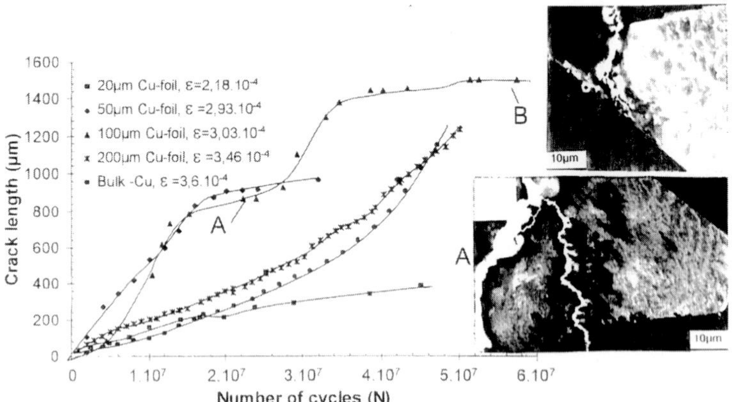

Figure 2: Crac growth-curves of Cu-foils of different thickness and ECCI micrographs

As presented in Fig. 2 by ECCI-micrographs the intermediate crack-arrest are attributed to a strong interaction of the growing crack with grain-boundaries and the onset of crack bifurcation. From this a behavior very similar to that of short cracks may be deduced [6]. This behavior may be explained by the assumption of a plane stress state for foils thinner than 200μm and a state of plane strain for foils with thicknesses exceeding 100μm. For bulk material the entire crack propagation is governed by plane strain. From the findings presented above it is obvious that crack growth in foils is characterized by the transition from plane stress to plane strain with increasing thickness. This assumption has been made firm by an investigation of the extent of the plastic zone in the vicinity of the growing crack. The radius of the plastic zone r_p was calculated from LEFM-considerations ($r_p = 1/2\pi.(K_{max}/\sigma_y)^2$).

For a selected crack length (a) of 1200μm the r_p values are listed in Tab. 2 for various foil thicknesses. The yield stress σ_y was determined after fatigue loading indicating cyclic hardening. A direct observation of the plastic zone is given in Fig. 2 for a foil of thickness 100μm. If the size of the plastic zone is defined by the extent of characteristic slip bands, the experimental findings indicate good agreement with the calculated values.

The added column of r_p/t-values should allow to distinguish between the state of plane stress and plane strain. Due to LEFM-considerations rp/t-values may be regarded as criteria [7]: $r_p/t < 0{,}025$ ... cracking in plane strain , $r_p/t > 1$ ... cracking in plane stress.

From the results given in Tab. 2 it may be deduced that cracking in thin foils is a process controlled mainly by plane stress; this is especially true for foil-thicknesses up to 100μm. The state of plane stress for thin foils is the reason for the decreasing crack propagation rates with increasing number of loading cycles. This may be explained by the fact that crack resistance in plane stress is increasing with increments of crack length so that an increase of stress is required to maintain crack growth.

Table 2 Radii of plastic zones for foils of different thicknesses for a selected crack length of 1200µm (a was selected to reduce the notch influence)

Thickness t (µm)	σ_y cyclic (MPa)	r_p (µm)	r_p / t
20	60	34	1,70
50	70	32	0,64
100	65	42	0,42
200	60	82	0,42
6000 (bulk)	60	87	0,019

The findings presented above are emphasized by the results of electrodeposited foils. In Fig. 3 these results are presented as a fatigue crack growth curve at a selected constant strain amplitude. The behavior is very similar to that of bulk material. As reason for this behavior the microstructure of these foils may be regarded. The average grain size is about 5µm, this fine grained structure is also responsible for the high σ_y-values of approximately 100MPa in the tensile test which raises to about 140MPa after cycling due to hardening. On the other hand this fine grained structure is also the reason for the absence of intermediate crack arrest during crack growth. r_p/t-values for crack growth range from 0,02 (200µm crack length) up to 0,21 for a 1800µm long crack. A predominant state of plane strain is obvious.

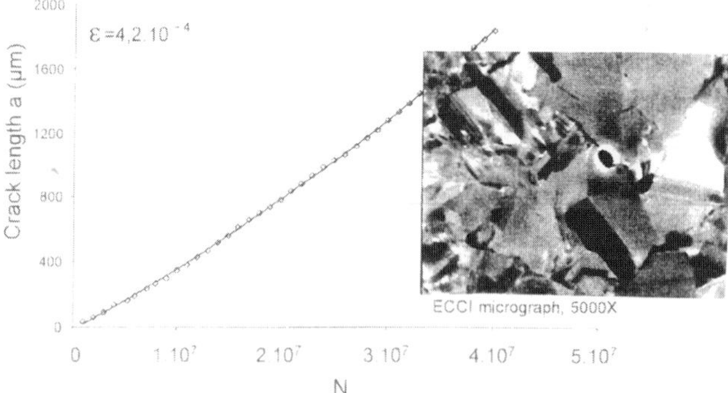

Figure 3:Crack growth-curve of a 105µm thick electrodeposited Cu-foil and ECCI migrographs

3.2 Fracture topography

In a first attempt to reveal size effects on the fracture behavior tests were performed under tension . SEM micrographs of recrystallized Cu-foils reveal two types of fracture topography after rupturing. For foils with thicknesses <100µm ductile failure with typical knife edge rupture without voids and dimples appears. With increasing foil thickness dimples and voids can be observed. These dimples are very small (about 5 µm). Similar findings have been reported by [8]. For thin foils in the thickness range of about 1µm Arzt [8] reports the lack of dimples on the fracture surfaces. Their minimum size of approx. 5µm makes it impossible to fit into a foil thinner than this size. In the present investigation, however, this minimum thickness can be extended to about 100µm, if twin boundaries are neglected.

3.3 Fatigue thresholds

In Tab. 3 a selection of fatigue threshold stress intensity values for foils of different thicknesses is given. The experimental error is about +/- 5%.

Table 3 Fatigue threshold of foils of different thickness

Thickness (μm)	Heat treatment	ΔK_{th} (MPa√m)	note
19	850 °C/ 6h	2,8	⊥
		2,0 – 2,9	texture effect
50	900 °C/ 2h	2,1 – 2,2	⊥
100	600 °C/ 4h	3,1 – 3,2	⊥
	900 °C/ 2h	3,1 – 3,6	⊥
	900 °C/ 2h	2,7 – 3,0	45 °
200	600 °C/ 4h	3,3 – 3,7	⊥
6000 (bulk)		2,0 – 2,3	⊥

⊥ crack direction normal to the rolling direction
45 ° crack direction 45° to the rolling direction

The given results reveal the following facts:

- The scatter of ΔK_{th} values for a given thickness is considerably large
- The angle between the direction of crack growth and the rolling direction influences the ΔK_{th}-values severely.
- A trend of increasing ΔK_{th}-values with increasing foil thickness can be observed.
- In comparison to the data of bulk material the ΔK_{th}-values are surprisingly high.

The large scatter of ΔK_{th} may be related to the pronounced interaction of the crack path with grain boundaries. This is especially true for the case of foils which thickness is in the order of the average grain size, so that the whole foil may be regarded as a two-dimensional array of grains. Copper foils often appear as highly textured. It is known that texture may be the reason of the variation of the Young`s modulus. For bulk material commonly a value of 126GPa is used, so that taking into account texture-effects results in ΔK_{th} which may be 46% lower or 20% higher than their anticipated value.

The trend of increasing threshold values with increasing thickness may be assumed only on a first rough view. A more detailed evaluation yields that the ratio of grain size to thickness and the grain size distribution (mostly influenced by predeformation) is to be considered. Generally the threshold values increase when the grain size becomes larger than the thickness so that the foil consists of a single grain across its thickness. High threshold values in thicker foils (consisting of a few layers of grains) are mainly attributed to an irregular distribution of grains of different shapes, sizes and amounts of deformation.

da/dN vs. N-curves of foils consisting of a single or double layer of grains through the thickness resemble the behavior of short cracks indicating a similar constraint effect. In the case of thin foils the dimensional influence is dominant.

4 Summary and Conclusions

i. A special experimental set-up has been presented, which allows fatigue testing of free-standing thin foils. Specimens are of a simple rectangular shape so that the present testing is more suitable to the determination of basic material properties than to the testing of given structures which will be of primary interest in engineering applications. *The high testing frequency corresponds to the high frequency operation of MEMS-devices. It seems feasible to use the test set-up for cycling at low testing frequencies and different load ratios in conventional testing-machines.*

ii. The unexpected crack growth behavior (decreasing crack propagation rate with increasing crack length and temporary thresholds) for foil-thicknesses up to approx. 100μm was explained by a transition of plane stress state to plane strain state with increasing foil thickness. This was confirmed by using the ratio of the plastic zone size to the foil thickness as a criteria. Intermediate crack arrests are due to a strong interaction of the growing crack with grain boundaries.

iii. Fatigue threshold stress intensity values have been determined and calculated under the assumption of the validity of LEFM. Generally the values are high which may be due to the ratio grain-size to foil-thickness being larger than 1, resulting in a reduction of the number of glide systems. The considerable large scatter of ΔK_{th} may be attributed to interaction of the crack tip with grain boundaries.

iv. The crack topography (under tension loading) is changed markedly with foil-thickness characterized by the absence of dimples for foil thicknesses up to 100μm. The formation of dimples is influenced by the dimensional constraint and the prevailing microstructure.

v. The application of LEFM seems questionable, when the dislocation motion is constrained by the dimensions of the investigated structure which is supposed to be in the 50 nm regime [8].

5 Acknowledgements

The authors thank the Austrian Bundesminsterium für Wissenschaft und Verkehr (Cost 510) and Fonds zur Förderung der wissenschaftlichen Forschung, Wien for financial support. For stimulating discussions and aid during experiments Prof. R. Stickler is gratefully acknowledged.

6 References

[1] W.N. Sharpe, Jr. and K.T. Turner in: „Fatigue '99", EMAS, Vol.3/4, 1837-1844

[2] S.B. Brown, et al. in: „Proc. Transducers '97", Chicago, 1997, 607-610

[3] D.T. Read, Int. J. Fatigue, 1998, 20, 203-209

[4] M. Anwander, B. Weiss, B. Zagar, H. Weiss in: „Experimental Mechanics", (Ed. I.M. Allison), 1998, Balkema, Rotterdam, 697-702

[5] J. Weerts, Zeitschrift für Metallkunde, 1933, Heft 5

[6] „The Behaviour of Short Fatigue Cracks", (Ed. K.J. Miller, E.R. de los Rios), MEP, London, 1986

[7] D. Broek, „Elem. Eng. Fracture Mech.", 4.ed, Martinus Nijhoff, Den Haag, 1986

[8] E. Arzt, Acta Mater., 1998, Vol. 46, No. 16, 5611-5626

Smart Fluids and Their Applications – Perspectives for New Actuators

Holger Böse, Alexandra Trendler
Fraunhofer-Institut für Silicatforschung, Würzburg, Germany

1 Abstract

The rheological behaviour of electrorheological and magnetorheological fluids is drastically and reversibly changed by strong electric or magnetic fields, respectively. The fact that the rheological properties are controllable can be used for many technical applications. Both kinds of smart fluids can be synthesized from various combinations of materials, which opens wide possibilities to adapt the fluid properties and the design of the actuatoric devices in which the fluid is used. An overview of the most relevant properties and a description of some of the possible applications of smart fluids are given.

2 Introduction

Electrorheological (ER) fluids are suspensions of electrically polarizable particles in nonconducting liquids which show drastic changes of their rheological properties, as for example a remarkable increase of the viscosity in strong electric fields [1], [2]. The polarized particles attract each other due to dipole forces and form particle chains between the electrodes resulting in a transition of the material from the liquid to a geleous state within some milliseconds. This transition is reversible, i. e. after removal of the field the ER fluid returns to its original liquid state. Correspondingly, magnetorheological (MR) fluids are suspensions of magnetically polarizable particles in various base liquids revealing a strong viscosity increase in a magnetic field [3]. Such smart fluids have a huge potential for a large number of applications where the ER or MR activity of the fluid can be usefully applied.

The ER or MR fluid, respectively, is generally confined in an electromechanical device in which the electric signal is transferred to a mechanical response like friction. This device is part of a mechatronic system in which the properties of the fluid and the mechanical response can be regulated by sensors and an electronic control. The development of such mechatronic systems requires an interdisciplinary approach, i. e. the cooperation of materials scientists and chemists as well as of mechanical, electrical and software engineers. Especially the fluid and the mechanical construction of the device containing the fluid should be adapted with respect to each other. In this paper an overview of the outstanding properties of ER and MR fluids and examples of the most promising applications of these smart fluids are given.

3 Properties of ER and MR Fluids

Although the electrorheological effect has been known for about 50 years ER fluids with practical use have been available only for a decade. The activity of early fluids has been attributed to the occurence of water at the surface of the particles which supports their polarization. If the water evaporizes at elevated temperatures the activity decreases or even vanishes. Present ER fluids contain particles with intrinsic polarization and do not need water to become active. Today there are materials of very different chemical origin, like polymers, ceramics or zeolites, which can be used for ER fluids. Base fluids in which the particles are dispersed are silicone or mineral oils as well as halogenated hydrocarbons. An advantage of the latter is their partly higher density which opens the possibility to match the densities of the particulate and liquid material in order to increase the sedimentation stability of the suspension. Furthermore, chemical additives are used for dispersing the particles and improving the stability. In Table 1 some of the most relevant properties of ER and MR fluids are summarized.

Table 1. Comparison of ER and MR fluids

ER fluids	MR fluids
Composition	
Dielectric particles (polymers, ceramics, zeolites, etc.) Nonconducting base fluid (silicone oils, mineral oils, halogenated hydrocarbons) Chemical additives for stabilization	Magnetic particles (iron, iron-cobalt alloys, ferrites, etc.) Base fluid (silicone oils, mineral oils, water, etc.) Chemical additives for stabilization
Main properties	
Base viscosity 50...200 mPas Shear stress in field 1...5 kPa Electric fields of 1...5 kV/mm necessary High voltage (kV) – low current (mA)	Base viscosity 200...1000 mPas Shear stress in field 10...50 kPa Magnetic fields of 0.2...1 T necessary Low voltage (V) – high current (A)
Advantages	
Lower base viscosity Higher sedimentation stability Smaller and lighter actuators possible	Much higher shear stress Lower temperature dependence Not sensitive to humidity

The base viscosity of ER fluids without electric field is generally in a range between 50 and 200 mPas. In the field a shear stress in the order of magnitude of 5 kPa can be reached. For this grade of stiffness fields of some kV/mm are necessary. The current density in the ER fluid sensitively depends on the field strength, signal shape (AC or DC), field frequency and temperature, but is normally in a range of up to 100 $\mu A/cm^2$. This means that for devices with electrode distances of 1 mm and surfaces of some 10 cm^2 the voltage is in the kilovolt and the current in the milliampere range giving a power in the watt range.

As an example Fig. 1 shows the rheological flow curves of an ER fluid containing zeolite particles in silicone oil at different electric field strengths. Without field the flow curve is nearly Newtonian but when the field is applied a yield stress arises which increases with field strength. Correspondingly, the relative increase of shear stress or apparent viscosity with field strength is larger at lower than at higher shear rates (see Fig. 1) and can reach values of a hundred and more. The zeolite powder used in this ER fluid has been thermally activated and contains essentially no water. It has to be emphasized that ER fluids based on different kinds of zeolites show a strongly different behaviour concerning shear stress and current density which depend on the zeolite type and its cation composition [4], [5]. For the ER fluid in Fig. 1 zeolite A containing potassium and sodium cations has been used.

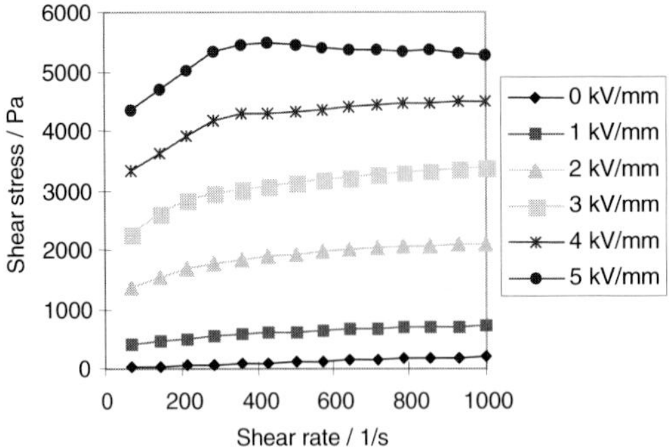

Figure 1. Flow curves of an ER fluid containing zeolite particles at different field strengths

For the preparation of MR fluids a lower variety of particulate materials can be applied than for ER fluids. The most commonly used material is iron due to its high magnetization saturation and low prize, but also Fe-Co alloys and ferrites have already been applied. The choice of the base fluid is less restricted than in case of ER fluids where the conductivity must be low due to the high field strength. For MR fluids different kinds of oils and also water can be used. Chemical additives for the stabilization of the suspension are even more important for MR than for ER fluids since due to their higher particle density they tend to sedimentate faster.

Related to the larger effort of stabilization the base viscosity of MR fluids without field is higher than in case of ER fluids (see Table 1). But in contrast the shear stress which can be reached with iron based fluids is an order of magnitude larger than for ER fluids. The magnetic field strength which is necessary for such strong effects is in a range of up to 1 T. In comparison to ER fluids much lower voltages but correspondingly higher currents have to be applied, which results in a comparable power consumption.

Both kinds of smart fluids have their specific advantages and disadvantages. The base viscosity of ER fluids is lower resulting in a lower streaming resistance but the shear stress in the field is much higher in case of MR fluids. With ER fluids a better stabilization against sedimentation is possible but most ER fluids strongly change their rheological and electric properties with temperature. This dependence is less pronounced for MR fluids. Due to the more flexible possibilities to generate electric fields in a restricted space by electrodes than to generate corresponding magnetic fields smaller and lighter devices working with ER fluids can be designed. On the other hand, MR fluids have the advantage not to be sensitive to humidity which changes the electric properties in ER fluids. Resuming the differences between the two kinds of smart fluids it has to be stated that the choice of the fluid depends on the special application.

4 Possible Applications of ER and MR Fluids

A large number of possible actuators has been discussed where the increase of the viscosity of ER or MR fluids by electric or magnetic fields is used. Table 2 gives an overview of the most relevant applications of ER and MR technology. Some of the applications can be realized with both kinds of fluids whereas for other devices a preference for ER or MR fluids exists. However, in each case the type of fluid, its profile of properties and the design of the actuator should be carefully adapted.

Table 2. Possible applications of ER and MR fluids

Application	ER fluids	MR fluids
Shock absorber	X	X
Vibration damper	X	X
Clutches	X	X
Brakes	X	X
Force feedback	X	X
Tactile devices	X	
Flexible grippers	X	X
Hydraulic systems	X	
Positioning control	X	
Polishing	X	X

The most prominent examples of ER and MR fluid devices are shock absorbers and vibration dampers [6] which can be used for automotive applications or in fastly rotating machines. Since the viscosity of the fluid in the damper can be controlled by the electric or magnetic field the strength of damping is adaptable to the working state of the machine. For instance, the damping should be strong near the resonance frequency. Other obvious applications of smart fluids are clutches and brakes in which the torque is transferred via the fluid without solid friction and the corresponding wearing. The transferrable torque can be limited by the field strength. For clutches and brakes MR fluids are more convenient than ER fluids due to the significantly larger shear stress. The scheme of a very simply designed MR clutch which basically consists of rotating disks is shown in Fig. 2. Since the MR fluid based on iron particles in silicone oil is located between the inner disk connected to the motor and the surrounding disks fixed to the other shaft the transferred torque is determined by the magnetic flux penetrating the clutch.

Promising perspectives arise from the force feedback capabilities of smart fluid devices. Fig. 3 shows a haptic device similar to a joystick which exerts resistance forces against the movement of the stick [7]. The stick position is detected by sensors and displayed by the cursor on a computer screen. If the cursor enters special fields on the screen the user perceives an increased resistance force on the stick. The resistance force is generated by an ER fluid based on zeolite particles which is located between two concentric spheres where the inner sphere can be rotated by the stick. By this way the user feels the motion of the cursor as a haptic response. This response can support applications in virtual reality (VR) or help blind persons working with a computer.

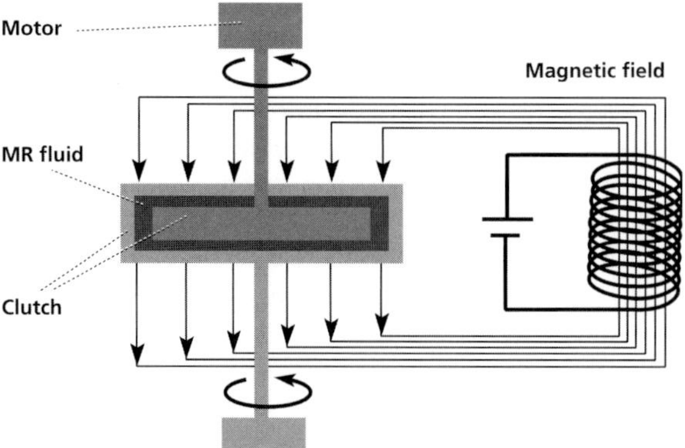

Figure 2. Scheme of a clutch working with an MR fluid

Another possibility is to apply smart fluids in tactile devices where the stiffness or shape of a flexible membrane which is touched by the user can be locally changed. For this purpose ER fluids are more advantageous than MR fluids due to the patterning of the surface which is more easily achieved with electrodes than with magnetic flux guides. Such tactile devices could also help blind or other persons to receive information nonvisually [8]. The concept of surfaces temporarily stiffened by smart fluids can also be used for flexible grippers in robot systems in order to grip sensitive pieces with complex shapes.

If the fluid in a hydraulic system is substituted by an ER fluid the flow is controllable by electric valves without mechanical parts because the flow of the ER fluid in a channel between the electrodes can be stopped by the electric field. Due to the fast response of ER fluids complex systems for positioning control are possible. If the fluidic motion in the hydraulic system is fast the abrasiveness of the ER fluid must be very low. The opposite requirement exists for applications like polishing where the particles in the fluid should be particularly abrasive. In such polishing systems the stiffening of the fluid is used to concentrate the abrasive effect on the spot where the field is active. The polishing effect is stronger with MR than with ER fluids due to the higher yield stress.

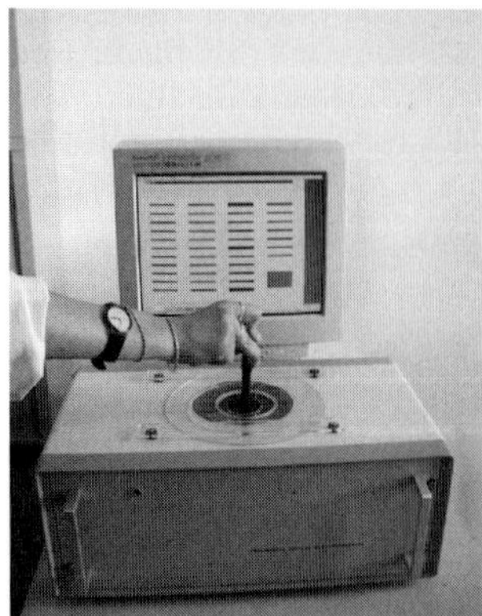

Figure 3. Haptic device based on an ER fluid

5 Conclusions

As described above smart fluids which respond to strong electric or magnetic fields possess a large potential for many technical applications. Since the requirements from various applications are different the type of the fluid should be carefully selected. Especially ER fluids but also MR fluids can be synthesized with very different profiles of properties. This opens the possibility to adapt the design of the actuatoric device and the smart fluid with respect to each other. In order to develop mechatronic systems based on ER and MR fluids an interdisciplinary approach is necessary.

6 References

1. K. M. Blackwood, H. Block, Trends in Polymer Sci. 1993, 1, 98-105
2. H. Conrad, MRS Bull. 1998, 23, 35-42
3. P. P. Phule, MRS Bull. 1998, 23, 23-25
4. H. Böse, Proc. 6th Int. Conf. on ER Fluids, MR Suspensions and Their Applications (M. Nakano, K. Koyama, Eds.), World Scientific, 1998, 240-247
5. H. Böse, A. Trendler, Proc. 7th Int. Conf on. ER Fluids and MR Suspensions 1999, in press
6. D. A. Brooks, Proc. 6th Int. Conf. on ER Fluids, MR Suspensions and Their Applications (M. Nakano, K. Koyama, Eds.), World Scientific, 1998, 689-696
7. H. Böse, H.-J. Berkemeier, Proc. 7th Int. Conf. on ER Fluids and MR Suspensions 1999, in press
8. P. M. Taylor, A. Hosseini-Sianaki, C. J. Varley, Proc. 5th Int. Conf. on ER Fluids and MR Suspensions (W. A. Bullough, Ed.), World Scientific, 1996, 184-191

Ag-Migration in La-doped PZT-Multilayers

Bernd Fischer, Michael J. Hoffmann
Institute for Ceramics in Mechanical Engineering, University of Karlsruhe, Haid-und-Neu-Str. 7, D-76131 Karlsruhe, Germany

1 Introduction

Lead zirconate-titanate (PZT)-multilayer ceramics are of increasing interest as actuators due to their low operating voltage which can be realized by ceramic layers less than 100μm thick. However, to achieve such thin layers and to reduce the costs, the preparation of PZT-multilayer components requires cofiring of a ceramic with internal electrodes. For a sufficient densification of the PZT-ceramic layers, sintering must be carried out at temperatures near 1100°C. Because of this, there is a need for an electrode material with a liquidus temperature above 1100°C. The mostly used material which combines these requirements is a Ag/Pd alloy with 30 atom-% Pd. However, it has been observed that silver can diffuse into the ceramic during cofiring, leading to an increasing conductivity and a change in dielectric and structural properties of the PZT ceramic layers. Although the problem has been identified, only less data can be found in the literature about the Ag-migration behaviour and kinetics. The present study shows that a considerable amount of silver can migrate from the electrodes into the ceramic and influences the microstructural development as well as the electromechanical properties. It will be shown that the diffusion is strongly correlated to the initial doping of the PZT, so that a microstructure adjusted to the requirements of application can be attained.

2 Experimental Procedure

PZT powder with the stoichiometric composition $Pb_{0.985}La_{0.01}(Zr_{0.54}Ti_{0.46})O_3$ and $Pb_{0.97}La_{0.02}(Zr_{0.54}Ti_{0.46})O_3$ has been prepared by the mixed oxide route. The composition is located at the morphotropic phase boundary (MPB) for 2mol% La-doping and its microstructure is well known for the presently used sintering conditions. The calcined and milled powders were consolidated into pellets by uniaxial pressing in a stainless steel die at 200MPa. Further densification has been achieved by subsequent cold isostatic pressing (CIP) at 400 MPa. A commercial Ag/Pd ink with the starting composition of 70Ag/30Pd was used to sceen print an electrode on one side of the pellet. XRD investigations revealed a completely alloyed state of the AgPd ink. The prepared samples were cofired at 1125°C and 1165°C for 2h, in an closed alumina crucible in oxygen atmosphere. Further sintering experiments were carried out at 1000°C for 10h in order to achieve a higher silver diffusion. A constant PbO vapor pressure during sintering was established by a $PbZrO_3/ZrO_2$ powder bed.

The microstructure was analyzed by SEM and digital image processing. Grain size was characterized by equivalent circle diameter (ECD). The electrode was carefully removed after cofiring to avoid artefacts during the preparation of a cross section perpendicular to the electrode. The cross sections were grinded and polished to determine the Ag content in the ceramic by microprobe analysis (WDX). The measurement error was less than 2% relative. In addition, tape casting have been performed to prepare green tapes with 100 μm thickness with screen printed Ag/Pd electrodes of 5 μm thickness. Multilayer ceramics were then prepared by laminating and subsequent cofiring. The multilayer ceramics were finally electrically characterized in order to study the changes in material properties by Ag-migration.

3 Results and Discussion

3.1 Microstructure

The microstructure of the PZT pellet with only one electrode on top strongly depends on the distance from the electrode. Figure 1 indicates that the average grain size is different from that of the opposite side (without electrode) almost 2500μm away from the electrode. A reference pellet prepared with the same PZT powder but without an electrode revealed a constant average grain diameter over the whole cross section and the same grain size distribution as it has been determined for samples with an electrode on the opposite side of the electrode. The data indicate a high reproducability of the microstructure especially for areas which are not influenced by silver diffusion. The average grain size at the electrode free side is approximately the same as predicted by Hammer and Hoffmann [1] for the same composition and under similar sintering conditions.

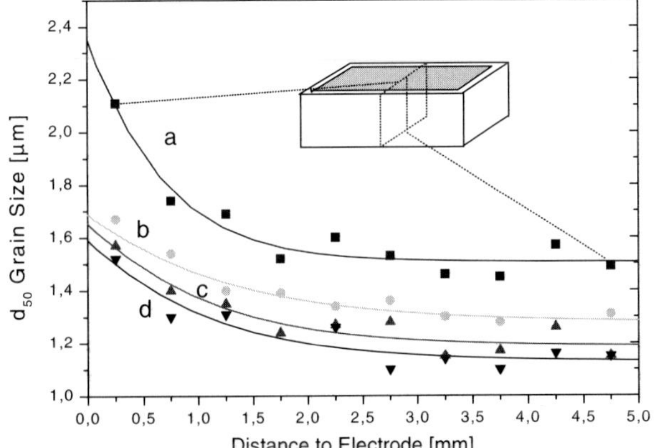

Figure 1. Average grain size of specimens a) sintered at 1165°C for 2h in PbO controlled atmosphere, b) same temperature, but without PbO control, c) 1125°C for 2h with powder bed d) 1125°C without powder bed.

A comparison of the present results with earlier published data [1] for PZT (54/46) with different La contents (Fig 2) shows a similar grain growth behaviour for the region just beneath the electrode (2.11 μm) as reported for PZT with 1.5 mol% La. Therefore it is assumed that Ag reduces the grain growth inhibitor effect of the lanthanum by occupying A-sites of the perovskite structure and thus reducing the number of lead vacancies. It seems that 25% of the initially existing Pb-vacancies are eliminated. No secondary phases could be detected by optical as well as SEM investigations which gives a further evidence for the incorporation of Ag into the perovskite structure. Similar observations have been reported by Schuh et al. [2] in Nd doped PZT ceramics.

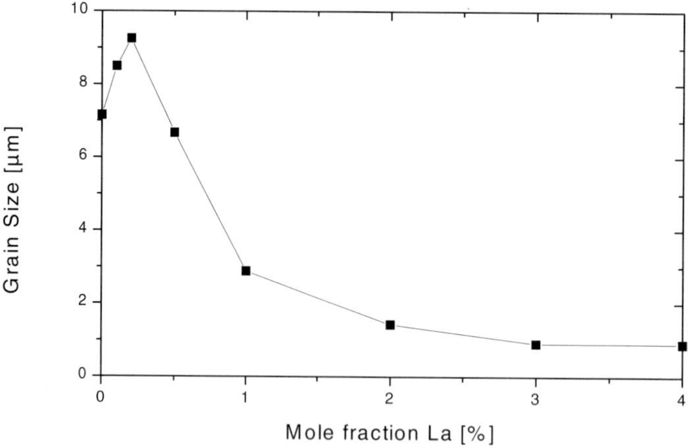

Figure 2. Average grain sizes as a function of La-content for PZT with a Zr/Ti ratio of 54/46 after sintering at 1170°C for 2h in a PbO controlled atmosphere [1].

3.2 Silver distribution

Figure 3 shows a WDX-linescan along the cross section of a sample with one top electrode. Although Pd undergoes an oxidation to PdO at 450°C and a subsequent reduction to Pd at 700°C [3], no Pd could be detected in the perovskite lattice. However, considerable amounts of silver could be found in the PZT-grains. This means that silver must be oxidized despite its higher stability against oxidation compared to Pd. The silver distribution shown in Fig. 3 can be divided into three regions:

- At a distance from 0 to 200µm from the top electrode, a constant silver content of 0.5 mol% has been determined for a 2 mol% La-doped material. The Ag content decreases to 0.25 mol% in case of a 1 mol% La-doped PZT. The magnitude of the observed saturation level depends obviously on the donor doping content. Such a saturation has been already observed by others authors [4]. From the WDX measurements it could be concluded that silver occupies half of the initial A-site vacancies, but charge compensation requires the evaporation of a Pb-ion which leads to an overall reduction of A-vacancies of 25%. These considerations are in full agreement with the results of the grain size measurements described in the previous chapter. Caballero et al [5] sintered an undoped PZT-ceramic multilayer with internal Ag/Pd-electrodes and found no silver solubility. This result also correlates well with our findings that the maximum Ag-solubility will be determined by the number of the initially available Pb-vacancies which are controlled by the donor doping content.
- The second range from 200 to 2500µm beneath the electrode is characterized by a linear decrease of the Ag-content with an increasing distance from the electrode. The data indicate a very fast transport mechanism. Due to the high final density of > 99% th. d., open porosity seems not to play a significant role as major transport mechanism as pointed out by Wen et al. [6].

- No silver could be detected in the third range at distances greater than 2500μm away from the electrode. This means that the properties of PZT ceramic will not be influenced by silver migration. The data are again in a good agreement with the measured grain size distributions shown in Fig. 1.

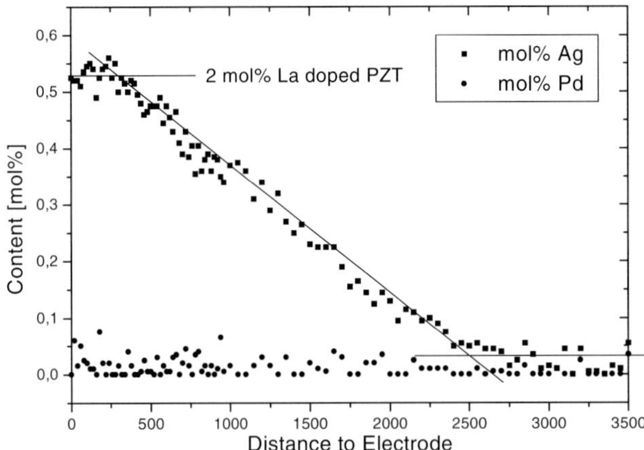

Figure 3. Ag distribution in 2mol% La-doped PZT after 10h sintering at 1000°C. The Ag-saturation in the PZT-lattice is obtained at 0.5mol%. The Pd content is under the dectection limit.

Previous investigations with a Ag_2O containing PZT-powder show a reduction of Ag_2O to metallic Ag under usual sintering conditions of PZT-ceramics [7,8]. These results are an evidence for the thermodynamic stability of silver during firing of PZT in air. The authors reported also an increase of the Curie-temperature of the Ag_2O containing PZT in comparison to a reference material without Ag_2O. This clearly indicates that a certain amount of the initially present Ag_2O is not reduced, but dissolved in the PZT-lattice. Therefore it can be concluded that an incorporation of Ag-ions in the perovskite lattice would be only possible in the presence of Ag_2O, but the oxidation mechanism is still not clear.

3.3 Electromechanical Behavior

Figure 4 shows the hysteresis loop of a cofired multilayer ceramic with 20 internal electrodes in comparison to a bulk ceramic of the same PZT-composition. The incorporation of Ag into the perovskite lattice clearly reduces the total strain. The polarisation is also reduced compared to the bulk ceramic. Both effects can be attributed to a reduction of the Pb-vacancy concentration which causes a more "hard" materials behavior. Nevertheless, the coercive field remained nearly constant. This could be explained by the different grain sizes of the multilayer and bulk ceramic: the grain size of the Ag-containing material is increased by a factor of 1.5.

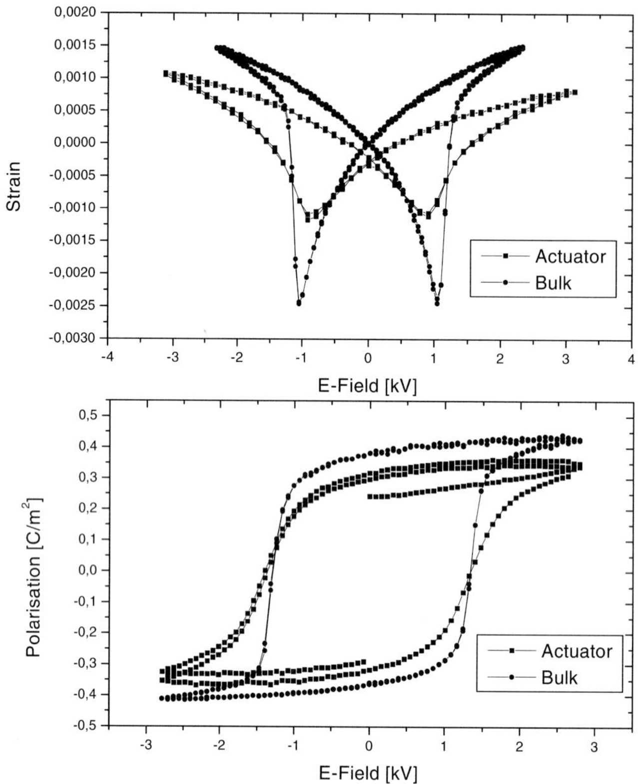

Figure 4. Electromechanical response of a multilayer ceramic with Ag/Pd electrodes in comparison to a bulk ceramic of the same composition.

4 Conclusions

La-doped PZT-ceramics with Ag/Pd-electrodes have been cofired below 1165°C for 2 h in order to study the silver migration during densification. WDX-linescans revealed a Ag-migration over a distance of 2.5 mm. The regions up to 200 μm away from the electrode showed a saturation with silver, whereas the Ag-content decreases linearly for larger distances. The extended Ag-migration was also supported by quantitative microstructural analysis which reveals a decreasing grain size with an increasing distance from the electrode. The grain size effect strongly depends on the sintering temperature as well as the sintering atmosphere. The proposed mechanism for the silver migration is based on a partial oxidation of silver during sintering. The Ag-ions formed during this oxidation could then diffuse into the perovskite lattice and occupy A-site vacancies which are created by doping with lanthanum. It seems that the number of the initial A-vacancies could be reduced by maximal 25%. Measurements of the electromechanical properties give further evidence for an incorporation of Ag-ions in the PZT-lattice. A Ag-saturated multilayer ceramic shows a shift towards a more hard ferroelectric behavior in comparison to a bulk ceramic of the same PZT-composition.

5 References

1. M. Hammer and M.J. Hoffmann, A Sintering Model for Mixed Oxide Derived Lead Zirconate Titanate (PZT) Ceramics, J. Am. Ceram. Soc., **81** [12] (1998) 3277.
2. C. Schuh, M. Kulig, K. Lubitz, Ag doping of rare earth modified PZT, Electroceramics V, S. 201
3. S.C. Sanford, Oxidation and Reduction of Palladium in the Presence of Silver, J. Am. Ceram. Soc., 68 [4] (1985), 106-107
4. M.V. Slinkina, G.I. Dontsov, V.M. Zhukovski, Diffusional Penetration of Silver from Electrodes into PZT Ceramics, J. Mater. Sci., 5 (1993), 5189-5192
5. A.C. Caballero, E. Nieto, P. Duran, C. Moure, Ceramic-Electrode interaction in PZT and PNN-PZT multilayer piezoelectric ceramics with Ag/Pd 70/30 inner electrode, J. Mater. Sci. 32 (1997), 3257-3262
6. J. Wen, H. Hellebrand, D. Cramer, K. Lubitz, G. Tomandl, Grain Growth in Multilayer PZT, Proceedings Electroceramics IV, Aachen (1994), 247-252
7. H.J. Hwang, K. Watari, M. Sando, M. Toriyama, K. Niihara, Low-Temperature-Sintering and High-Strength $Pb(Zr,Ti)O_3$ Matrix Composites Incorporating Silver Particles, J. Am. Ceram. Soc. 80[3] (1997), 791-793
8. H.J- Hwang, T. Nagai, T. Ohji, M. Sando, M. Toriyama, K. Niihara, Curie Temperature Anomaly in Lead Zirconate Titanate/Silver Composites, J. Am. Ceram. Soc. 81 [3] (1998), 709-712

Ferromagnetic Films for Temperature Measurement

C. Willeke, H. Klümper-Westkamp, F. Hoffmann, P. Mayr
Stiftung Institut für Werkstofftechnik (IWT), Bremen, Germany

1 Introduction

The temperature is in many processes an important parameter. For that reason a lot of methods of measurement were developed [1]. These methods can be diveded into contact and noncontact methods of temperature measurement. In the industrial application the resistance and the thermocouple thermometer as contact method have established. Among noncontact method of temperature measurement only the pyrometer has obtained a great importance up to now. Another noncontact method is the eddy-current thermometer [2] [3]. This work describes a sensor which indicates if the temperature exceeds or falls below a threshold temperature.

2 Measurement Principle

The sensor consists of a measuring coil and a ferromagnetic film. The film apply to the measured object. Fig. 1 shows the alignment of the system coil and film. The magnetization and the eddy currents induced by an applied electromagnetic field in the film cause a change of the measuring coil impedance. If the temperature of the film exceeds the Curie point, the film experiences a magnetic transformation from ferromagnetic to paramagnetic. This magnetic transformation leads to a significant decrease of the permeability of the film. This decrease is used to detect a threshold temperature.

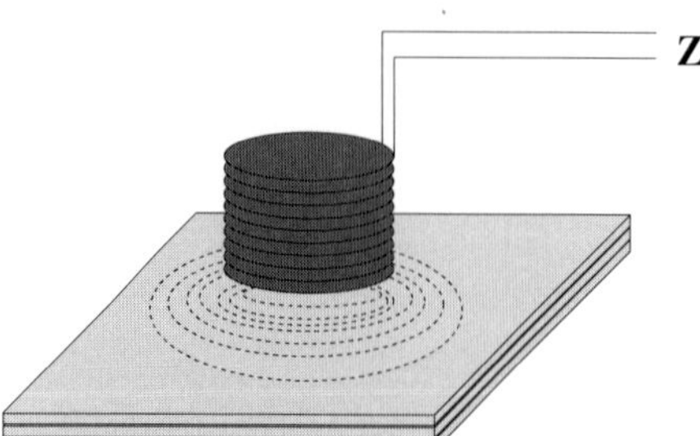

Figure 1. Alignment of the system: measuring coil - ferromagnetic film.

Alloy systems whose Curie point increase continuously by increase of one of the alloying constituent can be used for the production of the necessary ferromagnetic films. For example within the alloy system nickel-cobalt every curie temperatures between 361 °C and 1121 °C can be produced. Fig. 2 shows the phase diagram of nickel-cobalt.

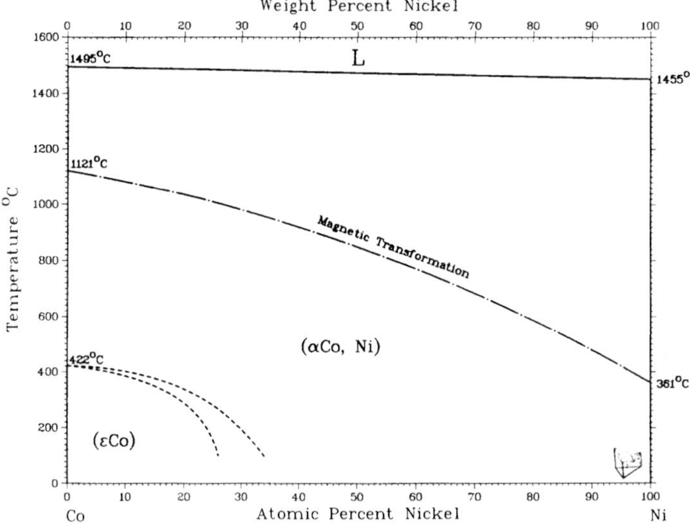

Figure 2. Phase diagram of nickel-cobalt.

The investigations in this work were performed for pure nickel films and nickel-cobalt films. This films were electrodeposited from chloride solutions. Copper substrates were used as the base material.

The sensitivity of the sensor is dependent on many parameters. The important parameters are:

- Alignment of system: coil - film.
- Surface area and thickness of the film.
- Coil data.
- Measurement frequency.

3 Investigations

The dependence of the impedance on the surface area and thickness of the film and on the distance between the film and the measurement coil was investigated. The measurement frequency was 40 kHz and the outside diameter of the coil was 1,3 cm. Fig. 3 shows the influence of the film thickness on the temperature dependence of the imaginary component of the impedance. The size of the film was 2,25 cm^2 and the lift-off distance was zero. The temperature was measured with a thermocouple thermometer which is placed on the film surface. The sensitivity of the sensor raises with increasing film thickness. The sensitivity can't be raised infinitely as the result of penetration of the electromagnetic field which depends on the measurement frequency. The influence of the film surface area is shown in Fig. 4. The thickness of the films was 100 µm and the lift-off distance was zero. Fig. 5 shows the influence of the lift-off distance. The thickness of the films was 100 µm and the surface area was 2,25 cm^2. The sensitivity of the sensor raises with increasing surface area and with decreasing lift-off distance.

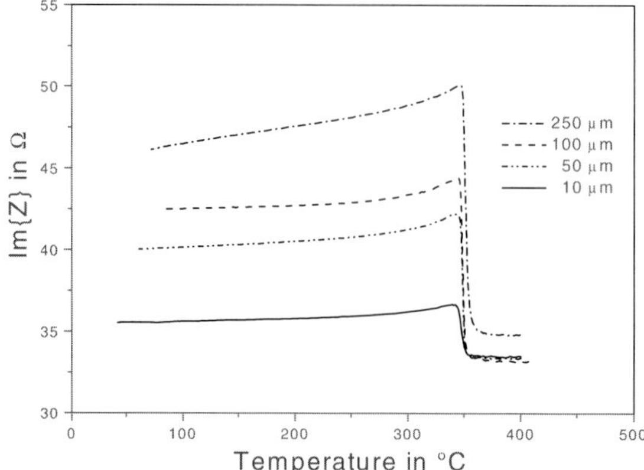

Figure 3. Temperature dependence of the imaginary part at various film thickness.

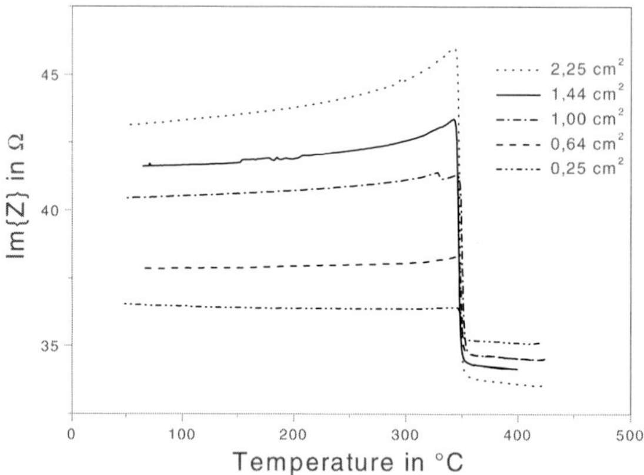

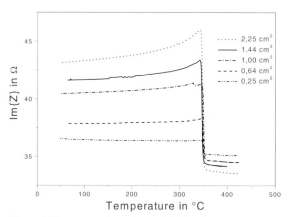

Figure 4. Temperature dependence of the imaginary part at various film surface areas.

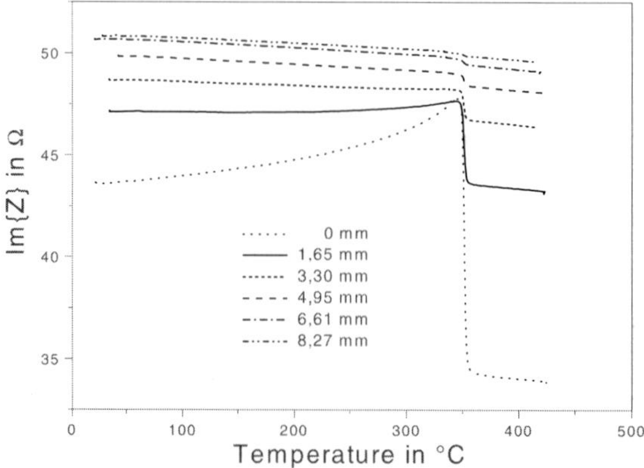

Figure 5. Temperature dependence of the imaginary part at various coil distances.

With nickel films a threshold temperature from about 361 °C can be detected. Higher threshold temperatures were obtained with nickel-cobalt alloys. Fig. 7 shows the temperature dependence of the imaginary part for various nickel-cobalt alloys. The percentage content of cobalt was determined by Energy-Dispersive X-Ray Spectroscopy (EDX). The Curie point increases to higher temperatures if the percentage content of cobalt increases. The temperature interval, within which the significant decrease of the imaginary part occurs, enlarges as well.

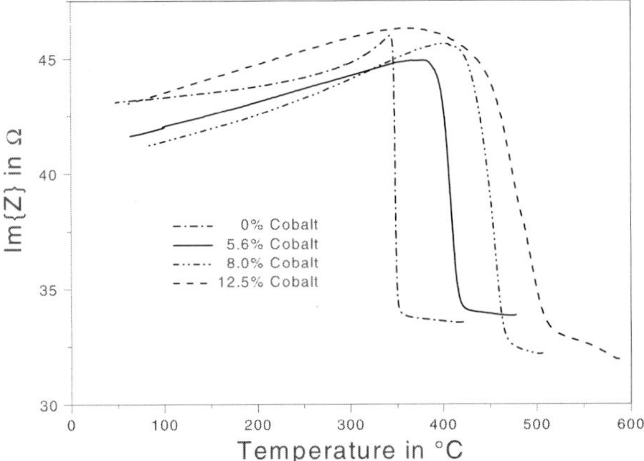

Figure 6. Temperature dependence of the imaginary part for various nickel-cobalt alloys.

4 Summary

In this work the influence of the thickness and surface area of the films and of the coil distance on the sensor signal were investigated for pure nickel films. The dependence of the sensitivity of the sensor on these parameters was showed. Furthermore it was showed how the threshold temperature can be changed in terms of alloying with cobalt. The enlargement of the temperature interval with increasing cobalt content reduces the accuracy of the threshold temperature.

5 Acknowledgement

This work was supported by Materials Science and Technology Cooperation (MATEC).

6 References

1] L. Weichert, Temperaturmessung in der Technik, 3th ed., expert verlag, 1981

[2] P. Keller, Iron and Steel Engineer, 1980, 42-44

[3] D. Schwarzennau, Untersuchung von Wirbelstromverfahren zur berührungslosen Temperaturmessung, Dissertation, Universität Hannover, 1989

Material Requirements in Hot Embossing of Microstructures

K. Feit, K.D. Müller, M. Heckele
Forschungszentrum Karlsruhe

1 Abstract

Plastic microstructures may be produced by thermoplastic deformation. Very good results in terms of highly complex LIGA microstructures with high aspect ratios and small roughness of the lateral walls are achieved by the hot embossing process. Selection of adequate process parameters requires an analysis of the thermal and thermomechanical material behavior of the semi-finished plastic products used as well as equipment-related technical studies. Using an amorphous plastic material (PMMA) and a semi-crystalline plastic material (POM), basic studies are performed with regard to the molding behavior of thermoplastic semi-finished products during hot embossing in microstructure technology.

2 Hot Embossing[1]

In the past years, hot embossing proved to be successful besides injection molding when manufacturing high-quality microstructures of a high aspect ratio from plastics by means of the LIGA method. Hot embossing is understood to be the process of pressing a mold insert under vacuum into a pre-fabricated semi-finished plastic product that is located on a substrate. The process takes place at a temperature that ensures sufficient flowability of the plastic material (Figure 1).

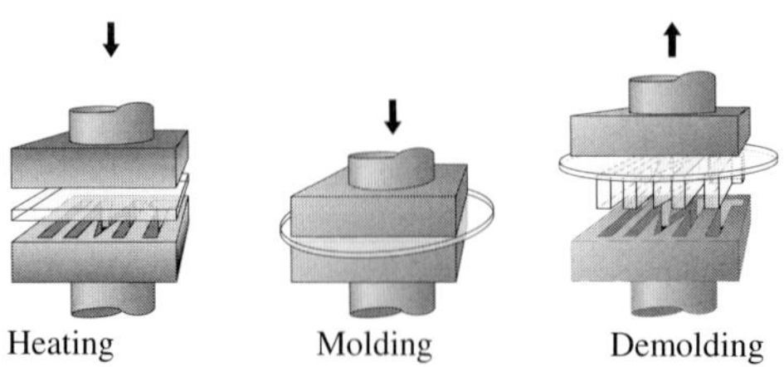

Figure 1. Process steps of hot embossing

After filling the microcavities of the mold insert, the plastic material is cooled down to a temperature which provides for a sufficient strength, and the microstructured plastic material can be demolded. As compared to injection molding, hot embossing has several significant advantages. The small deformation rates arising during molding result in small flow velocities which facilitate the molding of very fine structures, such as free columns with a small base area or long, thin walls in the mold insert and, thus, long, narrow channels are obtained in the embossed part without any deformation occurring. Due to a process conduct close to the thermal equilibrium and small flow paths of the order of structural height only, internal stresses in the molded plastic material can be minimized. Precise control of the process parameters, such as molding forces and temperature, allows to manufacture microstructures with very thin, unstructured carrier layers.

3 Material Requirements in Hot Embossing

3.1 The Embossing Process

As mentioned above, the hot embossing process serves to bring the semi-finished plastic product into a flowable state by an increase in temperature. Hence, only thermoplastic amorphous or semi-crystalline plastics are suited for embossing. The embossing parameters are determined decisively by the change of the mechanical behavior in the softening or melting range. The variable mechanical behavior of amorphous and semi-crystalline plastics is illustrated by thermomechanical diagrams, with Young's modulus being plotted as a function of temperature. Such diagrams can be obtained by means of the Dynamic-Mechanical Analysis (DMA) (Figure 2).

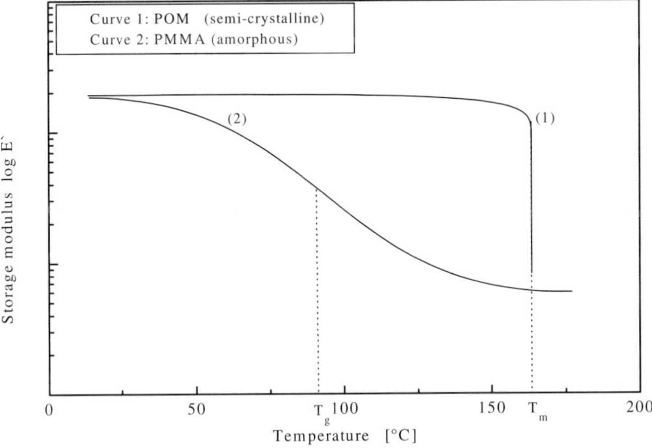

Figure 2. Thermomechanical behavior of an amorphous thermoplastic material (PMMA) and a semi-crystalline thermoplastic material (POM), as determined by means of the Dynamic-Mechanical Analysis (DMA)

Amorphous plastics exhibit softening over a wider temperature range (characterized by the glass temperature T_g), which is identical with the phase transition from the glassy state to the rubber-elastic and subsequent viscous state. Accordingly, semi-crystalline plastics show a sharp decrease of strength in the melting range (characterized by the crystallite melting temperature T_m). When filling the microcavities in the embossing process and setting a certain carrier layer thickness, a certain flowability must be reached. In case of the semi-crystalline plastic material, it may be achieved in a narrow temperature range above the crystallite melting temperature T_m. As far as amorphous polymers are concerned, the embossing temperature mostly is far above the glass temperature T_g.

3.2 Demolding

Demolding of plastic microstructures takes place by transferring the plastic mass into a state of a mechanical strength that allows to apply forces to the plastic mass, by means of which the plastic microstructure is separated from the mold insert. This increase in the mechanical strength of the plastic material is achieved by reducing the process temperature to below the glass temperature in case of amorphous plastics and below the melting temperature for semi-crystalline plastics. In analogy to embossing, the required material strength of semi-crystalline plastics is reached already somewhat below the melting temperature (or to be more precise, the crystallization temperature T_c). In case of amorphous plastics, a demolding temperature far below T_g must be reached due to the larger softening range. Hence, considerably smaller cycle times can be attained for semi-crystalline plastics as compared to amorphous plastics (Table 1).

Table 1. Embossing and demolding temperatures for an amorphous plastic material (PMMA) and a semi-crystalline plastic material (POM)

	Demolding temperature [°C]	Glass temperature T_g [°C]	Melting temperature T_m [°C]	Embossing temperature [°C]
POM	150		156	180
PMMA	70 - 80	90 - 105		170 - 210

During demolding, demolding stresses are generated, because adhesion and friction between the plastic material and the mold insert have to be overcome for demolding from the molding tool. If the demolding stresses exceed the yield stress of the plastic material during demolding, tear off and shearing as well as distortions and deformations may occur in the demolding direction of the microstructures. The mechanisms contributing to the generation of demolding stresses are of variable nature. By the dwell pressure applied during embossing, compressive stresses are generated in the plastic mass. They cause shear stresses between the plastic mass and the mold insert at the boundaries of the vertical microstructure walls. The compressive stresses are superimposed by tensile stresses that result from the variable expansion coefficients of the plastic mass, the mold insert and the substrate. The materials of the mold insert and the substrate have a Young's modulus that is higher by two orders of magnitude and a thermal expansion coefficient which is smaller by about one order of magnitude. During the temperature cycle of the molding process, these materials may be subject to free shrinking, while shrinking of the plastic material is impeded. In addition to these thermally induced demolding stresses, they may also be generated in both tension and compression direction by e.g. various designs, existing deformations of the mold inserts and adhesion problems between the plastic mass and the mold insert. Optimum demolding results would be achieved, if tensile and compressive stresses compensate each other, i.e. when the demolding stresses tend to zero. This state, however, is not reached such that it becomes necessary to minimize the existing demolding stresses. Possibilities to minimize the demolding stresses can be derived from studying the solidification behavior of the plastics. Taking into account the temperature influence on the material data, the developing demolding stresses can be described as follows (Equation 1)[2]:

$$\sigma = \sigma_D + \int_{Te}^{Tg} \frac{E_P(T)}{1-2\nu_P(T)} \left(\alpha_{th,P}(T) - \alpha_{th,M}(T)\right) dT \qquad (1)$$

demolding stress E_P: Young`s modulus (plastic)
ν_P: Poisson`s ratio (plastic) $\alpha_{th,P}$: thermal expansion coefficient (plastic)
$\alpha_{th,M}$: thermal expansion coefficient (mold insert) T_e: demolding temperature
T_g: glass temperature

Demolding stresses are generated above all by the strong temperature dependencies of the expansion behavior (Figure 3) and Young's modulus (Figure 4) of the plastics.

The p, v, T diagram (Figure 3) describes the expansion behavior of the plastics as a function of temperature. The thermal expansion coefficient can be obtained from the slope of the curves. The high pressures reflect the embossing conditions. In case of the semi-crystalline plastic material POM, a sudden change of expansion is found in the melting range, while expansion of the amorphous plastic material PMMA is nearly linear in the softening range. To minimize the demolding stresses that may cause structural distortions, the demolding temperature of POM must be set precisely in contrast to that of PMMA. It must not be below 150°C. It is also obvious from the temperature dependence of Young's modulus (Figure 4) that the high mechanical strength of the semi-crystalline POM is maintained up to about 5°C below its crystallite melting point T_m. Hence, it has a sufficiently high mechanical strength for applying the necessary demolding forces to the plastic material. For PMMA, the same mechanical strength is reached at a temperature of about 70°C to 80°C only, i.e. at a demolding temperature which is about 20°C to 30°C below the glass temperature T_g.

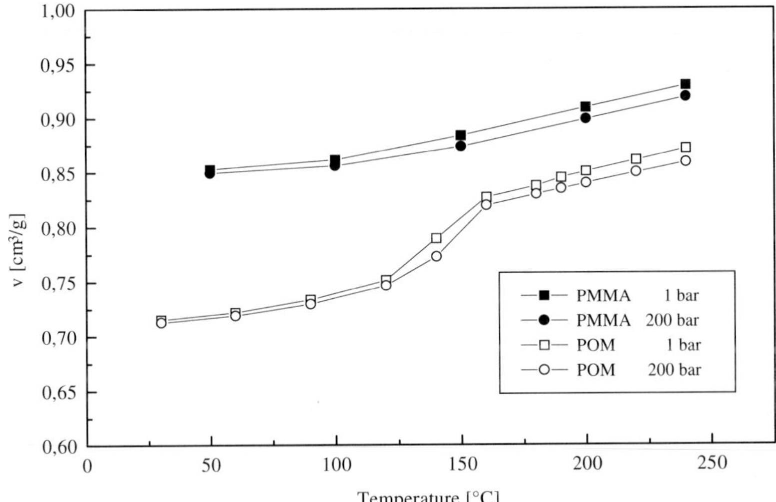

Figure 3. Thermal expansion behavior of an amorphous plastic material (PMMA) and a semi-crystalline plastic material (POM) under embossing conditions (200 bar) in the p, v, T diagram

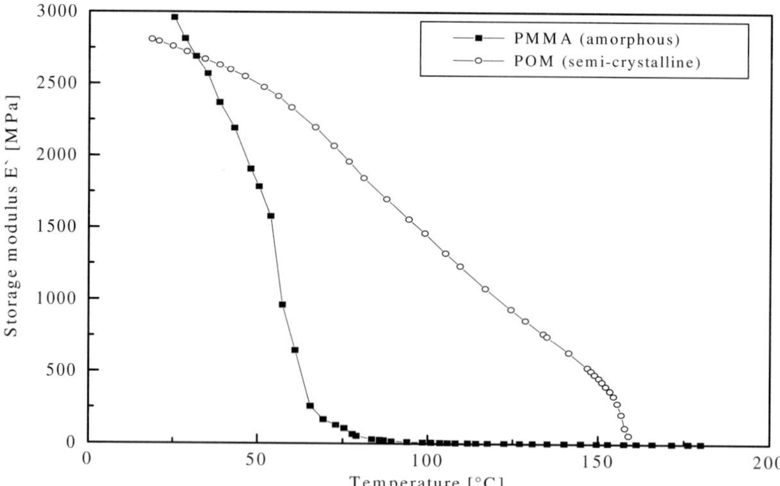

Figure 4. Young's modulus as a function of temperature for an amorphous plastic material (PMMA) and a semi-crystalline plastic material (POM)

Use of the data represented in Figure 3 and 4 to calculate the demolding stresses according to equation 1 and plotting them as a function of temperature (Figure 5) and Young's modulus (Figure 6) allows to select practically relevant methods for the reduction of the demolding stresses and, hence, to determine the appropriate demolding temperature for optimum molding of microstructures.

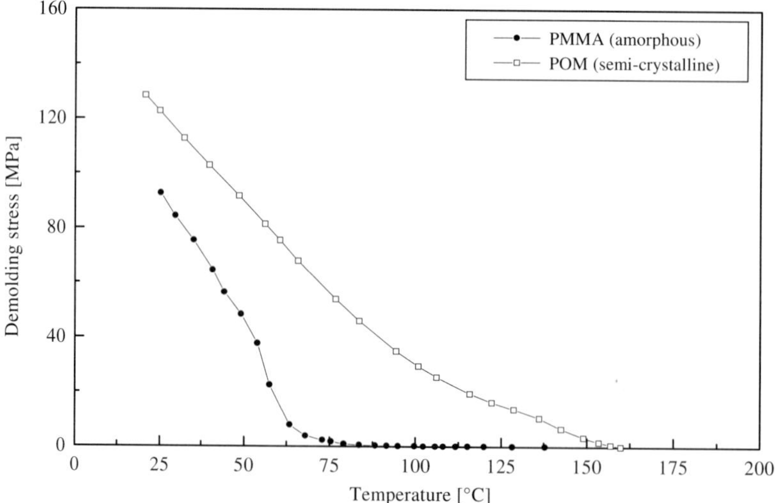

Figure 5. Demolding stresses as a function of temperature for an amorphous plastic material (PMMA) and a semi-crystalline plastic material (POM)

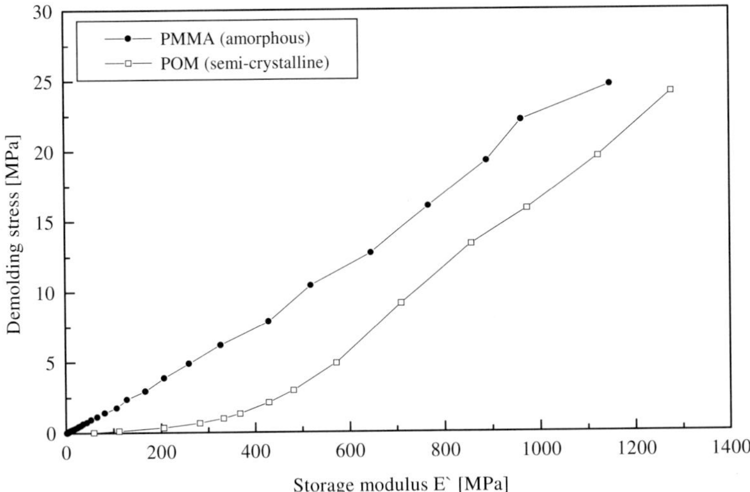

Figure 6. Demolding stresses as a function of Young's modulus for an amorphous plastic material (PMMA) and a semi-crystalline plastic material (POM)

The increase in the demolding stresses of POM at some °C below its crystallite melting temperature T_m already results from the significant change of its thermal expansion coefficient in this temperature range. Consequently, demolding temperatures below 150°C will result in considerable deformations of the microstructure. On the other hand, POM has a very high Young's modulus at these small thermal demolding stresses. As tensile strength of many plastics behaves linearly to Young's modulus, POM already has a sufficiently high mechanical strength such that demolding may take place some °C below the crystallite melting temperature T_m already.

4 References

1. M. Heckele, W. Bacher, K.D. Müller, Microsystem Technologies 1998, 4, 122-124
2. K.D. Müller, W. Bacher, M. Heckele, Proceedings IEEE, MEMS 98, 263-267

Structural and Spectroscopic Studies of Heteropoly Acid in Gel Electrolyte

U.Lavrenčič Štangar, N.Grošelj, U.Opara Krašovec, M.Gaberšček, B.Orel
National Institute of Chemistry, Slovenia

1 Introduction

The development of organically modified electrolytes (so-called ORMOLYTES) (1, 2) as a fast growing branch of sol-gel hybrid materials was oriented towards the application in electrochromic devices (3, 4), fuel cells (5) and solid-state lithium batteries (6).

Ormolytes in which organic and inorganic part are linked by chemical bonds, were often constructed by using reactive silicon alkoxide like 3-isocyanatopropyltriethoxysilane in combination with amino groups of poly(propylene glycol)bis(2-aminopropyl ether) (1) or with hydroxyl group of poly(ethylene glycol) methyl ether (2, 6). For this work, we have used the hybrid precursor with urea connecting groups (noted ICS-PPG) following the preparation procedure of Dahmouche et al. (1).

Phosphotungstic acid (PWA) is a good candidate for a functional molecule to be incorporated in sol-gel matrices. It belongs to a large group of polyoxometalates (7), exhibiting beside high protonic conductivity (0.02-0.1 S/cm at room temperature (8)), also electrochromic and photochromic properties. Relatively small size (~10 Å) of Keggin's ions and high solubility of PWA in water and many organic solvents facilitate the incorporation of heteropolyacids in various matrices.

Recently, it was suggested that PWA as aqueous solution could represent an efficient electrolyte for low temperature fuel cells (9) although the risk of its continuous leakage during cell operation is high. In order to overcome the stability problems and to increase the lifetime of the cell, phosphotungstic acid and silicotungstic acid (SiWA) immobilised in silica gel were prepared and the proton conductivity studied (10). It was found that up to 30 wt.% of PWA and up to 45 wt.% of SiWA can be entrapped inside the gel structure. The proton conductivity of the solid materials at room temperature was in the range 10^{-3}-10^{-4} S/cm, depending on the heteropoly acid loading and relative humidity. The interactions between Keggin units of PWA or SiWA with the silica network, studied by IR spectroscopy, were found to be stronger in the case of SiWA than in the case of PWA (10).

The aim of our work was to show that organic-inorganic gel matrix derived from ICS-PPG hybrid precursor, i.e. ethoxy silane functionalized by polypropylene glycol, is a suitable host for the incorporation of high amounts of heteropoly acids. IR spectroscopic measurements were used to reveal the preservation, entrapment, aggregation and the interactions of PWA and SiWA with the sol-gel derived network. The proton conductivity of gel samples were measured by impedance spectroscopy. X-ray diffractionanalyses were performed to see to what extent PWA and SiWA incorporation changes the amorphous character of the silica host.

2 Experimental

2.1 Preparation of proton-conducting gel samples

Unhydrolysed hybrid silicon precursor ICS-PPG was prepared according to (1). For further sol-gel synthesis of protonic conductor a portion of ICS-PPG was used, depending on the desired weight concentration of PWA in the gel: wt.% PWA = $m_{PWA(anhydrous)}/(m_{PWA}+m_{ICS-PPG})$. For example, to prepare a gel electrolyte containing 75 wt.% of PWA, 6.3g of PWA which contained 14% of the crystalline water (Aldrich) in ethanol (3 ml) was mixed with 1g of ICS-PPG. A homogeneous mixture was obtained after few minutes of vigorous stirring. Gelation was completed within few minutes to few days, depending on the PWA/ICS-PPG ratio.

2.2 Instrumental and measuring techniques

X-ray diffraction (XRD) measurements of PWA/ICS-PPG membranes, obtained by ageing of the gel samples in teflon dish, were made on a Philips PW1710 automated X-ray diffractometer. Thermogravimmetric (TG) measurements for the determination of water content in PWA were done on a TG951 MODUL TA2000, TA Instrument INC Castle Point, DE, USA.

FTIR spectra were measured on a FT-IR Perkin Elmer 2000 system spectrometer (spectral range used was 4000-400 cm^{-1}). Spectra of the PWA/ICS-PPG composites were recorded after depositing the sols directly on CdTe supporting plates and leaving them on the plates to dry. The resolution of all the measured spectra was 4 cm^{-1}.

Impedance measurements were made using a 1286 Solartron Electrochemical Interface and a 1250 Solartron Frequency Response Analyser. Impedance spectra were recorded between 65000 to 0.001 Hz. For the gels, two parallel $F{:}SnO_2$ covered glass plates served as electrodes. To achieve a good contact, the electrodes were immersed into the sol prior to gelation. The distance between the electrodes was 0.3 mm and their surface area was approximately 1 cm^2 (the exact value was determined individually for each sample).

3 Results

3.1 Structural analysis

As a first step in studying the PWA/ICS-PPG and SiWA/ICS-PPG compounds, we performed XRD analysis of the corresponding mixtures in the concentration range 30-80 wt.% (Fig. 1). As expected, XRD spectra of the composites with PWA or SiWA concentrations below 30 wt.% are amorphous. This trend remains the same up to 70 wt.% of heteropoly acid in the composite while the composites containing more than 70 wt.% of PWA or SiWA start to exhibit the diffraction peaks. The attained concentrations (i.e. ~ 70 wt.%) outrange those achieved when SiWA and PWA were incorporated in silica derived from TEOS sols indicating also that the specific interactions in our hybrids are quite different with respect to PWA/ and SiWA/silica composites (10). Accordingly, IR spectra of PWA/ICS-PPG and SIWA/ICS-PPG composites were measured and spectra analysed.

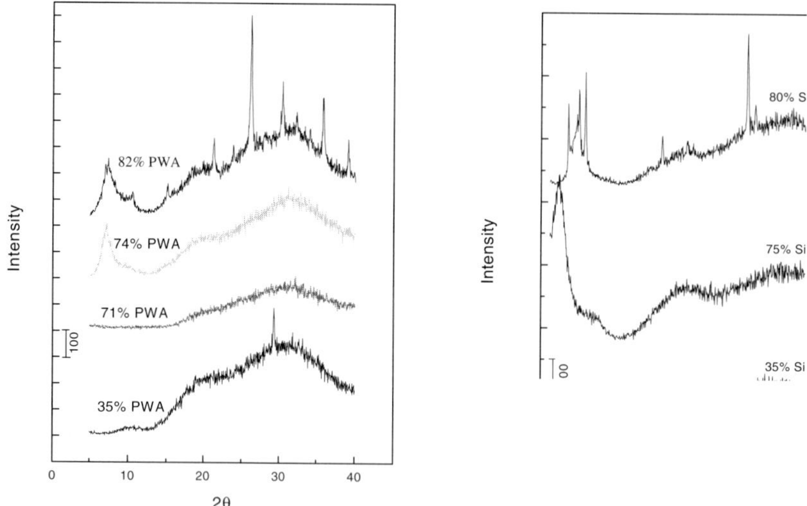

Figure 1. XRD spectra of PWA/ICS-PPG and SiWA/ICS-PPG composites.

IR spectra of PWA and SiWA show characteristic bands which have been assigned according to Deltcheff (11) in terms of W-O-W vibrations of edge- and corner-sharing W-O_6 units linked to the central P-O_4 or Si-O_4 tetrahedra (see Fig. 2). The stretching modes of edge-sharing (W-O_b-W) and corner-sharing (W-O_c-W) units appear in the range 795-890 cm^{-1} while the Si-O_d and P-O_d modes are at 924 cm^{-1} and 1017 cm^{-1} (SiWA) and 1078 cm^{-1} (PWA). The stretching modes of the terminal W=O_d groups do not differ so much and could be found in the range 979-982 cm^{-1} for both heteropolyacids. All the W-O modes are sensitive to the change of water content and shifts also when electrostatic interactions are changed when various ions (K^+, Na^+, Cs^+, Tetramethylamonium ion (TBA) or Tetramethylammonium ions (TMA)) replaced the smaller ones (11).

When ICS-PPG is added to SiWA the W=O_d mode shifts from 982 cm^{-1} (SiWA) to 971 cm^{-1} (SiWA/ICS-PPG = 10 wt%) while the W-O_c-W mode is shifted from 795 cm^{-1} to 800 cm^{-1}. This suggests that the dipolar interactions between the neighbouring SiWA molecules are diminished due to dispersion of anions in the composite which results in the decreasing of their mutual electrostatic interactions.

Contrary to expectations, in the spectrum of PWA the W=O_d mode at 979 cm^{-1} slightly shifts to higher frequencies with the decreasing concentration of the Keggin's ions in the ICS-PPG matrix and in addition, the blue frequency shift of the W-O_c-W stretching mode at 798 cm^{-1} is considerably greater as in the case of SiWA ($\Delta\nu = 825\ cm^{-1} - 798\ cm^{-1} = 27\ cm^{-1}$).

Since this mode does not exhibit any systematic shifts in frequencies when H^+ is replaced with bulkier cations (11), this means that the influence of the dipolar interactions on this mode is negligible in the crystalline PWA. Obviously, other type of interactions must be invoked to describe the observed frequency shifts of PWA in the composite.

It is very likely that the observed shift of the W-O_c-W mode is due to the H-bond interactions which exist between the edge- and corner-sharing oxygens of the PWA and the ICS-PPG host. On the base of the results we propose that Keggin' s ions in ICS-PPG matrix are solvated with remaining water which was introduced in ICS-PPG during the mixing with

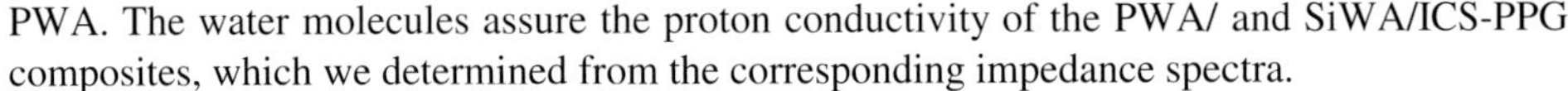
PWA. The water molecules assure the proton conductivity of the PWA/ and SiWA/ICS-PPG composites, which we determined from the corresponding impedance spectra.

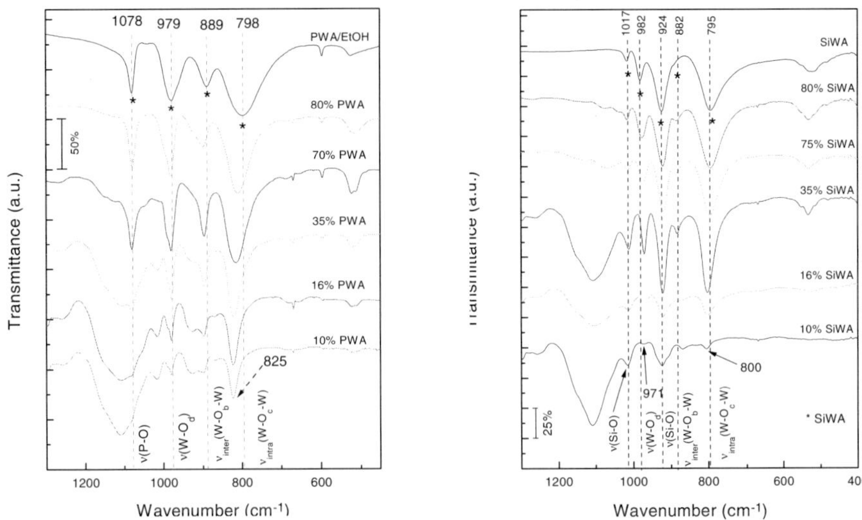

Figure 2. FT-IR transmission spectra of PWA/ICS-PPG and SiWA/ICS-PPG composites deposited on CdTe plates.

3.2 Electrical properties.

Table 1: Electronic (σ_{el}) and protonic (σ_{pr}) conductivities of gels

Sample	σ_{el}/Scm^{-1}	σ_{pr}/Scm^{-1}
Gels (wt.% of the acid in The ICS-PPG network)		
3% SiWA		$1.45 \cdot 10^{-6}$
16% SiWA		$5.90 \cdot 10^{-6}$
35% SiWA		$1.41 \cdot 10^{-4}$
70% SiWA		$1.55 \cdot 10^{-3}$
75% SiWA		$2.33 \cdot 10^{-3}$
6% PWA	$6.1 \cdot 10^{-12}$	$6.27 \cdot 10^{-6}$
16% PWA	$2.0 \cdot 10^{-9}$	$1.25 \cdot 10^{-4}$
35% PWA		$3.85 \cdot 10^{-4}$
60% PWA		$6.11 \cdot 10^{-4}$
70% PWA	$1.04 \cdot 10^{-7}$	$5.24 \cdot 10^{-4}$
75% PWA	$3.92 \cdot 10^{-7}$	$8.33 \cdot 10^{-4}$
80% PWA	$1.72 \cdot 10^{-7}$	$1.12 \cdot 10^{-3}$
0.036% HCl		$2.8 \bullet 10^{-6}$
0.036% HCl+10% $LiClO_4$		$1.1 \cdot 10^{-5}$
0.036% HCl+18% $LiClO_4$		$4.0 \cdot 10^{-5}$

Proton conductivity (σ_{pr}) and electronic conductivity (σ_{el}) of the SiWA and PWA/ICS-PPG composites have been evaluated from the measured impedance spectra according to procedure described in (12). Electrical conductivities presented in Table I revealed that the composites are characterised by a low σ_{el} (10^{-8} - 10^{-12} S/cm) irrespective of the amount of added ICS-PPG while the proton conductivities gradually increase with the increasing amount of PWA and SiWA in the composites.

The increase of the σ_{pr} with heteropolyacid concentration from 10^{-6} S/cm (<10 wt.% of PWA or SiWA) to 10^{-3} S/cm (>70 wt.% of PWA or SiWA) in the ICS-PPG host could be explained with the uniform distribution of heteropolyacids in the sol-gel matrix. The presence of water entrapped in the sol-gel matrix undoubtedly enhances the σ_{pr}, however, this does not explain relatively high proton conductivities of those composites in which the concentration of the heteropolyacid is below 10 wt.%. In this case, a small amount of remaining solvent contributes to higher ionic conductivity.

4 Acknowledgement

This work was supported by the Ministry of Science and Technology of Slovenia. U. Opara Krašovec and N.Grošelj wish to thank the Ministry for providing Ph.D funding.

5 References

1. K. Dahmouche, M. Atik, N.C. Mello, T.J. Bonagamba, H. Panepucci and M.A. Aegerter, P. Judeinstein, J. Sol-Gel Sci. Technol. 1997, **8,** 711-715.
2. M.-H. Lee, S. Taeko, K.S. Lee and S.B. Rhee, Mol. Cryst. Liq. Cryst. 1997, **294,** 229-232.
3. B. Orel, U. Opara Krašovec, U. Lavrenčič Štangar, P. Judeinstein, J. Sol-Gel Sci. Technol. 1998, **11**, 87-104.
4. B. Orel, U. Opara Krašovec, M. Maček, F. Švegl, U. Lavrenčič Štangar, Sol. Energy Mater. Sol. Cells 1999, **56,** 343.
5. O. Savadogo, J. New Mater. Electrochem. Systems 1998, **1,** 47-66.
6. V. de Zea Bermudez, L. Alcácer, J.L. Acosta, E. Morales, Solid State Ionics 1999, **116,** 197-209.
7. E. Coronado and C.J. Gómez-García, Chem. Rev. 1998, **98,** 273-296.
8. U. Mioć, M. Davidović, N. Tjapkin, Ph. Colomban and A. Novak, Solid State Ionics 1991, **46**, 103-109.
9. P. Staiti, A.S. Aricò, S. Hočevar and V. Antonucci, J. New Mater. Electrochem. Systems 1998, **1,** 1-6.
10. P. Staiti, S. Freni, S. Hočevar, J. Power Sources 1999, **79**, 250-255.
11. C.R. Deltcheff, M. Fournier, R. Franck and R.Thouvenot, Inorganic Chemistry, 1983, **22**, 207-216.
12. N. Grošelj, M. Gaberšček, U. Opara Krašovec, B. Orel, G. Dražič and P. Judeinstein, Solid State Ionics (accepted for publication).

Development of Materials and Processes for High-Performance Microtechnology Components

K.-H. Zum Gahr[1)], R. Knitter[1)], R. Ruprecht[1)], K. Seemann[1)], J. Aktaa[1)], W. Brocks[4)], H.P. Buchkremer[2)], R. Gerling[4)], P. Grünberg[2)], M. Peters[5)], W. Reimers[3)]
Forschungszentrum Karlsruhe[1)]; Forschungszentrum Jülich[2)]; Hahn-Meitner-Institut[3)], Berlin; GKSS-Forschungszentrum Geesthacht[4)]; Deutsches Zentrum für Luft- und Raumfahrt[5)], Köln

1 Introduction

Constructing high performance microcomponents requires development of new materials, processing and manufacturing, modeling and characterization of material properties in the micro range as well as integrating the component into the microsystem and testing of the system performance. Development of new materials and processes for microcomponents can open up a wide range of new applications, e.g. in communication technology, in households, medical technology, chemical process engineering or transport technology. The objective of this research project is the development of materials and processes for microtechnological demonstration objects such as microcomponents for functions in electronics, mechanics, and in chemical process engineering. The research programme is divided in three parts: (A) *microinductors* for use as matching networks in the range of 1 to 4 GHz for mobile communications, or as microtransformers, (B) technologies for processing materials required to make three-dimensional *wear microcomponents* of metals and ceramics with and without coatings, especially components for annular gear micropumps and micropatterned injection molding inserts, and (C) modular chemical *microreactors* designed with internal structural details in the submillimeter range to be made of chemically resistant ceramic materials. As the aim of the project is the realization of functioning demonstration units, industry is included right at the beginning of the definition phase with own contributions and will accompany research at the involved HGF-centers for the duration of the research activities.

2 Microinductors

Thin ferromagnetic films [1] applied in microinductors [2] for high frequency applications are of increasing interest. In the project part A different types of thin film materials are developed and optimized for high frequency applications in telecommunication. The intention is to develop microinductors with high quality factors Q and low losses at frequencies up to 4 GHz.

2.1 Objectives

In order to obtain high frequency ability the magnetic thin films must be developed in view of their magnetic properties like DC magnetization as well as their AC permeability. The magnetic anisotropy field H_a, saturation magnetization M_s and consequently the ferromagnetic resonance f_{FMR} play a decisive role in high frequency applications. Eddy current losses depend on the resistivity and thickness of the magnetic films and must be minimized. To obtain microinductors with a higher efficiency the magnetic films with the highest frequency ability have to be integrated in microcoils like toroids or solenoids with different windings and lateral dimensions of 100 to 250 μm approximately. Like the magnetic films the conducting lines of the microinductors are fabricated by thin film technology.

2.2 Methods

The ferromagnetic films are deposited on silicon substrates by magnetron- and ion beam sputtering. A high saturation magnetization as well as a high anisotropy field are required to attain magnetic materials with high cut-off frequencies. Various anisotropy fields can be achieved in FeCo, NiFe (crystalline), FeTaN (granular) and CoB (amorphous) alloys when depositing them on silicon substrates under different sputtering conditions and in-plane magnetic bias fields. To suppress frequency limiting eddy current losses a multilayer arrangement is realized by sputtering insulating SiO_2 and thin magnetic layers of certain thickness alternately.

Various methods are used to characterize the properties of the thin films. Residual stresses affecting the magnetic properties are investigated by x-ray diffraction and by measuring the curvature of a cantilever. Microstructure of the films is investigated by transmission electron microscopy and chemical composition is analyzed by Auger electron spectroscopy and electron microprobe analysis. The magnetization curves are obtained by vibrating sample magnetometer measurements. The AC permeability of the films is measured using an impedance analyzer up to frequencies of 1.8 GHz. An experimental set-up for characterizing the films up to 4 GHz could be successfully constructed. The Finite Element method is applied for optimizing the inductance of the microinductors.

2.3 Results

Up to now different crystalline (FeCo, NiFe), granular (FeTaN) and amorphous (CoB) single layer ferromagnetic films, as well as a multilayer system (FeCo/ SiO_2) were prepared and optimized for their use in microinductors. FeCo films, magnetron-sputtered from different targets ($Fe_{60}Co_{40}$, $Fe_{50}Co_{50}$) grow polycrystalline with a strong (110) texture normal to the substrate. The deposited films exhibit high tensile stress which is strongly affecting the magnetic properties due to the high positive magnetostriction of FeCo. By the variation of the Ar-pressure and the sputtering power the residual film stress and the in-plane magnetization of the films can be changed. Optimized single layer $Fe_{50}Co_{50}$ films reach cut-off frequencies up to 0.5 GHz, $Fe_{50}Co_{50}$/ SiO_2 multilayer films show a permeability breakdown in the 2 GHz range. Thin $Ni_{81}Fe_{19}$ (Permalloy) films, ion beam-sputtered under an in-plane magnetic bias field of 55 Oe show good soft magnetic properties with low coercivity fields and a saturation magnetization of about 1 T. The measured in-plane anisotropy field reaches values up to 1 Oe. High frequency measurements of these permalloy films exhibit cut-off frequencies of about 1 GHz. Granular FeTaN films prepared by magnetron-sputtering from a $Fe_{95}Ta_5$ target in Ar/ N_2 plasma with N_2 concentrations up to 50 vol.-% exhibit up to ten times higher resistivities as the FeCo films. By optimizing the sputtering conditions 1 μm thick single layer films could be obtained with cut-off frequencies of about 2.5 GHz. Negative magnetostrictive amorphous $Co_{85}B_{15}$ films, magnetron-sputtered from a $Co_{79}B_{21}$ target, show soft magnetic properties with coercivity fields less than 2 Oe and saturation magnetizations up to 1.3 T. Anisotropy fields up to 25 Oe could be realized by applying an in-plane magnetic bias field of

60 Oe during deposition. The high frequency permeability measurements show a breakdown at 0.3 GHz, higher cut-off frequencies should be reached by realizing CoB/ SiO_2 multilayer films.

FEM calculations are performed to optimize the design and geometry of the microinductors. The simulations on toroid shaped inductors with different numbers of windings and diameters yield an inductance in the low nH range. The computation also shows that the magnetic films have to be deposited on flat, planarized conducting lines to yield maximum inductance.

3 Wear Microcomponents

The increasing tendency of miniaturization in mechanical systems gives rise to an increasing demand of wear resistant microparts. However, both the knowledge and understanding of tribology and the manufacturing techniques of metallic and ceramic materials in microdimensions are still lacking.

3.1 Objectives

The tasks of the project part B are the development of materials, process technologies, testing methods and simulation tools for high-performance microcomponents. As a result the high potential of microtechnology shall be demonstrated for applications such as annular gear micropumps and micropatterned mold inserts for injection molding with filled plastics as examples for microcomponents subjected to wear.

3.2 Methods

The development of materials and process technologies is focussed on micro powder injection molding (μPIM) of hardenable steels or advanced ceramics, which includes the fabrication and conditioning of adequate powder or molding masses (feedstocks) for steels, as well as on the hot forging of metallic bulk glasses and suitable PVD coatings for wear protection. Microstructure of the selected materials depends with others on the process technology and has to be optimized with regard to wear behavior of the microparts in the applications considered as well as thermal and mechanical properties. In addition to the experimental work, computer based simulations of the μPIM process and the microcomponent behavior under tribological loading are carried out.

3.3 Results

Microparts fabricated by the μPIM process have been subjected to a dimensional control as a function of process parameters and powders (Fe, 316L, WC-Co, Al_2O_3, Y-ZrO_2, AlN) [3]. The obtained shrinkage data were considered by the industrial partners at the design of the mold inserts. New molding masses have been developed for metallic microparts which contain fine powders fabricated by the Electrode Induction Melting Gas Atomization (EIGA) technique or air separated from purchased powders [4]. Fine powders of the austenitic steels 316L and 17-4PH were gas classified using a 25 μm mesh and showed 3.6 wt.% grits of sizes < 10 μm. By adding a binder, a novel feedstock containing powder of 17-4PH has been produced, which is presently being checked for the μPIM process. In addition, a hot forging process of metallic glasses for microcomponents is developed, whereby the glasses have been manufactured using the alloys $Zr_{46.75}Ti_{8.25}Cu_{7.5}Ni_{10}Be_{27.5}$ and $Pd_{40}Cu_{30}Ni_{10}P_{20}$. Current investigations are focussed to the interaction between the metallic glasses and the tool materials (e.g. Cu, Ta, Mo, quartz glass). The aim is to avoid primary crystallites and oxide layers, which can deteriorate the surface quality of the forged microparts [5]. PVD coatings on the microparts may be required for improving wear properties. Glow discharge process was optimized to avoid undesired temperature increases during purification of microparts. Tribological characterization is carried out using a microtribometer running under conditions related to the demonstrators considered. Further facilities were set up for measuring mechanical properties in tension and bending in microdimensions as well as thermal properties using photoacoustic spectroscopy. First specimens produced by the μPIM process with Al_2O_3 and Y-ZrO_2 powders were tested. The Weibull parameter measured was greater than 16 on the green compacts and less than 8 on the sintered bodies. As expected, analyses with synchrotron radiation revealed marked internal stress gradients below the surfaces. At present the fabrication of microspecimens for mechanical and tribological characterization is improved by using novel mold inserts. Useful qualitative and some quantitative results related to process parameters could be obtained by using the software MOLDFLOW for simulating the μPIM process.

4 Ceramic Microreactors

In the last five years, the new field of microreaction technology attracted much attention with its goal to improve the conversion and the selectivity of chemical reactions. These key problems of chemical reaction technology are expected to be minimized by carrying out chemical reactions in reactors with inner dimensions in the micrometer to millimeter range. Up to now, however, the application of the microreactors available frequently has been limited by the properties of the used materials like glass, metals, silicon or polymers. To carry out chemical reactions at high temperatures or reactions with corrosive reactants or products, ceramics often are the only suitable materials.

4.1 Objectives

The project part C is aimed at developing ceramic microreactors of modular design for the application in chemical technology. Therefore, material optimization and process technology developments are carried out and joining and assembling techniques are employed. As a model reaction for the ceramic microreactor the oxidative coupling of methane to ethene was chosen. This reaction is well investigated in macroscopic reactors in the temperature range between 700°C and 1000°C with a wide variety of catalysts, but the attained yields did not exceed 25% [6]. For this exothermic reaction the gas has to be preheated to a defined temperature before it enters the reaction zone that is coated with a catalyst. In order to vary the catalyst material or to examine the effectiveness of the catalyst, the microreactor should have an exchangeable reaction zone.

4.2 Methods

Considering the requirements for the model reaction, a modular reactor system is designed. It is fabricated by low pressure injection molding, also called hot molding. In this process, paraffins and low viscosity waxes with low melting temperatures are used instead of the standard injection molding binders applied for the better known high pressure injection molding. Due to reduced working temperatures and lower injection pressures of the hot molding process, there are lower requirements made on the molds and it is possible to use rapid manufacturing techniques especially in the product development cycle [7]. The chosen designs of the different reactor parts are first generated by computer aided design (CAD), transferred to polymer molds by stereolithography and afterwards copied into silicone. These silicone molds are then used directly as tools for the hot molding process. By the combination of stereolithography and hot molding, modifications of the design can be realized in one week.

4.3 Results

A reactor design was chosen for the conversion of methane to ethene, which offers the possibility to feed and preheat the reactant gases - methane and oxygen - separately until they enter the reaction zone. The reaction zone itself consists of at least 2 catalyst carrier plates that are micropatterned with an array of about 20 mm long and 450μm wide channels. Alumina was chosen as material for the reactor housing, whereas the micropatterned reaction zone may consist of either catalyst-coated alumina or the pure catalyst material. The alumina feedstock was optimized to fulfill the special requirements of the hot molding process of the different reactor parts. After demolding, the ceramic parts were dewaxed carefully and sintered at 1700°C to a density of about 95%. The preheating of the reactants and the heating of the reaction zone will be realized by integrating ceramic heaters into the alumina reactor housing. For the joining and assembling of the reactor parts as well as for the catalyst coating, different techniques are under investigation.

5 Summary

The research project includes innovative concepts for development of system oriented new materials, processing and manufacturing, modeling and as well testing of materials on microscale as testing of microsystems. Combining electronic, mechanical and ceramic systems and participating of five HGF research centers in one interdisciplinary programme offer synergetic effects. It is expected that the results can be applied to a lot of other applications of microsystems beyond the topic of this project. Realization of demonstration units and the partnership with industry just from the beginning of the research programme should promote the transfer of microsystems technology and using of its benefits for advanced products in industrial applications.

6 Acknowledgement

The research project is closely related to one of the main activities "New Materials and Technology" of the Herrmann von Helmholtz Association of German Research Centers (HGF) and has been supported since July 1998 by its strategy fund and by the Federal Ministry for Education and Research (BMBF). In this project the following Helmholtz Centers are involved: Forschungszentrum Karlsruhe, Forschungszentrum Jülich, Hahn-Meitner-Institut Berlin, GKSS-Forschungszentrum Geesthacht and Deutsches Zentrum für Luft- und Raumfahrt Köln.

We gratefully acknowledge the contributions made by the companies TEMIC Semiconductors Heilbronn, Hydraulik Nord Parchim, Rolla Synthetics Grenchen, Männer Bahlingen, Schunk Sintermetalltechnik Gießen, Metallico Thale, FRIATEC Mannheim, and Akzo Nobel Chemicals Research Obernburg, and the support by our colleagues of the Central Experimental Department of the Forschungszentrum Karlsruhe.

7 References

[1] M. Senda, O. Ishii, IEEE Transactions on Magnetics 1995, 31, No. 2, 960-965.

[2] M. Yamaguchi, K. Suezawa, K. I. Arai, Y. Takahashi, S. Kikuchi, Y. Shimada, W. D. Li, S. Tabanabe, K. Ito, J. of Appl. Phys. 1999, 85, No. 11, 7919-7922.

[3] V. Piotter, T. Benzler, T. Gietzelt, R. Ruprecht, J. Hausselt: Micro Powder Injection Molding; Proc. EUROMAT, München 1999.

[4] R. Gerling; K.W. Liu; F.-P. Schimansky: in Sprühkompaktieren; Bd 4,: K. Bauckhage, V. Uhlenwinkel (eds.), Universität Bremen 1999, S. 105-127.

[5] M. Seidel, M.-P Macht, S. Melcher, N. Wanderka: Suitability of $Zr_{46.8}Ti_{8.2}Cu_{7.5}Ni_{10}Be_{27.5}$-bulk glasses for forging above glass transition; Proc. EUROMAT, München 1999.

[6] G. J. Hutchings, M. S. Scurrell in: Methane Conversion by Oxidative Processes, E. E. Wolf (ed.), Van Nostrand Reinhold, New York 1992, Chapter 7.

[7] R. Knitter, W. Bauer, B. Linner-Krcmar, E. Hansjosten, Rapid Manufacturing of Ceramic Microcomponents, Proc. EUROMAT, München 1999.

Silicon Moulds for Micro-Injection Moulding Techniques

Andreas Menz[1], Wolfgang Benecke[1], Thanh Vinh Duong[2], Hans-Dieter Kunze[2], Ralf Höper[3]
[1] Institute for Microsensors, -Actuators and -Systems, Postfach 330440, 28334 Bremen, e-mail: amenz@imsas.uni-bremen.de
[2] Fraunhofer-Institute for Manufacturing and Advanced Materials, Bremen
[3] Materials Science and Technology Cooperation, Bremen

1 Introduction

The fabrication of silicon moulds for micro injection moulding techniques is described. These are made by means of a silicon deep etching process and can be used for the moulding of many materials, such as polymers, ceramics, and metals. This paper shows new results of metallic (WC-Co) microparts with superior thermal and mechanical properties.

2 Fabrication of Moulds

The demand for silicon as bulk material in microsystem technologies has lead to a significant improvement of etching technologies. One of these is the Advanced Silicon Etch (ASE) process developed by Robert Bosch GmbH and Surface Technology Systems Ltd. [1,2]. The principle of ASE is an alternation of silicon etching and passivation. During the passivation step the wafer is isotropically coated by a polymer. Within the etch step the passivation layer is etched slower at the perpendicular sidewalls of the wafer than in the plane and thus protects the sidewalls from reactive attack. The exposed silicon at the bottom of the grooves is etched until the sidewall passivation layer is removed. At this point another passivation step is initiated (figure 1).

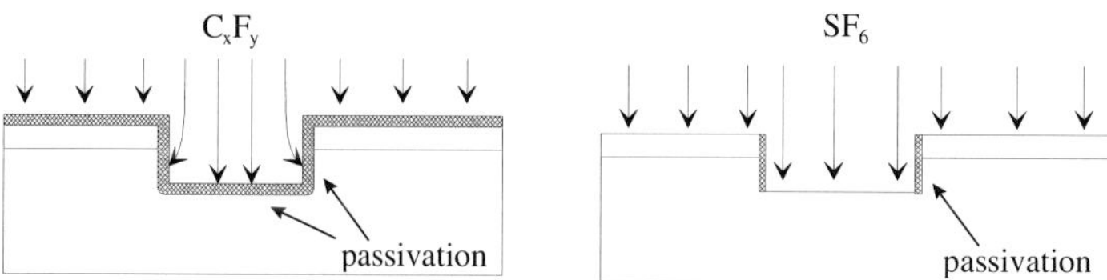

Figure 1. Principle of ASE. On the left: passivation by C_xF_y -plasma-polymerisation. On the right: reactive ion etching.

The features of ASE are:

etch rate	~ 3 μm/min
selectivity to resist	> 75:1
selectivity to silicon oxide	> 150:1
sidewall angle	~ 90°±1°
aspect ratios	> 30:1

Figure 2. Silicon mould

The etch depth is limited by the selectivity of the masking material and the wafer thickness. This qualifies the process for the fabrication of moulds for hot embossing, casting, and micro-injection moulding of polymers, ceramics, and metal. First experiments with mould structures in the micrometer range showed the necessity for very smooth sidewalls [3]. Therefore, a special treatment of the sidewalls has been developed powders (figure 2). The durability could be dramatically increased. Figure 3 shows a silicon mould after 100 injection cycles with hard metal powder.

Figure 3. Silicon mould after 100 injection moulding cycles with WC-10%Co

3 Metal Injection Moulding

Metal injection moulding (MIM) is a recent development in powder metallurgy and allows mass production of 3D-shaped parts of a wide variety of materials [4,5,6]. The basic steps of the MIM process are: powder preparation, mixing of powder and binder in a suitable feedstock processor to a highly uniform suspension, injection moulding to produce green bodies of the desired shape, careful removal of the binder and, finally, sintering (figure 4).

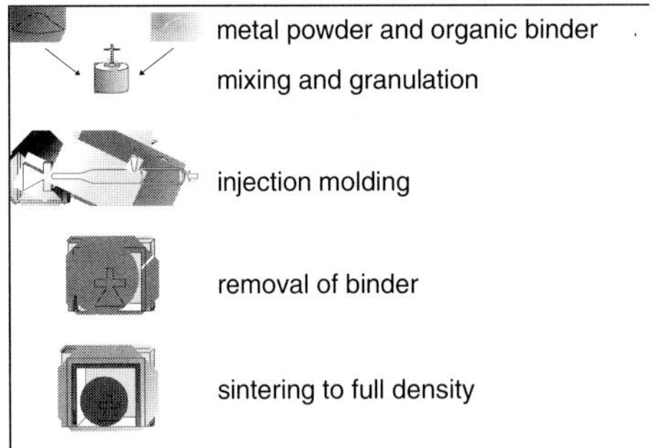

Figure 4. Flow diagram of the MIM process.

Figure 5. Sintered WC-10%Co sample (detail on the right).

Micro metal injection moulding with silicon moulds has been investigated using 316L stainless steel powder, iron powder and WC-Co powder. Very small structures were obtained with all materials (figure 5). Test structures with a width of 20 µm and a depth of 150 µm showed good results (figure 6). After sintering, these parts have a density of 98 % of solid state. The properties are comparable to bulk material. The linear shrinkage during sintering is roughly 17 %.

4 Conclusion

The ASE process has been successfully utilised for the manufacturing of silicon micro moulds. A special treatment of the mould sidewalls has dramatically increased the durability.

Therefore, a low-cost alternative to the expensive LIGA process has been generated. It has been shown that the combination of silicon moulds and MIM is a suitable technique for the reproduction of small structures and parts which may have applications in:

- micro-reaction and -mixing technology
- optical information technology
- micro tools

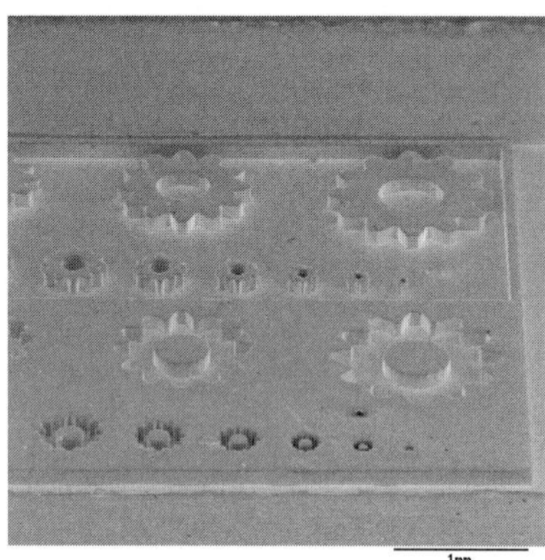

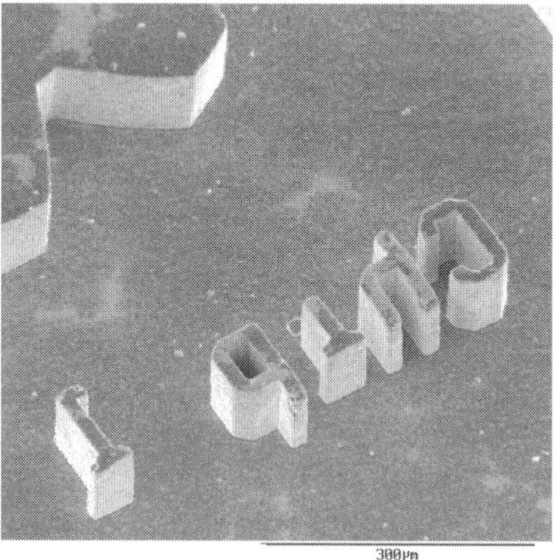

Figure 6. MIM reproduction of a silicon test structure in WC-10%Co

5 Acknowledgement

This project is supported by the federal state Bremen within the scope of the Materials Science and Technology Cooperation (MATEC Bremen).

6 References

[1] J.K. Bhardwaj, H. Ashraf: "Advanced Silicon Etching Using High Density Plasmas," SPIE Proc. Series 2639 (1995) 224-233.

[2] O. Lüdtke, A. Menz, V. Biefeld, A. Buhrdorf, J. Binder: "Wheel Bearing Integrated Accelerometers for Anti-Skid Control Systems," Proc. SENSORS EXPO, Chicago, fall 98.

[3] R. Höper, A. Menz, W. Benecke, T.V. Duong, T. Hartwig, H.-D. Kunze: "Entwicklung von Abformtechniken für metallische Mikrobauteile," Werkstoffwoche 98 Band 1 (1998) 303-307.

[4] R.M. German: "Powder Injection Molding," Metal Powder Industries Federation, New Jersey, 1990.

[5] F. Petzold, H. Eifert, T. Hartwig, G. Veltl: "Neuere Entwicklungen beim Metallpulverspritzguß," Tagungsband des Hagener Symposiums Pulvermetallurgie, Hagen, November 1995

[6] H. Eifert, G. Veltl: "Metallpulverspritzguß - wirtschaflich für komlizierte Bauteile," Metallhandwerk & Technik, September 1994, 118ff.

Influence of the Electrical and Structural Properties of Tin Oxide on the Performances of Combustible Gas Sensors

*Andreia M. Lopes, *Patrícia Nunes, ♠Paula Vilarinho, *Regina Monteiro, *Elvira Fortunato, *Rodrigo Martins

*Department of Materials Science, CENIMAT, Faculty of Sciences and Technology, New University of Lisbon and CENIMAT, 2825-114 Caparica, Portugal

♠Department of Ceramics and Glass Engineering, UIMC, University of Aveiro, 3810-193 Aveiro, Portugal

1 Introduction

Semiconducting metal oxides gained attention for potential application as gas sensors due to their property of varying their resistance when exposed to oxidizing or reducing gases, giving rise to a measurable signal. The well-known theory of operation of these sensors involves adsorption/desorption phenomena at the surface of the material [1, 2]. Tin oxide (SnO_2) is the most widely used inorganic sensor material with respect to practical applications due to its low cost (raw material and technological process), long-life and requires very simple electronics, so that little maintenance is involved. Up to now, most of the sensors known are based on sintering thick SnO_2. Nevertheless, the use of this material in the form of thin film starts paying a wide range of attention [3]. In this case, several methods are available to produce SnO_2 and the most common ones involve processes such as sputtering [4], ion-assisted deposition [5], chemical vapour deposition [6], sol-gel [7] and spray pyrolysis [8].

Many authors [2, 9, 10] wrote about this subject mentioning the SnO_2 properties as active material and as sensing element, but usually the intrinsic properties of the material and its correlation with the performances of the devices are not described. In this paper we present preliminary results on the influence of the electrical and structural characteristics of undoped SnO_2 thin films prepared by spray pyrolysis, on the performances presented by combustible gas (methane) sensors.

2 Experimental Details

Undoped SnO_2 thin films were deposited by spray pyrolysis technique. A solution of $SnCl_4 \cdot 5H_2O$ (Riedel-de-Haën, ref. 14550 with 98% purity) dissolved in isopropilic alcohol (Pronalab with 99.7% purity) with concentrations varying from 0.1 to 0.4 M was sprayed onto glass substrates pre-heated at temperatures between 300 and 400 ºC using nitrogen as carrier gas at a constant flow rate of 8 l/min. The substrate-nozzle distance was kept constant at 25 cm.

The films thickness (d) was measured with a Sloan Dektak 3D profilometer. The morphological properties of the films produced were studied by scanning electron microscope (SEM), through a Hitachi S-4100. Electrical properties of the material [Hall mobility, (μ), carrier concentration (n) and resistivity (ρ)] were determined through Hall effect measurements employing the Van der Pauw´s method using a HL5500PC system.

Finally, SnO_2 thin films were characterised [current-voltage I(V) curves] in air, and in an atmosphere containing 2000 ppm of methane at a temperature (T) of 200 °C. These measurements were performed in a test gas chamber, as represented in figure 1.

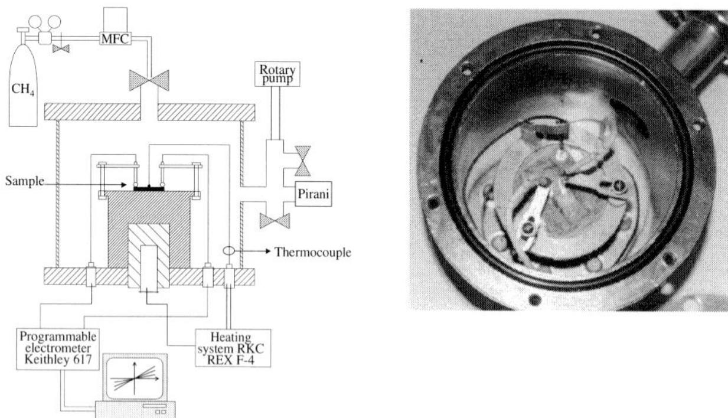

Figure 1. Schematic diagram of the test gas chamber and corresponding photograph (top view).

3 Results and Discussion

The electrical properties of undoped SnO_2 thin films are drastically influenced by the substrate temperature. Figure 2 shows the variation of ρ with T. The data show that for all the concentrations of $SnCl_4$ used there is a decrease on ρ.

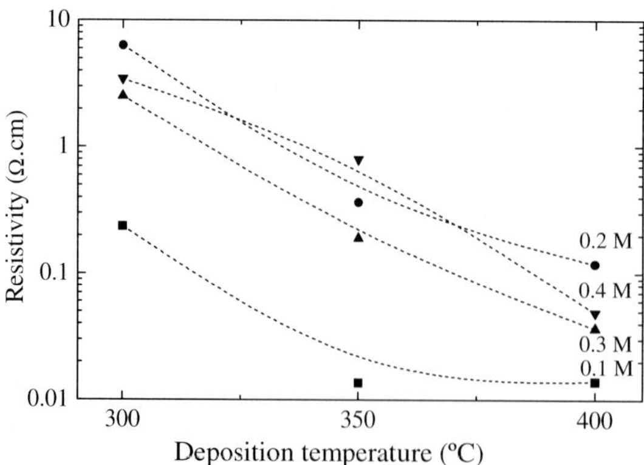

Figure 2. Dependence of ρ on T of SnO_2 films deposited with various concentrations of $SnCl_4$.

These observations suggest that the contribution of the potential associated with grain boundary decreases as T increases, due to an increase on grain size. This behaviour agrees with the results depicted in figure 3. The decrease in grain boundary potential is also responsible for an increase in n with T as shown in figure 4.

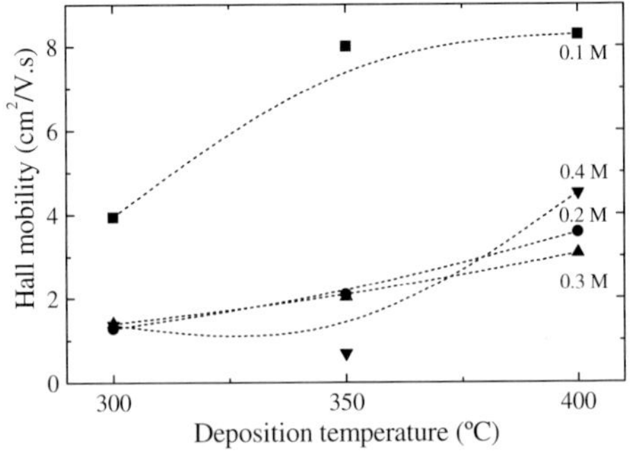

Figure 3. Dependence of μ on T of SnO_2 films deposited with various concentrations of $SnCl_4$.

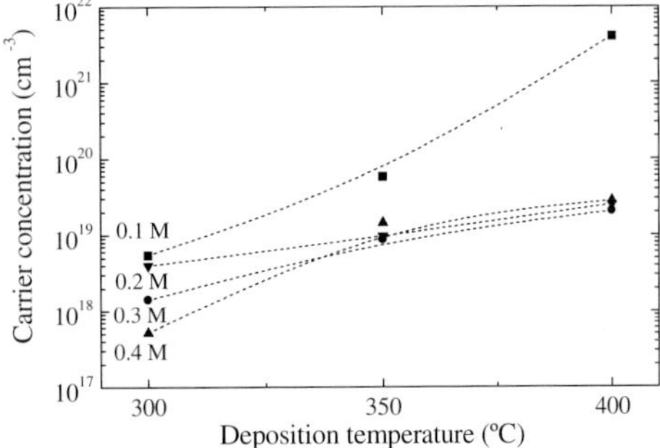

Figure 4. Dependence of n on T of SnO_2 films deposited with various concentrations of $SnCl_4$.

These results have been confirmed through SEM observation, as shown on figure 5. It's possible to see that the surface morphology and film's uniformity increase with T [from figure 5a) to 5c)]. There, we notice that at T=300 °C the film is amorphous (figure 5a). As T increases, some aggregates of nanocrystals start appearing, not possible to be clearly detected by XRD experiments.

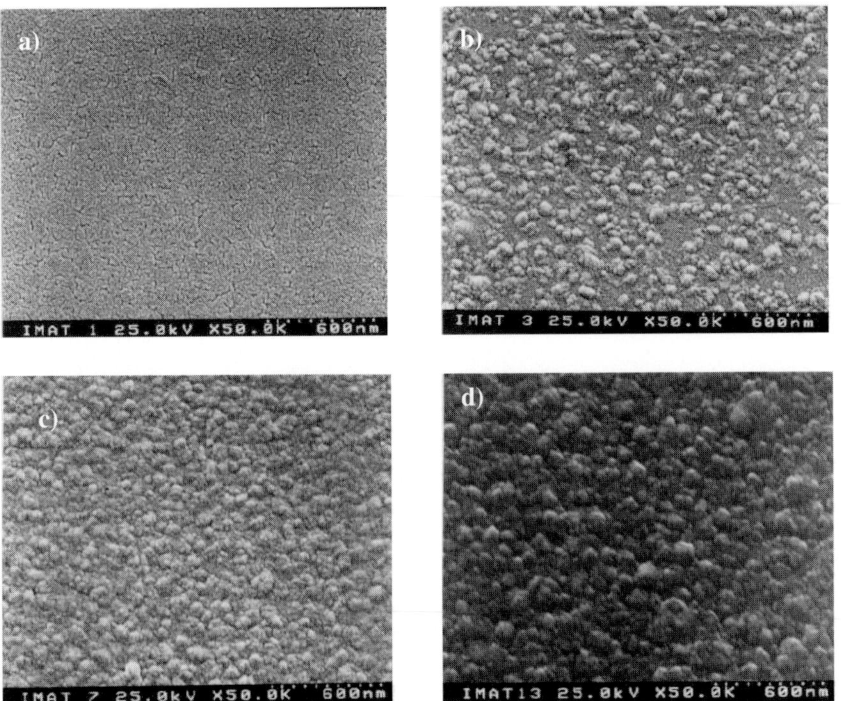

Figure 5. SEM micrographs of SnO_2 films produced at different T and concentration, of $SnCl_4$. (a) T=300 °C, solution concentration 0.1 M; (b) T=350 °C, solution concentration 0.1 M; (c) T=400 °C, solution concentration 0.1 M; (d) T=400 °C, solution concentration 0.3 M.

Besides the influence of the T was also study the dependence of the electrical and structural properties on the concentration of $SnCl_4$ used. Regarding the role of the $SnCl_4$ solution concentration used, figures 6 to 8 show that this behaviour is opposite to the one obtained with T: μ and n decreases while ρ increases, as the $SnCl_4$ solution concentration increases, up to 0.2 M. Above this value of concentration, within the experimental error of measurement, the values of ρ, μ and n tend to level off, more pronounced at high T.

This behaviour could be related to an increase of the surface states associated with the grain boundaries, which can trap or scatter free carriers, generating potential barriers or severely deplete the grains of free carriers.

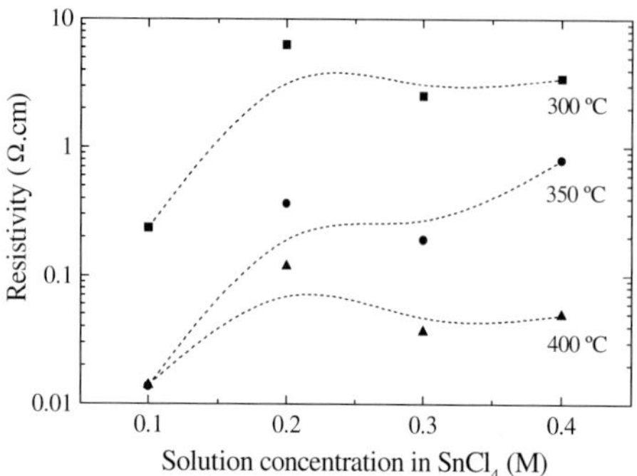

Figure 6. Dependence of ρ on $SnCl_4$ concentration used for SnO_2 films deposited at different T.

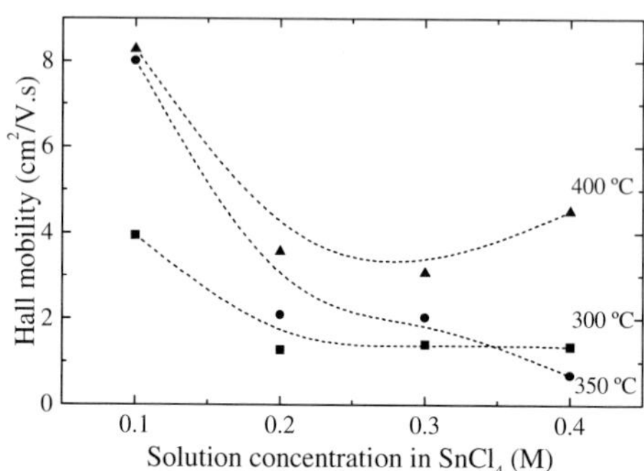

Figure 7. Dependence of μ on $SnCl_4$ concentration used for SnO_2 films deposited at different T.

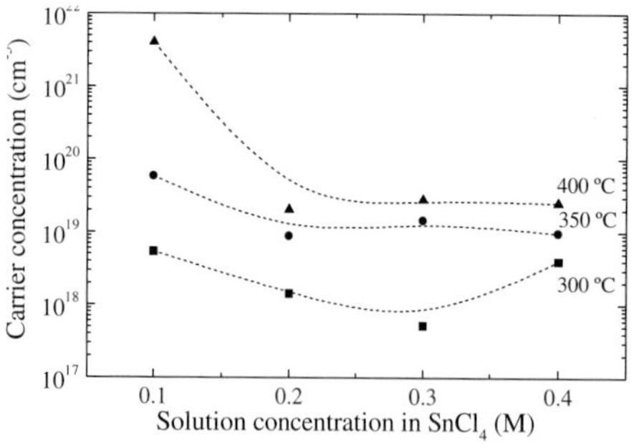

Figure 8. Dependence of n on $SnCl_4$ concentration used for SnO_2 films deposited at different T.

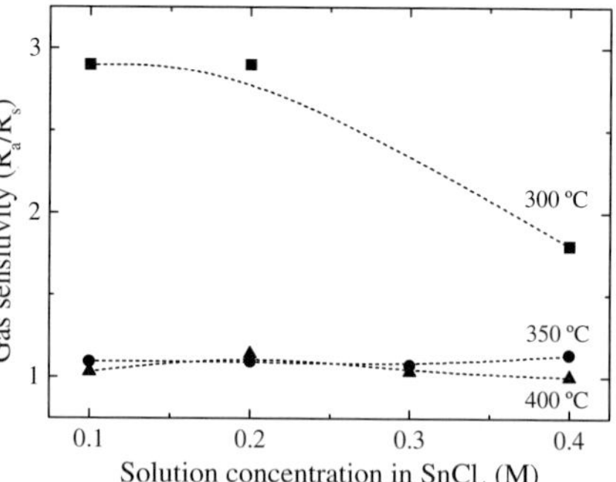

Figure 9. Dependence of S_G on the $SnCl_4$ concentration used for different T.

To measure the sensor's sensitivity (S_G), I(V) curves were plotted at the test gas chamber depicted in figure 1. S_G was calculated using the expression [3]: $S_G = R_a/R_s$, where R_a represents the sensor resistance in air and R_s the sensor resistance in the presence of gas. The tested gas was methane and the measurements were done at 200 °C. Figure 9 shows the dependence of S_G as a function of solution concentration for the different T used. Only for films produced at T=300 °C a clear dependence of S_G on the solution concentration is observed.

Figures 10 and 11 show the dependence of S_G on ρ and thickness of the SnO_2 film used, respectively. The data depicted show that $S_G > 1$ only when $\rho > 1$ Ωcm or $d < 1000$ Å. That is, for gas sensor applications it is required SnO_2 films with a disordered structure, presenting high ρ and low d.

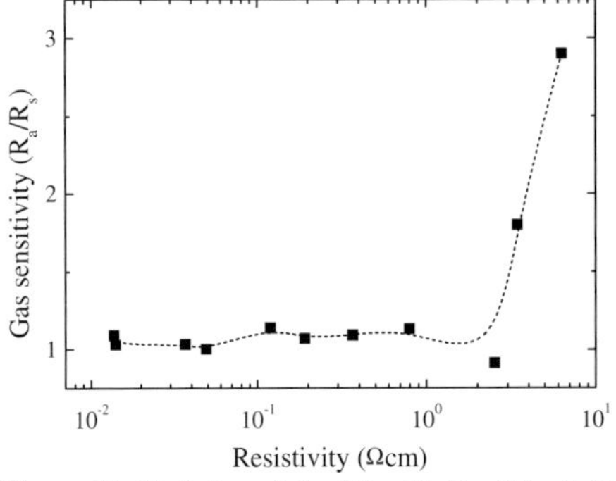

Figure 10. Variation of the SG with □ of the SnO2 films analysed.

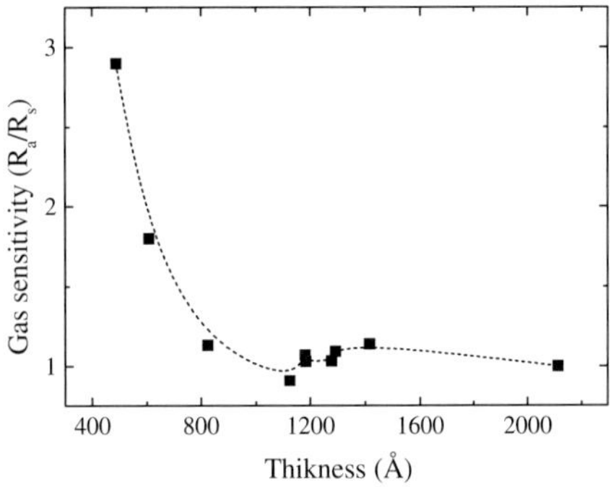

Figure 11. Variation of the SG with the SnO2 thickness, for the films analysed.

4 Conclusions

The set of results achieved show that SnO_2 films produced at T> 300 ºC start being nanostructured and presenting an enhanced surface roughness that increases with T, as can be seen from the SEM micrographs taken. The electrical results obtained are coherent with this feature in the sense that μ decreases with crystallite size. The results achieved also show that the $SnCl_4$ solution concentrations to be used should be ≤ 0.2 M and T ≤ 300 ºC, to produce films with high S_G. This corresponds to SnO_2 films with low μ and n and high ρ.

Apart from that, we also noticed that required d is below 1000 Å, showing that most of the processes conducting to gas sensing take place at the surface and not in the bulk of the material. Therefore, highly sensitive and low cost gas sensors could be produced using very thin SnO_2 films that could be conveniently integrated with a micro-heating element, aiming not only to improve the range of sensor's detectibility, but also to allow the proper and adequate sensor's regeneration and range of work.

5 Acknowledgements

The authors thank to João Grossinho for drawing the test gas chamber in Autocad R14.

This work was supported by the Fundação para a Ciência e a Tecnologia through Pluriannual Contracts with CENIMAT and project ECOCLIMAT.

6 References

1. C. Nayral, T. Ould-Ely, A. Maissonnat, B. Chaudret, P. Fau, L. Lescouzères, A. Peyre-Lavigne, Adv. Mater., 1999, 11, 61-66.
2. S. Capone, G. Leo, R. Rella, P. Siciliano, A. Ortiz, J. C. Alonso, V. Pankov, EurosensorsXII 1998, 681-684.
3. K. Ihokura, J. Watson, The Stannic Oxide Gas Sensor–Principles & Applications, CRC Press, Boca Raton, 1994, p. 35.
4. M. Ferroni, D. Gnani, V. Guidi, G. Martinelli, P. Nelli, G. Sberveglieri, EurosensorsXII, 1998, 617-620.
5. R. Lal, R. Grover, R. Vispute, R. Viswanathan, V. Godbole, S. Ogale, Thin Solid Films, 1991, 206, 88-93.
6. K. Kim, C. Park, Journal of the Electrochemical Society, 1991, 138, 2408-2412.
7. S. Capone, R. Rella, P. Siciliano, L. Vasanelli, M. Epifani, A. Licciulli, EurosensorsXII 1998, 613-616.
8. G. Blandenet, M. Court, Y. Lagarde, Thin Solid Films, 1981, 77, 81-90.
9. S. Semancik, R. E. Cavicchi, Thin Solid Films, 1991, 206, 81-87.
10. S. Song, J. Cho, W. Choi, H. Jung, D. Choi, J. Lee, H. Baik, S. Koh, Sensors & Actuators B, 1998, 46, 42-49.

V Adaptronics

Smart Structures (Adaptronics) – State of the Art and Future Outlook

E. J. Breitbach[1]
[1]German Aerospace Center (DLR), Institute of Structural Mechanics, Lilienthalplatz 7, 38108 Braunschweig, Germany

1 Abstract

In the first half of the eighties the development of new kinds of structural systems, called „smart", „intelligent", or „adaptive", was initiated first in the USA and soon after in Japan. Apart from suitable sensors and controllers, emphasis was placed on actuator systems of thermally, electrically or magnetically activaitable materials (piezoelectrics, electrostrictives, magnetostrictives, electro- and magnetorheological fluids, etc.) capable of being integrated into selected structural components, thus enabling the structure to adapt automatically to changing environmental and operational conditions. Approximately five years later similar efforts were also started in Europe, especially in Germany, leading to the development of „adaptronic" structures.

Due to its wide range of potential applications, the concept of adaptronics represents a typical multi-disciplinary key technology. Realizations achieved to date tend to focus on applications which are aimed at vibration / noise reduction, high precision positioning, and shape control (Ref. 1).

The goal of this paper is to demonstrate the current state of the art with some examples, to point out existing deficits, and to identify new application areas, in order to prepare the ground for a rapid transfer of adaptronic concepts into industrial practice.

2 Introduction

As mentioned above, adaptronic structural systems are characterized through the interplay of several related disciplines, of which structural mechanics (including material aspects and lightweight design), and controls concepts, sensor and actuator technology, are of central importance. The mastery of other, system-dependent disciplines such as the dynamics of rail and road vehicles, aerodynamics and flight mechanics for aircraft, and thermomechanics for spacecraft is also indispensable.

In contrast to standard structural systems in which controller-supported active elements are added subsequently as so-called „repair solutions", adaptronic structural systems are characterized by the requirement that from the beginning of the conceptual phase, all sub-systems are integrated into the design and development process with the aim of a fully optimized system. The success of this approach requires an unusually high degree of conceptual imagination, system experience, and systems orientation.

Such an approach leads to structural/mechanical constraints whose observance is especially critical in highly-stressed and safety-critical adaptronic structures. This applies, above all, for aerospace structures where a compromise of safety due to the integration of adaptronic elements is unacceptable. Further constraints such as

- neutral mass and space balance;
- positive energy balance;
- at least neutral balance of construction and service costs;
- reduction of total energy consumption and, thereby, environmental pollution;

are of great significance for the acceptability of adaptronics in the aerospace industry.

The mastery of the safety issue inevitably leads to the concept of *multifunctionality,* i.e., adaptronic actuator systems are to be structurally integrated such that they serve not only as actuators but also as load-carrying members of the structure itself. With respect to the *fail-safe* principle prevalent in aircraft design, this implies that the actuator elements ought to be operated parallel to passive load-carrying structural elements, allowing not only a sensible distribution of operationally induced loads on active and passive structural elements, but sufficient residual strength after a failure of the actuation function as well. The requirements for the realization of this principle lead necessarily to a new class of structural systems. A few typical examples in which this principle is realized are presented below.

3 Concepts and Applications

Aerospace technology is a trendsetter and a driving force for adaptronics in the USA, in Japan and in Europe. This trend is demonstrated by several conceptual developments and concrete project examples. Beyond aerospace, however, a growing demand is evident in other areas of technology as well.

3.1 Applications in Aeronautics

In the following, four main applications will be detailed in the field of aviation which can be addressed with adaptronic technology.

3.1.1 Adaptive Wing

In the major project ADIF (Adaptiver Flügel - Adaptive Wing) DaimlerChrysler Aerospace Airbus, DaimlerChrysler Research and Technology, and the German Aerospace Center are partners in developing new technologies in order to significantly improve the aerodynamic and structural efficiency of transonic civil transport aircraft (**Fig. 1a-e**).

In order to have optimal operating conditions during the take-off and landing phase flaps are generally used in transonic commercial airplanes. However, these flaps are retracted during cruise flight which means that during most of the flight time the wing remains invariable. The major disadvantage of the fixed geometry is that it can be optimized for only one design point, characterized by the parameters altitude, mach number and aircraft weight. Since these parameters vary continuously during flight the wing geometry is seldom optimal. The development of wings with fixed geometry is therefore always the best compromise between the design point and off-design points, where a better performance at the design point entails a worse off-design performance. With the demand for even more economical airplanes due to the increasing competition between the aircraft manufacturers the necessity of an additional adaptation of the airfoil geometry is a primary focus of interest.

Aerodynamic computations and wind tunnel tests have shown that the wave drag of an airfoil can be reduced when a bump is positioned at the location of the transonic shock. However the bump maximum has to be adjusted according to the variations of the shock positions during flight. The *tube spring concept* developed at the Institute of Structural Mechanics (ISM) allows to realize a structure, able to cope with these complex requirements.

Further aerodynamic investigations have shown that a chordwise and spanwise differential variation of the wing camber at the trailing edge reduces structural loading and improves the aerodynamic behavior. The *finger concept* and the *belt rib concept* of the ISM are capable to realize this innovative cambering optimization. Furthermore a shape varying design of a Fowler flap allows to save one track and simultaneously to gain better control over the gap shape.

With highly integrated actuators on the basis of new multifunctional materials, with improved sensor technologies, and with powerful, high-speed computers it is possible to perform the required contour optimization of the airfoil and in this way to achieve the required increase of the overall efficiency.

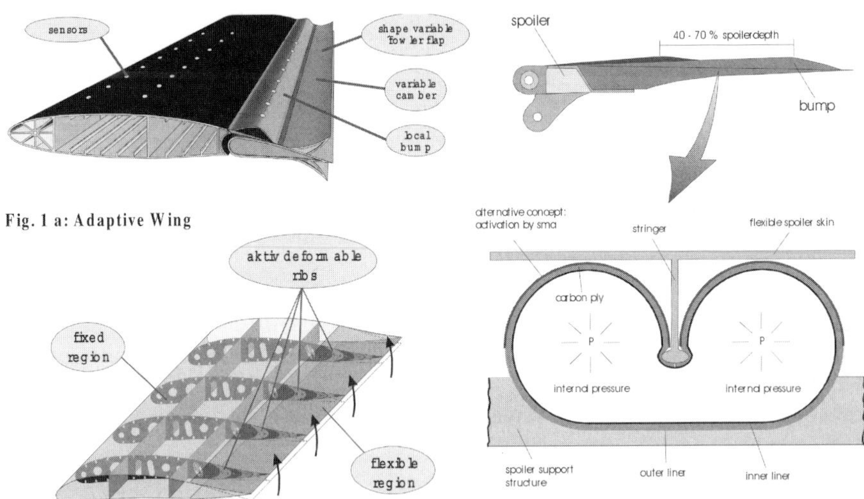

Fig. 1 a: Adaptive Wing

Fig. 1c: Adaptive Wing

Fig. 1b: Bump

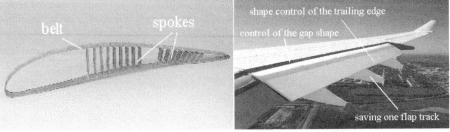

Fig. 1d: Adaptive Wing

Fig. 1e: Adaptive Wing

3.1.2 Adaptive Rotor

Present helicopter research mainly focuses on the improvement of the aerodynamic efficiency and on the reduction of vibrations and acoustic emission. Applying adaptronics to helicopters has a high potential to achieve this goals.

The *Adaptive Blade Twist* (ABT) concept allows to directly control the twist of the helicopter blades by smart adaptive elements and through this to positively influence the main rotor area which is the primary source for helicopter noise and vibrations. Since the interaction of non-stationary helicopter aerodynamics and elastomechanical structural characteristics of the helicopter blades causes flight envelope limitations, vibration and noise, a good comprehension of the aerodynamics is essential for the development of structural solutions to effectively influence the local airflow conditions and finally develop the structural blade concept.

The rotor blade twist, especially at the outer part of the rotor, can be structurally achieved by mainly three actuator principles:

- Torsion caused by tension/compression forces with 45° fiber orientation

In this concept, torsional moments caused by tension forces are utilised. Thin-walled actuator materials like piezoceramic plates or active fibers have to be implemented in the skin of the rotor blade. The advantage of this simple concept is the good control characteristics. One disadvantage of this concept is the insufficient damage tolerance behavior.

- Torsion due to torsion-warping-coupling

The twist deformations of the rotor blade are caused by warping forces. In comparison to the previously mentioned concept, cylindrical actuators, for example piezoelectric elongators (piezo-stacks) can be used to induce warping. However, it is necessary to change the geometry of the rotor blade cross section to realize this warping-torsion-coupling effect. The locally restricted effect of the warping forces, the changes in the geometry, and the installation space of the actuators may cause problems by implementing this concept into a rotor blade.

- Torsion due to torsion-tension-coupling

In general, torsion-tension-coupling is a specific behaviour appearing in anisotropic structural components. The anisotropic material behaviour must be distinguished from the anisotropic structure behaviour resulting from structure elements like ribs or stringers. In this concept anisotropic material behaviour caused by helical winding is illustrated in **Fig. 2**.

For practical realisation, cylindrical actuators like piezoelectric elongators (piezo-stacks) can also be employed. A disadvantage of this concept is the high spanwise stiffness of the rotor blade. Thus, an uncoupling layer between the spar and the skin is needed. An actuator supported at the rotor blade spar generates the axial forces. The principles of the above mentioned actuator concepts are presently being developed at DLR.

The ABT concept bases on such an actively controlled tension-torsion-coupling of the structure.

For this, an actuator is integrated within a helicopter blade that is made of anisotropic material based on fiber composites. The actuator is supported at the rotor blade spar and generates axial forces at the blade tip. **Fig. 4** shows the active rotor blade segment with adaptive blade twist.

Driving the actuator results in a local twist of the blade tip in such a way that the blade can be considered a torsional actuator. Influencing the blade twist distribution finally results in a higher aerodynamic efficiency.

In **Fig. 3** the first torsion mode at 113 Hz is shown. At the resonance frequency a deformation of ±1.9 degrees is possible. For the first flapwise mode at 9 Hz deflections of ±1.2 mm were measured.

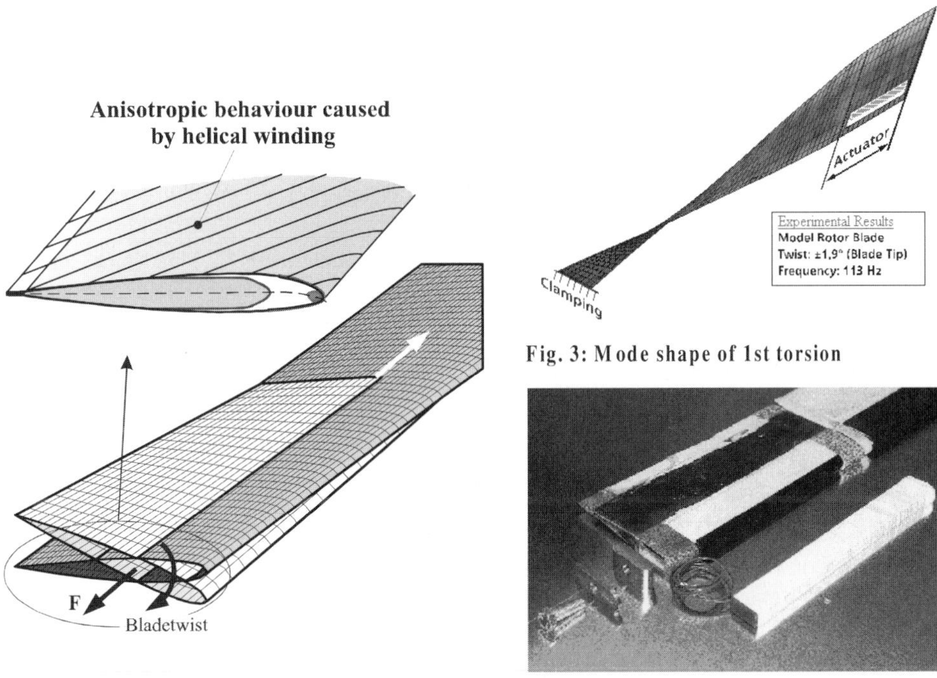

Fig. 3: Mode shape of 1st torsion

Fig. 2: Adaptive blade twist

Fig. 4: Active rotor blade segment

With these experimental results it could be shown that:

- an adaptive fibre composite rotor blade based on tension-torsion-coupling can be manufactured.
- the uncoupling layer between skin and spar is suitable to be used for tension-torsion-coupling in rotor blades and
- a piezoelectric stack actuator is suitable to twist the blade. Near the first torsion mode a deformation of ±1.9 degrees is possible. The actuated blade section must be specially designed for this purpose.

It could be determined that an adaptive blade twist in the outer part of the rotor is feasable with comparatively small effort and in its range of application, depending on the form of excitation, it shows a very strong potential. The realisation of such a control concept, covering static and controlled dynamic operation modes, depends on the proper choice of the actuator. In addition, the installation space, the power to mass ratio, and the duration of life are further criteria determining functionality and efficiency of such a device. Moreover, the variety of applications, the small torsional stiffness and the small external forces (inertia force, rotor moment and aerodynamic force) make it attractive to integrate the actuator in the aerodynamically efficient outer part of the rotor. Based on the underlying physics there is a number of other advantages:

- It is possible to influence the aerodynamic forces at the outer part of the rotor. Disturbances induced by the flowfield can be compensated for at the source.
- For vibration reduction the influence on special blade modes is important. The adaptive blade twist allows active modal vibration reduction.
- Active influence on blade deflections makes it possible to reduce dynamic stall at the retreating blade.
- Using special controllers leads to adaptive aeroelastic systems with no instabilities.
- There is no additional aerodynamic drag, the actuator is completely integrated in the rotor blade and causes controlled changes of the blade twist.

This work has to be complemented

- by investigations of alternative actuator concepts, based on fully integrated piezoceramic stacks, plates, films, and fibres.
- by additional experimental investigations with aerodynamic loads in wind tunnel tests on nonrotating and rotating models
- by investigation in full-scale flight tests.

3.1.3 Vibration Reduction in Propeller Aircraft

Adaptronic solutions can also be applied to the suppression of vibrations in propeller-driven airplanes. The vibrations creating disturbances are induced mostly aerodynamically, and their frequencies depend on rate of revolution and the number of propeller blades.

Apart from acoustical and mechanical absorbers (anti-resonators), and isolating shock mounts between the power plant and the wing structure, a highly attractive possibility for the suppression of vibrations is an engine/wing pylon structure mainly consisting of strut/truss elements with some of them being replaced by activatable struts, as shown in **Fig. 5**.

Without going to deep into technical/physical details, it seems possible to compensate vibrations which lead to strains or, respectively, forces in a load-carrying truss-type structure through the introduction of directed counter-forces into the most highly strained elements. This can be accomplished by replacing normal passive struts with active truss elements (Ref. 7 and 15).

Fig. 5: Active engine/wing pylon interface

Fig. 6: Active Suppression of fin vibration

3.1.4 Vibration Reduction of Military Aircraft

Military aircraft are increasingly endangered by severe buffet loading on the fin structure when flying at high angles of attack. Vortices shedding from the wing leading edge/fuselage interface hit the fin and excite structural vibrations as high as the structural limit loads or even higher. To reduce these severe vibrations an adaptive interface has been developed. The operational dynamic loads are reduced by a special piezoelectric actuator system, integrated directly into the bending support of the fin structure. To validate the stability and the static and dynamic strength of the interface FE calculations and extensive measurements on piezoceramic stack actuators were performed. The manufactured interface was integrated in an existing test structure and realistically loaded. The results are even better than anticipated. Further investigations are focusing on a different design, characterised by actuators working in parallel to the main load path. This solution promises an efficiency increase by about one order of magnitude.

3.2 Space Applications

Adaptive Structural Systems show an enormous potential to a vast variety of modern space mission scenarios. This is due to the fact that conventional structural technologies are currently reaching their limits with respect to fulfilling extreme structural requirements defined by the continuously increasing scientific, commercial and economic demands for both civil and military space applications. Major emphasis is presently placed on problems such as

- vibration reduction in satellite payloads during launch into orbit
- control of microvibrations in sensitive payload components in orbit
- shape control of components working under high orbital thermomechanical loading
- positioning and pointing control of antenna, reflector, mirror, and concentrator structures

3.2.1 Adaptive Launcher Interface

The structural design of spacecraft payloads (satellites, space probes, etc.) is largely dependent on the launch loads resulting among others from static and dynamic effects during engine ignition and shut-down and stage separation. Especially the dynamic forces can lead to excessive dynamic responses in the payload components which are conventionally accommodated by stiffening tructural elements or through additional passive damping measures, entailing in any case weight penalties.

As shown in **Fig. 7**, this problem can be solved by isolating the payload from any dynamic excitations through an adaptive interface between rocket and payload (Ref. 9). A controller drives the interface actuators such that the payload, ideally, vibrates only in its rigid-body modes without any elastic deformations.

Design and implementation of such a high load adaptive interface is a real challenge. Apart from its main task to actively compensate for dynamic forces all the interface components must be capable to sustain all static and dynamic loads during the rockets ascent. Due to the specific characteristics of piezoelectric materials it therefore is generally necessary to mechanically pre-stress all active members. This can be achieved by means of highly elastic bolts and a cylindrical load-carrying FRP structure, both acting in parallel to the actuator elements. This concept requires a tailored anisotropy in the FRP element with relatively high flexibility in the direction the actuator forces, but very high stiffness in the shear direction of the actuator.

Detailed information on a representative high load adaptive interface is detailed in Ref.9. Special focus of this device is the vibration reduction of the model support sting in wind tunnels, but the technical principle and design are generally applicable to a great variaty of other technical problems.

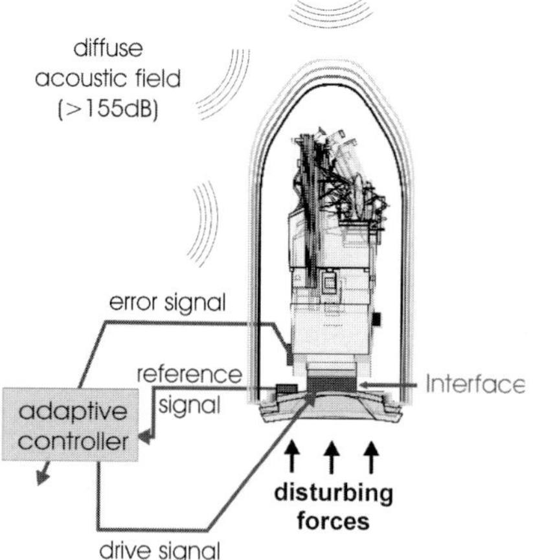

Fig. 7: General set-up of an adaptive interface including signal flow and control system

3.2.2 Adaptive Reflectors

Modern lightweight reflectors for use in the harsh thermal space environment have to fulfill the requirement of shape stability with ever increasing accuracy demands. Conventional design concepts are mainly based on the selection of material systems with extremely low coefficients of thermal expansion like CFRP or C/SiC, additionally stiffened by rigid and complex backup structures. These passive structures are developed to withstand the estimated worst case conditions and are highly over-dimensioned for most of their service life in orbit. This actually contradicts to common space mission demands for minimum weight and minimum volume. Moreover, any further passive optimization of the contour stability can hardly be realized, with respect to technical limitations as well as for manufacture and launch cost reasons. As an example, the advanced reflector of the FIRST mission marks the end of this conservative structural technology.

To overcome these limitations, adaptive quasi-static and dynamic shape control by proper activation of structurally integrated multifunctional components offer a promising option.

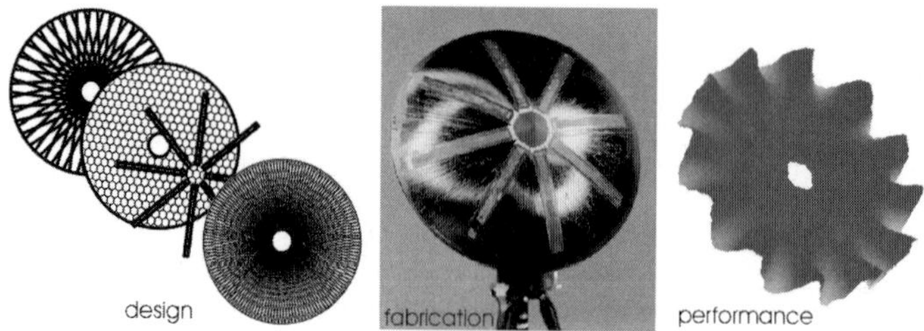

Fig. 8: General setup for quasi-static shape control for adaptive reflectors.

An example of this kind is shown in **Fig. 8** where the reflector's shape stability is achieved not by means of a passive design but by artificially activated deformations counteracting e.g. thermomechanical distortions. These artificial deformations are locally induced into the reflector structure by discretely distributed actuators, here through integrated piezoceramic patches and resulting in a global structural response. The actuator regulating variables are determined by specific sensors, transformed in an appropriate signal conditioning system and then transferred to the actuators (Ref. 12).

This and similar approaches have strong potential for use in large deployable antennas, for solar dynamic concentrators as well as for optical and infrared telescopes.

3.2.3 Vibration Absorber

In many present and future infrared telescope missions it is imperative to generate a cryogenic environment for specific equipment. Due to severe weight and volume restrictions of most space projects, the use of cryocoolers is required to ensure the necessary operating temperatures. Especially linear Stirling cryocoolers that have been successfully space qualified and implemented in a variety of flight missions offer quite suitable and efficient performance characteristics. Unfortunately, they produce significant uniaxial vibrations induced through the interfaces into the satellite and into sensitive components like infrared sensors, thus entailing significant quality deficits.

To overcome this problem by means of an adaptive structures technology approach, counteracting forces are generated in an active absorber to compensate for these disturbances. Such a device, consisting of a mass/spring system with an active piezoelectric interface between the absorber spring and the cooler has successfully been qualified for a small satellite application. A vibration reduction of 53dB at the critical cooler's drive frequency was achieved. Moreover, the system has demonstrated the capability of broadband and transient vibration reduction as well. The active absorber system, though lacking any launch locking device, was designed to withstand all launch loads though being highly flexible and efficient in the direction of the actuator action by implementation of specially designed spring mounts (Ref. 13). The system was built to comply with space mission demands. The system has been fully space proved with respect to the static and dynamic launch loads, outgassing, vacuum and thermal loads.

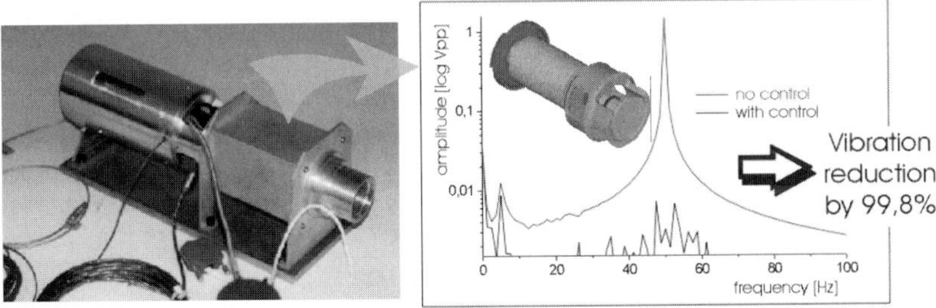

Fig. 9: Adaptive vibration compensation for cryocooler induced disturbances at IR-camera

3.3 Terrestrial Applications

Spinoff transfered from aerospace into terrestrical applications is beginning to increase. A detailed treatment of what has already been achieved and a discussion of future potential applications beyond aerospace exceeds the scope of this presentation. Nevertheless, some exemplary applications shall be presented below.

As skeched in **Fig. 10**, in passenger cars a variety of prospective adaptronic applications are subject to present and future investigations. Related to typical car components the following research issues can be derived.

- active engine suspension (vibration suppression)
- active splash board (cabin noise reduction)
- active roof sheet (cabin noise reduction)
- active interface between shock absorber and car body frame (vibration and noise reduction, noise source: wheel/road contact)

Quite similar problems arise in train systems as well as in ships. Specially related to electric trains is the dynamic contact problem between the trolley wire and the current collector. Through these highly flexible components electric energy is transmitted into the train motor. A continuous contact to the trolley wire has to be guaranteed by a constant contact force and also wear resistance must be monitored to maintain transmission quality and avoid flash over. The most relevant requirements for this component are constant contact forces, light weight, low inertial forces and minimal noise emission.

By integrating adaptive structure components in the current collector, the dynamic properties can be improved such that unwanted effects like disconnections and flash over are reduced (**Fig. 11**).

In addition to the above mentioned applications, a broad range of further terrestrial applications can be efficiently improved by means of adaptronic measures. Examples are:

- wind turbines
- machine tooling
- precision measurement devices
- precision optics
- vibration sensitive measurement equipment
- storage and transportation of sensitive goods, and
- rescue equipment / transport of the critically injured and sick

Some of these are focus applications in the framework of a huge German key technology project, funded by the Federal Ministry of Science and Technology with special emphasis on research and further development of Adaptronics.

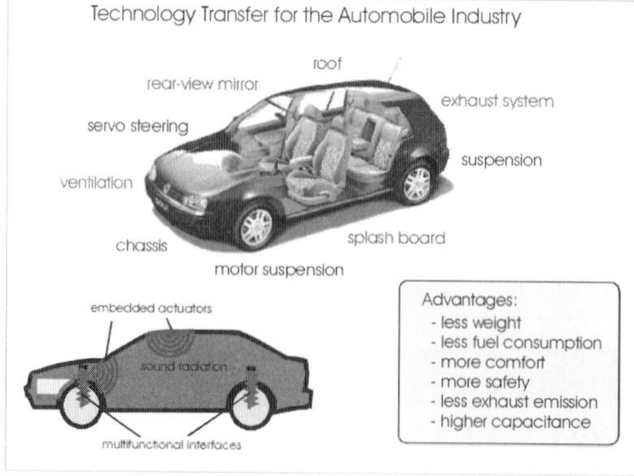

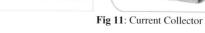

Fig. 10: Passenger Car

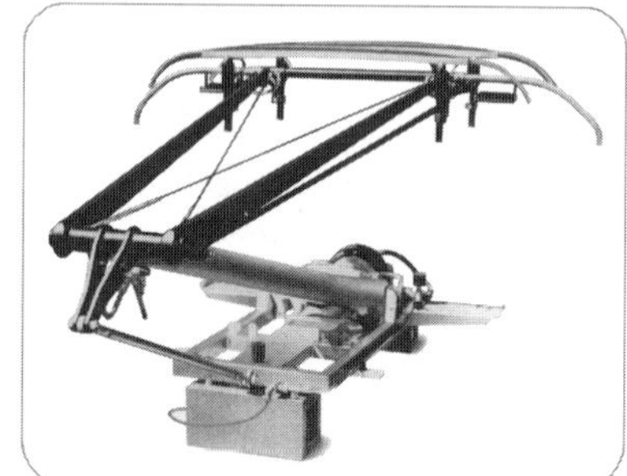

Fig 11: Current Collector

4 Conclusions

The technical and political range of adaptronics as well as examples of its application, mostly in the aerospace field, have been presented. It is shown that the added value potential does not lie only in the field of adaptronic materials, though smart materials with better performance (active elongation/compression, forces, frequency range of operation, etc.) are badly needed. All present major advances can be achieved with off-the-shelve materials, when system functions, construction materials, and design methods are combined in new ways. The multi-disciplinary problems call for a focused effort of technicians and engineers combined with fantasy, system knowledge and the ability to think in interdisciplinary terms to take advantage of the high innovative potential of adaptronics.

Controversies corresponding to the correct description of adaptronic systems or rigidly clinging to definitions interferes with the way to success. Thus, a pragmatic procedure which focuses on the technical and economical success is needed to ensure the acceptance and growth of this new technology concept.

5 References

1. Breitbach, E.: Research Status on Adaptive Structure in Europe, Proc. of the 2nd Joint Japan-USA Conference on Adaptive Structures, Nov. 12-14, 1991, Nagoya Int. Center, Japan.
2. Tönshoff, H.K.; Laux, Th.: Piezo-Aktuatoren für den mm-Bereich, VDI-Z Special Antriebstechnik, Apr. 1994.
3. Bein, Th.; Pabsch, A.; Piening, M.: Linienförmiger Aktuator zur Erzeugung großer Stellkräfte bei große Stellwegen. (Patent submitted).
4. Breitbach, E.; Büter, A.: The main Source of Helicopter Vibration and Noise Emission and Adaptive Concepts to Reduce Them, Journal of Structural Control, Dec. 1995.
5. Büter, A.; Piening, M.: Verdrehbares Rotorblatt aus faserverstärktem Kunst-harz. (Patent 1995).
6. Breitbach, E.; Büter, A.: Adaptiver Schwenkantrieb in Steckbauweise. (Patent 1997).
7. Schütze, R.: Aktiver Stab. Patent P4310825.3-25.
8. Breitbach, E.; Lammering, R.; Melcher, J.; Wimmel, R.: Adaptive Structures, Concepts and Prospects, DLR-Sonderdruck (1993).
9. Breitbach, E.: Hochbelastete adaptive Schnittstelle, Patent 1997.
10. Breitbach, E.: Research Status on Adaptive Structures in Europe, Proceedings of the 2.nd Int. Conf. On Adaptive Structures, Nagoya, Japan, 1991
11. Breitbach, E.: Adaptive Structural Concepts: State of the Art and Future Outlook, Proceeding CEAS International Forum on Elasticity and Structural Dynamics, Rome, June 17th – 20th, 1997
12. Hanselka, H.; Melz, T.: An Adaptive Reflector for a Satellite Antenna, Madrid, Spain, 1996, IB 131-96/24
13. Breitbach, E. und H., ERAS GmbH,'Elastisches Element mit Blattfedern'; Deutsches Patent, No. 19723515
14. Melz, T.; Sachau, D.; Melcher, J.; Breitbach, E.: Active Vibration Suppression on Small Satellite by Means of Adaptive Structures Technology, Adaptronic Congress, Potsdam, 1999
15. Breitbach, E.; Grützmacher: Europäische Patentanmeldung 98 104 863.0, Göttingen, 1999

Active Fiber Composites Actuation Materials and Applications

Nesbitt W. Hagood, Mauro J. Atalla, Benon Z. Janos, Mads Schmidt, Viresh K. Wickramasinghe
Massachusetts Institute of Technology, Cambridge, MA, USA

1 Introduction

In recent past many applications have been realized using intelligent materials systems. Most commercialized examples utilize piezoelectric and shape memory materials and to a lesser extent magnetostrictive, electrostrictive, and rheological materials. The main advantages for piezoelectrics compared to other smart materials are availability, material costs, material knowledge and a high bandwidth allowing operations up to the kHz-range resulting in a high work output. The high bandwidth and large specific actuation stress are some of the key reasons why piezoelectric ceramics are commonly used for structural actuation despite its limited strain capabilities. Relatively large and expensive power amplifiers are needed however to actuate piezoelectrics adding to the total cost and weight of the total system. One of the most critical limitations for monolithic piezoceramic materials is its fracture behavior and inherent brittleness leading to unexpected catastrophical failure without any signs of warning.

Monolithic piezoelectric actuators are typically thin sheets of piezoceramic with uniform surface electrodes. The distance between the electrodes is relatively small and high electric fields are easy to realize resulting in acceptable driving voltages. This design has two shortcomings: 1) the strain used for actuation is the in-plane strain, it relies on the transverse piezoelectric effect and is only about half the value for the out-of-plane strain and 2) the in-plane strains are the same for orthogonal directions. The latter adds complexity to the design of control systems, as one cannot independently address different directions.

2 Overview of Active Fiber Composites

Active Fiber Composites (AFCs) have been successfully developed at MIT to address the most critical properties restricting designers and engineers from using piezoelectric materials: brittleness, limited robustness, limited conformability and in-plane actuation isotropy.[1,2,3] As illustrated in Fig. 1, AFCs consist of small semi-continuous piezoelectric fibers uniaxially aligned in a polymer matrix between two interdigitated electrodes. The electric field and electric potential distribution are also schematically illustrated. This configuration exhibits in-plane actuation anisotropy and increased performance over monolithic actuators as it is designed to utilize the larger longitudinal piezoelectric effect. AFCs are much more robust than monolithic ceramics. Each fiber is surrounded and protected by the polymer matrix. In case of fiber cracking the system does not fail catastrophically but degrades continuously. Several studies on robustness and fatigue of AFCs supported by material models have been carried out at MIT. The studies showed that AFCs have good and predictable residual actuation properties allowing engineers to plan for it in their designs.[4]

Currently, Active Fiber Composites consist of commercially available extruded piezoceramic PZT5A fibers with a circular cross section of approximately 140 microns in diameter[5], photolithography etched copper or silver ink screen-printed interdigitated electrodes and commercially available epoxy resin. The different components are assembled in a lay-up process.[1, 2, 3]

3 Modeling and Characterization of Active Fiber Composites

Active fiber composites belong to the class of deformable actuators/sensors and can be well modeled within the theory of electroelastic continua. Two modeling issues specific to AFC devices are consideration of nonlinear constitutive equations that capture the hysteretic material response of the piezoelectric ceramic, and consideration of weak electric currents in the highly electrically insulating polymer matrix and electro-active ceramic materials. The nonlinear material modeling was addressed using a 27 function scalar expansion (of tensor invariants from the set of stress tensor, electric displacement vector, and a "memory" vector) that captures electromechanical coupling, and an evolution rule for the "memory" vector that captures the hysteresis of piezoelectric material in the nonlinear response regime.[6] The nonlinearity of piezoelectric ceramics was characterized in terms of this material model for a typical range of compressive stresses. Equations for electroelastic continua with electric conduction were used to account for the effect of weak electric currents present in AFCs, and finite element solutions demonstrated significant electrical transients under large electric voltage loading (nonlinear response regime).[7] The results illustrate the significant cumulative effects of weak electric currents present during "poling" of AFCs and actuation using DC voltage offset, which is commonly used to maximize the effective linear loading regime of the device.

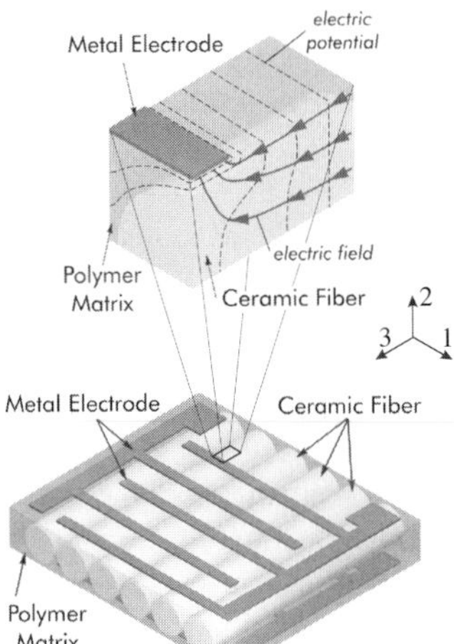

Figure 1: Illustration of Active Fiber Composites (bottom) with representative volume element (top) showing schematically electric field and potential distribution

Detailed studies have also been performed on the characterization of AFCs.[3,4,8,9] The characterization tests were designed to understand the electrical, mechanical and coupled electromechanical behavior and properties of AFCs under various loading and laminate configurations. Recently, an extensive series of characterization tests were conducted simulating realistic helicopter rotor blade conditions in order to validate the "screen-printed" AFC actuator material system for the active twist rotor blade application. Screen-printed AFC material system includes 140-micron diameter PZT5A fibers embedded in epoxy resin matrix (Shell Epon 9405/9470). Interdigitated electrodes were screen-printed on Kapton sheets with silver ink and the mean total thickness of the AFCs was 221micron.

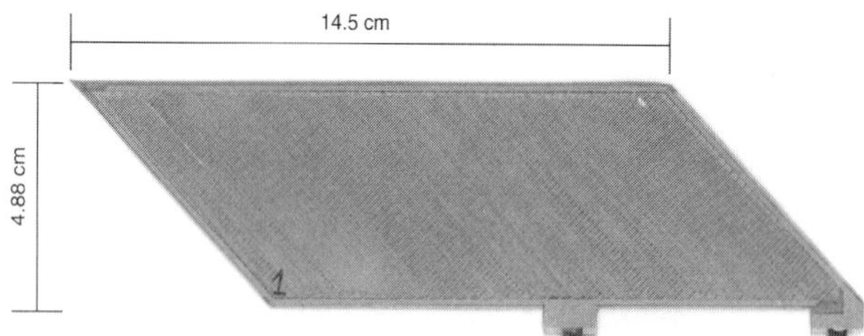

Figure 2: AFC actuator as used in active rotor blade

The primary metric of performance for AFCs is the peak-to-peak free strain produced in defined representative electrical cycles.[9] The average actuation performance at the nominal electrical cycle of 3000Vpp with 600Vdc at 1 Hz was 1083ppm. When AFCs were sandwiched with a single 0°/90° E-Glass ply on either side, the performance decreased by ~56%. Experiments also indicated that the actuation of unlaminated AFCs reduced by ~10% when the frequency increased from 1 Hz to 100 Hz. However, the frequency dependence on performance was not significant on E-Glass laminated actuators.

Stress-strain tests on unlaminated as well as 0°/90° E-Glass laminated AFCs were conducted to determine the fiber direction elastic modulus and the tensile strength. The tensile strength was 28.2MPa and 204MPa for unlaminated and E-Glass laminated AFCs respectively. Fiber direction moduli were measured for various strain ranges. Measured fiber direction modulus for unlaminated AFCs decreased from 20.8GPa for the 100-1000ppm range to 16.6GPa for the 100-3000ppm range. For 0°/90° E-Glass laminated actuators, the modulus decreased from 22.8GPa for the 100-1000ppm range to 17.2GPa for the 100-60000ppm range. The decrease of the elastic modulus with mechanical strain is mainly attributed to fiber breakage.[4]

Performance of AFCs under mechanical tensile loads was investigated to determine robustness of the actuators at various maximum mechanical strain levels. Figure 3 illustrates the gradual degradation in performance as the mechanical strain is increased. At the maximum mechanical strain of 4000ppm, the performance was decreased to 60%. It is seen that in contrast to monolithic actuators, AFC actuators do not fail catastrophically under mechanical load. Therefore AFC actuators can be integrated in structural designs with confidence. Successful correlation for predicting actuation capability and mechanical properties under high tensile loads using Fiber Cracking Model has been demonstrated.[4]

Electrical fatigue tests were conducted on both 0°/90° E-Glass laminated and unlaminated actuators to evaluate the robustness of AFCs. All specimens tested (3 unlaminated and 2 laminated) showed no degradation in actuation performance on the fiber direction elastic modulus up to 10^7 electrical cycles. Specimen temperature monitored by thermocouples indicated that high frequency electrical cycling does not cause measurable heat generation in the AFC actuators.

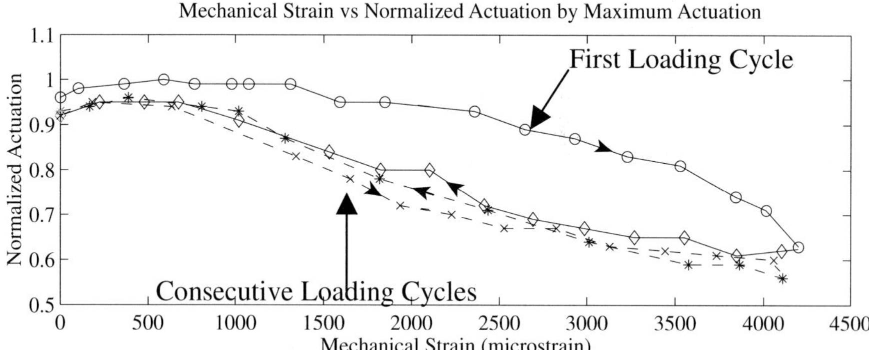

Figure 3: Performance under mechanical tensile load (4000ppm max. strain)

4 Applications

The advent of active fiber composites has resulted in increased design flexibility due to its unidirectional actuation characteristics and easy of embedding into composite structures. The Active Materials and Structures Lab at MIT has been actively involved in the development of new applications that benefit from this new technology. Two applications are addressed here: active rotor blade control and active structural-acoustic control.

4.1 Actively Controlled Rotor Blade using Active Fiber Composites

Helicopter rotor blades experience significant vibration as a result of the unsteady aerodynamics of rotating blades. Several methods for damping vibration have been developed, among them higher harmonic control (HHC) and individual blade control (IBC).[8] IBC uses blade mounted actuators and sensors to control each blade separately. By controlling the pitch of each blade independently, unsteady aerodynamic disturbances can be reduced. A reduction in vibration will improve ride quality, decrease maintenance costs, increase payload capability, and increase cruising speed.[8]

MIT, in conjunction with Boeing Helicopters, Philadelphia, is designing and manufacturing actively twisting helicopter blades. AFCs are embedded in the spar of the blade at 45° and -45° angles with respect to the span wise direction. By actuating these plies out of phase, shear stresses are induced in the walls of the spar, creating a twist along the span of the blade.[10]

In 1998, the first-generation active integral twist blade was developed, built and hover tested at MIT. This blade contained three active plies, two at 45° and one at -45°. In total, 42 AFC actuators were embedded in the spar skin. The blade achieved a 0.8°/m twist rate. The limited performance of the full blade can be attributed to numerous AFC pack failures, due to material failure and electrical connection failure, and to a reduced level of actuation voltage. After several actuators failed during initial testing, the voltage cycle was reduced to 50% of the design voltage cycle to prevent further damage to the blade.[8] In earlier tests of a half-span blade section, a twist rate of 2.06°/m was achieved.[10]

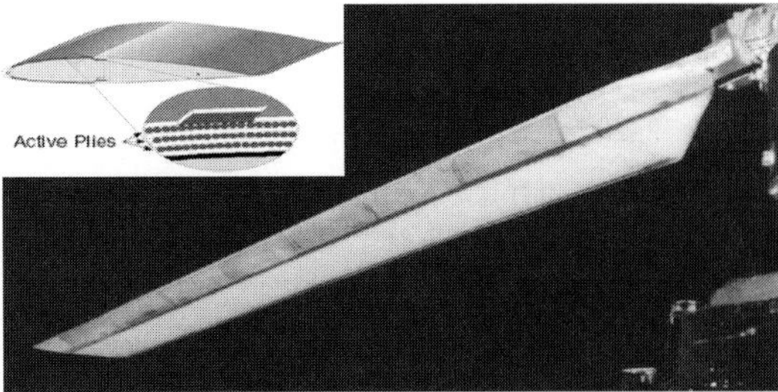

Figure 4: Actively twisting blade. Insert shows concept.

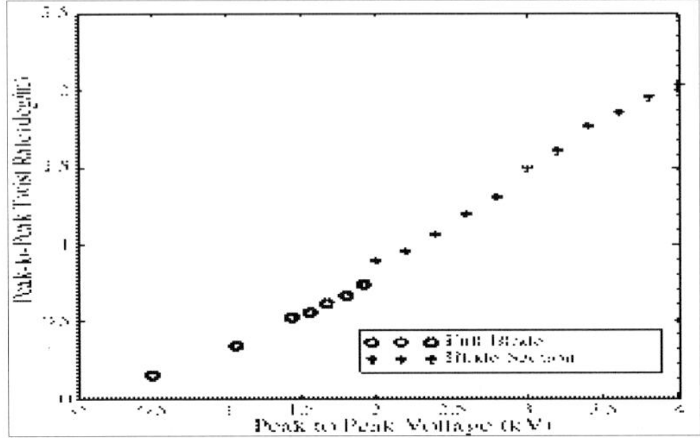

Figure 5: Performance data of the full active blade and a half blade section.

A new generation active blade is currently under development by MIT and Boeing under DARPA funding. One of the main goals of the second-generation blade is to eliminate the problems that were found in the first generation blade. Manufacturing has been improved to solve the problems that led to the reduced performance in the first generation blade. Core voids that were found in the first generation blade were found to be caused by an adhesive used in the instrumentation of the blade. New material systems are being investigated to make the actuators more robust and therefore prevent the actuator pack failures that occurred in the first generation blade. Electrical connections to the AFC actuators have been redesigned to prevent disconnection during the blade cure and testing. The new connection design places the electrical connections near the surface of the blade to improve manufacturability and to eliminate undesirable sharp bends in the electrode lead-outs.

Early tests have shown the new design to be effective. In a blade section with 6 actuators and 12 electrical connections, 100% of the connections survived the blade cure and the subsequent testing. The section was tested at 3000Vpp, up to 100Hz, which is representative of the operating cycle expected in future hover and forward flight testing.

Another blade section is being built to verify the electrical connection integrity. This section will also be used to demonstrate the robustness of the new material system. Actuation tests under load will be performed by placing the blade section under tension while actuating the AFCs. These tests will also serve as a structural test of the blade. Following testing, the blade design will be finalized and a full set of model scale active blades will be manufactured. A full set of blades will be tested in hover and forward flight.

4.2 Active Structural-Acoustic Control

The recently application of piezoceramic elements to the problem of acoustic control has resulted in a new area of research termed Active Structural-Acoustic Control (ASAC). ASAC uses structural-based sensors and actuators to control the acoustically radiating structural vibrations. Current research at MIT focuses in three areas: i) development of techniques to embed sensor, actuator and electronic components into composite structures, ii) active control of interior acoustics, and iii) control of radiated noise from thick-walled cylindrical shells.

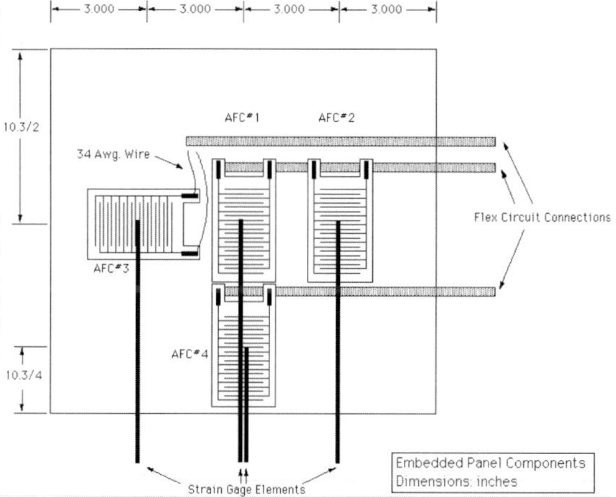

Figure 6: Layout of composite panel

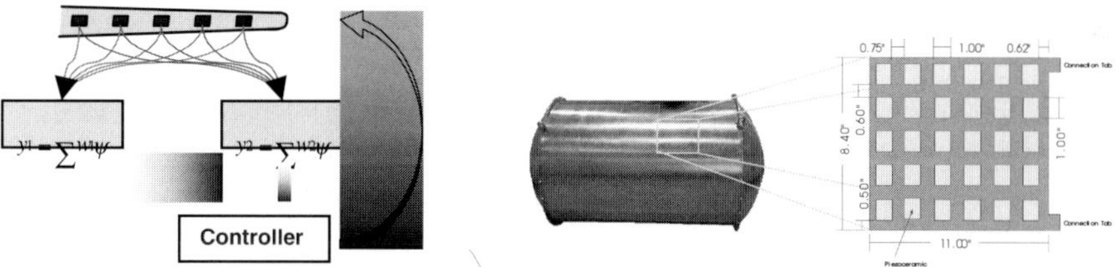

Figure 7: Reconfigurable array concept and application to an aircraft fuselage test-bed.

Embedding sensors, actuators and electronic components into composite structures: In order to increase structural robustness it is desirable to embed sensor and actuators into the host structure. This is particularly attractive in composite structures used in modern aircraft. In order to embed sensor, actuators and electronic components, small thickness and robustness to the curing environment is required. A flat composite panel with embedded active fiber composites and strain gauges has been manufactured.[11] The dimensions of the panel were chosen to have dynamic response similar to a single panel of a typical commercial aircraft (fig 6). To accommodate the embedded sensors and actuators and provide a measure of electrical isolation, three e-glass host plies were incorporated into the lay-up for each active ply. This composite panel has been used to investigate the performance of the embedded sensor and actuators, and also to compare different control techniques applied to the problem of active structural-acoustic noise radiation control.[11]

Active control of interior acoustics: Reconfigurable arrays of sensors and actuators have been developed to allow the design of reduced-order controllers for complex structures with high modal density and complex dynamics that are time varying. These arrays are able to isolate modes in a manner that is insensitive to assembly and modeling errors, by weighting and summing the sensor measurements to detect only the desired vibration modes. The weights are computed from experimental data, not relying on a mathematical model of the structure, and can be easily adapted in real-time since it has been shown that the adaptation process is stable and yields a globally optimal solution.[12] Many modes can be individually targeted for control since a different set of weights can be computed for any desired mode. These weighted signals are fed into reduced-order controllers and the output signals are then redistributed to the array of actuators.

An important aspect is sensor-actuator collocation. Since the ultimate objective is to manufacture composite panels with embedded arrays, PVDF-PZT sensor-actuator elements have been developed (fig. 7). These elements have shown collocated behavior past 10KHz and are now being used in closed-loop control experiments on a helicopter fuselage test-bed.[13]

Control of radiated noise from thick-walled cylindrical shells: Two approaches are being investigated for controlling the noise radiated from thick-walled cylindrical shells in water. In cooperation with the U.S. Naval Undersea Warfare Center active fiber composites have been placed on the inside of cylindrical shells with the objective of reducing shell vibrations, and therefore the radiated sound. Numerical models developed for this problem indicate that the active fiber composites have enough control authority to influence the dynamics of thick-walled cylindrical shells, and a test-bed is currently being instrumented to validate the numerical model in broadband closed-loop experiments.

An alternative approach is being investigated in cooperation with the Northrop Grumman Corporation. This approach makes use of 1-3 piezo-composite panels developed by Material Systems Inc. mounted on the surface of a thick cylindrical shell as a conformal array. These piezo-composite panels have injection-molded accelerometers mounted on its inner and outer surfaces, allowing the measurement of both the shell and interface surfaces' accelerations. The concept calls for a hybrid analog-digital controller that uses the acceleration signals as inputs and commands a voltage to the panel with the objective of maintaining the interface surface quiet.

5 Summary

This paper presented the current state-of-the-art in active fiber composite technologies along with new characterization results and status of its application to active control problems. The nonlinear material model and the governing equations and finite element formulation for electroelastic continua with electric conduction provide an engineering analysis tool useful for analyzing AFC devices. The research being conducted in the field of active fiber composites currently focuses on further development of the manufacturing procedure, on extensive electro-mechanical characterization, and on the application of this technology to the problems of active rotor blade control and active structural acoustic control.

6 Acknowledgments

The authors would like to acknowledge the support of DARPA under grants F49620-97-2-0527 and NDA972-98-3-0001, the Army Research Office under grant DAAH04-95-1-0104, and of the Office of Naval Research under grants 87-JJ-0J-9148 and N00-14-92-J-4067.

7 References

1. Bent, N.W. Hagood, J.P. Rodgers, J. of Int. Mat. Sys. and Struct., 1995, 6, 3, 338-349
2. B.Z. Janos, N.W. Hagood in Proc. of 6th Intern. Conf. on New Actuators, ACTUATOR98, Bremen, Germany, June 1998
3. A.A. Bent, Active Fiber Composites for Structural Actuation, Ph.D. Thesis, Massachusetts Institute of Technology, 1997
4. N.W. Hagood, A. Pizzochero, J. of Int. Mat. Sys. and Struct., 1997, 8, 9, 724-737
5. H.B. Strock, M.R. Pascucci, M.V. Parish, Proc. of SPIE, 1999, 3675, 22-31
6. K. Ghandi, Nonlinear Modeling and Characterization Techniques for Phase Transitions in Electro-Mechanically Coupled Devices, PhD Thesis, Massachusetts Institute of Technology, 1998
7. J.E. Harper, Analysis of Nonlinear Electroelastic Continua with Electric Conduction, SM Thesis, Massachusetts Institute of Technology, 1999
8. J. P. Rogers, Development of an Integral Twist-Actuated Rotor Blade for Individual Blade Control, Ph.D. Thesis, Massachusetts Institute of Technology, 1998, 73-128
9. J. P. Rogers, A. A. Bent, N. W. Hagood, Proc. of SPIE, 1996, 2717, 642-69
10. J. P. Rodgers, N. W. Hagood, Proc. of SPIE, 1998, 3329, 291-308
11. B.S. Bingham, N.W. Hagood, M.J. Atalla, Proc. of SPIE, 1999, 3668, 222-234.
12. M.L. Fripp, M.J. Atalla, N. W. Hagood, S. Tistaert, C. Savran, Reconfigurable arrays for broadband feedback control of aircraft fuselage vibration, Proc. 9th Int. Conf. Adaptive Structures, Paris, France, October 1999, to be published
13. M.L. Fripp, D.Q. O'Sullivan, S.R. Hall, N.W. Hagood, K. Lilienkamp, Proc. of SPIE, 1997, 3041, 88-99

In-Situ Structure Control of Ceramic Composite with Improved Mechanical Properties and Remote Sensing for Crack Propagation

Shin-ichi Hirano, Toshinobu Yogo, Wataru Sakamoto, Takashi Kojima and *Takashi Fujii
Department of Applied Chemistry, Graduate School of Engineering Nagoya University, Furo-cho, Chikusa-ku, Nagoya 464-8603, Japan, *Technical Research Center, Nisshin Flour Milling Co., Ltd., 5-3-1 Tsurugaoka, Oi-machi, Iruma-gun, Saitama 354, Japan

1 Abstracts

Novel Ce-TZP based composites were successfully developed by the microstructure control of in-situ formed plate-like Lanthanum-β-aluminate or La M-type Hexaferrite crystals via conventional powder processing and sintering procedure. Formation of the anisotropic crystals was effective to improve the mechanical properties of the composite. Composite with in-situ formed hexaferrite particles exhibits improved mechanical properties as well as magnetism, which can be applied for remote sensing of crack propagation utilizing the inverse magnetostrictive effect.

2 Introduction

Tetragonal Zirconia Polycrystal (TZP) ceramics have excellent mechanical properties with high fracture strength and fracture toughness. Ceria-stabilized tetragonal zirconia polycrystal (Ce-TZP) is superior in fracture toughness and hydrothermal stability to yttria-stabilized tetragonal zirconia polycrystal (Y-TZP). However, the use of Ce-TZP is limited in the specific field of engineering ceramics because its strength is relatively low. Recent studies have shown that the mechanical properties of TZP ceramics can be improved by the incorporation of the second phases. The second phase most examined for Ce-TZP is alumina. The dispersion of spherical alumina particles in Ce-TZP matrix is effective in increasing the fracture strength of the composite. However, the incorporation of spherical alumina particles also generates the problem of reduced fracture toughness. Although whisker or platelets addition is effective for improving the fracture toughness, the additives often give rise to the reduction in fracture strength due to the difficulty of sintering. The authors found that plate-like lanthanum-β-aluminate ($LaAl_{11}O_{18}$; LBA) particles are in-situ grown in Ce-TZP matrix during sintering.[1)-3)] Coexistence of LBA and alumina in Ce-TZP matrix is quite effective to improve both the fracture strength and fracture toughness. In addition, Mn^{2+} addition to Ce-TZP/Al_2O_3/LBA composite remarkably improved the microstructure and mechanical properties of the composite.[4)]

$LaFe_{12}O_{19}$ (La M-type hexaferrite; LaM), one of the hexaferrites, is ferromagnetic and has a plate-like hexagonal shape. Therefore, $LaFe_{12}O_{19}$ is quite suitable as a second phase of composites to improve the mechanical properties of structural ceramics. In addition, ferrite particles add magnetism to structural ceramics, and can be utilized for stress or fracture sensing based upon the inverse magnetostrictive effect. However, the narrow stable temperature range of LaM makes the in-situ formation of LaM in Ce-TZP matrix quite difficult. To improve the thermal stability of the LaM, a solid solution of LaM and LBA, $La(Fe,Al)_{12}O_{19}$, was synthesized.[5), 6)] Furthermore, the substitution of Fe^{2+} ion for other divalent ions is known to effective to stabilize LaM. Therefore, Co substitution for Fe^{2+} ions in $La(Fe,Al)_{12}O_{19}$ was investigated during in-situ processing of in Ce-TZP-based composites.[7)] This paper reviews the concept and scheme of Ce-TZP/LBA, Ce-TZP/$La(Fe,Al)_{12}O_{19}$ and Ce-TZP/$La\{Co(Fe_{0.9}Al_{0.1})_{11}\}O_{19}$ in-situ composites with the combination of the fracture sensing ability.

3 Experimental Procedure

3.1 Preparation of starting powder mixtures and green compacts

Reagents used for starting materials are commercially available powders as follows: Ce-TZP (12 mol% CeO_2-ZrO_2; Tosoh, TZ-12CE), lanthanum oxalate ($La_2(C_2O_4)_3$?$10H_2O$), a-Fe_2O_3, γ-Al_2O_3 (Sumitomo, AKP-G), α-Al_2O_3 (Sumitomo,AKP-30 or Taimei chemical, TM-DAR) , CoO and MnO_2. The preliminary experiments revealed that LBA and LaM were formed at two steps via $LaAlO_3$ and $La(Al_{1-x}Fe_x)O_3$, respectively. First, $La(Al_{1-x}Fe_x)O_3$ powder was prepared by the reaction of lanthanum oxalate, γ-Al_2O_3 and Fe_2O_3 powders. The appropriate amounts of the starting powder materials were mixed with ethanol in a plastic ball mill for 24h. After drying at 50°C, the mixture powder was calcined at 1400? for 4h. The calcined $La(Al_{1-x}Fe_x)O_3$ powder thus prepared was mixed with α-Al_2O_3, Fe_2O_3 and CoO powder by ball milling in a similar way. The mixture powder and Ce-TZP powder were weighed to be the composition of LBA or $La(Fe, Al)_{12}O_{19}$ or $La\{Co(Fe, Al)_{11}\}O_{19}$ in Ce-TZP, and then ball milled using ethanol. The powder was cold isostatic pressed (CIP) at 200 MPa yielding green compacts. The compacts were heat-treated under various conditions.

3.2 Characterization

The density was measured by the Archimedes method using water. The crystalline phases were identified by X-Ray diffractometry (XRD) using CuKα radiation with a monochromator. The composites were observed by scanning electron microscopy (SEM). The fracture strength of composites were evaluated by the three point bending method on sintered samples cut into 3×4×35 mm bars. The fracture toughness was measured by the single-edge-precracked-beam (SEPB) method using the same test piece after the bending test. Magnetic properties of composites were analyzed using a vibrating sample magnetometer (VSM) at room temperature. The distribution of anisotropic residual strain in the composite was examined by the neutron time-of-flight diffraction method. The magnetic flux of the composites was measured using a scanning superconducting quantum interference device (SQUID) gradiometer.[7)] The composite (15mm in length, 3mm in width and 4mm in height) placed on a scanning table was analyzed using a pick-up coil of the SQUID gradiometer.

4 Results and Discussion

4.1 Ce-TZP/Al_2O_3/LBA composite

Single-phase LBA particles were in-situ formed in Ce-TZP matrix after sintering around 1600°C for 4h. In Ce-TZP/Al_2O_3/LBA composite, plate-like LBA crystals and spherical Al_2O_3 particles were well dispersed in Ce-TZP matrix. 0.2wt% MnO_2 addition to Ce-TZP/Al_2O_3/LBA composite was found to decrease the particle size of LBA in Ce-TZP matrix.

Table 1 shows the mechanical properties of the composites. Al_2O_3 addition to the Ce-TZP was effective to improve the bending strength, however, fracture toughness of the composite decreased. The in-situ formation of anisotropic plate-like LBA crystals increased the fracture toughness of the composite. Mn addition to the starting mixture of the composite attained the pronounced improvement of the mechanical properties. From the neutron diffraction analysis, plate-like LBA crystals were under anisotropic compressive stress.[8] This results was derived from the crack deflection, the bridging effect by thermal residual stresses and the direct contact to the LBA crystals. The LBA crystals of Mn doped composite were smaller in size than that of non-doped composite. The smaller size of LBA crystals increased the number of crack deflection site, leading to the increase in bending strength and fracture toughness.

Table 1 Mechanical properties of 12Ce-TZP/Al_2O_3/LBA composites

	Al_2O_3 (wt%)	LBA (wt%)	MnO_2 (wt%)	Bending strength (MPa)	Fracture toughness (MPa?m$^{1/2}$)
12Ce-TZP	0	0	0	650	8.1
12Ce-TZP/Al_2O_3	20	0	0	930	6.4
12Ce-TZP/Al_2O_3/LBA	15	5	0	910	11.2
12Ce-TZP/Al_2O_3/LBA	15	5	0.2	1020	12.1

4.2 Ce-TZP/La M-type Hexaferrite composite

Single-phase LaM could not be formed in Ce-TZP matrix due to the low reaction rate and poor thermal stability of the LaM, which was overcome by forming a solid solution with LBA to $La(Fe,Al)_{12}O_{19}$. LBA was found to form a stable solid solution with LaM below 95mol% of LaM. In-situ formation of $La(Fe,Al)_{12}O_{19}$ composite was successfully performed by controlling the solid solution of LaM and LBA. Fig. 1 shows the microstructure of Ce-TZP/$La(Fe,Al)_{12}O_{19}$ (90% LaM in $La(Fe,Al)_{12}O_{19}$) composite sintered at 1350°C for 12h. Plate-like $La(Fe,Al)_{12}O_{19}$ crystals with hexagonal shape are well distributed in the Ce-TZP matrix.

The Ce-TZP/$La(Fe,Al)_{12}O_{19}$ composite exhibited typical magnetism. Since the highest magnetization of the composite was obtained when the composition of the LaM in $La(Fe,Al)_{12}O_{19}$ was 90mol% (10% LBA), the composition of the $La(Fe,Al)_{12}O_{19}$ was fixed at 90 mol% LaM. The saturation magnetization and the coercive force of the composites including $La(Fe,Al)_{12}O_{19}$ (90mol% LaM) were 7.6 emu/g and 2.6 kOe, respectively. However, when the composite was sintered blow or above 1350°C, single-phase $La(Fe,Al)_{12}O_{19}$ could not be formed in the Ce-TZP matrix and the saturation magnetization of the composite decreased steeply. In order to improve the thermal stability of the composite, Fe^{2+} ions in $La(Fe,Al)_{12}O_{19}$ were found to be substituted for Co. From the XRD analysis, single-phase La{Co($Fe_{0.9}Al_{0.1}$)$_{11}$}O_{19} was formed when the composite was sintered from 1200 to 1325°C. The stable temperature range of the second phase hexaferrite was much more extended. Figure 2 shows the magnetic properties of the Ce-TZP/20wt% La{Co($Fe_{0.9}Al_{0.1}$)$_{11}$}O_{19} composite sintered at various temperatures. The saturation magnetization of Ce-TZP/La{Co($Fe_{0.9}Al_{0.1}$)$_{11}$}O_{19} composites shows a constant value after sintering between 1200°C and 1350°C, which is much wider than that of Ce-TZP/La(Fe,Al)$_{12}$$O_{19}$ composite. At the temperatures, the main second phase in the composites was La{Co($Fe_{0.9}Al_{0.1}$)$_{11}$}O_{19}. The highest value of the coercivity at 1350°C reflects that the particle of magnetic phase approaches the size of magnetic single-domain.

Fig. 1 Microstructure of Ce-TZP/20wt% La(Fe,Al)$_{12}$$O_{19}$ composite (LaM : LBA=90:10)

Table 2 shows the bending strength of Ce-TZP/LBA, La(Fe,Al)$_{12}$$O_{19}$ and La{Co($Fe_{0.9}Al_{0.1}$)$_{11}$}O_{19} composite. The bending strength of these composites are much improved from Ce-TZP ceramics. The highest value of the bending strength of Ce-TZP/La{Co($Fe_{0.9}Al_{0.1}$)$_{11}$}O_{19} composite was attained at 1350°C, and the value was higher than that of Ce-TZP/20wt% La(Fe,Al)$_{12}$$O_{19}$ composite. The reason for the relatively high strength is the same as that for Mn added Ce-TZP/LBA composite.

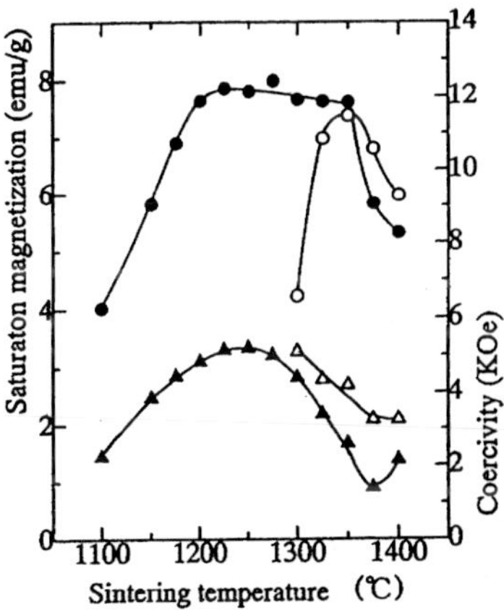

Fig. 2 Dependence of magnetic properties on sintering temperature for Ce-TZP/20wt% La{Co($Fe_{0.9}Al_{0.1}$)$_{11}$}O_{19} composites and Ce-TZP/20wt% La(Fe,Al)$_{12}$$O_{19}$ composites (The ratio of LaM to LBA in the second phase is 90 to 10). ?,?Ce-TZP/20wt% La{Co($Fe_{0.9}Al_{0.1}$)$_{11}$}O_{19} composites, ?,?Ce-TZP/20wt% La(Fe,Al)$_{12}$$O_{19}$ composites.

Table 2 Bending strength of Ce-TZP/LBA and Ce-TZP/La M-type hexaferrite composites

	Sintering condition	Bending strength (MPa)
12Ce-TZP/20wt% LBA	1600°C, 4h	780
12Ce-TZP/20wt% La(Fe,Al)$_{12}$O$_{19}$ (90% LaM)	1350°C, 12h	780
12Ce-TZP/20wt% La{Co(Fe$_{0.9}$Al$_{0.1}$)$_{11}$}O$_{19}$	1350°C, 12h	880

4.3 Remote sensing of crack propagation

In order to examine the fracture sensing ability of Ce-TZP/La M-type hexaferrite composite, the magnetic field around the crack was measured using a scanning SQUID gradiometer. Figure 3 shows the distribution of magnetic flux gradient of the composite before and after the introduction of a crack in the composite. The difference in magnetic flux gradient along the vertical direction to the top plane is revealed as positive and negative signs in Gauss in Fig. 3. The minimum square in the upper is 1 mm × 1 mm in size. The blank rectangles in the bottom correspond to the top surface of the composites. The distribution of magnetic flux after the introduction of the crack is clearly different from that before the crack indentation. The change in magnetic flux gradient is considered to result from the inverse magnetostrictive effect. The stress applied to a magnetic body affects the direction of domain magnetization through magnetostriction. This result reveled that this concept can be used not only for the detection of crack propagation but also for the remote fracture sensing system.

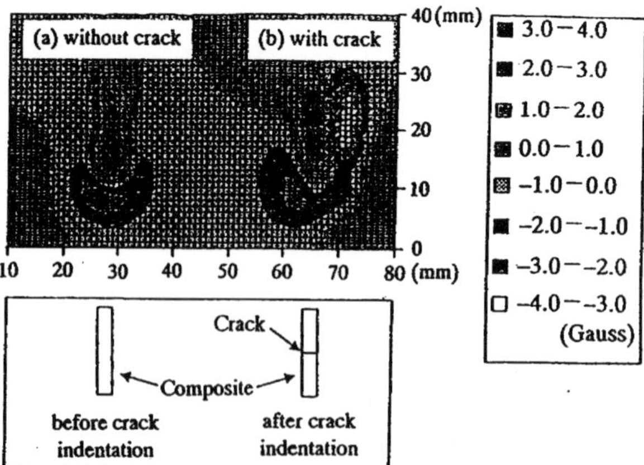

Fig. 3 Change in magnetic gradient with indentation of crack on Ce-TZP/La(Fe,Al)$_{12}$O$_{19}$ (LaM : LBA=90 : 10); (a) magnetic field gradient of the composites without crack and (b) magnetic field gradient of the composites with crack.

5 Conclusions

Processing of novel Ce-TZP-based in-situ composites were reviewed. Composites with well-controlled microstructures were successfully synthesized through in-situ powder processing. The in-situ formation of plate-like crystals found to improved markedly the mechanical properties of Ce-TZP. In-situ formation of ferromagnetic $La(Fe,Al)_{12}O_{19}$ and $La\{Co(Fe_{0.9}Al_{0.1})_{11}\}O_{19}$ crystals in Ce-TZP matrix creates the synergetic effect of improved mechanical properties as well as magnetism, which can be utilized for detection of crack propagation.

6 References

1) S. Hirano, H. Muragaki, H. Hatano and T. Fujii, Ceramic. Transaction., 1991, 22, 693-698.
2) S. Hirano, T. Fujii, H. Muragaki and H. Hatano, , Mat. Res. Soc. Symp. Proc., 1992, 274, 141-147.
3) S. Hirano, T. Yogo, M. Miura, H. Hongoh and T. Fujii, , J. Mat. Sci., 1994, 29, 262-268.
3) S. Hirano, T. Yogo, M. Miura, H. Hongoh, S. Kawai and T. Fujii, to be published.
4) T. Kojima, W. Sakamoto, T. Yogo, T. Fujii, S. Hirano, J. Am. Ceram. Soc., 1998, 81, 2965-70.
5) T. Kojima, W. Sakamoto, T. Yogo, T. Fujii, S. Hirano, J. Ceram. Soc. Japan, 1999, 107, 215-21
6) T. Kojima, W. Sakamoto, T. Yogo, T. Fujii, S. Hirano, J. Am. Ceram. Soc., in press.
7) T. Morooka, S. Nakayama, A. Odagawa, N. Shimizu, K. Chinone, T. Ataka and N. Kasai, Jpn. J. Appl. Phys., 1996, 35 Part 2, 4B, L486-L488.
8) S. Hirano, T. Yogo, M. Miura, T. Kojima, T. Fujii, J. W. Richedson. Jr, and C. K. Loong, to be published.

An Overview to the BMBF-Leitprojekt ADAPTRONIK

Holger Hanselka
Institut für Mechanik, Lehrstuhl für Adaptronik, Otto-von-Guericke-Universität Magdeburg

1 Abstract

In 1997 BMBF, within the framework of an idea competition for future-oriented key technologies and their industrial utilization, called for project proposals from industries and research for so-called „Leitprojekts". An independent group of experts selected few project proposals from the many submitted, and proposed them to BMBF for promotion. One of these projects is the BMBF-Leitprojekt ADAPTRONIK which is introduced in this paper. The Leitprojekt ADAPTRONIK which consists of 24 partners from industry and research institutes and is conducted under the responsibility of Deutsches Zentrum für Luft- und Raumfahrt e.V. (DLR) in Brunswick, focuses on the structure-conforming integration of piezoelectric fibers and patches in structures for lightweight construction. It is aimed at active vibration and noise reduction, contour deformation and micro-positioning in the very sense of adaptronics in various industrial applications. The project targets are prototype assemblies from the fields of automotive industry, rail vehicles, mechanical engineering, medical engineering, and aerospace.

2 Objectives

Adaptronics describes the field of technology focusing on the development of a new class of so-called *smart structures*. This concept is based on the development of adaptive systems which, through autonomous (i.e. self-controlled) mechanisms, are able to self-adapt to different operating conditions. This approach requires optimum integration of sensors and actuators within the system on the basis of new functional materials, such as piezoceramic fibers and patches with adaptive controllers. These elements satisfy both load-bearing and actuatoric/sensoric requirements at a time and, hence, are *multifunctional*. On the one hand, this process requires new approaches by engineers as, apart from classical stiffness and strength considerations, they also need to include in their design and calculations 'virtual' properties such as changing dynamic stiffness, damping, or mass distribution. In this way, structural elements can be developed which are <u>not</u> subject to any deformation as a result of external forces and, consequently, exhibit an apparently 'infinite' stiffness. On the other hand, an interdisciplinary system approach is necessary to implement such adaptive structures. With this approach it becomes possible to equip optimized mechanical structural systems with structure-conforming integrated actuators and sensors as well as adaptive control systems offering real-time capabi-lity. The technical potentials resulting from this development will considerably broaden the range of new products in the various branches of industry. New developments will focus on *Noise and vibration reduction, Contour deformation and stabilization, Micro-positioning, Sensor technology in the ultrasonic range* which, in terms of a typical cross-sectional techno-logy, will lead to quantum leaps in the product development in branches such as *Traffic engineering, Mechanical engineering, Optical engineering, Medical engineering and Aerospace.*

However, full advantage from adaptronics as to lightweight construction can only be drawn if the actuator and sensor components are successfully integrated in the structure such that the material and/or structural properties are only negligibly affected if at all (*structural conformity and multifunctionality*). The ADAPTRONIK „Leitprojekt" mainly aims at providing the scientific and technological basis for industrial applications in as many branches as practical. It pursues the strategy of establishing a synergetic network between the advanced scientific potential in world-wide competition and industrial partners from the various branches with the aim of preparing a new generation of products with properties that are unique on the international scale. In this process the small and medium-sized companies will play a special role as technological centers. To accomplish this ambitious task a *network* of industrial and research partners has been established which purposefully combines and manages the individual specialist disciplines of the highly complex field of adaptronics.

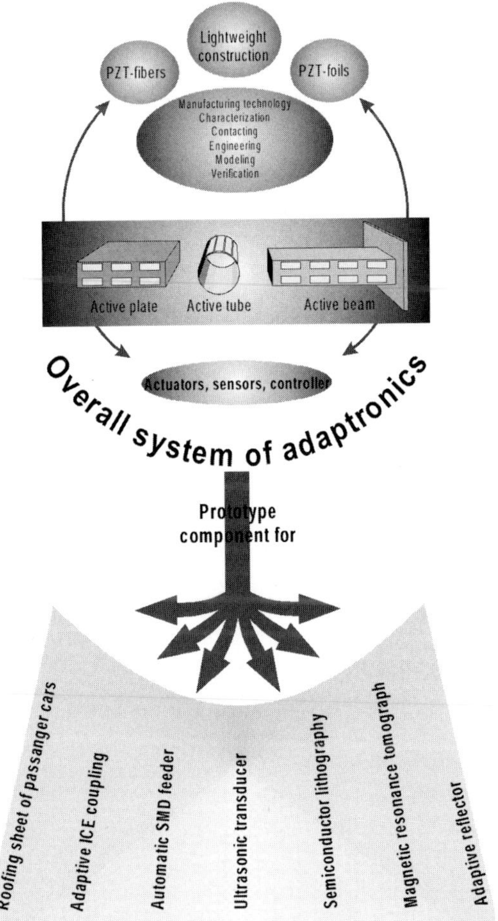

Figure 1: Scientific and industrial implementation of technical contents

The project aims at developing as many options for an industrial application as possible. Scientific and industrial implementation first requires the development of adaptive structures with a simple geometry, i.e. so-called *prototype specimens*, which can be used for a multiple development and utilization also in very specific product fields within specific constraints. Based on these prototype specimens industrial applications up to *prototype assemblies* can be developed and implemented under the overall concept of adaptronics. The overall system of adaptronics comprises a *multifunctional structure* with integrated actuator and sensor systems, the electronic equipment required, and a real-time adaptive control system tailored to the specific application.

The system incorporates the results of material development and of the related process metrology for piezoelectric fibers and patches, of the structure-conforming integration in lightweight materials and the development of modeling and optimization strategies, in particular, in the macromechanical field. As to the time factor, the project is designed such that some of the industrial users will be able to start very early to implement and introduce their products on the markets whereas users with a higher integration density may require more time depending on activities in process and potential experience in the technological environment.

3 Workprogram and Essential Results

The industrial and research partners involved perform the required development work along the value-added chain (material development → composite technology → adaptive overall system → prototype assemblies). The technological bases (AP 1 and AP 2) are incorporated in the overall system of adaptronics (AP 3) which, through its specifically selected prototype specimens, will form the basis for implementing prototype assemblies (AP 4). This requires parallel development work within the individual work packages. For this very complex and interdisciplinary approach, the establishment of synergies and exchange of results between the individual disciplines are absolutely essential. Most of the partners find themselves within the various sections of the value-added chain. This reflects the complexity and interdisciplinarity as the individual sections of the value-added chain are closely interlinked and, necessarily, depend on each other. Aiming at *Noise and vibration reduction, Contour deformation and stabilization, Micro-positioning and Ultrasonic technology*, the prototype assemblies are designed for use in the following branches of industry with their special targets:

- In *traffic engineering* multifunctional adaptive elements will be used in passenger car construction to reduce vibrations through the roofing sheet, and in rail car construction to reduce vibrations through skin fields in order to exert an active influence on the noise level inside the cabin.
- The application in *mechanical engineering* will focus on the active influence on machine vibrations to obtain a higher level of accuracy at reduced attenuation times and new ultrasonic transducers.
- The aim in the *optical industry* are a correction of mirrors and deformation-free carriers of lenses, both for use in objectives for semiconductor lithography.
- In *medical engineering* they are intended to give patients 'acoustic' relief in magnetic resonance tomography (MRT).
- In the branch of *aerospace* they will help to improve the accuracy of antenna and satellite structures.

The work packages are structured on the basis of the following scheme:

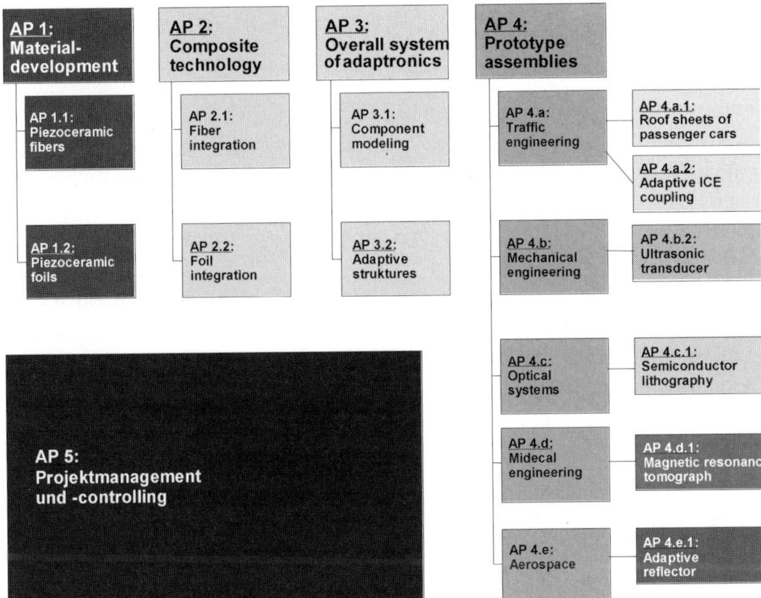

Figure 2: Structural scheme

The following sections provide a short description of the scientific and technical objectives of the individual work packages (AP). *Material development (AP 1)* PZT Fibers and PZT patches are applied as activatable materials with sensor and actuator properties.

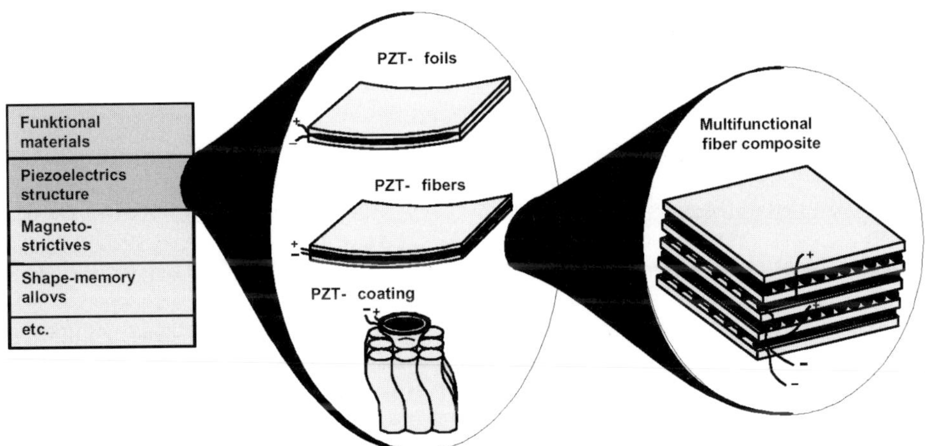

Figure 3: Multifunctional material system on the basis of PZT fibers and patches

Referring to piezoceramic fibers (*AP 1.1*) the Fraunhofer Institut für Silicatforschung (ISC) has gathered many years of experience in the field of fiber development. Under this project they have been able to provide first experimental specimens of piezoelectric fibers. This work package focuses on the following aims:

- Optimization of the piezoelectric properties of fibers (chemical composition, structure, dia-meter).
- Technology for small-scale continuous fiber production.
- Technology transfer for fiber production to a small company (HTM, Berlin),
- Development and testing of a hybrid, i.e. carbon-fiber supported, piezoelectric fiber.
- Analysis and evaluation of degradation and failure of piezoelectric fibers used under load composite materials.

The scientific and technical objectives of the work package of *piezoceramic patches (AP 1.2)* include:

- Development and provision of piezoceramic patches with tailored optimized performance parameters (area, bending, useful life, blocking force, etc.
- Technology and material development for electroding piezoceramic components.
- Material characterization and determination of material parameters for an optimized material design and component modeling.
- Analysis of ceramic-induced failure of and degradation effects in prototype specimens and prototype components, and deriving material-specific loading limits.

Composite technology (AP 2) To obtain a multifunctional material system which is able to satisfy actuatoric, sensoric and load-bearing properties ceramics, which are brittle by nature, are required to be integrated into lightweight construction materials or applied to lightweight construction. The core of the work package of *fiber integration (AP 2.1)* is the development of suitable manufacturing processes for producing lightweight structures with actuatorically and sensorically effective PZT fibers. First material specimens have been produced.

Figure 4: Material specimen on the basis of PZT fibers

To this end, it is necessary to develop equipment and methods for producing semifinished products with integrated and contacting PZT fibers, i.e. so-called fiber-based microsystems. Hence, the scientific and technical objectives of this work package mainly focus on disciplines relevant to production technology:

- Adhesives: Develop and optimize adhesives for fixing and contacting piezoceramic fibers.
- Microsystem technology: Develop and provide microassembly and microbonding processes for producing fiber-based microsystems with integrated, contacting PZT fibers.
- Provide geometrically simple and prototype fiber-based micro-systems.

- Integration and fiber composite processes: Develop processes for integrating microsystems into composite and lightweight construction materials with the aim of ensuring the highest level of structural conformity and piezoelectric efficiency.
- Produce geometrically simple prototype specimens (beam, plate, tube) with integral fiber-based microsystems.
- Characterize fiber-based microsystems and prototype specimens to ensure the primary componen function.
- Provide material parameters for numerical modeling (AP 3.1).

Foil integration (AP 2.2) mainly focuses on integrating PZT patches into fiber composite materials such that they conform to the particular structure, or apply them to lightweight construction materials. Figure 5 shows an example of such active composites.

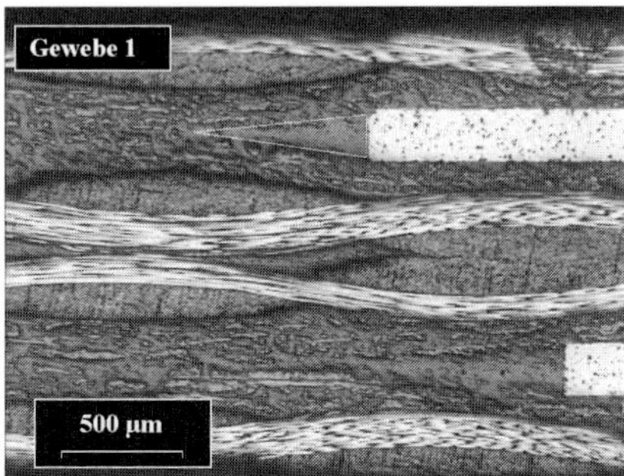

Figure 5: Picture of an active PZT fiber composite

The scientific and technical objectives of this work package include:

- Apply and integrate plain and curved PZT patches to/into lightweight construction materials, such as aluminum and carbon-fiber reinforced plastics, and fiber composite materials.
- Characterize multifunctional material systems and provide material parameters for component modeling (AP 3.1).
- Provide a cost-efficient manufacturing process for efficient and reliable production of lightweight structures using PZT patches.
- Manufacture the structural-mechanical component of the prototype specimen with PZT patches as actuator/sensor material.
- Develop, qualify and validate suitable methods of non-destructive material diagnosis.

Overall system of adaptronics (AP 3) This work package aims at employing the materials engineering basis in the form of prototype specimens to develop an overall adaptronic system by means of the structural-mechanical component, power electronics and adaptive control engineering.

Component modeling (AP 3.1) The scientific and technical objectives of the work package of component modeling (AP 3.1) include

- Develop and improve analytical and numerical methods for modeling and computer-aided simulation of the electrical thermomechanical behavior of adaptive composite structures with fibers and patches integrated such that they conform to the structure.
- Develop modeling strategies for adaptive lightweight structures under the special aspect of predicting the shape, the number and the physical location(s) of sensors and actuators in the structure.
- Develop design guidelines for adaptive composite structures.

- Develop appropriate application software and provide it in the form of design and modeling tools for implementing optimized adaptive lightweight structures.

Hence, this part project establishes the prerequisites for computer-aided modeling, design and optimization of adaptronic structural systems the capability of which to self-adapt is based on piezoelectric fibers and patches integrated such that they conform to a given structure. Therefore, close cooperation with work package 3.2 and the work packages 4.x permit rapid and efficient use of the technology of adaptronics for developing and designing new products and optimizing their properties. All scientific results are combined in the work package of *adaptive structures (AP 3.2)* and processed for industrial-scale use. The first example is the prototype specimen of an actuatorically and sensorically effective plate with PZT fibers. The scientific and technical objectives of this work package include

- Ensure structure-mechanical implementation of prototype specimen with integrated/
- appliedmultifunctional elements.
- Develop/Provide suitable control electronics.
- Conduct control engineering modeling of prototype specimens.
- Develop suitable adaptive control algorithms and implement these for appropriate hardware.
- Implement the overall adaptronic system.
- Verify the operability of the overall adaptronic system with the aim of industrial implementation.
- Compare the potential of piezoelectric fibers and patches.

Prototype assemblies (AP 4) Prototype assemblies are understood to be the results of product-related development work of the overall adaptronic system, and serve to provide more comprehensive evidence of their operability, performance potentials for their respective application, design solutions, control aspects, etc.. Outlined below are the prototype assemblies by their branches of industry:

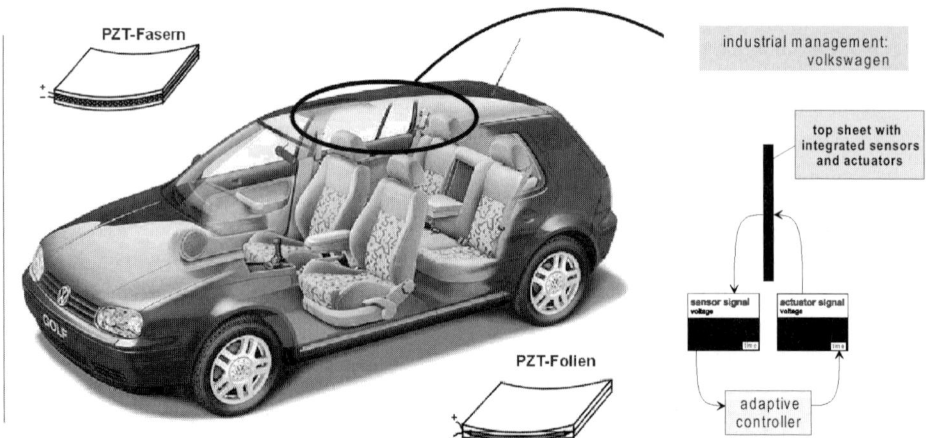

Figure 6: Roof sheet of a passenger car with distributed PZT actuators/sensors

Traffic engineering (AP 4.a) In traffic engineering, structural vibrations and the associated acoustic emissions as well as fatigue of materials form a major problem. Adopting passive measures such as local stiffness modifications, oscillation dampers, anti-noise coatings, damping etc., attempts have been made to minimize the effects of vibrations occurring. The adaptronic approach to be developed under this „Leitprojekt“ is aimed at preventing such vibrations from the very beginning. This would also render the weight-increasing passive measures unnecessary. The work *project of roof sheeting of passenger cars (AP 4.1.1)* focuses on influencing the structural vibrations of the roof sheet by means of distributed PZT actuators/sensors such that the noise in the car interior is reduced.

The scientific and technical objectives of this work package include:

- Evaluate the acoustic behavior of the plate prototype specimen within car-related constraints.
- Adapt the adaptive control to the requirements of the car roof sheet.
- Modeling of the overall adaptronic system comprising the roof sheet of the passenger car, electronic control equipment, and the adaptive controller.
- Study structural vibrations and acoustic emissions on unfinished roof body without and activated piezoelements applied to the roof structure.
- Investigate acoustics in the interior of the entire vehicle without and with activated piezoelements applied to the roof as well as the control on piezoelements and microphone.
- Conduct final analysis of the performance capabilities of adaptive systems versus passive measures with regard to vibration reduction and acoustic efficiency.

Adaptive ICE coupling (AP 4.a.2) focuses on the improvement of comfort in modern traffic and transport systems. A factor exerting a decisive influence on comfort is the transmission of vibrations and noise from outside into the interior of the vehicle. Here, adaptive technology offers advantages over conventional damping measures when used as weight and space saving alternatives to reduce vibrations and structure-borne noise. To this end, piezoceramic actuator/sensor elements offer promising approaches. A typical application is the transmission of structure-borne noise from the bogie into the interior of the superstructure of high-speed rail vehicles (e.g. ICE) via specific coupling elements. These so-called 'anti-rolling' devices are low-frequency structure-borne noise bridges (frequency range of 100 Hz) which are subjected to high loads (superstructure weight of 40 tonnes). With regard to this problem work package AP 4.a.2 focuses on the following scientific and technical aims:

- Develop basic concepts for adaptive vibration and structure-borne noise reduction in the 100 Hz range for typical coupling elements used for rail vehicles.
- Model vibration and structure-borne noise transmission for typical coupling elements as well as optimal actuator/sensor configuration by using piezoceramic actuators.
- Design and implement a hybrid system comprising conventional elements of vibration engineering (dampers for low-frequency vibrations < 30 Hz) and a piezo-highload actuator sys-tem for the higher-frequency portion (> 30 Hz).
- Verify the models for vibration and structure-borne noise transmission and optimal actuator/sensor configuration by means of a lab set-up resembling the real situation as closely as practical.
- Design/Adapt and verify of the adaptive control method.
- Make final evaluation of the performance of the adaptive system in reducing vibrations and structure-borne noise and assess its applicability to more complex systems.

Mechanical engineering (AP 4.b) Similar to traffic engineering, structural vibrations also play an important role in mechanical engineering, even though in the light of (i) improving the position accuracy, (ii) reducing the idle time for bridging attenuations, and (iii) similar phenomena rather than for user's convenience. Hence, it is the primary objective to improve the performance of machines and installations. Another aspect of mechanical engineering is attri-butable to the fact that in the industrial societies quality that can be demonstrated has become a major factor in competition. Companies will be enabled to perform complete and direct quality tests on their own premises and should also benefit from real-time measurements, in particular in quality-assuring monitoring and control of processes. In this connection sensor technology in the ultrasonic range plays an important role which will be intensified by developing novel ultrasonic transducers in future. The scientific and technical objectives of the work package of *ultrasonic transducers (AP 4.b.2)* include

- Develop a new generation of ultrasonic transducers, which cannot be implemented with the presently available piezoelectric materials, to be used for non-destructive testing of industrial components and installations. They mainly include 2D arrays with a high number of elements (matrix and segmented ring arrays) for three-dimensional electronic sound field control (sound beam orienting and focusing), ultrasonic endoscopes for pipe testing (circular arrays), extremely broad-banded transducers for testing coarse grained materials, such as austenites, high-frequency transducers providing very high axial and lateral resolution as well as broad-banded and, at the same time, sensitive airborne-noise transducer suitable for frequencies between 100 kHz and 2 MHz. The availability of such new ultrasonic transducers marks an essential improvement of the testing results and, hence, testing quality in many applications of non-destructive materials testing.
- Develop ultrasonic transducers for measuring flowrates of gases (200 kHz) and liquids (2 MHz) featuring improved properties (lower voltage and power requirements, stronger radiation than transmitting transducers and good receive performance, improved pulse form, generation of less structure-borne noise) while the manufacturing costs are much reduced.

Optical systems (AP 4.c) Optical system require maximum accuracy, whether in dynamic or in quasistatic loading. Employing adaptive structural systems is aimed at optimizing optical high-precision instruments. The scientific and technical objectives of the work package of *semiconductor topography* (*AP 4.c.1*) include

- Model elements on the basis of fibers or patches integrated to conform to the given structure so as to implement active mirrors or adaptronic carriers of lenses.
- Model control systems and implement the actuator, control electronics, sensor and the adaptive controller to satisfy highly accurate deformations.
- Produce test specimens with simple optics for initial experimental verification.
- Implement high-precision interferometer units to prove the optical performance of the active mirrors or the adaptronically carried lenses.
- Implement demonstrators which meet the requirements for use in objectives for semiconductor lithography.

Medical technology (AP 4.d) Medical engineering calls for extremely high technological performance to meet the requirements of future therapeutic and diagnostic methods. Efforts focus on protecting man and improving precision. As an example, magnetic resonance tomography is investigated in this project (AP 4.d.1). A central problem of magnetic resonance tomographs (MRT) is noise development during the operating time. Under extremely unfavorable conditions sound pressure levels exceeding 120 dB may occur. These disturbing noise levels affect various groups of people: patients, operators, people living in the vicinity of indoor or outdoor hospitals. Hence, this work package is aimed at developing quieter installations requiring the following work to be done:

- Investigate the influence of integrated piezoelectric materials by and on the environment of the MR tomograph.

- Demonstrate the capability of structure-integrated fiber or foil technologies in the given application.
- Draft a concept of an overall adaptronic system under conditions which are as close to practice as possible.
- Build a prototype (either active enclosure or active coil structure, or both) including extensive testing.

Aerospace (AP 4.e) Product quality requirements are especially high in aerospace. Orbital structures, for instance, are subjected to temperature cycles from +150 °C to -150°C which must not affect their performance. This means for reflectors that their contour must remain stable under any conditions. Today, this is achieved by employing highly stiff structures which is somehow disadvantageous in terms of lightweight construction. During transport in aerospace or even in passenger aircraft, systems are subjected to high dynamic loads. The adaptive vibration isolation of systems, such as the *adaptive reflector (AP 4.e.1),* is of utmost importance to the user. The scientific and technical objectives of this work package include

- Research the international state of the art as to active satellite structures with piezoceramic actuators and sensors.
- Draft a prototype component to exemplify a satellite reflector so as to test active technologies on the basis of piezoceramic elements. A major point will be adaptive shape control; active vibration damping and damage detection will also be investigated.
- Design a suitable component for active shape control of satellite reflectors to improve the contour accuracy.
- Implement the developed fiber and foil technology in the prototype component.
- Demonstrate the function and assess against analytical predictions.
- Derive operational and cost-effective constructions for implementing the developed technologies in future satellite structures.

4 References

1. D. Sporn and A. Schönecker: Composites with piezoelectric thin fibers - first evidence of piezoelectric behavior. Eingereicht bei: Mat. Res. Innovat., Springer-Verlag (1999)
2. E. Breitbach, H. Hanselka, D. Sachau: Adaptronik – Einsatz multifunktionaler Werkstoffsysteme in einem interdisziplinären Forschungsfeld.Werkstoffwoche „Werkstoffe u. Materialica“ München, 14.10.1998
3. D. Sporn and A. Schönecker: Composites with piezoelectric thin fibers - first evidence of piezoelectric behavior. Innovations in Materials Conference IMC, Washington, DC, USA, 19.-22.7.1998
4. H. Scholz, W. Watzka, P. Vierhaus: Keramische und insbesondere piezoelektrische Mono- oder Multifilamentfasern und Verfahren zu deren Herstellung. DE 19635748 A, offengelegt 5.3.98

Smart Structures based on Thin Piezoceramic Plates

D. Sachau*, P. Wierach*, H. P. Monner*, A. Schönecker**
* German Aerospace Center (DLR), Braunschweig; ** FhG IKTS, Dresden

1 Introduction

The rapid development of industrial products in the fields of traffic technology, mechanical engineering, etc. leads to the demand for higher efficiency of materials and structures. Thus, lightweight design has become very important, mainly to reduce the effects of accelerated mass. However lightweight structures often suffer from vibrational sensitivity, tendency to buckling, and low damage tolerance. A promising way to solve these problems is the use of multifunctional materials [1]. Critical deformations, accelerations or other physical quantities can be detected by integrated sensors, which in combination with suitable real-time controllers can be eliminated with structural conformable embedded or applied actuators and sensors [2]. In this paper thin monolithic piezoceramic wafers are used as structural actuators and sensors. Significant differences between mechanical and thermal properties of the ceramic plates as well as the extreme brittleness of the piezoceramic material demand sophisticated manufacturing techniques to attach these plates on metal substrates or to integrate them into carbon fiber composite structures. A substantial part of the "Leitprojekt Adaptronik" is the development of manufacturing techniques for adaptive structures with regard to industrial requirements like reliability and long term properties. With structures of elementary shape (beam, plate, tube) investigations are made to compare different material systems and control concepts for vibration-, noise- and shape control.

2 Piezoceramic Materials

Because of excellent properties (low energy consumption in quasi static applications, high efficiency, fast response, etc.) lead zirconate-titante piezoceramic materials are of increasing interest. In this paper it will be focused on thin monolithic piezoceramic plates manufactured by CeramTec (Sonox P53, 50x25x0.2mm) that are used as structural actuators and sensors for adaptive structures. Provided with uniformly electroded surfaces these plates are operating in the lateral d_{31} mode (Fig. 1.a).

The performance data of piezoceramic materials given by manufactures are produced with specimens of a special size and shape. To provide a reliable database for the modeling of adaptive structures, it is necessary to measure the parameters of the piezoceramic material with specimens of a similar configuration which will be used in later applications. The shape and size of the specimens have a significant influence on the performance of the piezoceramic material. Fig. 1.b shows the difference between manufacturer data and measured data [3].

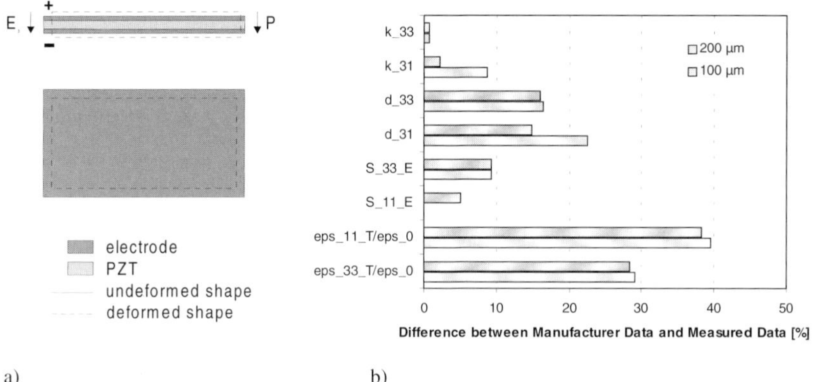

a) b)

Figure 1. Characterization of piezoceramic plates

3 Development of Lightweight Structures with Piezoceramic Plates

Some principle designs of active composites are shown in Fig.2. In Fig. 2a the ceramic plate is embedded between thin non-conductive fiber material (e.g. glass- or polyester fiber) which serve as insulation. The load carrying CFRP-layers above and below the ceramic are used to electrically contact the piezoceramic, therefore the insulation layers are required to avoid short circuits. As the usual disturbing wires are not needed, the manufacturing of the active composite has become easier and there is no additional weakening of the structure.

In structures with arbitrary distributed ceramics carbon fiber bundles are used as a structural conformable way to contact the piezoceramic. Fig 2.c shows a beam structure with two piezoceramics. Each electrode in the active beam is contacted separately. In this way it can be used as strain actuator as well as a bimorph.

In these examples the piezoceramic plates are directly integrated into the fiber composite during the manufacturing process. This procedure has some essential drawbacks. In structures with a great number of piezoceramic elements the danger is high that some ceramic get damaged during the manufacturing process. The result is a failure of the whole structure.

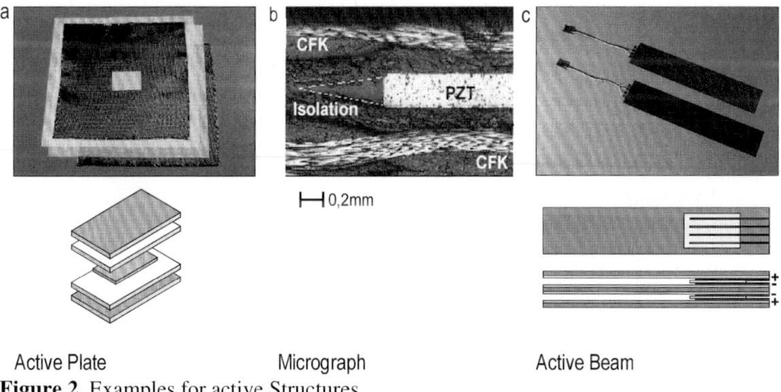

Figure 2. Examples for active Structures

An appropriate solution for this problem is to pre-encapsulate the brittle piezoceramic plates before processing. The PZT can be provided with an mechanical stabilization, electrical insulation and electrical contacts.

Even though an additional manufacturing step is necessary the advantages predominate. To fulfill the different requirements of the industrial partners in this project it was not possible to use existing solutions. Including the entire process of development starting with the manufacturing of the piezoceramic material the goal is to develop new elements with the desired characteristics. Fig. 3 shows the first generation of these elements and their performance as strain actuators. The piezoceramic is embedded between a thin layer of an insulating fiber material (<0,05mm) and a layer of metal mesh to contact the electrodes of the ceramic. These elements are especially optimized with regard to the integration into fiber composite structures and can also be attached on any surface.

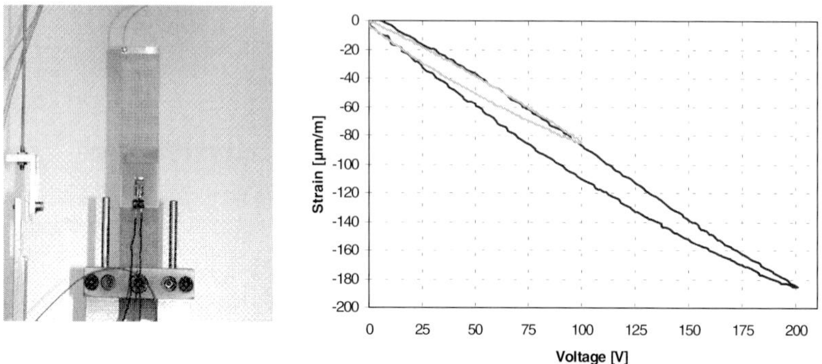

Figure 3. Design and performance of the active elements

4 Manufacturing Technology

Significant differences between mechanical and thermal properties of the ceramic plates as well as the extreme brittleness of the piezoceramic material demands sophisticated techniques for the manufacturing of the active elements. Very good results have been produced with an improved RTM technology, the so-called DP-RTM (Differential Pressure – Resin Transfer Moulding) [4]. This guarantees an extreme high quality and reproducibility of the components. The fiber material is laid out in dry state which facilitates the positioning of the mesh and ceramics. The DP-RTM procedure becomes especially interesting as it is not necessary to provide massive moulds since the clamping forces in the autoclave are created by the differential pressure. Thus, a simple sheet plate can serve as sub mould while a vacuum foil is applied for the upper mould. Fiber volume content and fluid rate can directly be controlled by adjusting differential pressure during the stages of injection. In order to minimize the weight of the active fiber composite a high fiber volume content is required. This can be achieved by increasing the differential pressure. Simultaneously, the increasing mechanical load on the ceramics has to be considered since it might result in mechanical damage of the brittle actuators. The different process parameters like autoclave pressure, autoclave temperature and differential pressure have been optimized during several test procedures.

5 Integration into Adaptive Structures

The multifunctional components described above have both excellent sensor and actuator capabilities. When they are integrated in or applied to a load carrying structure and combined with an appropriate adaptive controller the new class of adaptive structures is created.

In order to make a reliable statement about the performance of these multifunctional components they have to be investigated in theoretical and experimental way in combination with different load carrying structures, controller hardware, and adaptive control algorithms.

The simple geometry of the beam, plate and tube have been chosen as the basic load carrying structures because they allow to compare the different theoretical formulations for the structural model and the control algorithms with the experimental results in a very sufficient way. But also the manufacturing technologies dealing with how to integrate or apply the multifunctional components can be investigated with these basic structures which are demonstrated in Fig. 4.

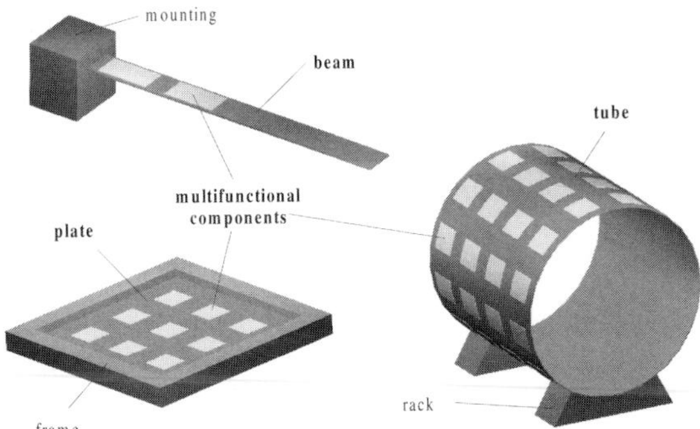

Figure 4. Basic structures

These basic structures have also been chosen because they are fundamental for many industrial applications: Rail vehicles have a tube like structure where significant problems with interior noise occurs. Active noise control with smart structures have the potential to reduce this effect. Problems with interior noise occurs in cars, too. By integrating multifunctional components into the car roof a considerable reduction of noise is expected. This effect can be investigated on plate like structures. The beam represents many 2D problems like a manipulator of automation systems where the accuracy is expected to be improved. These examples only represent a small part of possible applications that can profit from the use of smart structures being investigated on the basis of beams, plates, and tubes.

From these basic structures first the beam structure will be investigated in an extensive test program. For this reason the following descriptions will concentrate on the beam. As can be seen in Fig. 4 the beam has two multifunctional components applied on both upper and lower side an is rigidly bond on one end. The other end is not supported. Different types of static an dynamic tests will be performed in order to be able to quantify the sensor and actuator capabilities of the multifunctional components.

On the basis of the results of this test program new control algorithms will be developed an tested on these structures. The goal is to develop a controller that is capable of performing multiple input and multiple output (MI MO). Especially with regard to the next structures to be investigated, the plate and tube, this controller type is essential.

In order to reach this goal a step-by-step approach will be carried out. First the well known single input and single output controllers (SI SO) will be developed and tested (see Fig.5 (a) and (b)). After this step one more actuator will be added which is why the controller will have to be able to cope with a single input and multiple output (SI MO) as can be seen in Fig.5 (c). Finally one more sensor will be added so that the controller has to be able to perform the wanted MI MO. Similar test programs and approaches for developing controllers will be established for the plate and tube, too.

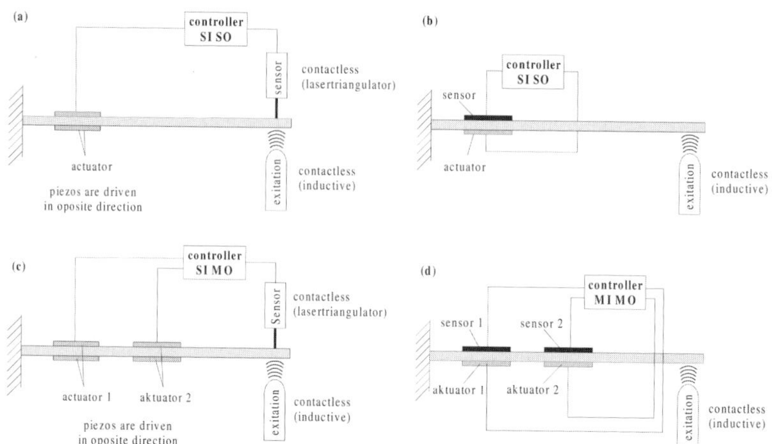

Figure 5. Different control algorithms

6 Conclusions

The new materials and material systems that can be used as multifunctional actuators and sensors with new real-time control concepts and adaptive signal processing, have helped to establish a new class of structures called adaptive structures. The results of recent investigations indicate that these concepts contain very promising potentials for future structural and acoustic purposes. With the research progress and the realization of functional demonstrators as described above by some examples, the intention is now to make this technology available for common industrial applications.

7 References

1. H. Hanselka, H.P. Monner, V. Krajenski, Hochintegreirte Materialsysteme für adaptive Leichtbaustrukturen, 1996, AC-Berlin
2. E.F. Crawley, E.H. Anderson, Detailed Models of Piezoceramic Actuation of Beams, 1989, AIAA Journal
3. Produktinformation CeramTec, Piezoelektrische Bauteile
4. M.Kleineberg, A. Pabsch, C. Siegle, Ermittlung und Charakterisierung der Prozeßparameter eines weiterentwickelten RTM-Verfahrens, DLR-Braunschweig, 1995, IB 131-95/34

Mathematical Modeling and Numerical Simulation of Smart Structures Controlled by Piezoelectric Wafers and Fibers

U. Gabbert[#], W. Kreher*, H. Köppe[#]
[#]Institut für Mechanik, Otto-von-Guericke Universität Magdeburg
*Institut für Werkstoffwissenschaft, Technische Universität Dresden

1 Introduction

Smart structures are characterized by a synergistic integration of active materials into a passive structure connected by a control system to enable an automatic adaptation to changing environmental conditions[1]. The increasing engineering activities in the development and industrial applications of structronic (structure + electronic) systems require effective and reliable simulation and design tools in this field[2,3].

In vibration and noise control as well as in shape control piezoelectric wafers and fibers embedded in composite materials are very common smart structural components in structronic systems. The global behavior of such high integrated complex engineering smart structures can be simulated and designed with sufficient accuracy on the basis of the finite element method extended by control and optimization strategies. In piezoelectric controlled smart composite structures the coupled electro-thermo-mechanical field equations have to be solved, where the parameters of the material tensors are needed as input parameters[9,10]. These macroscopic (homogenized) material tensors of such composites are non-linearly dependent on the properties as well as the arrangements of the constituents in the composite. The measurement of these data is time consuming and expensive. Alternatively, analytical (e.g. based on the Mori-Tanaka-type mean field approach) as well as numerical (e.g. based on the finite element analysis of a representative volume element) methods can be used to calculate homogenized material tensors of a heterogeneous material. The paper first presents the application of the Effective Field Approximation (EFA) for the analytical determination of homogenized material tensors of electro-thermo-mechanical material systems, consisting of a matrix material with piezoelectric inclusions[7] (e.g. in form of thin piezoelectric fibers[4]). Then a brief description of a general purpose finite element package for the simulation of electro-thermo-mechanical coupled field problems is given[9,10]. The presented analytical homogenization strategy can be simply incorporated in the finite element simulation tool. Finally, as a simple test example an actively controlled beam structure is given, where piezoelectric fiber wafers are used as actuators. The homogenized material tensors are calculated first based on the EFA. These data are than used in the finite element simulation.

2 Constitutive Equations

The general linear relationship between the field quantities stress σ, strain γ, electric field E and dielectric displacement D are written here in the form

$$\sigma_{ik} = c^E_{iklm}\gamma_{lm} - e_{lik}E_l - \lambda^E_{ik}\theta \qquad i,k,l,m = 1,2,3$$
$$D_i = e_{ilm}\gamma_{lm} + \varepsilon^\gamma_{il}E_l + p^\gamma_i\theta \tag{1}$$

where we have introduced the standard notation for the linear material property tensors (θ denotes the temperature change). These constitutive equations can be written in the compact form

$$j_p = L_{pq}f_q + j^s_p \qquad p,q = 1,2,\ldots,9 \tag{2}$$

with the 9×9 material property matrix

$$L = \begin{pmatrix} c^E_{\alpha\beta} & e^t_{\alpha\ j} \\ e_{i\beta} & -\varepsilon^\gamma_{ij} \end{pmatrix} \tag{3}$$

This type of constitutive equations can be applied to describe the material behavior of single components as well as of piezoelectric composites.

3 Homogenized Properties of Composites

In order to derive homogenized properties of fiber reinforced materials, we apply here the effective field method. This method is especially useful if we are concerned with a matrix material $\boldsymbol{L}^{(0)}$ which contains inclusions $\boldsymbol{L}^{(k)}$ ($k = 1,2,\ldots,n$). In the theoretical scheme every inclusion is embedded in an infinite matrix medium, and the interaction between different inclusions is projected into an effective field which acts on the considered inclusion. This effective field is then determined in a self-consistent manner[5,6,7]. The result for the effective (homogenized) material properties (here denoted by an upper asterisk) can be written into the following concise form

$$\boldsymbol{L}^* = \left\langle \boldsymbol{A}(\boldsymbol{L},\boldsymbol{L}^{(0)})\right\rangle^{-1}\left\langle \boldsymbol{A}(\boldsymbol{L},\boldsymbol{L}^{(0)})\,\boldsymbol{L}\right\rangle \tag{4}$$

$$\boldsymbol{j}^* = \left\langle \boldsymbol{A}(\boldsymbol{L},\boldsymbol{L}^{(0)})\right\rangle^{-1}\left\langle \boldsymbol{A}(\boldsymbol{L},\boldsymbol{L}^{(0)})\,\boldsymbol{j}^s\right\rangle \tag{5}$$

where the 9×9 matrix

$$\boldsymbol{A}(\boldsymbol{L},\boldsymbol{L}^{(0)}) = \left(\boldsymbol{I} + \left(\boldsymbol{L}-\boldsymbol{L}^{(0)}\right)\boldsymbol{P}^0\right)^{-1} \tag{6}$$

depends on the matrix properties $\boldsymbol{L}^{(0)}$, the inclusion properties $\boldsymbol{L}$, and on $\boldsymbol{P}^0$ which is the generalized Eshelby Tensor formed with the matrix properties (see e.g. Huang and Kuo[8]). This Eshelby tensor depends also on the shape of inclusions which are approximated here as spheroids, where the aspect ratio is used to model different inclusion geometry. The angular brackets denote averaging over all types of components.

4 Finite Element Analysis for Coupled Problems

The basic equation for the finite element analysis of an coupled electro-mechanical field problem can be obtained as virtual energy formulation that combines the Cauchy's equation of motion

$$\sigma_{ij|,j} + \rho\, B_i = \rho\, a_i \tag{7}$$

with the electrical balance equation

$$D_{i|,i} = 0 \tag{8}$$

in form of

$$G = \int_V \left(\sigma_{ij|,j} + \rho B_i - \rho a_i\right)\delta v_k dV + \int_V \left(D_{i|,i}\right)\delta\phi_k dV = 0 . \qquad (9)$$

With the constitutive equations (1), the strain displacement relation $\gamma_{ij} = \frac{1}{2}(u_{i|j} + u_{j|i})$ and the relation between the electric field and the electric potential $E_i = -\phi_{|i}$ we can express equation (9) by the unknown field variables u_i and ϕ. We approximate these field variables in a finite element by shape functions

$$u_i = \sum_{(L)} N_L^{(u)} u_{iL} , \qquad \phi = \sum_{(L)} N_L^{(\phi)} \phi_L \qquad (10)$$

where $N^{(u)}$, $N^{(\phi)}$ are the shape function for the displacements and the electric potential, respectively. Following the standard procedure for the development of finite element equations we get the semi-discrete form of the equations of motion for a coupled electro-mechanical problem in the form[9]

$$\begin{bmatrix} M_{uu} & 0 \\ 0 & 0 \end{bmatrix}\begin{bmatrix} \ddot{u} \\ \ddot{\phi} \end{bmatrix} + \begin{bmatrix} R_{uu} & 0 \\ 0 & 0 \end{bmatrix}\begin{bmatrix} \dot{u} \\ \dot{\phi} \end{bmatrix} + \begin{bmatrix} K_{uu} & K_{u\phi} \\ K_{\phi u} & -K_{\phi\phi} \end{bmatrix}\begin{bmatrix} u \\ \phi \end{bmatrix} = \begin{bmatrix} f_u \\ f_\phi \end{bmatrix} \qquad (11)$$

Based on the above given theoretical background, a library of piezoelectric finite elements has been developed and tested. This library includes solid elements, plane elements, axisymmetric elements, rod elements as well as special multilayer composite shell elements[9,11] and has been implemented in the general purpose finite element system COSAR. For static solutions a specially optimized sub-matrix oriented Cholesky solver can be used. For the solution of eigenvalue problems the subspace iteration method by Mc Cormick&Noe is used. In transient problems modal based techniques as well as time integration schemes such as the Newmark-Method, the Wilson-Method, or the Central-Difference-Method can be applied. The finite element code has the capability to use a substructure technique. This can be utilized as an excellent precondition to solve optimization problems[12] and nonlinear problems effectively. The finite element software can also be used to simulate controlled structures. A data interface between the finite element software and control design tools such as MATLAB/SIMULINK has been developed to design controller for real engineering applications.

5 Numerical Results

First, homogenized material properties of composites made of piezoelectric fibers embedded in a polymer matrix are calculated on the basis of the effective field method presented. The electromechanical material properties of each of the two constituents of the composite as well as the calculated homogenized data for a special composite consisting of a 30% fiber volume fraction and a fiber aspect ratio of 50 are given in the Table 1.

Table 1. Electromechanical properties of components and a special composite (fiber volume fraction 30%, aspect ratio 50)

	c^E_{11} GPa	c^E_{12} C/m^2	c^E_{13} nF/m	c^E_{33}	c^E_{44}	e_{31}	e_{33}	e_{15}	ε^γ_{11}	ε^γ_{33}
PZT-fibers	106.0	58.2	57.4	90.2	20.0	-4.1	12.1	8.6	14.0	13.0
Polymer	6.4	3.5	3.5	6.4	1.5	0	0	0	0.04	0.04
Composite	9.8	5.0	5.2	22.1	2.6	-0.1	3.3	0.0	0.08	3.1

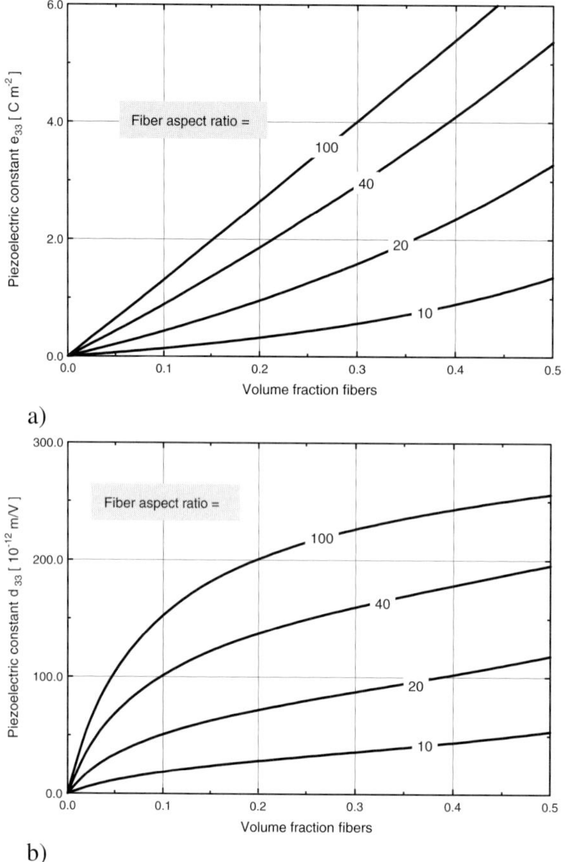

Figure 1. Predicted piezoelectric constants e_{33} (a) and d_{33} (b) for a composite made of piezoelectric fibers in a polymer matrix

In Figures 2a and 2b results are presented for different volume fractions and fiber aspect ratios which show that a high fiber volume fraction and a large aspect ratio is required to get the desired performance of the composite. This is especially important when the e_{33}-effect has to be utilized, but is not so critical for the d_{33}-effect. Note that all calculations have been performed with an identical poling state of the fibers as shown in Table 1. If the poling of the composite is done "in-situ", the resulting poled fiber properties may vary in the same manner as the effective properties of the composite, which can be also calculated within the same theoretical approximation scheme. The main advantages of fiber composites are the better structural conformity in comparison with PZT wafers (see the low stiffness properties in Table 1) as well as the an-isotropic sensing and actuation behavior.

The homogenized material properties were used for a numerical simulation. As test case a beam is used with two attached piezoelectric fiber composite layers at top and bottom of the beam (see Figure 2), where these layers consist of the homogenized material properties given in Table 1. The actuators are controlled with a time depending electric potential $\phi(t) = \phi_0 \sin \Omega t$ with $\phi_0 = 100V, \Omega = 100 s^{-1}$. In Figure 3 the displacement response of a point at the free end of the beam is shown.

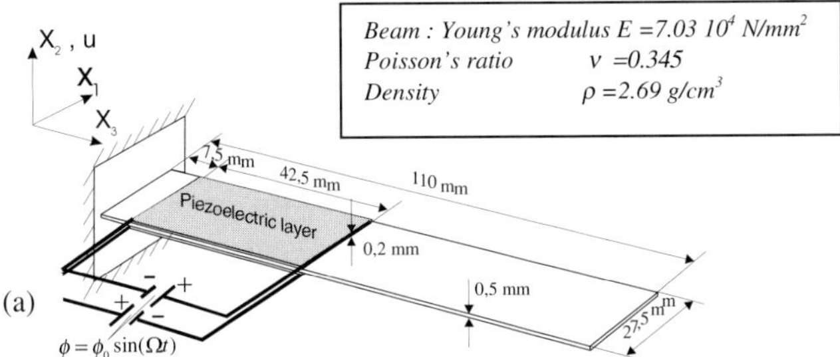

Figure 2. Elastic beam controlled by piezoelectric composite fiber paches

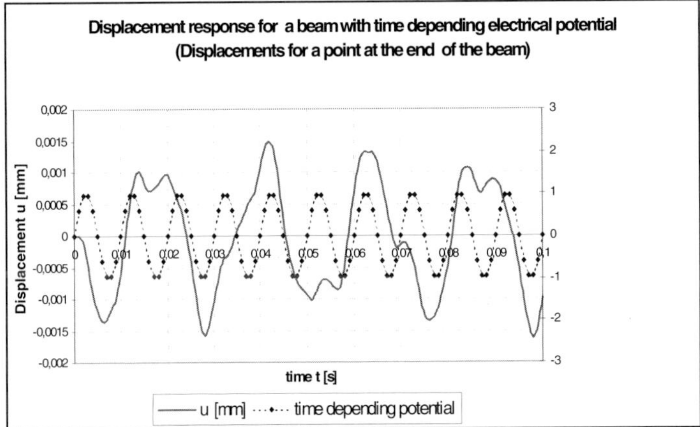

Figure 3. Response of the beam in the time domain

6 Conclusions

A simulation concept for piezoelectric controlled smart structures has been presented, where in the first step a homogenization method based on the Effective Field Approximation is given. In the second step the homogenized material tensors are used in the finite element simulation of the global structural behavior. I the near future this solution will be used to optimize smart material systems with respect to the overall performance of structronic systems.

7 Acknowledgment

This work has been supported by the German Ministry for Science and Technology (BMBF) and the German Research Society (DFG). This support is gratefully acknowledged.

8 References

1. H. S. Tzou, G. L. Anderson (Eds.), Intelligent Structural Systems, Kluwer 1993
2. Tzou, H.-S., Guran, A. (Eds.), Structronic Systems: Smart Structures, Devices and Systems, Part 1: Materials and Structures, Part 2: Systems and Control, World Scientific, 1998
3. U. Gabbert (Ed.), Modelling and Control of Adaptive Mechanical Structures, Fortschr.-Ber. VDI Reihe 11, Nr. 268, Düsseldorf, VDI Verlag 1998
4. D. Sporn, A. Schoenecker, Composites with piezoelectric thin fibers – first evidence of piezoelectric behaviour, Mat Res Innovat (1999) 2:303-308
5. S. K. Kanaun, V. M. Levin, Effective field methods in mechanics of matrix composite materials, in: Advances in mathematical modelling of composite materials (ed: K. Z. Markov), Series on Advances in Mathematics for Applied Sciences, vol. 15, pp. 1-58, World Scientific Publ. Comp., Singapore, 1994.
6. V. M. Levin, The overall properties of piezoactive matrix composite materials, in: Continuum Models and Discrete Systems (ed: K. Z. Markov), Proc. of the 8th Int.Symposium, Varna 1995, pp. 225-232, World Scientific Publ. Comp., Singapore, 1996.
7. V. M. Levin, M. I. Rakovskaja, W. S. Kreher, The effective thermoelectroelastic properties of microinhomogeneous materials, Int. J. Solids & Structures, 36 (1999) 2683-2705.
8. J. H. Huang, W.-S. Kuo, Micromechanics determination of the effective properties of piezoelectric composites containing spatially oriented short fibers, Acta mater., 44 (1996) 12, 4889-4898
9. H. Berger, X. Cao, H. Köppe, U. Gabbert, Finite Element Based Analysis of Adaptive Structures, in Fortschr.-Ber. VDI Reihe 11, Nr. 268, Düsseldorf, VDI Verlag 1998, pp. 103-114
10. H. Köppe, U. Gabbert, H. S. Tzou, On Three-Dimensional Layered Shell Elements for the Simulation of Adaptive Structures, in Fortschr.-Ber. VDI Reihe 11, Nr. 268, Düsseldorf, VDI Verlag 1998, pp. 103-114
11. U. Gabbert, A. Görnandt, H. Köppe, F. Seeger, Benchmark Problems for the Analysis of Piezoelectric Controlled Smart Structures, ASME 1999, Design Engineering Technical Conferences DETC'99, Las Vegas, Nevada, September 12-16, 1999, paper: DETC99/VIB-8391
12. U. Gabbert, C-T. Weber, Analysis and Optimal Design of Piezoelectric Smart Structures by the Finite Element Method, Proceedings of the European Conference on Computational Mechanics, August 31 – September 3, Munich, 1999, 14 pages

An On-line Design and Simulation Procedure for the Development of Instationary or Nonlinear or Adaptive Systems

J. Melcher and A. Büter
German Aerospace Center (DLR), Institute of Structural Mechanics, Germany

1 Abstract

This paper describes an on-line design and a simulation procedure that allows rapid prototyping of time-variant systems, instationary processes and non-linear structures like adaptives systems.

The proposed procedure is mainly based on digital MX filters whose coefficients vary with time in such a way that even structures can be modelled whose eigenvalues change very fast. In this paper two examples are carried out:

Firstly an on-line or real-time system identification of activatable materials that are implemented into mechanical structures and secondly a brand-new on-line helicopter model that considers the structural dynamics, the aeroelasticity and the aerodynamics of the rotor blades.

Using these MX filter based models a decisive progress in R&D can be achieved:

Materials with time variant properties can be modelled, e.g. parameters and form of the constitutive equations can be derived.

Wind tunnel test of helicopters can be confirmed. Instationary processes can be interpreted. There is the possibility to get more understanding of the physics of non-linear and adaptive systems. The description of the proposed procedure is analytical, numerical and experimental.

2 Adaptive Modelling with Adaptive MX Filters

The concept of the proposed online identification tool is based on the procedure of adaptive modelling, as shown in Fig. 1. Both the system to be identified and the model realized by a digital MX filter system are fed identically by external forces. A corresponding adaptation algorithm, the MX-LMS filter system, calculates the best model parameters (in this case filter coefficients) to minimize the error signal, which is the difference between the measured and adapted response signals. A block of transformations contains few mathematical operations that calculate the modal parameters from the adapted coefficients.

The two filter systems, the MX, and the MX-LMS, are the main ideas of the new identification method. The MX filter system is based on digital lattice-typed filterings. It represents a linear dynamic system with any modal density. At each time step it is iteratively corrected by the MX-LMS filter system. The latter needs data from the model in order to copy new or better coefficients to the MX filter. The adaptation algorithm yields the information as to whether or not the model is already operating perfectly from the error signal.

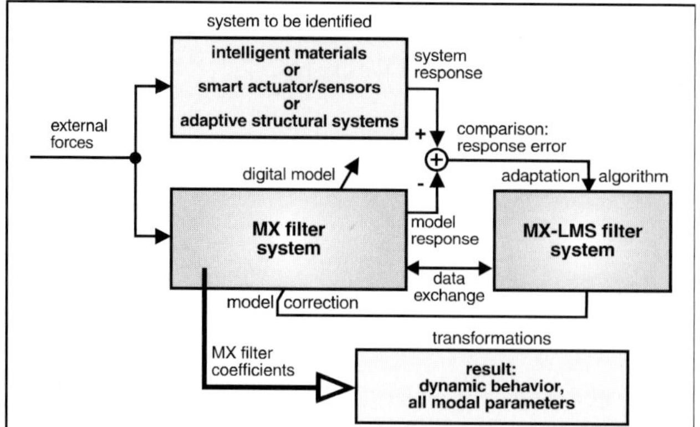

Figure 1. Adaptive modelling with MX filters for online system identification.

The signal flow and form of the MX filter system are shown in Fig. 2 (Melcher, 1994). Calling it after its property (modally formulated due to the parallel form) and its appearance (cross-coupled structure), it was named MX.

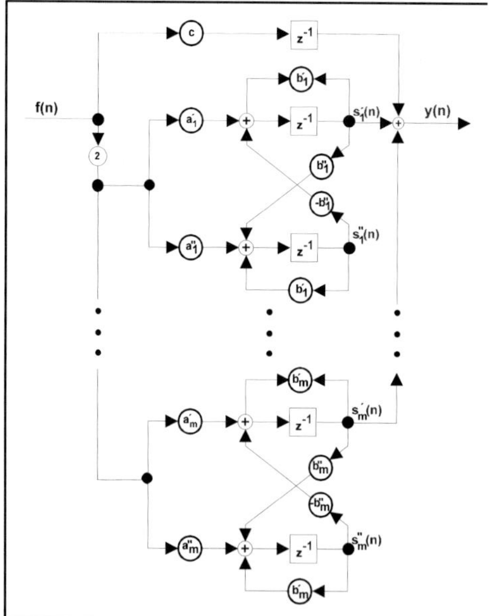

Figure 2. Signal flow of the MX filter system representing m modes.

In the discrete form presented here, the filter is non-canonical (the number of delays z^{-1} is higher than the order of the represented transfer function). In cases of m modes it has (1+4m) real coefficients: the transversal coefficients c, a_i', a_i'' and the recursive coefficients b_i' and b_i'', whereas $i \in [1,m]$. The functional relationship between the input samples f(n), (n is the discrete integer time step) and the resulting output samples y(n) can be described in different ways. The filter difference equations are

$$
\begin{aligned}
s_0(n) &= c(n-1)\, f(n-1)\\
s_i'(n) &= 2a_i'(n-1)\;\; f(n-1) + b_i'(n-1)\, s_i'(n-1) - b_i''(n-1)\, s_i''(n-1)\\
s_i''(n) &= 2a_i''(n-1)\;\; f(n-1) + b_i'(n-1)\, s_i''(n-1) + b_i''(n-1)\, s_i'(n-1)\\
y(n) &= s_0(n) + \sum_{i=1}^{m} s_i'(n)
\end{aligned}
\tag{1}
$$

The corresponding transfer function is

$$
H(z) = \left[c(n) + \sum_{i=1}^{m} \left(\frac{a_i(n)}{1-b_i(n)z^{-1}} + \frac{a^*{}_i(n)}{1-b^*{}_i(n)z^{-1}} \right) \right] z^{-1} \quad . \tag{2}
$$

with the substitutions $a_i(n) := a_i'(n)+j\, a_i''(n)$ and $b_i(n) := b_i'(n)+j\, b_i''(n)$ for all $i \in [1,m]$, where the sign * denotes complex conjugation. Eq. 2 clearly demonstrates three essential properties: The MX filter system represents only systems that can be formulated mathematically with conjugate complex pairs of residues a_i and poles b_i. Furthermore, in practice, the stability of this filter system can be proved easily by taking the conditions $b_i'^2(n)+b_i''^2(n) < 1$ for all $i \in \left[1,m\right]$ at each time step n. Finally, H(z) is modelling a structural transfer function multiplied by z^{-1}. This means that the structural response must be time-delayed with z^{-1} for the calculation of the error signal. The coefficients c(n) only play an important role if accelerations are measured to yield the structural responses. The most often used cost function in adaptive signal procesing, and the one to be minimized here, is the mean square error function $\psi(n) = E\left[e^2(n)\right]$ with E[...] as the statistical expectation operator and e(n) as the measured discrete error signal. This performance function defines a transformation from a vector space spanned by the MX filter coefficients into the space of a real scalar. Fig. 3 examplarily shows the surface of a typical mean square error function using MX filters.

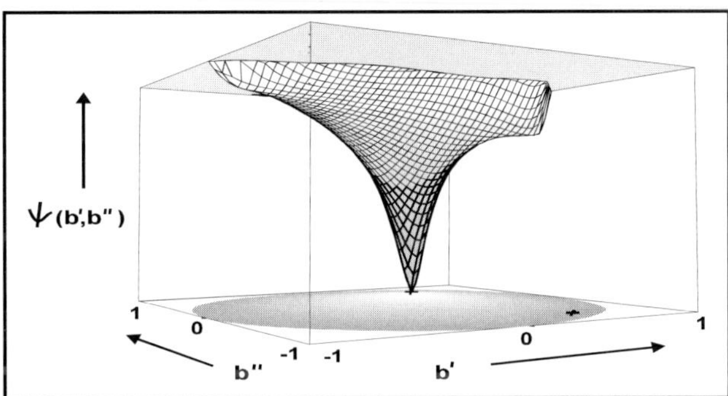

Figure 3. Performance function inside of the stability circle.

The MX filter system is optimized by finding the coefficient vector

$$
w(n) = \left[c(n) \;\; a_1'(n) \;\; a_1''(n) \;\; b_1'(n) \;\; b_1''(n) \;\; \ldots \;\; a_m'(n) \;\; a_m''(n) \;\; b_m'(n) \;\; b_m''(n) \right]^t
$$

that minimizes the mean square error criterion. The search for optimal values that is proposed here is the iterative method of the negative gradients $w(n+1) = w(n) - \mu \cdot \nabla\psi(n)$. In this adaptation formula, the variables are

$$\mu = \mathrm{diag}\left[\mu_c \quad \mu_{a_1'} \quad \mu_{a_1''} \quad \mu_{b_1'} \quad \mu_{b_1''} \quad \cdots \quad \mu_{a_m'} \quad \mu_{a_m''} \quad \mu_{b_m'} \quad \mu_{b_m''}\right]^t$$

as the diagonal matrix of convergence parameters which define the compromise between convergence accuracy and convergency rate, and $\nabla\psi(n) = \nabla e^2(n) = -2\,e(n)\cdot\nabla y(n)$ as the estimated spatial gradient of the coefficient-dependent performance function. The derivatives of the MX filter system output y(n) are

$$\nabla y(n) = \left[\frac{\partial y(n)}{\partial c} \quad \frac{\partial y(n)}{\partial a_1'} \quad \frac{\partial y(n)}{\partial a_1''} \quad \frac{\partial y(n)}{\partial b_1'} \quad \frac{\partial y(n)}{\partial b_1''} \quad \cdots \quad \frac{\partial y(n)}{\partial a_m'} \quad \frac{\partial y(n)}{\partial a_m''} \quad \frac{\partial y(n)}{\partial b_m'} \quad \frac{\partial y(n)}{\partial b_m''}\right]^t \tag{3}$$

$$= : \left[\gamma(n) \quad \alpha_1'(n) \quad \alpha_1''(n) \quad \beta_1'(n) \quad \beta_1''(n) \quad \ldots \quad \alpha_m'(n) \quad \alpha_m''(n) \quad \beta_m'(n) \quad \beta_m''(n)\right]^t .$$

With the further substitutions

$$\begin{aligned} \alpha_{0,i}'(n) &:= \frac{\partial s_i''(n)}{\partial a_i'} \quad \text{and} \quad \alpha_{0,i}''(n) := \frac{\partial s_i''(n)}{\partial a_i''} \\ \beta_{0,i}'(n) &:= \frac{\partial s_i''(n)}{\partial b_i'} \quad \text{and} \quad \beta_{0,i}''(n) := \frac{\partial s_i''(n)}{\partial b_i''} \end{aligned} \tag{4}$$

for all $i \in [1,m]$, the MX-LMS algorithm becomes

$$\begin{aligned}
&1) \quad \gamma(n) = f(n-1) \\
&2) \quad \alpha_i'(n) = 2\,f(n-1) + b_i'(n-1)\,\alpha_i'(n-1) - b_i''(n-1)\,\alpha_{0,i}'(n-1) \\
&\qquad \alpha_{0,i}'(n) = b_i'(n-1)\,\alpha_{0,i}'(n-1) + b_i''(n-1)\,\alpha_i'(n-1) \\
&3) \quad \alpha_i''(n) = b_i'(n-1)\,\alpha_i''(n-1) - b_i''(n-1)\,\alpha_{0,i}''(n-1) \\
&\qquad \alpha_{0,i}''(n) = 2\,f(n-1) + b_i'(n-1)\,\alpha_{0,i}''(n-1) + b_i''(n-1)\,\alpha_i''(n-1) \\
&4) \quad \beta_i'(n) = s_i'(n-1) + b_i'(n-1)\,\beta_i'(n-1) - b_i''(n-1)\,\beta_{0,i}'(n-1) \\
&\qquad \beta_{0,i}'(n) = s_i''(n-1) + b_i'(n-1)\,\beta_{0,i}'(n-1) + b_i''(n-1)\,\beta_i'(n-1) \\
&5) \quad \beta_i''(n) = -s_i''(n-1) + b_i'(n-1)\,\beta_i''(n-1) - b_i''(n-1)\,\beta_{0,i}''(n-1) \\
&\qquad \beta_{0,i}''(n) = s_i'(n-1) + b_i'(n-1)\,\beta_{0,i}''(n-1) + b_i''(n-1)\,\beta_i''(n-1) \; .
\end{aligned} \tag{5}$$

These terms are the difference equations of further MX filters with the required derivations of gradient components as the outputs. The performance of the combination of both filter systems MX as well as MX-LMS is illustrated in Fig. 4 and 5. The contours correspond to the performance function of Fig. 3. In the simulation of Fig. 4 it is assumed that all transversal coefficients are already optimized and that they can be held constant. Remarkably, the estimated and the analytical track clearly show how proper the desired gradients are estimated. In a further simulation, see Fig. 5, all MX filter coefficients are adaptive using a constant convergence matrix μ.. For a better orientation, the contours implemented in this plot are the same as those in the prior test, even though they correspond to only two dimensions. Thus, the curves of both the adapted and the desired track do not perpendicularly cross the contours, but both curves should be approximately equally apart from the noisy characteristics. In this typical case the fastest performance has nearly 200 iterations using the analytical gradient and nearly 600 iterations for the MX-LMS algorithm. The parameter of the unknown structure are identified within one second.

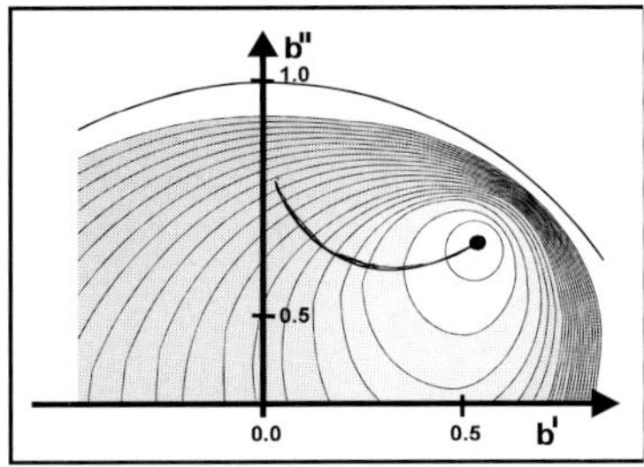

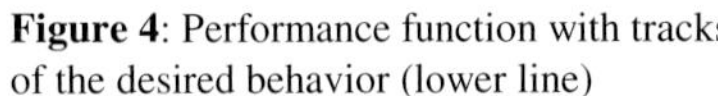
Figure 4: Performance function with tracks of the desired behavior (lower line)

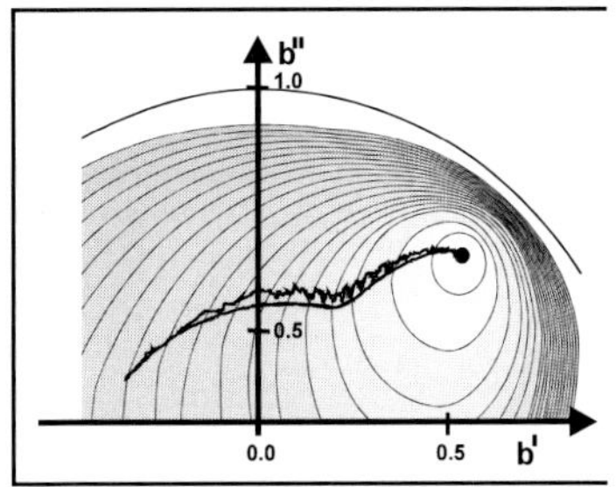

Figure 5: MX-LMS performance when all coefficients are adaptive.

A comparison between the modal description of the dynamic system to be identified and the model description leads to the time-dependent relation between the system parameters as eigenfrequencies f in Hz and damping factors:

$$f_i(n) = \frac{1}{2\pi\tau}\sqrt{\frac{1}{4}\ln^2\left(b_i'^2(n)+b_i''^2(n)\right) + \arctan^2\left(\frac{b_i''^2(n)}{b_i'^2(n)}\right)}$$

$$\delta_i(n) = -\frac{\ln\left(b_i'^2(n)+b_i''^2(n)\right)}{2\sqrt{\frac{1}{4}\ln^2\left(b_i'^2(n)+b_i''^2(n)\right) + \arctan^2\left(\frac{b_i''^2(n)}{b_i'^2(n)}\right)}} \tag{6}$$

for all $i\in[1,m]$, τ is the sampling rate in seconds. These are all neccessary mathematical calculations summarized in the transformation block of Fig. 1.

3 Experiments

To cope with the practical aspects, special consideration is devoted to experimental investigations of the application of the proposed tool to a dynamic structure with time-varying parameters, see Fig. 6. The structural system, whose eigenfrequencies and damping factors should be identified, consist of a simple carbon-fiber-composite beam with 3 PZT patches embedded in the neutral plane. The length of the beam in the clamped-clamped configuration is 47cm, its thickness is 0.72mm,and its width is 25mm. The dimensions of PZT is 25mm $\times$50mm $\times$0.2mm.

For transfer function measurements, a usual electrodynamic shaker is attached near the middle of the beam. An accelerometer is located on the left half of the beam, observing bending eigenfrequencies. When the PZTs which embedded in the beam are applied different voltages, the modal frequencies and modal damping of the beam are different. The Fig. 8 shows that the modal frequencies are changed, when the PZTs are applied 0V and 300V. The external force *f(n)* was zero-mean white noise. The beam's power spectrum density magnitude and modal frequencies are changed with the sine angle of the voltage. Fig. 8 shows the time varying modal frequencies and modal damping, when the PZTs are applied sinoidal voltage.

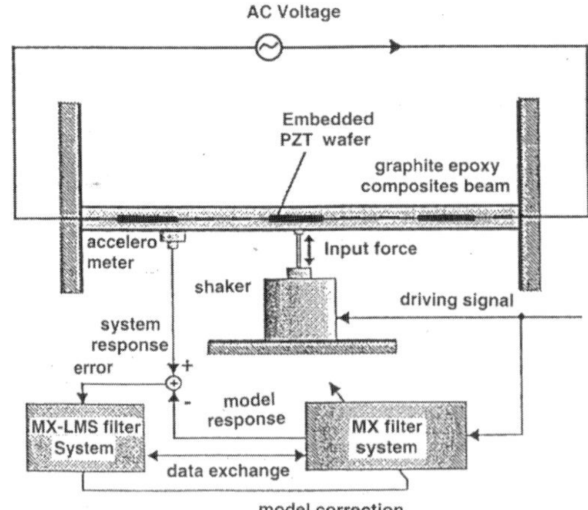

Figure 6. Experimental set-up for online system identification of a "smart" beam.

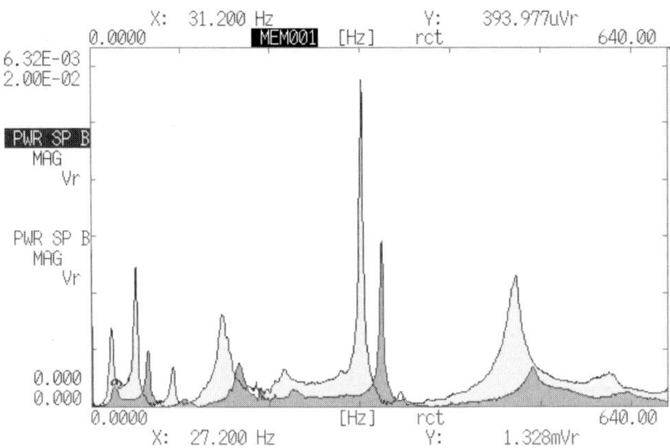

Figure 7: The Beam's Modal Frequencies changed with Voltage.

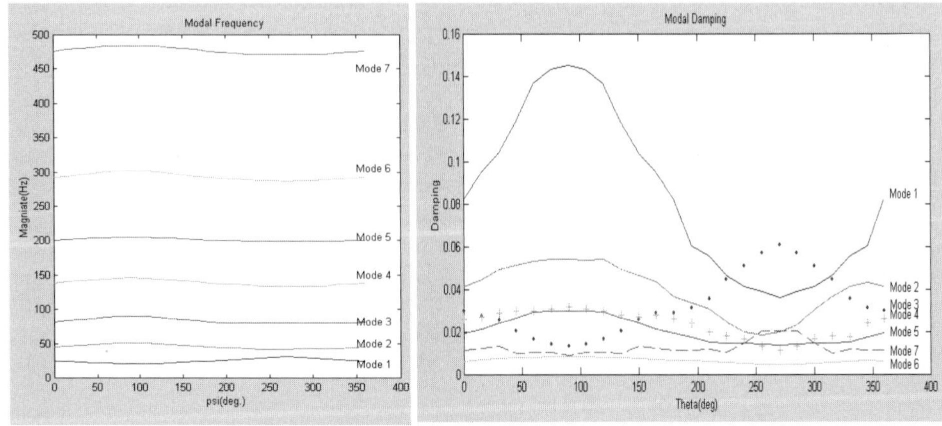

Figure 8: Modal frequencies (left) and damping factors (right)

4 Simulations

Matlab/Simulink was used for modeling the dynamic system. It can be found that the coefficients of each section of the MX-filter can be determined by each modal feature (modal frequency, modal damping ratio, modal mass and shapes) of the beam. The rotating beam has two bending motions and a torsion yielding six MX filter sections. Assuming that the relationship between azimuth angle with eigenfrequencies and damping of rotating beam are shown in Fig. 9.

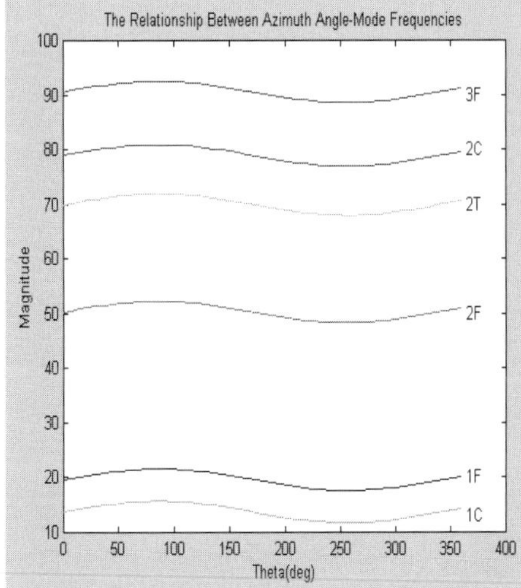

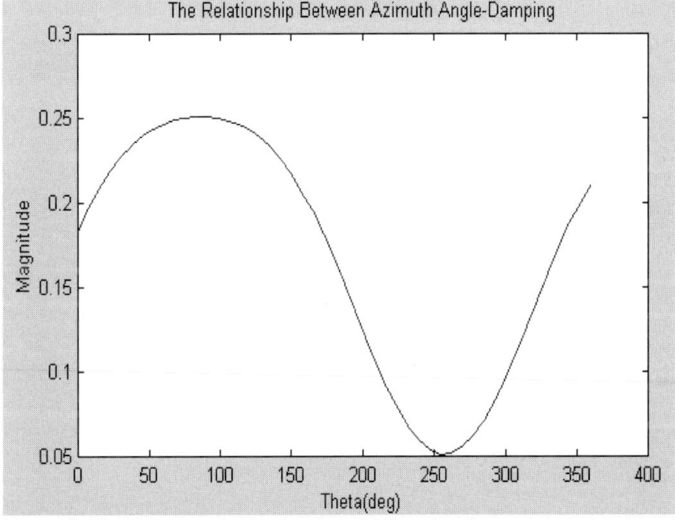

Figure 9: Modal frequencies (top) and modal damping (bottom) vs. azimuth angle.

Fig. 10 shows the response and power spectrum density of the rotor mode excited by dual-frequency, the frequencies are 70Hz, and 140Hz. From the Fig. 10, it can be seen that many eigenfrequencies have been excited except for the frequencies 70Hz and 140Hz, so the rotating beam should be controlled during the operation.

The experiments demonstrate that the results of using MX-filters to model and simulate a dynamic system are same as that of using a FFT analyzer. It verifies that using MX-filters can model and simulate the nonlinear, time-vary structure dynamic system, such as rotating beam.

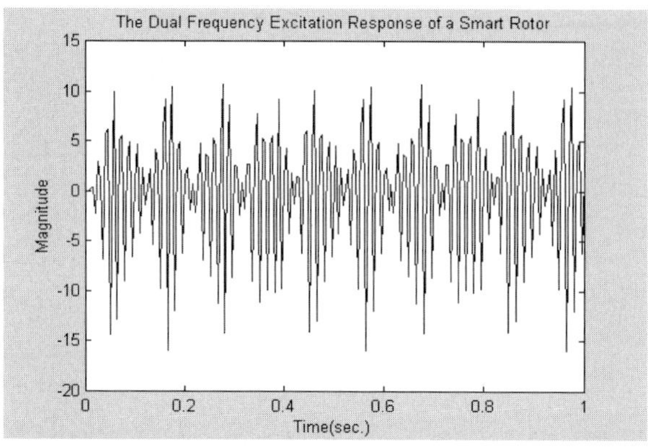

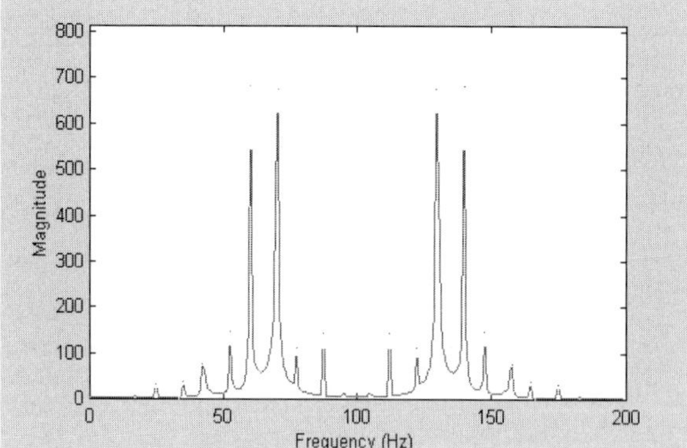

Figure 10: FRF and PSD of the smart rotor mode excited by dual frequencies.

5 Reference

1. Melcher, J. 1994, "MX Filters: A New Tool For Performance Tests of Adaptive Structural Systems ", Journal of Int. Mat. Syst. And Struct., Vol. 5, Nov., 1994.

On the Application of Smart Materials in Automotive Structures

Holger Manz, Knut Schmidt
Volkswagen AG Wolfsburg

1 Abstract

The demand in the automobile sector for greater comfort in the vehicle is of a high importance alongside the requirements for a low emission of pollutants. This applies in particular to the vehicle's acoustics as perceived in the first instance by the passengers in the vehicle. Measures presently employed to reduce the noise levels by using additional damping material are associated with a higher structural weight. Alternatives for the future are therefore being searched for in combination with the secondary condition of realizing a low structural weight. It is for this reason that the application of so-called intelligent materials is appropriate since these can be used to realize an overall adaptive system.

The term adaptronics is generally understood as being the process of creating "more intelligent" structures on the basis of "multi-functional" elements. Multi-functional elements are thereby characterized by the property that these contain at least two functional components (actuator, sensor or controlling mechanism). Included in such multi-functional elements are for instance, shape-memory alloys, bimetals, piezo-electric components, etc. There are many uses for adaptronic elements in automotive applications. Automatic inside day/night mirrors and driving-dynamics control systems are just two examples mentioned here. There are diverse possibilities available for acoustics applications in the interior of the vehicle. Alongside adaptronic systems for the chassis, the use of active vibration compensators for the roof-panel constitutes a very promising approach for improving the acoustics in the interior of the vehicle.

2 Introduction

Alongside the price aspects like fuel consumption, weight, safety, comfort and emission of pollutants will more and more influence the market of a new car in the future. The purchase decision is going to be vital at the fuel consumption and comfort. In order to realize a lower fuel consumption, both, the development of engines with a high efficiency and the lightweight design have to be considered as well. Beside this a higher comfort is mostly coupled with additional aggregates and isolation materials and consequently a higher weight. Adding material is contrary to lightweight design. A solution could be the use of intelligent materials because of their multifunctional properties. They are characterized by their ability to be used in shape with the structure as an actuator as well as a sensor.

Beside other smart materials like shape-memory-alloys the piezoelectic ceramics seem to be the most promisingly ones for the use in automobile structures because of their high manipulating speed combined with a sufficient accuracy. The piezo-electric effect concerns the property of the material that this can set up an electric field by the polarization caused by an applied deformation. An inverse effect of this is given when an electric field is applied to the piezo-ceramic material. The material deforms as a result of applying the electric field. The following figure provides various designs of piezoelectric actuators.

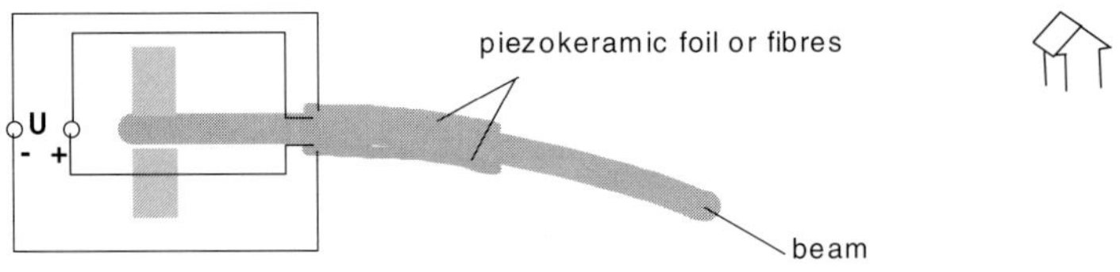

laminar-design stacked design (Ceramtec)

Figure 1. Design of piezoceramic actuators

As shown in figure 2 the use of piezoelectric actuators can be very diverse.

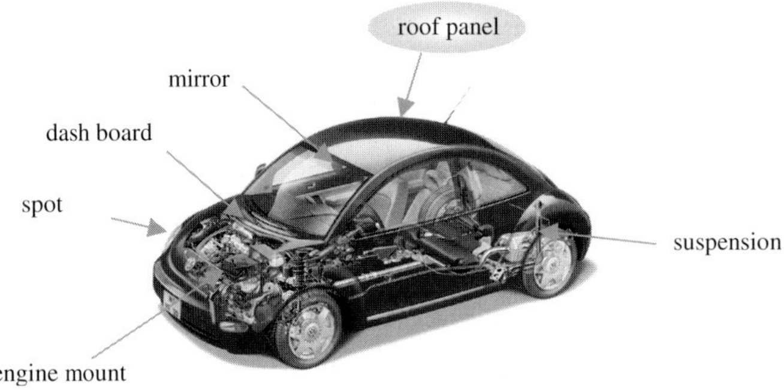

Figure 2. Application of piezoelectric actuators in the automobile

Looking at figure 2 two main principles can be carried out. On the one hand there is the application of piezoceramics at the enginge mount or the suspension. The aim of this configuration would be to decouple the body of the engine or suspension, preventing any vibrations from the engine the suspension to be transferred to the body of the car. The aim of the remaining applications is to reduce the vibrations of the body itself by active vibration control of the body. The interface between body an engine for example is not discussed here.

3 Active Vibration Control of the Roof Panel

3.1 Previous concepts of interior noise reduction

With regard to the roof panel of an automobile the active vibration control is explained in the following sections.

Noise reduction is one of the main tasks an acoustical engineer is concerned with today. Until today the use of isolation materials, e.g. at the roof, has proved to be evective enough.

Besides the advantage of a passive system, to be inexpensive and with no need of any maintenance this concept becomes less effective for applications in higher class automobiles because of its mass and unsufficient noise reduction. Considerations have been made in the past to use anti-noise to reduce the unwanted sound pressure. Anti-noise in this case means the targeted use of loudspeakers to generate a noise field that is phase-shifted by 180° from the original noise field, such that the amplitudes are mutually canceled out. In the case of a permanently located listener, that is to say a passenger retaining his position in the vehicle for example, these considerations could lead to improvements in the acoustics. The use of anti-noise from loudspeakers is not however purposeful since there are areas in the vehicle where the amplitudes cannot be eliminated but rather are amplified. It therefore appears all the more important to reduce the development of the interfering noise waves or in other words, to intervene in the oscillation behavior of the structure. By the way controlling the loudspeakers costs much effort and consequently a lot of money.

3.2 Active vibration control of the roof panel

Concerning the effectiveness, weight and costs most of the concepts up to now have proved to be unsuitable. That is why the use of piezoceramic actuators promises more success. Their size and electrical control capabilities provide a simple possibility of active vibration control. The roof panel of the automobile proves suitable for preliminary investigations since this constitutes an oscillating membrane and acts as a loudspeaker when excited by structural components oscillating in the immediate vicinity.

For the roof-application the so-called piezoceramic foils will be used. Figure 2 illustrates the basic concept for the adaptronic roof panel.

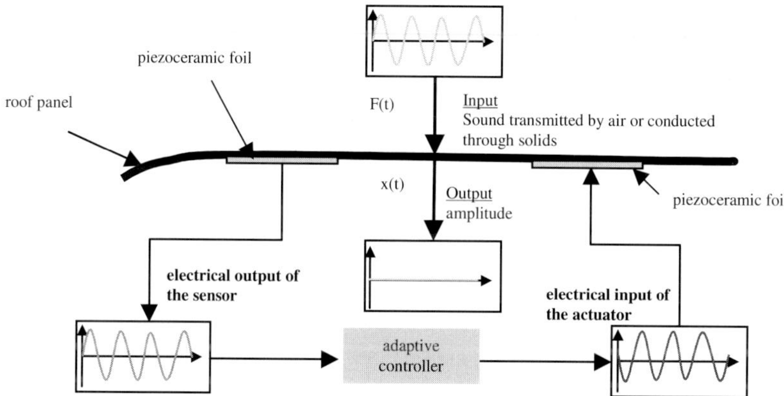

Figure 3. Basic concept for the adaptronic roof panel

The aim of the application is the vibration control of the roof panel, which is excited by the engine, suspension or the air. The sensor measures the vibration and sends an electrical signal to the controller. The control output is then impressed on the actuator.

The size of the piezoceramic foils is given by 25 mm·50 mm·0,2 mm. By the size of the foils it already becomes clear that one aim should be to place the foils as patches at suitable locations on the roof panel.

Therefore the knowledge of the best locations for the patches is necessary. In order to reach this aim the magnitude of the vibration has to be assessed and it has to be observed whether the vibrations correlate with the interior noise or not. For this a measurement of a Passat Variant had been carried out. Depending on the engine speed, which had been determined by a rise of the interior sound pressure, the transverse wave velocity of the roof panel had been measured by the use of a scanning vibrometer. Due to the size of the scanning area, the measurement of the roof panel had to be divided into three sections.

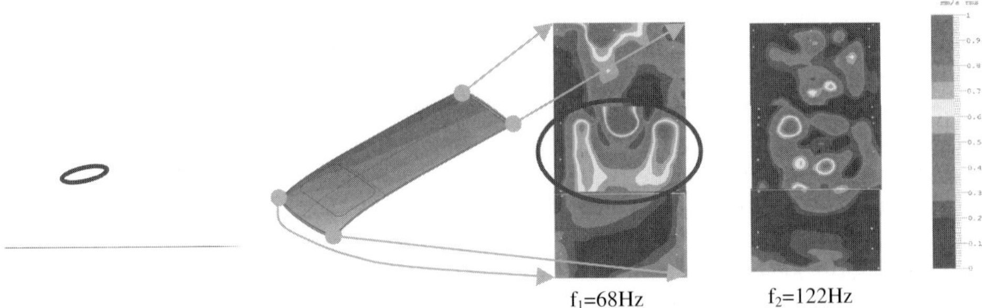

Figure 4. Transverse wave velocity of the Passat Variant

Looking at figure 4 the sections, which do not exactly fit, become evident. A reason for this may be the influence of the temperature of the car. Different temperatures may cause thermal stresses which then modify the vibrational characteristics of the roof panel. As seen in Fig. 4 the section with the largest deflections can be found at the section of roof panel, which is near the passengers ear in the back of the car.

3.3 Feasibility study

The determination of a suitable location for the piezoceramic foils is one major task. On the other hand the assessment of the required power, weight and size of the adaptronic system is vital at further investigations. In order to get the desired information a study of a flat 1m·2m·0,9mm steel plate was carried out. 6 actuators were bonded only on one side of the roof panel. The foils were controlled individually using the conventional car power supply. The frequency of excitation was set to be at 50Hz with a maximum amplitude of ±0,2 mm.

With these boundary conditions the required power of the piezoceramics foils was 69,25 mW, including the controller it was 6 W.

This result shows small power requirement. One should keep in mind that this value is based on a flat plate. A curved plate would lead to much higher power inputs. Nevertheless in conclusion the required power fits in a range which might be acceptable for the application in an automobile.

3.4 Material properties and requirements on piezo-materials

In Figure 1 the design of piezoceramics had already been carried out. The ceramic is made of lead-zirkonate-titanate. The piezoelectric effect is created in piezo-ceramic materials above a higher by polarizing the ceramic. The piezo-material has an Young´s modulus of 70000 N/mm², which is comparable to that of aluminium. Its density is given by 7800 kg/m³, which is same as the density of steel. The maximum strain to be actuated at an electric field-strength of 2 kV/mm is about 0,1 %. As seen at the density the ceramic is characterized by an important disadvantage, namely its weight. Therefore the number of patches to be used must be well assessed. Furthermore the maximum strain seems to be very small. With regard to the bonding to the roof panel, this strain is sufficient enough because of the distance between the median plane of the foil and the neutral plane of the sandwich.

Apart from the material properties several boundary conditions for the use of piezoelectric actuators have to be determined. One of the most important boundary conditions is use of the piezocereamic as a multifunctional element, which means to be used as an actuator, sensor and may be controlling mechanism as well. This will keep the size as small as possible. The desired size of the patches has already been pointed out in chapter 2.2. Due to the possibility of bonding the ceramics to a curved shape the realization of a radius of curvature of 1000 mm should be guaranteed. Furthermore the piezo-material should cover a specified frequency band. In the function as a sensor the frequency ranges from 1 to 3000 Hz. The frequencies concerned with the actuation can be divided in three sections. The excitation of the suspension reaches frequencies about 1-50 Hz. The engine is characterized by a frequency band of 20-200 Hz and the excitation by air reaches frequencies about 20-3000 Hz. Summing up the actuator should be able to manage the same frequencies as the sensor.

Another boundary condition is given by the operating temperature from –40°C to +100°C at atmospheric pressure and hygroscopic moisture <90 %. Looking at the temperature, the maximum of 100°C seems to be very high, but thinking of a dark coloured car under a high solar radiation this value might be realistic. The endurable temperature during the car production process should be about 200°C, which means that the polarization of the ceramic should not be influenced during the process of production.

Further requirements are given by the transverse forces of the roof panel, which are supposed to be about 200 N, combined with maximum accelerations of 130 g for the active actuator and 65 g for the passive actuator. In order to keep the ceramic away from any harm the ceramic should endure an impact of 5J. Accordingly the maximum strain is given by 0,3 %. The number of maximum load cycles should exceed 10^9. In order to guarantee the application in an automobile the power supply of 12/14 V respectively 36/42 V in the future must be considered as well. Furthermore the adaptronic system should fulfill the electromagnetic compatibility. Apart from this there are a lot of further requirements, which are concerned with the application of piezo-materials in an automotive structure, e.g. suitability for mass production, possibility of health monitoring, handling and at least the price.

4 Open Questions

Growing up from the foregoing section there are a multitude of questions apart from the boundary conditions which govern the material properties of the piezo-material. According to fig. 4 one might draw the conclusion that thermal stresses may influence the vibrational characteristics of the roof panel. Consequently the manufacturing tolerances, which may influence the location of the piezo-patches, must be evaluated. Apart from the foregoing aspects another question is formed by the configuration of the bonding. For the safety of the ceramic the adhesive coat should be ductile. In contrast with a ductile characteristic the adhesive coat has to guarantee a high transfer of force which means an adhesive coat with a high stiffness.

Another very important question is the control strategy. Therefore it should be investigated whether a stochastic or a harmonic excitation is of importance. Equally it should be asked if a proportional feedback of the sensor-signal is suitable for a sufficient vibration control. As a consequence the control of the vibration would cost less effort.

The most important question aims at the efficiency of the arrangement. Does the use of the presented principle of active vibration control properly improve the interior acoustics under all conditions? Due to the dependence of various components of the car this question can not be answered yet.

5 Conclusions and Outlook

The previous sections explained which structural components of an automobile might be proper for the use of smart materials. Exemplary the operating characteristics of these materials were presented for the roof-panel application. According to the fact, that hitherto concepts of improving the interior noise are either expensive or weighty the use of piezoceramic patches at the roof-panel might result in a success.

In conclusion the goal will be to improve the interior acoustic behaviour without using any isolating materials at the roof-panel of the automobile. By the way this may reduce the mass of the car as well. Nevertheless this application is concerned with lots of questions, which are not answered yet. In case of a succesful concept the dash board will be another application which is similar to the roof-panel. The problems arising from this application grow out of the holes inside the dash board, which enable an easy sound transmission. That is why the roof panel was investigated first.

Apart from the application of piezoceramic foils a general aim will be to use applicated piezo fibres in order to increase the efficiency of the arrangement.

Possibilities of Adaptronics in Optics

Frank Höller, M.Roß-Meßemer, T.Möller, A.Menck
Carl Zeiss / Oberkochen (D)

1 Abstract

Adaptronics in optics is a rather new field if smart materials are considered but it is a rather traditional field if the general concept of improvement by adaptive mechanical alignment is seen. Adaptronics in optics is therefore a rather promising field. Adaptronic as the innovative combination of sensor and actor within one material offers new possibilities namely in the field of high precision optics. This allows to realize new mechanical systems for fine positioning and alignment of optical elements with sub μm-accuracy. Furthermore adaptronical structures can be used for deforming optical elements like lenses and mirrors and hence reduce optical aberrations to zero. Vibrations cannot only be actively reduced by adaptronics but can also be actively introduced into optical surfaces for the realization of completely new illumination elements. This contribution will show the numerous potential applications of adaptronics in optics.

2 Introduction

Optics is one of the central enabling technologies of our time as has been pointed out in the harnessing light study, which has been recently published[1]. Photolithography and it's applications in semiconductor industry are the central design drivers for maximum optical and mechanical precision in imaging optics. Due to the continously increasing demand for higher integration density optical systems have to be improved accordingly.

These improvements can no more be solved by completely conventional techniques. Therefore investigations on new concepts for optical as well as mechanical elements are a central research topique in advanced optics. Adaptronical systems and system components offer the chance to solve some of the fundamental problems in optomechanical design and further more offer the chance to realize optical elements which cannot be realized by means of conventional approaches.

3 Active Vibration Control

is one of the best known applications of adaptronical systems and may become necessary in the future. At the moment there is no urgent need for vibration control in optics, since the

[1] Harnessing Light -Optical Science and Engineering for the 21st Century / National Research Council / National Academy Press / Washington DC 1998

existing systems are used in vibration isolated structures. There may arise a necessity for active vibration control in the future, if non rotation symmetrical systems like catadioptric systems for 157 nm or off axes mirror systems for 13 nm light will dominate. This may become important for amplitudes < 10 nm and frequencies below 300 Hz.

4 Alignment by Adaptronical Means

Alignment in optical systems usually requires 1-3d movements over typically 0.1 to 1 mm with ppm accuracy and hence µm to sub µm accuracy. These movements are traditionally performed by screws or other mechanical means. Adaptronics offers not only the chance to perform these movements by more advanced means but also the possibility to automatically compensate effects due to temperature, air pressure and humidity change during system operation. Alignment can be splitted up into vertical and horizontal alignment. Lithography lenses are usually built up with a vertical optical axes in order to avoid any disturbance of rotational symmetry.

Vertical alignment requires the movement of lenses with a weight of some kg over 50 µm and more. This requires energies of some mJ which cannot be achieved by piezoceramic elements without bi- or trimorph structures. Horizontal alignment seems to be much more promising, since energy is not the limiting factor. Energy is only needed in order to overcome friction. Under these conditions the lateral alignment movement is mainly limited by the applicable electric field.

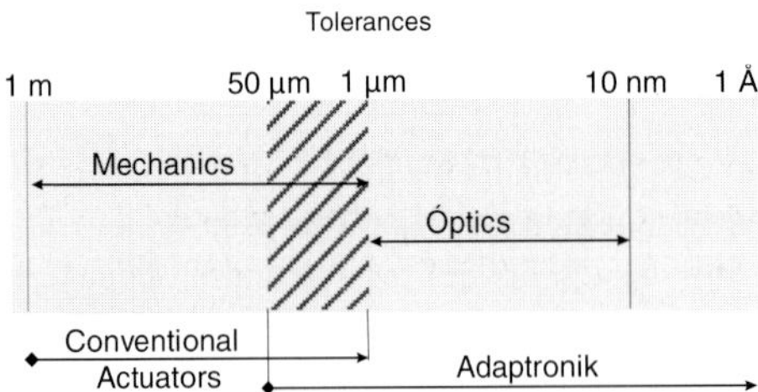

Figure 1: Tolerances

Alignment accuracies and methods can be potentially improved by application of adaptronic methods. There is a strong synergy between the needs for ultraprecise alignment and the increasing application of computer based alignment methods during production and operation on the one hand and the development of adaptronic actuator/sensor concepts on the other hand.

5 Surface Control by Adaptronical Means

Whereas the methods discussed in the first part of this paper are based on the acceleration of parts of the system or the complete system (hopefully) without changing the geometry of the system, the following methods are based on the user intended deformation of the system.

Optical elements in imaging and illumination systems can be categorized in 4 types of function :

- Mirrors
- Gratings (reflective as well as refractive)
- Diffusors (reflective as well as refractive)
- Lenses

The function of all these elements is **defined by their surface** as long as gradient lenses etc are not considered. Traditionally all these elements cannot be influenced in their function directly by changing their surface except for uncontrollable effects due to the influences like temperature, humidity, air pressure ... Modern optical lithography systems inhibit typically 60 surfaces. The sum of the surface deviations nevertheless shall not exceed a small amount of the wavelength of the operating light (i.e. 365nm, 248 nm, 193nm and in future 157 nm and 13 nm). Surface quality during production as well as during lifetime of optical systems is therefore of enormous importance.

6 Macro-Adaptronical Deformation of Mirrors

Residual aberrations of optical systems cannot only be reduced by movement of optical elements relative to the mechanical structure of the optical system but also by adaptive control of the surface of the involved optical elements. This method was originally developed for astronomical mirrors like the NTT = New Technology Telescope from Carl Zeiss[2]. Most of the following telescopes like the 8 m diameter VLT = Very Large Telescope and the american GEMINI-telescope are based on this adaptive concept. Mirrors are the most obvious candidates for adaptive surface control, since on the one hand they have a „refractive index“ of 2, which means that surface deformations of mirrors are approximately four times more critical than surface deformations of lenses, and on the other hand there is no limitation to fixate actuators and sensors on their backside. Nevertheless practical experiments have shown, that print through effects limitate the achievable accuracy severely. This was not that critical for astronomical mirrors since their optical performance is limited by the turbulence of the atmosphere and the typical wavelengths are in the visible range.

Lithography optics do not have these limitations and hence need more advanced mirror designs. Adaptronical structures offer the chance to overcome these problems with bimorph mirrors since there is no print through effect by principle as long as the number of actuators is high enough. The adaptronic principle on the other hand offers the possibility to control the deformation directly and hence introduce not only simple aberations like focus but also complicated aberations like astigmatism, coma and higher order Zernike polynomials. Especially astigmatism is of central interest since there is no possibility to elliminate astigmatism in optical systems by simple alignment of optical elements.

The decreasing wavelength in lithography systems now urgently requires the application of mirrors instead of lenses in either catadioptric systems for 157 nm[3] or fully off axes mirror systems for 13 nm. There is no doubt that mirrors will become more important in optics than they are now. Especially 13 nm light can only be influenced by either mirrors or gratings. The

[2] Wilson R.N. et alia , From passive Support Systems the NTT Active Support, Proc. IAU,Vol. 79, 1984, 23-40

[3] McLay J.A. & McIntyre A.S.L., 157 nm optical lithography: the accomplishments and the challenges, Solid State Technology, June 1999, 57-64

main drivers for mirrors instead of lenses are perfect chromatic behavior, reduced costs of material, no homogeneity problems and last but not least reduced degradation effects. These obvious advantages have to be „paid“ by aspherical instead of spherical surfaces and increased sensitivity to surface deformations.

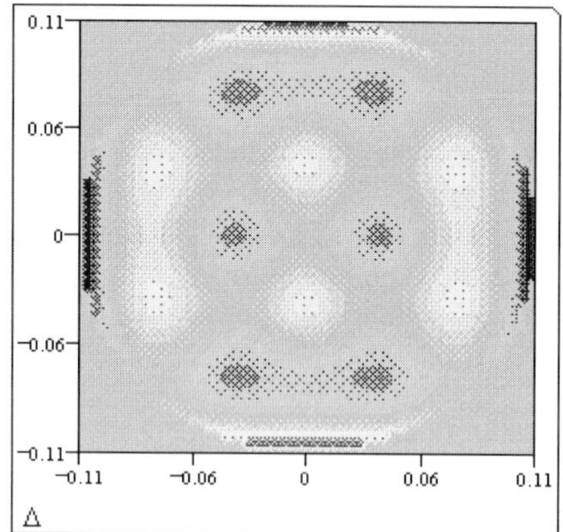

Figure 2: Print Through Effect on a conventional mirror

Figure 3: Hydraulical Support Structure of NTT

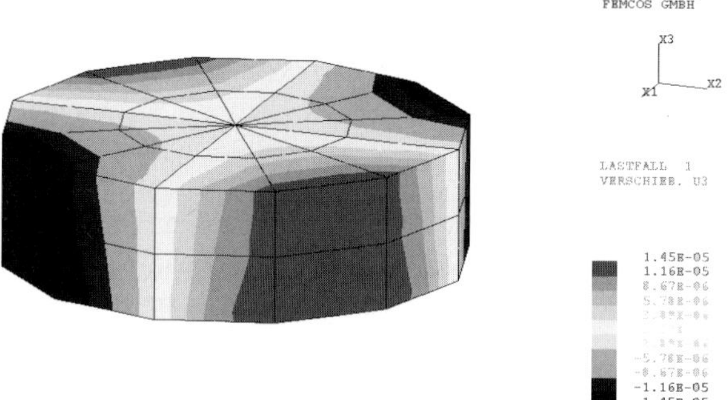

Figure 4: Adaptronic bimorph mirror

7 Micro-Adaptronical Deformation = Gratings and Diffusors

Usually only deformations with a spatial resolution of 1 to 0.01 mm^{-1} are considered if the deformation of optical elements is discussed. But there is another interesting spatial spectrum for optical elements in the region of << 100 mm^{-1}. If these deformations are periodical and the amplitude of the deformation is in the same order of size as the wavelength of light we can observe the effect of diffraction. If these deformations are broadband with respect to spatial frequency and the amplitude of deformation is in the order of size as the wavelength of light we can observe diffusion. Neither adaptive gratings nor adaptive diffusors up to now were possible, since there was no possibility to realize a grating structure with variable period except for solutions based on the Moire effect, i.e. the overlapping of two or more gratings.

8 Lenses

Lenses can be deformed in order to compensate effects like astigmatism and or anisotropy of the optical material. Main problems are the unfavourable conditions for the design of the support structure, since there is no free backside.

9 Summary

Adaptronic systems potentially have various applications in optics. Especially in semiconductor lithography nearly all relevant mechanooptical problems require new technical solutions. PZT and $LiNbO_3$ seem to be the most promising materials. Besides quasistatic **macroadaptronic** problems there seems to be an increasing chance for applications of **microadaptronic** methods.

Industrial Application of Adaptronics

Roger Wimmel
ERAS GmbH, Göttingen, Germany

1 Abstract

The innovation aims in machine building with regard to product quality and throughput are to increase the accuracy, the velocity and the size in the first place.

However, the accuracy of a process is questioned by an enlargement of the velocity if one considers that the excitation amplitudes increase for all vibration phenomena with increasing velocity. The demand to increase throughput by means of larger system widths normally leads to a mass increase and a stiffness reduction at the same time. Therefore, it must be an important aim to make use of lightweight construction designs for these applications. However, the potential of lightweight construction technology is limited by lower damping referring to vibration problems

Summarizingly it can be stated that for the future the vibration behavior of machines has to be moved into the centre of dimensioning since the innovation aims are currently not achieved due to vibration problems. The useful way to overcome the design dilemma with active systems will be shown herein at the example of an aggregate of the worlds largest wood-free coated paper production line.

2 Phenomenon

In a paper machine, there is a great number of rotating rolls. Pairs of rotating rolls are used to press the paper or to apply additives (color, glue etc.).

Eigenmodes with the rolls vibrating against each other (contact vibrations) can be observed as shown here in *figure 1* for example.

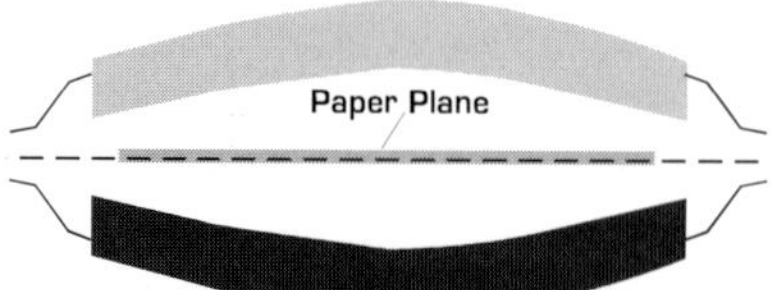

Figure 1: Vibration eigenmode of a pair of rolls (lateral view)

Furthermore, roundness errors or polygons are formed especially on elastomer coated rolls (*figure 2*) if multiples of the rotating frequencies might coincide with a natural frequency of the system given by the pair of rolls.

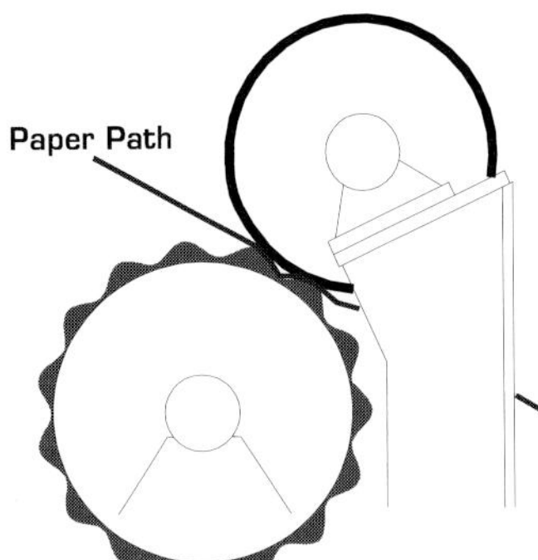

Figure 2: Formation of a polygonity on a rubber-lined roll due to contact vibrations (side view).

3 Remedy Possibilities

Possibilities to influence the vibration behavior of a machine are basically given through:

- changes of the system parameters,
- changes of the operating parameters,
- a reconstruction or another conception,
- supplementary aggregates.

However, it can be stated that up to now all attempts in the above mentioned directions did not reach the goal that production is damaged by vibration problems no more. For this reason, the concept and the realization of an active system for reduction of the contact vibrations should be described in the following.

4 Concept

In cooperation with the machine supplier, ERAS developped a system consisting of a displacement sensor, an adaptive controller and actuator between the roll bearings. The displacement sensor has to be fitted to a support arm over the mid-position of the rolls. This secured the highest resolution of the signals, however, also meant to install a costly traverse device not to reduce the operational- and service conditional space for the movement of the rolls

The controller with a digital signal processor (DSP) as a heart as well as the electronics were to be integrated into a console which was placed in greater distance of the machine. For the cabling as well as for all electronic components, the relevant standards and determinations of the system operator were to be kept.

Only the bearing plane was available to place the actuator (*figure 3*). However, the necessary actuator principle could not clearly be determined with the prior knowledge.

Into consideration were taken:

- hydraulic,
- piezo electric or
- electromagnetic

actuators. In particular the extreme limiting conditions

- 24h operation,
- permanent moisture and
- high pressure cleaning etc.

had to be taken into account here.

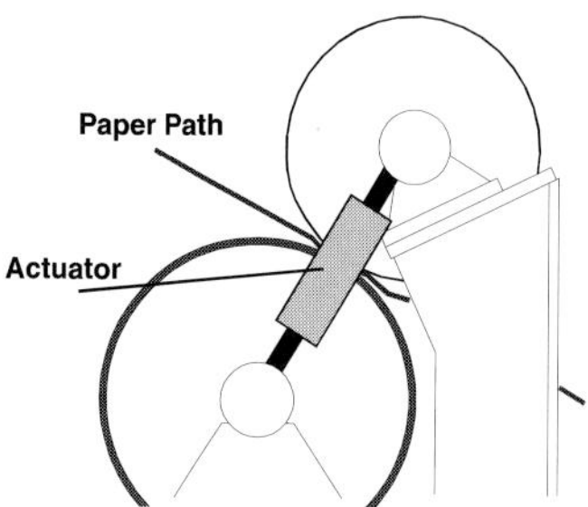

Figure 3: Positioning of the actuator

5 Preliminary Investigations

An vibration analysis in operational state was carried out on the appropriate production machine. By this, the problematic eigenmode could be determined and the connection between the multiples of the rotating frequency could be proved. The measured vibration amplitudes were used for comparison with finite element analysises which were done together with the customer.

Based on the results of the finite element analysises, the selection and dimensioning the actuator could be carried out. Considering all limiting conditions such as expenditure, price and delivery period, an electro-magnetic actuator in the form of a linear motor with a force amplitude of 5kN proved itself as optimal.

6 Realization

The realization of the design for the production machine was started directly after finishing of the preliminary investigations. The work contents were

- the construction, the building and/or the procurement of all components for the actuator, the sensor arm and the control console as well as
- the roughcut of all modules,
- the assembly,
- the installation on site,
- the startup and
- the installation of an interface to the paper machine process control.

Figure 4 shows a photo of the paper machine component with the actuator in the final state.

Figure 4: The actuator in operation

The functional verification could be shown at the machine one week after installation of the active system.

Figure 5 shows the corresponding results (behavior of the system with and without operation of the active system).

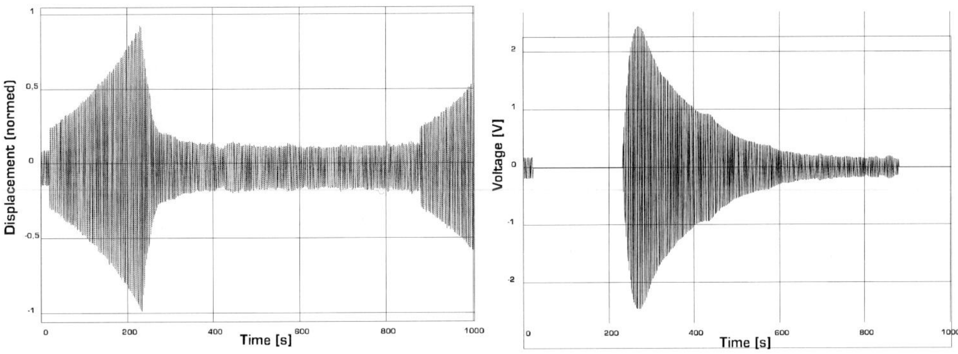

Figure 5: Displacement sensor signal (above) and drive signal of the actuator (below)

The active system works in permanent operation since June 1998.

7 Outlook

In order to take advantage from adaptive structures technology (adaptronics) further progress of such systems can be imagined by putting sensor and actuator systems into the rolls. As already stated, any additional component near the roll surface or the bearing can diminish the operational space. But if once a complete active system has been installed, a step towards higher integration is not too far.

Piezoceramic Materials – Potential of a new Actuator Technology

Peter Jänker, Frank Hermle, Thomas Lorkowski, Stefan Storm, Markus Christmann
DaimlerChrysler AG Research and Technology, D-81663 Munich

1 Introduction

Over the past few years, above all microelectronics has shaped and accelerated technical developments. Modern electronics makes it possible to realize "smart" adaptive systems. These systems cover operating conditions with sensors, further process the information to actuator commands and, by means of actuators, control mechanical vehicle subfunctions. Developments in the area of actuator systems have not kept pace at all with microelectronics and sensor systems. Actuators based on mechanically active materials (smart materials) are a new approach toward closing this technological gap. Of extraordinary technical importance are electrically controlled actuators that can be integrated into electronic control systems and that represent the core modules of mechatronic systems. Within this major group, piezoelectric ceramics (PZT) offer a high potential compared to the electromagnetic actuators that are currently used most[1]. The industrial application technology of piezoelectrical actuators is currently in an early development stage. A topical example for potential industrial application is vibration isolation of high-speed trains. Other examples for application presented in this paper are adaptive systems for future helicopters.

2 Metrological Characterization of Piezoactuators

The aim of metrological investigations is to support product development with application-oriented data. In order to describe the material behavior, a model of the electro-mechanical behavior is required. In order to ensure characterization of piezoactuators that is as complete as possible, a measurement stand with an active measuring head was developed (Fig. 1) at DaimlerChrysler. The measurement stand comprises an extremely stiff frame with axial optical access on both sides to the object to be measured, hydraulics to preset a static force component, a differential optical length-measuring system, force measurement and a temperature chamber. The active test head is equipped with piezoelectric actuators. Electronic control permits representing various external load situations by controlling the active test head. The measurement stand covers the determination of blocking force (infinitely stiff clamping), free stroke (constant force) as well as further processes such as variable spring loads.

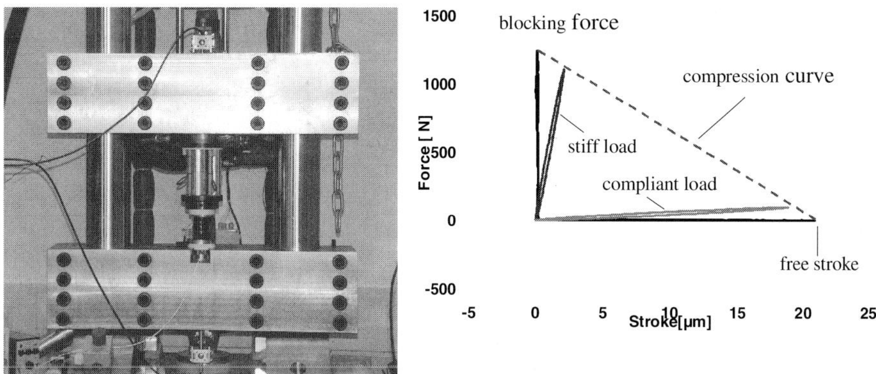

Fig. 1: Dynamic measurement stand with active test head. Determination of the behavior X(F,U) and Q(F,U) at a constant temperature. Results of measurements on the dynamic test stand.

3 Application Examples

Piezoactuators are high-force, but low stroke devices. They have high specific work capability, excellent dynamics, good controllability and resolution of travel, but mechanical properties are poor. Open questions are concerning lifetime, reliability, costs and availability. There are manifold application possibilities mainly in the field of positioning and control. In the field of automobiles systems piezoactuators are for example extremely interesting for fuel injection systems.

Applications in aviation, with relatively small quantities, are for the emerging piezoelectric actuator technology of initial interest. Helicopters represent a platform of this type. Helicopters are vehicles with unique application possibilities. A number of advanced technical developments are still required in order to open up further market segments. Urgent development tasks are to reduce noise pollution and vibrations and to enhance flight performance. Examples of solution-offering technologies that have been promoted by ECD and DaimlerChrysler Research and Technology comprise the Active Rotor Control, Active Noise as well as Vibration Control.

3.1 Active Rotor Control

The key element of a future helicopter is active rotor control. The active rotor control has three main technical objectives:

- Reduction of main-rotor noise and cabin vibrations
- Improvement of the rotor aerodynamics
- Stabilization of the rotor dynamics.

3.1.1 Servo-Flap Technology

An approach to realize rotor active control is to install servoflaps in the outer part of the rotor blades which are driven by piezoelectrical actuators[2 3]. Controlling each rotor blade is effected during a rotor revolution with high speed and precision

The flap system comprises a flap with hinged supports that is driven by linear actuators. In order not to unacceptably change the nominal position of the c.g. line at approx. 25% of the blade depth, the actuator is mounted near the blade leading edge. A control rod is used for force transmission.

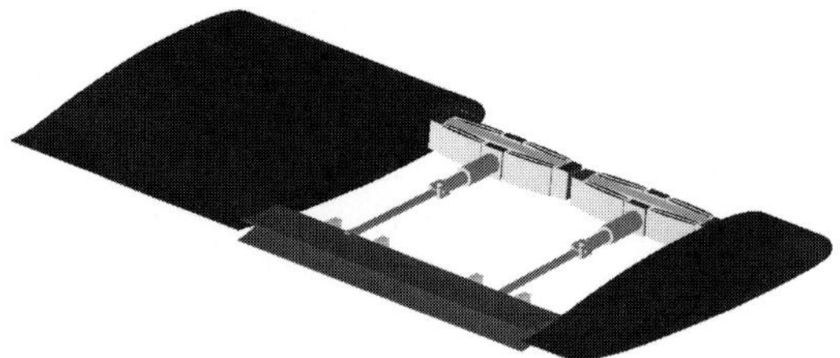

Fig. 2: Rotorblade segment with integrated actuator flap unit.

3.1.2 Leading-Edge-Flap Technology

The performance of current helicopters is also limited by the occurrence of dynamic stall on the retreating side of the rotor. This situation may be remedied by the use of dynamic high-lift devices such as the leading-edge flap [4] integrated in the rotor blades. General aspects for the structural implementation are shown in figure 3. This approach allows to redevelop the entire rotor system using reduced airfoil thickness in the outer blade area. This has a great potential for power saving and possible reduction of high-impulsive noise. Combined use with a servoflap would also be conceivable.

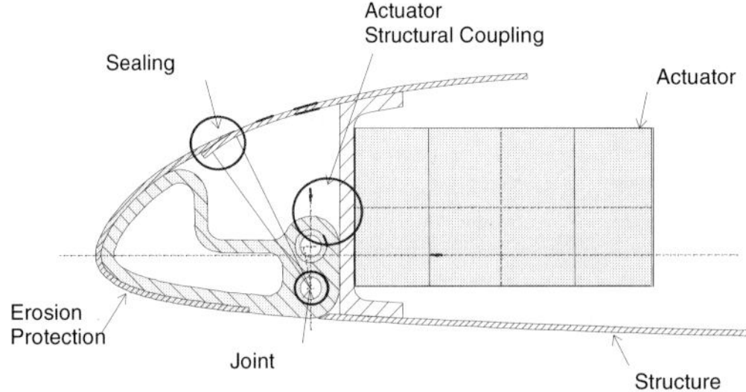

Fig. 3: General design aspects for implementation of the leading-edge flap for adaptive airfoil variation.

3.1.3 Large Displacement, High Force Actuators

Controlling a flap integrated in the outer area demands a lot of the actuator system:

- Only little installation space is available in the rotor blade.
- The high centrifugal acceleration (typ. 1,000 g) results in large mass forces.
- The actuator must cope with the high air loads. The actuator dynamics must be better than 5/rev. This corresponds with the EC135 helicopter to 35 Hz.
- The actuator system must generate sufficient stroke. In the concrete case of an EC135, a stroke of approx. 1 mm and a blocking force of 2,000 N is necessary.

The decisive technical hurdle in developing the piezoelectrically controlled flap system was the low elongation capacity of the PZT materials. A suitable method of construction that was developed by Daimler-Benz Research in recent years is the hybrid actuator comprising a piezostack with a mechanical step-up gear. The step-up gear must meet the following requirements:

- No mechanical play, no/slight friction
- High gear stiffness
- High energy efficiency: low elastic energy losses, low weight

An important criterion of the quality of a design as regards the working capacity is the mechanical efficiency η_{MECH}, which is defined as the ratio of the working capacity of the hybrid actuator (W_H) to the working capacity of the piezo (W_P): $\eta_{MECH} = W_H / W_P$. A further important criterion of the design quality is therefore applied, namely the mass efficiency η_{MASS}, the ratio of the specific working capacity of hybrid actuator and piezostack: $\eta_{MASS} = \eta_{MECH}\ m_H/m_P$. An integrated design is considered to be the optimum construction method for a weight- and volume-optimized hybrid actuator. The stiff mounting frame is designed through the integration of joints as a gear. In order to ensure freedom from play and wear, flexures are used as joints. Fig. 4 shows the diamond-shaped geometry of the gear.

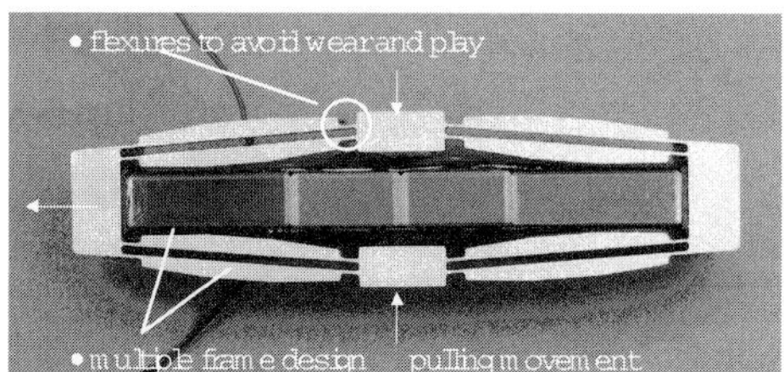

Fig. 4: DWARF

Due to the geometric arrangement, an expansion movement of the piezostack is transformed into a pulling movement. The effectivity of the gear was optimized decisively by being designed as a multiple frame. The dynamics of the hybrid actuator is very good, the first resonance frequency lies at 220 Hz. The DWARF generates a blocking force of 720 N and a free stroke of 1.1 mm.

3.2 Active Vibration Control

The noise and vibration situation in modern helicopters is still unacceptable compared to modern aircraft cabins. Oscillations of the rotor gear box unit reaches the passenger cabin via the connecting struts. To solve this problems, DaimlerChrysler have developed together with the project partner ECD two active strut technologies.

The first strut technology (Fig. 6 left) is designed for acoustic control. For this, conventional gearbox struts were modified by bonding an active layer of PZT directly onto the surface. The piezoelectric layer allows to generate dynamic forces of ±250N up to 2 kHZ. The advanced noise reduction capabilities of the smart strut system developed were demonstrated on a BK117 helicopter [5].

A second active strut technology is to control vibrations around 26 Hz. Force actuators are being developed replacing conventional gearbox struts (Fig. 6 right) [6]. The force actuators were used as struts between the fuselage and the gearbox. They were controlled as to isolate the gearbox isolations form the fuselage. The actuators are hydraulically amplified stacks generating an output of ±8 kN and dynamic amplitude of ±0.5 mm.

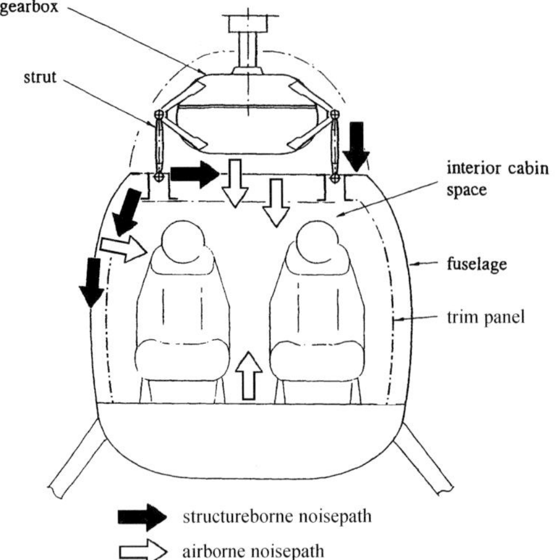

Fig. 5 Noise transmission paths.

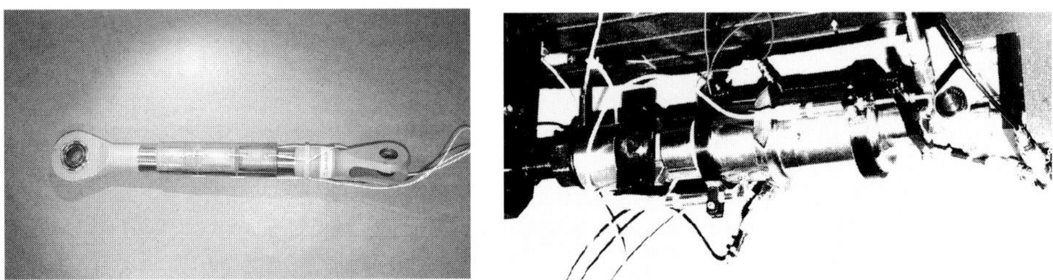

Fig. 6 Smart Strut (left) for acoustic control. Isolator (bottom right) for vibration contol.

4 Conclusions

Actuators based on piezoelectrical materials are extremely promising candidates for implementing adaptive mechanical systems in vehicle construction. In this paper a whole host of applications developed for future helicopters are presented. This includes effective construction methods for high-performance piezoactuators.

5 References

1. P. Jänker, W. Martin, "Performance and Characteristics of Actuator Materials", Fourth International Conference on Adaptive Structures, 2-4 November 1993, Cologne, Germany
2. D. Schimke, P. Jänker, A. Blaas, R. Kube, and Ch. "Keßler, Individual Blade Control by Servo-Flap and Blade Root Control A Collaborative Research and Development Programme", 23rd European Rotorcraft Forum in Dresden, Germany, September 16-18, 1997

3. P. Jänker et al., Development and Evaluation of Advanced Flap Control Technology Utilizing Piezoelectric Actuators, 25rd European Rotorcraft Forum in Rom, Italy, September 14 – 16, 1999
4. T. Lorkowski et al., 25rd European Rotorcraft Forum in Rom, Italy, September 14 – 16, 1999
5. W. Gembler, et al., Smart Struts -The Solution for Helicopter Interior Noise Problems, 25rd European Rotorcraft Forum in Rom, Italy, September 14 – 16, 1999
6. H. Rottmayr et al., Application of Modern Vibration Control Techniques on EC135 and Future Trends, 25rd European Rotorcraft Forum in Dresden, September 16 – 18, 1997

Author Index*

* The page numbers refer to the first page of the respecting article

Subject Index*

* The page numbers refer to the first page of the respecting article

N

O

P